LARGE-SCALE DISASTERS

Extreme events—climatic events such as hurricanes, tornadoes, and drought, as well as events not related to the climate such as earthquakes, wars, and transportation-related disasters—can cause massive disruption to society, including large death tolls and property damage in the billions of dollars. Events in recent years have shown the importance of being prepared and that countries need to work together to help alleviate the resulting pain and suffering. This volume presents an integrated review of the broad research field of large-scale disasters. It establishes a common framework for predicting, controlling, and managing both manmade and natural disasters. There is a particular focus on events caused by weather and climate change. Other topics include air pollution, tsunamis, disaster modeling, the use of remote sensing, and the logistics of disaster management. It will appeal to scientists, engineers, first responders, and health care professionals, in addition to graduate students and researchers who have an interest in the prediction, prevention, or mitigation of large-scale disasters.

MOHAMED GAD-EL-HAK is the Inez Caudill Eminent Professor of Biomedical Engineering and Chair of Mechanical Engineering at Virginia Commonwealth University in Richmond. He is a Fellow of the American Academy of Mechanics, the American Physical Society, and the American Society of Mechanical Engineers.

LARGE-SCALE DISASTERS
Prediction, Control, and Mitigation

Edited by

MOHAMED GAD-EL-HAK
Virginia Commonwealth University

CAMBRIDGE UNIVERSITY PRESS
Cambridge, New York, Melbourne, Madrid, Cape Town, Singapore, São Paulo, Delhi

Cambridge University Press
32 Avenue of the Americas, New York, NY 10013-2473, USA

www.cambridge.org
Information on this title: www.cambridge.org/9780521872935

© Cambridge University Press 2008

This publication is in copyright. Subject to statutory exception
and to the provisions of relevant collective licensing agreements,
no reproduction of any part may take place without
the written permission of Cambridge University Press.

First published 2008

Printed in Hong Kong by Golden Cup

A catalog record for this publication is available from the British Library.

Library of Congress Cataloging in Publication Data
Large-scale disasters : prediction, control, and mitigation / edited by Mohamed Gad-el-Hak.
p. cm.
Includes bibliographical references and index.
ISBN 978-0-521-87293-5 (hardback)
1. Natural disasters. 2. Natural disaster warning systems. 3. Natural disasters—Risk assessment.
4. Hazardous geographic enviornments—Risk assessment. I. Gad-el-Hak, Mohamed, 1945– II. Title.
GB5014.L365 2008
363.34–dc22   2007035502

ISBN   978-0-521-87293-5 hardback

Cambridge University Press has no responsibility for
the persistence or accuracy of URLs for external or
third-party Internet Web sites referred to in this publication
and does not guarantee that any content on such
Web sites is, or will remain, accurate or appropriate.

On the cover: An earthquake of magnitude 7.6 struck the Pakistan-administered region of Kashmir on 8 October 2005. Most of the affected areas are in mountainous regions and access was impeded by landslides that blocked the roads. More than 80,000 people died, and 3.3 million were left homeless in this calamitous event.

Back cover: A satellite image of category 5 Hurricane Katrina, taken over the Gulf of Mexico on 29 August 2005. The sheer physical size of Katrina caused devastation far from the eye of the storm; it was possibly the largest hurricane of its strength ever recorded.

Dedicated to all those who succumbed to any natural or manmade disaster, and to all those who through their kindness, generosity, courage, resourcefulness, dedication, and hard work helped alleviate the pain and suffering.

We start life with a bag full of luck and an empty bag of experience. Aim of life is to fill the bag of experience before the bag of luck runs out.

(From a Hindu fable)

Contents

Preface			page xv
About the editor			xix
List of contributors			xxi
1	Introduction		1
	Mohamed Gad-el-Hak		
	1.1	What is a large-scale disaster?	1
	1.2	Book contents	2
	References		3
2	The art and science of large-scale disasters		5
	Mohamed Gad-el-Hak		
	2.1	Are disasters a modern curse?	5
	2.2	Disaster scope	6
	2.3	Facets of large-scale disasters	9
	2.4	The science of disaster prediction and control	10
		2.4.1 Modeling the disaster's dynamics	12
		2.4.2 The fundamental transport equations	13
		2.4.3 Closing the equations	14
		2.4.4 Compressibility	17
		2.4.5 Prandtl's breakthrough	20
		2.4.6 Turbulent flows	21
		2.4.7 Numerical solutions	23
		2.4.8 Other complexities	23
		2.4.9 Earthquakes	25
		2.4.10 The butterfly effect	26
	2.5	Global Earth Observation System of Systems	30
	2.6	The art of disaster management	31
	2.7	A bit of sociology	32
	2.8	Few recent disasters	34
		2.8.1 San Francisco Earthquake	34
		2.8.2 Hyatt Regency walkway collapse	38
		2.8.3 Izmit Earthquake	41
		2.8.4 September 11	42
		2.8.5 Pacific Tsunami	43
		2.8.6 Hurricane Katrina	45
		2.8.7 Kashmir Earthquake	49
		2.8.8 Hurricane Wilma	49

		2.8.9	Hajj stampede of 2006	54
		2.8.10	*Al-Salam Boccaccio 98*	56
		2.8.11	Bird flu	59
		2.8.12	Energy crisis	61
	2.9	Concluding remarks		64
	References			64
3	Multiscale modeling for large-scale disaster applications			69
	Ramana M. Pidaparti			
	3.1	Introduction		69
	3.2	Definition and modeling of scales in climate and weather		71
		3.2.1	Global climate modeling	72
		3.2.2	Long-term climate simulation	72
		3.2.3	Limits to predictability	72
		3.2.4	Global and regional climate models	74
	3.3	Definition and modeling of scales during accidental release of toxic agents in urban environments		77
	3.4	Multiscale modeling methods		81
		3.4.1	Key challenges	81
		3.4.2	Application of modeling methods to large-scale disasters	81
		3.4.3	Multiscale modeling techniques	82
		3.4.4	Molecular dynamics method	83
		3.4.5	Coarse-grained methods	84
		3.4.6	Monte Carlo methods	84
		3.4.7	Cellular automata	85
		3.4.8	Neural networks	86
		3.4.9	Mathematical homogenization	86
		3.4.10	Quasi-continuum method	87
		3.4.11	Heterogeneous multiscale method	87
		3.4.12	Continuum methods	88
		3.4.13	Domain decomposition method and parallel computations	89
		3.4.14	Lattice Boltzmann method	90
	3.5	Summary and outlook		91
	Acknowledgments			91
	References			91
4	Addressing the root causes of large-scale disasters			94
	Ilan Kelman			
	4.1	Definitions and context		94
		4.1.1	Defining disasters	94
		4.1.2	Do natural disasters exist?	97
	4.2	Root causes of disaster		98
		4.2.1	Case studies	98
		4.2.2	Root cause: vulnerability	101
		4.2.3	Root causes of vulnerability	105
	4.3	Tackling root causes of disaster		108
		4.3.1	Principles	108
		4.3.2	Illustrative case studies	109
	4.4	Conclusions		113
	References			114

5	Issues in disaster relief logistics		120
	Nezih Altay		
	5.1	Introduction	120
	5.2	Disaster relief issues identified in literature	122
	5.3	Supply chain issues	123
		5.3.1 Funding issues	126
		5.3.2 Needs assessment and procurement	127
		5.3.3 Management of information	129
		5.3.4 Coordination issues	130
		5.3.5 Transportation infrastructure and network design	132
		5.3.6 Standardization of relief	132
	5.4	Operational issues	132
		5.4.1 Personnel issues	134
		5.4.2 Availability of technology	136
		5.4.3 Local resources	137
	5.5	Ethical issues	139
		5.5.1 Discrimination	139
		5.5.2 Corruption	139
	5.6	Political issues	140
		5.6.1 Military use in disaster relief	140
	5.7	Conclusions and future research directions	142
	References		143
6	Large-scale disasters: perspectives on medical response		147
	Jehan Elkholy and Mostafa Gad-el-Hak		
	6.1	Introduction	147
	6.2	Characteristics of disasters	148
	6.3	Classification of disasters	148
	6.4	Disaster management	149
	6.5	Phases of a disaster	150
		6.5.1 Phase I: disaster preparedness	150
		6.5.2 Phase II: medical response	152
		6.5.3 Phase III: recovery	156
	6.6	Role of specialists	156
	6.7	Disaster evaluation	158
	6.8	Failure of disaster response and problems encountered during disaster management	159
	6.9	Conclusions	159
	References		159
7	Augmentation of health care capacity in large-scale disasters		161
	Atef M. Radwan		
	7.1	Introduction	161
	7.2	Definitions	162
	7.3	Capacity augmentation of health care facility	163
		7.3.1 Variables of health care capacity	163
		7.3.2 Triage priorities	164
		7.3.3 Capacities in the medical assistance chain	164
		7.3.4 Mass trauma casualty predictor	166
		7.3.5 Increasing hospital bed capacity	168

			7.3.6	Adaptation of existing capacity	168
			7.3.7	Staff calling in and staff augmentation plan	169
			7.3.8	Modification of the standards of care	170
			7.3.9	Triage of patients in mass critical care	171
		7.4	Cooperative regional capacity augmentation		171
		7.5	Off-site patient care		173
		7.6	Role of government		173
		7.7	Community involvement		174
		7.8	Summary		174
		References			175
8	Energy, climate change, and how to avoid a manmade disaster				177
	Ahmed F. Ghoniem				
		8.1	Introduction		177
		8.2	Energy consumption—now and then		179
			8.2.1	How much we use	179
			8.2.2	Energy and how we live	179
			8.2.3	How much we will use	181
		8.3	Carbon dioxide		183
			8.3.1	Greenhouse gases	184
			8.3.2	Energy balance	184
			8.3.3	Climate modeling	186
			8.3.4	Global warming and climate change	189
		8.4	CO_2 emission mitigation		194
			8.4.1	Implementing multiple solutions	195
			8.4.2	The "wedges"	196
		8.5	Low-carbon fossil conversion technologies		198
			8.5.1	Chemical energy	199
			8.5.2	CO_2 capture	200
			8.5.3	Electrochemical separation	203
			8.5.4	Synfuel production	203
		8.6	Zero-carbon conversion technologies: nuclear and renewable sources		203
			8.6.1	Nuclear energy	204
			8.6.2	Hydraulic power	204
			8.6.3	Geothermal energy	205
			8.6.4	Wind energy	205
			8.6.5	Solar energy	206
			8.6.6	Biomass energy	206
			8.6.7	Renewable sources and storage	207
		8.7	Transportation		208
		8.8	Conclusions		209
		References			210
9	Seawater agriculture for energy, warming, food, land, and water				212
	Dennis M. Bushnell				
		9.1	Introduction		212
		9.2	Biomass and the Sahara		213
		9.3	Saline/salt water agriculture		214
		9.4	Additional impacts/benefits of saline/seawater agriculture		215
		9.5	Summary		216
		References			217

10	Natural and anthropogenic aerosol-related hazards affecting megacities		218
	Hesham El-Askary and Menas Kafatos		
	10.1	Introduction	218
	10.2	Aerosol properties	221
	10.3	Sand and dust storms	221
		10.3.1 Remote sensing of sand and dust storms	222
		10.3.2 Egypt case study	223
		10.3.3 India case study	230
		10.3.4 Modeling of dust storms (dust cycle model)	236
	10.4	Air pollution	239
		10.4.1 Cairo air pollution case study	239
		10.4.2 Pollution effects forcing on large-scale vegetation in India	242
	10.5	Forcing component	245
		10.5.1 Egypt case study	248
		10.5.2 China case study	249
	10.6	Conclusions	252
	Acknowledgments		254
	References		254
11	Tsunamis: manifestation and aftermath		258
	Harindra J. S. Fernando, Alexander Braun, Ranjit Galappatti,		
	Janaka Ruwanpura, and S. Chan Wirasinghe		
	11.1	Introduction	258
	11.2	Causes of tsunamis: a general overview	262
	11.3	Hydrodynamics of tsunamis	263
	11.4	Ecological impacts of tsunamis—a general overview	265
	11.5	The Sumatra Earthquake and Tsunami	267
		11.5.1 The Sumatra–Andaman Island Earthquake, 26 December 2004	267
		11.5.2 The Sumatra Tsunami in Sri Lanka	269
		11.5.3 Wave observations and impacts on Sri Lanka	270
		11.5.4 The impact on Sri Lanka	271
	11.6	Tsunami warning systems	282
	11.7	Planning for tsunamis	283
		11.7.1 Components of the stochastic scheduling network	284
	11.8	Conclusions	289
	Acknowledgments		289
	References		290
12	Intermediate-scale dynamics of the upper troposphere and stratosphere		293
	James J. Riley		
	12.1	Background	293
	12.2	More recent interpretation of data	295
	12.3	Results from numerical simulations	296
	12.4	Implications	298
	12.5	Summary	300
	References		300
13	Coupled weather–chemistry modeling		302
	Georg A. Grell		
	13.1	Introduction	302
	13.2	Fully coupled online modeling	302
		13.2.1 Grid scale transport of species	303

		13.2.2	Subgrid scale transport	303
		13.2.3	Dry deposition	304
		13.2.4	Gas-phase chemistry	304
		13.2.5	Parameterization of aerosols	305
		13.2.6	Photolysis frequencies	306
	13.3	Online versus offline modeling		306
	13.4	Application in global change research		312
	13.5	Concluding remarks		314
	References			315
14	Seasonal-to-decadal prediction using climate models: successes and challenges			318
	Ramalingam Saravanan			
	14.1	Introduction		318
	14.2	Potentially predictable phenomena		321
	14.3	Successes in dynamical climate prediction		322
	14.4	Challenges that remain		325
	14.5	Summary		326
	References			327
15	Climate change and related disasters			329
	Ashraf S. Zakey, Filippo Giorgi, and Jeremy Pal			
	15.1	Introduction		329
		15.1.1	Definitions of climate parameters	330
	15.2	A brief review of regional climate modeling		332
	15.3	ICTP regional climate model		335
	15.4	Climate change and extreme events		337
		15.4.1	Defining changes of extremes	337
	15.5	Extremes and climate variability		341
	15.6	Regional impact studies		346
		15.6.1	Severe summertime flooding in Europe	346
		15.6.2	Warming and heat wave	347
		15.6.3	Wind storms (hurricanes)	351
	15.7	Summary		352
	Acknowledgments			357
	References			357
16	Impact of climate change on precipitation			363
	Roy Rasmussen, Aiguo Dai, and Kevin E. Trenberth			
	16.1	Introduction		363
	16.2	Precipitation processes in observations and models		364
		16.2.1	Evaluation of model simulated changes in precipitation by examination of the diurnal cycle	366
		16.2.2	Observed trends in moisture and extreme precipitation events	368
	16.3	How should precipitation change as the climate changes?		371
	16.4	Questions and issues		373
	16.5	Summary		374
	Acknowledgments			374
	References			374
17	Weather-related disasters in arid lands			377
	Thomas T. Warner			
	17.1	Introduction		377

	17.2	Severe weather in arid lands	378
		17.2.1 Dust storms and sand storms	378
		17.2.2 Rainstorms, floods, and debris flows	391
	17.3	Desertification	396
		17.3.1 What is desertification?	396
		17.3.2 Extent of desertification	399
		17.3.3 Anthropogenic contributions to desertification	401
		17.3.4 Natural contributions to desertification	409
		17.3.5 Additional selected case studies and examples of desertification	409
		17.3.6 Physical process feedbacks that may affect desertification	414
		17.3.7 Satellite-based methods for detecting and mapping desertification	417
	17.4	Summary	418
		References	420
18	The first hundred years of numerical weather prediction		427
	Janusz Pudykiewicz and Gilbert Brunet		
	18.1	Forecasting before equations	427
	18.2	The birth of theoretical meteorology	429
	18.3	Initial attempts of scientifically based weather prediction	432
	18.4	Bergen school of meteorology	433
	18.5	First numerical integration of the primitive meteorological equations	434
	18.6	Weather forecasting after Richardson	436
	18.7	Richardson's experiment revisited and the birth of forecasting based on primitive equations	439
	18.8	Expansion of the scope of traditional meteorological prediction	439
	18.9	Development of the modern atmospheric prediction systems	440
	18.10	From weather prediction to environmental engineering and climate control	441
	18.11	Conclusions	444
		References	444
19	Fundamental issues in numerical weather prediction		447
	Jimy Dudhia		
	19.1	Introduction	447
	19.2	Disaster-related weather	447
	19.3	Disaster prediction strategies	448
		19.3.1 Medium-range prediction (5–10 days)	448
		19.3.2 Short-range prediction (3–5 days)	448
		19.3.3 Day-to-day prediction (1–3 days)	449
		19.3.4 Very short-range prediction (<1 day)	449
	19.4	Fundamental issues: atmospheric predictability	449
	19.5	The model	450
		19.5.1 Model physics	450
		19.5.2 Model dynamics	451
		19.5.3 Model numerics	451
	19.6	Model data	451
	19.7	Conclusions	452
		References	452

20	Space measurements for disaster response: the International Charter		453
	Ahmed Mahmood and Mohammed Shokr		
	20.1	Introduction	453
	20.2	Space remote sensing and disaster management	454
	20.3	General principles of remote sensing	459
		20.3.1 Optical, thermal, and microwave imaging	461
		20.3.2 Image processing, information contents, and interpretation	470
		20.3.3 Geophysical parameter retrieval and value adding	475
		20.3.4 Image classification and change detection	477
	20.4	Space-based initiatives for disaster management	479
	20.5	About the charter	482
		20.5.1 History and operations	484
		20.5.2 A constellation of sensors and satellites	491
		20.5.3 Mission summaries	500
		20.5.4 Applicable policies	504
		20.5.5 Performance update	511
	20.6	Disaster coverage	515
		20.6.1 Activation criteria	515
		20.6.2 Data acquisition planning	516
		20.6.3 Reporting and user feedback	518
	20.7	Case histories	519
		20.7.1 Nyiragongo volcanic eruption	519
		20.7.2 Southern Manitoba flood	521
		20.7.3 Galicia oil spill	522
		20.7.4 South Asian Tsunami	525
		20.7.5 French forest fires	527
		20.7.6 Hurricane Katrina	528
		20.7.7 Kashmir Earthquake	531
		20.7.8 Philippines landslide	532
		20.7.9 Central Europe floods	534
	20.8	Concluding remarks	539
	Acknowledgments		539
	References		540
21	Weather satellite measurements: their use for prediction		542
	William L. Smith		
	21.1	Introduction	542
	21.2	Weather satellite measurements	542
	21.3	Global Earth Observation System of Systems	544
	21.4	The current and planned space component	545
	21.5	Vegetation index	547
	21.6	Flash floods	549
	21.7	Severe thunderstorms and hurricanes	551
	21.8	Improvements in the satellite observing system	556
	21.9	Summary	565
	Acknowledgments		566
	References		566
Epilogue			569
Index			573

Preface

> I am the daughter of Earth and Water
>   And the nursling of the Sky;
> I pass through the pores of the ocean and shores
>   I change, but I cannot die.
> For after the rain with never a strain
>   The pavilion of Heaven is bare,
> And the winds and sunbeams with their convex gleams
>   Build up the blue dome of air,
> I silently laugh at my own cenotaph,
>   And out of the caverns of rain,
> Like a child from the womb, like a ghost from the tomb
>   I arise and unbuild again.
>   *(From* The Cloud *by Percy Bysshe Shelley)*

> Blow, winds, and crack your cheeks! rage! blow!
> You cataracts and hurricanoes, spout
> Till you have drench'd our steeples, drown'd the cocks!
> You sulphurous and thought-executing fires,
> Vaunt-couriers to oak-cleaving thunderbolts,
> Singe my white head! And thou, all-shaking thunder,
> Smite flat the thick rotundity o' the world!
> Crack nature's moulds, an germens spill at once,
> That make ingrateful man!
>   *(From William Shakespeare's* King Lear*)*

This book is a collection of review-type chapters that covers the broad research field of large-scale disasters, particularly their prediction, prevention, control, and mitigation. Both natural and manmade disasters are considered. The seed for the project was a meeting organized by the book's editor, the U.S.–Egypt Workshop on Predictive Methodologies for Global Weather-Related Disasters, held in Cairo, Egypt, 13–15 March 2006. Sponsored by the U.S. State Department and its National Science Foundation, the meeting organizers invited fifty American and Egyptian scientists, engineers, meteorologists, and medical personnel. Thirty formal presentations were made, and plenty of both formal and informal discussions were carried out. The three-day conference concluded with two panel discussions. Despite its more limited title, the workshop's scope expanded considerably beyond predictive methodologies for weather-related disasters to include other types of disasters and their prediction, control, and management. This book reflects that expansion.

Although the book covers considerable territories, breadth does not come at the expense of depth. *Large-Scale Disasters: Prediction, Control, and Mitigation* ties together the disparate topics encompassed by its title, and a few of those topics are covered in greater detail. The extreme event could be natural, manmade, or a combination of the two. Examples of naturally occurring disasters include earthquakes; wildfires; pandemics; volcano eruptions; mudslides; floods; droughts; and extreme weather phenomena, such as ice ages, hurricanes, tornadoes, and sandstorms. Human's folly, meanness, mismanagement, gluttony, or unchecked consumption of resources may cause war, energy crises, fire, global warming, famine, air/water pollution, urban sprawl, desertification, bus/train/airplane/ship accidents, or terrorist acts. The book attempts to establish a common framework for predicting, controlling, and managing manmade and natural disasters, thus delivering a more integrated review of a coherent subject, in contrast to a mere collection of disparate articles around a loose theme.

The laws of nature, reflected in the science portion of any particular chapter, and even crisis management, reflected in the art portion, should be the same, or at least quite similar, regardless of where or what type of disaster strikes. Humanity should benefit from both the science and the art of predicting, controlling, and managing large-scale disasters, as extensively and thoroughly discussed in the book.

This book is timely and will hopefully be favorably received by professionals around the globe. The last *annus horribilis*, in particular, has shown the importance of being prepared for large-scale disasters and how the world can get together to help alleviate the resulting pain and suffering. In its own small way, the book will better prepare scientists, engineers, first responders, emergency room professionals and other health care providers, and, above all, political leaders to deal with manmade and natural disasters.

The book is intended for engineers, physicists, first responders, physicians, educators, and graduate students at universities and research laboratories in government and industry who have an interest in the prediction, prevention, or mitigation of large-scale disasters. The different chapters here are written in a pedagogic style and are designed to attract newcomers to the field.

Large-Scale Disasters: Prediction, Control, and Mitigation is organized into twenty-one chapters as follows. Following a brief introduction, the art and science of large-scale disasters are broadly outlined, including a discussion of a proposed disaster metric. Chapter 3 discusses multiscale modeling for large-scale disasters, and Chapter 4 focuses on the root causes of the same. Issues in disaster relief logistics, medical response, and health care capacity are then covered in Chapters 5 to 7. Chapters 8 to 10 discuss global warming, energy crisis, seawater irrigation, and anthropogenic aerosol-related hazards. Chapter 11 is devoted to tsunamis. Chapter 12 is concerned with the fundamentals of intermediate-scale dynamics of the upper troposphere and stratosphere, and Chapter 13 briefly covers coupled weather–chemistry modeling. Chapters 14 to 17 focus on climate prediction, climate change, impact on precipitation, and arid lands. Chapters 18 and 19, respectively, discuss the history and the present of numerical weather predictions. Finally, Chapters 20 and 21 introduce the International Charter and weather satellite measurements.

I want to thank my colleague at Virginia Commonwealth University, Professor Ramana M. Pidaparti, who conceived the original idea of the workshop. My sincere gratitude is extended to Dr. Basman El Hadidi of Cairo University, who worked tirelessly for many months and well beyond the call of duty to ensure the success of the meeting. Professor Atef O. Sherif, chairman of the Egyptian National Authority for Remote Sensing and Space Sciences, hosted the meeting in Cairo and invited the Egyptian speakers. Ms. Joan Mahoney, the coordinator of the Science and Technology Program within the American Embassy in Cairo, attended every presentation. Her calm demeanor, even in the face of chaos, crisis, and potential disaster, was an inspiration to us all. Mr. John P. Desrocher,

Counselor for Economic and Political Affairs in the American Embassy, kindly offered perceptive remarks during the opening ceremony. My sincere gratitude is extended to all who attended the meeting and made it a resounding success. I am indebted to the other thirty-one contributing authors for their dedication to this endeavor and selfless, generous giving of their time with no material reward other than the knowledge that their hard work may one day make the difference in someone else's life. Special thanks to the authors of Chapters 2, 3, 10, 11, and 20, and their respective institutions—Virginia Commonwealth University, Center for Earth Observing and Space Research at George Mason University, Arizona State University, University of Calgary, Canadian Space Agency, and Environment Canada—for subsidizing the color production throughout the entire book. Last but not least, I acknowledge the financial support of the U.S. National Science Foundation (grant NSF-OISE 0541963) and its capable senior program manager, Dr. Osman A. Shinaishin.

Mohamed Gad-el-Hak
Richmond, Virginia

About the editor

Mohamed Gad-el-Hak received his B.Sc. (summa cum laude) in mechanical engineering from Ain Shams University in 1966 and his Ph.D. in fluid mechanics from The Johns Hopkins University in 1973, where he worked with Professor Stanley Corrsin. Gad-el-Hak has since taught and conducted research at the University of Southern California, University of Virginia, University of Notre Dame, Institut National Polytechnique de Grenoble, Université de Poitiers, Friedrich-Alexander-Universität Erlangen-Nürnberg, Technische Universität München, and Technische Universität Berlin, and has lectured extensively at seminars in the United States and overseas. Dr. Gad-el-Hak is currently the Inez Caudill Eminent Professor of Biomedical Engineering and Chair of Mechanical Engineering at Virginia Commonwealth University in Richmond. Prior to his Notre Dame appointment as Professor of Aerospace and Mechanical Engineering, Gad-el-Hak was Senior Research Scientist and Program Manager at Flow Research Company in Seattle, Washington, where he managed a variety of aerodynamic and hydrodynamic research projects.

Professor Gad-el-Hak is world renowned for advancing several novel diagnostic tools for turbulent flows, including the laser-induced fluorescence technique for flow visualization; for discovering the efficient mechanism via which a turbulent region rapidly grows by destabilizing a surrounding laminar flow; for conducting the seminal experiments that detailed the fluid–compliant surface interactions in turbulent boundary layers; for introducing the concept of targeted control to achieve drag reduction, lift enhancement, and mixing augmentation in wall-bounded flows; and for developing a novel viscous pump suited for microelectromechanical systems applications. Gad-el-Hak's work on Reynolds number effects in turbulent boundary layers, published in 1994, marked a significant paradigm shift in the subject. His 1999 paper on the fluid mechanics of microdevices established the fledgling field on firm physical grounds and is one of the most cited articles of the 1990s.

Gad-el-Hak holds two patents: one for a drag-reducing method for airplanes and underwater vehicles and the other for a lift control device for delta wings. Dr. Gad-el-Hak has published more than 460 articles; authored/edited 18 books and conference proceedings;

and presented 260 invited lectures in the basic and applied research areas of isotropic turbulence, boundary layer flows, stratified flows, fluid–structure interactions, compliant coatings, unsteady aerodynamics, biological flows, non-Newtonian fluids, hard and soft computing including genetic algorithms, reactive flow control, and microelectromechanical systems. Gad-el-Hak's papers have been cited more than 1,500 times in the technical literature. He is the author of the book *Flow Control: Passive, Active, and Reactive Flow Management* and editor of the books *Frontiers in Experimental Fluid Mechanics, Advances in Fluid Mechanics Measurements, Flow Control: Fundamentals and Practices, The MEMS Handbook* (first and second editions), *Transition and Turbulence Control*, and *Large-Scale Disasters: Prediction, Control, and Mitigation*.

Professor Gad-el-Hak is a Fellow of the American Academy of Mechanics, a Fellow and life member of the American Physical Society, a Fellow of the American Society of Mechanical Engineers, an associate Fellow of the American Institute of Aeronautics and Astronautics, and a member of the European Mechanics Society. He has recently been inducted as an eminent engineer in Tau Beta Pi, an honorary member in Sigma Gamma Tau and Pi Tau Sigma, and a member-at-large in Sigma Xi. From 1988 to 1991, Dr. Gad-el-Hak served as Associate Editor of the *AIAA Journal*. He is currently serving as Editor-in-Chief of *e-MicroNano.com*; Associate Editor of *Applied Mechanics Reviews* and *e-Fluids*; editorial advisor to the *Bulletin of the Polish Academy of Sciences*; and Contributing Editor of Springer-Verlag's *Lecture Notes in Engineering* and *Lecture Notes in Physics*, McGraw-Hill's *Year Book of Science and Technology* and *Encyclopedia of Science & Technology*, and CRC Press's *Mechanical Engineering Series*.

Dr. Gad-el-Hak serves as consultant to the governments of Egypt, France, Germany, Italy, Poland, Singapore, Sweden, and the United States; the United Nations; and numerous industrial organizations. Professor Gad-el-Hak has been a member of several advisory panels for the U.S. Department of Defense, the National Aeronautics and Space Administration, and the National Science Foundation. During the 1991/1992 academic year, he was a visiting professor at Institut de Mécanique de Grenoble, France. During the summers of 1993, 1994, and 1997, Dr. Gad-el-Hak was, respectively, a distinguished faculty Fellow at Naval Undersea Warfare Center, Newport, Rhode Island; a visiting exceptional professor at Université de Poitiers, France; and a Gastwissenschaftler (guest scientist) at Forschungszentrum Rossendorf, Dresden, Germany. In 1998, Professor Gad-el-Hak was named the fourteenth ASME Freeman Scholar. In 1999, Gad-el-Hak was awarded the prestigious Alexander von Humboldt Prize—Germany's highest research award for senior U.S. scientists and scholars in all disciplines—as well as the Japanese Government Research Award for Foreign Scholars. In 2002, Gad-el-Hak was named ASME Distinguished Lecturer, as well as inducted into The Johns Hopkins University Society of Scholars.

List of contributors

Dr. Nezih Altay
Robins School of Business
University of Richmond
Richmond, VA 23173-0001
U.S.A.
E-mail: naltay@richmond.edu

Dr. Alexander Braun
Department of Geomatics Engineering
Schulich School of Engineering
University of Calgary
Calgary, Alberta T2N 1N4
CANADA
E-mail: braun@ucalgary.ca

Dr. Gilbert Brunet
Environment Canada
2121 Trans-Canada Highway
Dorval, Quebec H9P 1J3
CANADA
E-mail: Gilbert.Brunet@ec.gc.ca

Mr. Dennis M. Bushnell
NASA Langley Research Center
Hampton, VA 23681-0001
U.S.A.
E-mail: d.m.bushnell@larc.nasa.gov

Dr. Aiguo Dai
National Center for Atmospheric Research
P.O. Box 3000
Boulder, CO 80307-3000
U.S.A.
E-mail: adai@ucar.edu

Dr. Jimy Dudhia
National Center for Atmospheric Research
P.O. Box 3000
Boulder, CO 80307-3000
U.S.A.
E-mail: dudhia@ucar.edu

Dr. Hesham El-Askary
Department of Environmental Studies
Faculty of Science
Alexandria University
Alexandria, EGYPT
E-mail: helaskar@gmu.edu

Dr. Jehan Elkholy
Department of Anesthesia
Faculty of Medicine
Cairo University
Cairo, EGYPT
E-mail: jehanelkholy04@yahoo.com

Professor Harindra J. S. (Joe) Fernando
Department of Mechanical and Aerospace Engineering
Arizona State University
Tempe, Arizona 85287-6106
U.S.A.
E-mail: j.fernando@asu.edu

Professor Mohamed Gad-el-Hak
Department of Mechanical Engineering
Virginia Commonwealth University
Richmond, VA 23284-3015
U.S.A.
E-mail: gadelhak@vcu.edu

List of contributors

Professor Mostafa Gad-el-Hak
Department of Anesthesia
Faculty of Medicine
Cairo University
Cairo, EGYPT
E-mail: m_gadelhak@hotmail.com

Dr. Ranjit Galappatti
25/4 Barnes Place
Colombo 7
SRI LANKA
E-mail: ranjitgalappatti@yahoo.co.uk

Professor Ahmed F. Ghoniem
Department of Mechanical Engineering
Massachusetts Institute of Technology
Cambridge, MA 02139-4301
U.S.A.
E-mail: ghoniem@mit.edu

Professor Filippo Giorgi
International Centre for Theoretical
 Physics
Strada Costiera 11
34014 Trieste
ITALY
E-mail: giorgi@ictp.it

Dr. Georg A. Grell
National Oceanic and Atmospheric
 Administration
325 Broadway Street
Boulder, CO 80305-3337
U.S.A.
E-mail: Georg.A.Grell@noaa.gov

Professor Menas Kafatos
School of Computational Sciences
George Mason University
Fairfax, VA 22030-4444
U.S.A.
E-mail: mkafatos@gmu.edu

Dr. Ilan Kelman
Center for Capacity Building
National Center for Atmospheric Research
P.O. Box 3000
Boulder, CO 80307-3000
U.S.A.
E-Mail: ilan_kelman@hotmail.com

Dr. Ahmed Mahmood
Canadian Space Agency
6767 Route de l'Aéroport
Longueuil, Quebec J3Y 8Y9
CANADA
E-mail: Ahmed.Mahmood@space.gc.ca

Dr. Jeremy Pal
International Centre for Theoretical
 Physics
Strada Costiera 11
34014 Trieste
ITALY
E-mail: jpal@ictp.it

Professor Ramana M. Pidaparti
Department of Mechanical Engineering
Virginia Commonwealth University
Richmond, VA 23284-3015
U.S.A.
E-mail: rmpidaparti@vcu.edu

Dr. Janusz Pudykiewicz
Environment Canada
2121 Trans-Canada Highway
Dorval, Quebec H9P 1J3
CANADA
E-mail: Janusz.Pudykiewicz@ec.gc.ca

Professor Atef M. Radwan
Department of Anesthesia
Faculty of Medicine
Zagazig University
Zagazig, Sharkia Governorate
EGYPT
E-mail: a_m_radwan@yahoo.com

Dr. Roy Rasmussen
National Center for Atmospheric
 Research
P.O. Box 3000
Boulder, CO 80307-3000
U.S.A.
E-mail: rasmus@ucar.edu

Professor James J. Riley
Department of Mechanical Engineering
University of Washington
Seattle, WA 98195-2600
U.S.A.
E-mail: rileyj@u.washington.edu

Dr. Janaka Ruwanpura
Department of Civil Engineering
Schulich School of Engineering
University of Calgary
Calgary, Alberta T2N 1N4
CANADA
E-mail: janaka@ucalgary.ca

Professor Ramalingam Saravanan
Department of Atmospheric Sciences
Texas A&M University
College Station, TX 77843-3150
U.S.A.
E-mail: sarava@tamu.edu

Dr. Mohammed Shokr
Environment Canada
4905 Dufferin Street
Toronto, Ontario M3H 5T4
CANADA
E-mail: mohammed.shokr@ec.gc.ca

Professor William L. Smith
Center for Atmospheric Sciences
Hampton University
Hampton, VA 23668-0001
U.S.A.
E-mail: bill.smith@hamptonu.edu

Dr. Kevin E. Trenberth
National Center for Atmospheric Research
P.O. Box 3000
Boulder, CO 80307-3000
U.S.A.
E-mail: trenbert@ucar.edu

Dr. Thomas T. Warner
Research Applications Laboratory
National Center for Atmospheric Research
P.O. Box 3000
Boulder, CO 80307-3000
U.S.A.
E-mail: warner@ucar.edu

Professor S. Chan Wirasinghe
Department of Civil Engineering
Schulich School of Engineering
University of Calgary
Calgary, Alberta T2N 1N4
CANADA
E-mail: wirasing@ucalgary.ca

Dr. Ashraf S. Zakey
International Centre for Theoretical
  Physics
Strada Costiera 11
34014 Trieste
ITALY
E-mail: azakey@ictp.it

1
Introduction

Mohamed Gad-el-Hak

IT WAS the best of times, it was the worst of times, it was the age of wisdom, it was the age of foolishness, it was the epoch of belief, it was the epoch of incredulity, it was the season of Light, it was the season of Darkness, it was the spring of hope, it was the winter of despair, we had everything before us, we had nothing before us, we were all going direct to Heaven, we were all going direct the other way—in short, the period was so far like the present period, that some of its noisiest authorities insisted on its being received, for good or for evil, in the superlative degree of comparison only.

(From A Tale of Two Cities *by Charles Dickens)*

Rumble thy bellyful! Spit, fire! spout, rain!
Nor rain, wind, thunder, fire, are my daughters:
I tax not you, you elements, with unkindness;
I never gave you kingdom, call'd you children,
You owe me no subscription: then let fall
Your horrible pleasure: here I stand, your slave,
A poor, infirm, weak, and despised old man:
But yet I call you servile ministers,
That have with two pernicious daughters join'd
Your high engender'd battles 'gainst a head
So old and white as this. O! O! 'tis foul!

(From William Shakespeare's King Lear*)*

This book is a collection of review-type chapters that cover the broad research field of large-scale disasters, particularly their prediction, prevention, control, and mitigation. Both natural and manmade disasters are covered. The seed for the project is the *U.S.–Egypt Workshop on Predictive Methodologies for Global Weather-Related Disasters*, held in Cairo, Egypt, 13–15 March 2006. Sponsored by the U.S. State Department and its National Science Foundation, the meeting organizers invited fifty American and Egyptian scientists, engineers, meteorologists, and medical personnel. Thirty formal presentations were made, and plenty of both formal and informal discussions were carried out. The 3-day conference concluded with two panel discussions, and its proceedings have been subsequently published (Gad-el-Hak, 2006). Despite its more limited title, the workshop's scope expanded considerably beyond predictive methodologies for weather-related disasters to include other types of natural and manmade disasters and their prediction, control, and management. This book reflects that expansion.

1.1 What is a large-scale disaster?

There is no absolute answer to this question. The mild injury of one person may be perceived as catastrophic by that person or by his or her loved ones. What we consider herein, however, is the adverse effects of an event on a community or an ecosystem. What makes a large-scale disaster is the number of people affected by it and/or the extent of the geographic area involved. Such disaster taxes the resources of local communities and central governments and leads those communities to diverge substantially from their normal social structure. The extreme event could be natural, manmade, or a combination of the two. Examples of naturally occurring disasters include earthquakes, wildfires, pandemics, volcanic eruptions, floods, droughts, and extreme weather phenomena such as ice ages, hurricanes, tornadoes, and sandstorms. Humans' foolishness, folly, cruelty, mismanagement, gluttony, unchecked consumption of resources, or simply sheer misfortune may cause war, energy crisis, fire, global warming, famine, air/water pollution, urban sprawl, desertification, bus/train/airplane/ship accident, or terrorist act.

In addition to the degree or scope of the disaster, there is also the issue of the rapidity of the calamity. Earthquakes, for example, occur over extremely short time periods measured in seconds, whereas air or water pollution and global warming are slowly evolving disasters, their duration measured in years and even decades, although their devastation, over the long term, can be worse than that of a rapid, intense calamity.

For the disaster's magnitude, how large is large? Herein, we propose a metric by which disasters are measured in terms of the number of people affected and/or the extent of the geographic area involved. The suggested scale is nonlinear, logarithmic in fact, much the same as the Richter scale used to judge the severity of an earthquake. The *scope* of a disaster is determined if at least one of two criteria is met, relating to either the number of displaced/injured/killed people or the adversely affected area of the event. We classify disaster types as being of Scopes I to V, according to the following scale:

Scope I	Small disaster	<10 persons	or	<1 km²
Scope II	Medium disaster	10–100 persons	or	1–10 km²
Scope III	Large disaster	100–1,000 persons	or	10–100 km²
Scope IV	Enormous disaster	1,000–10^4 persons	or	100–1,000 km²
Scope V	Gargantuan disaster	>10^4 persons	or	>1,000 km²

We elaborate on this classification in Chapter 2.

1.2 Book contents

There are several recent books on natural disasters on the market, but less available on manmade disasters. Most books are written from either a sociologist's or a tactician's point of view, in contrast to a scientist's viewpoint.[1] A sample is listed in the Bibliography at the end of this chapter. Numerous journals deal at least in part with one aspect or another of large-scale disasters, whether it is technological, scientific, logistical, medical, economical, social, or political. Few archival publications are exclusively dedicated to disasters, for example, *Crisis Response Journal*; *Disaster Prevention and Management*; *Disaster Recovery Journal*; *Disasters: The Journal of Disaster Studies, Policy and Management*; *Journal of Homeland Security and Emergency Management*; *Journal of Emergency Management*; *Journal of Prehospital and Disaster Medicine*; *International Journal of Emergency Management*; *International Journal of Mass Emergencies and Disasters*; and *Natural Hazards*

---

[1] There are a few popular science or high school–level books on disasters—Engelbert et al. (2001) and Allen (2005)—and even fewer more advanced science books— Bunde et al. (2002).

Review. Numerous resources are available on the Internet. A Google search on the word "disaster" yielded 404,000,000 links, most of them of course irrelevant to the study of large-scale disasters. However, three portals and two university research centers, in particular, are worth listing herein because they lead to many useful sites in a well-organized fashion:

www.disastercenter.com/;
www.disaster.net/;
www.gwu.edu/~guides/sciences/crisis.html#use;
www.udel.edu/DRC/; and
www.colorado.edu/hazards/.

This book is divided into twenty-one chapters, covering many aspects of natural and manmade disasters, including their prediction, control, mitigation, and management. Use of scientific principles to improve prediction is emphasized. Following this introduction, the art and science of large-scale disasters are broadly described, including elaborating on the disaster classification scheme just introduced. Chapter 3 discusses multiscale modeling for large-scale disasters, and Chapter 4 focuses on the root causes of the same. Issues in disaster relief logistics, medical response, and health care capacity are then covered in Chapters 5 to 7. Chapters 8 to 10 discuss global warming, energy crisis, seawater irrigation, and anthropogenic aerosol-related hazards. Chapter 11 is devoted to tsunamis. Chapter 12 is concerned with the fundamentals of intermediate-scale dynamics of the upper troposphere and stratosphere, and Chapter 13 briefly covers coupled weather–chemistry modeling. Chapters 14 to 17 focus on climate prediction, climate change, impact on precipitation, and arid lands. Chapters 18 and 19, respectively, discuss the history and the present of numerical weather predictions. Finally, Chapters 20 and 21 introduce the International Charter and weather satellite measurements. *Large-Scale Disasters: Prediction, Control, and Mitigation* ties together the disparate topics encompassed by its title, and attempts to establish a common framework for predicting, controlling, and managing manmade and natural disasters, thus delivering a more integrated review of a coherent subject, in contrast to a mere collection of disparate chapters around a loose theme.

References

Allen, J. (2005) *Predicting Natural Disasters*, Thomson Gale, Farmington Hills, Michigan.
Bunde, A., Kropp, J., and Schellnhuber, H. J. (2002) *The Science of Disasters: Climate Disruptions, Heart Attacks, and Market Crashes*, Springer, Berlin, Germany.
Engelbert, P., Deschenes, B., Nagel, R., and Sawinski, D. M. (2001) *Dangerous Planet—The Science of Natural Disasters*, Thomson Gale, Farmington Hills, Michigan.
Gad-el-Hak, M. (2006) *Proceedings of the U.S.–Egypt Workshop on Predictive Methodologies for Global Weather-Related Disasters*, Cairo, Egypt, 13–15 March 2006, CD Publication, Virginia Commonwealth University, Richmond, Virginia.

Bibliography

Abbott, P. (2005) *Natural Disaster*, McGraw-Hill, San Diego, California.
Alexander, D. (1993) *Natural Disaster*, Springer, Berlin, Germany.
Alexander, D. (2000) *Confronting Catastrophe: New Perspectives on Natural Disasters*, Oxford University Press, London, United Kingdom.
Bankoo, G., Frerks, G., and Hilhorst, D. (2004) *Mapping Vulnerability: Disasters, Development, and People*, Earthscan Publications, Gateshead, United Kingdom.

Burby, R. J. (1998) *Cooperating with Nature: Confronting Natural Hazards with Land-Use Planning for Sustainable Communities*, John Henry Press, Washington, DC.

Childs, D. R., and Dietrich, S. (2002) *Contingency Planning and Disaster Recovery: A Small Business Guide*, Wiley, New York, New York.

Cooper, C., and Block, R. (2006) *Disaster: Hurricane Katrina and the Failure of Homeland Security*, Times Books, New York, New York.

Cutter, S. L. (2001) *American Hazardscapes: The Regionalization of Hazards and Disasters*, John Henry Press, Washington, DC.

de Boer, J., and van Remmen, J. (2003) *Order in Chaos: Modelling Medical Disaster Management Using Emergo Metrics*, LiberChem Publication Solution, Culemborg, The Netherlands.

der Heide, E. A. (1989) *Disaster Response: Principles of Preparation and Coordination*, C.V. Mosby, St. Louis, Missouri.

Dilley, M., Chen, R. S., Deichmann, U., Lerner-Lam, A. L., and Arnold, M. (2005) *Natural Disaster Hotspots: A Global Risk Analysis*, World Bank Publications, Washington, DC.

Fischer, H. W., III (1998) *Response to Disaster: Fact Versus Fiction & Its Perpetuation: The Sociology of Disaster*, second edition, University Press of America, Lanham, Maryland.

Gist, R., and Lubin, B. (1999) *Response to Disaster: Psychosocial, Community, and Ecological Approaches*, Bruner/Mazel, Philadelphia, Pennsylvania.

Kunreuther, H., and Roth, R. J., Sr. (1998) *The Status and Role of Insurance Against Natural Disasters in the United States*, John Henry Press, Washington, DC.

McKee, K., and Guthridge, L. (2006) *Leading People Through Disasters: An Action Guide*, Berrett-Koehler, San Francisco, California.

Mileti, D. S. (1999) *Disasters by Design: A Reassessment of Natural Hazards in the United States*, Joseph Henry Press, Washington, DC.

Office of the United Nations Disaster Recovery (1991) *Mitigating Natural Disasters: Phenomena, Effects and Options*, Sales No. E.90.Iii.M.1, United Nations, New York, New York.

Olasky, M. (2006) *The Politics of Disaster: Katrina, Big Government, and a New Strategy for Future Crises*, W Publishing Group, Nashville, Tennessee.

Pelling, M. (2003) *The Vulnerability of Cities: Natural Disaster and Social Resilience*, Earthscan Publications, Gateshead, United Kingdom.

Pickett, S. T. A., and White, P. S. (1985) *The Ecology of Natural Disturbance and Patch Dynamics*, Academic Press, San Diego, California.

Posner, R. A. (2004) *Catastrophe: Risk and Response*, Oxford University Press, Oxford, United Kingdom.

Quarantelli, E. L. (1998) *What Is a Disaster? Perspectives on the Question*, Routledge, London, United Kingdom.

Smith, K. (1991) *Environmental Hazards: Assessing Risk and Reducing Disaster*, Routledge, London, United Kingdom.

Stein, S., and Wysession, M. (2003) *An Introduction to Seismology, Earthquakes, and Earth Structure*, Blackwell, Boston, Massachusetts.

Steinberg, T. (2000) *Acts of God: The Unnatural History of Natural Disasters in America*, Oxford University Press, London, United Kingdom.

Tierney, K. J., Lindell, M. K., and Perry, R. W. (2001) *Facing the Unexpected: Disaster Preparedness and Response in the United States*, John Henry Press, Washington, DC.

Tobin, G. A. (1997) *Natural Hazards: Explanation and Integration*, Guilford Press, New York, New York.

Vale, L. J., and Campanella, T. J. (2005) *The Resilient City: How Modern Cities Recover from Disaster*, Oxford University Press, London, United Kingdom.

Wallace, M., and Webber, L. (2004) *The Disaster Recovery Handbook*, Amacom, New York, New York.

2
The art and science of large-scale disasters
Mohamed Gad-el-Hak

There was no pause, no pity, no peace, no interval of relenting rest, no measurement of time. Though days and nights circled as regularly as when time was young, and the evening and morning were the first day, other count of time there was none. Hold of it was lost in the raging fever of a nation, as it is in the fever of one patient. Now, breaking the unnatural silence of a whole city, the executioner showed the people the head of the king—and now, it seemed almost in the same breath, the head of his fair wife which had had eight weary months of imprisoned widowhood and misery, to turn it grey.
(From A Tale of Two Cities *by Charles Dickens)*

> Alas, sir, are you here? things that love night
> Love not such nights as these; the wrathful skies
> Gallow the very wanderers of the dark,
> And make them keep their caves: since I was man,
> Such sheets of fire, such bursts of horrid thunder,
> Such groans of roaring wind and rain, I never
> Remember to have heard: man's nature cannot carry
> The affliction nor the fear.
> *(From William Shakespeare's* King Lear*)*

The subject of large-scale disasters is broadly introduced in this chapter, leaving much of the details to subsequent chapters. Both the art and the science of predicting, preventing, and mitigating natural and manmade disasters are discussed. The laws of nature govern the evolution of any disaster. In some cases, such as weather-related disasters, those first principles laws could be written in the form of field equations, but exact solutions of these often nonlinear differential equations are impossible to obtain, particulary for turbulent flows, and heuristic models together with intensive use of supercomputers are necessary to proceed to a reasonably accurate forecast. In other cases, such as earthquakes, the precise laws are not even known, and prediction becomes more or less a black art. Management of any type of disaster is more art than science. Nevertheless, much can be done to alleviate the resulting pain and suffering.

2.1 Are disasters a modern curse?

Although it appears that way when the past few years are considered, large-scale disasters have been with us since *Homo sapiens* set foot on this third planet from the Sun. Frequent disasters struck the Earth even before then, as far back as the time of its formation around 4.5 billion years ago. In fact, the geological Earth that we know today is believed to be the result of agglomeration of the so-called planetesimals and subsequent impacts of bodies of similar mass (Huppert, 2000). The planet was left molten after each giant impact, and its outer crust was formed upon radiative cooling to space. Those were the "good" disasters perhaps. On the bad side, there have been several mass extinctions throughout the Earth's history. The dinosaurs, along with about 70% of all species existing at the time, became extinct because a large meteorite struck the Earth 65 million years ago and the resulting airborne dust partially blocked the Sun, thus making it impossible for cold-blooded animals to survive. However, if we concern ourselves with our own warm-blooded species, then starting 200,000 years ago, ice ages, famines, infections, and attacks from rival groups and animals were constant reminders of humans' vulnerability. On average, there are about three large-scale disasters that strike the Earth every day, but only a few of these natural or manmade calamities make it to the news. Humans have survived because we were programmed to do so. We return to this point in Section 2.7.

This book is a collection of review-type chapters that cover the broad research field of large-scale disasters, particularly their prediction, prevention, control, and mitigation. Technological, scientific, medical, logistical, sociological, economical, and political aspects of both natural and manmade disasters are covered, some aspects to greater extent than others. The seed for the project is the *U.S.–Egypt Workshop on Predictive Methodologies for Global Weather-Related Disasters*, held in Cairo, Egypt, 13–15 March 2006. Sponsored by the U.S. State Department and its National Science Foundation, the meeting organizers invited fifty American and Egyptian scientists, engineers, meteorologists, and medical personnel. Thirty formal presentations were made and plenty of both formal and informal discussions were carried out. The 3-day conference concluded with two panel discussions, and its proceedings have been subsequently published (Gad-el-Hak, 2006a). Despite its more limited title, the workshop's scope expanded considerably beyond predictive methodologies for weather-related disasters to include other types of natural and manmade disasters and their prediction, control, and management. This book reflects that expansion. The subject of large-scale disasters is broadly introduced in this chapter, leaving many of the details to the subsequent chapters of this book, each focusing on a narrow aspect of the bigger scope. Both the art and the science of predicting, preventing, and mitigating natural and manmade disasters are discussed. We begin by proposing a metric by which disasters are sized in terms of the number of people affected and/or the extent of the geographic area involved.

2.2 Disaster scope

There is no easy answer to the question of whether a particular disaster is large or small. The mild injury of one person may be perceived as catastrophic by that person or by his or her loved ones. What we consider herein, however, is the adverse effects of an event on a community or an ecosystem. What makes a disaster a large-scale one is the number of people affected by it and/or the extent of the geographic area involved. Such disaster taxes the resources of local communities and central governments. Under the weight of a large-scale disaster, a community diverges substantially from its normal social structure. Return to *normalcy* is typically a slow process that depends on the severity, but not the duration, of the antecedent calamity as well as the resources and efficiency of the recovery process.

The extreme event could be natural, manmade, or a combination of the two in the sense of a natural disaster made worse by human's past actions. Examples of naturally occurring disasters include earthquakes, wildfires, pandemics, volcanic eruptions, mudslides, floods, droughts, and extreme weather phenomena such as ice ages, hurricanes, tornadoes, and sandstorms. Humans' foolishness, folly, meanness, mismanagement, gluttony, unchecked consumption of resources, or simply sheer misfortune may cause war, energy crisis, economic collapse of a nation or corporation, market crash, fire, global warming, famine, air/water pollution, urban sprawl, desertification, deforestation, bus/train/airplane/ship accident, oil slick, or terrorist act. Citizens suffering under the tyranny of a despot or a dictator can also be considered a disaster, and, of course, genocide, ethnic cleansing, and other types of mass murder are gargantuan disasters that often test the belief in our own humanity. Although technological advances exponentially increased human prosperity, they also provided humans with more destructive power. Manmade disasters have caused the death of at least 200 million people during the twentieth century, a cruel age without equal in the history of man (de Boer & van Remmen, 2003).

In addition to the degree or scope of a disaster, there is also the issue of the rapidity of the calamity. Earthquakes, for example, occur over extremely short time periods measured in seconds, whereas anthropogenic catastrophes such as global warming and air and water pollution are often slowly evolving disasters, their duration measured in years and even decades or centuries, although their devastation, over the long term, can be worse than that of a rapid, intense calamity (McFedries, 2006). The painful, slow death of a cancer patient who contracted the dreadful disease as a result of pollution is just as tragic as the split-second demise of a human at the hands of a crazed suicide bomber. The latter type of disaster makes the news, but the former does not. This is quite unsettling because the death of many spread over years goes largely unnoticed. The fact that 100 persons die in a week in a particular country as a result of starvation is not a typical news story. However, 100 humans perishing in an airplane crash will make CNN all day.

For the disaster's magnitude, how large is large? Much the same as is done to individually size hurricanes, tornadoes, earthquakes, and, very recently, winter storms, we propose herein a universal metric by which all types of disaster are sized in terms of the number of people affected and/or the extent of the geographic area involved. This quantitative scale applies to both natural and manmade disasters. The suggested scale is nonlinear, logarithmic in fact, much the same as the Richter scale used to measure the severity of an earthquake. Thus, moving up the scale requires an order of magnitude increase in the severity of the disaster as it adversely affects people or an ecosystem. Note that a disaster may affect only a geographic area without any direct and immediate impact on humans. For example, a wildfire in an uninhabited forest may have long-term adverse effects on the local and global ecosystem, although no human is immediately killed, injured, or dislocated as a result of the event.

The *scope* of a disaster is determined if at least one of two criteria is met, relating to either the number of displaced/tormented/injured/killed people or the adversely affected area of the event. We classify disaster types being of Scope I to V, according to the scale depicted in Table 2.1.

Table 2.1. *Disaster scope according to number of casualties and/or geographic area affected*

Scope I	Small disaster	<10 persons	or	<1 km^2
Scope II	Medium disaster	10–100 persons	or	1–10 km^2
Scope III	Large disaster	100–1,000 persons	or	10–100 km^2
Scope IV	Enormous disaster	1000–10^4 persons	or	100–1,000 km^2
Scope V	Gargantuan disaster	>10^4 persons	or	>1,000 km^2

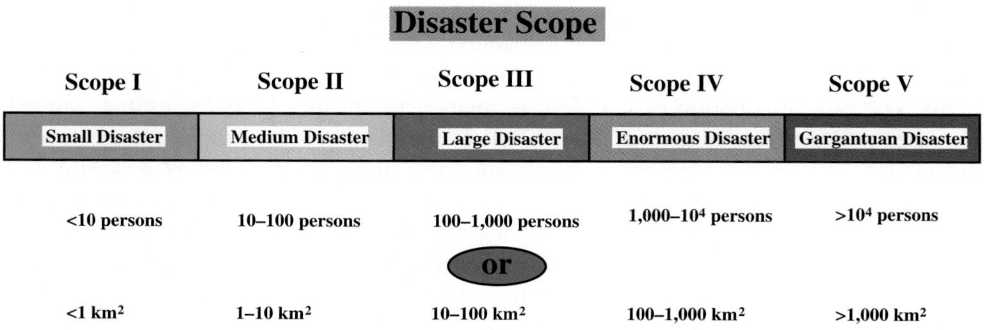

Disaster Scope

Scope I	Scope II	Scope III	Scope IV	Scope V
Small Disaster	Medium Disaster	Large Disaster	Enormous Disaster	Gargantuan Disaster
<10 persons	10–100 persons	100–1,000 persons	1,000–10^4 persons	>10^4 persons

or

<1 km²	1–10 km²	10–100 km²	100–1,000 km²	>1,000 km²

Figure 2.1 Classification of disaster severity.

These classifications are pictorially illustrated in Figure 2.1. For example, if 70 persons were injured as a result of a wildfire that covered 20 km², this would be considered Scope III, large disaster (the larger of the two categories II and III). However, if 70 persons were killed as a result of a wildfire that covered 2 km², this would be considered Scope II, medium disaster. An unusual example, at least in the sense of even attempting to classify it, is the close to 80 million citizens of Egypt (area slightly larger than 1 million km²) who have been tormented for more than a half-century[1] by a virtual police state. This manmade cataclysm is readily stigmatized by the highest classification, Scope V, gargantuan disaster.

The quantitative metric introduced herein is contrasted to the conceptual scale devised by Fischer (2003a, 2003b), which is based on the degree of social disruption resulting from an actual or potential disaster. His ten disaster categories are based on the scale, duration, and scope of disruption and adjustment of a normal social structure, but those categories are purely qualitative. For example, Disaster Category (DC)-3 is indicated if the event partially strikes a small town (major scale, major duration, partial scope), whereas DC-8 is reserved for a calamity massively striking a large city (major scale, major duration, major scope).

The primary advantage of having a universal classification scheme such as the one proposed herein is that it gives officials a quantitative measure of the magnitude of the disaster so that proper response can be mobilized and adjusted as warranted. The metric suggested applies to *all* types of disaster. It puts them on a common scale, which is more informative than the variety of scales currently used for different disaster types; the Saffir-Simpson scale for hurricanes, the Fujita scale for tornadoes, the Richter scale for earthquakes, and the recently introduced Northeast Snowfall Impact Scale (notable, significant, major, crippling, extreme) for the winter storms that occasionally strike the northeastern region of the United States. Of course, the individual scales also have their utility; for example, knowing the range of wind speeds in a hurricane as provided by the Saffir-Simpson scale is a crucial piece of information to complement the number of casualties the proposed scale supplies. In fact, a prediction of wind speed allows estimation of potential damage to people and property. The proposed metric also applies to disasters such as terrorist acts or droughts, where no quantitative scales are currently available to measure their severity.

In formulating all scales, including the proposed one, a certain degree of arbitrariness is unavoidable. In other words, none of the scales is totally objective. The range of 10 to 100 persons associated with a Scope II disaster, for example, could very well be 20

---

[1] Of course, the number of residents of Egypt was far less than 80 million when the disaster commenced in 1952.

to 80, or some other range. What is important is the relative comparison among various disaster degrees; a Scope IV disaster causes an order of magnitude more damage than a Scope III disaster, and so on. One could arbitrarily continue beyond five categories, always increasing the influenced number of people and geographic area by an order of magnitude, but it seems that any calamity adversely affecting more than 10,000 persons or 1,000 km^2 is so catastrophic that a single Scope V is adequate to classify it as a gargantuan disaster. The book *Catastrophe* is devoted to analyzing the risk of and response to unimaginable but not impossible calamities that have the potential of wiping out the human race (Posner, 2004). Curiously, its author, Richard A. Posner, is a judge in the U.S. Seventh Circuit Court of Appeals.

In the case of certain disasters, the scope can be predicted in advance to a certain degree of accuracy; otherwise, the scope can be estimated shortly after the calamity strikes with frequent updates as warranted. The magnitude of the disaster should determine the size of the first-responder contingency to be deployed: which hospitals to mobilize and to what extent; whether the military forces should be involved; what resources, such as food, water, medicine, and shelter, should be stockpiled and delivered to the stricken area; and so on. Predicting the scope should facilitate the subsequent recovery and accelerate the return to normalcy.

2.3 Facets of large-scale disasters

A large-scale disaster is an event that adversely affects a large number of people, devastates a large geographic area, and taxes the resources of local communities and central governments. Although disasters can naturally occur, humans can cause their share of devastation. There is also the possibility of human actions causing a natural disaster to become more damaging than it would otherwise. An example of such an anthropogenic calamity is the intense coral reef mining off the Sri Lankan coast, which removed the sort of natural barrier that could mitigate the force of waves. As a result of such mining, the 2004 Pacific Tsunami devastated Sri Lanka much more than it would have otherwise (Chapter 11). A second example is the soil erosion caused by overgrazing, farming, and deforestation. In April 2006, wind from the Gobi Desert dumped 300,000 tons of sand and dust on Beijing, China. Such gigantic dust tempests—exasperated by soil erosion—blow around the globe, making people sick, killing coral reefs, and melting mountain snow packs continents away. Examples such as this incited the 1995 Nobel laureate and Dutch chemist Paul J. Crutzen to coin the present geological period as *anthropocene* to characterize humanity's adverse effects on global climate and ecology (www.mpch-mainz.mpg.de/ air/anthropocene/).

What could make the best of a bad situation is to be able to predict the disaster's occurrence, location, and severity. This can help prepare for the calamity and evacuating large segments of the population out of harm's way. For certain disaster types, their evolution equations can be formulated. Predictions can then be made to different degrees of success using heuristic models, empirical observations, and giant computers. Once formed, the path and intensity of a hurricane, for example, can be predicted to a reasonable degree of accuracy up to 1 week in the future. This provides sufficient warning to evacuate several medium or large cities in the path of the extreme event. However, smaller-scale severe weather such as tornadoes can only be predicted up to 15 minutes in the future, giving very little window for action. Earthquakes cannot be predicted beyond stating that there is a certain probability of occurrence of a certain magnitude earthquake at a certain geographic location during the next 50 years. Such predictions are almost as useless as stating that the Sun will burn out in a few billion years.

Once disaster strikes, mitigating its adverse effects becomes the primary concern: how to save lives, take care of the survivors' needs, and protect properties from any further damage. Dislocated people need shelter, water, food, and medicine. Both the physical and the mental health of the survivors, as well as relatives of the deceased, can be severely jeopardized. Looting, price gouging, and other law-breaking activities need to be contained, minimized, or eliminated. Hospitals need to prioritize and even ration treatments, especially in the face of the practical fact that the less seriously injured tend to arrive at emergency rooms first, perhaps because they transported themselves there. Roads need to be operable and free of landslides, debris, and traffic jams for the unhindered flow of first responders and supplies to the stricken area, and evacuees and ambulances from the same. This is not always the case, especially if the antecedent disaster damages most if not all roads, as occurred after the 2005 Kashmir Earthquake (Section 2.8.7). Buildings, bridges, and roads need to be rebuilt or repaired, and power, potable water, and sewage need to be restored. Chapters 5 to 7 cover some of the logistical and medical aspects of disasters.

Figure 2.2 depicts the different facets of large-scale disasters. The important thing is to judiciously employ the finite resources available to improve the science of disaster prediction, and to artfully manage the resulting mess to minimize loss of life and property.

2.4 The science of disaster prediction and control

Science can help predict the course of certain types of disaster. When, where, and how intense would a severe weather phenomena strike? Are the weather conditions favorable for extinguishing a particular wildfire? What is the probability of a particular volcano erupting? How about an earthquake striking a population center? How much air and water pollution is going to be caused by the addition of a factory cluster to a community? How would a toxic chemical or biological substance disperse in the atmosphere or in a body of water? Below a certain concentration, certain danger substances are harmless, and "safe" and "dangerous" zones could be established based on the dispersion forecast. The degree of success in answering these and similar questions varies dramatically. Once formed, the course and intensity of a hurricane (tropical cyclone), which typically lasts from inception to dissipation for a few weeks, can be predicted about 1 week in advance. The path of the much smaller and short-lived, albeit more deadly, tornado can be predicted only about 15 minutes in advance, although weather conditions favoring its formation can be predicted a few hours ahead.

Earthquake prediction is far from satisfactory but is seriously attempted nevertheless. The accuracy of predicting volcanic eruptions is somewhere in between those of earthquakes and severe weather. Patanè et al. (2006) report on the ability of scientists' to "see" inside Italy's Mount Etna and forecast its eruption using seismic tomography, a technique similar to that used in computed tomography scans in the medical field. The method yields time photographs of the three-dimensional movement of rocks to detect their internal changes. The success of the technique is in no small part due to the fact that Europe's largest volcano is equipped with a high-quality monitoring system and seismic network, tools that are not readily available for most volcanoes.

Science and technology can also help control the severity of a disaster, but here the achievements to date are much less spectacular than those in the prediction arena. Cloud seeding to avert drought is still far from being a practical tool, but still a notch more rational than the then-Governor of Texas George W. Bush's 1999 call in the midst of a dry period to "pray for rain." Slinging a nuclear device toward an asteroid or a meteor to avert its imminent collision with Earth remains solidly in the realm of science fiction

2.4 The science of disaster prediction and control

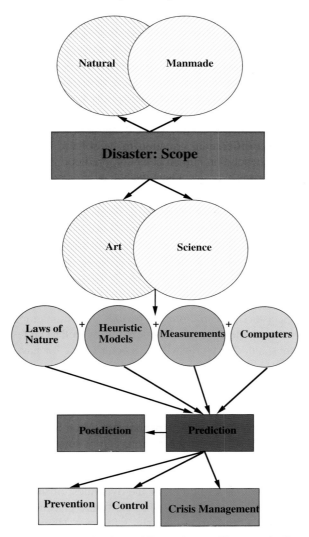

Figure 2.2 Schematic of the different facets of large-scale disasters.

(in the 1998 film *Armageddon*, a Texas-size asteroid was courageously nuked from its interior!). In contrast, employing scientific principles to combat a wildfire is doable, as is the development of scientifically based strategies to reduce air and water pollution; moderate urban sprawl; evacuate a large city; and minimize the probability of accident for air, land, and water vehicles. Structures could be designed to with-stand an earthquake of a given magnitude, wind of a given speed, and so on. Dams could be constructed to moderate the flood–drought cycles of rivers, and levees/dikes could be erected to protect land below sea level from the vagaries of the weather. Storm drains; fire hydrants; fire-retardant materials; sprinkler systems; pollution control; simple hygiene; strict building codes; traffic rules and regulations in air, land, and sea; and many other examples are the measures a society should take to mitigate or even eliminate the adverse effects of certain natural and manmade disasters. Of course, there are limits to what we can do. Although much better fire safety will be achieved if a firehouse is erected, equipped, and manned around every city block, and less earthquake casualties will occur if every structure is built to withstand the strongest possible tremor, the prohibitive cost of such efforts clearly cannot be justified or even afforded.

In contrast to natural disasters, manmade ones are generally somewhat easier to control but more difficult to predict. The war on terrorism is a case in point. Who could predict the behavior of a crazed suicide bomber? A civilized society spends its valuable resources on intelligence gathering, internal security, border control, and selective/mandatory screening to prevent (control) such devious behavior, whose dynamics (i.e., time evolution) obviously cannot be distilled into a differential equation to be solved. However, even in certain disastrous situations that depend on human behavior, predictions can sometimes be made; crowd dynamics being a prime example where the behavior of a crowd in an emergency can to some degree be modeled and anticipated so that adequate escape or evacuation routes can be properly designed (Adamatzky, 2005). Helbing et al. (2002) write on simulation of panic situations and other crowd disasters modeled as nonlinear dynamical systems.

The tragedy of the numerous manmade disasters is that they are all preventable, at least in principle. We cannot prevent a hurricane, at least not yet, but global warming trends could at least be slowed down by using less fossil fuel and seeking alternative energy sources (Chapters 8 and 9). Conflict resolution strategies can be employed between nations to avert wars. Speaking of wars, the Iraqi–American poet Dunya Mikhail, lamenting on the many manmade disasters, calls the present period "The Tsunamical Age." A bit more humanity, common sense, selflessness, and moderation, as well as bit less greed, meanness, selfishness, and zealotry, and the world will be a better place for having fewer manmade disasters.

2.4.1 Modeling the disaster's dynamics

For disasters that involve (fluid) transport phenomena, such as severe weather, fire and release of toxic substance, the governing equations can be formulated subject to some assumptions, the less the better. Modeling is usually in the form of nonlinear partial differential equations with an appropriate number of initial and boundary conditions. Integrating those field equations leads to the time evolution, or the dynamics, of the disaster. In principle, marching from the present (initial conditions) to the future gives the potent predictability of mechanics and ultimately leads to the disaster's forecast. However, the first principles equations are typically impossible to solve analytically, particularly if the fluid flow is turbulent, which unfortunately is the norm for the high Reynolds number flows encountered in the atmosphere and oceans. Furthermore, initial and boundary conditions are required for both analytical and numerical solutions, and massive amounts of data need to be collected to determine those conditions with sufficient resolution and accuracy. Computers are not big enough either, so numerical integration of the instantaneous equations (direct numerical simulations) for high Reynolds number natural flows is computationally prohibitively expensive if not outright impossible at least for now and the foreseeable future. Heuristic modeling then comes to the rescue but at a price. Large-eddy simulations, spectral methods, probability density function models, and the more classical Reynolds stress models are examples of such closure schemes that are not as computationally intensive as direct numerical simulations and not as reliable. This type of second-tier modeling is phenomenological in nature and does not stem from first principles. The more heuristic the modeling is, the less accurate the expected results.

Together with massive ground, sea, and sky data to provide at least in part the initial and boundary conditions, the models are entered into supercomputers that come out with a forecast, whether it is a prediction of a severe thunderstorm that is yet to form, the future path and strength of an existing hurricane, or the impending concentration of a toxic gas that was released in a faraway location some time in the past. For other types of disasters such as earthquakes, the precise laws are not even known mostly because proper constitutive relations are lacking. In addition, deep underground data are difficult to gather to say the least. Predictions in those cases become more or less a black art. The issue of nonintegrability of certain dynamical systems is an additional challenge and opportunity that is revisited in Section 2.4.10.

In the next seven subsections, we focus on the prediction of disasters involving fluid transport. Several subsequent chapters (e.g., Chapters 12–21) revisit this important subject, which has spectacular successes within the past few decades, for example, in being able to predict the weather a few days in advance. The accuracy of today's 5-day forecast is the same as the 3-day and 1.5-day ones in 1976 and 1955, respectively. The 3-day forecast of a hurricane's strike position is accurate to within 100 km, about a 1-hour drive on the highway (Gall & Parsons, 2006). The painstaking advances made in fluid mechanics in general and turbulence research in particular, together with the exponential growth of computer memory and speed, undoubtedly contributed immeasurably to those successes.

The British physicist Lewis Fry Richardson was perhaps the first to make a scientifically based weather forecast. Based on data taken at 7:00 am, 20 May 1910, he made a 6-hour "forecast" that took him 6 weeks to compute using a slide rule. The belated results[2] were totally wrong as well! In his remarkable book, Richardson (1922) wrote, *"Perhaps some day in the dim future it will be possible to advance the computations faster than the weather advances and at a cost less than the saving to mankind due to the information gained. But that is a dream (p. vii of the original 1922 edition)."* We are happy to report that Richardson's dream is one of the few that came true. A generation ago, the next day's weather was hard to predict. Today, the 10-day forecast is available 24/7 on www.weather.com for almost any city in the world. Not very accurate perhaps, but far better than the pioneering Richardson's 6-hour forecast. Chapter 18 discusses this and other aspects of the first 100 years of numerical weather prediction.

The important issue in this section is to precisely state the assumptions needed to write the evolution equations for transport, which are basically statements of the conservation of mass, momentum, and energy, in a certain form. The resulting equations and their eventual analytical or numerical solutions are only valid under those assumptions. This seemingly straightforward fact is often overlooked, and wrong answers readily result when the situation we are trying to model is different from that assumed.

2.4.2 The fundamental transport equations

Each fundamental law of fluid mechanics and heat transfer—conservation of mass, momentum, and energy—are listed first in their *raw form* (i.e., assuming only that the speeds involved are non-relativistic and that the fluid is a continuum). In nonrelativistic situations, mass and energy are conserved separately and are not interchangeable. This is the case for all normal fluid velocities that we deal with in everyday situations—far below the speed of light. The continuum assumption ignores the grainy (microscopic) structure of matter. It implies that the derivatives of the dependent variables exist in some reasonable sense. In other words, local properties such as density and velocity are defined as averages over

---

[2] Actually delayed by a few years due to of World War I and relocation to France. Richardson chose that particular time and date because upper air and other measurements were available to him some years before.

large elements compared with the microscopic structure of the fluid but small enough in comparison with the scale of the macroscopic phenomena to permit the use of differential calculus to describe them. The resulting equations therefore cover a broad range of situations, the exception being flows with spatial scales that are not much larger than the mean distance between the fluid molecules, as, for example, in the case of rarefied gas dynamics, shock waves that are thin relative to the mean free path, or flows in micro- and nanodevices. Thus, at every point in space–time in an inertial (i.e., nonaccelerating/nonrotating), Eulerian frame of reference, the three conservation laws for nonchemically reacting fluids, respectively, read in Cartesian tensor notations

$$\frac{\partial \rho}{\partial t} + \frac{\partial}{\partial x_k}(\rho u_k) = 0 \tag{2.1}$$

$$\rho \left(\frac{\partial u_i}{\partial t} + u_k \frac{\partial u_i}{\partial x_k} \right) = \frac{\partial \Sigma_{ki}}{\partial x_k} + \rho g_i \tag{2.2}$$

$$\rho \left(\frac{\partial e}{\partial t} + u_k \frac{\partial e}{\partial x_k} \right) = -\frac{\partial q_k}{\partial x_k} + \Sigma_{ki} \frac{\partial u_i}{\partial x_k} \tag{2.3}$$

where ρ is the fluid density, u_k is an instantaneous velocity component (u, v, w), Σ_{ki} is the second-order stress tensor (surface force per unit area), g_i is the body force per unit mass, e is the internal energy per unit mass, and q_k is the sum of heat flux vectors due to conduction and radiation. The independent variables are time t, and the three spatial coordinates x_1, x_2, and x_3 or (x, y, z). Finally, the Einstein's summation convention applies to all repeated indices. Gad-el-Hak (2000) provides a succinct derivation of the previous conservation laws for a continuum, nonrelativistic fluid.

2.4.3 Closing the equations

Equations (2.1), (2.2), and (2.3) constitute five differential equations for the seventeen unknowns ρ, u_i, Σ_{ki}, e, and q_k. Absent any body couples, the stress tensor is symmetric, having only six independent components, which reduces the number of unknowns to fourteen. To close the conservation equations, relation between the stress tensor and deformation rate, relation between the heat flux vector and the temperature field, and appropriate equations of state relating the different thermodynamic properties are needed. Thermodynamic equilibrium implies that the macroscopic quantities have sufficient time to adjust to their changing surroundings. In motion, exact thermodynamic equilibrium is impossible because each fluid particle is continuously having volume, momentum, or energy added or removed, and so in fluid dynamics and heat transfer we speak of quasiequilibrium. The second law of thermodynamics imposes a tendency to revert to equilibrium state, and the defining issue here is whether the flow quantities are adjusting fast enough. The reversion rate will be very high if the molecular time and length scales are very small as compared to the corresponding macroscopic flow scales. This will guarantee that numerous molecular collisions will occur in sufficiently short time to equilibrate fluid particles whose properties vary little over distances comparable to the molecular length scales. Gas flows are considered in a state of quasiequilibrium if the Knudsen number—the ratio of the mean free path to a characteristic length of the flow—is less than 0.1. In such flows, the stress is linearly related to the strain rate, and the (conductive) heat flux is linearly related to the temperature gradient. Empirically, common liquids such as water follow the same laws under most flow conditions. Gad-el-Hak (1999) provides extensive discussion of situations in which the quasiequilibrium assumption is violated. These may include gas flows at great altitudes, flows of complex liquids such as long-chain molecules, and even ordinary gas and liquid flows when confined in micro- and nanodevices.

For a Newtonian, isotropic, Fourier,[3] ideal gas, for example, those constitutive relations read

$$\Sigma_{ki} = -p\,\delta_{ki} + \mu\left(\frac{\partial u_i}{\partial x_k} + \frac{\partial u_k}{\partial x_i}\right) + \lambda\left(\frac{\partial u_j}{\partial x_j}\right)\delta_{ki} \tag{2.4}$$

$$q_i = -\kappa\,\frac{\partial T}{\partial x_i} + \text{Heat flux due to radiation} \tag{2.5}$$

$$de = c_v\,dT \quad \text{and} \quad p = \rho\,\mathcal{R}\,T \tag{2.6}$$

where p is the thermodynamic pressure, μ and λ are the first and second coefficients of viscosity, respectively, δ_{ki} is the unit second-order tensor (Kronecker delta), κ is the thermal conductivity, T is the temperature field, c_v is the specific heat at constant volume, and $\mathcal{R}$ is the gas constant. The Stokes' hypothesis relates the first and second coefficients of viscosity, $\lambda + \frac{2}{3}\mu = 0$, although the validity of this assumption has occasionally been questioned (Gad-el-Hak, 1995). With the previous constitutive relations and neglecting radiative heat transfer,[4] Equations (2.1), (2.2), and (2.3), respectively, read

$$\frac{\partial \rho}{\partial t} + \frac{\partial}{\partial x_k}(\rho\,u_k) = 0 \tag{2.7}$$

$$\rho\left(\frac{\partial u_i}{\partial t} + u_k\frac{\partial u_i}{\partial x_k}\right) = -\frac{\partial p}{\partial x_i} + \rho g_i$$
$$+ \frac{\partial}{\partial x_k}\left[\mu\left(\frac{\partial u_i}{\partial x_k} + \frac{\partial u_k}{\partial x_i}\right) + \delta_{ki}\,\lambda\,\frac{\partial u_j}{\partial x_j}\right] \tag{2.8}$$

$$\rho c_v\left(\frac{\partial T}{\partial t} + u_k\frac{\partial T}{\partial x_k}\right) = \frac{\partial}{\partial x_k}\left(\kappa\,\frac{\partial T}{\partial x_k}\right) - p\,\frac{\partial u_k}{\partial x_k} + \phi \tag{2.9}$$

The three components of the vector equation (2.8) are the Navier–Stokes equations expressing the conservation of momentum (or, more precisely, stating that the rate of change of momentum is equal to the sum of all forces) for a Newtonian fluid. In the thermal energy equation (2.9), ϕ is the always positive (as required by the Second Law of Thermodynamics) dissipation function expressing the irreversible conversion of mechanical energy to internal energy as a result of the deformation of a fluid element. The second term on the right-hand side of Equation (2.9) is the reversible work done (per unit time) by the pressure as the volume of a fluid material element changes. For a Newtonian, isotropic fluid, the viscous dissipation rate is given by

$$\phi = \frac{1}{2}\mu\left(\frac{\partial u_i}{\partial x_k} + \frac{\partial u_k}{\partial x_i}\right)^2 + \lambda\left(\frac{\partial u_j}{\partial x_j}\right)^2 \tag{2.10}$$

---

[3] Newtonian implies a linear relation between the stress tensor and the symmetric part of the deformation tensor (rate of strain tensor). The isotropy assumption reduces the 81 constants of proportionality in that linear relation to two constants. Fourier fluid is that for which the conduction part of the heat flux vector is linearly related to the temperature gradient, and again isotropy implies that the constant of proportionality in this relation is a single scalar.

[4] An assumption that obviously needs to be relaxed for most atmospheric flows, where radiation from the Sun during the day and to outer space during the night plays a crucial rule in weather dynamics. Estimating radiation in the presence of significant cloud cover is one of the major challenges in atmospheric science.

There are now six unknowns, ρ, u_i, p, and T, and the five coupled Equations (2.7), (2.8), and (2.9) plus the equation of state relating pressure, density, and temperature. These six equations, together with sufficient number of initial and boundary conditions, constitute a well-posed, albeit formidable, problem. The system of Equations (2.7) to (2.9) is an excellent model for the laminar or turbulent flow of most fluids, such as air and water under most circumstances, including high-speed gas flows for which the shock waves are thick relative to the mean free path of the molecules.

Polymers, rarefied gases, and flows in micro- and nanodevices are not equilibrium flows and have to be modeled differently. In those cases, higher-order relations between the stress tensor and rate of strain tensor, and between the heat flux vector and temperature gradient, are used. In some cases, the continuum approximation is abandoned altogether, and the fluid is modeled as it really is—a collection of molecules. The molecular-based models used for those unconventional situations include molecular dynamics simulations, direct simulation Monte Carlo methods, and the analytical Boltzmann equation (Gad-el-Hak, 2006b). Under certain circumstances, hybrid molecular–continuum formulation is required.

Returning to the continuum, quasiequilibrium equations, considerable simplification is achieved if the flow is assumed incompressible, usually a reasonable assumption provided that the characteristic flow speed is less than 0.3 of the speed of sound and other conditions are satisfied. The incompressibility assumption, discussed in greater detail in Section 2.4.4, is readily satisfied for almost all liquid flows and for many gas flows. In such cases, the density is assumed either a constant or a given function of temperature (or species concentration). The governing equations for such flows are

$$\frac{\partial u_k}{\partial x_k} = 0 \tag{2.11}$$

$$\rho \left(\frac{\partial u_i}{\partial t} + u_k \frac{\partial u_i}{\partial x_k} \right) = -\frac{\partial p}{\partial x_i} + \frac{\partial}{\partial x_k} \left[\mu \left(\frac{\partial u_i}{\partial x_k} + \frac{\partial u_k}{\partial x_i} \right) \right] + \rho g_i \tag{2.12}$$

$$\rho c_p \left(\frac{\partial T}{\partial t} + u_k \frac{\partial T}{\partial x_k} \right) = \frac{\partial}{\partial x_k} \left(\kappa \frac{\partial T}{\partial x_k} \right) + \phi^* \tag{2.13}$$

These are five equations for the five dependent variables u_i, p, and T. Note that the left-hand side of Equation (2.13) has the specific heat at constant pressure c_p and not c_v. This is the correct incompressible flow limit—of a compressible fluid—as discussed in detail in Section 10.9 of Panton (1996); a subtle point perhaps, but one that is frequently missed in textbooks. The system of Equations (2.11) to (2.13) is coupled if either the viscosity or density depends on temperature; otherwise, the energy equation is uncoupled from the continuity and momentum equations, and can therefore be solved *after* the velocity and pressure fields are determined from solving Equations (2.11) and (2.12). For most geophysical flows, the density depends on temperature and/or species concentration, and the previous system of five equations is coupled.

In nondimensional form, the incompressible flow equations read

$$\frac{\partial u_k}{\partial x_k} = 0 \tag{2.14}$$

$$\left(\frac{\partial u_i}{\partial t} + u_k \frac{\partial u_i}{\partial x_k} \right) = -\frac{\partial p}{\partial x_i} + \frac{Gr}{Re^2} T \delta_{i3} + \frac{\partial}{\partial x_k} \left[\frac{F_v(T)}{Re} \left(\frac{\partial u_i}{\partial x_k} + \frac{\partial u_k}{\partial x_i} \right) \right] \tag{2.15}$$

$$\left(\frac{\partial T}{\partial t} + u_k \frac{\partial T}{\partial x_k} \right) = \frac{\partial}{\partial x_k} \left(\frac{1}{Pe} \frac{\partial T}{\partial x_k} \right) + \frac{Ec}{Re} F_v(T) \phi_{\text{incomp}} \tag{2.16}$$

where $F_\nu(T)$ is a dimensionless function that characterizes the viscosity variation with temperature, and Re, Gr, Pe, and Ec are, respectively, the Reynolds, Grashof, Péclet, and Eckert numbers. These dimensionless parameters determine the relative importance of the different terms in the equations.

For both the compressible and the incompressible equations of motion, the transport terms are neglected away from solid walls in the limit of infinite Reynolds number (i.e., zero Knudsen number). The flow is then approximated as inviscid, nonconducting, and nondissipative; in other words, it is considered in perfect thermodynamic equilibrium. The corresponding equations in this case read (for the compressible case):

$$\frac{\partial \rho}{\partial t} + \frac{\partial}{\partial x_k}(\rho u_k) = 0 \tag{2.17}$$

$$\rho\left(\frac{\partial u_i}{\partial t} + u_k \frac{\partial u_i}{\partial x_k}\right) = -\frac{\partial p}{\partial x_i} + \rho g_i \tag{2.18}$$

$$\rho c_v \left(\frac{\partial T}{\partial t} + u_k \frac{\partial T}{\partial x_k}\right) = -p \frac{\partial u_k}{\partial x_k} \tag{2.19}$$

The Euler equation (2.18) can be integrated along a streamline, and the resulting Bernoulli's equation provides a direct relation between the velocity and the pressure.

2.4.4 Compressibility

The issue of whether to consider the continuum flow compressible or incompressible seems to be rather straightforward, but is in fact full of potential pitfalls. If the local Mach number is less than 0.3, then the flow of a compressible fluid such as air—according to the conventional wisdom—can be treated as incompressible. However, the well-known $Ma < 0.3$ criterion is only a necessary, not a sufficient, condition to allow a treatment of the flow as approximately incompressible. In other words, there are situations where the Mach number can be exceedingly small while the flow is compressible. As is well documented in advanced heat transfer textbooks, strong wall heating or cooling may cause the density to change sufficiently and the incompressible approximation to break down, even at low speeds. Less known is the situation encountered in some microdevices, where the pressure may strongly change due to viscous effects, even though the speeds may not be high enough for the Mach number to go above the traditional threshold of 0.3. Corresponding to the pressure changes would be strong density changes that must be taken into account when writing the continuum equations of motion. In this section, we systematically explain all situations where compressibility effects must be considered.

Let us rewrite the full continuity equation (2.1) as follows

$$\frac{D\rho}{Dt} + \rho \frac{\partial u_k}{\partial x_k} = 0 \tag{2.20}$$

where $\frac{D}{Dt}$ is the substantial derivative $\left(\frac{\partial}{\partial t} + u_k \frac{\partial}{\partial x_k}\right)$, expressing changes following a fluid element. The proper criterion for the incompressible approximation to hold is that $\left(\frac{1}{\rho}\frac{D\rho}{Dt}\right)$ is vanishingly small. In other words, if density changes following a fluid particle are small, the flow is approximately incompressible. Density may change arbitrarily from one particle to another without violating the incompressible flow assumption. This is the case, for example, in the stratified atmosphere and ocean, where the variable density/temperature/salinity flow is often treated as incompressible.

From the state principle of thermodynamics, we can express the density changes of a simple system in terms of changes in pressure and temperature,

$$\rho = \rho(p, T) \tag{2.21}$$

Using the chain rule of calculus,

$$\frac{1}{\rho} \frac{D\rho}{Dt} = \alpha \frac{Dp}{Dt} - \beta \frac{DT}{Dt} \tag{2.22}$$

where α and β are, respectively, the isothermal compressibility coefficient and the bulk expansion coefficient—two thermodynamic variables that characterize the fluid susceptibility to change of volume—defined by the following relations

$$\alpha(p, T) \equiv \frac{1}{\rho} \left. \frac{\partial \rho}{\partial p} \right|_T \tag{2.23}$$

$$\beta(p, T) \equiv -\frac{1}{\rho} \left. \frac{\partial \rho}{\partial T} \right|_p \tag{2.24}$$

For ideal gases, $\alpha = 1/p$ and $\beta = 1/T$. Note, however, that in the following arguments it will not be necessary to invoke the ideal gas assumption.

The flow must be treated as compressible if pressure and/or temperature changes are sufficiently strong. Equation (2.22) must, of course, be properly nondimensionalized before deciding whether a term is large or small. In here, we closely follow the procedure detailed in Panton (1996).

Consider first the case of adiabatic walls. Density is normalized with a reference value ρ_o; velocities with a reference speed v_o; spatial coordinates and time with, respectively, L and L/v_o; and the isothermal compressibility coefficient and bulk expansion coefficient with reference values α_o and β_o. The pressure is nondimensionalized with the inertial pressure-scale $\rho_o v_o^2$. This scale is twice the dynamic pressure (i.e., the pressure change as an inviscid fluid moving at the reference speed is brought to rest).

For an adiabatic wall, temperature changes result from the irreversible conversion of mechanical energy into internal energy via viscous dissipation. Temperature is therefore nondimensionalized as follows

$$T^* = \frac{T - T_o}{\left(\frac{\mu_o v_o^2}{\kappa_o}\right)} = \frac{T - T_o}{Pr \left(\frac{v_o^2}{c_{p_o}}\right)} \tag{2.25}$$

where T_o is a reference temperature; μ_o, κ_o, and c_{p_o} are, respectively, reference viscosity, thermal conductivity, and specific heat at constant pressure; and Pr is the reference Prandtl number, $(\mu_o c_{p_o})/\kappa_o$.

The scaling used previously for pressure is based on Bernoulli's equation, and therefore neglects viscous effects. This particular scaling guarantees that the pressure term in the momentum equation will be of the same order as the inertia term. The temperature scaling assumes that the conduction, convection, and dissipation terms in the energy equation have the same order of magnitude. The resulting dimensionless form of Equation (2.22) reads

$$\frac{1}{\rho^*} \frac{D\rho^*}{Dt^*} = \gamma_o Ma^2 \left\{ \alpha^* \frac{Dp^*}{Dt^*} - \frac{Pr B \beta^*}{A} \frac{DT^*}{Dt^*} \right\} \tag{2.26}$$

where the superscript $*$ indicates a nondimensional quantity, Ma is the reference Mach number ($Ma \equiv v_o/a_o$, where a_o is the reference speed of sound), and A and B are dimensionless constants defined by $A \equiv \alpha_o \rho_o c_{p_o} T_o$ and $B \equiv \beta_o T_o$. If the scaling is properly chosen, the terms having the $*$ superscript in the right-hand side should be of order one, and the relative importance of such terms in the equations of motion is determined by the magnitude of the dimensionless parameter(s) appearing to their left (e.g., Ma, Pr). Therefore, as $Ma^2 \to 0$, temperature changes due to viscous dissipation are neglected (unless Pr is very large, as in the case of highly viscous polymers and oils). Within the same order of approximation, all thermodynamic properties of the fluid are assumed constant.

Pressure changes are also neglected in the limit of zero Mach number. Hence, for $Ma < 0.3$ (i.e., $Ma^2 < 0.09$), density changes following a fluid particle can be neglected, and the flow can then be approximated as incompressible.[5] However, there is a caveat in this argument. Pressure changes due to inertia can indeed be neglected at small Mach numbers, and this is consistent with the way we nondimensionalized the pressure term previously. If, however, pressure changes are mostly due to viscous effects, as is the case in a long microduct or microgas bearing, pressure changes may be significant even at low speeds (low Ma). In that case, the term $\frac{Dp^*}{Dt^*}$ in Equation (2.26) is no longer of order one and may be large regardless of the value of Ma. Density then may change significantly, and the flow must be treated as compressible. Had pressure been nondimensionalized using the viscous scale $\left(\frac{\mu_o v_o}{L}\right)$ instead of the inertial one $\left(\rho_o v_o^2\right)$, the revised Equation (2.26) would have Re^{-1} appearing explicitly in the first term in the right-hand side, accentuating the importance of this term when viscous forces dominate.

A similar result can be gleaned when the Mach number is interpreted as follows

$$Ma^2 = \frac{v_o^2}{a_o^2} = v_o^2 \left.\frac{\partial \rho}{\partial p}\right|_s = \frac{\rho_o v_o^2}{\rho_o}\left.\frac{\partial \rho}{\partial p}\right|_s$$

$$\approx \frac{\Delta p}{\rho_o}\frac{\Delta \rho}{\Delta p} = \frac{\Delta \rho}{\rho_o} \qquad (2.27)$$

where s is the entropy. Again, the previous equation assumes that pressure changes are inviscid; therefore, small Mach number means negligible pressure and density changes. In a flow dominated by viscous effects—such as that inside a microduct—density changes may be significant even in the limit of zero Mach number.

Identical arguments can be made in the case of isothermal walls. Here, strong temperature changes may be the result of wall heating or cooling, even if viscous dissipation is negligible. The proper temperature scale in this case is given in terms of the wall temperature T_w and the reference temperature T_o as follows

$$\hat{T} = \frac{T - T_o}{T_w - T_o} \qquad (2.28)$$

where $\hat{T}$ is the new dimensionless temperature. The nondimensional form of Equation (2.22) now reads

$$\frac{1}{\rho^*}\frac{D\rho^*}{Dt^*} = \gamma_o Ma^2 \alpha^* \frac{Dp^*}{Dt^*} - \beta^* B \left(\frac{T_w - T_o}{T_o}\right)\frac{D\hat{T}}{Dt^*} \qquad (2.29)$$

---

[5] With an error of about 10% at $Ma = 0.3$, 4% at $Ma = 0.2$, 1% at $Ma = 0.1$, and so on.

Here, we notice that the temperature term is different from that in Equation (2.26). Ma is no longer appearing in this term, and strong temperature changes (i.e., large $(T_w - T_o)/T_o$) may cause strong density changes, regardless of the value of the Mach number. In addition, the thermodynamic properties of the fluid are not constant but depend on temperature, and, as a result, the continuity, momentum, and energy equations all couple. The pressure term in Equation (2.29), however, is exactly as it was in the adiabatic case, and the same arguments made previously apply: the flow should be considered compressible if $Ma > 0.3$, or if pressure changes due to viscous forces are sufficiently large.

There are three additional scenarios in which significant pressure and density changes may occur without inertial, viscous, or thermal effects. First is the case of quasistatic compression/expansion of a gas in, for example, a piston-cylinder arrangement. The resulting compressibility effects are, however, compressibility of the fluid and not of the flow. Two other situations where compressibility effects must also be considered are problems with length scales comparable to the scale height of the atmosphere and rapidly varying flows as in sound propagation (see Lighthill, 1963).

2.4.5 Prandtl's breakthrough

Even with the simplification accorded by the incompressibility assumption, the viscous system of equations is formidable and has no general solution. Usual further simplifications—applicable only to laminar flows—include geometries for which the nonlinear terms in the (instantaneous) momentum equation are identically zero, low Reynolds number creeping flows for which the nonlinear terms are approximately zero, and high Reynolds number inviscid flows for which the continuity and momentum equations can be shown to metamorphose into the linear Laplace equation. The latter assumption spawned the great advances in perfect flow theory that occurred during the second half of the nineteenth century. However, neglecting viscosity gives the totally erroneous result of zero drag for moving bodies and zero pressure drop in pipes. Moreover, none of those simplifications apply to the rotational, (instantaneously) time-dependent, and three-dimensional turbulent flows.

Not surprisingly, hydraulic engineers of the time showed little interest in the elegant theories of hydrodynamics and relied instead on their own collection of totally empirical equations, charts, and tables to compute drag, pressure losses, and other practically important quantities. Consistent with that pragmatic approach, engineering students then and for many decades to follow were taught the art of hydraulics. The science of hydrodynamics was relegated, if at all, to mathematics and physics curricula.

In lamenting the status of fluid mechanics at the dawn of the twentieth century, the British chemist and Nobel laureate Sir Cyril Norman Hinshelwood (1897–1967) jested that fluid dynamists were divided into hydraulic engineers who observed things that could not be explained and mathematicians who explained things that could not be observed.

In an epoch-making presentation to the third International Congress of Mathematicians held in Heidelberg, the German engineer Ludwig Prandtl resolved, to a large extent, the previous dilemma. Prandtl (1904) introduced the concept of a fluid boundary layer, adjacent to a moving body, where viscous forces are important and outside of which the flow is more or less inviscid. At sufficiently high Reynolds number, the boundary layer is thin relative to the longitudinal length scale and, as a result, velocity derivatives in the streamwise direction are small compared to normal derivatives. For the first time, that single simplification made it possible to obtain viscous flow solutions, even in the presence of nonlinear terms, at least in the case of laminar flow. Both the momentum and the energy equations are parabolic under such circumstances, and are therefore amenable to similarity solutions and marching numerical techniques. From then on, viscous flow theory was in vogue for both scientists and engineers. Practical quantities such as skin friction drag could be computed from first principles, even for noncreeping flows. Experiments in wind tunnels, and their cousins water tunnels and towing tanks, provided valuable data for problems too complex to submit to analysis.

2.4.6 Turbulent flows

All the transport equations listed thus far are valid for nonturbulent and turbulent flows. However, in the latter case, the dependent variables are generally random functions of space and time. No straightforward method exists for obtaining stochastic solutions of these nonlinear partial differential equations, and this is the primary reason why turbulence remains as the last great unsolved problem of classical physics. Dimensional analysis can be used to obtain crude results for a few cases, but first principles analytical solutions are not possible even for the simplest conceivable turbulent flow.

The contemporary attempts to use dynamical systems theory to study turbulent flows have not yet reached fruition, especially at Reynolds numbers far above transition (Aubry et al., 1988), although advances in this theory have helped with reducing and displaying the massive bulk of data resulting from numerical and experimental simulations (Sen, 1989). The book by Holmes et al. (1996) provides a useful, readable introduction to the emerging field. It details a strategy by which knowledge of coherent structures, finite-dimensional dynamical systems theory, and the Karhunen-Loève or proper orthogonal decomposition could be combined to create low-dimensional models of turbulence that resolve only the organized motion, and describes their dynamical interactions. The utility of the dynamical systems approach as an additional arsenal to tackle the turbulence conundrum has been demonstrated only for turbulence near transition or near a wall, so that the flow would be relatively simple, and a relatively small number of degrees of freedom would be excited. Holmes et al. summarize the (partial) successes that have been achieved thus far using relatively small sets of ordinary differential equations and suggest a broad strategy for modeling turbulent flows and other spatiotemporal complex systems.

A turbulent flow is described by a set of nonlinear partial differential equations and is characterized by an infinite number of degrees of freedom. This makes it rather difficult to model the turbulence using a dynamical systems approximation. The notion that a complex, infinite-dimensional flow can be decomposed into several low-dimensional subunits is, however, a natural consequence of the realization that quasiperiodic coherent structures dominate the dynamics of seemingly random turbulent shear flows. This implies that low-dimensional, localized dynamics can exist in formally infinite-dimensional extended systems, such as open turbulent flows. Reducing the flow physics to finite-dimensional dynamical systems enables a study of its behavior through an examination of the fixed points and the topology of their stable and unstable manifolds. From the dynamical systems theory viewpoint, the meandering of low-speed streaks is interpreted as hovering of the flow state near an unstable fixed point in the low-dimensional state space. An intermittent event that produces high wall stress—a burst—is interpreted as a jump along a heteroclinic cycle to a different unstable fixed point that occurs when the state has wandered too far from the first unstable fixed point. Delaying this jump by holding the system near the first fixed point should lead to lower momentum transport in the wall region and, therefore, to lower skin friction drag. Reactive control means sensing the current local state and, through appropriate manipulation, keeping the state close to a given unstable fixed point, thereby preventing further production of turbulence. Reducing the bursting frequency by

50%, for example, may lead to a comparable reduction in skin friction drag. For a jet, relaminarization may lead to a quiet flow and very significant noise reduction. We return to the two described facets of nonlinear dynamical systems—predictability and control—in Section 2.4.10.

Direct numerical simulations (DNS) of a turbulent flow, the brute force numerical integration of the instantaneous equations using the supercomputer, is prohibitively expensive—if not impossible—at practical Reynolds numbers (Moin and Mahesh, 1998). For the present at least, a statistical approach, where a temporal, spatial, or ensemble average is defined and the equations of motion are written for the various moments of the fluctuations about this mean, is the only route available to obtain meaningful engineering results. Unfortunately, the nonlinearity of the Navier–Stokes equations guarantees that the process of averaging to obtain moments results in an open system of equations, where the number of unknowns is always greater than the number of equations, and more or less heuristic modeling is used to close the equations. This is known as the closure problem, and again makes obtaining first principles solutions to the (averaged) equations of motion impossible.

To illustrate the closure problem, consider the (instantaneous) continuity and momentum equations for a Newtonian, incompressible, constant density, constant viscosity, turbulent flow. In this uncoupled version of Equations (2.11) and (2.12) for the four random unknowns u_i and p, no general stochastic solution is known to exist. However, would it be feasible to obtain solutions for the nonstochastic mean flow quantities? As was first demonstrated by Osborne Reynolds (1895) more than a century ago, all the field variables are decomposed into a mean and a fluctuation. Let $u_i = \overline{U}_i + u'_i$ and $p = \overline{P} + p'$, where $\overline{U}_i$ and $\overline{P}$ are ensemble averages for the velocity and pressure, respectively, and u'_i and p' are the velocity and pressure fluctuations about the respective averages. Note that temporal or spatial averages could be used in place of ensemble average if the flow field is stationary or homogeneous, respectively. In the former case, the time derivative of any statistical quantity vanishes. In the latter, averaged functions are independent of position. Substituting the decomposed pressure and velocity into Equations (2.11) and (2.12), the equations governing the mean velocity and mean pressure for an incompressible, constant viscosity, turbulent flow becomes

$$\frac{\partial \overline{U}_k}{\partial x_k} = 0 \tag{2.30}$$

$$\rho \left(\frac{\partial \overline{U}_i}{\partial t} + \overline{U}_k \frac{\partial \overline{U}_i}{\partial x_k} \right) = -\frac{\partial \overline{P}}{\partial x_i} + \frac{\partial}{\partial x_k} \left(\mu \frac{\partial \overline{U}_i}{\partial x_k} - \rho \overline{u_i u_k} \right) + \rho \overline{g}_i \tag{2.31}$$

where, for clarity, the primes have been dropped from the fluctuating velocity components u_i and u_k.

This is now a system of four equations for the ten unknowns $\overline{U}_i$, $\overline{P}$, and $\overline{u_i u_k}$.[6] The momentum equation (2.31) is written in a form that facilitates the physical interpretation of the turbulence stress tensor (Reynolds stresses), $-\rho \overline{u_i u_k}$, because additional stresses on a fluid element are to be considered along with the conventional viscous stresses and pressure. An equation for the components of this tensor may be derived, but it will contain third-order moments such as $\overline{u_i u_j u_k}$. The equations are (heuristically) closed by expressing the second- or third-order quantities in terms of the first or second moments, respectively. For comprehensive reviews of these first- and second-order closure schemes, see Lumley

---

[6] The second-order tensor $\overline{u_i u_k}$ is obviously a symmetric one with only six independent components.

(1983, 1987), Speziale (1991), and Wilcox (1993). A concise summary of the turbulence problem in general is provided by Jiménez (2000).

2.4.7 Numerical solutions

Leaving aside for a moment less conventional, albeit just as important, problems in fluid mechanics such as those involving non-Newtonian fluids, multiphase flows, hypersonic flows, and chemically reacting flows, in principle almost any laminar flow problem can presently be solved, at least numerically. Turbulence, in contrast, remains largely an enigma, analytically unapproachable, yet practically very important. The statistical approach to solving the Navier–Stokes equations always leads to more unknowns than equations (the closure problem), and solutions based on first principles are again not possible. The heuristic modeling used to close the Reynolds-averaged equations has to be validated case by case, and does not, therefore, offer much of an advantage over the old-fashioned empirical approach.

Thus, turbulence is a conundrum that appears to yield its secrets only to physical and numerical experiments, provided that the wide band of relevant scales is fully resolved—a far-from-trivial task at high Reynolds numbers (Gad-el-Hak and Bandyopadhyay, 1994). Until recently, direct numerical simulations of the canonical turbulent boundary layer have been carried out, at great cost despite a bit of improvising, up to a very modest momentum-thickness Reynolds number of 1,410 (Spalart, 1988).

In a turbulent flow, the ratio of the large eddies (at which the energy maintaining the flow is inputed) to the Kolmogorov microscale (the flow smallest length scale) is proportional to $Re^{3/4}$ (Tennekes and Lumley, 1972). Each excited eddy requires at least one grid point to describe it. Therefore, to adequately resolve, via DNS, a three-dimensional flow, the required number of modes would be proportional to $(Re^{3/4})^3$. To describe the motion of small eddies as they are swept around by large ones, the time step must not be larger than the ratio of the Kolmogorov length scale to the characteristic root-mean-square (rms) velocity. The large eddies, however, evolve on a time scale proportional to their size divided by their rms velocity. Thus, the number of time steps required is again proportional to $Re^{3/4}$. Finally, the computational work requirement is the number of modes × the number of time steps, which scales with Re^3 (i.e., an order of magnitude increase in computer power is needed as the Reynolds number is doubled) (Karniadakis and Orszag, 1993). Because the computational resource required varies as the cube of the Reynolds number, it may not be possible to directly simulate very high Reynolds number turbulent flows any time soon.

2.4.8 Other complexities

Despite their already complicated nature, the transport equations introduced previously could be further entangled by other effects. We list herein a few examples. Geophysical flows occur at such large length scales as to invalidate the inertial frame assumption made previously. The Earth's rotation affects these flows, and such things as centrifugal and Coriolis forces enter into the equations rewritten in a noninertial frame of reference fixed with the rotating Earth. Oceanic and atmospheric flows are more often than not turbulent flows that span the enormous range of length scales of nine decades, from a few millimeters to thousands of kilometers (Garrett, 2000; McIntyre, 2000).

Density stratification is important for many atmospheric and oceanic phenomena. Buoyancy forces are produced by density variations in a gravitational field, and those forces drive significant convection in natural flows (Linden, 2000). In the ocean, those forces are

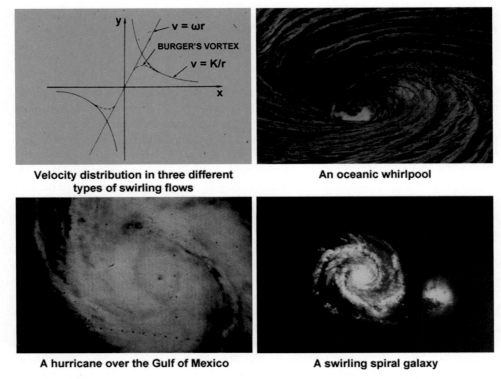

Figure 2.3 Simple modeling of an oceanic whirlpool, a hurricane, and a spiral galaxy.

further complicated by the competing influences of temperature and salt (Garrett, 2000). The competition affects the large-scale global ocean circulation and, in turn, climate variability. For weak density variations, the Bousinessq approximation permits the use of the coupled incompressible flow equations, but more complexities are introduced in situations with strong density stratification, such as when strong heating and cooling is present. Complex topography further complicates convective flows in the ocean and atmosphere.

Air–sea interface governs many of the important transport phenomena in the ocean and atmosphere, and plays a crucial role in determining the climate. The location of that interface is itself not known a priori and thus is the source of further complexity in the problem. Even worse, the free boundary nature of the liquid–gas interface, in addition to the possibility of breaking that interface and forming bubbles and droplets, introduces new nonlinearities that augment or compete with the customary convective nonlinearity (Davis, 2000). Chemical reactions are obviously important in fires and are even present in some atmospheric transport problems (Chapter 13). When liquid water or ice is present in the air, two-phase treatment of the equations of motion may need to be considered, again complicating even the relevant numerical solutions.

However, even in those complex situations described previously, simplifying assumptions can be rationally made to facilitate solving the problem. Any spatial symmetries in the problem must be exploited. If the mean quantities are time independent, then that too can be exploited.

An extreme example of simplification that surprisingly yields reasonable results includes the swirling giants depicted in Figure 2.3. Here, an oceanic whirlpool, a hurricane, and a spiral galaxy are simply modeled as a rotating, axisymmetric viscous core and an external inviscid vortex joined by a Burger's vortex. The viscous core leads to a circumferential velocity proportional to the radius, and the inviscid vortex leads to a velocity proportional to $1/r$. This model leads to surprisingly good results in some narrow sense for those exceedingly complex flows.

A cyclone's pressure is the best indicator of its intensity because it can be precisely measured, whereas winds have to be estimated. The previous simple model yields the maximum wind speed from measurements of the center pressure, the ambient pressure, and the size of the eye of the storm. It is still important to note that it is the difference in the hurricane's pressure and that of its environment that actually give it its strength. This difference in pressure is known as the "pressure gradient," and it is this change in pressure over a distance that causes wind. The bigger the gradient, the faster will be the winds generated. If two cyclones have the same minimum pressure, but one is in an area of higher ambient pressure than the other, that one is in fact stronger. The cyclone must be more intense to get its pressure commensurately lower, and its larger pressure gradient would make its winds faster.

2.4.9 Earthquakes

Thus far in this section, we discuss prediction of the type of disaster involving fluid transport phenomena, weather-related disasters being the most rampant. Predictions are possible on those cases, and improvements in forecast's accuracy and extent are continually being made as a result of enhanced understanding of flow physics, increased accuracy and resolution of global measurements, and exponentially expanded computer power. Other types of disaster do not fare as well, earthquakes being calamities that thus far cannot be accurately predicted. Prediction of weather storms is possible in part because the atmosphere is optically transparent, which facilitates measurements that in turn provide not only the initial and boundary conditions necessary for integrating the governing equations, but also a deeper understanding of the physics. The oceans are not as accessible, but measurements there are possible as well, and scientists learned a great deal in the past few decades about the dynamics of both the atmosphere and the ocean (Garrett, 2000; McIntyre, 2000). Our knowledge of *terra firma*, in contrast, does not fare as well mostly because of its inaccessibility to direct observation (Huppert, 2000). What we know about the Earth's solid inner core, liquid outer core, mantle, and lithosphere comes mainly from inferences drawn from observations at or near the planet's surface, which include the study of propagation, reflection, and scatter of seismic waves. Deep underground measurements are not very practical, and the exact constitutive equations of the different constituents of the "solid" Earth are not known. All that inhibits us from writing down and solving the precise equations, and their initial and boundary conditions, for the dynamics of the Earth's solid part. That portion of the planet contains three orders of magnitude more volume than all the oceans combined and six orders of magnitude more mass than the entire atmosphere, and it is a true pity that we know relatively little about the solid Earth.

The science of earthquake basically began shortly after the infamous rupture of the San Andreas fault that, devastated San Francisco a little more than a century ago (Section 2.8.1). Before then, geologists had examined seismic faults and even devised primitive seismometers to measure shaking. However, they had no idea what caused the ground to heave without warning. A few days after the Great Earthquake struck on 18 April 1906, Governor George C. Pardee of California charged the state's leading scientists with investigating how and why the Earth's crust had ruptured for hundreds of miles with such terrifying violence. The foundation for much of what is known today about earthquakes was laid 2 years later, and the resulting report carried the name of the famed geologist Andrew C. Lawson (Lawson, 1908).

Earthquakes are caused by stresses in the Earth's crust that build up deep inside a fault—until it ruptures with a jolt. Prior to the Lawson Report, many scientists believed earthquakes created the Faults instead of the other way around. The San Andreas Fault system marks the boundary between two huge moving slabs of the Earth's crust: the Pacific Plate and the North American Plate. As the plates grind constantly past each other, strain builds until it is released periodically in a full-scale earthquake. A few small sections of the San Andreas Fault had been mapped by scientists years before 1906, but Lawson and his team discovered that the entire zone stretched for more than 950 km along the length of California. By measuring land movements on either side of the fault, the team learned that the earthquake's motion had moved the ground horizontally, from side to side, rather than just vertically as scientists had previously believed.

A century after the Lawson Report, its conclusions remain valid, but it has stimulated modern earthquake science to move far beyond. Modern scientists have learned that major earthquakes are not random events—they apparently come in cycles. Although pinpoint prediction remains impossible, research on faults throughout the San Francisco Bay Area and other fault locations enables scientists to estimate the probability that strong quakes will jolt a region within the coming decades. Sophisticated broadband seismometers can measure the magnitude of earthquakes within a minute or two of an event and determine where and how deeply on a fault the rupture started. Orbiting satellites now measure within fractions of an inch how the Earth's surface moves as strain builds up along fault lines, and again how the land is distorted after a quake has struck. "Shakemaps," available on the Internet and by e-mail immediately after every earthquake, can swiftly tell disaster workers, utility companies and residents where damage may be greatest. Supercomputers, simulating ground motion from past earthquakes, can show where shaking might be heaviest when new earthquakes strike. The information can then be relayed to the public and to emergency workers.

One of the latest and most important ventures in understanding earthquake behavior is the borehole drilling project at Parkfield in southern Monterey County, California, where the San Andreas Fault has been heavily instrumented for many years. The hole is about 3.2 km deep and crosses the San Andreas underground. For the first time, sensors can actually be inside the earthquake machine to catch and record the earthquakes right where and when they are occurring.

The seismic safety of any structure depends on the strength of its construction and the geology of the ground on which it stands—a conclusion reflected in all of today's building codes in the United States. Tragically, the codes in some earthquakeprone countries are just as strict as those in the United states, but are not enforceable for the most part. In other nations, building codes are not sufficiently strict or nonexistent altogether.

2.4.10 *The butterfly effect*

There are two additional issues to ponder for all disasters that could be modeled as nonlinear dynamical systems. The volume edited by Bunde et al. (2002) is devoted to this topic, and is one of very few books to tackle large-scale disasters purely as a problem to be posed and solved using scientific principles. The modeling could be in the form of a number of algebraic equations or, more likely, ordinary or partial differential equations, with nonlinear term(s) appearing somewhere within the finite number of equations. First, we examine the bad news. Nonlinear dynamical systems are capable of producing chaotic solutions, which limit the ability to predict too far into the future, even if infinitely powerful computers are available. Second, we examine the (potentially) good news. Chaotic systems can be controlled, in the sense that a very small perturbation can lead to a significant change in the future state of the system. In this subsection, we elaborate on both issues.

In the theory of dynamical systems, the so-called "butterfly effect" (a lowly *diurnal lepidopteran* flapping its wings in Brazil may set off a future tornado in Texas) denotes sensitive dependence of nonlinear differential equations on initial conditions, with phase-space solutions initially very close together and separating exponentially. Massachusetts Institute of Technology's atmospheric scientist Edward Lorenz originally used seagull's wings for the metaphor in a paper for the New York Academy of Sciences (Lorenz, 1963), but in subsequent speeches and papers, he used the more poetic butterfly. For a complex system such as the weather, initial conditions of infinite resolution and infinite accuracy are clearly never going to be available, thus further making certain that precise long-term predictions are never achievable.

The solution of nonlinear dynamical systems of three or more degrees of freedom[7] may be in the form of a strange attractor whose intrinsic structure contains a well-defined mechanism to produce a chaotic behavior without requiring random forcing (Ott, 1993). Chaotic behavior is complex, aperiodic, and, although deterministic, appears to be random. The dynamical system in that case is nonintegrable,[8] and our ability for long-term forecast is severely hindered because of the extreme sensitivity to initial conditions. One can predict the most probable weather, for example, a week from the present, with a narrow standard deviation to indicate all other possible outcomes. We speak of a 30% chance of rain 7 days from now, and so on. That ability to provide reasonably accurate prediction diminishes as time progresses because the sensitivity to initial conditions intensifies exponentially, and Lorenz (1967) proposes a 20-days theoretical limit for predicting weather. This means that regardless of how massive future computers will become, weather prediction beyond 20 days will always be meaningless. Nevertheless, we still have a way to go to double the extent of the current 10-day forecast.

Weather and climate should not be confused, however. The latter describes the long-term variability of the climate system whose components comprise the atmosphere, hydrosphere, cryosphere, pedosphere, lithosphere and biosphere. Climatologists apply models to compute the evolution of the climate a hundred years or more into the future (Fraedrich & Schönwiese, 2002; Hasselmann, 2002). Seemingly paradoxical, meteorologists use similar models but have difficulties forecasting the weather beyond just a few days. Both weather and climate are nonlinear dynamical systems, but the former concerns the evolution of the system as a function of the initial conditions with fixed boundary conditions, whereas the latter, especially as influenced by human misdeeds, concerns the response of the system to changes in boundary conditions with fixed initial conditions. For long time periods, the

---

[7] The number of first-order ordinary differential equations, each of the form $\frac{dx_i}{dt} = f_i(x_1, x_2, \ldots, x_N)$, which completely describe the autonomous system's evolution, is in general equal to the number of degrees of freedom N. The later number is in principle infinite for a dynamical system whose state is described by partial differential equation(s). As examples, a planar pendulum has two degrees of freedom, a double planar pendulum has three, a single pendulum that is free to oscillate in three dimensions has four, and a turbulent flow has infinite degrees of freedom. The single pendulum is incapable of producing chaotic motion in a plane, the double pendulum does if its oscillations have sufficiently large (nonlinear) amplitude; the single, nonplanar, nonlinear pendulum is also capable of producing chaos, and turbulence is spatiotemporal chaos whose infinite degrees of freedom can be reduced to a finite but large number under certain circumstances.
[8] Meaning analytical solutions of the differential equations governing the dynamics are not obtainable, and numerical integrations of the same lead to chaotic solutions.

dependence of the time-evolving climate state on the initial conditions becomes negligible asymptotically.

Now for the good news. A question arises naturally: just as small disturbances can radically grow within a deterministic system to yield rich, unpredictable behavior, can minute adjustments to a system parameter be used to reverse the process and control (i.e., regularize) the behavior of a chaotic system? This question was answered in the affirmative both theoretically and experimentally, at least for system orbits that reside on low-dimensional strange attractors (see the review by Lindner and Ditto, 1995).

There is another question of greater relevance here. Given a dynamical system in the chaotic regime, is it possible to stabilize its behavior through some kind of active control? Although other alternatives have been devised (e.g., Fowler, 1989; Huberman, 1990; Huberman and Lumer, 1990; Hübler and Lüscher, 1989), the recent method proposed by workers at the University of Maryland (Ott et al., 1990a, 1990b; Romeiras et al., 1992; Shinbrot et al., 1990, 1992a, 1992b, 1992c, 1998) promises to be a significant breakthrough. Comprehensive reviews and bibliographies of the emerging field of chaos control can be found in the articles by Lindner and Ditto (1995), Shinbrot (1993, 1995, 1998), and Shinbrot et al. (1993).

Ott et al. (1990a) demonstrate through numerical experiments with the Hénon map, that it is possible to stabilize a chaotic motion about any prechosen, unstable orbit through the use of relatively small perturbations. The procedure consists of applying minute time-dependent perturbations to one of the system parameters to control the chaotic system around one of its many unstable periodic orbits. In this context, targeting refers to the process whereby an arbitrary initial condition on a chaotic attractor is steered toward a prescribed point (target) on this attractor. The goal is to reach the target as quickly as possible using a sequence of small perturbations (Kostelich et al., 1993a).

The success of the Ott-Grebogi-Yorke's (OGY) strategy for controlling chaos hinges on the fact that beneath the apparent unpredictability of a chaotic system lies an intricate but highly ordered structure. Left to its own recourse, such a system continually shifts from one periodic pattern to another, creating the appearance of randomness. An appropriately controlled system, however, is locked into one particular type of repeating motion. With such reactive control, the dynamical system becomes one with a stable behavior.

The OGY method can be simply illustrated by the schematic in Figure 2.4. The state of the system is represented as the intersection of a stable manifold and an unstable one. The control is applied intermittently whenever the system departs from the stable manifold by a prescribed tolerance; otherwise, the control is shut off. The control attempts to put the system back onto the stable manifold so that the state converges toward the desired trajectory. Unmodeled dynamics cause noise in the system and a tendency for the state to wander off in the unstable direction. The intermittent control prevents this and the desired trajectory is achieved. This efficient control is not unlike trying to balance a ball in the center of a horse saddle (Moin and Bewley, 1994). There is one stable direction (front/back) and one unstable direction (left/right). The restless horse is the unmodeled dynamics, intermittently causing the ball to move in the wrong direction. The OGY control need only be applied, in the most direct manner possible, whenever the ball wanders off in the left/right direction.

The OGY method has been successfully applied in a relatively simple experiment by Ditto et al. (1990) and Ditto and Pecora (1993) at the Naval Surface Warfare Center, in which reverse chaos was obtained in a parametrically driven, gravitationally buckled, amorphous magnetoelastic ribbon. Garfinkel et al. (1992) apply the same control strategy to stabilize drug-induced cardiac arrhythmias in sections of a rabbit ventricle. Other

2.4 The science of disaster prediction and control

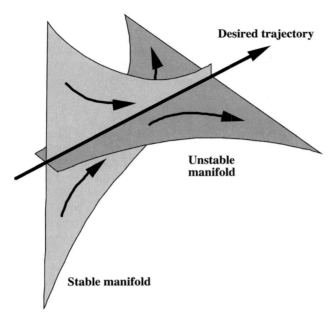

Figure 2.4 OGY method for controlling chaos.

extensions, improvements, and applications of the OGY strategy include higher-dimensional targeting (Auerbach et al., 1992; Kostelich et al., 1993b); controlling chaotic scattering in Hamiltonian (i.e., nondissipative, area conservative) systems (Lai et al., 1993a, 1993b); synchronization of identical chaotic systems that govern communication, neural, or biological processes (Lai and Grebogi, 1993); use of chaos to transmit information (Hayes et al., 1994a, 1994b); control of transient chaos (Lai et al., 1994); and taming spatiotemporal chaos using a sparse array of controllers (Auerbach, 1994; Chen et al., 1993; Qin et al., 1994).

In a more complex system, such as a turbulent boundary layer, numerous interdependent modes, as well as many stable and unstable manifolds (directions), exist. The flow can then be modeled as coherent structures plus a parameterized turbulent background. The proper orthogonal decomposition (POD) is used to model the coherent part because POD guarantees the minimum number of degrees of freedom for a given model accuracy. Factors that make turbulence control a challenging task are the potentially quite large perturbations caused by the unmodeled dynamics of the flow, the nonstationary nature of the desired dynamics, and the complexity of the saddle shape describing the dynamics of the different modes. Nevertheless, the OGY control strategy has several advantages that are of special interest in the control of turbulence: (1) the mathematical model for the dynamical system need not be known, (2) only *small* changes in the control parameter are required, and (3) noise can be tolerated (with appropriate penalty).

How does all this apply to large-scale disasters? Suppose, for example, global warming is the disaster under consideration. Suppose further that we know how to model this complex phenomena as a nonlinear dynamical system. What if we can ever so gently manipulate the present state to greatly, and hopefully beneficially, affect future outcome? A quintessential butterfly effect. For example, what if we cover a modest-size desert with reflective material that reduces the absorption of radiation from the Sun? If it is done right, that manipulation of a microclimate may result in a macroclimate change in the future.

However, is it the desired change? What if it is not done right? This, of course, is the trillion-dollar question! Other examples may include prevention of future severe storms, droughts, famines, and earthquakes. More far-fetched examples include being able to control, via small perturbations, unfavorable human behaviors such as mass hysteria, panic, and stampedes. Intensive theoretical, numerical, and experimental research is required to investigate the proposed idea.

2.5 Global Earth Observation System of Systems

To predict weather-related disasters, computers use the best available models, together with massive data. Those data are gathered from satellites and manned as well as unmanned aircraft in the sky, water-based sensors, and sensors on the ground and even beneath the ground. Hurricanes, droughts, climate systems, and the planet's natural resources could all be better predicted with improved data and observations. Coastal mapping, nautical charting, ecosystem, hydrological and oceanic monitoring, fisheries surveillance, and ozone concentration can all be measured and assessed. In this subsection, we briefly describe the political steps that led to the recent formation of a global Earth observation system, a gigantic endeavor that is a prime example of the need for international cooperation. More details are forthcoming in Chapters 20 and 21.

Producing and managing better information about the environment has become a top priority for nations around the globe. In July 2003, the Earth Observation Summit brought together thirty-three nations as well as the European Commission and many international organizations, to adopt a declaration that signified a political commitment toward the development of a comprehensive, coordinated, and sustained Earth observation system to collect and disseminate improved data, information, and models to stakeholders and decision makers.

Earth observation systems consist of measurements of air, water and land made on the ground, from the air, or in space. Historically observed in isolation, the current effort is to look at these elements together and to study their interactions. An ad hoc group of senior officials from all participating countries and organizations, named the Group on Earth Observations (GEO), was formed to undertake this global effort. GEO was charged to develop a "framework document" as well as a more comprehensive report, to describe how the collective effort could be organized to continuously monitor the state of our environment, increase understanding of dynamic Earth processes, and enhance forecasts on our environmental conditions. Furthermore, it was to address potential societal benefits if timely, high-quality, and long-term data and models were available to aid decision makers at every level, from intergovernmental organizations to local governments to individuals. Through four meetings of GEO, from late 2003 to April 2004, the required documents were prepared for ministerial review and adoption.

In April 2004, U.S. Environmental Protection Agency Administrator Michael Leavitt and other senior cabinet members met in Japan with environmental ministers from more than fifty nations. They adopted the framework document for a 10-year implementation plan for the Global Earth Observation System of Systems (GEOSS).

As of 16 February 2005, 18 months after the first-ever Earth Observation Summit, the number of participating countries has nearly doubled, and interest has accelerated since the recent tsunami tragedy devastated parts of Asia and Africa. Sixty-one countries agreed to a 10-year plan that will revolutionize the understanding of Earth and how it works. Agreement for a 10-year implementation plan for GEOSS was reached by member countries of the GEO at the Third Observation Summit held in Brussels. Nearly forty international organizations also support the emerging global network. The GEOSS project will help all nations involved produce and manage their information in a way that benefits both the environment and humanity by taking the planet's "pulse." In the coming months, more countries and global organizations are expected to join the historic initiative.

GEOSS is envisioned as a large national and international cooperative effort to bring together existing and new hardware and software, making it all compatible in order to supply data and information at no cost. The United States and developed nations have a unique role in developing and maintaining the system, collecting data, enhancing data distribution, and providing models to help the world's nations. Outcomes and benefits of a global informational system will include:

- Disaster reduction
- Integrated water resource management
- Ocean and marine resource monitoring and management
- Weather and air quality monitoring, forecasting, and advisories
- Biodiversity conservation
- Sustainable land use and management
- Public understanding of environmental factors affecting human health and well-being
- Better development of energy resources
- Adaptation to climate variability and change

The quality and quantity of data collected through GEOSS should help improve the prediction, control, and mitigation of many future manmade and natural disasters. Chapters 20 and 21 further discuss GEOSS.

2.6 The art of disaster management

The laws of nature are the same regardless of what type of disaster is considered. A combination of first principles laws, heuristic modeling, data collection, and computers may help, to different degrees of success, the prediction and control of natural and manmade disasters, as discussed in Section 2.4 and several subsequent chapters. Once a disaster strikes, mitigating its adverse effects becomes the primary concern. Disaster management is more art than science, but the management principles are similar for most types of disaster, especially those that strike suddenly and intensely. The organizational skills and resources needed to mitigate the adverse effects of a hurricane are not much different from those required in the aftermath of an earthquake. The scope of the disaster (Section 2.2) determines the extent of the required response. Slowly evolving disasters such as global warming or air pollution are different, and their management requires a different set of skills, response, and political will. Although millions of people may be adversely affected by global warming, the fact that that harm may be spread over decades and thus diluted in time does not provide immediacy to the problem and its potential mitigation. Political will to solve long-range problems—not affecting the next election—is typically nonexistent, except in the case of the rare visionary leader.

In his book, der Heide (1989) states that disasters are the ultimate test of emergency response capability. Once a large-scale disaster strikes, mitigating its adverse effects becomes the primary concern. There are concerns about how to save lives, take care of the survivors' needs, and protect property from any further damage. Dislocated people need shelter, water, food, and medicine. Both the physical and the mental health of the survivors, as well as relatives of the deceased, can be severely jeopardized. Looting, price gouging, and other law-breaking activities need to be contained, minimized, or eliminated. Hospitals need to prioritize and even ration treatments, especially in the face of the practical fact that the less seriously injured tend to arrive at emergency rooms first, perhaps because they transported themselves there. Roads need to be operable and free of landslides, debris, and traffic jams for the unhindered flow of first responders and supplies to the stricken area, and evacuees and ambulances from the same. This is not always the case, especially if the antecedent disaster damages most if not all roads, as occurred after the 2005 Kashmir Earthquake (Section 2.8.7). Buildings, bridges, and roads need to be rebuilt or repaired, and power, potable water, and sewage need to be restored. Chapters 5 to 7 cover some of the logistical, engineering, and medical aspects of disasters.

Lessons learned from one calamity can be applied to improve the response to subsequent ones (Cooper & Block, 2006; Olasky, 2006). Disaster mitigation is not a trial-and-error process, however. Operations research (operational research in Britain) is the discipline that uses the scientific approach to decision making, which seeks to determine how best to design and operate a system, usually under conditions requiring the allocation of scarce resources (Winston, 1994). Churchman et al. (1957) similarly define the genre as the application of scientific methods, techniques, and tools to problems involving the operations of systems so as to provide those in control of the operations with optimum solutions to the problems. Operations research and engineering optimization principles are skillfully used to facilitate recovery and return to normalcy following a large-scale disaster (Chapter 5, this book; Altay & Green, 2006). The always finite resources available must be used so as to maximize their beneficial impact. A lot of uncoordinated, incoherent activities are obviously not a good use of scarce resources. For example, sending huge amounts of perishable food to a stricken area that has no electricity makes little sense. Although it seems silly, it is not difficult to find such examples that were made in the heat of the moment.

Most books on large-scale disasters are written from either a sociologist's or a tactician's point of view, in contrast to the scientist's viewpoint of this book. There are few popular science or high school–level books on disasters, such as Allen (2005) and Engelbert et al. (2001), and even fewer more advanced science books, such as Bunde et al. (2002). The other books deal, for the most part, with the behavioral response to disasters and the art of mitigating their aftermath. Current topics of research include disaster preparedness and behavioral and organizational responses to disasters. A small sample of recent books includes Abbott (2005), Alexander (1993, 2000), Bankoff et al. (2004), Burby (1998), Childs and Dietrich (2002), Cooper and Block (2006), Cutter (2001), de Boer and van Remmen (2003), der Heide (1989), Dilley et al. (2005), Fischer (1998), Gist and Lubin (1999), Kunreuther and Roth (1998), McKee and Guthridge (2006), Mileti (1999), Olasky (2006), Pelling (2003), Pickett and White (1985), Posner (2004), Quarantelli (1998), Smith (1991), Stein and Wysession (2003), Steinberg (2000), Tierney et al. (2001), Tobin (1997), Vale and Campanella (2005), and Wallace and Webber (2004). Chapters 5 to 7 of this book cover some of the logistical, engineering, and medical aspects of disasters, but the rest focus on the modeling and prediction of disasters, particularly weather-related ones.

2.7 A bit of sociology

Although it appears that large-scale disasters are more recent when the past few years are considered, they have actually been with us since *Homo sapiens* set foot on Earth. Frequent disasters struck the planet as far back as the time of its formation. The dinosaur went extinct because a meteorite struck the Earth 65 million years ago. However, if we concern ourselves with humans, then starting 200,000 years ago, ice ages, famines, attacks from rival groups or animals, and infections were constant reminders of human's vulnerability. We survived because we were programmed to do so.

Humans deal with natural and manmade disasters with an uncanny mix of dread, trepidation, curiosity, and resignation, but they often rise to the challenge with acts of resourcefulness, courage, and unselfishness. Disasters are common occurrences in classical and modern literature. William Shakespeare's comedy *The Tempest* opens with a storm that becomes the driving force of the plot and tells of reconciliation after strife. Extreme weather forms the backdrop to three of the Bard's greatest tragedies: *Macbeth*, *Julius Caesar*, and *King Lear*. In *Macbeth*, the tempest is presented as unnatural and is preceded by "portentious things." Men enveloped in fire walked the streets, lions became tame, and night birds howled in the midday sun. Order is inverted, man acts against man, and the gods and elements turn against humanity and mark their outrage with "a tempest dropping fire." In *Julius Caesar*, humanity's abominable actions are accompanied through violent weather. Caesar's murder is plotted while the sea swells, rages, and foams, and "All the sway of earth shakes like a thing unfirm." In *King Lear*—quoted at the beginning of the parts of this book authored by its editor—extreme weather conditions mirror acts of human depravity. The great storm that appears in Act 2, Scene 4, plays a crucial part in aiding Lear's tragic decline deeper into insanity.

On the popular culture front, disaster movies flourish in Hollywood, particularly in times of tribulation. Witness the following sample of the movie genre: *San Francisco* (1936), *A Night to Remember* (1958), *Airport* (1970), *The Poseidon Adventure* (1972), *Earthquake* (1974), *Towering Inferno* (1974), *The Hindenburg* (1975), *The Swarm* (1978), *Meteor* (1979), *Runaway Train* (1985), *The Abyss* (1985), *Outbreak* (1995), *Twister* (1996), *Titanic* (1997), *Volcano* (1997), *Armageddon* (1998), *Deep Impact* (1998), *Flight 93* (2006), *United 93* (2006), and *World Trade Center* (2006).

Does disaster bring out the worst in people? Thomas Glass, professor of epidemiology at The Johns Hopkins University, argues the opposite (Glass, 2001; Glass & Schoch-Spana, 2002). From an evolutionary viewpoint, disasters bring out the best in us. It almost has to be that way. Humans survived ice ages, famines, and infections, not because we were strong or fast, but because in the state of extreme calamity, we tend to be resourceful and cooperative, except when there is a profound sense of injustice—that is, when some group has been mistreated or the system has failed. In such events, greed, selfishness, and violence do occur. A sense of breach of fairness can trigger the worst in people. Examples of those negative connotations include distributing the bird flu vaccine to the rich and mighty first, and the captain and crew escaping a sinking ferry before the passengers. The first of these two examples has not yet occurred, but the second is a real tragedy that recently took place (Section 2.8.10).

If reading history amazes you, you will find that the bird flu pandemic (or similar flu) wiped out a lot of the European population in the seventeenth century, before they cleaned it up. The bright side of a disaster is the reconstruction phase. Disasters are not always bad, even if we think they are. We need to look at what we learn and how we grow to become stronger after a disaster. For example, it is certain that the local, state, and federal officials in the United States are now learning painful lessons from Hurricane Katrina (Section 2.8.6), and that they will try to avoid the same mistakes again. It is up to us humans to learn from mistakes and to not forget them. However, human nature is forgetful, political leaders are not historians, and facts are buried.

The sociologist Henry Fischer (1988) argues that certain human depravities commonly perceived to emerge during disasters (e.g., mob hysteria, panic, shock, looting) are the exception not the rule. The community of individuals does not break down, and the norms that we tend to follow during normal times hold during emergency times. Emergencies bring out the best in us, and we become much more altruistic. Proving his views using several case studies, Fischer writes about people who pulled through a disaster: "Survivors share their tools, their food, their equipment, and especially their time. Groups of survivors tend to emerge to begin automatically responding to the needs of one another. They search for the injured, the dead, and they begin cleanup activities. Police and fire personnel stay on the job, putting the needs of victims and the duty they have sworn to uphold before their own personal needs and concern. The commonly held view of behavior is incorrect (pp. 18–19)." Fischer's observations are commonly accepted among modern sociologists. Indeed, as stated previously, we survived the numerous disasters encountered throughout the ages because we were programmed to do so.

2.8 Few recent disasters

It is always useful to learn from past disasters and to prepare better for the next one. Losses of lives and property from recent years are staggering. Not counting the manmade disasters that were tallied in Section 2.2, some frightening numbers from natural calamities alone, as quoted in Chapter 5, are

- Seven hundred natural disasters in 2003, which caused 75,000 deaths (almost seven times the number in 2002), 213 million people adversely affected to some degree, and $65 billion in economic losses
- In 2004, 244,577 persons killed globally as a result of natural disasters
- In 2005, $150 billion in economic losses, with hurricanes Katrina and Rita, which ravaged the Gulf Coast of the United States, responsible for 88% of that amount
- Within the first half of 2006, natural disasters already caused 12,718 deaths and $2.3 billion in economic damages

In the following twelve subsections we briefly recall a few manmade and natural disasters. The information herein and the accompanying photographs are mostly as reported in the online encyclopedia *Wikipedia* (http://en.wikipedia.org/wiki/Main_Page). The numerical data were cross-checked using archival media reports from such sources as *The New York Times* and ABC News. The numbers did not always match, and more than one source was consulted to reach the most reliable results. Absolute accuracy is not guaranteed, however. The dozen or so disasters sampled herein are not by any stretch of the imagination comprehensive, merely a few examples that may present important lessons for future calamities. Remember, they strike Earth at the average rate of three per day!

2.8.1 San Francisco Earthquake

A major earthquake of magnitude 7.8 on the Richter scale struck the city of San Francisco, California, at around 5:12 am, Wednesday, 18 April 1906. The *Great Earthquake*, as it became known, was along the San Andreas Fault with its epicenter close to the city. Its violent shocks were felt from Oregon to Los Angeles and inland as far as central Nevada. The earthquake and resulting fires would go down in history as one of the worst natural disasters to hit a major U.S. city.

At the time only 478 deaths were reported, a figure concocted by government officials who believed that reporting the true death toll would hurt real estate prices and efforts to rebuild the city. This figure has been revised to today's conservative estimate of more than 3,000 victims. Most of the deaths occurred in San Francisco, but 189 were reported

Figure 2.5   San Francisco after the 1906 earthquake.

elsewhere across the San Francisco Bay Area. Other places in the Bay Area such as Santa Rosa, San Jose, and Stanford University also received severe damage.

Between 225,000 and 300,000 people were left homeless, out of a population of about 400,000. Half of these refugees fled across the bay to Oakland, in an evacuation similar to the Dunkirk Evacuation that would occur years later. Newspapers at the time described Golden Gate Park, the Panhandle, and the beaches between Ingleside and North Beach as covered with makeshift tents. The overall cost of the damage from the earthquake was estimated at the time to be around $400 million. The earthquake's notoriety rests in part on the fact that it was the first natural disaster of its magnitude to be captured by photography. Furthermore, it occurred at a time when the science of seismology was blossoming. Figures 2.5 and 2.6 depict the devastation.

Eight decades after the Great Earthquake, another big one struck the region. This became known as the Loma Prieta Earthquake. At 5:04 pm on 17 October 1989, a magnitude 7.1 earthquake on the Richter scale severely shook the San Francisco and Monterey Bay regions. The epicenter was located at 37.04°N latitude, 121.88°W longitude near Loma Prieta peak in the Santa Cruz Mountains, approximately 14 km northeast of Santa Cruz and 96 km south–southeast of San Francisco. The tremor lasted for 15 seconds and occurred when the crustal rocks comprising the Pacific and North American Plates abruptly slipped as much as 2 m along their common boundary—the San Andreas Fault system (Section 2.4.9). The rupture initiated at a depth of 18 km and extended 35 km along the fault, but it did not break the surface of the Earth.

This major earthquake caused severe damage as far as 110 km away; most notably in San Francisco, Oakland, the San Francisco Peninsula, and in areas closer to the epicenter in the communities of Santa Cruz, the Monterey Bay, Watsonville, and Los Gatos. Most of the major property damage in the more distant areas resulted from liquefaction of soil used over the years to fill in the waterfront and then built on. The magnitude and distance of the earthquake from the severe damage to the north were surprising to geotechnologists. Subsequent analysis indicates that the damage was likely due to reflected seismic waves—the reflection from well-known deep discontinuities in the Earth's gross structure, about 25 km below the surface.

Figure 2.6   Fires erupted over many parts of the city shortly after the tremor.

There were at least 66 deaths and 3,757 injuries as a result of this earthquake. The highest concentration of fatalities, 42, occurred in the collapse of the Cypress structure on the Nimitz Freeway (Interstate 880), where a double-decker portion of the freeway collapsed, crushing the cars on the lower deck. One 15-m section of the San Francisco–Oakland Bay Bridge also collapsed, causing two cars to fall to the deck below and leading to a single fatality. The bridge was closed for repairs for 1 month.

Because this earthquake occurred during the evening rush hour, there could have been a large number of cars on the freeways at the time, which on the Cypress structure could have endangered many hundreds of commuters. Very fortunately, and in an unusual convergence of events, the two local Major League Baseball teams, the Oakland Athletics and the San Francisco Giants, were about to start their third game of the World Series, which was scheduled to start shortly after 5:30 pm. Many people had left work early or were participating in early afterwork group viewings and parties. As a consequence, the usually crowded highways were experiencing exceptionally light traffic at the time.

Extensive damage also occurred in San Francisco's Marina District, where many expensive homes built on filled ground collapsed. Fires raged in some sections of the city as water mains broke. The San Francisco's fireboat Phoenix was used to pump salt water from San Francisco Bay using hoses dragged through the streets by citizen volunteers. Power was cut to most of San Francisco and was not fully restored for several days. Deaths in Santa Cruz occurred when brick storefronts and sidewalls in the historic downtown, which was then called the Pacific Garden Mall, tumbled down on people exiting the buildings. A sample of the devastation is shown in the three photographs in Figures 2.7 to 2.9. The earthquake also caused an estimated $6 billion in property damage, the costliest natural disaster in U.S. history at the time. It was the largest earthquake to occur on the San Andreas Fault since the Great Earthquake. Private donations poured in to aid relief efforts, and on 26 October 1986, President George H. W. Bush signed a $3.45-billion earthquake relief package for California.

2.8 Few recent disasters

Figure 2.7 Land break near the Loma Prieta Earthquake's epicenter. The man taking a photograph provides a length scale for the width of the hole.

Figure 2.8 A car is crushed by a collapsed rowhouse.

Figure 2.9 Elegant homes were not exempt from the devastation.

2.8.2 Hyatt Regency walkway collapse

The Hyatt Regency Hotel was built in Kansas City, Missouri, in 1978. A state-of-the-art facility, this hotel boasted a forty-story hotel tower and conference facilities. These two components were connected by an open-concept atrium, within which three suspended walkways connected the hotel and conference facilities on the second, third, and fourth levels. Due to their suspension, these walkways were referred to as "floating walkways" or "skyways." The atrium boasted 1,580 m^2 and was 15 m high. It seemed incredulous that such an architectural masterpiece could be involved in the United States' most devastating structural failure (not caused by earthquake, explosion, or airplane crash) in terms of loss of life and injuries.

It was 17 July 1981, when the guests at Kansas City Hyatt Regency Hotel witnessed the catastrophe depicted in Figure 2.10. Approximately 2,000 people were gathered to watch a dance contest in the hotel lobby. Although the majority of the guests were on the ground level, some were dancing on the floating walkways on the second, third, and fourth levels. At about 7:05 pm, a loud crack was heard as the second- and fourth-level walkways collapsed onto the ground level. This disaster took the lives of 114 people and left more than 200 injured.

What did we learn from this manmade disaster? The project for constructing this particular hotel began in 1976 with Gillum–Colaco International, Inc., as the consulting structural engineering firm. Gillum–Colaco Engineering (G.C.E.) provided input into various plans that were being made by the architect and owner, and were contracted in 1978 to provide "all structural engineering services for a 750-room hotel project." Construction began in the spring of 1978. In the winter of 1978, Havens Steel Company entered the contract to fabricate and erect the atrium steel for the project under the standards of the American

Figure 2.10 Collapsed walkway in the Kansas City Hyatt Regency Hotel.

Institute of Steel Construction for steel fabricators. During construction in October 1979, part of the atrium roof collapsed. An inspection team was brought in to investigate the collapse, and G.C.E. vowed to review all steel connections in the structure, including that of the roof.

The proposed structure details of the three walkways were as follows:

- Wide-flange beams were to be used on either side of the walkway, which was hung from a box beam.
- A clip angle was welded to the top of the box beam, which connected to the flange beams with bolts.
- One end of the walkway was welded to a fixed plate, while the other end was supported by a sliding bearing.

- Each box beam of the walkway was supported by a washer and a nut that were threaded onto the supporting rod. Because the bolt connection to the wide flange had virtually no movement, it was modeled as a hinge. The fixed end of the walkway was also modeled as a hinge, while the bearing end was modeled as a roller.

Due to disputes between the G.C.E. and Havens, design changes from a single- to a double-hanger, rod-box beam connection were implemented. Havens did not want to have to thread the entire rod in order to install the washer and nut. This revised design consisted of the following:

- One end of each support rod was attached to the atrium's roof cross-beams.
- The bottom end went through the box beam where a washer and nut were threaded onto the supporting rod.
- The second rod was attached to the box beam 10 cm from the first rod.
- Additional rods suspended downward to support the second level in a similar manner.

Why did the design fail? Due to the addition of another rod in the actual design, the load on the nut connecting the fourth-floor segment was increased. The original load for each hanger rod was to be 90 kN, but with the design alteration the load was doubled to 181 kN for the fourth-floor box beam. Because the box beams were longitudinally welded, as proposed in the original design, they could not hold the weight of the two walkways. During the collapse, the box beam split and the support rod pulled through the box beam, resulting in the fourth- and second-level walkways falling to the ground level.

The following paradigm clarifies the design failure of the walkways quite well. Suppose a long rope is hanging from a tree, and two people are holding onto the rope, one at the top and one near the bottom. Under the conditions that each person can hold their own body weight and that the tree and rope can hold both people, the structure would be stable. However, if one person was to hold onto the rope, and the other person was hanging onto the legs of the first, then the first person's hands must hold both people's body weights, and thus the grip of the top person would be more likely to fail. The initial design is similar to the two people hanging onto the rope, while the actual design is similar to the second person hanging from the first person's legs. The first person's grip is comparable to the fourth-level hanger-rod connection. The failure of this grip caused the walkway collapse.

Who was responsible? One of the major problems with the Hyatt Regency project was lack of communication between parties. In particular, the drawings prepared by G.C.E. were only preliminary sketches but were interpreted by Havens as finalized drawings. These drawings were then used to create the components of the structure. Another large error was G.C.E.'s failure to review the final design, which would have allowed them to catch the error in increasing the load on the connections. As a result, the engineers employed by G.C.E., who affixed their seals to the drawings, lost their engineering licenses in the states of Missouri and Texas. G.C.E. also lost its ability to be an engineering firm.

An engineer has a responsibility to his or her employer and, most important, to society. In the Hyatt Regency case, the lives of the public were hinged on G.C.E.'s ability to design a structurally sound walkway system. Their insufficient review of the final design lead to the failure of the design and a massive loss of life. Cases such as the Hyatt Regency walkway collapse are a constant reminder of how an error in judgment can create a catastrophe. It is important that events in the past are remembered so that engineers will always fulfill their responsibility to society.

Figure 2.11 Following the Izmit Earthquake, a mosque stood with few other structures amid the rubble of collapsed buildings in the town of Gölcük.

2.8.3 Izmit Earthquake

On 17 August 1999, the Izmit Earthquake with a magnitude of 7.4 struck northwestern Turkey. It lasted 45 seconds and killed more than 17,000 people according to the government report. Unofficial albeit credible reports of more than 35,000 deaths were also made. Within 2 hours, 130 aftershocks were recorded and two tsunamis were observed.

The earthquake had a rupture length of 150 km from the city of Düzce to the Sea of Marmara along the Gulf of Izmit. Movements along the rupture were as large as 5.7 m. The rupture passed through major cities that are among the most industrialized and urban areas of Turkey, including oil refineries, several car companies, and the navy headquarters and arsenal in Gölcük, thus increasing the severity of the life and property loss.

This earthquake occurred in the North Anatolian Fault Zone (NAFZ). The Anatolian Plate, which consists primarily of Turkey, is being pushed west by about 2 to 2.5 cm/yr because it is squeezed between the Eurasian Plate on the north, and both the African Plate and the Arabian Plate on the south. Most of the large earthquakes in Turkey result as slip occurs along the NAFZ or a second fault to the east, the Eastern Anatolian Fault.

Impacts of the earthquake were vast. These included in the short term, 4,000 buildings destroyed, including an army barracks, an ice skating rink, and refrigerated lorries used as mortuaries; cholera, typhoid, and dysentery were spread; homelessness and posttraumatic stress disorder were experieced in around 25% of those living in the tent city set up by officials for the homeless. An oil refinery leaked into the water supply and Izmit Bay and, subsequently, caught fire. Because of the leak and the fire, the already highly polluted bay saw a two- to threefold increase in polycyclic aromatic hydrocarbon levels compared to 1984 samples. Dissolved oxygen and chlorophyll reached their lowest levels in 15 years. Economic development was set back 15 years, and the direct damage of property was estimated at $18 billion, a huge sum for a developing country. A curious scene of the damage in the town of Gölcük, 100 km east of Istanbul, is depicted in Figure 2.11.

Figure 2.12 United Airlines Flight 175 goes through the southern tower of the World Trade Center. This scene was captured on live TV because filming crews were already photographing the northern tower, which was attacked 18 minutes earlier.

2.8.4 September 11

A series of coordinated suicide attacks upon the United States were carried out on Tuesday, 11 September, 2001, in which nineteen hijackers took control of four domestic commercial airliners. The terrorists crashed two planes into the World Trade Center in Manhattan, New York City, one into each of the two tallest towers, about 18 minutes apart (Figure 2.12). Within 2 hours, both towers had collapsed. The hijackers crashed the third aircraft into the

Pentagon, the U.S. Department of Defense headquarters, in Arlington County, Virginia. The fourth plane crashed into a rural field in Somerset County, Pennsylvania, 129 km east of Pittsburgh, following passenger resistance. The official count records 2,986 deaths in the attacks, including the hijackers, the worst act of war against the United States on its own soil.[9]

The National Commission on Terrorist Attacks Upon the United States (9/11 Commission) states in its final report that the nineteen hijackers who carried out the attack were terrorists, affiliated with the Islamic Al-Qaeda organization. The report named Osama bin Laden, a Saudi national, as the leader of Al-Qaeda, and as the person ultimately suspected as being responsible for the attacks, with the actual planning being undertaken by Khalid Shaikh Mohammed. Bin Laden categorically denied involvement in two 2001 statements, before admitting a direct link to the attacks in a subsequent taped statement.

The 9/11 Commission reported that these hijackers turned the planes into the largest suicide bombs in history. The 9/11 attacks are among the most significant events to have occurred so far in the twenty-first century in terms of the profound economic, social, political, cultural, psychological, and military effects that followed in the United States and many other parts of the world.

Following the September 11 disaster, the Global War on Terrorism was launched by the United States, enlisting the support of NATO members and other allies, with the stated goal of ending international terrorism and state sponsorship of the same. The difficulty of the war on terrorism, now raging for more than 6 years, is that it is mostly a struggle between a super power and a nebulously defined enemy: thousands of stateless, loosely connected, disorganized, undisciplined religion fanatics scattered around the globe, but particularly in Africa, Middle East, South Asia, and Southeast Asia.

2.8.5 *Pacific Tsunami*

A tsunami is a series of waves generated when water in a lake or a sea is rapidly displaced on a massive scale. Earthquakes, landslides, volcanic eruptions, and large meteorite impacts all have the potential to generate a tsunami. The effects of a tsunami can range from unnoticeable to devastating. The Japanese term "tsunami" means harbor and wave. The term was created by fishermen who returned to port to find the area surrounding the harbor devastated, although they had not been aware of any wave in the open water. A tsunami is not a subsurface event in the deep ocean; it simply has a much smaller amplitude offshore and a very long wavelength (often hundreds of kilometers long), which is why it generally passes unnoticed at sea, forming only a passing "hump" in the ocean.

Tsunamis have been historically referred to as tidal waves because as they approach land, they take on the characteristics of a violent onrushing tide rather than the more familiar cresting waves that are formed by wind action on the ocean. However, because tsunamis are not actually related to tides, the term is considered misleading and its usage is discouraged by oceanographers. Chapter 11 focuses on the powerful phenomenon.

The 2004 Indian Ocean Earthquake, known by the scientific community as the Sumatra–Andaman Earthquake, was an undersea earthquake that occurred at 00:58:53 UTC (07:58:53 local time) on 26 December 2004. According to the U.S. Geological Survey (USGS), the earthquake and its tsunami killed more than 283,100 people, making it one of the deadliest disasters in modern history. Indonesia suffered the worse loss of life at more than 168,000.

---

[9] The Imperial Japanese Navy's surprise attack on Pearl Harbor, Oahu, Hawaii, on the morning of 7 December 1941, was aimed at the Pacific Fleet and killed 2,403 American servicemen and 68 civilians.

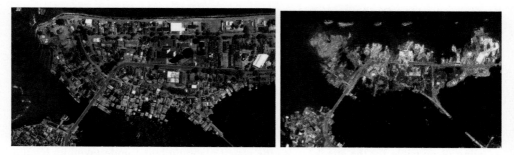

Figure 2.13   Satellite photographs of an island before and after the Pacific Tsunami. The total devastation of the entire area is clear.

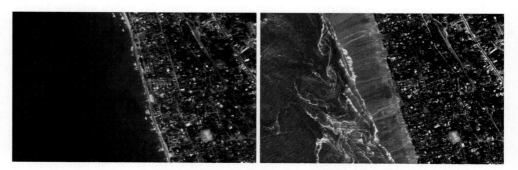

Figure 2.14   Satellite photographs of a coastal area. The receding tsunami wave is shown in the bottom photograph.

The disaster is known in Asia and the media as the Asian Tsunami; in Australia, New Zealand, Canada and the United Kingdom, it is known as the Boxing Day Tsunami because it took place on Boxing Day, although it was still Christmas Day in the Western Hemisphere when the disaster struck.

The earthquake originated in the Indian Ocean just north of Simeulue Island, off the western coast of northern Sumatra, Indonesia. Various values were given for the magnitude of the earthquake that triggered the giant wave, ranging from 9.0 to 9.3 (which would make it the second largest earthquake ever recorded on a seismograph), although authoritative estimates now put the magnitude at 9.15. In May 2005, scientists reported that the earthquake itself lasted close to 10 minutes, even though most major earthquakes last no more than a few seconds; it caused the entire planet to vibrate at least a few centimeters. It also triggered earthquakes elsewhere, as far away as Alaska.

The resulting tsunami devastated the shores of Indonesia, Sri Lanka, South India, Thailand, and other countries with waves up to 30 m high. The tsunami caused serious damage and death as far as the east coast of Africa, with the furthest recorded death due to the tsunami occurring at Port Elizabeth in South Africa, 8,000 km away from the epicentre. Figures 2.13 and 2.14 show examples of the devastation caused by one of the deadliest calamities of the twenty-first century. The plight of the many affected people and countries prompted a widespread humanitarian response.

Unlike in the Pacific Ocean, there is no organized alert service covering the Indian Ocean. This is partly due to the absence of major tsunami events between 1883 (the Krakatoa eruption, which killed 36,000 people) and 2004. In light of the 2004 Indian Ocean Tsunami, UNESCO and other world bodies have called for a global tsunami monitoring system.

Human's actions caused this particular natural disaster to become more damaging than it would otherwise. The intense coral reef mining off the Sri Lankan coast, which removed the sort of natural barrier that could mitigate the force of waves, amplified the disastrous effects of the tsunami. As a result of such mining, the 2004 Pacific Tsunami devastated Sri Lanka much more than it would have otherwise (Chapter 11).

2.8.6 Hurricane Katrina

Hurricane Katrina was the eleventh named tropical storm, fourth hurricane, third major hurricane, and first category 5 hurricane of the 2005 Atlantic hurricane season. It was the third most powerful storm of the season, behind Hurricane Wilma and Hurricane Rita, and the sixth strongest storm ever recorded in the Atlantic basin. It first made landfall as a category 1 hurricane just north of Miami, Florida, on 25 August 2005, resulting in a dozen deaths in South Florida and spawning several tornadoes, which fortunately did not strike any dwellings. In the Gulf of Mexico, Katrina strengthened into a formidable category 5 hurricane with maximum winds of 280 km/h and minimum central pressure of 902 mbar. It weakened considerably as it was approaching land, making its second landfall on the morning of 29 August along the Central Gulf Coast near Buras-Triumph, Louisiana, with 200 km/h winds and 920 mbar central pressure, a strong category 3 storm, having just weakened from category 4 as it was making landfall.

The sheer physical size of Katrina caused devastation far from the eye of the hurricane; it was possibly the largest hurricane of its strength ever recorded, but estimating the size of storms from before the presatellite 1960s era is difficult to impossible. On 29 August, Katrina's storm surge breached the levee system that protected New Orleans from Lake Pontchartrain and the Mississippi River. Most of the city was subsequently flooded, mainly by water from the lake. Heavy damage was also inflicted onto the coasts of Mississippi and Alabama, making Katrina the most destructive and costliest natural disaster in the history of the United States and the deadliest since the 1928 Okeechobee Hurricane.

The official combined direct and indirect death toll now stands at 1,836, the fourth highest in U.S. history, behind the Galveston Hurricane of 1900, the 1893 Sea Islands Hurricane, and possibly the 1893 Chenier Caminanda Hurricane, and ahead of the Okeechobee Hurricane of 1928.

More than 1.2 million people were under an evacuation order before landfall. In Louisiana, the hurricane's eye made landfall at 6:10 am CDT on Monday, 29 August. After 11:00 am CDT, several sections of the levee system in New Orleans collapsed. By early September, people were being forcibly evacuated, mostly by bus to neighboring states. More than 1.5 million people were displaced—a humanitarian crisis on a scale unseen in the United States since the Great Depression. The damage is now estimated to be about $81.2 billion (2005 U.S. dollars) more than double the previously most expensive Hurricane Andrew, making Katrina the most expensive natural disaster in U.S. history.

Federal disaster declarations blanketed 233,000 km^2 of the United States, an area almost as large as the United Kingdom. The hurricane left an estimated 3 million people without electricity, taking some places several weeks for power to be restored (but faster than the 4 months originally predicted). Referring to the hurricane itself plus the flooding of New Orleans, Homeland Security Secretary Michael Chertoff described on 3 September that the aftermath of Hurricane Katrina as "probably the worst catastrophe, or set of catastrophes" in U.S. history.

The devastation of Katrina is depicted in Figures 2.15 to 2.20. The aftermath of the hurricane produced the perfect political storm whose winds lasted long after the hurricane.

Figure 2.15 Aerial photograph of flooded New Orleans.

Figure 2.16 Wind damage.

2.8 Few recent disasters

Figure 2.17 Flooded homes due to the levee's breakdown.

Figure 2.18 Wreckage of cars and homes are mingled in a heap of destruction.

Figure 2.19 Boats washed ashore.

Figure 2.20 This boat was found many kilometers from the water.

Congressional investigations reaffirmed what many have suspected: Governments at all levels failed. The city of New Orleans, the state of Louisiana, and the United States let the citizenry down. The whole episode was a study in ineptitude—and in buck passing that fooled no one. The then-director of the Federal Emergency Management Agency (FEMA), Michael Brown, did not know that thousands of New Orleanians were trapped in the Superdome with subhuman conditions. In the middle of the bungled response, President George W. Bush uttered his infamous phrase "Brownie, you're doin' a heckuva job." Several books were published in the aftermath of the calamity, mostly offering scathing criticism of the government as well as more sensible strategies to handle future crises (e.g., Cooper & Block, 2006; Olasky, 2006).

On 23 October 2007, slightly more than two years after the Katrina debacle, the new FEMA Deputy Administrator, Vice Admiral Harvey E. Johnson, held a news conference as wildfires raged in California. The briefing went very well and was carried out live on several news outlets. Only problem, all present were FEMA staffers playing reporters! FEMA yet once agian became the subject of national ridicule. In a Washington Post column entitled "FEMA Meets the Press, Which Happens to Be ... FEMA" (26 October 2007, p. A19), Al Kamen derided the notorious government agency, "FEMA has truly learned the lessons of Katrina. Even its handling of the media has improved dramatically."

2.8.7 Kashmir Earthquake

The Kashmir Earthquake—aka the Northern Pakistan Earthquake or South Asia Earthquake—of 2005 was a major seismological disturbance that occurred at 08:50:38 Pakistan Standard Time (03:50:38 UTC, 09:20:38 India Standard Time, 08:50:38 local time at epicenter) on 8 October 2005, with the epicenter in the Pakistan-administered region of the disputed territory of Kashmir in South Asia. It registered 7.6 on the Richter scale, making it a major earthquake similar in intensity to the 1935 Quetta Earthquake, the 2001 Gujarat Earthquake, and the 1906 San Francisco Earthquake.

Most of the casualties from the earthquake were in Pakistan where the official death toll is 73,276, putting it higher than one massive scale of destruction of the Quetta earthquake of 31 May 1935. Most of the affected areas are in mountainous regions and access is impeded by landslides that have blocked the roads. An estimated 3.3 million people were left homeless in Pakistan. According to Indian officials, nearly 1,400 people died in the Indian-administered Kashmire. The United Nations (UN) reported that more than 4 million people were directly affected. Many of them were at risk of dying from cold and the spread of disease as winter began. Pakistan Prime Minister Shaukat Aziz made an appeal to survivors on 26 October to come down to valleys and cities for relief. It has been estimated that damages incurred are well more than $5 billion US. Three of the five crossing points have been opened on the line of control between India and Pakistan. Figure 2.21 depicts a small sample of the utter devastation.

2.8.8 Hurricane Wilma

In the second week of October 2005, a large and complex area of low pressure developed over the western Atlantic and eastern Caribbean with several centers of thunderstorm activity. This area of disturbed weather southwest of Jamaica slowly organized on 15 October 2005 into tropical depression number 24. It reached tropical storm strength at 5:00 am EDT on 17 October, making it the first storm ever to use a "W" name since alphabetical naming began in 1950, and tying the 1933 record for most storms in a season. Moving

Figure 2.21 Homes crumbled under the intense Kashmir Earthquake.

slowly over warm water with little wind shear, tropical storm Wilma strengthened steadily and became a hurricane on 18 October. This made it the twelfth hurricane of the season, tying the record set in 1969.

Hurricane Wilma was the sixth major hurricane of the record-breaking 2005 Atlantic hurricane season. Wilma set numerous records for both strength and seasonal activity. At its peak, it was the most intense tropical cyclone ever recorded in the Atlantic Basin. It was the third category 5 hurricane of the season (the other two being hurricanes Katrina and Rita), the only time this has occurred in the Atlantic, and only the third category 5 to develop in October. Wilma was the second twenty-first storm in any season and the earliest-forming twenty-first storm by nearly a month.

Wilma made several landfalls, with the most destructive effects experienced in the Yucatán Peninsula of Mexico, Cuba, and the U.S. state of Florida. At least 60 deaths were reported, and damage is estimated at between $18 billion and $22 billion, with $14.4 billion in the United States alone, ranking Wilma among the top ten costliest hurricanes ever recorded in the Atlantic and the fifth costliest storm in U.S. history.

Figures 2.22 to 2.25 show different aspects of Hurricane Wilma. Around 4:00 pm EDT on 18 October 2005, the storm began to intensify rapidly. During a 10-hour period, Hurricane Hunter aircraft measured a 78-mbar pressure drop. In a 24-hour period from 8:00 am EDT 18 October to the following morning, the pressure fell 90 mbar. In this same 24-hour period, Wilma strengthened from a strong tropical storm with 110 km/h winds to a powerful category 5 hurricane with 280 km/h winds. In comparison, Hurricane Gilbert of 1988—the previous recordholder for lowest Atlantic pressure—recorded a 78-mbar pressure drop in

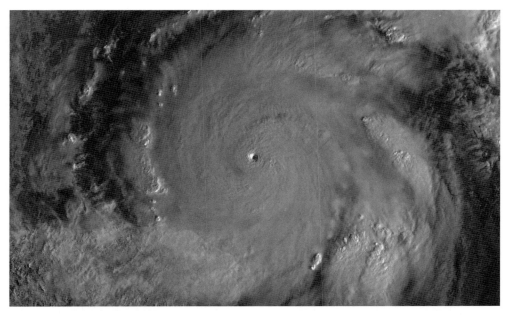

Figure 2.22 Visible image of Hurricane Wilma near record intensity with a central pressure of 882 mbars. Image captured by satellite at 9:15 EDT on 19 October 2005.

Figure 2.23 Hurricane Wilma as it is ready to leave the Atlantic side of Florida. Satellite image courtesy of The Weather Channel (www.weather.com).

Figure 2.24 The destruction of Wilma.

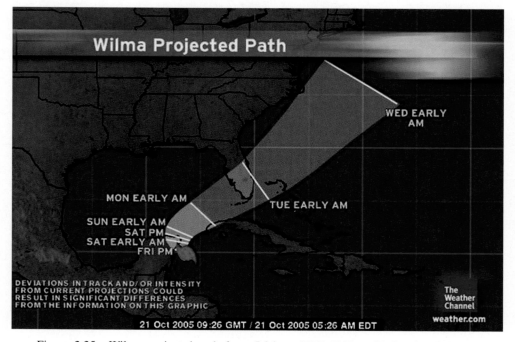

Figure 2.25 Wilma projected path from 5:26 am EDT, Friday, 21 October 2005, to early morning Wednesday, 26 October 2005. The 5-day forecast is reasonably accurate. Photograph courtesy of The Weather Channel (www.weather.com).

a 24-hour period for a 3 mbar/h pressure drop. This is a record for the Atlantic Basin and is one of the most rapid deepening phases ever undergone by a tropical cyclone anywhere on Earth—the record holder is 100 mbar by Super Typhoon Forrest in 1983.

During its intensification on 19 October 19, the eye's diameter shrank to 3 km—one of the smallest eyes ever seen in a tropical cyclone. Quickly thereafter, Wilma set a record for the lowest pressure ever recorded in an Atlantic hurricane when its central pressure dropped to 884 mbar at 8:00 am EDT and then dropped again to 882 mbar 3 hours later before rising slowly in the afternoon, while remaining a category 5 hurricane. In addition, at 11:00 pm EDT that day, Wilma's pressure dropped again to 894 mbar, as the storm weakened to a category 4 with winds of 250 km/h. Wilma was the first hurricane ever in the Atlantic Basin, and possibly the first tropical cyclone in any basin, to have a central pressure below 900 mbar while at category 4 intensity. In fact, only two other recorded Atlantic hurricanes have ever had lower pressures at this intensity; these two storms being previous Atlantic record holder Hurricane Gilbert of 1988 and the Labor Day Hurricane of 1935.

Although Wilma was the most intense hurricane (i.e., a tropical cyclone in the Atlantic, Central Pacific, or Eastern Pacific) ever recorded, there have been many more intense typhoons in the Pacific. Super Typhoon Tip is the most intense tropical cyclone on record at 870 mbar. Hurricane Wilma existed within an area of ambient pressure that was unusually low for the Atlantic Basin, with ambient pressures below 1,010 mbar. These are closer to ambient pressures in the northwest Pacific Basin. Indeed, under normal circumstances, the Dvorak matrix would equate an 890-mbar storm in the Atlantic basin—a Current Intensity (CI) number of 8—with an 858-mbar storm in the Pacific. Such a conversion, if normal considerations were in play, would suggest that Wilma was more intense than Tip. However, Wilma's winds were much slower than the 315 km/h implied by an 8 on the Dvorak scale. A speed of 280+ km/h may seem incredibly fast, but for an 882-mbar hurricane it is actually quite slow. In comparison, Hurricane Gilbert had a pressure of 888 mbar but winds of 300 km/h. In fact, at one point after Wilma's period of peak intensity, it had a pressure of 894 mbar, but was actually not even a category 5, with winds of just 250 km/h. Before Wilma, it had been unheard of for a storm to go under 900 mbar and not be a category 5. These wind speeds indicate that the low ambient pressure surrounding Wilma caused the 882-mbar pressure to be less significant than under normal circumstances, involving a lesser pressure gradient. By the gradient standard, it is entirely possible that Hurricane Gilbert, and not Wilma, is still the strongest North Atlantic hurricane on record.

Hurricane Wilma's southeast eyewall passed the greater Key West area in the lower Florida Keys in the early morning hours of 24 October 2005. At this point, the storm's eye was approximately 56 km in diameter, and the north end of the eye wall crossed into the south and central section of Palm Beach County as the system cut a diagonal swath across the southern portion of the Florida peninsula. Several cities in the South Florida Metropolitan Area, which includes Palm Beach, Fort Lauderdale, and Miami, suffered severe damage as a result of the intense winds of the rapidly moving system. The center of the eye was directly over the South Florida Metropolitan Area at 10:30 am on Monday, 24 October. After the hurricane had already passed, there was a 3-m storm surge from the Gulf of Mexico that completely inundated a large portion of the lower Keys. Most of the streets in and near Key West were flooded with at least 1 m of salt water, causing the destruction of tens of thousands of vehicles. Many houses were also flooded with 0.5 m of seawater.

Despite significant wind shear in the Gulf, Hurricane Wilma regained some strength before making a third landfall just north of Everglades City, Florida, near Cape Romano, at 6:30 am EDT, 24 October 2005, as a category 3 hurricane. The reintensification of Hurricane Wilma was due to its interaction with the Gulf Loop Current. At landfall, Wilma had sustained winds of 200 km/h. Over the Florida peninsula, Wilma weakened slightly to a category 2 hurricane, and exited Florida and entered the Atlantic at that strength about 6 hours later. Unexpectedly, Wilma regained strength over the Gulf Stream and once again became a category 3 hurricane north of the Bahamas, regaining all the strength it lost within 12 hours. However, on 25 October, the storm gradually began weakening and became extratropical late that afternoon south of Nova Scotia, although it still maintained hurricane strength and affected a large area of land and water with stormy conditions.

2.8.9 Hajj stampede of 2006

There have been many serious incidents during the Hajj that have led to the loss of hundreds of lives. The Hajj is the Islamic annual pilgrimage to the city of Mecca, Saudi Arabia. There are an estimated 1.3 billion Muslims living today, and during the month of the Hajj, the city of Mecca must cope with as many as 4 million pilgrims. The Muslim world follows a lunar calendar, and therefore, the Hajj month shifts from year to year relative to the Western, solar calendar.

Jet travel also makes Mecca and the Hajj more accessible to pilgrims from all over the world. As a consequence, the Hajj has become increasingly crowded. City officials are consequently required to control large crowds and provide food, shelter, and sanitation for millions. Unfortunately, they have not always been able to prevent disasters, which are hard to avoid with so many people. The worst of the incidents has occurred during the ritual stoning of the devil, an event near the tail end of the Hajj. Saudi authorities had replaced the pillar, which had represented the devil in the past, with an oval wall with padding around the edges to protect the crush of pilgrims. The officials had also installed cameras and dispatched about 60,000 security personnel to monitor the crowds.

On 12 January 2006, a stampede during the ritual stoning of the devil on the last day of the Hajj in Mina, Saudi Arabia, killed at least 346 pilgrims and injured at least 289 more. The stoning ritual is the most dangerous part of the pilgrimage because the ritual can cause people to be crushed, particularly as they traverse the massive two-layer flyover-style Jamarat Bridge (Figure 2.26) that affords access to the pillars. The incident occurred shortly after 1:00 pm local time, when a passenger bus shed its load of travelers at the eastern access ramps to the bridge. This caused pilgrims to trip, rapidly resulting in a lethal crush. An estimated 2 million people were performing the ritual at the time. Figure 2.27 shows the mobile morgue the Saudi officials moved to the scene of the stampede. Tragically, the stampede was the second fatal tragedy of the Islamic month of Dhu al-Hijjah in 2006. On 5 January 2006, the Al Ghaza Hotel had collapsed. The death toll was seventy-six, and the number of injured was sixty-four.

There is a long and tragic history for the Hajj stampede. The surging crowds, trekking from one station of the pilgrimage to the next, cause a stampede. Panic spreads, pilgrims jostle to avoid being trampled, and hundreds of deaths can result. A list of stampede and other accidents during the Hajj season follows.

- In December 1975, an exploding gas cylinder caused a fire in a tent colony; 200 pilgrims were killed.
- On 20 November 1979, a group of approximately 200 militant Muslims occupied Mecca's Grand Mosque. They were driven out by special commandos—allowed into the city under these special circumstances despite their being non-Muslims—after bloody fighting that left 250 people dead and 600 wounded.
- On 31 July 1987, Iranian pilgrims rioted, causing the deaths of more than 400 people.

2.8 *Few recent disasters* 55

Figure 2.26 The massive two-layer flyover-style Jamarat Bridge.

Figure 2.27 Mobile morgue for victims of the stampede.

- On 9 July 1989, two bombs exploded, killing one pilgrim and wounding sixteen. Saudi authorities beheaded sixteen Kuwaiti Shiite Muslims for the bombings after originally suspecting Iranian terrorists.
- On 15 April 1997, 343 pilgrims were killed and 1,500 injured in a tent fire.
- On 2 July 1990, a stampede inside a pedestrian tunnel—Al-Ma'aisim tunnel—leading out from Mecca toward Mina and the Plains of Arafat led to the deaths of 1,426 pilgrims.
- On 23 May 1994, a stampede killed at least 270 pilgrims at the stoning of the devil ritual.
- On 9 April 1998, at least 118 pilgrims were trampled to death and 180 injured in an incident on Jamarat Bridge.
- On 5 March 2001, 35 pilgrims were trampled in a stampede during the stoning of the devil ritual.
- On 11 February 2003, the stoning of the devil ritual claimed 14 pilgrims' lives.
- On 1 February 2004, 251 pilgrims were killed and another 244 injured in a stampede during the stoning ritual in Mina.
- A concrete multistory building located in Mecca close to the Grand Mosque collapsed on 5 January 2006. The building—Al Ghaza Hotel—is said to have housed a restaurant, a convenience store, and a hostel. The hostel was reported to have been housing pilgrims to the 2006 Hajj. It is not clear how many pilgrims were in the hotel at the time of the collapse. As of the latest reports, the death toll is 76, and the number of injured is 64.

Critics say that the Saudi government should have done more to prevent such tragedies. The Saudi government insists that any such mass gatherings are inherently dangerous and difficult to handle, and that they have taken a number of steps to prevent problems.

One of the biggest steps, that is also controversial is a new system of registrations, passports, and travel visas to control the flow of pilgrims. This system is designed to encourage and accommodate first-time visitors to Mecca, while imposing restrictions on those who have already embarked on the trip multiple times. Pilgrims who have the means and desire to perform the Hajj several times have protested what they see as discrimination, but the Hajj Commission has stated that they see no alternative if further tragedies are to be prevented.

Following the 2004 stampede, Saudi authorities embarked on major construction work in and around the Jamarat Bridge area. Additional accessways, footbridges, and emergency exits were built, and the three cylindrical pillars were replaced with longer and taller oblong walls of concrete to enable more pilgrims simultaneous access to them without the jostling and fighting for position of recent years. The government has also announced a multimillion-dollar project to expand the bridge to five levels; the project is planned for completion in time for the 1427 AH Hajj (December 2006–January 2007).

Smith and Dickie's (1993) book is about engineering for crowd safety, and they list dozens of crowd disasters, including the recurring Hajj stampedes. Helbing et al. (2002) discuss simulation of panic situations from the point of view of nonlinear dynamical systems theory.

2.8.10 Al-Salam Boccaccio 98

Al-Salam Boccaccio 98 was an Egyptian ROPAX (passenger roll on–roll off) ferry, operated by al-Salam Maritime Transport, that sank on 3 February 2006 in the Red Sea en route from Duba, Saudi Arabia, to Safaga in southern Egypt. Its last known position was 100 km from Duba, when it lost contact with the shore at about 22:00 EET (20:00 UTC).

The vessel was built by the Italian company Italcantieri in 1970 with IMO number 6921282 and named the Boccaccio at Castellammare di Stabia, Italy. It was originally intended for Italian domestic service. Its dimensions included 130.99-m length overall, with

Figure 2.28   The ferry *Al-Salam Boccaccio 98* prior to the disaster.

23.60-m beam and 5.57-m draft. The main engines are rated at 16,560 kW for a maximum speed of 19 knots (35 km/h). The vessel had an original capacity of 200 automobiles and 500 passengers. Five sister ships were built.

The vessel was rebuilt in 1991 by INMA at La Spezia, maintaining the same outer dimensions albeit with a higher superstructure, changing the draught to 5.90 m. At the same time, its automobile capacity was increased to 320, and the passenger capacity was increased to 1,300. The most recent gross registered tonnage was 11,799.

The *Boccaccio* was purchased in 1999 by al-Salam Maritime Transport, headquartered in Cairo, the largest private shipping company in Egypt and the Middle East, and renamed *al-Salam Boccaccio 98*; the registered owner is Pacific Sunlight Marine, Inc., of Panama. The ferry is also referred to as *Salam 98*.

At the doomed voyage, the ship was carrying 1,312 passengers and 96 crew members, according to Mamdouh Ismail, head of al-Salaam Maritime Transport. Originally, an Egyptian embassy spokesman in London had mentioned 1,310 passengers and 105 crew, while the Egyptian presidential spokesman mentioned 98 crew and the Transport Minister said 104. The majority of passengers are believed to have been Egyptians working in Saudi Arabia. Passengers also included pilgrims returning from the Hajj in Mecca. The ship, pictured in Figure 2.28, was also carrying about 220 vehicles.

First reports of statements by survivors indicated that smoke from the engine room was followed by a fire that continued for some time. There were also reports of the ship listing soon after leaving port, and that after continuing for some hours the list became severe and the ship capsized within 10 minutes as the crew fought the fire. In a BBC radio news broadcast, an Egyptian ministerial spokesman said that the fire had started in a storage area, was controlled, but then started again. The ship turned around and as it turned the capsize occurred. The significance of the fire was supported by statements attributed to crew members who were reported to claim that the firefighters essentially sank the ship when sea-water they used to battle the fire collected in the hull because drainage pumps were not working.

The Red Sea is known for its strong winds and tricky local currents, not to mention killer sharks. The region had been experiencing high winds and dust storms for several days at the time of the sinking. These winds may have contributed to the disaster and may have complicated rescue efforts.

There are several theories expressed about possible causes of the sinking:

- *Fire*: Some survivors dragged from the water reported that there was a large fire on board before the ship sank, and there were eyewitness accounts of thick black smoke coming from the engine rooms.
- *Design flaws*: The *al-Salam Boccaccio 98* was a roll on–roll off (ro–ro) ferry. This is a design that allows vehicles to drive on one end and drive off the other. This means that neither the ship nor any of the vehicles need to turn around at any point. It also means that the cargo hold is one long chamber going through the ship. To enable this to work, the vehicle bay doors must be very near the waterline, so if these are sealed improperly, water may leak through. Even a small amount of water moving about inside can gain momentum and capsize the ship, in what is known as the free surface effect.
- *Modifications*: In the 1980s, the ship was reported to have had several modifications, including the addition of two passenger decks, and the widening of cargo decks. This would have made the ship less stable than it was designed to be, particularly as its draught was only 5.9 m. Combined with high winds, the tall ship could have been toppled easily.
- *Vehicle movement*: Another theory is that the rolling ship could have caused one or more of the 220 vehicles in its hold to break loose and theoretically be able to puncture a hole in the side of the ship.

At 23:58 UTC on 2 February 2006, the air–sea rescue control room at RAF Kinloss in Scotland detected an automatic distress signal relayed by satellite from the ship's position. The alert was passed on via France to the Egyptian authorities, but almost 12 hours passed before a rescue attempt was launched. As of 3 February 2006, some lifeboats and bodies were seen in the water. It was then believed that there were still survivors. At least 314 survivors and around 185 dead bodies have been recovered. Reuters reported that "dozens" of bodies were floating in the Red Sea.

Rescue boats and helicopters, including four Egyptian frigates, searched the area. Britain diverted the warship *HMS Bulwark* that would have arrived in a day and a half, but reports conflict as to whether the ship was indeed recalled. Israeli sources report that an offer of search-and-rescue assistance from the Israeli Navy was declined. Egyptian authorities did, however, accept a United States offer of a P-3 Orion maritime naval patrol aircraft after initially having said that the help was not needed.

The sinking of *al-Salam Boccaccio 98* is being compared to that of the 1987 *M/S Herald of Free Enterprise* disaster, which killed 193 passengers, and also to other incidents. In 1991, another Egyptian ferry, the *Salem Express*, sunk off the coast of Egypt after hitting a small habili reef; 464 Egyptians lost their lives. The ship is now a landmark shipwreck for SCUBA divers along with the *SS Thistlegorm*. In 1994, the *M/S Estonia* sank, claiming 852 lives. On 26 September 2002, the *M/S Joola*, a Senegalese government-owned ferry, capsized off the coast of Gambia, resulting in the deaths of at least 1,863 people. On 17 October 2005, the *Pride of al-Salam 95*, a sister ship of the *al-Salam Boccaccio 98*, also sank in the Red Sea, after being struck by the Cypriot-registered cargo ship *Jebal Ali*. In that accident, 2 people were killed and another 40 injured, some perhaps during a stampede to leave the sinking ship. After evacuating all the ferry passengers and crew, the *Jebal Ali* went astern and the *Pride of al-Salam 95* sank in about 3 minutes.

What is most tragic about the *al-Salam Boccaccio 98*'s incident is the utter ineptness, corruption, and collusion of both the Egyptian authorities and the holding company staff, particularly its owner, a member of upper chamber of parliament and a close friend to an even more powerful politician in the inner circle of the president. The 35-year-old ferry was not fit for sailing, and was in fact prevented from doing so in European waters, yet licensed to ferry passengers despite past violations and other mishaps by this and other ships owned by the same company. The captain of the doomed ferry refused to turn the ship around to its nearer point of origin despite the fire on board, and a passing ship owned by the same company ignored the call for help from the sinking ferry. Rescue attempts by the government did not start for almost 12 hours after the sinking, despite a distress signal from the ship that went around the globe and was reported back to the Egyptian authorities. Many officials failed to react promptly because an "important" soccer game was being televised. Rescued passengers told tales of the ship's crew, including the captain, taking the few lifeboats available to themselves before attempting to help the helpless passengers. The company's owner and his family were allowed to flea the country shortly after the disaster, despite a court order forbidding them from leaving Egypt. Local news media provided inaccurate reporting and then ignored the story altogether within a few weeks to focus on another important soccer event. Victims and their relatives were left to fend for themselves, all because they were the poorest of the poor, insignificant to the rich, powerful and mighty. Disasters occur everywhere, but in a civilized country, inept response as occurred in Egypt would have meant the fall of the government, the punishment of a few criminals and, most important, less tragic loss of life.

2.8.11 Bird flu

A pandemic is a global disease outbreak. A flu pandemic occurs when a new influenza virus emerges for which people have little or no immunity, and for which there is no vaccine. The disease spreads easily from person to person, causes serious illness, and can sweep across countries and around the world in a very short time. It is difficult to predict when the next influenza pandemic will occur or how severe it will be. Wherever and whenever a pandemic starts, everyone around the world is at risk. Countries might, through measures such as border closures and travel restrictions, delay arrival of the virus, but they cannot prevent or stop it.

The highly pathogenic avian H5N1 avian flu is caused by influenza A viruses that occur naturally among birds. There are different subtypes of these viruses because of changes in certain proteins (hemagglutinin [HA] and neuraminidase [NA]) on the surface of the influenza A virus and the way the proteins combine. Each combination represents a different subtype. All known subtypes of influenza A viruses can be found in birds. The avian flu currently of concern is the H5N1 subtype.

Wild birds worldwide carry avian influenza viruses in their intestines, but they usually do not get sick from them. Avian influenza is very contagious among birds and can make some domesticated birds, including chickens, ducks, and turkeys, very sick and even kill them. Infected birds shed influenza virus in their saliva, nasal secretions, and feces. Domesticated birds may become infected with avian influenza virus through direct contact with infected

Figure 2.29   Animals killed by the bird flu were left in the muddy streets of a village in Egypt.

waterfowl or other infected poultry, or through contact with surfaces (e.g., dirt or cages) or materials (e.g., water or feed) that have been contaminated with the virus.

Avian influenza infection in domestic poultry causes two main forms of disease that are distinguished by low and high extremes of virulence. The "low pathogenic" form may go undetected and usually causes only mild symptoms such as ruffled feathers and a drop in egg production. However, the highly pathogenic form spreads more rapidly through flocks of poultry. This form may cause disease that affects multiple internal organs and has a mortality rate that can reach 90% to 100%, often within 48 hours.

Human influenza virus usually refers to those subtypes that spread widely among humans. There are only three known A subtypes of influenza viruses (H1N1, H1N2, and H3N2) currently circulating among humans. It is likely that some genetic parts of current human influenza A viruses originally came from birds. Influenza A viruses are constantly changing, and other strains might adapt over time to infect and spread among humans. The risk from avian influenza is generally low to most people because the viruses do not usually infect humans. H5N1 is one of the few avian influenza viruses to have crossed the species barrier to infect humans, and it is the most deadly of those that have crossed the barrier.

Since 2003, a growing number of human H5N1 cases have been reported in Azerbaijan, Cambodia, China, Egypt, Indonesia, Iraq, Thailand, Turkey, and Vietnam. More than half of the people infected with the H5N1 virus have died. Most of these cases are all believed to have been caused by exposure to infected poultry (e.g., domesticated chicken, ducks, and turkeys) or surfaces contaminated with secretion/excretions from infected birds. There has been no sustained human-to-human transmission of the disease, but the concern is that H5N1 will evolve into a virus capable of human-to-human transmission. The virus has raised concerns about a potential human pandemic because it is especially virulent; it is being spread by migratory birds; it can be transmitted from birds to mammals and, in some limited circumstances, to humans; and similar to other influenza viruses, it continues to evolve.

In 2005, animals perished by the bird flu were left in the muddy streets of a village in Egypt as seen in Figure 2.29, exasperating an already dire situation. Rumors were rampant about contaminating the entire water supply in Egypt, which comes from the Nile River. Cases of the deadly H5N1 bird flu virus have been reported in at least fifteen governorates, and widespread panic among Egyptians has been reported. The Egyptian government has ordered the slaughter of all poultry kept in homes as part of an effort to stop the spread of bird flu in the country. A ban on the movement of poultry between governorates is in place. Measures already announced include a ban on the import of live birds, and officials say there have been no human cases of the disease. The government has called on Egyptians to stay calm, and not to dispose of slaughtered or dead birds in the roads, irrigation canals, or the Nile River.

Symptoms of avian influenza in humans have ranged from typical human influenza like symptoms (e.g., fever, cough, sore throat, muscle aches) to eye infections, pneumonia, severe respiratory diseases such as acute respiratory distress, and other severe and life-threatening complications. The symptoms of avian influenza may depend on which virus caused the infection.

A pandemic may come and go in waves, each of which can last for 6 to 8 weeks. An especially severe influenza pandemic could lead to high levels of illness, death, social disruption, and economic loss. Everyday life would be disrupted because so many people in so many places would become seriously ill at the same time. Impacts can range from school and business closings to the interruption of basic services such as public transportation and food delivery.

If a pandemic erupts, a substantial percentage of the world's population will require some form of medical care. Health care facilities can be overwhelmed, creating a shortage of hospital staff, beds, ventilators, and other supplies. Surge capacity at nontraditional sites such as schools may need to be created to cope with demand. The need for vaccine is likely to outstrip supply, and the supply of antiviral drugs is also likely to be inadequate early in a pandemic. Difficult decisions will need to be made regarding who gets antiviral drugs and vaccines. Death rates are determined by four factors: the number of people who become infected, the virulence of the virus, the underlying characteristics and vulnerability of affected populations, and the availability and effectiveness of preventive measures.

The U.S. government site (http://www.pandemicflu.gov/general/) lists the following pandemic death tolls since 1900:

- 1918–1919; United States 675,000+; worldwide 50 million+
- 1957–1958; United States 70,000+; worldwide 1–2 million
- 1968–1969; United States 34,000+; worldwide 700,000+

The United States is collaborating closely with eight international organizations, including the UN's World Health Organization (WHO), the Food and Agriculture Organization also of the UN, the World Organization for Animal Health, and eighty-eight foreign governments to address the situation through planning, greater monitoring, and full transparency in reporting and investigating avian influenza occurrences. The United States and its international partners have led global efforts to encourage countries to heighten surveillance for outbreaks in poultry and significant numbers of deaths in migratory birds, and to rapidly introduce containment measures. The U.S. Agency for International Development and the U.S. Department of State, Department of Health and Human Services, and Department of Agriculture are coordinating future international response measures on behalf of the White House with departments and agencies across the federal government. Together, steps are being taken to minimize the risk of further spread in animal populations, reduce the risk of human infections, and further support pandemic planning and preparedness. Ongoing detailed mutually coordinated onsite surveillance and analysis of human and animal H5N1 avian flu outbreaks are being conducted and reported by the USGS National Wildlife Health Center, the Centers for Disease Control and Prevention, the WHO, the European Commission, and others.

2.8.12 Energy crisis

The energy crisis is one of those slowly evolving disasters that does not get the attention it deserves. It is defined as any great shortfall (or price rise) in the supply of energy resources to an economy. There is no immediacy to this type of calamity, despite the adverse effects on the health, economic, and social well-being of billions of people around the globe. Chapters 8 and 9 touch on this subject, and the former addresses the related topic of global warming. For now, I offer a few personal reflections on energy and its looming crisis, with the United States in mind. The arguments made, however, may apply with equal intensity to many other countries.

Nothing can move let alone survive without it. Yet, the word was rarely uttered during the 2004, U.S. presidential campaign. Promises to effect somehow a lower price of gas at the pump are at best a short-term Band-Aid to what should be a much broader and longer-term national debate. Much like company executives, politicians mind, envision, and articulate issues in terms of years, not decades. A 4-year horizon is about right because this is the term for a president, twice that for a representative, and two-thirds of a senator's term. A chief executive officer's tenure is typically shorter than that for a senator. But the debate on energy should ideally be framed in terms of a human life span, currently about 75 years. The reason is twofold. First, fossil fuel, such as oil, gas, and coal, is being consumed at a much faster rate than nature can make it. These are not renewable resources. Considering the anticipated population growth (with a conservative albeit unrealistic assumption of no increase in the per capita demand) and the known reserves of this type of energy sources, the world supply of oil is estimated to be exhausted in 0.5 life span, gas in one life span, and coal in 1.5 life spans. Second, alternative energy sources must be developed to prevent a colossal disruption of our way of life. However, barring miracles, those cannot be found overnight, but rather over several decades of intensive research and development. The clock is ticking, and few people seem to be listening to the current whisper and, inevitably, the future thunder.

Uranium fission power plants currently supply less than 8% of the U.S. energy need. Even at this modest rate of consumption, the country will exhaust its supply of uranium in about two life spans. Real and imagined concerns about the safety of nuclear energy and depositions of their spent fuel have brought all new construction to a halt. Controlled nuclear fusion has the potential to supply inexhaustibly all of our energy needs, but, even in the laboratory, we are far from achieving the breakeven point (meaning getting more energy from the reactor than needed to sustain the reaction).

With 5% of the world population, the United States consumes 25% of the world's annual energy usage, generating in the process a proportional amount of greenhouse gases, which in turn is suspected of causing another type of disaster: global warming. Conservation alone is not going to solve the problem; it will merely relegate the anticipated crisis to a later date. A whopping 20% conservation effort this year will be wiped out by a 1% annual population increase over the next 20 years. However, this does not mean that conservation efforts should not be carried out. Without conservation, the situation will be that much worse.

The energy crisis exemplified by the 1973 Arab oil embargo brought about a noticeable shift of attitudes toward energy conservation. During the 1970s and 1980s, governments, corporations and citizens around the world, but particularly in the industrialized countries, invested valuable resources in searching for methods to conserve energy. Dwellings and other buildings became better insulated, and automobiles and other modes of transportation became more energy efficient. Plentiful fuel supplies during the 1990s and the typical short memory of the long gas lines during 1973 have, unfortunately, somewhat dulled the urgency and enthusiasm for energy conservation research and practice. Witness—at least in the United States—the awakening of the long-hibernated gas-guzzler automobile and the recent run on house-size sport utility vehicles, aka land barges. The $75 plus barrel of crude oil in 2006 has reignited interest in conservation. But in my opinion, the gas at the pump needs to skyrocket to $10 per gallon to have the required shock value. The cost is already half that much in Europe, and the difference in attitudes between the two continents is apparent.

Conservation or not, talk of energy independence is just that, unless alternative energy sources are developed. The United States simply does not have traditional energy sources in sufficient quantities to become independent. In fact, our energy dependence has increased steadily since the 1973 oil crisis. The nontraditional sources are currently either nonexistent or too expensive to compete with the $3 per gallon at the pump. However, a $10 price tag will do the trick, one day.

How do we go from here to there? We need to work on both the supply side and the demand side. On the latter, consumers need to moderate their insatiable appetite for energy. Homes and workplaces do not have to be as warm in the winter as a crowded bus or as cold in the summer as a refrigerator. A car with a 300-horsepower engine (equivalent to 300 live horses, really) is not needed to take one person to work via city roads. In addition, new technology can provide even more efficient air, land, and sea vehicles than exist today, with better insulated buildings; less wasteful energy conversion, storage, and transmission systems; and so on.

On the supply side, we need to develop the technology to deliver nontraditional energy sources inexpensively, safely, and with minimum impact on the environment. The United States and many other countries are already searching for those alternative energy sources. But are we searching with sufficient intensity? Enough urgency? I think not, simply because the problem does not affect the next presidential election, but rather the fifth or tenth one down the road. Who is willing to pay more taxes now for something that will benefit the next generation? Witness the unceremonious demise of former President Carter's Energy Security Corporation, which was supposed to kick off with the issuance of $5-billion worth of energy bonds. One way to assuage the energy problem is to increase usage taxes to help curb demands and to use the proceeds to develop new supplies.

Herein, we recite some alarming statistics provided in a recent article by the chair of the U.S. Senate Energy and Natural Resources Committee (Domenici, 2006). Federal funding for energy Research and Development (R&D) has been declining for years, and it is not being made up by increased private-sector R&D expenditure. Over the 25-year period from 1978 to 2004, federal appropriations fell from $6.4 billion to $2.75 billion in constant 2000 dollars, nearly 60% reduction. Private sector investment fell from about $4 billion to $2 billion during the period from 1990 to 2006. Compared to high-technology industries, energy R&D expenditure is the least intensive. For example, the private sector R&D investment is about 12% of sales in the pharmaceuticals industry and 15% in the airline industry, while the combined federal and private-sector energy R&D expenditure is less than 1% of total energy sales.

Let us briefly appraise the nontraditional sources known or even (sparingly) used today. The listing herein is not exhaustive, and other technologies unforeseen today may be developed in the future. Shale oil comes from sedimentary rock containing dilute amounts of near-solid fossil fuel. The cost, in dollars as well as in energy, of extracting and refining that last drop of oil is currently prohibitive.

There are also the so-called renewable energy sources. Although the term is a misnomer because once energy is used it is gone forever, those sources are inexhaustible in the sense that they cannot be used faster than nature makes them. The Sun is the source of all energy on Earth, providing heat, light, photosynthesis, winds, waves, life, and its eventual albeit very slow decay into fossil fuel. Renewable energy sources will always be here as long as the Sun stays alight, hopefully for more than a few billion years.

Using the Sun's radiation, when available, to generate either heat or electricity is limited by the available area, the cost of the heat collector or the photovoltaic cell, and the number of years of constant operation it takes the particular device to recover the energy used in its manufacturing. The United States is blessed with its enormous land and can, in principle, generate all its energy needs via solar cells using less than 3% of available land area. Belgium, in contrast, requires an unrealistic 25% of its land area to supply its energy need using the same technology. Solar cells are presently inefficient and expensive. They also require about 5 years of constant operation just to recover the energy spent on their manufacturing. Extensive R&D is needed to improve on these fronts.

Wind energy, although not constant, is also inexhaustible, but has similar limitations to those of solar cells. Windmills currently cannot compete with fossil fuel without tax subsidies. Other types of renewable energy sources include hydroelectric power, biomass, geophysical and oceanic thermal energy, and ocean waves and tides. Hydrogen provides clean energy, but has to be made using a different source of energy, such as photovoltaic cells. Despite all the hype, the hydrogen economy is not a net energy saver, but has other advantages nevertheless.

In his "malaise" speech of 15 July 1979, Jimmy Carter lamented, "Why have we not been able to get together as a nation to resolve our serious energy problem?" Why not indeed Mr. President.

2.9 Concluding remarks

The prediction, control, and mitigation of both natural and manmade disasters is a vast field of research that no one book or book chapter can cover in any meaningful detail. In this chapter, we defined what constitutes a large-scale disaster and introduced a metric to evaluate its scope. Basically, any natural or manmade event that adversely affects many humans or an expanded ecosystem is a large-scale disaster. The number of people tormented, displaced, injured, or killed and the size of the area adversely affected determine the disaster's scope. Large-scale disasters tax the resources of local communities and central governments and disrupt social order.

In this chapter, we showed how science can help predicting different types of disaster and reducing their resulting adverse effects. We listed a number of recent disasters to provide few examples of what can go right or wrong with managing the mess left behind every large-scale disaster.

The laws of nature, reflected in the science portion of any particular calamity, and even crisis management, reflected in the art portion, should be the same, or at least quite similar, no matter where or what type of disaster strikes. Humanity should benefit from the science and the art of predicting, controlling, and managing large-scale disasters, as extensively and thoroughly discussed in this book.

The last *annus horribilis*, in particular, has shown the importance of being prepared for large-scale disasters, and how the world can get together to help alleviate the resulting pain and suffering. In its own small way, this and subsequent chapters better prepare scientists, engineers, first responders, and, above all, politicians to deal with manmade and natural disasters.

References

Abbott, P. (2005) *Natural Disaster*, McGraw-Hill, San Diego, California.
Adamatzky, A. (2005) *Dynamics of Crowd-Minds: Patterns of Irrationality in Emotions, Beliefs and Actions*, World Scientific, London, United Kingdom.
Alexander, D. (1993) *Natural Disaster*, Springer, Berlin, Germany.
Alexander, D. (2000) *Confronting Catastrophe: New Perspectives on Natural Disasters*, Oxford University Press, London, United Kingdom.

References

Allen, J. (2005) *Predicting Natural Disasters*, Thomson Gale, Farmington Hills, Michigan.
Altay, N., and Green, W. G., III (2006) "OR/MS Research in Disaster Operations Management," *European J. Operational Res.* **175**, pp. 475–493.
Aubry, N., Holmes, P., Lumley, J. L., and Stone, E. (1988) "The Dynamics of Coherent Structures in the Wall Region of a Turbulent Boundary Layer," *J. Fluid Mech.* **192**, pp. 115–173.
Auerbach, D. (1994) "Controlling Extended Systems of Chaotic Elements," *Phys. Rev. Lett.* **72**, pp. 1184–1187.
Auerbach, D., Grebogi, C., Ott, E., and Yorke, J. A. (1992) "Controlling Chaos in High Dimensional Systems," *Phys. Rev. Lett.* **69**, pp. 3479–3482.
Bankoo, G., Frerks, G., and Hilhorst, D. (2004) *Mapping Vulnerability: Disasters, Development, and People*, Earthscan Publications, Gateshead, United Kingdom.
Bunde, A., Kropp, J., and Schellnhuber, H. J. (2002) *The Science of Disasters: Climate Disruptions, Heart Attacks, and Market Crashes*, Springer, Berlin, Germany.
Burby, R. J. (1998) *Cooperating with Nature: Confronting Natural Hazards with Land-Use Planning for Sustainable Communities*, John Henry Press, Washington, DC.
Chen, C. -C., Wolf, E. E., and Chang, H. -C. (1993) "Low-Dimensional Spatiotemporal Thermal Dynamics on Nonuniform Catalytic Surfaces," *J. Phys. Chemistry* **97**, pp. 1055–1064.
Childs, D. R., and Dietrich, S. (2002) "Contingency Planning and Disaster Recovery: A Small Business Guide," Wiley, New York.
Churchman, C. W., Ackoff, R. L., and Arnoff, E. L. (1957) *Introduction to Operations Research*, Wiley, New York.
Cooper, C., and Block, R. (2006) *Disaster: Hurricane Katrina and the Failure of Homeland Security*, Times Books, New York, New York.
Cutter, S. L. (2001) *American Hazardscapes: The Regionalization of Hazards and Disasters*, John Henry Press, Washington, DC.
Davis, S. H. (2000) "Interfacial Fluid Dynamics," in *Perspectives in Fluid Dynamics: A Collective Introduction to Current Research*, eds. G. K. Batchelor, H. K. Moffatt, and M. G. Worster, pp. 1–51, Cambridge University Press, London, United Kingdom.
de Boer, J., and van Remmen, J. (2003) *Order in Chaos: Modelling Medical Disaster Management Using Emergo Metrics*, LiberChem Publication Solution, Culemborg, The Netherlands.
der Heide, E. A. (1989) *Disaster Response: Principles of Preparation and Coordination*, C. V. Mosby, St. Louis, Missouri.
Dilley, M., Chen, R. S., Deichmann, U., Lerner-Lam, A. L., and Arnold, M. (2005) *Natural Disaster Hotspots: A Global Risk Analysis*, World Bank Publications, Washington, DC.
Ditto, W. L., and Pecora, L. M. (1993) "Mastering Chaos," *Scientific American* **269**, August, pp. 78–84.
Ditto, W. L., Rauseo, S. N., and Spano, M. L. (1990) "Experimental Control of Chaos," *Phys. Rev. Lett.* **65**, pp. 3211–3214.
Domenici, P. V. (2006) "Meeting Our Long-Term Energy Needs Through Federal R&D," *APS News* **15**, no. 9, p. 8.
Engelbert, P., Deschenes, B., Nagel, R., and Sawinski, D. M. (2001) *Dangerous Planet–The Science of Natural Disasters*, Thomson Gale, Farmington Hills, Michigan.
Fischer, H. W., III (1998) *Response to Disaster: Fact Versus Fiction & Its Perpetuation: The Sociology of Disaster*, second edition, University Press of America, Lanham, Maryland.
Fischer, H. W., III (2003a) "The Sociology of Disaster: Definition, Research Questions, Measurements in a Post–September 11, 2001 Environment," presented at the 98th Annual Meeting of the American Sociological Association, Atlanta, Georgia, 16–19 August 2003.
Fischer, H. W., III (2003b) "The Sociology of Disaster: Definition, Research Questions, and Measurements. Continuation of the Discussion in a Post–September 11 Environment," *Int. J. Mass Emergencies & Disasters* **21**, pp. 91–107.
Fowler, T. B. (1989) "Application of Stochastic Control Techniques to Chaotic Nonlinear Systems," *IEEE Trans. Autom. Control* **34**, pp. 201–205.
Fraedrich, K., and Schönwiese, C. D. (2002) "Space–Time Variability of the European Climate," in *The Science of Disasters: Climate Disruptions, Heart Attacks, and Market Crashes*, eds. A. Bunde, J. Kropp, and H. J. Schellnhuber, pp. 105–139, Springer, Berlin, Germany.
Gad-el-Hak, M. (1995) "Questions in Fluid Mechanics: Stokes' Hypothesis for a Newtonian, Isotropic Fluid," *J. Fluids Engin.* **117**, pp. 3–5.

Gad-el-Hak, M. (1999) "The Fluid Mechanics of Microdevices—The Freeman Scholar Lecture," *J. Fluids Engin.* **121**, pp. 5–33.

Gad-el-Hak, M. (2000) *Flow Control: Passive, Active, and Reactive Flow Management*, Cambridge University Press, London, United Kingdom.

Gad-el-Hak, M. (2006a) *Proceedings of the U.S.–Egypt Workshop on Predictive Methodologies for Global Weather-Related Disasters*, Cairo, Egypt, 13–15 March 2006, CD Publication, Virginia Commonwealth University, Richmond, Virginia.

Gad-el-Hak, M. (2006b) *The MEMS Handbook*, second edition, vols. I–IIII, CRC Taylor & Francis, Boca Raton, Florida.

Gad-el-Hak, M., and Bandyopadhyay, P.R. (1994) "Reynolds Number Effects in Wall-Bounded Flows," *Appl. Mech. Rev.* **47**, pp. 307–365.

Gall, R., and Parsons, D. (2006) "It's Hurricane Season: Do You Know Where Your Storm Is?" *IEEE Spectrum* **43**, pp. 27–32.

Garfinkel, A., Spano, M. L., Ditto, W. L., and Weiss, J. N. (1992) "Controlling Cardiac Chaos," *Science* **257**, pp. 1230–1235.

Garrett, C. (2000) "The Dynamic Ocean," in *Perspectives in Fluid Dynamics: A Collective Introduction to Current Research*, eds. G. K. Batchelor, H. K. Moffatt, and M. G. Worster, pp. 507–556, Cambridge University Press, London, United Kingdom.

Gist, R., and Lubin, B. (1999) *Response to Disaster: Psychosocial, Community, and Ecological Approaches*, Bruner/Mazel, Philadelphia, Pennsylvania.

Glass, T. A. (2001) "Understanding Public Response to Disasters," *Public Health Reports* **116**, suppl. 2, pp. 69–73.

Glass, T. A., and Schoch-Spana, M. (2002) "Bioterrorism and the People: How to Vaccinate a City Against Panic," *Clinical Infectious Diseases* **34**, pp. 217–223.

Hasselmann, K. (2002) "Is Climate Predictable?," in *The Science of Disasters: Climate Disruptions, Heart Attacks, and Market Crashes*, eds. A. Bunde, J. Kropp, and H. J. Schellnhuber, pp. 141–169, Springer, Berlin, Germany.

Hayes, S., Grebogi, C., and Ott, E. (1994a) "Communicating with Chaos," *Phys. Rev. Lett.* **70**, pp. 3031–3040.

Hayes, S., Grebogi, C., Ott, E., and Mark, A. (1994b) "Experimental Control of Chaos for Communication," *Phys. Rev. Lett.* **73**, pp. 1781–1784.

Helbing, D., Farkas, I.J., and Vicsek, T. (2002) "Crowd Disasters and Simulation of Panic Situations," in *The Science of Disasters: Climate Disruptions, Heart Attacks, and Market Crashes*, eds. A. Bunde, J. Kropp, and H. J. Schellnhuber, pp. 331–350, Springer, Berlin, Germany.

Holmes, P., Lumley, J. L., and Berkooz, G. (1996) *Turbulence, Coherent Structures, Dynamical Systems and Symmetry*, Cambridge University Press, Cambridge, Great Britain.

Huberman, B. (1990) "The Control of Chaos," *Proc. Workshop on Applications of Chaos*, 4–7 December, San Francisco, California.

Huberman, B. A., and Lumer, E. (1990) "Dynamics of Adaptive Systems," *IEEE Trans. Circuits Syst.* **37**, pp. 547–550.

Hübler, A., and Lüscher, E. (1989) "Resonant Stimulation and Control of Nonlinear Oscillators," *Naturwissenschaften* **76**, pp. 67–69.

Huppert, H. E. (2000) "Geological Fluid Mechanics," in *Perspectives in Fluid Dynamics: A Collective Introduction to Current Research*, eds. G. K. Batchelor, H. K. Moffatt, and M. G. Worster, pp. 447–506, Cambridge University Press, London, United Kingdom.

Jiménez, J. (2000) "Turbulence," in *Perspectives in Fluid Dynamics: A Collective Introduction to Current Research*, eds. G. K. Batchelor, H. K. Moffatt, and M. G. Worster, pp. 231–288, Cambridge University Press, London, United Kingdom.

Karniadakis, G. E., and Orszag, S. A. (1993) "Nodes, Modes and Flow Codes," *Physics Today* **46**, no. 3, pp. 34–42.

Kostelich, E. J., Grebogi, C., Ott, E., and Yorke, J. A. (1993a) "Targeting from Time Series," *Bul. Am. Phys. Soc.* **38**, p. 2194.

Kostelich, E. J., Grebogi, C., Ott, E., and Yorke, J. A. (1993b) "Higher-Dimensional Targeting," *Phys. Rev. E* **47**, pp. 305–310.

Kunreuther, H., and Roth, R. J., Sr. (1998) *The Status and Role of Insurance Against Natural Disasters in the United States*, John Henry Press, Washington, DC.

Lai, Y. -C., Deng, M., and Grebogi, C. (1993a) "Controlling Hamiltonian Chaos," *Phys. Rev. E* **47**, pp. 86–92.

Lai, Y.-C., and Grebogi, C. (1993) "Synchronization of Chaotic Trajectories Using Control," *Phys. Rev. E* **47**, pp. 2357–2360.

Lai, Y.-C., Grebogi, C., and Tél, T. (1994) "Controlling Transient Chaos in Dynamical Systems," in *Towards the Harnessing of Chaos*, ed. M. Yamaguchi, Elsevier, Amsterdam, The Netherlands.

Lai, Y.-C., Tél, T., and Grebogi, C. (1993b) "Stabilizing Chaotic-Scattering Trajectories Using Control," *Phys. Rev. E* **48**, pp. 709–717.

Lawson, A. C. (1908) *The California Earthquake of April 18, 1906: Report of the State Earthquake Investigation Commission*, vols. I & II, Carnegie Institution of Washington Publication 87, Washington, DC.

Lighthill, M. J. (1963) "Introduction. Real and Ideal Fluids," in *Laminar Boundary Layers*, ed. L. Rosenhead, pp. 1–45, Clarendon Press, Oxford, Great Britain.

Linden, P. F. (2000) "Convection in the Environment," in *Perspectives in Fluid Dynamics: A Collective Introduction to Current Research*, eds. G. K. Batchelor, H. K. Moffatt, and M. G. Worster, pp. 289–345, Cambridge University Press, London, United Kingdom.

Lindner, J. F., and Ditto, W. L. (1995) "Removal, Suppression and Control of Chaos by Nonlinear Design," *Appl. Mech. Rev.* **48**, pp. 795–808.

Lorenz, E. N. (1963) "Deterministic Nonperiodic Flow," *J. Atmos. Sci.* **20**, pp. 130–141.

Lorenz, E. N. (1967) *The Nature and Theory of the General Circulation of the Atmosphere*, World Meteorological Organization, Geneva, Switzerland.

Lumley, J. L. (1983) "Turbulence Modeling," *J. Appl. Mech.* **50**, pp. 1097–1103.

Lumley, J. L. (1987) "Turbulence Modeling," *Proc. Tenth U.S. National Cong. of Applied Mechanics*, ed. J. P. Lamb, pp. 33–39, ASME, New York.

McFedries, P. (2006) "Changing Climate, Changing Language," *IEEE Spectrum* **43**, August, p. 60.

McIntyre, M. E. (2000) "On Global-Scale Atmospheric Circulations," in *Perspectives in Fluid Dynamics: A Collective Introduction to Current Research*, eds. G. K. Batchelor, H. K. Moffatt, and M. G. Worster, pp. 557–624, Cambridge University Press, London, United Kingdom.

McKee, K., and Guthridge, L. (2006) *Leading People Through Disasters: An Action Guide*, Berrett-Koehler, San Francisco, California.

Mileti, D. S. (1999) *Disasters by Design: A Reassessment of Natural Hazards in the United States*, Joseph Henry Press, Washington, DC.

Moin, P., and Bewley, T. (1994) "Feedback Control of Turbulence," *Appl. Mech. Rev.* **47**, no. 6, part 2, pp. S3–S13.

Moin, P., and Mahesh, K. (1998) "Direct Numerical Simulation: A Tool in Turbulence Research," *Annu. Rev. Fluid Mech.* **30**, pp. 539–578.

Olasky, M. (2006) *The Politics of Disaster: Katrina, Big Government, and a New Strategy for Future Crises*, W Publishing Group, Nashville, Tennessee.

Ott, E. (1993) *Chaos in Dynamical Systems*, Cambridge University Press, Cambridge, Great Britain.

Ott, E., Grebogi, C., and Yorke, J. A. (1990a) "Controlling Chaos," *Phys. Rev. Lett.* **64**, pp. 1196–1199.

Ott, E., Grebogi, C., and Yorke, J. A. (1990b) "Controlling Chaotic Dynamical Systems," in *Chaos: Soviet–American Perspectives on Nonlinear Science*, ed. D. K. Campbell, pp. 153–172, American Institute of Physics, New York, New York.

Panton, R. L. (1996) *Incompressible Flow*, second edition, Wiley-Interscience, New York, New York.

Patanè, D., de Gori, P., Chiarabba, C., and Bonaccorso, A. (2006) "Magma Ascent and the Pressurization of Mount Etna's Volcanic System," *Science* **299**, pp. 2061–2063.

Pelling, M. (2003) *The Vulnerability of Cities: Natural Disaster and Social Resilience*, Earthscan Publications, Gateshead, United Kingdom.

Pickett, S. T. A., and White, P. S. (1985) *The Ecology of Natural Disturbance and Patch Dynamics*, Academic Press, San Diego, California.

Posner, R. A. (2004) *Catastrophe: Risk and Response*, Oxford University Press, Oxford, United Kingdom.

Prandtl, L. (1904) "Über Flüssigkeitsbewegung bei sehr kleiner Reibung," *Proc. Third Int. Math. Cong.*, pp. 484–491, Heidelberg, Germany.

Qin, F., Wolf, E. E., and Chang, H.-C. (1994) "Controlling Spatiotemporal Patterns on a Catalytic Wafer," *Phys. Rev. Lett.* **72**, pp. 1459–1462.

Quarantelli, E. L. (1998) *What Is a Disaster: Perspectives on the Question*, Routledge, London, United Kingdom.

Reynolds, O. (1895) "On the Dynamical Theory of Incompressible Viscous Fluids and the Determination of the Criterion," *Phil. Trans. Roy. Soc. Lond. A* **186**, pp. 123–164.

Richardson, L. F. (1922) *Weather Prediction by Numerical Process*, reissued in 1965 by Dover, New York.
Romeiras, F. J., Grebogi, C., Ott, E., and Dayawansa, W. P. (1992) "Controlling Chaotic Dynamical Systems," *Physica D* **58**, pp. 165–192.
Sen, M. (1989) "The Influence of Developments in Dynamical Systems Theory on Experimental Fluid Mechanics," in *Frontiers in Experimental Fluid Mechanics*, ed. M. Gad-el-Hak, pp. 1–24, Springer-Verlag, New York.
Shinbrot, T. (1993) "Chaos: Unpredictable Yet Controllable?" *Nonlinear Science Today* **3**, pp. 1–8.
Shinbrot, T. (1995) "Progress in the Control of Chaos," *Adv. Physics* **44**, pp. 73–111.
Shinbrot, T. (1998) "Chaos, Coherence and Control," in *Flow Control: Fundamentals and Practices*, eds. M. Gad-el-Hak, A. Pollard, and J. -P. Bonnet, pp. 501–527, Springer-Verlag, Berlin.
Shinbrot, T., Bresler, L., and Ottino, J. M. (1998) "Manipulation of Isolated Structures in Experimental Chaotic Fluid Flows," *Exp. Thermal & Fluid Sci.* **16**, pp. 76–83.
Shinbrot, T., Ditto, W., Grebogi, C., Ott, E., Spano, M., and Yorke, J. A. (1992a) "Using the Sensitive Dependence of Chaos (the "Butterfly Effect") to Direct Trajectories in an Experimental Chaotic System," *Phys. Rev. Lett.* **68**, pp. 2863–2866.
Shinbrot, T., Grebogi, C., Ott, E., and Yorke, J. A. (1992b) "Using Chaos to Target Stationary States of Flows," *Phys. Lett. A* **169**, pp. 349–354.
Shinbrot, T., Grebogi, C., Ott, E., and Yorke, J. A. (1993) "Using Small Perturbations to Control Chaos," *Nature* **363**, pp. 411–417.
Shinbrot, T., Ott, E., Grebogi, C., and Yorke, J. A. (1990) "Using Chaos to Direct Trajectories to Targets," *Phys. Rev. Lett.* **65**, pp. 3215–3218.
Shinbrot, T., Ott, E., Grebogi, C., and Yorke, J. A. (1992c) "Using Chaos to Direct Orbits to Targets in Systems Describable by a One-Dimensional Map," *Phys. Rev. A* **45**, pp. 4165–4168.
Shinbrot, T., and Ottino, J. M. (1993a) "Geometric Method to Create Coherent Structures in Chaotic Flows," *Phys. Rev. Lett.* **71**, pp. 843–846.
Shinbrot, T., and Ottino, J. M. (1993b) "Using Horseshoes to Create Coherent Structures in Chaotic Fluid Flows," *Bul. Am. Phys. Soc.* **38**, p. 2194.
Smith, K. (1991) *Environmental Hazards: Assessing Risk and Reducing Disaster*, Routledge, London, United Kingdom.
Smith, R. A., and Dickie, J. F. (1993) *Engineering for Crowd Safety*, Elsevier, Amsterdam, The Netherlands.
Spalart, P. R. (1988) "Direct Simulation of a Turbulent Boundary Layer up to $R_\theta = 1410$," *J. Fluid Mech.* **187**, pp. 61–98.
Speziale, C. G. (1991) "Analytical Methods for the Development of Reynolds-Stress Closures in Turbulence," *Annu. Rev. Fluid Mech.* **23**, pp. 107–157.
Stein, S., and Wysession, M. (2003) *An Introduction to Seismology, Earthquakes, and Earth Structure*, Blackwell, Boston, Massachusetts.
Steinberg, T. (2000) *Acts of God: The Unnatural History of Natural Disasters in America*, Oxford University Press, London, United Kingdom.
Tennekes, H., and Lumley, J. L. (1972) *A First Course in Turbulence*, MIT Press, Cambridge, Massachusetts.
Tierney, K. J., Lindell, M. K., and Perry, R. W. (2001) *Facing the Unexpected: Disaster Preparedness and Response in the United States*, John Henry Press, Washington, DC.
Tobin, G. A. (1997) *Natural Hazards: Explanation and Integration*, Guilford Press, New York, New York.
Vale, L. J., and Campanella, T. J. (2005) *The Resilient City: How Modern Cities Recover from Disaster*, Oxford University Press, London, United Kingdom.
Wallace, M., and Webber, L. (2004) *The Disaster Recovery Handbook*, Amacom, New York, New York.
Wilcox, D. C. (1993) *Turbulence Modeling for CFD*, DCW Industries, Los Angeles, California.
Winston, W. L. (1994) *Operations Research: Applications and Algorithms*, third edition, Duxbury Press, Belmont, California.

3
Multiscale modeling for large-scale disaster applications

Ramana M. Pidaparti

This chapter discusses various examples of large-scale disasters to illustrate the governing phenomena and the wide range of spatial and temporal scales involved. Due to the multiscale nature of disasters, an overview of applicable multiscale modeling techniques for investigating and predicting the behavior of large-scale disasters is presented. It is hoped that multiscale modeling techniques may help predict disasters before they occur; monitor and recover property, infrastructures, and environment; and provide response and quick relief in terms of damage mapping, communications, logistics, and medical facilities.

3.1 Introduction

Large-scale disasters can be classified into two groups: manmade and natural, as shown in Figure 3.1. Manmade disasters can be classified into technological disasters and human-related disasters. Some examples of natural disasters include avalanches, tornadoes, hurricanes, earthquakes, tsunamis, thunderstorms, and wildfires. Transportation accidents, nuclear and radiation mishaps, nano/microparticle emissions from transportation vehicles, air and water pollution, structural collapses, and power and utility failures are examples of technological disasters. Human-related disasters include war, terrorism, sabotage, arson, civil unrest, and mass hysteria (see the U.S. Occupational Safety and Health Administration document no. OR-OSHA 212, at www.cbs.state.or.us/external/osha/pdf/workshops/212i.pdf). Additional examples, as well as a detailed overview of the science of disasters, can be found in Bunde et al. (2002).

Large-scale disasters cause great suffering to people and society. Since the mid-1990s, almost 2 billion people have been adversely affected by disasters worldwide, with natural disasters alone causing more than 900,000 deaths. Hurricanes and tropical storms resulted in a US $100 billion damage in North America and Central America. The 2004 'tsunami,' and the 2005 hurricanes Rita and Katrina are among the more recent and major natural disasters. See Chapter 2 for additional examples of recent disasters. The impact of natural disasters on victims, personal property, and societal infrastructure is shown in Figure 3.2. Emergency response services, including medical personnel and resources, need to be mobilized quickly and effectively in the wake of a disaster. Chapters 6 and 7 describe various phases involved in coordinating a medical response to disasters. Due to the large-scale impact of natural disasters on essential services in society, world scientists and engineers are being challenged to improve disaster prediction capabilities and improve understanding of global climate changes. The challenges in global climate prediction include improving prediction of climate change at regional, national, and global levels, and identifying the means to adapt, mitigate, and prevent these extreme changes. The United Nations sponsored a "Disaster Conference" in 2004 to

70   *Multiscale modeling for large-scale disaster applications*

Natural Disasters	Technological	Human
Avalanche	Aircraft Crash	Arson
Biological	Structural Collapse	Civil Unrest
Drought	Business Interruption	Economic
Dust/Sand Storms	Communication	Enemy Attack
Earthquakes	Dam/Levee Failure	General Strike
Extreme Heat/Cold	Explosions/Fire	Hostage Situation
Fire	Extreme Air Pollution	Mass Hysteria
Flood	Financial Collapse	Sabotage
Hurricane/Tsunami	Fuel/Resource Shortage	Special Events
Landslide/Mudslide	Hazardous Materials Release	Terrorism
Lightning	Power Utility Failure	War
Snow/Ice/Hail	Radiological/Nuclear Accidents	Workplace Violence
Tornado	Strikes	
Volcanic Eruption	Transportations Accidents	
Windstorm		

Figure 3.1   Examples of large-scale disasters. Redrawn from U.S. Occupational Safety and Health Administration document no. OR-OSHA 212, Washington, DC.

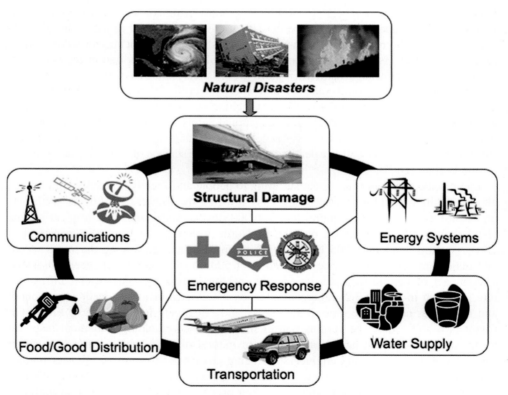

Figure 3.2   Effect of large-scale disasters on various systems affecting people and societal infrastructure.

address the impact of hurricanes in Florida and other hurricane-prone areas in the world. Scientists around the world, including Tokyo and Singapore, came together to share knowledge and skills on such topics as global weather conditions and climate change. Robert Gall and David Parsons (2006) from the U.S. National Center for Atmospheric Research (NCAR) in Boulder, CO, suggest that prediction of hurricanes and other weather-related disasters will significantly improve due to availability of supercomputers, sophisticated sensors, weather-specific satellites, and advances in modeling techniques. See Chapters 20 and 21 for a discussion of the role of space and weather satellite measurements in disaster response.

To protect our environment from severe climate changes and natural disasters, and preserve it for future generations, we must develop modeling and simulation methods that lead to accurate predictions of future climate and other natural disasters. Heaney et al. (2000) discuss the need for more research on the engineering aspects of natural disasters, specifically how engineering practice has evolved over the years in the United States, particularly the standards of codes and practice related to hazards management. Disaster management includes assessing the weather patterns and hazards, as well as a response phase and a recovery phase. An overview of the issues in disaster relief logistics is presented in Chapter 5. New response and recovery strategies often emerge from reviewing lessons learned from past events and experiences, and estimating/predicting likelihood of future events. This requires the use of physical models combined with disaster assessment data and observations. Multiscale modeling techniques can be applied at regional, national, and global levels, and offer a new approach to disaster management by helping monitor and recover property, infrastructures, environment, and provide response and quick relief in terms of damage mapping, communications, logistics, and medical facilities.

Disasters vary due to temporal duration, specific locations, surroundings, and geologic extent (neighborhood, village, city, and region). Disasters also have specific spatial properties. Examples include affected zone around earthquake epicenter, toxic spills in rivers or close to buildings, plumes for windbound pollutants and contaminants. In this chapter, two disasters, one resulting from climate and weather conditions and another due to the accidental release of chemical or biological agents into an urban environment, are discussed to illustrate the governing phenomena and the spatial and temporal scales associated with them. These are described in the next section.

3.2 Definition and modeling of scales in climate and weather

Climate describes the relatively long-term behavior of the climate system. The climate system consists of five major components. These are the atmosphere (time scale of days); the hydrosphere consisting of ocean and freshwater areas (time scale of months to years); the cryosphere consisting of land and sea ice (time scale of 10–1,000 years); the pedo- and lithospheres, which represent the land surface; and the biosphere, representing the vegetation. Chapter 12 presents a discussion of the dynamics of upper troposphere and stratosphere simulations. The parameters characterizing the atmosphere subsystem include the air temperature, precipitation, humidity, pressure, wind, and cloudiness, all of which vary in time and space. Climate is usually described by using statistics relevant across all space–time scales. The scale range of atmospheric phenomena spans from raindrops to planetary waves in space. The scales involved in environmental sciences range spatially from the distance between molecules to roughly the diameter of the Earth and temporally from fast chemical reactions to the age of the Earth. Operative at these scales are diverse physical, chemical, and biological processes, many of which are not well understood.

State-of-the-art climate models require the integration of atmosphere, ocean, sea ice, land, possible carbon cycle and geochemistry data, and much more. Additional information about climate and weather predictions can be found in Bunde et al. (2002) and Neugebauer and Simmer (2003), among others. In the following section, issues related to modeling global climate are presented. The terms weather and climate are defined to highlight implications for modeling and discuss limits to predictability.

3.2.1 Global climate modeling

To understand the scope of global climate modeling, one must first define the terms "climate" and "weather". Climate is the "envelope of possibilities" within which the weather bounces around, and is defined in terms of the statistics of weather. Climate is determined by the properties of the boundary conditions of the Earth itself, whereas weather depends very sensitively on the evolution of the system from moment to moment. Weather is the temporal distortion and precipitation that occurs within the Earth's atmospheres.

A sample global climate modeling problem is presented here to define the approach, computational requirements, and possible applications. Let us assume the global climate modeling problem involves computing such parameters as the latitude, longitude, elevation, time, temperature, pressure, humidity, or wind velocity. The approach, computational requirements, and applications are presented. A long-term stable representation of the Earth's climate is presented next.

- Approach
  1. Model the fluid flow in the atmosphere.
  2. Discretize the domains, and solve the Navier–Stokes problem.
  3. Devise an algorithm to predict weather at a given time.
- Computational Requirements
  1. Weather Predictions—60 Gflops/s (7 days in 24 hours)
  2. Climate predictions—5 Tflops/s (50 years in 30 days)
- Applications
  1. Predict major events, (e.g., hurricane, El Nino).
  2. Use in setting standards.
  3. Negotiate policy.

3.2.2 Long-term climate simulation

Figure 3.3 presents a 1,000-year climate simulation of temperature change using the CCSM2 (community climate model) software (developed by NCAR). This simulation involved 760,000 process hours and shows a long-term, stable representation of the Earth's climate. Additional information and discussion of long-term climate change and weather prediction can be found in Chapters 14 to 16. Similarly, Figure 3.4 shows the simulation of projected sea ice during 2010 to 2080, affecting the global warming in the Arctic region.

3.2.3 Limits to predictability

There are basic differences between forecasting weather and predicting climate. Both regard the problem as initial and boundary conditions with the relevant time scales. Weather predictions are concerned with the time-dependent evolution as a function of initial conditions with fixed boundary conditions. However, the prediction of climate is concerned with the response of a system to changes in boundary conditions with fixed initial conditions.

3.2 Definition and modeling of scales in climate and weather

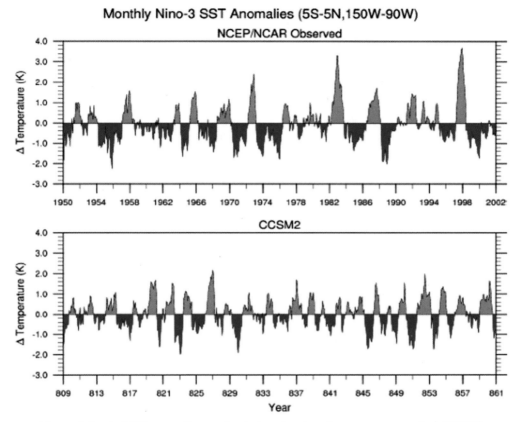

Figure 3.3 A 1,000-year climate simulation showing the temperature using CCSM2 software. *Source:* Warren Washington and Jerry Meehl, National Center for Atmospheric Research; Bert Semtner, Naval Postgraduate School; John Weatherly, U.S. Army Cold Regions Research and Engineering Laboratory.

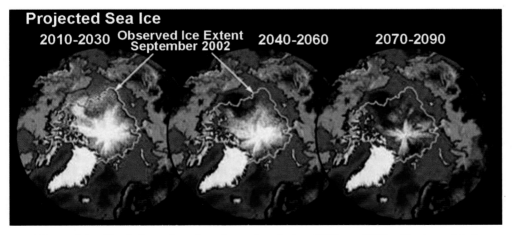

Figure 3.4 Simulation of projected sea ice during 2010 to 2090, affecting global warming in the Arctic region. *Source:* www.amap.no/acia/files/ProjIceExt-5ModelAvgS%2379058.jpg.

The dynamic equations that govern the motion of the atmosphere and the oceans are strongly nonlinear. This makes these equations very sensitively dependent on their initial conditions. Errors in the initial conditions, no matter how trivial or how small on a spatial scale, quickly grow in magnitude and propagate to larger spatial scales. The flow around an airplane wing is governed by the same strong nonlinear Navier–Stokes equations which govern the atmosphere. For these same reasons, we cannot forecast the weather a month in advance. Also, we can never predict the instantaneous, three-dimensional, turbulent flow around a wing. However, with given boundary values and parameters, we can predict and analyze with confidence the statistics of this flow. Otherwise, modern flight would have been impossible!

Given the limitations to predictability, let us now review methods commonly used to model climate change at the global and regional levels. The following section presents information on global and regional climate models, particularly their advantages and disadvantages in modeling and simulating large-scale disasters.

3.2.4 Global and regional climate models

In general, the climate change process at both the regional and the global level involves multiscale physics and chemistry from the molecular to a continuum scale. The physical and chemical mechanisms occurring at various length and time scales play an important role in accurate global weather predictions. The challenge in global weather modeling and simulations is to bridge the wide range of length and time scales with new and improved methods.

Global climate models (GCMs) run at resolutions of 200 to 500 km on a horizontal grid, and represent the atmosphere/land/surface/oceans and sea ice at a global level. The GCM are usually simulated for few years to a few centuries. In contrast, regional climate models (RCMs) run at resolutions of about 50 km represent only the land surface, and do not include oceans. The results of both RCMs and GCMs are strongly dependent on parameterizations of various phenomena (cloud processes, radiation, turbulence, and land-surface processes). The modeling approaches between RCMs and GCMs should consider the compatibility in parameterizations, and at the same time, they should be physically consistent. Due to the complex process of the weather phenomena, many parameterized processes are expected to be behaving differently at different scales both in RCMs and in GCMs. A fundamental aspect of climate change mechanisms is that they usually initiate at multiple levels, and it is important to study the relationship between various levels in terms of grids scaling, etc., in RCMs and GCMs. Figure 3.5 shows the simulation results by Duffy et al. from the Lawrence Livermore National Laboratory (LLNL). This figure shows that as the model's resolution becomes finer, results converge toward observations from data. Similarly, Figure 3.6 shows the urban to regional scale predictions of anthropogenic particulates as they are transported from urban sources and mixed into the regional environment with precursor trace gases, natural particulates, and particulates from other anthropogenic sources (Fast 2005). More information related to natural and anthropogenic aerosol-related hazards may be found in Chapter 10. Chapter 17 presents a discussion of weather-related disasters in arid lands.

There is a vast amount of literature on global and regional climate research. In this chapter, the focus is on work related to regional climate research with respect to modeling approaches, and their prospects and challenges. Many investigators have reviewed climate research for the past decade (Houghton et al., 2001). A few summaries and review articles have been published based on special conferences or workshops (Easterling et al., 2000;

3.2 *Definition and modeling of scales in climate and weather* 75

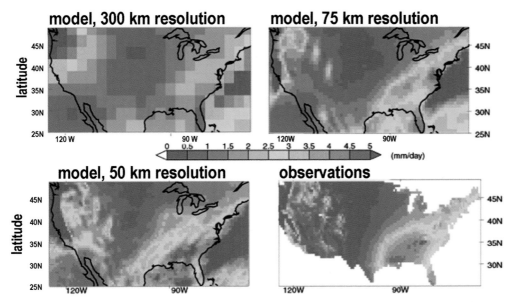

Figure 3.5 Effect of model refinement in climate modeling for wintertime precipitation. *Source:* Philip P. Duffy, Atmospheric Science Division, Lawrence Livermore National Laboratory, Livermore, California.

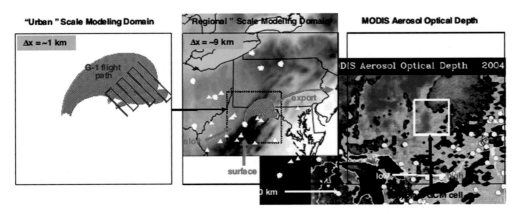

Figure 3.6 Effect of model refinement in predicting anthropogenic particulates as they are transported from urban sources. *Source:* Jerome Fast, presentation at the Science Team Meeting of the U.S. Department of Energy Atmospheric Science Program, 30 October–1 November 2005, Alexandria, Virginia, www.asp.bnl.gov/Agenda-ASP-STMtg-FY2006.html.

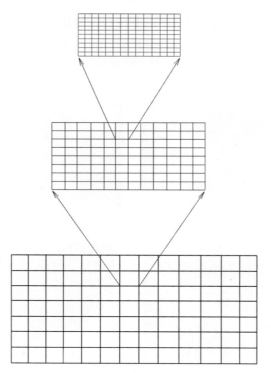

Figure 3.7 Upscaling procedure from RCMs to GCMs in the climate prediction process.

Asrar et al., 2000; Lueng et al., 2003; Wang et al., 2004). This chapter summarizes the salient features from each of these articles. Easterling et al. (2000) discuss the effect of climate extremes based on observations, impacts, and modeling. They observe that climate extremes are modeled through two broad groups, one based on simple climate statistics, and the other based on complex event-driven extremes such droughts, floods, and hurricanes. Societal impacts, as well as impact on natural systems, are discussed. Asrar et al. (2000 discuss National Aeronautics and Space Administration's (NASA's) research strategy for earth system science, especially the climate component. Their research foci for climate research for the next 5 years include documenting climate variability and trends in relevant climate forcing factors, investigating the key climate response and feedback mechanisms, and addressing climate prediction issues. Lueng et al. (2003) summarize the regional climate research needs and opportunities, which include GCMs and RCMs, statistical downscaling, data diagnostics/validation, and applications of upscaling from RCMs to GCMs (Figure 3.7). They argue that dynamic downscaling methods may help address issues such as urban air quality, heat island effects, lake effects, and storm surges. Wang et al. (2004) review regional climate modeling studies, including RCM development, and their application to dynamical downscaling, and climate assessment and predictability. See Chapter 15 for a detailed review of the regional climate modeling approach and its use in predicting weather extremes and climate variability.

Based on these review articles, it appears that existing RCMs do a good job of capturing regional climate changes and predict extreme events to some extent. RCMs have been shown to be useful for improving our understanding of various climate processes (land–atmosphere interaction, cloud-radiation feedback, and topographic forcing). Regional climate modeling is also integral to biogeochemical modeling and measurement studies, as well as research on greenhouse gases and aerosols. There are, however, several aspects that RCMs do not yet capture effectively, including understanding the model behaviors at high-spatial resolutions due to physics-based parameterizations, examining the relationships at various temporal and spatial scales in depth in order to capture certain climate features, and accurately predicting the regional climate conditions. There is still a need for downscaling to coarser/fine grids, and increasing computational efficiency and linking to GCMs with nesting approaches and others.

Enhanced RCMs based on multiscale approaches will help in better understanding the regional climate processes and improve accuracy of weather predictions. Incorporating various physical/chemical process parameters at multiple scales in the weather prediction modules is a fundamental step in accurately predicting regional climate, and nesting the RCM model to GCMs may provide more accurate information on global weather conditions and disasters. Due to the complexity of climate dynamics and their sensitivity to many parameters, it will be of interest to approach the RCM with modern multiscale computational methods. Krasnopolsky and Schiller (2003a, b) apply neural network (NN) methods to environmental science applications, including the atmospheric chemistry, and wave models.

A variety of research packages exist in the climate modeling community that have open sources and can be easily used (Wang et al., 2004). For example, NASA's Seasonal-to-Interannual Prediction Program, NCAR Community Atmospheric Model, and Penn State/NCAR MM5 mesoscale model have been used by most weather prediction agencies and scientists. These sources support most of the physics involved in weather prediction, for example, convection modeling, precipitation modeling, cloud parameterizations, ice/snow modeling, and basic land surface models. Also, they support grid nesting, which is essential in numerical weather prediction. See Chapters 12, 13, and 19 for additional discussion of the physics and dynamics of numerical weather prediction and coupled weather–chemistry modeling.

The regional modeling programs mentioned previously have limitations in grid spacing at regional and global levels, and capturing the outer boundary information. To alleviate problems associated with grid spacing and boundary domains, alternate approaches may be of interest. At NCAR, a multiscale modeling project titled "Predicting the Earth System Across Scales," was initiated by Greg Holland Mesoscale and Microscale Meteorology division and Hurrell Climate and Global Dynamics division with partners from Pacific Northwest National Lab (PNNL) and other researchers from three NCAR divisions. More recently, Rasmussen (2005) showed a comparison of winter precipitation as observed in Figure 3.8 and simulated by the nested grid (*center*) and global scale (*bottom*) models during the first phase of the multiscale modeling project. The results pointed to the improved accuracy of the nested model.

3.3 Definition and modeling of scales during accidental release of toxic agents in urban environments

Accidental releases from chemical and biological agents/materials can create a civil emergency and pose an immediate hazard to human health and life. In 1984, a chemical accident in Bhopal, India released a toxic cloud over a short time and killed more than 5,000 people. There was an accidental release of sarin at several locations in the Tokyo subway in 1995, and another incident in Moscow involved the threat of chemical release from a pipe bomb in a popular park. Estimating the atmospheric dispersion of accidental release of chemical, biological, and radiological agents and pollutants into the air involves a review of variety of physical processes, including release, advection, diffusion, rainout, or washout (Bacon

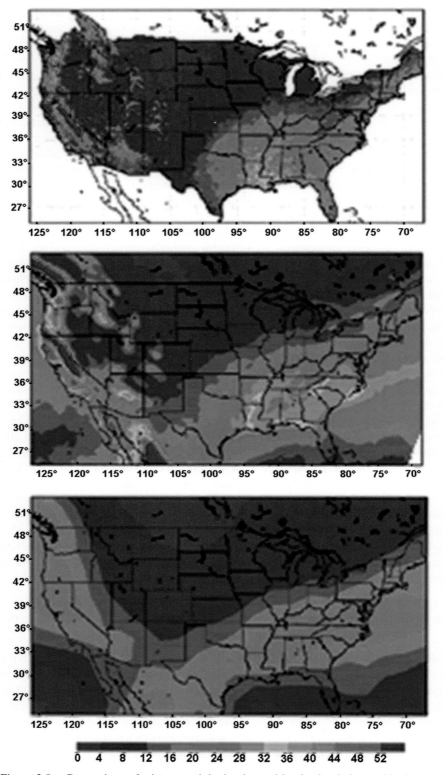

Figure 3.8 Comparison of winter precipitation by multiscale simulations with observations. From Rasmussen (2005).

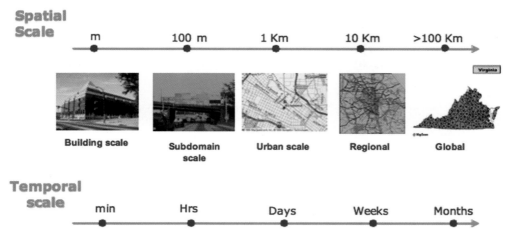

Figure 3.9 Spatial and temporal scales in a typical urban environment modeling system.

et al., 2005). In general, atmospheric dispersion covers many different spatial (10 m–1000 km) and temporal scales (minutes to months). The physical phenomena of atmospheric dispersion is an interdisciplinary problem involving physics (fluid dynamics, thermodynamics, and radiation transport), meteorology (atmospheric microphysics, cloud and mesoscale dynamics, planetary and surface physics), atmospheric chemistry (absorption, adsorption, homogeneous and heterogeneous reactions), and mathematics (numerical methods such as computational fluid dynamics, computational structural mechanics, turbulence, and chaos).

It is essential to develop tools and techniques that can model urbanization as it relates to the meteorological environment, particularly in situations where there is deployment of biological, chemical, or radiological weapons of mass destruction or others. These tools and techniques may also improve communication and transportation networks during such disasters. Urban landscapes have complex geometry over large areas with varying spatial and temporal scales, as shown in Figure 3.9. For example, in describing the regional characteristics of an urban area, the effects of streets and buildings should be considered. For modeling urban characteristics, data obtained from aerial photos, laser surveys, geographic information systems, and satellite methods, and computer-aided-design programs (SolidWorks, ProE, Autodesk, and others) can be used. The critical factors for dispersion studies include the injection altitude, particle size, and reactivity. Also, to address terrorism in urban settings, all these details need to be included in modeling and prediction methods. Long-range dispersion studies may involve modeling at meso- and global scales related to climatic issues due to releases of radioactivity from nuclear accidents, biological agents, and greenhouse gases. Radioactivity is invisible in comparison to smoke from fires, which is clearly visible for meso-regional scale dispersion studies (Bacon et al., 2000). Modeling dispersion at mesoscale is more complex than global scale due to nonuniformity of the surface (topology, roughness, and canopy). Therefore, the urban environment modeling system involves several spatial/time scales as follows:

- Single to few buildings scale (10–100 m, time scale: 10 min)
- Urban subdomain scale (km, few hours, building clusters)
- Urban scale (1 km-10 km, 1–10 hr, buildings are parameterized)
- Meso-regional scale (10–100 km, 10–100 hr)
- Global scale (100–1,000 km, 100–1,000 hr)

Figure 3.10 Results of a large urban release of CBR agent simulated through CT-Analyst, developed by G. Patnaik, J. P. Boris, and F. F. Grinstein, Laboratory for Computational Physics & Fluid Dynamics, U.S. Naval Research Laboratory, Washington, DC.

In multiscale methods, city and urban characteristics should be well indicated with respect to their individual features and gross features to capture the spreading of the phenomena throughout the scales (Figure 3.9). The whole system describing the appropriate scale dependent models, urban characteristics, and meteorology data should have high precision and contain accurate geographic information (e.g., terrain height, land type, building distribution, energy sources consumption). Bacon et al. (2000) developed an atmospheric simulation system based on an adaptive unstructured grid called OMEGA (Operational Multiscale Environment model with Grid Adaptivity). This system has been applied to real-time aerosol and gas hazard prediction (Bacon et al., 2000), as well as to hurricane forecasting (Gopalakrishnan et al., 2002). Fang et al. (2004) developed a customized multiscale modeling system based on three spatial scales to simulate the urban meteorological environment. Several researchers at the LLNL have been developing multiscale methods for high-performance global climate modeling and urban meteorological environment (www.llnl.gov/CASC/).

Multiscale methodology and simulations can be directly used in emergency assessment of industrial spills, transportation accidents, or terrorist attacks. Figure 3.10 shows the simulation of spread of smoke around a large city through the code CT-Analyst developed by the Naval Research Laboratory (Patnaik et al., 2005; Pullen et al., 2005). However, the multiscale methodology should be fast and accurate enough so that practical chemical, biological or radiological (CBR) scenarios can be simulated and used to avoid catastrophes, and protect and save human lives (Boris, 2005). The next section introduces the multiscaling modeling approach and reviews issues in applying multiscale modeling methods to large-scale disasters.

3.4 Multiscale modeling methods

3.4.1 Key challenges

As discussed in previous sections, large-scale disasters generally involve processes that operate over a wide range of length and time scales, and provide compelling challenges. To effectively model and prevent large-scale disasters, we need to describe detailed scientific phenomena occurring at multiple scales (nano-, micro-, meso-, macro-, and megascales) by capturing information at small scales and examining their effects on the megascale level. Many problems will remain unresolved without the capability to bridge these scales for modeling and simulation of large-scale disasters.

Historically, the study of large-scale disasters has been driven by the need for information to guide restoration, policy, and logistics because they are vital for human safety. However, the challenge is to develop more effective modeling and simulation tools that can be used for prediction and disaster management. This will require a systematic, multidisciplinary approach consisting of basic science, mathematical descriptions, and computational techniques that can address large-scale disasters across time and space scales, involving multiple orders of magnitude.

3.4.2 Application of modeling methods to large-scale disasters

Multiscale modeling and simulation is concerned with the methods for computing, manipulating, and analyzing information and data at different spatial and time resolution levels. The techniques of multiscale modeling in various fields have undergone tremendous advances during the past decade because of the cost effectiveness of the hardware environment. Advances in computational methods and the distributed hardware have enabled the development of new mathematical and computational methods that enable multiple simulations. To establish the validity of a model and its application in simulations, both verification and validation aspects need to be addressed (Dolling, 1996). Verification involves checking the model for correctly solving the governing equations mathematically, whereas the validation involves checking that the right equations are solved from the perspectives of the model applications. Validations require advances in experimental measurement and data gathering, as well as advances in instrumentations. Chapters 20 and 21 provide information on space and weather satellite measurements in disaster response, and underscore the importance of instrumentation. It is therefore important and necessary that the multiscale simulation methods use experimental data to establish the range of capability of the models and to validate predicted behavior at macro- and megascales. The potential impact of applying multiscale methods to large-scale disasters has been realized by a number of government labs and agencies, including NCAR, National Science Foundation, U.S. Environmental Protection Agency, LLNL, U.S. Department of Energy, NASA, and others.

Figure 3.11 shows an overview of the multiscale methods (discrete and continuum) over multiple spatial and temporal scales that can be used in developing a multiscale methodology for investigating large-scale disaster applications. Discrete modeling methods employ reduced models and integrate stochastic equations of motion. The advantages of these models include explicit atomistic detail and incorporation of stochastic phenomena.

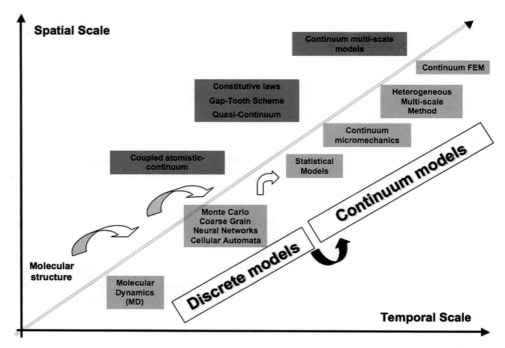

Figure 3.11 Overview of multiscale modeling methods for investigating large-scale disaster applications.

The disadvantages include slow convergence and difficulty in scaling. Continuum modeling methods are deterministic and offer the advantages of good convergence, are highly scalable, and provide connections to other continuum mechanics behaviors.

3.4.3 Multiscale modeling techniques

In recent years, several researchers from various fields have developed multiscale modeling techniques by taking into account various phenomena at multiple scales. For example, these include computational materials science (Gates et al., 2005; Nieminen, 2002), computational mechanics (Liu et al., 2004), biomedical engineering (Ayati et al., 2006), and nanotechnology (Fish, 2006). Traditional multiscale techniques, such as the multigrid method, domain decomposition, adaptive mesh refinement, the fast multipole method, and the conjugate gradient method (Reddy, 2006), have focused on efficiently resolving the fine scale. Recent techniques reduce computational complexity by adopting different computational approaches and different laws of physics on various space and time scales. For example, on the macroscale (> millimeters), fluids are accurately described by density, temperature, and velocity fields that obey continuum Navier–Stokes equations. However, on the scale of the mean free path of the fluid particles, it is necessary to use kinetic theory (Boltzmann's equations) to get a more detailed description. Averaging, where the leading order behavior of a slow time-varying variable is replaced by its time-average value, and homogenization (Fish et al., 2005), where approximate equations are obtained to leading order in the ratio of fine and coarse spatial scales, are examples of powerful analytical techniques. The quasi-continuum (QC) method (Knap and Ortiz, 2001), gap-tooth technique (Gear et al., 2003), and the heterogeneous multiscale method (HMM; Weinan et al., 2003a, b) are examples of recently developed methods.

Multiscale methods offer powerful modeling and simulation techniques for the characterization and prediction of large-scale disasters. In this chapter, some of the recent development in multiscale methods from other research areas are reviewed for application to large-scale disasters.

3.4.4 Molecular dynamics method

Alder and Wainwright (1959) developed the molecular dynamics method (MD), which describes the behavior of a collection of atoms by their position and momentum. In this framework, the macroscale process is the molecular dynamics of the nuclei, and the microscale process is the state of the electrons that determines the potential energy of the system. The MD simulations provide the results of structural information, transport phenomena, and time dependence of physical properties. There are various MD simulations addressing specific issues related to thermodynamics of biological processes, polymer chemistry, and material crystal structures. Most molecular dynamics simulations are performed under conditions of constant number of atoms (N), volume (V) and energy (E) or constant number of atoms (N), temperature (T), and pressure (P) to better simulate experimental conditions.

The basic steps in the MD simulation include (1) establishing initial coordinates of existing atoms in the minimized structure and assigning them initial velocities, (2) establishing thermal dynamics conditions and performing equilibration dynamics to rescale the velocities and checking the temperature, and (3) performing dynamic analysis of trajectories using Newton's second law. The result of the MD simulation is a time series of conformations or the path followed by each atom.

In general, MD simulations generate information about atomic positions and velocities at the nanolevel. The conversion of this position and velocity information to macroscopic quantities (pressure, energy, heat) that can be observed requires the use of statistical mechanics. Usually, an experiment is carried out on a representative macroscopic unit that contains an extremely large number of atoms or molecules, representing an enormous number of conformations. Averages corresponding to experimental measurements are defined in terms of ensemble averages in statistical mechanics. To ensure a proper average, an MD simulation must account for a large number of representative conformations. For example, the total energy (E) of a particle (atom, molecule, etc.) can be written as

$$E = T + V \tag{3.1}$$

where T is the kinetic energy and V is the potential energy. For example, the average potential energy of the system is defined (Gates et al., 2005) as

$$V = \frac{1}{M} \sum_{i=1}^{M} V_i \tag{3.2}$$

where M is the number of configurations in the molecular dynamics trajectory and V_i is the potential energy of each configuration. Similarly, the average kinetic energy (K) is given by

$$K = \frac{1}{M} \sum_{j=1}^{M} \left\{ \sum_{i=1}^{N} \frac{m_i}{2} v_i \bullet v_i \right\}_j \tag{3.3}$$

where M is the number of configurations in the simulation, N is the number of atoms in the system, m_i is the mass of the particle i, and v_i is the velocity of particle i.

Once the total energy is calculated, the potential can be calculated as the difference of the energies of different particles, and the molecular forces can be calculated from the derivatives of the potentials.

3.4.5 Coarse-grained methods

Due to inherent difficulties in numerical and computational boundaries in MD simulations, the size and time scales of the model may be limited. Even though, MD methods may provide the details necessary to resolve molecular structure and localized interactions, they are computationally expensive. However, coarse-grained methods may overcome these limitations by representing molecular chains as simpler, bead–spring models (Rudd, 2004). Although the coarse-grained models lack the atomistic details, they preserve many of the important aspects of the structural and chemical information. The connection to the more detailed atomistic model can be made directly through an atomistic-to-coarse-grained mapping procedure that when reversed allows one to model well-equilibrated atomistic structures by performing this equilibration by using the coarse-grained model. This mapping helps overcome the time scale upper limits of MD simulations.

Several approaches to coarse graining have been proposed for both continuous and lattice models. The continuous models seem to be preferable for dynamic problems such as might occur when considering dynamic changes in volume. The systematic development of the coarse-grain models requires determining the degree of coarse graining and the geometry of the model, choosing the form of the intra- and interchain potentials, and optimizing the free parameters (Hahn et al., 2001). Coarse-grained models are usually constructed using Hamilton's equations from MD under fixed thermodynamic conditions. To preserve the average position and momentum of the fine scale atoms, representative atoms are enforced in the model. Coarse-grained models have shown a four orders of magnitude decrease in CPU time in comparison to MD simulations (Lopez et al., 2002).

3.4.6 Monte Carlo methods

Due to the time scales (femtoseconds to microseconds) involved in the various physical and chemical phenomena in large-scale disasters, linking diverse time scales is very challenging. The kinetic Monte Carlo (KMC)-based methods can be used to address the time scales (Binder, 1995). Also, the coarse-grain models are often linked to Monte Carlo (MC) simulations to provide a solution in time. The KMC method is used to simulate stochastic events and provide statistical approaches to numerical integration.

The integration scheme in the KMC method is simply implemented to integrate a function over a complicated domain D by picking randomly selected points over some simple domain D', which is a superset of D. The area of D is estimated as the area of D' multiplied by the fraction of points within domain D. The integral of a function f in a multidimensional volume V is determined by picking N randomly distributed points $X_1, \ldots, X_N$ as follows,

$$\int f dV \approx V \langle f \rangle \pm V \sqrt{\frac{\langle f^2 \rangle - \langle f \rangle^2}{N}} \tag{3.4}$$

where

$$\langle f \rangle \equiv \frac{1}{N} \sum_{i=1}^{N} f(x_i) \tag{3.5}$$

and

$$\langle f^2 \rangle \equiv \frac{1}{N} \sum_{i=1}^{N} f^2(x_i) \tag{3.6}$$

There are three important characteristic steps in the MC simulation: (1) translating the physical problem into an analogous probabilistic or statistical model, (2) solving the probabilistic model by a numerical sampling experiment, and (3) analyzing the resultant data by using statistical methods.

To deal with stochastics in modeling micro- and macroscale processes (discrete or continuous), MC methods can be used. Traditional MC methods are robust, but slow to converge, and require many trials to compute high-order moments with adequate resolution. Coupling discrete-stochastic models with continuous stochastic or deterministic models, where appropriate, would enable the simulation of many complex disaster-related problems.

3.4.7 Cellular automata

Cellular automata (CA)-based modeling techniques are powerful methods to describe, simulate, and understand the behavior of complex physical systems (Chopard and Droz, 1998). The original CA model proposed by Von Neumann (Wolfram, 1986) is a two-dimensional square lattice in which each square is called a *cell*. Each cell can be in a different state at any given time. The evolution of each cell and the updating of the internal states of each cell occurs synchronously and is governed by a set of *rules*. The cellular space thus created is a complete discrete dynamic system. Earlier work by Wolfram (1986, 1994) showed that the CA as a discrete dynamic system exhibits many of the properties of a continuous dynamic system, yet CA provide a simpler framework.

A CA is an array (1D string or 2D grid or 3D solid) of identically programmed "cells" that interact with one another. The cells are usually arranged as a rectangular grid, honeycomb, or other form. The essential features of a CA are the State, its Neighborhood, and its Program. The State is a variable that takes a separate value for each cell, and the State can be either a number or a property. The Neighborhood is the set of cells with which it interacts. The Program is the set of rules that define how its state changes in response to its current state, and that of its neighbors. If we consider a three-dimensional space, then the CA on a cubic lattice over a period of time would occur as follows. In the lattice, each cell position is labeled as $\vec{r} = (i, j, k)$, where i, j, and k are the indices in three directions, respectively. A function $f_t(\vec{r})$ is applied to the cubic lattice to describe the state of each cell at iteration, t. This value can be Boolean 0 or 1, or it can be a continuous value by using a probabilistic function. The rule R specifies how the state changes are to be computed from an initial state configuration of $f_0(\vec{r})$ starting at $t = 0$. The state of the lattice at $t = 1$ is obtained by applying the rule to each cell in the entire lattice.

CA has been successfully applied to model various physical phenomena from forest fires, game of life, and diffusion to coalescence and self-organizing systems (Chopard and Droz, 1998). The CA-based modeling techniques can be used to model for urban land-use simulations (Lau and Kam, 2005) and also to describe, simulate, and understand the spread of CBR agents into building surroundings and estimate the damage, as shown in Figure 3.12.

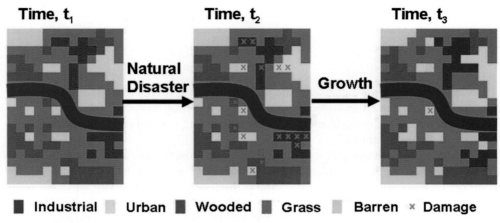

Figure 3.12 Progression of damage and spread of CBR agents through CA modeling.

3.4.8 Neural networks

Neural networks (NNs) are intelligent arithmetic computing elements that can represent, by learning from examples, complex functions with continuous valued and discrete outputs, as well as a large number of noisy inputs (Russell and Norvig, 1995). These networks imitate the learning process in the brain and can be thought of as mathematical models for the operation of the brain. The simple arithmetic elements correspond to the neurons—the cells that process information inside the brain. The network as a whole corresponds to the collection of interconnected neurons. Each link has a numeric weight associated with it. Weights are the primary means of long-term storage in neural networks, and learning usually takes place by updating these weights. The weights are adjusted so as to bring the network's input/output behavior more in line with that of the phenomena being modeled by the network. Each node has a set of input links from other nodes, a set of output links to other nodes, a current activation level, and a means of computing the activation level at the next step, given its inputs and weights. The computation of activation level is based on the values of each input signal received from a neighboring node and the weights on each input link.

The most popular method for learning in multilayer networks is called back-propagation, first invented by Bryson and Ho (1969). In such a network, learning starts with presenting the examples to the network and comparing the output vector computed by the feedforward network with the target vector (known outcomes for the given examples). If the network output and the target vector match, nothing is done. However, if there is an error (a difference between the outputs and the target), then the weights are adjusted to reduce this error. The trick here is to assess the blame for an error and divide it among the contributing weights, thereby minimizing the error between each target output and the output computed by the network. More details can be found in Russell and Norvig (1995). Artificial NNs are capable of realizing a variety of nonlinear relationships of considerable complexity and have been applied to solve many engineering problems. The prediction of climate change parameters with varying input parameters can be modeled using NN, as shown in Figure 3.13.

3.4.9 Mathematical homogenization

Homogenization and averaging are the mathematical/computational processes by which local properties are obtained on a coarse space/time grid from the variables on fine scales. Several analytical studies of homogenization problems in random and periodic heterogeneous

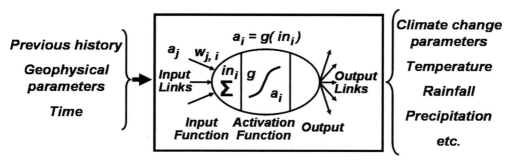

Figure 3.13 Neural network modeling and prediction of climate change parameters.

materials can be found in Schwab and Matache (2000). Homogenization was first developed for periodic structures involving boundary value problems in material sciences, continuum mechanics, quantum physics, and chemistry. Homogenization results in a coarse scale description of processes that occur on different space and time scales. Usually, the process is described by an initial boundary value problem for a partial differential equation, and the medium considered is periodical. For example, the transport process of water and solutes occurs at micro- and mesoscales, and depends on the aggregate and grain size distribution, porosity, and the porous media properties. The particles and pore sizes may range from subnano-, nano-, to micro- and macropores.

3.4.10 Quasi-continuum method

The Quasi-continuum (QC) method was originally developed for material science applications (Tadmor et al., 1996; 1999) to simulate isolated defects, such as dislocations and cracks in single crystals. In this method, constitutive equations describing the material behavior are constructed directly from atomistic information. This is different from a phenomenological model. The QC model is applicable as long as the continuum fields are slowly varying over a unit cell domain. The microscale model is based on molecular mechanics, whereas the macroscale model is based on nonlinear elasticity. An adaptive mesh refinement procedure allows QC method to identify the defects and define locally to the atomistic scale in order to resolve the full details around the defects. The QC method can also be used as a way of simulating the macroscopic deformation of a material without using empirical stored energy functional, but using directly the atomic potential.

3.4.11 Heterogeneous multiscale method

The Heterogeneous multiscale method (HMM) proposes a unified framework for linking models at different scales, emphasizing "multiphysics" applications of very different nature (Weinan et al., 2003a, b). It follows a "top-down strategy" starting from an incomplete macroscale model, with the microscale model as a supplement. The HMM essentially consists of two main components: the macroscale solver, which is knowledge based, uses the knowledge about the macroscale process, and a procedure for estimating the missing data from the microscale model. Several classes of problems are solved using this framework, including atomistic–continuum modeling for the dynamics of solids and fluids, interface dynamics, and homogenization problems. The details of this method and applications can be found in Weinan et al. (2003a, b). One distinct feature of HMM is its emphasis on selecting the right macroscale solver. This is especially important for problems whose macroscale behavior contains singularities or phase transitions. Therefore, the macroscale solver should be stable and facilitate coupling with the microscale model.

3.4.12 Continuum methods

At the continuum level, the macroscopic behavior is determined by assuming that the material is continuously distributed throughout its volume. This method disregards the discrete atomistic and molecular structure of the material. The continuum material is assumed to have an average density and can be subjected to body forces such as gravity and surface forces such as the contact between two bodies.

The continuum is assumed to obey several fundamental laws: continuity, equilibrium, and momentum. The law of continuity is derived from the conservation of mass, and the law of equilibrium is derived from momentum considerations and Newton's second law. The law of moment of momentum principle is based on the model that the time rate of change of angular momentum with respect to an arbitrary point is equal to the resultant moment. In addition to the previous laws, the conservation of energy and entropy, which are based on the first and second laws of thermodynamics, respectively, need to be satisfied. These laws provide the basis for the continuum model and must be coupled with the appropriate constitutive equations and equations of state to provide all equations necessary for solving a continuum problem. The state of a continuum system is described by several thermodynamic and kinematical state variables. The equations of state provide the relationships between the nonindependent state variables. The finite element method (FEM) is a popular method based on continuum mechanics, which is described next.

The simulation of various physical, biological, and chemical processes typically requires the numerical solution of various types of partial differential equations. The finite element method provides a numerical solution to initial value and boundary value problems for both space and time. The FEM has been very popular and can find applications in mechanical, aerospace, biological, and geological systems. The FEM is based on a variational technique for solving differential equations, and the details can be found in Reddy (2006). In general, in the FEM, the physical domain or shape is broken down to simple subdomains (finite elements) that are interconnected and fill the entire domain without any overlaps. To this end, one can prescribe a domain to be modeled with selected finite elements and a uniform mesh size. Solving the governing equations produces an approximate solution. The finer the mesh, the better the solution. Discretizing the equations in three dimensions on realistic domains requires solving a very large but sparsely populated system of linear equations. In addition, the discretization on an adapted finite element mesh shown in Figure 3.14 can be used to improve the solution.

For problems involving fluid mechanics, the basic equations governing motion of fluid are the Navier–Stokes equations. The conservation of mass, momentum, and energy are, respectively,

$$\frac{\partial \rho}{\partial t} + \frac{\partial}{\partial x_k}(\rho u_k) = 0 \qquad (3.7)$$

$$\frac{\partial \rho u_i}{\partial t} + \frac{\partial}{\partial x_k}(\rho u_i u_k - \tau_{ik}) + \frac{\partial P}{\partial x_i} = S_i \qquad (3.8)$$

$$\frac{\partial (\rho E)}{\partial t} + \frac{\partial}{\partial x_k}[(\rho E + P)u_k + q_k - \tau_{tk}u_t)] = S_k u_k + Q_H \qquad (3.9)$$

3.4 Multiscale modeling methods

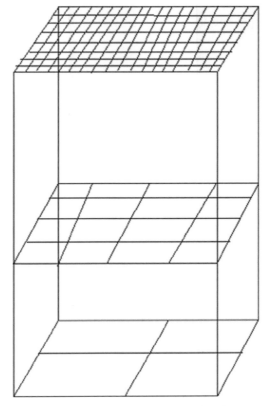

Figure 3.14 Adaptive mesh refinement in finite element method.

where u is the fluid velocity, S_i is the external volume force per unit mass, ρ is the fluid density, E is the total energy per unit mass, τ_{tk} is the viscous shear stress tensor, Q_H is the heat source per unit volume, and q_i is the diffusive heat flux.

3.4.13 Domain decomposition method and parallel computations

In the domain decomposition method, the domain of interest is split into as many partitions as processors are available, as shown in Figure 3.15. The definition of interface nodes must guarantee the correct connection between the partitions. In general, the parallel computations are carried out as follows:

- Given the problem data and P processors of a parallel machine
- Each processor $i = 1, \ldots, P$
- Obtains a coarse solution over the global domain
- Subdivides the global domain into P subdomains, each of which is assigned a processor
- Assigns boundary conditions to a fine discretization of its subdomain using the coarse global solution
- Solves the equation on its subdomain
- A master processor collects observable data from other processors and controls input/output

Parallel computers consisting of multiple processors and distributed network of computers (PCs and workstations) are well suited to solve large-scale disaster problems.

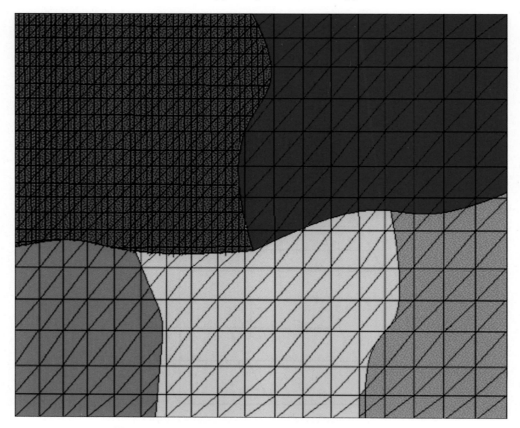

Figure 3.15 Multigrid modeling in parallel computations.

3.4.14 Lattice Boltzmann method

Lattice Boltzmann method (LBM) is a powerful alternative to the standard "top-down" and "bottom-up" approaches to simulate fluid flows. The top-down approaches begin with a continuum description of macroscopic phenomena provided by partial differential equations, such as Navier–Stokes equations. Numerical methods are then used to transform this continuum description into discrete points to obtain a solution numerically. The bottom-down approaches begin with molecular dynamics description of microscopic phenomena. The position and velocity of each particle in the system is obtained from Newton's equation of motion. Unlike these two schemes, LBM can be considered as a mesoscopic particle-based approach. Gases of particles are arranged on a node of a discrete lattice. Time and velocity are divided into discrete values. During each time step, particles move from one lattice node to the next based on their discrete velocity, and then scatter according to a specific collision rule that conserves mass, momentum, and energy. This rule is carefully chosen to capture the real microscopic phenomena and give a good approximation of the macroscopic phenomena.

The general form of the lattice Boltzmann equation is

$$f_i(x + \mathbf{e}_i \Delta t, t + \Delta t) = f_i(x, t) + \Omega_i(f(x, t))(t = 0, 1, \ldots, M) \qquad (3.10)$$

where f_i is a distribution of particles that travel with a discrete velocity $\mathbf{e}_i$ in the ith direction, Ω_i is a collision operator that determines the rate of change of f_i resulting from the collision, and M is the number of directions of particle velocities at each node. With this discrete velocity $\mathbf{e}_i$, the particle distribution moves to the next lattice node in one time step Δt. The collision operator Ω_i differs for many LBMs. See Chen and Doolen (1998) for more details. The LBM is very suitable for modeling flow fields in complex geometries, like porous media, because a no-slip boundary can be applied to any lattice nodes. This no-slip boundary condition that is based on a bounce-back scheme costs very little in terms of computational time.

3.5 Summary and outlook

Large-scale disasters, both natural and manmade, continue to cause intense suffering and damage to people and property around the world. Research and technological advances are focused on identifying more effective mechanisms for preventing, predicting, and responding to these disasters. Historically, the overriding concern for human safety and welfare led to the study of large-scale disasters from the perspective of restoration, policy, and logistics. However, the challenge is to develop more effective tools that can be used for prediction as well as disaster management. This will require a systematic, multidisciplinary approach consisting of basic science, mathematical descriptions, and computational techniques that can address large-scale disasters across time and space scales, involving multiple orders of magnitude. Due to the multiscale nature of large-scale disasters, multiscale modeling methods offer a promising methodology for the characterization and prediction of large-scale disasters. The idea of multiscale modeling is straightforward—one computes information at a smaller (finer) scale and passes it on to a model at a larger (coarser) scale, by leaving out degrees of freedom as one moves from finer to coarser scales.

In this chapter, the governing phenomena and spatial and temporal characteristics of two types of disasters, one natural and the second manmade/accidental, are discussed within a multiscale modeling approach. Some of the recent developments in multiscale methods from other research areas, such as nanotechnology, computational materials science, fluid dynamics, and artificial intelligence, were reviewed for application to large-scale disasters. Currently, multiscale methods are being developed and used by a number of research labs and scientists from various disciplines to study applications to large-scale disasters. However to establish the validity of modeling and simulation approaches, it is important and necessary that the multiscale simulation methods use experimental data to establish the range of capability of the models and to validate predicted behavior at multiple levels including micro- and megascales.

Acknowledgments

The author wants to thank Dr. M. Gad-el-Hak for his support, and graduate students Evan Neblett, Kittisak Koombua, Ihab Abdelsayed, and Gopakumar Sethuraman for their help with figures and word processing. Thanks also to Dr. Chitra Pidaparti for her help with reviewing the chapter. The author also thanks U.S. National Science Foundation for supporting this work through grant NSF-OISE 0541963, with Dr. Shinaishin as the program director.

References

Alder, B. J., and Wainwright, T. E. (1959) "Studies in Molecular Dynamics," *Chemical Physics* **31**, pp. 459–466.

Asrar, G., Kaye, J. A., and Morel, P. (2000) "NASA Research Strategy for Earth System Science: Climate Component," *Bulletin of the American Meteorological Society* **7**, pp. 1309–1329.

Ayati, B. P., Webb, G. F., and Anderson, A. R. A. (2006) "Computational Methods and Results for Structured Multi-Scale Models of Tumor Invasion," *Multiscale Model. Simul. Journal* **5**(1), pp. 1–20.

Bacon, D. P., Ahmad, N., Boybeyi, Z., Dunn, T. J., Hall, M. S., Lee, P. C. S., Sarma, R. A., Turner, M. D., Waight, K. T., Young, S. H., and Zack, J. W. (2000) "A Dynamically Adapting Weather and Dispersion Model: The Operational Multiscale Environment Model with Grid Adaptivity (OMEGA)," *Mon. Wea. Rev.* **128**, pp. 2044–2076.

Bacon, D. P., Garrett, R. M., Liotta, P. L., Mays, D. E., and Miller, T. E. (2005) "Industrial Hazards to Military Personnel," *J. Med. CBR Def.* **3**, pp. 1–20.

Binder, K. (1995) *Monte Carlo and Molecular Dynamics Simulations in Polymer Science*, Oxford University Press, London, United Kingdom.

Boris, J. P. (2005) "Dust in the Wind: Challenges for Urban Aerodynamics," *AIAA Journal*, pp. 1–63.

Bryson, A. E., and Ho, Y. C. (1969) *Applied Optimal Control*, Blaisdell, New York, New York.

Bunde, A., Kropp, J., and Schellnhuber, H. J. (2002) *The Science of Disasters*, Springer, Berlin, Germany.

Chen, S., and Doolen, G. D. (1998) "Lattice Boltzmann Methods for Fluid Flows," *Ann. Rev. Fluid Mech.* **30**, pp. 329–364.

Chopard, B., and Droz, M. (1998) *Cellular Automata Modeling of Physical Systems*, Cambridge University Press, Cambridge, United Kingdom.

Dolling, D. S. (1996) "Considerations in the Comparison of Experimental Data with Simulations—Consistency of Math Models and Flow Physics," *AIAA Fluid Dynamics Conference*, Paper No. 96-2030, Reston, Virginia.

Easterling, D. R., Meehl, G. A., Parmesan C., Changnon, C. A., Karl, T. R., and Mearns, L. O. (2000) "Climate Extremes: Observations, Modeling, and Impacts," *Science* **289**, pp. 2068–2074.

Fang, X., Jiang, W., Miao, S., Zhang, N., Xu, M., Ji, C., Chen, X., Wei, J., Wang, Z., and Wang, X. (2004) "The Multiscale Numerical Modeling System for Research on the Relationship between Urban Planning and Meterological Environment," *Advances in Atmospheric Sciences* **21**(1), pp. 103–112.

Fast, J. (2005) "Presentation at the Science Team Meeting" of The U.S. Department of Energy Atmospheric Science Program, Oct. 30–Nov. 1, 2005, Alexandria, Virginia.

Fish, J. (2006) "Bridging the Scales in Nano Engineering and Science," *Journal of Nanoparticle Research* **8**, pp. 577–594.

Fish, J., Chen, W., and Tang, Y. (2005) "Generalized Mathematical Homogenization of Atomistic Media at Finite Temperatures," *Int. J. Multiscale Computation Engineering* **3**, pp. 393–413.

Gall, R., and Parsons, D. (2006) "It's Hurricane Season: Do you know where your storm is?," *IEEE Spectrum*, August 2006, pp. 27–32.

Gates, T. S., Odegard, G. M., Frankland, S. J. V., and Clancy, T. C. (2005) "Computational Materials: Multiscale Modeling and Simulation of Nanostructured Materials," *Composites Science and Technology* **65**, pp. 2416–2434.

Gear, C. W., Li, J., and Kevrekidis, I. G. (2003) "The Gap–Tooth Scheme for Particle Simulations," *Phys. Lett. A* **316**, pp. 190–195.

Gopalakrishnan, S. G., Bacon, D. P., Ahmad, N. N., Boybeyi, Z., Dunn, T. J., Hall, M. S., Jin, Y., Lee, P. C. S., Mays, D. E., Madala, R. V., Sarma, A., Turner, M. D., and Wait, T. R. (2002) "An Operational Multiscale Hurricane Forecasting System," *American Meterology Society* **130**, pp. 1830–1848.

Hahn, O., Site, L. D., and Kremer, K. (2001) "Simulation of Polymer Melts: From Spherical to Ellipsoidal Beads," *Macromolecular Theory and Simulations Journal* **10**(4), pp. 288–303.

Heaney, J. P., Peterka, J., and Wright, L. T. (2000) "Research Needs for Engineering Aspects of Natural Disasters," *Journal of Infrastructure Systems* **6**(1), pp. 4–14.

Houghton, J. T., Ding, Y., and Noguer, M. (2001) *Climate Change 2001: The Scientific Basis*, Cambridge University Press, London, United Kingdom.

Knap, J., and Ortiz, M. (2001) "An Analysis of the Quasicontinuum Method," *J. Mech. Phys. Solids*. **49**(9) pp. 1899–1923.

Krasnopolsky, V. M., and Schiller, H. (2003a) "Some Neural Network Applications in Environmental Science. Part I: Forward and Inverse Problems in Geophysical Remote Measurements," *Neural Networks* **16**, pp. 321–334.

Krasnopolsky, V. M., and Schiller, H. (2003b) "Some Neural Network Applications in Environmental Science. Part II: Advancing Computational Efficiency of Environmental Numerical Models," *Neural Networks* **16**, pp. 334–348.

Lau, K. H., and Kam, B. H. (2005) "A Cellular Automata Model for Urban Land-use Simulation," *Environment and Planning B: Planning and Design* **32**, pp. 247–263.

Liu, W. K., Karpov, E. G., Zhang, S., and Park, H. S. (2004) "An Introduction to Computational Nanomechanics and Materials," *Computer Methods and Applied Mechanics in Engineering* **193**, pp. 1529–1578.

Lopez, C. F., Moore, P. B., Shelley, J. C., Shelley, M. Y., and Klein, M. L. (2002) "Computer Simulation Studies of Biomembranes Using a Coarse Grain Model," *Comput. Phys. Commun.* **147**, pp. 1–6.

Lueng, L. R., Mearns, L. O., Giorgi, F., and Wilby, R. L. (2003) "Regional Climate Research—Needs and Opportunities," *Bulletin of the American Metrological Society* **1**, pp. 89–95.

Neugebauer, H. J., and Simmer, C. (2003) *Dynamics of Multiscale Earth Systems*, Springer, Berlin, Germany.

Nieminen, R. M. (2002) "From Atomistic Simulation Towards Multiscale Modeling of Materials," *Journal of Physics: Condensed Matter* **14**, pp. 2859–2876.

Patnaik, G., Boris, J. P., Grinstein, F. F., and Iselin, J. (2005) "Large Scale Urban Simulations with Flux-Corrected Transport," Chapter 4 in High-Resolution Schemes for Convention-Dominated Flows: 30 years of FCT, Eds., D. Kuzmin, R. Lohner, and S. Turek, Springer, pp. 105–130.

Pullen, J., Boris, J. P., Young, T., Patnaik, G., and Iselin, J. (2005) "A Comparison of Contaminant Plume Statistics from a Gaussian Puff and Urban CFD Model for Two Large Cities," *Atmospheric Environment* **39**, pp. 1049–1068.

Rasmussen, R. (2005) "A Giant Step Down the Road to Multiscale Model," *UCAR Quarterly*, National Center for Atmospheric Research, Boulder, Colorado.

Reddy, J. N. (2006) *Finite Element Method*, McGraw-Hill, New York, New York.

Rudd, R. E. (2004) "Coarse Grained Molecular Dynamics for Computational Modeling of Nanomechanical Systems," *Int. J. Multiscale Comp. Engrg.* **2**(2), pp. 203–220.

Russell, P., and Norvig, P. (1995) *Artificial Intelligence: A Modern Approach*, Prentice Hall, Upper Saddle River, New Jersey.

Schwab, C., and Matache, A. M. (2000) "Generalized FEM for Homogenization Problems," *Lecture Notes in Comp. Sci. and Engrg.* **20**, pp. 197–237.

Tadmor, E. B., Ortiz, M., and Phillips, R. (1996) "Quasicontinuum Analysis of Defects in Solids," *Philosophical Magazine* **73**(6), pp. 1529–1563.

Tadmor, E. B., Smith, G. S., Bernstein, N., and Kaxiras, E. (1999) "Mixed Finite Element and Atomistic Formulation for Complex Crystals," *Phy. Rev. B.* **59**(1), pp. 235–245.

Wang, Y., Lueng, L. R., McGregor, J. L., Lee, D. K., Wang, W. C., Ding, Y., and Kimura, F. (2004) "Regional Climate Modeling: Progress, Challenges, and Prospects," *Journal of the Metrological Society of Japan* **82**(6), pp. 1599–1628.

Weinan, E., Engquist, B., and Huang, Z. (2003a) "Heterogeneous Multiscale Method: A General Methodology for Multiscale Modeling," *Physical Review B* **67**(9), pp. 1–4.

Weinan, E., Engquist, B., and Huang, Z. (2003b) "The Heterogeneous Multi-Scale Methods," *Comm. Math. Sci.* **1**, pp. 87–133.

Wolfram, S. (1986) *Theory and Application of Cellular Automata*, Addison Wesley, Reading, Massachusetts.

Wolfram, S. (1994) *Cellular Automata and Complexity*, Addison Wesley, Reading, Massachusetts.

4

Addressing the root causes of large-scale disasters

Ilan Kelman

This chapter explores the root causes of large-scale disasters and how those root causes might be addressed to reduce disaster risk. Definitions are given, indicating that the term "natural disaster" might be a misnomer because disasters tend to require human input to occur, making few disasters be truly "natural." Then, case studies of United Kingdom floods in 2000 and El Salvador earthquakes in 2001 are used to demonstrate vulnerability, the human input to disasters, as being the root cause of large-scale disasters. Vulnerability is shown to arise from population, economic, and political factors. As methods of dealing with the root cause of vulnerability, "localizing disaster risk reduction" and "living with risk" are described in theory and applied in practice to warning systems and education tools. To address root causes of large-scale disasters, opportunities need to be created for linking disaster risk reduction with wider livelihoods and sustainability activities.

4.1 Definitions and context

4.1.1 Defining disasters

Chapter 1 defines "a large-scale disaster" by the number of people affected by it and the extent of the geographic area involved: either 100 to 1,000 people need to be adversely affected, such as by being displaced, injured, or killed, or else the disaster's adverse effects must cover 10 to 100 km^2. Chapter 2 notes that "A large-scale disaster is an event that adversely affects a large number of people, devastates a large geographical area, and taxes the resources of local communities and central governments." The term adopted to explain the process by which root causes of disasters need to be identified and tackled is disaster risk reduction: "The conceptual framework of elements considered with the possibilities to minimize vulnerabilities and disaster risks throughout a society, to avoid (prevention) or to limit (mitigation and preparedness) the adverse impacts of hazards, within the broad context of sustainable development" (United Nations International Strategy for Disaster Reduction [UNISDR], 2006).

From these definitions, it is helpful to highlight two pedantic points. First, each event representing a potential hazard, peril, or danger must be separated from a disaster where casualties and/or damage are witnessed. Because the definitions begin with the assumption that a disaster adversely affects people, an environmental event that does not adversely affect people is not a disaster. A tornado must impact humans, such as by disrupting the food supply or killing people, in order to yield a disaster. A pyroclastic flow is a normal environmental process: volcanic activity. If people live near the volcano with the expectation that no pyroclastic flows will occur, then significant casualties and damage could occur: a disaster.

If society is not considered, then environmental events such as tornadoes, floods, and volcanic eruptions are considered to be "disturbances." In ecology, a disturbance is an event that removes organisms and opens up space for other individuals to colonize (Begon et al., 1996; Reice, 2001). "An event that removes organisms" implies killing the organisms or forcing them to migrate. The definition of disturbance is neutral—it removes organisms, but also opens up space for others—in contrast to the definition of "disaster," which is only about adverse consequences on society.

The second point is that the definition of "large-scale disaster" is not confined to disasters resulting from rapid-onset, relatively clearly defined events such as earthquakes and cyclones. Disasters resulting from events that are more diffuse in space and time are also incorporated, such as droughts and epidemics. Conditions that become disastrous but with less clear start and end points are also incorporated, for example, glaciation, glacial retreat, isostatic uplift, coastal subsidence, and climate change as per the Intergovernmental Panel on Climate Change's (2001) definition.

Different approaches have been adopted to differentiate among events with different onset times and with different clarity of start and end points. Pelling (2001) differentiates between "catastrophic" disasters that can be identified as specific events and "chronic" disasters that still overwhelm a community's ability to cope, yet are part of daily life. Examples of the latter are a low-quality water supply, energy overuse with dependence on nonrenewable supplies, and inadequate waste management. Glantz (1999, 2003a) uses the terms "creeping environmental changes" and "creeping environmental problems" to describe ongoing changes that overwhelm a community's ability to cope, such as desertification, salinization of water supplies, and sea-level rise. Usually, both environmental and human input are required to explain these creeping changes, although at times the environmental or human influence dominates. The longer-term processes could also be termed "disaster conditions" in contrast to "disaster events."

Expanding the definition of "disaster" leads to more political interpretations. Pelling's (2001) "chronic" disasters could be interpreted as including a lack of resources available for tackling outside interests that are exploiting local human and material resources alongside "the gendered and ethnic nature of social systems" (Pelling, 1999). Incompetence, ignorance, or corruption in failing to implement disaster risk reduction could be considered to be the disaster rather than the tornado or flood event in which people die. Figure 4.1 illustrates chronic disasters of coastal erosion and poor solid waste management.

Vocabulary, contexts, and interpretations regarding the definition of "disaster" are not fixed in the literature. Here, "large-scale disasters" are taken to mean both disaster events and disaster conditions, while further noting that this classification should not be binary but, rather, a continuum from rapid-onset, spatially and temporally well-defined events through to ongoing, poorly defined challenges without fixed start or end points in time or space.

These issues are not just an academic exercise to develop a baseline for subsequent discussion, but also have practical value and are needed in operational contexts. Insurance companies make payout decisions depending on whether the "event" or "occurrence" is covered in the policy, which includes the length of time over which the event occurred and whether the loss arose from single or multiple loss occurrences (e.g., Brun and Etkin, 1997; Martin et al., 2000). Currey (1992) asks "whether famine is a discrete event or an integral part of the process of development" as part of attempting to determine how to tackle famine: should famine be tackled as the fundamental problem or are there underlying development causes that should be addressed to avert famine? Therefore, understanding the disaster continuum, from events to conditions, has both academic and practical value. It also helps the following discussion on the term "natural disaster."

Figure 4.1 In the Faroese town of Sumba, a sea wall was built in an attempt to avert the chronic disaster of coastal erosion. The chronic disaster of poor solid waste management is also seen (July 2003).

4.1.2 Do natural disasters exist?

With the definition of disaster requiring human impacts, the term "natural disaster" is used to refer to a disaster in which the hazard or event originates in the environment. The term has led to connotations that the disaster is caused by nature or that these disasters are the natural state of affairs. In many belief systems, including Western thought, deities often cause "natural disasters" to punish humanity or to assert power.

On 1 November 1755, Lisbon, Portugal was devastated by an earthquake and a tsunami. Rousseau reacted by questioning the standard view of disaster as being natural or deific. In a letter to Voltaire, Rousseau (1756) noted that nature did not build the houses that collapsed and suggested that Lisbon's high population density contributed to the toll. He also asserted that unnecessary casualties resulted from people's inappropriate behavior following the initial shaking and that an earthquake occurring in wilderness would not be important to society.

Almost two centuries later, Gilbert White—a geographer regarded as the modern founder of disaster social science—viewed flood disasters from the perspective of people's, rather than nature's, behavior and proposed a range of "adjustments" to human behavior to be adopted for reducing flood damage, going beyond the standard government approach of seeking to control the water (White, 1942/1945). In a paper entitled "Taking the 'Naturalness' Out of Natural Disasters," O'Keefe et al. (1976) extended the focus on human behavior to all "natural disasters," identifying "the growing vulnerability of the population to extreme physical events," not changes in nature, as causing the observed increase in disasters.

The focus on human decisions leading to vulnerabilities that cause disasters, with the potential implication that disasters are never "natural," is now embedded in the disaster literature (e.g., Lewis, 1999; Mileti et al., 1999; Steinberg, 2000; Wisner et al., 2004). Smith (2005) summarizes: "It is generally accepted among environmental geographers that there is no such thing as a natural disaster. In every phase and aspect of a disaster—causes, vulnerability, preparedness, results and response, and reconstruction—the contours of disaster and the difference between who lives and who dies is to a greater or lesser extent a social calculus."

Practitioners also accept the notion that human input exists to all disasters. Abramovitz's (2001) report "Unnatural Disasters" describes the factors which make disasters with environmental phenomena unnatural: "undermining the health and resilience of nature, putting ourselves in harm's way, and delaying mitigation measures." Turcios (2001) asserts "Natural disasters do not exist; they are socially constructed." UNISDR (2002) notes that "Strictly speaking, there are no such things as natural disasters," while UNISDR's (2006) terminology of basic disaster risk reduction terms does not include "natural disaster."

The argument is that natural disasters do not exist because all disasters require human input. Nature provides input through what could be considered a normal and necessary environmental event, but human decisions have put people and property in harm's way without adequate mitigating measures. The conclusion is that those human decisions are the root causes of disasters, not the environmental phenomena.

This conclusion could be contentious. Should humanity be blamed if an astronomical object, such as a comet or asteroid, strikes Earth? The "no" perspective explains that any location in the universe has vulnerability to such objects and that humanity did not have a choice in evolving on Earth. The "yes" perspective contends that humanity has the ability to monitor for potential threats in order to provide enough lead-time to act to avert calamity, namely, by deflecting the object's trajectory or by breaking it up. Search efforts include

Figure 4.2 Flooded house in Yalding, England (October 2000).

the Lincoln Near-Earth Asteroid Research program (Stokes et al., 2000) and the Japanese Spaceguard Association (Isobe, 2000), which is now operational. Meanwhile, Cellino et al. (2006) and Price and Egan (2001) describe other monitoring possibilities that have not yet been implemented. Carusi et al. (2005) and Peter et al. (2004) discuss the feasibility of different options for taking action against threats.

Monitoring deep enough into space in all directions to provide enough lead-time for successful action, irrespective of the object's size, would be costly. The "yes" perspective contends that humanity makes the choice not to expend the resources. The "no" perspective suggests that it is impractical to demand that resources be used to monitor and counter every conceivable threat, among which are gamma-ray flares from stars. These flares, such as that reported by Palmer et al. (2005), occurring within several thousand light-years of Earth could potentially cause a mass extinction according to Mikhail V. Medvedev and Adrian L. Melott from the Department of Physics and Astronomy at the University of Kansas.

Without denying the importance of this debate, rather than trying to resolve it, the focus here is on issues with more immediate practical value. Here, the question "Do natural disasters exist?" is sidestepped by using the phrase "large-scale disaster" at the expense of "natural disaster" because, ultimately, the answer lies in the definition adopted. By definition, a disaster has human impacts. Whether all such human impacts result from human decisions depends on what would be defined as "root causes."

4.2 Root causes of disaster

4.2.1 Case studies

Two case studies of large-scale disasters are selected to examine root causes. First, the October–December 2000 flood disasters in the United Kingdom (relying on Kelman, 2001). Second, the January and February 2001 earthquake disasters in El Salvador.

From October to December 2000, England and Wales suffered devastating flooding, the worst episodes occurring during the approximate dates 10 to 15 October (Figure 4.2),

28 October to 12 November, and 8 to 14 December. At least fourteen people were killed (Jonkman and Kelman, 2005), and costs exceeded US $1.5 billion (2000 dollars; Environment Agency for England and Wales [EA], 2001).

An unusually high level of rainfall occurred over England and Wales throughout 2000 (EA, 2001). Should these meteorological, or potentially climatological, conditions be identified as the root cause of the flood disasters? The excessive rainfall tended to be regional, occurring over much of England and Wales, yet as with many flood disasters, damaged areas were highly localized.

In Uckfield, East Sussex, only a small urban area was inundated, a narrow strip along the river (Binnie Black & Veatch for the Environment Agency [BB&V], 2001). Specific areas of Bevendean, East Sussex, and Malton, Yorkshire, were damaged, although Norton, across the River Derwent from Malton, had more widespread flooding. Lewes, East Sussex, was also more extensively affected, but the 836 flooded properties represent a small proportion of buildings in the community (BB&V, 2001). Those small areas were affected immensely. Moreover, such small areas appeared in a large proportion of counties and in many locales within affected counties across England and Wales (EA, 2001).

Many of the small areas were affected because, once the rain had fallen, highly localized anthropogenic influences channeled and/or stored the water in such a way that flooding resulted. Addressing the impact of community design on observed depth and velocity of flood water, BB&V (2001) write "we believe that the extent, location and orientation of the various structures... on the flood plain, both in Uckfield and in Lewes, made the effect of the 12th October 2000 flood worse than it would have been otherwise." Residents in Partridge Green, West Sussex, explained that inadequate maintenance of drainage ditches left them filled with sediment and clogged by vegetation that stored rather than drained the October rainwater, yielding localized flooding.

Bevendean, East Sussex, lies at the bottom of a cultivated hill. The farmer, despite warnings against the practice and despite previous muddy floods, had been cultivating the fields with a technique that augments runoff even during nonextreme rainfall events (Boardman, 2001, 2003; Boardman et al., 2003). In October 2000, the intensive rain carved a gulley into the hill and, coupled with a sewer system unable to cope with the flow, deposited mud into dozens of homes at the slope's bottom. In Keighley, Yorkshire, the Stockbridge area was flooded by water from the River Aire. Structural flood defenses along the adjacent but nonflooding River Worth exacerbated the flooding, holding the water in the houses until the defenses were deliberately breached, permitting the water to drain (Figure 4.3). Similarly, floodwater in Lewes was trapped in properties by flood defenses on 13 October 2000 (BB&V, 2001).

These local factors are exacerbated by national trends. EA (2001) implicates "greater runoff from the continued expansion in urban areas" as one influence that augments inland flood vulnerability, along with property development in vulnerable areas without adequate precautions. Boardman (2003) notes that for the Bevendean flood, "the underlying cause... is economic policies that encourage, by means of high prices, the growing of cereal crops on all available ground." Across the United Kingdom, demographic changes and poor community planning over past decades, which continue today, increasingly place people and property in vulnerable areas without adequate understanding of how to alleviate the vulnerabilities (e.g., Baxter et al., 2001; Crichton, 2006; Office of the Deputy Prime Minister, 1999; Penning-Rowsell, 2001). For determining root causes, wider perspectives are necessary, beyond the day's or year's rainfall and beyond individuals' daily or yearly decisions.

The precipitation was extreme, yet human decisions over the long-term effectively dictated the flooding's locales and characteristics—such as duration, water velocity, and

Figure 4.3   Flood defenses along the River Worth in Keighley, England, under repair after being deliberately breached to permit floodwater to drain out of houses (November 2000).

mud content. Therefore, human decisions over the long-term effectively dictated the flood damage (i.e., the type of flood disaster).

As the political fallout from the 2000 floods was just starting in the United Kingdom, El Salvador was struck by earthquakes on 13 January 2001 killing 844 people and on 13 February 2001 killing 315 people (Wisner, 2001b). The death toll occurred despite the country knowing the seismic risk: an average of one destructive earthquake per decade has hit El Salvador in living memory (Bommer et al., 2002). The previous earthquake disaster in El Salvador, however, was in 1986, with the intervening years witnessing a civil war continuing until 1992 (it had started in 1980 and killed 75,000 people over the 12 years), major floods and landslides in 1988, and Hurricane Mitch in 1998.

During this period, earthquakes and wider disaster risk reduction had not been entirely sidelined. An urban renewal program after the 1986 earthquake included structural earthquake safety measures (de López and Castillo, 1997). The Inter-American Development Bank (IADB) (1999) set out to turn Hurricane Mitch into an opportunity for development, stating "the objective of Central America is not that of rebuilding the same type of society and productive structure so vulnerable to natural phenomena as the present one." El Salvador's National Reconstruction Program (IADB, 1999) acknowledged that concerns observed due to Hurricane Mitch were deep-rooted and that the rainfall experienced at the end of October 1998 simply exposed many of the chronic problems that the country faced.

Although such problems were seen yet again through the 2001 earthquakes, is it fair to complain that few improvements occurred in the short time frame between Hurricane Mitch and the earthquakes? In the case of 13 January 2001, the answer is yes because short-term action was taken that contributed to most of the deaths. Approximately 700 of the fatalities occurred in one neighborhood, buried by a landslide. The slope failed where deforestation had been completed in order to build luxury homes. A court case had been brought against the luxury home development by the municipality that suffered the landslide, but the case had failed, permitting the development to occur (Wisner, 2001a, 2001b). It is challenging to prove or disprove the assertion that trees would have prevented the landslide. It is legitimate to query whether good judgment was exercised in undertaking an activity likely to destabilize a slope in an earthquake-prone area when hundreds of people live at that slope's bottom—especially when those people actively opposed that activity.

In addition, the question of the short time frame between Hurricane Mitch and the earthquakes is misleading. As with earthquakes, hurricanes are not a new threat to El Salvador. As with earthquakes, lack of hurricane disaster risk reduction occurred for decades, even though that is highlighted mainly after a large-scale disaster, such as Hurricane Mitch. The rainfall and the earth shaking exposed longer-term issues to be discussed as the disasters' root causes. If proven, possibilities to include would be poverty, inequity before the courts, and the Cold War, which led to the civil war. Similarly, in the United Kingdom's 2000 floods, human decisions, values, and actions to be considered as root causes include the long-term choices of siting, designing, and constructing settlements and buildings, plus the land use around settlements. Does a model or identifiable root cause exist that describes such behavior overall, the weaknesses of which were then exposed when extreme rainfall or earth shaking was experienced?

4.2.2 Root cause: vulnerability

To answer the question that concludes the previous section with respect to disaster risk reduction, the origin of the "disaster risk" to be reduced is explored. Within the disaster context, various risk definitions appear in the literature, indicating that consensus does not exist. Examples are

- Alexander (1991) who cites the United Nations Educational, Scientific and Cultural Organization (UNESCO) and the Office of the United Nations Disaster Relief Coordinator: Total risk = Impact of hazard × Elements at risk × Vulnerability of elements at risk
- Blong (1996) who cites UNESCO: Risk = Hazard × Vulnerability
- Crichton (1999): "'Risk' is the probability of a loss, and this depends on three elements, hazard, vulnerability and exposure." If any of these three risk elements increases or decreases, then risk increases or decreases, respectively.
- De La Cruz-Reyna (1996): Risk = Hazard × Vulnerability × Value (of the threatened area)/ Preparedness
- Granger et al. (1999): "Risk is the outcome of the interaction between a hazard phenomenon and the elements at risk within the community (the people, buildings and infrastructure) that are vulnerable to such an impact. The relationship is expressed in pseudo-mathematical form as: $Risk_{(total)}$ = Hazard × Elements at Risk × Vulnerability."
- Helm (1996): Risk = Probability × Consequences
- Sayers et al. (2002): "Risk is a combination of the chance of a particular event, with the impact that the event would cause if it occurred. Risk therefore has two components—the chance (or probability) of an event occurring and the impact (or consequence) associated with that event. The consequence of an event may be either desirable or undesirable.... In some, but not all cases, therefore a convenient single measure of the importance of a risk is given by: Risk = Probability × Consequence."
- Smith (1996): "Risk is the actual exposure of something of human value to a hazard and is often regarded as the combination of probability and loss."
- Stenchion (1997): "Risk might be defined simply as the probability of the occurrence of an undesired event [but] be better described as the probability of a hazard contributing to a potential disaster... importantly, it involves consideration of vulnerability to the hazard."

- United Nations Department of Humanitarian Affairs (UN DHA) (1992): Risk is "Expected losses (of lives, persons injured, property damaged, and economic activity disrupted) due to a particular hazard for a given area and reference period. Based on mathematical calculations, risk is the product of hazard and vulnerability."
- UNISDR (2006): Risk is "The probability of harmful consequences, or expected losses (deaths, injuries, property, livelihoods, economic activity disrupted or environment damaged) resulting from interactions between natural or human-induced hazards and vulnerable conditions."

Some definitions are mathematical, some are prose, and some are both. Parallels emerge with "probability" representing "hazard" and "consequences" or "possible consequences" representing "vulnerability." Definitions of "vulnerability" or "possible consequences" could incorporate other terms used, such as "exposure," "value," "elements," or "preparedness," thereby aligning many of the definitions. The provisos beyond the equations and quotations are helpful, too. Helm (1996) notes that his "simple product is not sufficient in itself to fully describe the real risk, but ... it provides an adequate basis for comparing risks or making resource decisions." Sayers et al. (2002), in addition to their comment that their product is not universally applicable, write "Intuitively it may be assumed that risks with the same numerical value have equal 'significance' but this is often not the case ... low probability/high consequence events are treated very differently to high probability/low consequence events."

Lewis (1999) warns that "focus on risk of a given magnitude may cloud our perception of a reality which might in fact be lesser or partial" and suggests that vulnerability is the most important aspect in the risk debate. He observes "Consideration of vulnerability ... looks at the *processes at work* between the two factors of hazard and risk. It reverses the conventional approach, and focuses upon the location and condition of the element at risk and reasons for that location and condition.... Risk is static and hypothetical (though reassessable from period to period of time) but vulnerability is accretive, morphological and has a reality applicable to any hazard."

Vulnerability is not only about the present state, but also about what society has done to itself over the long term, why and how that has been done in order to reach the present state, and how the present state could be changed to improve for the future. Qualitative aspects of risk, as seen through vulnerability processes, describe more than the quantitative calculations. With disaster risk being the combination, mathematically and nonmathematically, of hazard (including probability) and vulnerability (including consequences), identifying and tackling vulnerability as the root cause of disaster risk becomes evident, even though society has the ability to alter both hazard and vulnerability. Floods and droughts are used to illustrate this aspect.

For altering hazard, long-term damaging effects can outweigh immediate gains. In the past, hydrological engineering frequently sought to control river flow including dampening out flow extremes, such as due to spring snow melt or the annual dry season. Riverside inhabitants thus tend to become inured to the absence of regular flood and drought cycles. Because few extremes occur, mitigation and preparedness activities for large-scale events tend to lapse. There is decreased awareness of the potential flood and drought hazards, decreased understanding of how to predict and react to floods and droughts, and decreased ability to psychologically cope with flood and drought events. The hazard is minimal in the short term, but vulnerability is immense over the long term. Eventually, a large-scale flood or drought must occur, yielding damage that is far greater than would have occurred if the affected community were used to regular, smaller-scale floods and droughts.

This phenomenon is termed risk transference (Etkin, 1999) in which risk is transferred onto future events. The consequences have been documented for major floods such as in the Mississippi basin, United States, in 1993 (Mileti et al., 1999) and in flood-vulnerable areas in England through "the investment... in the water-front Docklands area of London following the erection of the Thames barrier" (Ward and Smith, 1998). Fordham (1999), Kelman (2001), Pielke (1999), and Smith et al. (1996) further document how hydrological engineering for floods encourages settlement in flood-prone locations and discourages adequate precautions against floods.

The hydrological engineering, however, brought benefits. If a reservoir were constructed, then water would be available during droughts—except for extreme droughts. In extreme droughts, the community would have poor mitigation and preparedness because less extreme droughts were not experienced and because the perception would be that the reservoir has ended drought worries: risk transference again. In addition, the use of floodplain land reaped rewards for development—until the large-scale flood or drought.

Costs of the hydrological engineering also occur. Reservoir construction creates the risk of reservoir failure and potentially a flash flood level that could not have occurred without the reservoir. This extreme flood damages properties that would not have been in the floodplain without the hydrological engineering. Although hydrological engineering might be perceived as hazard control and flood control, it is actually hazard alteration and flood alteration. Criss and Shock (2001), for instance, document "flood enhancement through flood control" for the middle Mississippi River and the lower Missouri River in the United States. Costs and benefits occur with hazard alteration, even where short-term benefits are overemphasized and long-term costs are underemphasized.

For altering vulnerability, long-term analyses are also necessary. In Australia, the Bureau of Transport and Regional Economics (BTRE) (2002) details long-term flood vulnerability reduction measures that have been implemented. Land-use planning in Katherine has stopped structures from being built in vulnerable areas, while voluntary purchase of vulnerable properties in Bathurst is estimated to have saved AU $0.7 million (1998 dollars) in the 1998 flood. During a 1998 flood in Thuringowa, mandating and enforcing minimum heights above ground for property ground levels appears to have reduced inundation damage.

In Canada, Brown et al. (1997) examined similar 1986 floods in comparable locations in Michigan, United States, and Ontario, Canada. Ontario, with a vulnerability reduction approach to floodplain management since the Hurricane Hazel disaster in 1954, incurred economic losses less than 0.5% of Michigan's losses. The long-term vulnerability reduction approach required 32 years before an immense payback was witnessed. Meanwhile, other benefits accrued. Following torrential flow and fatalities alongside a river during Hurricane Hazel, Toronto, Ontario, Canada, prevented development in river valleys that now serve both as floodways and as a system of heavily used recreational greenways (Figure 4.4). Boulder, Colorado, United States, has adopted a similar approach, purchasing properties in the floodplain and constructing bicycle and walking paths along Boulder Creek, which has a large flash flood potential (Figure 4.5). The Association of State Floodplain Managers (ASFPM) (2002) documents many further American examples of flood vulnerability reduction and the resulting cost savings.

Altering vulnerability with a long-term plan therefore tends to be the most effective approach for disaster risk reduction. Nonetheless, hazard alteration provides a useful tool that should be implemented on occasion. When an asteroid heads toward the Earth, altering the asteroid's trajectory is reasonable rather than trying to reduce vulnerability to a planetwide cataclysm. Volcanic carbon dioxide release from Lake Nyos in Cameroon killed 1,746 people in 1986. A project to degas the lake slowly that does not affect the surrounding area has so far been effective, rather than waiting for the gas to build up and become another lethal release. Instead of moving the population from their livelihoods in the gas-vulnerable

Figure 4.4 A flood along the Don River in Toronto, Ontario, Canada, left debris and caused a fence to collapse along the biking and walking pathway. Prior decades of banning buildings from the floodplain prevented extensive property damage from this flood (October 2005).

Figure 4.5 Boulder Creek, Boulder, Colorado, United States, is a major flash flood hazard, but the recreation pathways and spaces alongside the creek provide some room for floodwater while keeping buildings farther away from the water (May 2006).

areas, which could expose them to other hazards, and instead of relying on a tight time frame for a warning system for gas release, a hazard-altering system has been found to be more effective. In these rare circumstances, the root vulnerability still exists, but altering the hazard has proved to be a useful tool for disaster risk reduction.

4.2.3 Root causes of vulnerability

Vulnerability arises from numerous factors, hence, addressing vulnerability must simultaneously consider multiple dimensions. In focusing on root causes, the summary here discusses societal aspects rather than considering how individuals' characteristics—for instance, gender, age, ethnicity, and state of health—affect each person's possibility for experiencing a disaster's adverse effects (e.g., Brenner and Noji, 1993; Carter et al., 1989; Jonkman and Kelman, 2005; Pearce, 1994).

For example, following the 26 December 2004 tsunamis around the Indian Ocean which killed approximately 300,000 people, Oxfam (2005) compared male and female mortality. In locations surveyed in Indonesia, India, and Sri Lanka, women consistently outnumbered men in the fatalities. In some villages, 80% of fatalities were female. Oxfam highlights that this imbalance results from gender roles within society; for instance, men are taught how to swim more often than women, and thus, men would be more comfortable being immersed in water than women. Similarly, in the 1991 Bangladesh cyclone, lack of swimming ability and/or lack of physical strength was likely a factor in the higher death rate for females than for males (see information given in Chowdhury et al., 1993). Swimming ability and physical strength are not guarantees of survival in coastal floods because debris and collisions with stationary objects are significant dangers, but vulnerability is augmented by lack of swimming ability, lack of comfort in water, and physical weakness.

Therefore, to focus on the individual characteristic of gender, and to claim that women are more inherently vulnerable than men in coastal floods, neglects the root cause of vulnerability. The root cause is the vulnerable role in which women are placed in society, which inhibits their swimming ability, their comfort in water, and their physical strength through factors such as restrictive clothes, less acceptance of women doing manual work, worse nourishment than males, and less acceptance of women appearing in water in public.

At the population scale, the predominant influences on vulnerability to large-scale disasters are increasing population numbers and, to some extent, increasing population densities. Increasing vulnerability from increasing population results from the definition of large-scale disaster: the more people affected, the larger a disaster. A given hazard in an area with more people will yield a larger disaster.

Increasing population densities, mainly through rural-to-urban migration, affects vulnerability in a complex fashion (Figure 4.6). If an equal probability of a hazard occurring exists over a large area, then a dense urban population will experience less frequent hazards than a dispersed rural population, but when a hazard does strike the urban area, more people will be impacted. In contrast, if the probability of a hazard occurring is unevenly distributed across an area, then the location of an urban area determines the effect of urbanization on vulnerability. For example, approximately two-thirds of all cities with a population over 2.5 million are coastal, where they are subjected to hazards including cyclones and tsunamis (Kullenberg, 2005). In the case of large cities, urbanization increases vulnerability to coastal hazards.

Economic structures frequently impose restrictions on addressing root causes of vulnerability. Despite their paybacks, mitigation investments can be deemed to be overly expensive. Harris et al. (1992) design schools and public buildings in tornado-prone areas

Figure 4.6 Increasing population density, mainly from rural-to-urban migration, is epitomized by Shanghai, China, which is vulnerable to typhoons, river flooding, subsidence, and epidemics, among other hazards (October 2003).

in the United States, yet they note that added immediate costs for tornado vulnerability reduction must be almost negligible or else they will not be awarded contracts, irrespective of the payback period. Only those who can afford the immediate investment have a choice, but that choice is often for increased vulnerability and increased long-term costs because short-term economic paradigms prevail.

Without a choice, vulnerability often increases. Fatalities increased during the 1995 Chicago heat wave because a legitimate fear of crime precluded many elderly from (1) responding to city workers and volunteers who were trying to check on them and (2) going outside to shaded locales or to common areas of apartment buildings where it was cooler and more airy than in their apartments (Klinenberg, 2002). These people perished from heat-related medical causes that resulted from the lack of affordability of indoor climate control and the lack of choice regarding their communities' poor security situation. Similarly, many deaths during cold waves are attributed to "fuel poverty," where residents cannot afford adequate insulation or heating (Howieson and Hogan, 2005; Rudge and Gilchrist, 2005).

Although exceptions exist, poverty is identified as a root cause contributing to vulnerability because poorer people tend to have fewer choices regarding (1) where they live, both the types of buildings and the locations relative to hazards; and (2) access to education, health care, insurance, political lobbying, and legal recourse. Three provisos are necessary when examining poverty's contribution to vulnerability.

First, poverty leads to vulnerability, but vulnerability can lead to poverty if disasters continually occur, precluding any opportunity to address vulnerability over the longer term. Following the overthrow of the Taleban government in Kabul in late 2001, the international humanitarian and development community faced state-building in Afghanistan. The country had faced decades of war, years of drought, and several earthquake disasters where immediate relief was hampered by downpours and local flooding. Continual disaster leaves little room for initiating and maintaining momentum for longer-term vulnerability reduction and poverty reduction programs.

Second, wealth can augment one aspect of vulnerability because more value exists to be damaged. A distinction is helpful between absolute impact, such as the total fatalities or the total monetary value of losses, and proportional impact, such as the percentage of a community killed or the percentage of assets lost. Prior to a volcanic eruption in 1995, the total population of Montserrat, a small island in the Caribbean, was approximately half the death toll from the 26 January 2001 earthquake in Gujarat, India. Considering absolute impact, the tragedy in India far surpasses any event that could afflict Montserrat. However, proportional impact illustrates that Montserrat has far greater vulnerability than India.

Since Montserrat's volcano started erupting in 1995 (see background in Clay, 1999; Davison, 2003; and Pattullo, 2000), every resident of Montserrat, 100% of the population, has been directly affected, and close to 100% of Montserrat's 1995 infrastructure has been severely damaged or destroyed. On 25 June 1997, pyroclastic flows killed at least nineteen people on Montserrat, a small number compared to the Gujarat earthquake, but which proportionally would be equivalent to more than 1 million people dying in India. Without downplaying disasters in India that are devastating in their own right, examining vulnerability as the root cause of disasters suggests giving disasters that are smaller by absolute metrics appropriate prominence alongside disasters that are larger in absolute impact but smaller in proportional impact. Thus, although more affluent countries appear to be more vulnerable than less affluent countries from a wealth perspective, the proportional cost of a disaster is often greater in less affluent countries (Lewis, 1999). Poverty breeds proportional vulnerability.

Third, wealth permits choices that can increase (or decrease or not affect) vulnerability. Living in mansions along Californian cliffs or on barrier islands along the United States' East Coast requires wealth, but increases vulnerability to storms. Recreational activities requiring wealth, such as skiing, expose people to hazards, such as avalanches, which they might not encounter otherwise. On some Pacific islands, wealth and "development" have led to replacing traditional wood dwellings with "modern" masonry dwellings that are often more vulnerable to earthquakes. Wealth has the potential for increasing vulnerability if certain options are selected, although poverty does not provide the luxury of making a choice regarding vulnerability.

Inadequate bottom-up and top-down leadership and political will are other root causes of vulnerability. The principles of good governance—participation, transparency, accountability, rule of law, effectiveness, and equity (United Nations Development Programme [UNDP], 1997)—at all political levels are identified as a necessary condition for vulnerability reduction (International Federation of Red Cross and Red Crescent Societies, 2004; UNDP, 2004). Transparency International (2005) implicates corruption in construction contractors and building inspectors, leading to shoddy building practices as being the root cause of many earthquake disaster deaths between 1990 and 2003.

Even when leadership and political will exist to promulgate procedures or regulations, more is required to monitor and enforce them. In 1992, Dade County, Florida, had one of the toughest building codes in the United States, but much of the damage caused that year by Hurricane Andrew occurred because buildings were not designed according to the code and because poor enforcement practices failed to uncover the problems (Coch, 1995). During various tornadoes in eastern Canada, "buildings in which well over 90% of the occupants were killed or seriously injured did not satisfy two key requirements of [Canada's] National Building Code" (Allen, 1992).

In summary, root causes of vulnerability and hence disasters are

- Increasing population and, to some degree, increasing population densities
- Inadequate economic structures, incorporating poverty
- Lack of leadership and political will

4.3 Tackling root causes of disaster

4.3.1 Principles

No fixed rules or optimal solutions exist for dealing with the root causes of vulnerability that also affect concerns other than vulnerability and large-scale disasters. Because this discussion is on large-scale disasters, this section summarizes two main principles that have emerged from disaster risk reduction with regard to reducing vulnerability: (1) localizing disaster risk reduction, and (2) living with risk. The next section provides examples of putting these general principles into practice.

Localizing refers to the evidence that disaster risk reduction, including predisaster activities such as preparedness and mitigation and postdisaster activities such as response and recovery, are best achieved at the local level with community involvement (e.g., Lewis, 1999; Twigg, 1999–2000; Wisner et al., 2004). Top-down guidance is frequently helpful, such as in a codified form (e.g., Australia's Workplace Relations Amendment (Protection for Emergency Management Volunteers) Act 2003), as guidelines and a plan (e.g., UNISDR, 2005a), for standardizing vocabulary (e.g., Thywissen, 2006; UNISDR, 2006) or for providing resources. Nonetheless, the most successful outcomes are seen with broad support and action from local residents, rather than relying only on external specialists, professionals, or interventions.

Examples of community involvement and leadership for disaster risk reduction are Townwatch (Ogawa et al., 2005), Community Fireguard (Boura, 1998), Future Search (Mitchell, 2006), and the Safe Living Program (Hennessy, 1998). Even for postdisaster activities, many manuals suggest that individuals and families must be prepared to take care of themselves for at least 72 hours without external assistance (e.g., Emergency Management Australia, 2003; Federal Emergency Management Agency [FEMA], 2004), although recent discussions have suggested 1 to 2 weeks or more. Community teams are increasingly being trained for such purposes, such as the Community Disaster Volunteer Training Program in Turkey (www.ahep.org/ev/egitim5_0e.htm) and Community Emergency Response Teams in the United States (Simpson, 2001; https://www.citizencorps.gov/cert).

Living with risk (UNISDR, 2004) means accepting that the hazard component of risk is a usual part of life and livelihoods. Hazards should not be ignored, and attempts to separate life and livelihoods from hazards frequently incurs long-term costs, as noted earlier for floods and droughts. Avoiding all hazards is also unlikely to be successful and could force movement away from areas with productive livelihoods, such as coastal fisheries or fertile floodplains or volcanic soils. For example, many cities in Asia's deserts are sited near earthquake faults because the tectonic activity makes freshwater available (Jackson, 2001). Rather than just surviving in the face of adversity or reacting to extreme events after they have occurred, living with risk means building and maintaining habitats and livelihoods by using available resources without destroying those resources.

The localized approach of living with risk helps communities create their future. Rather than feeling helpless and expecting external help whenever a crisis strikes, individuals and communities experience:

- Ownership. They can control their own fate, witness the impact of their actions on their own safety and security, and be motivated to help themselves.

- Relevance. Disaster risk reduction is seen to positively and tangibly impact day-to-day living, perhaps because it also provides improved choices, education, and, as noted before with ownership, control over one's life, livelihoods, and community.
- Savings. Disaster risk reduction is cost effective (ASFPM, 2002; Brown et al., 1997; BTRE, 2002; FEMA, c. 1997, 1998; Pan American Health Organization, 1998; Twigg, 2003). The resources gained can be quantified and, to continue momentum for disaster risk reduction, reinvested into the community.
- Continuity. Tackling vulnerability is not about single or one-off actions. Vulnerability is an ongoing process, hence, vulnerability reduction must be ongoing. It thus becomes part of usual, day-to-day life rather than being a special activity to be completed and then forgotten.

Different locations and different sectors have achieved different levels of success in implementing these principles. On a location basis, Abarquez and Murshed (2004) wrote a community-based disaster risk management handbook that applies to much of Asia. For Pacific small islands, the South Pacific Applied Geosciences Commission (SOPAC) produced a program on Comprehensive Hazard and Risk Management, which works to assimilate risk awareness into national planning processes (see SOPAC, 2005). In the programs mentioned previously, the United States' Community Emergency Response Teams were adapted to produce Turkey's Community Disaster Volunteer Training Program. On a sector basis, Corsellis and Vitale (2005) apply the longer-term vulnerability reduction principles to postdisaster settlement and shelter. For adobe construction, Blondet et al. (2003) produced a manual for making adobe construction safer, while GeoHazards International (2006) published a shorter public awareness pamphlet to introduce principles of adobe construction which is safer in earthquakes.

UNISDR (2004, 2005b) integrates many of these aspects by compiling successful case studies from around the world. Although caution is needed because transferability among locations and sectors might be limited, they provide insight into past achievements in tackling root causes of disaster.

4.3.2 Illustrative case studies

To illustrate the previous section's principles in practice, a helpful theme is warning systems. In July 2004, the Climate Forecast Applications in Bangladesh project—based at the Georgia Institute of Technology, United States, with international partners—forecasted that a significant probability existed of above danger-level flooding for the Brahmaputra River in Bangladesh (information on this case study is from Hopson and Webster, 2004, 2006). Government officials in Dhaka received and understood this information on forecast flood levels, but had not been funded to develop education or dissemination programs as part of a warning system. Work is currently underway with appropriate funding to raise awareness of Bangladeshis regarding river floods and to develop an effective information dissemination system for warnings. The river level forecasts are still being provided to Dhaka, and efforts are underway to further improve the skill of the 10-day forecast for the Ganges and Brahmaputra Rivers individually.

Yet, there is recognition that scientific data leading to accurate forecasts of river levels would not be enough. Information must reach vulnerable people in time to act, and they must believe that information and act appropriately. Education, awareness, and training that is started at the community level on their own initiative, with external support when requested, would achieve that and must be part of the warning system. A complete warning system therefore becomes a long-term endeavor, incorporating continual community-based education, awareness, and training.

Figure 4.7   The Pacific Tsunami Warning Center in Hawai'i, United States, is an essential part of international tsunami warning (see also Kelman, 2006), yet many local social components implemented over the long term in communities potentially affected by tsunamis must also be used for an effective warning system (October 2005).

This approach of using a long-term process to reduce vulnerability through warning systems is challenging (see also Glantz, 2003b, 2004; Gruntfest et al., 1978; Handmer, 2000; Kelman, 2006; Lewis, 1999; Mileti et al., 1999; Wisner et al., 2004), especially when a threat is imminent and should be addressed rapidly. Warning systems should be planned as integrated components of the communities to be warned rather than as top-down impositions from governments or scientists relying on technology (Figure 4.7). They should be incorporated into livelihood and development activities, and should be continually relevant to the people who will be warned. Again, concerted long-term effort would be required for success.

Specific principles to follow with examples where they have been implemented include

- Recognizing that people use many information sources to create their own warning and reaction contexts. Dow and Cutter (1998) document in the United States how North Carolinians find sources other than official warnings to be more relevant to their decision making regarding hurricane evacuation.
- Identifying reasons why people might not want to heed a warning and trying to allay those concerns before a warning must be issued. Two reasons why Bangladeshis have not used flood shelters despite receiving warnings were fears of looting and worries that they would be charged rent for using the shelters (Haque and Blair, 1992). Poverty impacted the warning system's effectiveness.
- Ensuring that elements of warning systems are part of people's day-to-day lives rather than being invoked and relevant only when an event threatens. Akhand (1998) describes such aspects for Bangladesh's cyclone warning and shelter system, including continued disaster awareness and the integration of the cyclone shelter and warning system with workplaces and schools, thereby also using the warning system to promote education. Wisner et al. (2004) explain how a flood warning system could and should be incorporated into improving water management in Guatemala

Figure 4.8 A child and adult in Suva, Fiji (November 2005): By teaching the Millennium Development Goals in schools, Fiji also reaches the students' parents and the wider community.

and Honduras. Johnston et al. (2005) note that preexisting beliefs must be accommodated to turn tsunami awareness into tsunami preparedness in the state of Washington, United States.
- Incorporating hazard, vulnerability, and warning knowledge into education systems. Anecdotes from the 26 December 2004 tsunamis around the Indian Ocean (e.g., Cyranoski, 2005) relate instances where an individual's observations of the sea's odd behavior led to a beach being evacuated just before inundation. Although nighttime events would limit use of this specific knowledge, one aspect of warning systems could be a set of simple observations on different environmental phenomena to which individuals would know how to react.

This operational experience with warning systems provides a basis for pursuing the long-term warning system process. Local context must always guide the specific implementation, but crossover to other locations and sectors can exist. In Fiji, the national stage show *Tadra Kahani* is developed by 6 to 19-year-olds and their teachers by taking one United Nations (UN) Millennium Development Goal (UN, 2000) and dramatizing or choreographing its meaning. This approach reaches, in order, educators, youth, the youth's families, their wider community, the national level, and—by making it an annual tourist event and advertizing it onboard the national airline Air Pacific—visitors to the country (Figure 4.8). This idea could be emulated for disaster risk reduction themes.

In Nepal, the National Society for Earthquake Technology (www.nset.org.np) uses a simple shake table to demonstrate in public that a scale-model standard house collapses in the same earthquake in which a scale-model earthquake-resistant house survives. This powerful visual display captures interest for explaining the principles of earthquake-resistant buildings and the low cost and relative simplicity of retrofitting. Witnesses then gossip to their neighbors and communities about what they saw. A collective will for action can develop. Further word-of-mouth promotion has occurred when a parent is proud of their child who is a mason and who has been trained in making buildings earthquake resistant. The parent tells a neighbor or colleague how wonderful their child is by saving lives through making schools safer from earthquakes around the country—which leads to questions and potentially action about making local homes and other infrastructure safer from earthquakes.

As with *Tadra Kahani*, an awareness chain, formal and informal, is created to percolate the information throughout the community. The distant form of disaster is made relevant to day-to-day activities. A relative's work highlights earthquake safety in Nepal. Tourists in Fiji learn how the Millennium Development Goals apply to a small island village. Tackling vulnerability as the root cause of large-scale disasters is integrated into ongoing environmental and community awareness and education. Education refers to any manner by which information is conveyed, both formally and informally. Examples commonly used around the world are chants, textbooks, dances, lectures, sagas, drawings, websites, interactive seminars, role-playing exercises, and sculptures. "Education" means getting information and ideas to people on their terms in their way.

Another useful tool is games that play a significant role in developing behavior, values, and attitudes for children and adults. The UNISDR develops useful disaster risk reduction education products, but often does not have the resources, especially the luxury of time, for full critique and resulting modification. For example, UNISDR deserves kudos for leading the production of a game for disaster risk reduction education, Riskland, but full consideration of tackling vulnerability would have improved the product.

First, dice are used, whereas a helpful message would be that disasters are seldom random, usually resulting from society's choices. Second, the game is competitive, whereas cooperative games promote a more helpful ethos. Accepting vulnerability as the root cause of disaster suggests that either everyone in a community wins or loses, that humanity (and not random events) has the choices for tackling vulnerability, and that one uncooperative individual can ruin the game for everyone.

These critiques of Riskland apply to the discussion in this section. In particular, highlighting education has advantages and matches other disaster risk reduction and sustainability goals (Figure 4.9), especially because education is labeled as a fundamental human right (UN, 1948; World Conference on Education for All [WCEA], 1990; World Education Forum [WEF], 2000), yet 14% of school age children do not attend school (United Nations Educational, Scientific and Cultural Organization [UNESCO], 2004). Although efforts are ongoing regarding "Education for All" (UNESCO, 2004; WCEA, 1990; WEF, 2000), tackling vulnerability must also incorporate the population without access to formal education. Wisner et al. (2007) further highlight that education should be safe. Otherwise, achieving "Education for All" could entail rushing to build vulnerable schools, adding to the long history of students and teachers killed in schools during disasters. Yet, demanding that schools be safe could reduce money for teachers and teaching materials if budget structures force such a dilemma and if political will does not exist to change this budget structure.

As with flood warnings in Bangladesh, the challenge in tackling vulnerability is the long-term, comprehensive nature of activities that must be integrated with day-to-day living at a community's initiative. An example combining warning systems and children's education arose from the rubble of the 2001 El Salvador earthquakes. Plan International and Northumbria University, United Kingdom, have been collaborating in El Salvador on

Figure 4.9 A school near Quito, Ecuador, educates its pupils to live with Cotopaxi volcano (January 2006).

children's roles in disaster risk reduction (Dr. Maureen Fordham, Northumbria University, personal communication, 2006). Children and youth have been taking an active role in warning dissemination through involvement with school- and/or community-based emergency planning or disaster management brigades. In one community built after the 2001 earthquakes, the brigades effected warning and evacuation just before Tropical Storm Stan struck in October 2005. This pattern was repeated elsewhere in El Salvador, with children and youth not only involved in pre-Stan warnings, but also involved in managing shelters as the storm passed over.

Reducing vulnerability to disasters means linking to education, livelihoods, and poverty reduction so that communities have the resources to make their own choices and to take their own initiatives on their own terms regarding localized living with risk.

4.4 Conclusions

Addressing the root causes of large-scale disasters means tackling vulnerability as part of the usual, day-to-day processes of living and pursuing livelihoods, whether that be for a business executive in Kuala Lumpur, a homeless child in New York, or a subsistence fisher near Kampala. The principal philosophical step is emphasizing the human contribution to disasters over the long term, through decisions and values—such as by recognizing that the term "natural disaster" is often a misnomer. The principal practical step is drawing on successes and positive experiences, from *Tadra Kahani* to village shake tables followed by village gossip, in order to overcome past decisions and to reshape values.

Some approaches to adopt, based on the discussion here, are as follows:

- Create opportunities to implement "localized living with risk" in order to explore alternatives to "protection from nature."
- Create opportunities for vulnerability reduction in order to explore alternatives to hazard control and hazard alteration.
- Create opportunities for supporting local decision making that do not neglect wider contexts and that use appropriate technology as one component of broad-based approaches, rather than relying on top-down, technology-only approaches.
- Create opportunities to link disaster risk reduction with wider livelihoods and sustainability activities.

The material covered in this chapter does not yield a complete set. Examples of other suggestions for addressing the root causes of large-scale disasters include researchers collaborating with decision makers to make their methods and results useable (e.g., Bailey, 2006; Glantz, 2003b, 2004); explicitly incorporating ethics into science-based decision making (e.g., Glantz, 2003a; Kelman, 2005); and melding traditional knowledge and techniques with Western scientific and technological knowledge and techniques (e.g., Ellemor, 2005; Few, 2003; Gaillard, 2006; Rautela, 2005; Schilderman, 2004).

Not every approach will work for every circumstance. Nature and deities may continue to be burdened with an excessive portion of the blame for large-scale disasters. By focusing on vulnerability and the human contribution to disasters, progress can continue to be made in addressing the root causes of large-scale disasters.

References

Abarquez, I. and Murshed, Z. (2004) *Community-Based Disaster Risk Management: Field Practitioners' Handbook*, Asia Disaster Preparedness Centre, Bangkok, Thailand.

Abramovitz, J. (2001) "Unnatural Disasters," Worldwatch Paper 158, Worldwatch Institute, Washington, DC.

Akhand, M. H. (1998) "Disaster Management and Cyclone Warning System in Bangladesh," Abstract presented at *EWC II—Second International Conference on Early Warning*, 11 September 1998, Potsdam, Germany.

Alexander, D. (1991) "Natural Disasters: A Framework for Research and Teaching," *Disasters* **15**(3), pp. 209–226.

Allen, D. E. (1992) "A Design Basis Tornado: Discussion," *Can. J. Civ. Eng.* **19**(2), p. 361.

Association of State Floodplain Managers (ASFPM). (2002) *Mitigation Success Stories in the United States*, fourth edition, ASFPM, Madison, Wisconsin.

Bailey, S. (2006) "Integrating Emergency Responders and Scientists into the Emergency Response Plan at Mount Rainier, Washington, USA," presentation in Symposium VI: Emergency Management at *Cities on Volcanoes 4*, 23–27 January 2006, Quito, Ecuador.

Baxter, P. J., Möller, I., Spencer, T., Spence, R., and Tapsell, S. (2001) "Flooding and Climate Change," in *Health Effects of Climate Change*, section 4.6, Document 22452.2P.1K.APR 01(WOR), Department of Health, United Kingdom Government, London, United Kingdom.

Begon, M., Harper, J. L., and Townsend, C. R. (1996) *Ecology*, third edition, Blackwell Science, Oxford, United Kingdom.

Binnie Black & Veatch for the Environment Agency (BB&V). (2001) *Sussex Ouse: 12th October 2000 Flood Report, Executive Summary*, BB&V, England and Wales, United Kingdom.

Blondet, M., Garcia M., G. V., and Brzev, S. (2003) *Earthquake-Resistant Construction of Adobe Buildings: A Tutorial*, Earthquake Engineering Research Institute, Oakland, California.

Blong, R. J. (1996) "Volcanic Hazards Risk Assessment," in *Monitoring and Mitigation of Volcano Hazards*, eds. R. Scarpa and R.I. Tilling, pp. 675–698, Springer-Verlag, Berlin, Germany.

Boardman, J. (2001) "Storms, Floods and Soil Erosion on the South Downs, East Sussex, Autumn and Winter 2000–01," *Geography* **84**(4), pp. 346–355.

Boardman, J. (2003) "Soil Erosion and Flooding on the South Downs, Southern England 1976–2001," *Trans. Inst. Brit. Geog.* **28**(2), pp. 176–196.

Boardman, J., Evans, R., and Ford, J. (2003) "Muddy Floods on the South Downs, Southern England: Problem and Response," *Environ. Sci. Pol.* **6**(1), pp. 69–83.

Bommer, J. J., Benito, M. B., Ciudad-Real, M., Lemoine, A., López-Menijvar, M. A., Madariaga, R., Mankelow, J., Méndez de Hasbun, P., Murphy, W., Nieto-Lovo, M., Rodriguez-Pineda, C. E., and Rosa, H. (2002) "The El Salvador Earthquakes of January and February 2001: Context, Characteristics and Implications for Seismic Risk," *Soil Dynamics and Earthquake Engineering* **22**(5), pp. 389–418.

Boura, J. (1998) "Community Fireguard: Creating Partnerships with the Community to Minimise the Impact of Bushfire," *Australian Journal of Emergency Management* **13**(3), pp. 59–64.

Brenner, S. A. and Noji, E. K. (1993) "Risk Factors for Death or Injury in Tornadoes: An Epidemiologic Approach," in *The Tornado: Its Structure, Dynamics, Prediction, and Hazards*, eds. C. Church, D. Burgess, C. Doswell, and R. Davies-Jones, pp. 543–544, Geophysical Monograph 79, American Geophysical Union, Washington, DC.

Brown, D. W., Moin, S. M. A., and Nicolson, M. L. (1997) "A Comparison of Flooding in Michigan and Ontario: 'Soft' Data to Support 'Soft' Water Management Approaches," *Can. Water Resour. J.* **22**(2), pp. 125–139.

Brun, S. E. and Etkin, D. (1997) "Occurrence Definition," in *Coping with Natural Hazards in Canada: Scientific, Government, and Insurance Industry Perspectives*, eds. S. E. Brun, D. Etkin, D. G. Law, L. Wallace, and R. White, pp. 111–119, Environmental Adaptation Research Group, Environment Canada and Institute for Environmental Studies, University of Toronto, Toronto, Ontario, Canada.

Bureau of Transport and Regional Economics (BTRE). (2002) *Benefits of Flood Mitigation in Australia (Report 106)*, BTRE, Department of Transport and Remedial Services, Canberra, Australia.

Carter, A. O., Millson, M. E., and Allen, D. E. (1989) "Epidemiological Study of Deaths and Injuries Due to Tornadoes," *Am. J. Epidemiol.* **130**(6), pp. 1209–1218.

Carusi, A., Perozzi, E., and Scholl, H. (2005) "Mitigation Strategy," *C. R. Phys.* **6**(3), pp. 367–374.

Cellino, A., Somma, R., Tommasi, L., Paolinetti, R., Muinonen, K., Virtanen, J., Tedesco, E. F., and Delbò, M. (2006) "NERO: General Concept of a Near-Earth Object Radiometric Observatory," *Adv. Space Res.* **37**, pp. 153–160.

Chowdhury, A. M. R., Bhuyia, A. U., Choudhury, A. Y., and Sen, R. (1993) "The Bangladesh Cyclone of 1991: Why So Many People Died," *Disasters* **17**(4), pp. 291–304.

Clay, E. (1999) *An Evaluation of HMG's Response to the Montserrat Volcanic Emergency*, Evaluation Report EV635, Department for International Development, London, United Kingdom.

Coch, N. K. (1995) *Geohazards: Natural and Human*, Prentice Hall, Upper Saddle River, New Jersey.

Corsellis, T. and Vitale, A. (2005) *Transitional Settlement: Displaced Populations*, Oxfam, Oxford, United Kingdom.

Crichton, D. (1999) "The Risk Triangle," in *Natural Disaster Management*, ed. J. Ingleton, pp. 102–103, Tudor Rose, United Kingdom.

Crichton, D. (2006) "UK Flood, Will Insurers Be Out of Their Depth?" paper presented at Session 2, UK Flood, *Insurance Times: Forces of Nature 2006*, 20 March 2006, London, United Kingdom.

Criss, R. E. and Shock, E. L. (2001) "Flood Enhancement Through Flood Control," *Geology* **29**(10), pp. 875–878.

Currey, B. (1992) "Is Famine a Discrete Event?" *Disasters* **16**(2), pp. 138–144.

Cyranoski, D. (2005) "Get Off the Beach—Now!" *Nature* **433**, p. 354.

Davison, P. (2003) *Volcano in Paradise*, Methuen, London, United Kingdom.

De La Cruz-Reyna, S. (1996) "Long-Term Probabilistic Analysis of Future Explosive Eruptions," in *Monitoring and Mitigation of Volcano Hazards*, eds. R. Scarpa and R.I. Tilling, pp. 599–629, Springer-Verlag, Berlin, Germany.

de López, E. M. and Castillo, L. (1997) "El Salvador: A Case of Urban Renovation and Rehabilitation of *Mesones*," *Environment and Urbanization* **9**(2), pp. 161–179.

Dow, K. and Cutter, S. L. (1998) "Crying Wolf: Repeat Responses to Hurricane Evacuation Orders," *Coastal Management* **26**, pp. 237–252.

Ellemor, H. (2005) "Reconsidering emergency management and indigenous communities in Australia," *Env. Haz.* **6**(1), pp. 1–7.

Emergency Management Australia (EMA). (2003) *Preparing for the Unexpected*, EMA, Canberra, Australia.

Environment Agency for England and Wales (EA). (2001) *Lessons learned: Autumn 2000 Floods*, EA, London, United Kingdom.

Etkin, D. (1999) "Risk Transference and Related Trends: Driving Forces Towards More Mega-Disasters," *Env. Haz.* **1**(2), pp. 69–75.

Federal Emergency Management Agency (FEMA). (c. 1997) *Report on Costs and Benefits of Natural Hazard Mitigation*, FEMA, Washington, DC, and developed under the Hazard Mitigation Technical Assistance Program (HMTAP), Contract Number 132, with Woodward-Clyde Federal Services, Gaithersburg, Maryland.

Federal Emergency Management Agency (FEMA). (1998) *Protecting Business Operations: Second Report on Costs and Benefits of Natural Hazard Mitigation*, FEMA, Washington, DC, and developed by the Mitigation Directorate, Program Assessment and Outreach Division under the Hazard Mitigation Technical Assistance Program (HMTAP), contract with Woodward-Clyde Federal Services, Gaithersburg, Maryland.

Federal Emergency Management Agency (FEMA). (2004) *Are You Ready? An In-depth Guide to Citizen Preparedness*, FEMA, Washington, DC.

Few, R. (2003) "Flooding, Vulnerability and Coping Strategies: Local Responses to a Global Threat," *Prog. Dev. Studies* **3**(1), pp. 43–58.

Fordham, M. (1999) "Participatory Planning for Flood Mitigation: Models and Approaches," *Australian Journal of Emergency Management* **13**(4), pp. 27–34.

Gaillard, J. -C. (2006) "Traditional Societies in the Face of Natural Hazards: The 1991 Mt. Pinatubo Eruption and the Aetas of the Philippines," *Int. J. Mass Emergencies and Disasters* **24**(1), pp. 5–43.

GeoHazards International (GHI). (2006) *Seismic Safety for Adobe Homes: What Everyone Should Know*, GHI, Palo Alto, California.

Glantz, M. H. (ed.) (1999) *Creeping Environmental Problems and Sustainable Development in the Aral Sea Basin*, Cambridge University Press, Cambridge, United Kingdom.

Glantz, M. H. (2003a) *Climate Affairs: A Primer*, Island Press, Covelo, California.

Glantz, M. H. (2003b) "Usable Science 8: Early Warning Systems: Do's and Don'ts," report from the Workshop held 20–23 October 2003 in Shanghai, China, by the National Center for Atmospheric Research, Boulder, Colorado, www.ccb.ucar.edu/warning/report.html.

Glantz, M. H. (2004) "Usable Science 9: El Niño Early Warning for Sustainable Development in Pacific Rim Countries and Islands," report from the Workshop held 13–16 September 2004 in the Galapagos Islands, Ecuador, by the National Center for Atmospheric Research, Boulder, Colorado, www.ccb.ucar.edu/galapagos/report.

Granger, K., Jones, T., Leiba, M., and Scott, G. (1999) *Community Risk in Cairns: A Multi-Hazard Risk Assessment*, Australian Geological Survey Organisation Cities Project, Department of Industry, Science and Resources, Canberra, Australia.

Gruntfest, E. C., Downing, T. E., and White, G. F. (1978) "Big Thompson Flood Exposes Need for Better Flood Reaction System to Save Lives," *Civ. Eng.* **February**, pp. 72–73.

Handmer, J. (2000) "Are Flood Warnings Futile? Risk Communication in Emergencies," *Australasian Journal of Disaster and Trauma Studies*, **2000-2**, www.massey.ac.nz/trauma.

Haque, C. E. and Blair, D. (1992) "Vulnerability to Tropical Cyclones: Evidence from the April 1991 Cyclone in Coastal Bangladesh," *Disasters* **16**(3), pp. 217–229.

Harris, H. W., Mehta, K. C., and McDonald, J. R. (1992) "Taming Tornado Alley," *Civ. Eng.* **62**(June), pp. 77–78.

Helm, P. (1996) "Integrated Risk Management for Natural and Technological Disasters," *Tephra* **15**(1), pp. 4–13.

Hennessy, M. (1998) "Effective Community Collaboration in Emergency Management," *Australian Journal of Emergency Management* **13**(2), pp. 12–13.

Hopson, T. M. and Webster, P. J. (2004) "Operational Short-Term Flood Forecasting for Bangladesh: Application of ECMWF Ensemble Precipitation Forecasts," presentation at the American Geophysical Union 2004 Fall Meeting, 13–17 December 2004, San Francisco, California.

Hopson, T. M. and Webster, P. J. (2006) "Operational Short-Term Flood Forecasting for Bangladesh: Application of ECMWF Ensemble Precipitation Forecasts," *Geophys. Res. Abs.*, **8**, SRef-ID 1607-7962/gra/EGU06-A-09912.

Howieson, S. G. and Hogan, M. (2005) "Multiple Deprivation and Excess Winter Deaths in Scotland," *Journal of The Royal Society for the Promotion of Health* **125**(1), pp. 18–22.

Inter-American Development Bank (IADB). (1999) *Central America after Hurricane Mitch: The Challenge of Turning a Disaster into an Opportunity*, IADB, Consultative Group for the Reconstruction and Transformation of Central America, Stockholm, Sweden.

Intergovernmental Panel on Climate Change (IPCC). (2001) *IPCC Third Assessment Report*, IPCC, Geneva, Switzerland.

International Federation of Red Cross and Red Crescent Societies (IFRC). (2004) *World Disasters Report*, IFRC, Geneva, Switzerland.

Isobe, S. (2000) "The Position of the Japanese Spaceguard Association with Regard to NEO Problems," *Planet. Space Sci.* **48**, pp. 793–795.

Jackson, J. (2001) "Living with Earthquakes: Know Your Faults," *J. Earthquake Eng.* **5**(special issue 1), pp. 5–123.

Johnston, D., Paton, D., Crawford, G. L., Ronan, K., Houghton, B., and Bürgelt, P. (2005) "Measuring Tsunami Preparedness in Coastal Washington, United States," *Nat. Haz.* **35**, pp. 173–184.

Jonkman, S. N. and Kelman, I. (2005) "An Analysis of Causes and Circumstances of Flood Disaster Deaths," *Disasters* **29**(1), pp. 75–97.

Kelman, I. (2001) "The Autumn 2000 Floods in England and Flood Management," *Weather* **56**(10), pp. 346–348, 353–360.

Kelman, I. (2005) "Operational Ethics for Disaster Research," *Int. J. Mass Emergencies and Disasters* **23**(3), pp. 141–158.

Kelman, I. (2006) "Warning for the 26 December 2004 Tsunamis," *Disaster Prev. Manag.* **15**(1), pp. 178–189.

Klinenberg, E. (2002) *Heat Wave: A Social Autopsy of Disaster in Chicago*, University of Chicago Press, Chicago, Illinois.

Kullenberg, G. (2005) "The Earth, the Wind and the Sea," in *Know Risk*, pp. 210–213, Tudor Rose and the United Nations Secretariat for the International Strategy for Disaster Reduction, United Kingdom/Switzerland.

Lewis, J. (1999) *Development in Disaster-Prone Places: Studies of Vulnerability*, Intermediate Technology Publications, London, United Kingdom.

Martin, R. J., Reza, A., and Anderson, L. W. (2000) "What Is an Explosion? A Case History of an Investigation for the Insurance Industry," *Journal of Loss Prevention in the Process Industries* **13**(6), pp. 491–497.

Mileti, D. and 136 contributing authors (1999) *Disasters by Design: A Reassessment of Natural Hazards in the United States*, Joseph Henry Press, Washington, DC.

Mitchell, T. (2006) *Building a Disaster Resilient Future: Lessons from Participatory Research on St. Kitts and Montserrat*, PhD dissertation, Department of Geography, University College London, London, United Kingdom.

Office of the Deputy Prime Minister (ODPM). (1999) *Projections of Households in England 2021*, ODPM, London, United Kingdom.

Ogawa, Y., Fernandez, A. L., and Yoshimura, T. (2005) "Town Watching as a Tool for Citizen Participation in Developing Countries: Applications in Disaster Training," *Int. J. Mass Emergencies and Disasters* **23**(2), pp. 5–36.

O'Keefe, P., Westgate, K., and Wisner, B. (1976) "Taking the Naturalness Out of Natural Disasters," *Nature* **260**, pp. 566–567.

Oxfam (2005) *Oxfam International Briefing: The Tsunami's Impact on Women*, Oxfam, Oxford, United Kingdom.

Palmer, D. M., Barthelmy, S., Gehrels, N., Kippen, R. M., Cayton, T., Kouveliotou, C., Eichler, D., Wijers, R. A. M. J., Woods, P. M., Granot, J., Lyubarsky, Y. E., Ramirez-Ruiz, E., Barbier, L., Chester, M., Cummings, J., Fenimore, E. E., Finger, M. H., Gaensler, B. M., Hullinger, D., Krimm, H., Markwardt, C. B., Nousek, J. A., Parsons, A., Patel, S., Sakamoto, T., Sato, G., Suzuki, M., and Tueller, J. (2005) "A Giant γ-Ray Flare from the Magnetar SGR 1806–20," *Nature* **434**, pp. 1107–1109.

Pan American Health Organization (PAHO). (1998) *Natural Disaster Mitigation in Drinking Water and Sewerage Systems: Guidelines for Vulnerability Analysis*, PAHO, Washington, DC.

Pattullo, P. (2000) *Fire from the Mountain: The Tragedy of Montserrat and the Betrayal of Its People*, Constable and Robinson, London, United Kingdom.

Pearce, L. (1994) "The Vulnerability of Society to Atmospheric Hazards," in *Proceedings of a Workshop on Improving Responses to Atmospheric Extremes: The Role of Insurance and Compensation*, 3–4 October 1994, Toronto, Onotario, Canada, eds. D. Etkin and J. McCulloch, pp. 2–12 to 2–28, Toronto, Ontario, Canada.

Pelling, M. (1999) "The Political Ecology of Flood Hazard in Urban Guyana," *Geoforum* **30**, pp. 249–261.

Pelling, M. (2001) "Natural Disasters?" in *Social Nature: Theory, Practice and Politics*, eds. N. Castree and B. Braun, pp. 170–188, Blackwell, Oxford, United Kingdom.

Penning-Rowsell, E. C. (2001) "Flooding and Planning—Conflict and Confusion," *Town & Country Planning* **70**(4), pp. 108–110.

Peter, N., Bartona, A., Robinson, D., and Salotti, J. -M. (2004) "Charting Response Options for Threatening Near-Earth Objects," *Acta Astronautica* **55**(3–9), pp. 325–334.

Pielke, R. A., Jr. (1999) "Nine Fallacies of Floods," *Climatic Change* **42**, pp. 413–438.

Price, S. D. and Egan, M. P. (2001) "Space Based Infrared Detection and Characterization of Near Earth Objects," *Adv. Space Res.* **28**(8), pp. 1117–1127.

Rautela, P. (2005) "Indigenous Technical Knowledge Inputs for Effective Disaster Management in the Fragile Himalayan Ecosystem," *Disaster Prev. Manag.* **14**(2), pp. 233–241.

Reice, S. R. (2001) *The Silver Lining: The Benefits of Natural Disasters*, Princeton University Press, Princeton, New Jersey.

Rousseau, J. J. (1756) "Rousseau à François-Marie Arouet de Voltaire" (Lettre 424, le 18 août 1756), in *Correspondance Complète de Jean Jacques Rousseau, Tome IV 1756–1757*, ed. R.A. Leigh, 1967, pp. 37–50, Institut et Musée Voltaire, Les Délices, Geneva, Switzerland.

Rudge, J. and Gilchrist, R. (2005) "Excess Winter Morbidity among Older People at Risk of Cold Homes: A Population-Based Study in a London Borough," *J. Public Health* **27**(4), pp. 353–358.

Sayers, P. B., Gouldby, B. P., Simm, J. D., Meadowcroft, I., and Hall, J. (2002) *Risk, Performance and Uncertainty in Flood and Coastal Defence—A Review*, R&D Technical Report FD2302/TR1 (HR Wallingford Report SR587), United Kingdom Government, London, United Kingdom.

Schilderman, T. (2004) "Adapting Traditional Shelter for Disaster Mitigation and Reconstruction: Experiences with Community-Based Approaches," *Build. Res. Inf.* **32**(5), pp. 414–426.

Simpson, D. (2001) "Community Emergency Response Training (CERTs): A Recent History and Review," *Nat. Haz. Rev.* **2**(2), pp. 54–63.

Smith, D. I., Handmer, J. W., McKay, J. M, Switzer, M. A. D., and Williams B. J. (1996) *Issues in Floodplain Management, A Discussion Paper, Volume 1*, Report to the National Landcare Program, Department of Primary Industries and Energy by the Centre for Resource and Environmental Studies, Australian National University, Canberra, Australia.

Smith, K. (1996) *Environmental Hazards: Assessing Risk and Reducing Disaster*, second edition, Routledge, London, United Kingdom.

Smith, N. (2005) "There's No Such Thing as a Natural Disaster," *Understanding Katrina: Perspectives from the Social Sciences*, http://understandingkatrina.ssrc.org/Smith.

South Pacific Applied Geoscience Commission (SOPAC). (2005) "Pacific CHARM," in *Know Risk*, p. 234, Tudor Rose and the United Nations Secretariat for the International Strategy for Disaster Reduction, United Kingdom/Switzerland.

Steinberg, T. (2000) *Acts of God: The Unnatural History of Natural Disaster in America*, Oxford University Press, New York, New York.

Stenchion, P. (1997) "Development and Disaster Management," *Australian Journal of Emergency Management* **12**(3), pp. 40–44.

Stokes, G. H., Evans, J. B., Viggh, H. E. M., Shelly, F. C., and Pearce, E. C. (2000) "Lincoln Near-Earth Asteroid Program (LINEAR)," *Icarus* **148**, pp. 21–28.

Thywissen, K. (2006) *Components of Risk: A Comparative Glossary*, United Nations University Institute for Environment and Human Security, Bonn, Germany.

Transparency International. (2005) *Global Corruption Report 2005*, Transparency International, Berlin, Germany.

Turcios, A. M. I. (2001) "Central America: A Region with Multiple Threats and High Vulnerability?" Norwegian Church Aid Occasional Paper Series Number 5/2001.

Twigg, J. (1999–2000) "The Age of Accountability: Future Community Involvement in Disaster Reduction," *Australian Journal of Emergency Management* **14**(4), pp. 51–58.

Twigg, J. (2003) "Lessons from Disaster Preparedness," Notes for presentation to "Workshop 3: It Pays to Prepare" at International Conference on Climate Change and Disaster Preparedness, 26–28 June 2002, The Hague, Netherlands.

United Nations (UN). (1948) *Universal Declaration of Human Rights*, United Nations, New York, New York.

United Nations (UN). (2000) *Millennium Development Goals*, United Nations, New York, New York, www.un.org/millenniumgoals.

United Nations Department of Humanitarian Affairs (UN DHA). (1992) *Internationally Agreed Glossary of Basic Terms Related to Disaster Management*, UN DHA, Geneva, Switzerland.

United Nations Development Programme (UNDP). (1997) *Governance for Sustainable Human Development: A UNDP Policy Document*, UNDP, New York, New York.

United Nations Development Programme (UNDP). (2004) *Reducing Disaster Risk: A Challenge for Development*, UNDP, Geneva, Switzerland.

United Nations Educational, Scientific and Cultural Organization (UNESCO). (2004) *Education for All (EFA): Global Monitoring Report 2003/2004*, UNESCO, Paris, France.

United Nations International Strategy for Disaster Reduction (UNISDR). (2002) *Disaster Reduction for Sustainable Mountain Development: 2002 United Nations World Disaster Reduction Campaign*, UNISDR, Geneva, Switzerland.

United Nations International Strategy for Disaster Reduction (UNISDR). (2004) *Living With Risk*, UNISDR, Geneva, Switzerland.

United Nations International Strategy for Disaster Reduction (UNISDR). (2005a) *Hyogo Framework for Action 2005–2015: Building the Resilience of Nations and Communities to Disasters*, UNISDR, Geneva, Switzerland.

United Nations International Strategy for Disaster Reduction (UNISDR). (2005b) *Know Risk*, Tudor Rose and the UNISDR, United Kingdom/Switzerland.

United Nations International Strategy for Disaster Reduction (UNISDR). (2006) *Terminology: Basic Terms of Disaster Risk Reduction*, UNISDR, Geneva, Switzerland, www.unisdr.org/eng/library/lib-terminology-eng%20home.htm.

Ward, R. and Smith, K. (1998) *Floods: Physical Processes and Human Impacts*, John Wiley & Sons, Chichester, United Kingdom.

White, G. F. (1942/1945) *Human Adjustment to Floods: A Geographical Approach to the Flood Problem in the United States*, Doctoral dissertation at the University of Chicago, Department of Geography (1942), republished as Research Paper No. 29 (1945), University of Chicago, Department of Geography, Chicago, Illinois.

Wisner, B. (2001a) *Disaster and Development: El Salvador 2001*, Quick Response Report no. 142, Natural Hazards Research and Applications Information Center, University of Colorado, Boulder, Colorado.

Wisner, B. (2001b) "Risk and the Neoliberal State: Why Post-Mitch Lessons Didn't Reduce El Salvador's Earthquake Losses," *Disasters* 25(3), pp. 251–268.

Wisner, B., Blaikie, P., Cannon, T., and Davis, I. (2004), *At Risk: Natural Hazards, People's Vulnerability and Disasters*, 2nd edition, Routledge, London, United Kingdom.

Wisner, B., Kelman, I., Monk, T., Bothara, J. K., Alexander, D., Dixit, A. M., Benouar, D., Cardona, O. D., Kandel, R. C., and Petal, M. (in press) "School Seismic Safety: Falling Between the Cracks?," in *Earthquakes*, eds. C. Rodrigué and E. Rovai, Routledge, New York, New York.

World Conference on Education for All (WCEA). (1990) *World Declaration on Education for All*, adopted at WCEA, 5–9 March 1990, Jomtien, Thailand.

World Education Forum (WEF). (2000) "Education For All: Meeting Our Collective Commitments," *Dakar Framework for Action*, text adopted at the WEF, 26–28 April 2000, Dakar, Senegal.

5
Issues in disaster relief logistics
Nezih Altay

> [T]he most deadly killer in any humanitarian emergency is not dehydration, measles, malnutrition or the weather, it is bad management ...
> *(John Telford, former senior emergency preparedness and response officer, United Nations High Commissioner for Refugees)*

This book covers scientific approaches and issues thereof to predicting, preparing, and responding to large-scale disasters. Even though scientific models may help tremendously in capturing most of the facets of each of these three stages, softer and hard-to-model issues surface at their interface. This chapter primarily looks at the soft issues at the interface of preparation and response which is logistics. Logistical problems in the private sector have been widely researched in the field of operations research and management science. This chapter reviews some of these modeling approaches that relate to disaster relief. However, the logistics of disaster relief present certain complex challenges that cannot be easily incorporated into mathematical models, yet directly affect the outcome of relief operations. Such challenges are the main interest of this chapter.

5.1 Introduction

In 2003, 700 natural events caused 75,000 deaths (almost seven times the number in 2002) and more than $65 billion in estimated economic losses, and affected 213 million people (United Nations Economic and Social Council [ECOSOC], 2004). In 2004, an estimated 244,577 people were killed in disasters globally. In 2005, natural disasters caused estimated economic losses in excess of $150 billion, with hurricanes Katrina and Rita, which ravaged the Gulf Coast of the United States, responsible for 88% of this amount. Within the first 5 months of 2006, natural disasters have already caused 12,718 deaths and $2.3 billion in economic damages.[1] This increasing trend in the occurrence of natural disasters is clearly presented in Figure 5.1.

Unlike our common inclination to believe that disasters are low-frequency high-consequence events, some disasters such as floods, cyclones, and droughts tend to affect the same regions repeatedly. For example, Vietnam and the Philippines have suffered serious floods every year between 1999 and 2005. Central America and the Caribbean are frequent targets for hurricanes with more than thirty hits in 2005. More than 1,000 tornados hit the United States every year, most occurring in the central plains states, also called "tornado alley."

---

[1] Data compiled from EM-DAT, an international emergency disasters database (www.em-dat.net).

5.1 Introduction

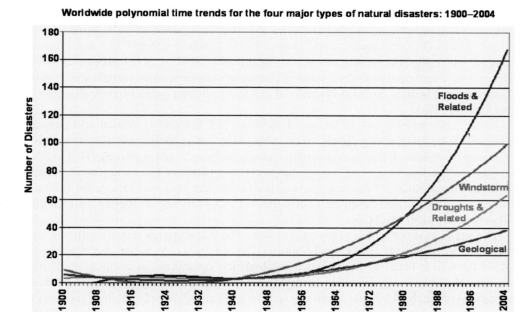

Figure 5.1 Natural disasters are occurring more frequently. *Source:* EM-DAT (2005).

The previous data suggests that as populations shift and their densities increase, and as supply networks grow and their interdependencies widen, concern about disasters becoming increasingly severe is very real, thus making better management of relief efforts critical. Disaster relief operations will need to increase in size and value, reinforcing the need to develop better preparedness programs. Preparedness is the key to a successful response to disasters, and logistics serves as a bridge between preparedness and response.

The Fritz Institute defines humanitarian relief logistics as "the process of planning, implementing and controlling the efficient, cost-effective flow and storage of goods and materials, as well as related information, from the point of origin to the point of consumption for the purpose of alleviating the suffering of vulnerable people" (Thomas and Kopczak, 2005, p. 2). This definition is parallel to the supply chain management concept in the private sector. Therefore, similar to the case with their corporate counterparts, success in humanitarian supply chains will depend on effective communication, coordination, and collaboration among supply chain partners.

Coordinating and collaborating during an emergency is not a trivial task. The mere size of the problem requires the involvement of international agencies, military forces, local authorities, and nongovernmental organizations (NGOs), creating bureaucratic and communication and collaboration difficulties. The response of the International Federation of Red Cross and Red Crescent Societies (IFRC) to the Gujarat Earthquake gives a good indication for the scale of the problem. Within 30 days of the incident, the Logistics and Resource Mobilization Department organized the delivery of forty-five charter planes full of 255,000 blankets, 34,000 tents, and 120,000 plastic sheets along with various other relief items for 300,000 people (Samii et al., 2002). Controlling the procurement and flow of goods into the region is only one side of that equation (Trunick, 2005). In most disasters, information is scarce, and coordination rarely exists (Long and Wood, 1995). Furthermore, disasters often occur in developing countries with poor transportation and communication infrastructures (Nollet, Leenders, and Diorio, 1994). Efforts are duplicated and agencies compete with each other for supplies, driving up prices. In addition, political constraints and domestic or international conflicts routinely make the situation more complicated. Often, corruption and bribery increase operational costs even during times of normalcy (Hecht and Morici, 1993).

Hence, coordination and collaboration between aid agencies is critical and requires investing in building strong supply chains. For many aid agencies, however, the design of high-performance logistics and supply chain operations is not a priority. This lack of investment into logistics, along with the unpredictable nature of emergency incidents and the intermittent nature of funding leads to high employee turnover, lack of institutional learning, and operations based on poorly defined processes and disjointed technologies (Thomas and Kopczak, 2005).

The situation is not as bad as it seems though. The humanitarian sector finally realized the criticality of logistics. The United Nations (UN) and other large humanitarian actors are taking the lead in bettering coordination and logistical operations. In addition, academics are now also interested in making contributions to this field. For example, the fields of operations research and management science are applying advanced analytical methods to help make better decisions. Optimization models have been developed for routing vehicles, allocating resources, and locating collection centers, and many more interesting problems in disaster relief logistics are being uncovered.

In this chapter, we first summarize some of the previously mentioned logistical challenges in the disaster relief sector. Then we attempt to provide a review of the important issues in this context keeping an eye on developments in operations research and management science. Finally, we identify key issues that will have the greatest impact if resolved.

5.2 Disaster relief issues identified in literature

In February 1995, the IFRC, in collaboration with the Danish Red Cross, Danish International Development Agency, and European Community Humanitarian Office, identified key factors to incorporate into the federation's disaster response methodology. The report took the lead in identifying some important issues still current for many aid agencies, such as standardization of services, transparency and accountability, building capacity, capabilities for preparedness, and decentralization (IFRC, 1996).

Later, the director of the Logistics and Resource Mobilization Department of IFRC, together with researchers from INSEAD in France, reiterated that obtaining funds dedicated to support preparedness and capacity building, identifying optimal structures for coordination, and clearly defining roles of involved agencies remain as supply chain challenges for most relief organizations (Chomilier et al., 2003). Lars Gustavsson, the director of Emergency Response and Disaster Mitigation at World Vision International, agreed with the funding problem, and added to this list the lack of depth in knowledge and lack of investment in technology and communication as factors hindering the ability of NGOs to incorporate best practices emerging in the private sector (Gustavsson, 2003).

Recently, based on several excellent case studies he authored or coauthored and his interaction with major humanitarian agencies, Van Wassenhove (2006) identified significant issues facing humanitarian logistics teams as complex operating conditions, safety and security, high staff turnover, uncertainty of demand and supply, time pressure, a need for robust equipment that can be set up and dismantled quickly, large number of stake holders, and the role of media.

Other research focused on observations of humanitarian activity in specific events. For example, after observing the response to Hurricane Georges in the Dominican Republic, McEntire (1999) listed inadequate preparedness, absence or unreliability of disaster-related information, difficulty of needs assessment, unjust distribution of aid, centralization of decision making, insufficient amount of aid, and distrust in emergency managers as perpetual problems of relief logistics. A Fritz Institute survey of humanitarian agencies after the Indian Ocean Tsunami revealed that assessments of needs and planning were inadequate, collaboration and coordination were limited, and supply chain processes were largely manual (Fritz Institute, 2005). Further research by the Fritz Institute identified lack of recognition of the importance of logistics, lack of professional staff, inadequate use of technology, and limited collaboration as some of the common challenges in humanitarian aid logistics (Thomas and Kopczak, 2005).

Logistical challenges in disaster relief highlighted in the literature are summarized as follows:

- Need for capacity building for preparedness
- Lack of funds dedicated to preparedness
- Centralized decision making slowing down response
- Lack of standardization of services
- Need for better coordination
- Tough operating conditions
- Safety and security problems
- High personnel turnover
- Uncertainty of demand and supply
- Need for affordable and robust equipment and technology
- Large number of stakeholders
- Need for more transparency and accountability
- Unreliable or incomplete influx of information
- Lack of recognition of the importance of logistics

Some of these issues are inherent to the very nature of the disasters. For example, disaster relief agencies will always be under time pressure with sudden-onset events such as earthquakes. The operating conditions will always be complex due to the chaotic nature of damage. Other issues mentioned previously, however, could be resolved or at least relaxed if we have a better understanding of the disaster relief business and its logistics function.

In this chapter, we divide logistical issues in disaster relief as supply chain issues and operational issues. With that said, the political and humanitarian nature of disaster relief cannot be ignored because these two areas provide unique challenges that indirectly and sometimes directly interact with relief operations and aid distribution. This chapter also reviews some of the critical political and ethical issues affecting disaster relief logistics.

5.3 Supply chain issues

The definition of disaster relief logistics mentioned in the first section of this chapter is reminiscent of corporate supply chains. However, the flow of physical goods, information, and cash in humanitarian supply chains differs from corporate as depicted in Figure 5.2. For example, shareholders invest in a company expecting financial returns. If returns are not realized, investments cease. Similarly, donors provide money to aid organizations and want

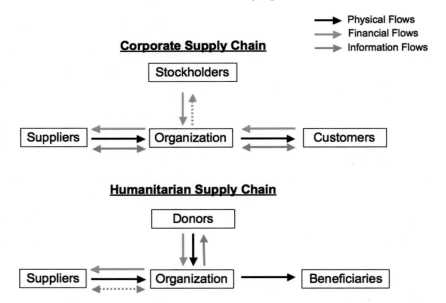

Figure 5.2 Physical, financial, and information flows in corporate versus humanitarian supply chains. *Source:* Blanco and Goentzel (2006).

to be well informed about the results of their philanthropy. If timely and clear feedback regarding the use of their donations is not available, donations may disappear or move to other agencies. With donors being the only significant income source for aid agencies, the quality of feedback becomes crucial for the future of operations.

Donors are not the only stakeholders of humanitarian supply chains. Media, governments, militaries, NGOs, suppliers, third-party logistics firms, and, of course, beneficiaries are all in some form involved in disaster aid operations. Similar to their corporate counterparts, humanitarian supply chains are essentially networks of organizations working for the same objective. Problems are usually strategic and involve all partners in the chain. A well-designed supply chain would thrive through its capabilities like physical and communications infrastructure, and coordination and collaboration among supply chain partners. IFRC's Logistics and Resource Mobilization Department (LRMD) is a good example of this. Recognizing that they cannot operate as an island, LRMD worked to expand the focus of their supply chain from merely purchasing of relief goods to include all activities such as planning, warehousing, training, and reporting (Chomilier et al., 2003).

The World Food Program (WFP) is another excellent example of a humanitarian organization investing in supply chain design. The WFP often establishes its own physical and communications infrastructure during relief operations by using prefabricated, prewired facilities, a logistical preparedness team called the Augmented Logistics Intervention Team for Emergencies, and an information and communications technology support team named the Fast Information Technology and Telecommunications Emergency Support Team. These teams build partnerships with other aid organizations so they can deploy them as standby partners. The main goal of WFP since 2000 has been to improve rapid response capability (Scott-Bowden, 2003). WFP is establishing warehouses in strategic locations to provide storage capacity and act as staging areas for response. Furthermore, WFP's efforts in building rapid local procurement capability allows them to reduce inventory levels significantly.

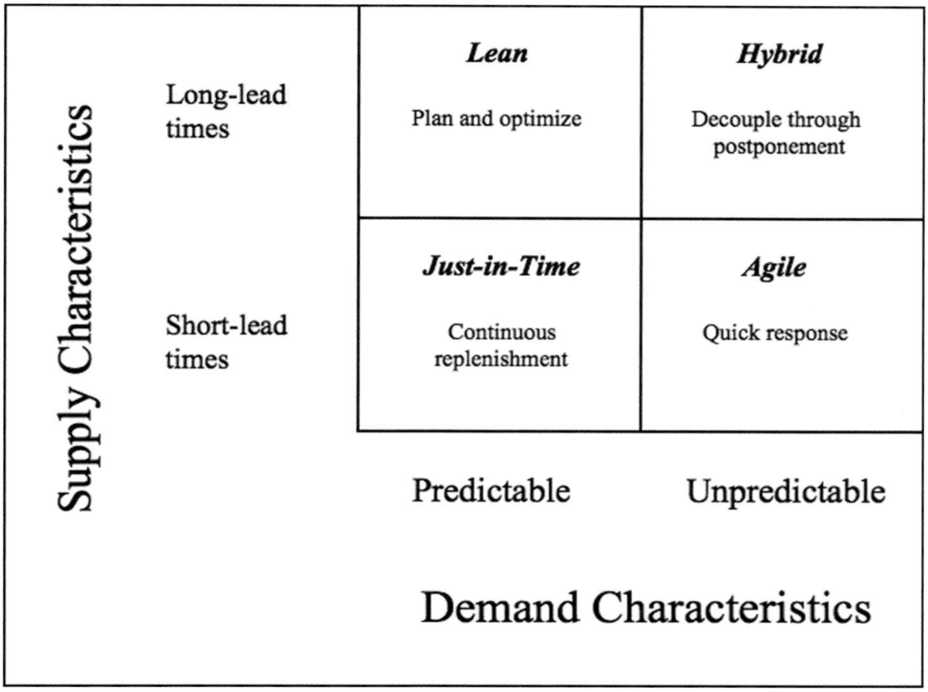

5.3 Supply chain issues

		Predictable	Unpredictable
Supply Characteristics	Long-lead times	**Lean** Plan and optimize	**Hybrid** Decouple through postponement
	Short-lead times	**Just-in-Time** Continuous replenishment	**Agile** Quick response

Demand Characteristics

Figure 5.3 Generic supply chain strategies in the private sector. *Source:* Christopher (2005).

Supply chains are usually designed to minimize cost, maximize throughput, or minimize response time. The private sector recognized long ago that different product lines require different supply chain designs. Figure 5.3 summarizes generic supply chain strategies based on one characteristic of supply and demand. Humanitarian aid supply chains face the challenge of minimizing response time and minimizing costs (McGuire, 2001). Hence, at first glance, disaster aid seems to fall to the lower right quadrant of Figure 5.3 because demand is usually considered unpredictable and short response times require short lead-times. However, similar to the private sector supply chains, different product families should be handled differently. For example, prepackaged food aid for emergencies is mostly standardized using staples, and could be procured and stocked well before the occurrence of a disaster, suggesting a lean supply chain that can be planned and optimized ahead of time. Portable decompression chambers, in contrast, are used to treat crush syndrome victims after earthquakes for which the demand is unpredictable and delivery lead-times must be short.

The previous example suggests that corporate and humanitarian supply chains have a lot to learn from each other. According to Hau Lee, the codirector of the Global Supply Chain Management Forum at Stanford University, supply chains need to possess three very different qualities to have sustainable competitive advantage: they need to be agile, adaptable, and align the interests of the firms in the supply network so that companies actually optimize the chain's performance when they maximize their interests (Lee, 2004). Agility and adaptability are natural skills that relief agencies have to use every day. However, the corporate world started paying attention to these skills only within the past two decades. For example, WFP's portable operations should be treated as a benchmark for agility and adaptability. In contrast, the humanitarian sector may learn alignment strategies from the corporate world because it does seem that many NGOs operate independently, causing the overall system to underperform.

Alignment requires standardization of tasks and products and a well-designed infrastructure that consequently will promote coordination through assessment, management, and dissemination of information. However, all this starts with money, and unfortunately, funding for building capacity and capabilities is not always available. These issues are discussed in the following sections.

5.3.1 Funding issues

Many people in this field believe that the main issue holding back many relief organizations from better preparing for disasters is the difficulty of finding funds for building capacity and capabilities for effective logistical operations (Van Wassenhove, 2006). Donors tag their donations with specific spending targets and want to see that their donation has been spent accordingly. Rarely are funds designated for infrastructure or system development. Furthermore, the public approaches the issue of spending with suspicion. In an interview in 2001, the president of World Vision Canada said, "I would say our biggest challenge in Canada has remained pretty much the same. The challenge, as shown by survey work we have done, is an estimated 80% of the Canadian public is still in a position where it questions whether the money and aid they give to the poor actually gets to them" (Moody, 2001, p. 1). Clearer reporting and demonstrable accountability are needed to keep donors satisfied. Therefore, keeping tabs of expenditures is critical for reporting purposes (ECOSOC, 2005).

During the response to the Indian Ocean Tsunami however, one thing that was abundant was donations. Some organizations received far more than they could spend in the response phase. For example, Doctors Without Borders, the American arm of the medical charity Medecins Sans Frontieres (MSM), announced that they received as much money as they can spend and what was needed was "supply managers without borders: people to sort goods, identify priorities, track deliveries and direct traffic of a relief effort in full gear" (Economist, 2005). Interestingly, when MSM asked if they could use some of the money they had raised for the tsunami disaster to address other crises, the government of Sri Lanka complained to the French government (Strom, 2006).

Although this availability of money allowed humanitarian aid agencies to operate without focusing on fundraising, it also put pressure on them to spend funds quickly. Driven by donor pressure for fast results, many of them used it inefficiently by launching simultaneous projects and executing them with large numbers of personnel (ECOSOC, 2005). Sometimes relief items were dumped in order to artificially raise the number of beneficiaries. Unfortunately, such behavior was not exclusive to the tsunami disaster. At the end of 2005, the health minister of Niger charged that some international aid groups had overstated the extent of the hunger crisis in his country as part of a strategy to raise money for their own purposes (Strom, 2006).

Not every disaster draws as much money as the tsunami. There is a close relationship between the influx of donations and the extent of media coverage of and interest in a disaster (Kim, 2005). During the response to the tsunami the media focus was on Indonesia, India, Thailand, and Sri Lanka. However, 4 months after the tsunami, the Indian Ocean state, Seychelles, received only $4.4 million in response to their appeal, despite their announcement that the estimated cost of repairing damages was $30 million. Similarly, only 3 % of the funding requested in the UN's 2005 Consolidated Appeal for Somalia had been pledged at the beginning of April of that year (Nyanduga, 2005). Not only is money not usually as plentiful as it was for the tsunami response, but donors frequently make financial pledges that are later not fulfilled. Ensuring that the promised funds are actually delivered may be a challenge. For example, only one-third of pledged funds for Darfur (Sudan) and Hurricane Mitch were actually delivered (Oloruntoba, 2005).

5.3.2 Needs assessment and procurement

The goal of logistics is to deliver the right product to the right location at the right time and at the right cost. Therefore, the correct assessment of need immediately after a disaster is crucial for sending the right products. The quality of assessment depends on the involvement of the local authorities in the process. For example, in 2002, Malawi, Lesotho, and Zimbabwe declared a state of emergency after the worst crop failure in recent history. The WFP started distributing food mostly donated by the U.S. government, but many African countries refused the food because it was genetically modified. They claimed that their economy depended on nongenetically modified produce, and they did not want to risk contamination. In another case, during the response to the Indian Ocean Tsunami, there were cases where food aid did not match the needs of the survivors. For example, wheat flower was distributed to rice-eating communities in Indonesia and Thailand, and the nutritional biscuits given as part of emergency food aid packages were not chewable by the elderly (Carballo and Heal, 2005).

Identification of the right products needed is important. Then these products need to be acquired, making procurement one of the key initial steps in a successful response to a disaster. A Fritz Institute (2005) report indicates that the NGOs responding to the Indian Ocean Tsunami had preestablished procurement processes, but nevertheless half of them experienced procurement delays because all organizations were simultaneously trying to purchase the same items.

Procurement is one of the areas where relief organizations may learn from the private sector. Some initial response items are common in every situation and could be secured or even purchased and prepositioned ahead of the time. For such items, continuous or periodic review inventory policies have already been developed (Wright, 1969; Whittmore and Saunders, 1977; Moinzadeh and Nahmias, 1988; Chiang and Gutierrez, 1996). These models assume two modes of demand, one for ordinary demand and one for emergency orders. Recently, Beamon and Kotleba (2006) developed such a model with two options for replenishment for humanitarian relief operations. An order size of $Q1$ is placed every time the inventory level reached $R1$, the regular reorder point. An emergency reorder option is an expedited order of size $Q2$ placed when the inventory reaches a position $R2$ (where $R1 > R2$). The lead times of emergency orders are assumed to be shorter than the regular orders. The emergency order items are also assumed to have higher ordering and purchasing costs than regular orders.

Not all relief items need to be purchased. Some are donated by the public. However, these donations, called gifts-in-kind (GIK), do not always match the assessed needs of the survivors. Management of the flow of incoming supplies and sorting out the essential items from the inappropriate donations becomes a resource draining challenge to the local humanitarian actors. We refer to GIK that are not sensitive to the local culture and norms and to unsolicited aid that is not useful as "donation pollution." Although the former undermines a successful aid operation, the latter takes up valuable time and resources. Disaster survivors need help, not pity. In Banda Aceh, a bundle of torn and stained clothes were dumped at the entrance of a camp. Camp residents found this degrading saying, "although we are in this situation we still have our pride" (Solvang, 2005, p. 29).

Figure 5.4 Tons of inappropriate drugs in Southeast Asia. *Source:* www.drugdonations.org.

Unless a relief organization already has a process in place for identifying and preventing unsolicited and inappropriate donations from entering their system, the extra effort of separating, prioritizing, transporting, and storing these items results in delays and increased logistics costs (Fritz Institute, 2005). During the response to the 1993 Bangladesh Earthquake, unwanted goods constituted 95% of all goods received (Chomilier et al., 2003). Diet pills and winter coats were sent to the Dominican Republic after Hurricane Georges (McEntire, 1999). After Hurricane Katrina, tons of donated clothing from all over the United States were sent to New Orleans. The donations exceeded warehouse capacities, and piles and piles of clothing were left to rot away. In the case of the tsunami relief, such gifts included high-heeled shoes and female swimsuits (Fraser, 2005). Wagonloads of quilts arrived from northern India but were of no use in hot and humid coastal regions (Banarjee and Chaudhury, 2005). Tons of inappropriate drugs were donated (Figure 5.4). The October 2005 newsletter of the Pharmaciens Sans Frontières (PSF) Comité International (www.psfci.org), on its assessment of medicine donations to Banda Aceh province in Indonesia after the tsunami, reported that 4,000 tons of medicine were received for a population of less than 2 million people. Nearly 60% were not on the National List of Essential Drugs. Ten percent had expired before they reached Banda Aceh, and another 30% were due to expire in less than 6 months or had missing expiration dates. About 345 tons of donated drugs have been identified for destruction, which was estimated to cost 1.4 million Euro. PSF-Germany has noted that the same problems have also been experienced in Pakistan.

5.3.3 Management of information

In disaster situations, there is a paucity of information. One does not really know how many survivors are present; what their immediate needs are; and how much food, medicine, and water is still good at their homes. This uncertainty reverberates along the supply chain and becomes exaggerated as we move upstream. It becomes very difficult to ship the correct type and quantity of supplies when the information at all levels within the supply chain is imprecise.

During the response to the tsunami, the problem was not absence of information but rather an absence of comprehensive, cross-functional information on the situation across the affected area (Hudspeth, 2005). During the tsunami response, with an unprecedented number of national and international NGOs (UNICEF alone had around 400 partners) the UN Humanitarian Information Center became critical for information consolidation, analysis, and dissemination (Hudspeth, 2005). However, the center did not become fully operational until several weeks after the tsunami (Volz, 2005).

Even though information may be readily available, most relief agencies do not necessarily have the capability to process that information. A 1997 World Disasters Report of IFRC points out that identification of the necessary information in providing effective relief, how it should be used, and the facilitating effect of technological tools on data collection and transfer has not been given much emphasis. As evidenced by Hurricane Katrina in August 2005, disaster response efforts are hindered by a lack of coordination, poor information flows, and the inability of disaster response managers to validate and process relevant information and make decisions in a timely fashion (Thompson et al., 2007). Even if we assume that information is readily and correctly available, the quantity of variables and parameters makes optimal decision making still difficult. As a point of reference, the IFRC catalog of relief items includes 6,000 stock-keeping units. The UNICEF warehouse in Copenhagen carries $22 million worth of relief supplies acquired from more than 1,000 vendors worldwide. Add to this the problem of donation pollution, and tracking and tracing of inventories becomes a taxing task without the use of information management technology. Outside the large international organizations, most humanitarian agencies use spreadsheets to accomplish this task, and reporting is still done manually (Thomas and Kopczak, 2005). Even if they wanted to develop such technology, they would not have the monetary resources to spend on such a project because donors mainly fund relief efforts rather than preparedness efforts. On average, of funds requested by the World Health Organization in 2002, the supplies component was 37%. Medical and other supplies (excluding vehicles) comprised 35%, and information technology approximately 2% (de Ville de Goyet, 2002). For this reason, the Fritz Institute recently developed the Humanitarian Logistics Software (HLS), which is free of charge. HLS is intended to assist with mobilization, procurement, transportation, and tracking decisions, and produce quick reports. It also connects to financial systems to provide real-time visibility of costs and to allow tracking of budgets (Lee and Zbinden, 2003).

Aid organizations with more resources developed their own technology. One such example is Pan American Health Organization's (PAHO's) Supply Management Software

System (SUMA). SUMA, also available free of charge, is mainly used to sort and identify relief supplies, rapidly identify and prioritize the distribution of supplies to affected population, maintain inventory and distribution control in warehouses, keep authorities and donors informed about items received, and keep managers informed about the availability of items (PAHO, 1999).

Other systems developed in-house include the United Nations' Internet-based operational alert system for earthquakes and sudden-onset emergencies called "virtual on-site operations coordination centers," WFP's International Food Aid Information System, which tracks movements of food aid; and IFRC's Disaster Management Information System to retain existing knowledge within its network (ECOSOC, 2004). Also, in 1991 the U.S. Army's Civil Affairs developed the Disaster Assistance Logistics Information System, a database used for tracking inventories (Long and Wood, 1995).

There have been information systems developed in academia and the private sector to assist emergency managers and relief agencies. For example, Science Applications International Corporation developed the Consequences Assessment Tool Set, which provides decision support with casualty and damage estimation and estimates the amount of resources required to mitigate the destruction and its aftermath. In the academic world, there have been many decision support systems developed, but these are mainly for specific disaster types or scenarios. Two of these systems are the Distributed Environmental Disaster Information and Control Systems developed to help during wildfires (Wybo and Kowalski, 1998) and the Information Management System for Hurricane Disasters, which offers support for preparedness and during and posthurricane response (Iakovou and Douligeris, 2001).

The development of such information systems by and for aid agencies is very encouraging. However, one potential future problem will be coordination using these individual systems, which at this point do not have the capability of communicating with each other. Businesses have already experienced this problem when they tried to coordinate decision making in their supply chains. One greedy solution was to create middleware. Middleware is a software application designed to enable effective information exchange between two systems that are otherwise incompatible. Whether it is middleware or some other solution, these individual decision support systems will need to be able to communicate with each other for effective coordination. This and other coordination challenges in disaster relief are discussed in the next section.

5.3.4 Coordination issues

The relief logistics business is a dynamic one. The actors are not always the same in every disaster. Nor do they have the same task assigned to them every time. Their capacities vary from nation to nation and even year to year within countries. Such uncertainties may easily hinder success of relief operations. Benini (1997) studied uncertainty management and information processing in WFP and UNICEF during their work with victims of the conflict in southern Sudan in 1995, and concluded that cooperation among agencies and public confidence in their work provide functional equivalents for certainty. This means effective coordination may help cope with uncertainties. An excellent review of coordination within social networks in the context of the interactions among public, private, and nonprofit organizations in response to the September 11 terrorist attacks is provided by Kapucu (2005). He showed that dynamic network theory and complex adaptive systems theory are good approaches to understanding social networks and coordination among their nodes.

In a nutshell, coordination is a big challenge given the very different origins, history, geographic, cultural, and political nature of humanitarian actors (Van Wassenhove, 2006).

A report to the UN's ECOSOC states that coordination was not always smooth during the response to the Indian Ocean Tsunami. Some communities were flooded with relief items that did not always match the needs. Miscommunication among aid agencies led to duplication of efforts, delays, and ad hoc plans (ECOSOC, 2005).

There are volumes of research regarding coordination and information sharing in the private sector. Chen (2003) reviewed information sharing and coordination in supply chains. When independent entities in a supply chain are sharing information and one entity has superior information, two things may occur. Information may be withheld to gain strategic advantage, or revealed to gain cooperation from others. If the former, the other (less informed) firms may try to provide incentives for the well-informed firm to reveal this private information; this is called *screening*. If the latter, *signaling may occur* (i.e., revealing information in a credible way). Sometimes it may not be possible to identify who has more or less information. In that case, a firm's willingness to share its information depends on if the others are going to share their information and how the revealed information will be used. Whether we look at information screening or signaling, the common assumption is that the firms in the supply chain are independent entities and information is decentralized. This seems to fit the nature of humanitarian supply chains. Although indeed NGOs, military, governments, and so on are independent units with decentralized information, they are not necessarily competing for beneficiaries' interests. This point of view is a better fit for the case with decentralized information but shared incentives. This seems to fit the team model introduced by Marschak and Radner (1972). Organizations only have access to partial information, but they share a common goal—to optimize the supply chain–wide costs or benefits. An organization acting as an information clearinghouse can simply remove the assumption of decentralized information in this scenario helping reach other organizations' goals. The following paragraph highlights one such organization.

The United Nations Joint Logistics Center (UNJLC) is taking on the challenge of coordinating large relief efforts by collaborating with various aid agencies, governments, and local authorities around the world. UNJLC was established during the humanitarian response to the 1996 Eastern Zaire crisis to deal specifically with logistics issues such as interagency coordination and asset management during a complex emergency. UNJLC views the disaster relief effort as a "modular" system and seeks to strengthen logistics of individual agencies. It accomplishes that by gathering, analyzing, and disseminating relevant information from and among humanitarian and nonhumanitarian actors (Kaatrud, Samii, and Van Wassenhove, 2003).

For effective coordination, it is important to understand the impediments to coordination in humanitarian settings. Oppenheim, Richardson, and Stendevad (2001) list the barriers to effective coordination within the relief aid supply chains as (1) involvement of large number of parties (in Rwanda, 200, and in Kosovo, 300 aid agencies were present), (2) locating and deploying appropriate skills and expertise, and (3) donor-induced constraints for allocating resources. Further research also found competition for donations, competition for media coverage, and cost of coordinating as legitimate issues hindering effective coordination (Stephenson, 2005).

Clearly, effective coordination is essential for successful relief operations. However, coordination overkill may have an adverse effect. For example, in Banda Aceh alone, UN agencies were holding seventy-two coordination meetings per week, but "most NGOs did not have the resources to attend even a small fraction of these meetings" (Volz, 2005, p. 26). Meetings were held in English without translation into the language of the host country. As a result, many local NGOs stopped attending the meetings.

5.3.5 Transportation infrastructure and network design

Network design is a critical element in building successful supply chains. For example, in the business world, a conventional rule of thumb for minimizing cost is for manufacturers to be located closer to suppliers and for retailers to be located closer to end-customers. However, Dell computers worked around this rule by selling directly to the end-customers. This required a different supply chain network design and gave them competitive advantage. Relief agencies have to think about two supply chains, one before a disaster occurs (preparedness stage) and one thereafter (response stage). The latter needs to be portable, agile, and adaptable, and depends on the local infrastructure much more than the former.

In most disaster scenarios, however, the local transportation infrastructure is often heavily damaged. Alternative transportation mediums and shipping routes need to be explored. For example, the Indian Ocean Tsunami destroyed 200 km of coastline in Banda Aceh, making it completely inaccessible. It washed away roads and bridges, and transformed the coastline with new sandbars and debris restricting the landing of large sea vessels (Figure 5.5). Consequently, access was only possible by helicopters and small boats (Hudspeth, 2005). One airport north of Banda Aceh had no lighting and no fuel available, so it could only operate during the day. Krueng Raya seaport had no handling equipment such as forklifts or cranes in place (Oloruntoba, 2005). In the case of the Maldives, its geography was the main challenge in distribution of aid. With 199 inhabited islands dispersed in a strip running 850 km north to south and with much of the transport infrastructure lost to the tsunami, delivery of water, food, and medical supplies was not a simple task (Brown, 2005). In the United States, Hurricane Katrina destroyed major highways and bridges that connect New Orleans and Biloxi to the rest of the country (Figure 5.6).

5.3.6 Standardization of relief

Coordination and collaboration among the providers of disaster relief would benefit from common standards for assessing emergencies and carrying out relief tasks (Oppenheim et al., 2001). In 1997, a group of humanitarian NGOs and IFRC launched the Sphere Project Initiative to identify minimum standards for relief in five key sectors: water supply and sanitation, nutrition, food aid, shelter, and health services (www.sphereproject.org). The Sphere Project aims to enhance the quality and accountability of humanitarian agencies to help them assist people more effectively. It is a very comprehensive undertaking and resulted in a 340-page handbook. IFRC also introduced standardization to procurement, transportation, and tracking of relief goods along with codes of conduct and frame agreements to improve collaboration and coordination with other agencies. This allowed them to keep donation pollution from diluting the essential relief operations.

5.4 Operational issues

Roughly speaking, operations could be explained as the execution side of a business. It is how strategies are implemented while keeping costs and use of resources as low as possible. Disaster relief agencies frequently find themselves helping the unfortunate while constrained with limited resources. Operations research and management science tools are good fits in solving constrained optimization problems. Altay and Green (2006) provided a review of operations research and management science approaches in disaster operations. Because most of the issues discussed in this chapter are soft issues (i.e., they are difficult to formulate mathematically), operations research approaches on relief logistics focus on

Figure 5.5 Devastated transportation infrastructure in Banda Aceh. *Source:* Associated Press.

Figure 5.6 Interstate 90 was completely destroyed during Hurricane Katrina.

resource allocation, transportation planning, scheduling, and routing. Kaplan (1973) formulated deployment of resources from one location to another as a transportation problem. Knott (1987) modeled the transportation of bulk food as a minimum-cost, single-commodity network flow problem with a single mode of transportation to minimize transportation costs. He used the same model to maximize the amount of food delivered. In a later article, Knott (1988) used linear programming along with expert knowledge to schedule delivery vehicles to distribute bulk relief of food. Similar to Knott's knowledge-based approach, Brown and Vasiliou (1993) developed a real-time decision support system that incorporates linear and integer programming models and simulation along with the judgment of the decision maker to assign resources to tasks after a disaster.

Haghani and Oh (1996) pointed out that the basic underlying problem in relief logistics is to move a number of different commodities using a number of different transportation modes from a number of origins to one or more destinations efficiently in a timely manner. Hence, they formulated a multicommodity, multimodal network flow problem with time windows for deliveries. The physical transportation network is converted to a time-space network and solved using a heuristic algorithm. Later modeling approaches to the transportation of relief were all modeled as realistic multicommodity, multimodal problems, while considering multiple conflicting objectives (Barbarosoglu et al., 2002), stochastic demands (Barbarosoglu and Arda, 2004), and dynamic time-dependent problem setups (Ozdamar et al., 2004).

Although the previously mentioned models are targeting to improve relief logistics by optimal allocation and movement of resources, it should not be forgotten that the most important resource in disaster relief logistics is people. Planning makes sense, and technology helps, but it is the people who execute. Therefore, it is crucial for any aid organization to retain their people, hence, retaining know-how and experience. The following subsection discusses personnel issues in disaster relief logistics. The availability and use of technology is another resource issue that is worth discussing. The last part of this section looks into the use of local resources during response to disasters.

5.4.1 Personnel issues

Locating and transporting experienced staff quickly every time a disaster hits is difficult. During the response to the Indian Ocean Tsunami, the humanitarian community was not able to quickly deploy and maintain enough experienced staff specialized in information management, communications, and civil–military liaisons (ECOSOC, 2005). For this reason, one NGO, World Vision, created a full-time division dedicated to predicting, preparing for, and responding to large-scale emergencies (Matthews, 2005). An advantage of having the same people tackling relief challenges is that they are building their own networks with other relief professionals and agencies facilitating effective coordination.

Responding to disasters is a high-pressure job with long hours under hazardous conditions. As if working in harsh environmental conditions were not stressful enough, lately threats from local militia and terrorist groups against foreign humanitarian workers have become very real, adding more stress to the lives of humanitarian actors. Such working conditions present real operational challenges to humanitarian agencies. In this section, we consider four personnel management challenges: employee turnover, safety and security of employees, management of volunteers, and logistics training of personnel.

Employee turnover is a common problem for all disaster relief agencies. High employee turnover impairs institutional knowledge, which is a valuable asset for organizations dealing

Table 5.1. *Disaster stressors*

Lack of warning	Sudden-onset events reduce reaction time
Event type	Psychological impact of natural disasters is relatively less than manmade disasters
Nature of destructive agent	Unseen or highly toxic threats trigger more intense reactions
Degree of uncertainty	Not knowing the duration of threat or possibility of recurrence
Time of occurrence	Responding at night tends to be more stressful
Presence of traumatic stimuli	Sights, sounds, and smells associated with a disaster
Lack of opportunity for effective action	Factors beyond the person's control
Knowing the victims or their families	A sense of shared fate associated with high levels of identification with the deceased
Intense media interest or public scrutiny	Media portraying aid workers in inappropriate ways or generating rumors
Higher than usual or expected responsibility	Making life or death decisions, often with incomplete or inaccurate information
Higher than usual physical, time, and emotional demands	Relief workers often believe that the success of the relief effort hinges on their personal involvement
Contact with victims	Increases traumatic stimuli and reduces workers' willingness to use support resources
Resource availability and adequacy	Insufficient equipment to perform particular tasks increases the sense of inadequacy
Coordination problems	Repetition of tasks and poor communication fuels frustration
Conflict between agencies	Causes additional stress and unrest
Inadequate and changing role definition	Relief workers frequently find themselves in different roles
Inappropriate leadership practices	An automatic management style is not appropriate for disaster work
Single versus multiple threats	Aftershocks, unstable buildings, water contamination, and fires are all threats during response to earthquakes

Adapted from Paton (1996).

with nonroutine problems. Staff members working in harsh conditions tend to get stressed and burned out rather quickly. Table 5.1 provides a comprehensive inventory of disaster stressors (Paton, 1996). Such an inventory may help identify high-risk situations, alert the organization to likely support requirements, and facilitate effective training programs and realistic simulations.

Lack of security is a major obstacle to the delivery of aid where the security of humanitarian personnel is threatened. Northern Uganda, Burundi, Somalia, Iraq, and Afghanistan are some of these locations. In Angola, despite the end of the war, landmine infestation and unexploded ordnance have hindered the delivery of disaster aid along key delivery routes. Separatist rebels in Banda Aceh, "Tamil Tigers" in Sri Lanka and Jemaa Islamiye, a militant Islamic group that warned the aid organizations that they will not tolerate long-term deployment of foreign aid workers, create very real security threats to humanitarian aid personnel (CNN, 2005). After the European Union added the group to its list of banned terrorists in May 2006, Liberation Tigers of Tamil Eelam demanded that European Union citizens leave the mission.

It is argued that aid workers everywhere in the world have become a chosen, deliberate, and direct target of terrorist groups. Between January 1992 and September 2002, 216 UN civilian staff lost their lives. An additional 265 were kidnapped or taken hostage.[2] Bombing of the UN headquarters in Iraq in August 2003 and the attack on the International Committee of the Red Cross the following October suggest that the emblematic protection traditionally afforded aid organizations is no longer recognized. Attacks on humanitarian workers in Afghanistan demonstrated that some belligerents perceive humanitarian organizations as taking sides (ECOSOC, 2004). In August 2006, UN observers were accidentally hit by Israeli bombs in Lebanon, and seventeen staff members of the French branch of the international aid agency, Action Against Hunger, were found killed execution style in Sri Lanka. Clearly, humanitarian agencies need to reorganize their security arrangements. Since the end of 2003, more than $100 million is estimated to have been spent on revamping the security arrangements of the UN and other aid agencies (Gassmann, 2005).

Work conditions may be harsh for humanitarian workers, but after each disaster volunteers offering help are aplenty. When it comes to extra hands, one might think the more the merrier. Indeed, individuals volunteering to help are great, but this may also create a management nightmare. Similar to the "donation pollution" issue discussed earlier in this chapter, volunteers in inappropriate clothing and gear simply cause delays in operations. After the 1999 Turkey Earthquake, many people showed up at the disaster site wearing open-toed shoes, shorts, and tank tops, and were angry and confused when their help offer was denied. Despite warnings, some still went ahead and started working in the rubble, causing medical personnel to use precious time to treat them when they got hurt. Which organizations need help, where will these volunteers be sent, what will they do, and who will supervise them and how, are important questions that will take time and resources to answer (Oloruntoba, 2005). Relief organizations should recognize that conduct and behavior of their staff must be sensitive to the local norms and practices of their duty stations (ECOSOC, 2004). Careless behavior of some volunteers may cause incidents that diminish hard-earned trust and reputation of relief organizations in a host country.

The last personnel issue we discuss is logistics training. Most people working for humanitarian aid agencies are social activists who are not professional logisticians. Donald Chaikin, the head of logistics at Oxfam, says that humanitarian agencies need logisticians with management experience and calls for "professionalism" in the sector (Chaikin, 2003). This shortage of logistics know-how affects the efficiency of distribution efforts (Long, 1997). Recently, in the United States, the humanitarian community started collaborating with professional organizations such as the Council of Supply Chain Management and the Association for Operations Management to provide logistics and operations training. In addition, the Fritz Institute now offers a Certification in Humanitarian Logistics. Several universities in Europe are also providing supply chain management training to humanitarian groups. Part of logistics training should include introduction of technology to the humanitarian sector. The following section focuses on new technology and discusses the importance of having low-tech backup systems.

5.4.2 Availability of technology

Stephenson and Anderson (1997) reviewed the developments in information technology likely to shape disaster planning, management, and research within the last decade. They focused on ultrabroadband networks, digital libraries, high-capacity data storage, cheap microsensors, "smart cards," mobile wireless PDAs, high-performance computing, and remote surveillance technology. Most of these technologies are yet to be found in use in

---

[2] Data from the September 2002 issue of WHO newsletter *Health in Emergencies*.

managing disasters. They foresaw that videoconferencing between field operations and coordination centers, real-time access to beneficiary information, wider use of electronic cash, and use of commercial satellite technology would redefine relief distribution.

International accessibility of satellite technology made remote sensing and use of geospatial technology (geographic information systems, global positioning systems, and related form of earth mapping) in disaster management a reality. Verjee (2005) provided an excellent review to geospatial technology and its use in complex humanitarian crises, and argued that most of the humanitarian community's use of this technology is simple and cartographic in nature, such as updating land use maps and creating transportation maps. According to Verjee, few relief agencies are exploiting geospatial technology to optimize the efficiency of relief distribution.

Even though technology increases the speed and quality of decision making, one thing to remember is that the more easily available the technology, the more dependent on it we become. The disaster management community knows that one should always have low-tech backup systems in place. For example, cell phones can become useless within hours due to their limited battery life. Cell phone towers also receive their power from large batteries or a generator, and after a catastrophic incident, they may not be accessible for replenishment. A robust communications infrastructure is a prerequisite for most of the upcoming information technology. However, in many poor or developing countries, such infrastructure does not exist. For example, doctors from Virginia Commonwealth University's Medical School had to develop a database to monitor hospital bed availability among nine hospitals in Ecuador, and nurses had to enter data by using the land phone line because using the Internet was not a viable solution.

5.4.3 Local resources

All disasters are essentially local. Therefore, a successful response depends heavily on local capabilities and on collaboration with the host government (Van Wassehove, 2006). During the critical initial hours, the speed and effectiveness of the response very much depends on the speed and effectiveness of the local response. In Iran, national authorities and the Iranian Red Crescent Society (IRCS) responded to the earthquake quickly and effectively. The IRCS managed to rapidly mobilize 8,500 relief workers for a massive rescue operation. The response to both, the Bam Earthquake and the earthquake in Morocco, demonstrated that investing in local capacities leads to a speedy response and a solid logistical network (ECOSOC, 2004). Furthermore, use of local groups in decision making and logistics of relief operations also eases the effects of sociocultural differences (Oloruntoba, 2005). In addition to local manpower, relief agencies also acquire part of their relief supplies locally.

EuronAID, an association of European NGOs concerned with food security, lists the benefits of supplying disaster aid locally as follows: first, it helps the damaged local economy. Procurement of food aid from the local farming community would have a positive development effect by improving rural livelihoods and would also catalyze food production for future seasons. A second advantage of using local suppliers is that the risk of delivering inappropriate relief supplies is minimized, and delivery lead-times are relatively shorter. Local procurement also improves the local food quality because international food aid agencies in control of the local procurement process can enforce more appropriate quality standards. Finally, local procurement of relief goods promotes regional trade and employment. Table 5.2 shows EuronAID purchases from local markets from 2000 to 2002 and gives an indication of the monetary impact on local markets. In 3 years, about 69 million Euros were spent locally.

Table 5.2. *Local purchases by EuronAID 2000–2002*

Year	Product	Quantity	Origin	Recipient	Purchase value (Euros)
2000	Cereals	47,091 tons	Ethiopia	Ethiopia	10,294,344
	Vegetable oil	1,077 tons	Sudan	Sudan	1,683,615
	Pulses	4,018 tons	Madagascar	Madagascar	10,687,225
	Other food products	6,320 tons	Ethiopia	Ethiopia	5,510,368
	Seeds & plants	549 tons	Nicaragua	Nicaragua	474,108
	Seeds & plants	1 lot	Sudan	Sudan	51,300
	Tools & inputs	4,000,000 pieces	Nicaragua	Nicaragua	18,390
	Tools & inputs	1,364 Euros	EU	Albania	16,913
	Tools & inputs	3,135 lt.	Honduras	Honduras	15,178
	Tools & inputs	127,172 pieces	Spain	Nicaragua	916,508
	Tools & inputs	1,376 tons	Nicaragua	Nicaragua	318,060
	Tools & inputs	15 lots	Ethiopia	Ethiopia	928,379
					Total: 30,914,388
2001	Cereals	93,031 tons	Sudan	Sudan	14,423,292
	Milk powder	220 tons	India	India	579,360
	Sugar	82 tons	Madagascar	Madagascar	72,721
	Vegetable oil	634 tons	Sudan	Sudan	663,054
	Pulses	1,533 tons	Sudan	Sudan	923,140
	Other food products	178,000 pieces	Rwanda	Rwanda	16,224
	Other food products	848 tons	Sudan	Sudan	630,717
	Seeds & plants	911,700 pieces	Nicaragua	Nicaragua	68,064
	Seeds & plants	2,444 tons	Sudan	Sudan	1,778,611
	Seeds & plants	1 lot	Sudan	Sudan	37,670
	Tools & inputs	11,176 meters	Ethiopia	Ethiopia	133
	Tools & inputs	4,500 lt.	El Salvador	El Salvador	46,507
	Tools & inputs	145 rolls	Ethiopia	Ethiopia	3,464
	Tools & inputs	165,379 pieces	Sudan	Sudan	493,848
	Tools & inputs	3,648 tons	China	N. Korea	768,057
	Tools & inputs	16 lots	Nicaragua	Nicaragua	372,878
					Total: 20,877,740
2002	Cereals	68,944 tons	Ethiopia	Ethiopia	11,906,211
	Milk powder	189 tons	India	India	262,485
	Sugar	204 tons	India	India	74,916
	Vegetable oil	776 tons	Sudan	Sudan	473,218
	Pulses	3,156 tons	Nicaragua	Nicaragua	1,387,710
	Other food products	1,225 tons	Ethiopia	Ethiopia	736,869
	Seeds & plants	2,848 tons	Eritrea	Eritrea	1,412,394
	Tools & inputs	2,745 Euros	Ethiopia	Ethiopia	2,057
	Tools & inputs	4,500 sets	S. Africa	Angola	48,825
	Tools & inputs	1,354 lt.	Nicaragua	Nicaragua	13,782
	Tools & inputs	324,049 pieces	Burkina Faso	Burkina Faso	792,185
	Tools & inputs	1,770 tons	Pakistan	Afghanistan	406,455
	Tools & inputs	1 lot	Sudan	Sudan	2,249
					Total: 17,519,356

Source: www.euronaid.net.

One negative issue with the local resources is the cost of merchandise and housing in the disaster area. The influx of hundreds and even thousands of international aid workers usually causes local housing prices to rise rapidly, sometimes as much as tenfold (Gustavsson, 2003). This artificial increase in prices also applies to food and sanitation supplies, and drains valuable funds that could be spent elsewhere in the relief process.

5.5 Ethical issues

Although the physical and psychological toll of disasters brings out the good in some people, it brings out the bad in others. There are many ethical issues in disasters; however, we only focus our attention on the issues of discrimination among aid recipients and the effects of corruption within local and state governments on relief logistics.

5.5.1 Discrimination

Internally displaced persons (IDPs) are those who have been forced to leave their home due to a natural disaster, political conflict, or war. IDPs bring about various human rights challenges to humanitarian aid operations. In complex emergencies such as wars, IDPs cause special security challenges because women and children can be used as negotiation weapons and for trafficking (Kalin, 2005). In other cases, IDPs are discriminated against and not given access to assistance by controlling governments based on their ethnic background. Some are discriminated against by other survivors based on class differences. After the Indian Ocean Tsunami, in a caste-based hierarchical social structure such as India, some survivors (dalits or "untouchables") were either reluctantly accepted to shelters or were driven away (Hedman, 2005).

According to the Refugee Studies Centre of University of Oxford, "there is evidence of de facto discrimination by local government authorities and Thai citizens against Burmese tsunami survivors" in the affected southern province of Thailand (Hedman, 2005, p. 4). Burmese migrant workers have been excluded in the distribution of emergency relief. Similar to this scenario was the discrimination against Haitian workers in the Dominican Republic during the response to Hurricane Georges (McEntire, 1999).

IDPs may also be difficult to track due to the loss of documentation and the difficulty of obtaining replacement documentation. After the tsunami, Hudspeth (2005) mentioned that "simply finding many of the IDPs in Aceh was difficult." Consequently, some IDPs staying with accepting families, instead of in the survivor camps, did not get aid.

5.5.2 Corruption

Corruption, or our perception of its existence, may affect relief logistics internally and externally. Internally, it affects the protection and distribution of relief supplies. In 1999, after the Kocaeli Earthquake in Turkey, cases were reported where warehouse personnel were sifting through donations to pick newer items for themselves or simply to sell. In such a scenario, fair distribution of aid becomes an issue. For example, after Hurricane Georges, it was reported that the public did not trust the government regarding even distribution of aid (McEntire, 1999).

External corruption, or the perception of it, affects the willingness of donors to contribute to the relief effort. Once again, after the 1999 earthquake in Turkey, monetary donations stalled immediately after the news that donation money may be being channeled from the Turkish Red Crescent Society accounts to the president of the society. Consequently, donors who perceived the Red Crescent Society as a government institution lost their trust in all other government organizations and scrambled to find small trustworthy local NGOs to give their donations.

Unfortunately, the influx of aid in a chaotic environment encourages corruption. Aid agencies need to take every precaution to make sure that distribution of relief is fair. Coordination with the local authorities and the public would help if aid agencies truly understand the power dynamics in the social fabric of the host country. Disregarding these dynamics will only reinforce inequitable social structures (Walker, 2005).

A report for Transparency International identifies transparency, and strong logistics and administrative systems within humanitarian organizations, as being among a set of variables affecting risk of corruption. The report shows establishing bogus NGOs or inflating budgets as examples of ways of moving funds for personal gain. It also identifies logistical risks under procurement and distribution as securing substandard, out-of-date, or below specification goods, giving undue preference to some suppliers, diversion of stock and equipment for private gain, and "taxation" of relief goods by local elites or authorities (Ewins et al., 2006).

5.6 Political issues

Ideally, the existence of developed national institutions and committed governments contributes greatly to the success of relief efforts (Couldrey and Morris, 2005). This does not always work out the way it should, however. The relief effort after the North Korean train explosion in April 2004 was severely hindered due to political reasons. Direct routing of relief by land from South Korea was not allowed by the North Korean government, forcing the diversion of aid materials to take longer, multimodal routes (Pettit and Beresford, 2005).

Aid to North Korea has been a highly politicized process. In the autumn of 2005, North Korean government stopped accepting humanitarian assistance in the form of food. Aid officials in North Korea were not given access even to basic data such as population and employment statistics. The government also refused to supply a complete list of institutions that received food from WFP. The WFP was banned from food markets and was required to provide prior notice of inspection visits (Fairclough, 2006).

The role or significance of political dynamics on humanitarian relief became more obvious during the response to the tsunami. In Indonesia, Sri Lanka, and Somalia, the tsunami occurred in areas with complex and protracted conflicts, which hindered the organization and delivery of relief. In Banda Aceh, for example, the identity of survivors who have been internally displaced before the tsunami due to an internal conflict has become politically sensitive. Sri Lanka asked the U.S. government to scale down its military deployment, India rejected international assistance to Tamil Nadu and Nicobar islands, and Indonesia allowed all foreign military groups only 3 months to operate in the Banda Aceh region (Oloruntoba, 2005). These examples show the critical role of military in disaster relief. We focus on this issue next.

5.6.1 Military use in disaster relief

It is almost impossible in large-scale disasters to provide aid without some kind of relationship with the military. Increasingly, some countries are including humanitarian activities in the mission mandates of their armed forces. In recent years, the United Nations has applied a multidimensional approach to peacekeeping operations, bringing together the

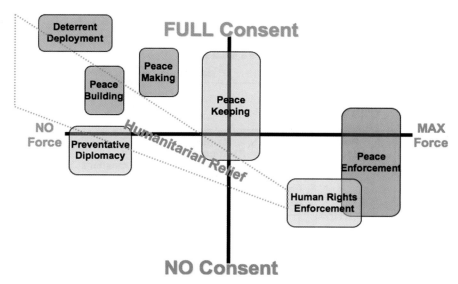

Figure 5.7 Humanitarian space. *Source:* Wolfson and Wright (1995).

peacekeeping humanitarian and development areas of the UN system (ECOSOC, 2004). Figure 5.7 presents the role of military in the humanitarian space as defined by the United Nations High Commissioner on Refugees (UNHCR). So, the question is not whether there should be a relationship with the military, but rather how to establish what the appropriate relationship should be and where the boundaries should lie.

In July 2000, a joint multiagency disaster relief exercise called "Strong Angel" (www.medibolt.com/strongangel) was carried out in the Asia-Pacific region. Military, United Nations, several NGOs, and civilians participated. The objective of the exercise was to establish a forum to exchange relevant information between relief organizations and the military. Military involvement in major natural disasters is not only acceptable, but also vital, because no other institution has the same means in terms of equipment and available personnel (Bredholt, 2005). However, for many NGOs, information sharing with military units to ensure fulfillment of needs through the use of available military assets such as aircraft, boats, vehicles, and personnel is poor at best (ECOSOC, 2005). The Civil–Military Operations Center (CMOC) formed during the Strong Angel exercise was intended to close this gap and coordinated relief operations between the military and civilian NGOs (Figure 5.8).

Not all humanitarian agencies have an adequate understanding of military command structures. For example, NGOs and military units have different organizational structures. Most NGOs are organized geographically, whereas military units are functionally organized. Despite the technical and organizational strength of the military, their main objective is to fight wars not provide disaster assistance. In addition, militaries are mission oriented, meaning that they identify the objective of a mission and carry out the necessary actions. The primary purpose of some humanitarian agencies such as Oxfam, in contrast, is rebuilding and development (Long, 1997).

Humanitarian organizations live by their principles of humanity, neutrality, and impartiality (Van Wassenhove, 2006). The involvement of the military in disaster relief operations is seen by some NGOs as being likely to compromise their neutrality. Neutrality is especially important to NGOs because they see it as their best defense (Pettit and Beresford,

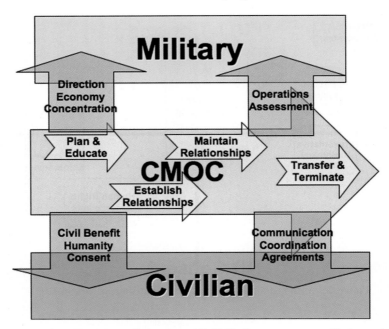

Figure 5.8 The civil–military operations center (CMOC). *Source:* www.medibolt.com/strongangel.

2005). However, a military presence is needed for security when the disaster area suffers from war, internal conflict, or looting.

5.7 Conclusions and future research directions

Disaster relief logistics is a complex task. Too many actors are working under uncertain and difficult conditions toward satisfying their objectives. Investing in relief logistics would make humanitarian organizations better prepared for responding to disasters and alleviate the pressure on field operations. Unlike the private sector, which enjoys the luxury of having flexibility of spending their cash, the humanitarian sector is constrained by the scarcity and tagging of funds. In contrast, in the private sector, money is invested into research and development, organizational design, and building of infrastructures that provide unique capabilities to their supply chains.

In this author's opinion, the one change that would have the largest impact on the success of relief operations would be the untagging of funds. The inflexibility of money is the main bottleneck in the humanitarian aid sector. If donors could be convinced to make their donations for the best use rather than for a self-defined specific purpose, relief organizations could build agility into their supply chains, retain more of their personnel, and afford investing in information and communications technology for better coordination.

The logistical issues discussed in this chapter may appear to be insurmountable. However, the private sector has been running global supply chains that are lean and agile over the past two decades and reacting to uncertainties posed by changing customer demands. Private sector supply chains adjust to environments with less uncertainty but with more variety of products, while optimizing cost, quality, and lead-times. There is much to learn from the private sector supply chain design for the relief industry. Lars Gustavsson, director of Emergency Response and Disaster Mitigation at World Vision International, agrees that "it is critical for NGOs to learn from the corporate and for-profit sector and incorporate emerging best practice" (Gustavsson, 2003, p. 7).

Surely, relief supply chains need to be agile and adaptable, but they also need to be efficient due to scarcity of funds. The interaction and boundary conditions between efficiency, agility, and adaptability need to be investigated to design well-balanced relief supply chains. Thus, more research is needed on disaster relief logistics. Beamon (2004) focused specifically on logistical issues in disaster relief and called for more research on appropriate supply chain structures and distribution network configurations, procurement and inventory control models specifically developed for humanitarian scenarios, and appropriate performance measures for disaster relief.

Many of the issues discussed in this chapter could also be interpreted as future research directions. In 1995, IFRC identified some specific research directions. They reported that new and practical methods were needed for analyzing capacity and vulnerability in specific disaster situations, and for evaluating the quality of the relief process. The IFRC called for the development of more holistic accountability systems and methodologies to evaluate the impact of international relief on local organizations. A point was also made regarding the importance of disseminating research results to the correct audience (IFRC, 1996).

References

Altay, N., and Green, W. G. (2006) "OR/MS Research in Disaster Operations Management," *European Journal of Operational Research* **175**, pp. 475–493.

Banerjee, P., and Chaudhury, S. B. R. (2005) "Indian Symposium Reviews Tsunami Response," *Forced Migration Review, Special Tsunami Issue*, pp. 42–43.

Barbarosoglu, G., Ozdamar, L., and Cevik, A. (2002) "An Interactive Approach for Hierarchical Analysis of Helicopter Logistics in Disaster Relief Operations," *European Journal of Operational Research* **140**, pp. 118–133.

Barbarosoglu, G., and Arda, Y. (2004) "A Two-Stage Stochastic Programming Framework for Transportation Planning in Disaster Response," *Journal of the Operational Research Society* **55**, pp. 43–53.

Beamon, B. M. (2004) "Humanitarian Relief Chains: Issues and Challenges," *Proceedings of the 34th International Conference on Computers and Industrial Engineering*, San Francisco, California, 14–16 November.

Beamon, B. M., and Kotleba, S. A. (2006) "Inventory Modeling for Complex Emergencies in Humanitarian Relief Operations," *International Journal of Logistics* **9**, pp. 1–18.

Benini, A. A. (1997) "Uncertainty and Information Flows in Humanitarian Agencies," *Disasters* **21**, pp. 335–353.

Blanco E., Goentzel, J. (2006) "Humanitarian Supply Chains: A Review," presentation given at the 17th Annual Conference of the Production and Operations Management Society, MIT Center for Transportation and Logistics.

Bredholt, N. (2005) "Some Southern Views on Relations with the Military in Humanitarian Aid," *Humanitarian Exchange* **30**, pp. 34–36.

Brown, H. (2005) "Lost Innocence: The Tsunami in the Maldives," *Forced Migration Review, Special Tsunami Issue*, pp. 48–50.

Brown, G. G., and Vassiliou, A. L. (1993) "Optimizing Disaster Relief: Real-Time Operational Tactical Decision Support," *Naval Research Logistics* **40**, pp. 1–23.

Carballo M., and Heal, B. (2005) "The Public Health Response to the Tsunami," *Forced Migration Review, Special Tsunami Issue*, pp. 12–14.

Chaikin, D. (2003) "Towards Improved Logistics: Challenges and Questions for Logisticians and Managers," *Forced Migration Review* **18**, p. 10.

Chen, F. (2003) "Information Sharing and Supply Chain Coordination," in *Handbook in Operations Research and Management Science: Supply Chain Management*, eds. T. de Kok and S. Graves, North-Holland, Amsterdam, The Netherlands.

Chiang, C., and Gutierrez, G. J. (1996) "A Periodic Review Inventory System with Two Supply Modes," *European Journal of Operational Research* **94**, pp. 527–547.

Chomilier, B., Samii, R., and Van Wassenhove, L. N. (2003) "The Central Role of Supply Chain Management at IFRC," *Forced Migration Review* **18**, pp. 15–16.

Christopher, M. (2005) *Logistics and Supply Chain Management: Creating Value-Adding Networks*, 3rd edition, Prentice Hall, New York, New York.

CNN. (2005) "Fear Grows for Aid Workers," www.cnn.com/2005/WORLD/asiapcf/01/11/asia.tsunami.security/index.html.

Couldrey M., and Morris, T. (2005) "UN Assesses Tsunami Response," *Forced Migration Review, Special Tsunami Issue*, pp. 6–9.

de Ville de Goyet, C. (2002) "Logistical Support Systems in Humanitarian Assistance," *WHO Newsletter Health in Emergencies* **14**, p. 1.

Economist. (2005) "Quality Over Quantity: Emergency Aid," Economist.com, The Global Agenda, 5 January.

Ewins, P., Harvey, P., Savage, K., and Jacobs, A. (2006) "Mapping the Risks of Corruption in Humanitarian Action," a report for Transparency International, www.transparency.org.

Fairclough, G. (2006) "In North Korea, a Plan to Solve Aid Dilemma," *Wall Street Journal*, February 22, p. B1.

Fraser, I. (2005) "Small Fish Trampled in Post-Tsunami Stampede," *Forced Migration Review, Special Tsunami Issue*, pp. 39–40.

Fritz Institute. (2005) "Logistics and the Effective Delivery of Humanitarian Relief," www.fritzinstitute.org/PDFs/Programs/TsunamiLogistics0605.pdf.

Gassmann, P. (2005) "Rethinking Humanitarian Security," *Humanitarian Exchange* **30**, pp. 32–34.

Gustavsson, L. (2003) "Humanitarian Logistics: Context and Challenges," *Forced Migration Review* **18**, pp. 6–8.

Haghani, A., and Oh, S. (1996) "Formulation and Solution of a Multi-Commodity, Multi-Modal Network Flow Model for Disaster Relief Operations," *Transportation Research A* **30**, pp. 231–250.

Hecht, L., and Morici, P. (1993) "Managing Risks in Mexico," *Harvard Business Review* **71**, pp. 32–34.

Hedman, E. (2005) "The Politics of the Tsunami Response," *Forced Migration Review, Special Tsunami Issue*, pp. 4–5.

Hudspeth, C. (2005) "Accessing IDPs in Post-Tsunami Aceh," *Forced Migration Review, Special Tsunami Issue*, pp. 19–21.

Iakovou, E., and Douligeris, C. (2001) "An Information Management System for the Emergency Management of Hurricane Disasters," *International Journal of Risk Assessment and Management* **2**, pp. 243–262.

International Federation of Red Cross and Red Crescent Societies (IFRC). (1996) "Key Factors for Developmental Relief," *International Review of the Red Cross* **310**, pp. 55–130.

Kaatrud, D. B., Samii, R., and Van Wassenhove, L. N. (2003) "UN Joint Logistics Center: A Coordinated Response to Common Humanitarian Logistics Concerns," *Forced Migration Review* **18**, pp. 11–14.

Kalin, W. (2005) "Natural Disasters and IDPs' Rights," *Forced Migration Review, Special Tsunami Issue*, pp. 10–11.

Kaplan, S. (1973) "Readiness and the Optimal Redeployment of Resources," *Naval Research Logistics Quarterly* **20**, pp. 625–638.

Kapucu, N. (2005) "Inter-Organizational Coordination in Dynamic Context: Networks in Emergency Response Management," *Connections* **26**, pp. 33–48.

Kim, J. S. (2005) "Media Coverage and Foreign Assistance," *Journal of Humanitarian Assistance*, www.jha.ac/articles/a176.pdf.

Knott, R. (1987) "The Logistics of Bulk Relief Supplies," *Disasters* **11**, pp. 113–115.

Knott, R. (1988) "Vehicle Scheduling for Emergency Relief Management: A Knowledge-Based Approach," *Disasters* **12**, pp. 285–293.

Lee, H. L. (2004) "The Triple-A Supply Chain," *Harvard Business Review* **82**, pp. 102–112.

Lee, H. W., and Zbinden, M. (2003) "Marrying Logistics and Technology for Effective Relief," *Forced Migration Review* **18**, pp. 34–35.

Long, D. C. (1997) "Logistics for Disaster Relief," *IEE Solutions* **29**, pp. 26–29.

Long, D. C., and Wood, D. F. (1995) "The Logistics of Famine Relief," *Journal of Business Logistics* **16**, pp. 231–229.

Marschak, J., and Radner, R. (1972) *Economic Theory of Teams*, Yale University Press, New Haven, Connecticut.

Matthews, S. (2005) "Logistical Challenges," *Forced Migration Review, Special Tsunami Issue*, p. 38.

McEntire, D. A. (1999) "Issues in Disaster Relief: Progress, Perpetual Problems, and Prospective Solutions," *Disaster Prevention and Management* **8**, pp. 351–361.

McGuire, G. (2001) "Supply Chain Management in the Context of International Humanitarian Assistance in Complex Emergencies—part 2," *Supply Chain Practice* **3**, pp. 4–18.

Moinzadeh, K., and Nahmias, S. (1988) "A Continuous Review Model for an Inventory System with Two Supply Modes," *Management Science* **34**, pp. 761–773.

Moody, F. (2001) "Emergency Relief Logistics—A Faster Way Across the Global Divide," *Logistics Quarterly* **7**, issue 2 pp. 1–5.

Nollet, J., Leenders, M. R., and Diorio, M. O. (1994) "Supply Challenges in Africa," *International Journal of Purchasing and Materials Management* **30**, pp. 51–56.

Nyanduga, B. T. (2005) "An African Perspective on the Tsunami," *Forced Migration Review, Special Tsunami Issue*, p. 50.

Oloruntoba, R. (2005) "A Wave of Destruction and the Waves of Relief: Issues, Challenges, and Strategies," *Disaster Prevention and Management* **14**, pp. 506–521.

Oppenheim, J. M., Richardson, B., and Stendevad, C. (2001) "A Standard for Relief," *McKinsey Quarterly, Nonprofit Anthology Issue*, pp. 91–99.

Ozdamar, L., Ekinci, E., and Kucukyazici, B. (2004) "Emergency Logistics Planning in Natural Disasters," *Annals of Operations Research* **129**, pp. 217–245.

Pan American Health Organization (PAHO). (1999) "Humanitarian Assistance in Disaster Situations: A Guide for Effective Aid," PAHO, Washington, DC, www.paho.org.

Paton, D. (1996) "Training Disaster Workers: Promoting Wellbeing and Operational Effectiveness," *Disaster Prevention and Management* **5**, pp. 11–18.

Pettit, S. J., and Beresford, A. K. C. (2005) "Emergency Relief Logistics: An Evaluation of Military, Non-Military and Composite Response Models," *International Journal of Logistics: Research and Applications* **8**, pp. 313–331.

Samii, R., Van Wassenhove, L. N., Kumar, K., and Becerra-Fernandez, I. (2002) "IFRC—Choreographer of Disaster Management: The Gujarat Earthquake," INSEAD Case, www.fritzinstitute.org/PDFs/Case-Studies/Gujarat%20Earthquake.pdf.

Scott-Bowden, P. (2003) "The World Food Program: Augmenting Logistics," *Forced Migration Review* **18**, pp. 17–19.

Solvang, I. (2005) "Powerless Victims or Strong Survivors," *Forced Migration Review, Special Tsunami Issue*, p. 29.

Stephenson, M. (2005) "Making Humanitarian Relief Networks More Effective: Operational Coordination, Trust and Sense Making," *Disasters* **29**, pp. 337–350.

Stephenson, R., and Anderson, P. S. (1997) "Disasters and the Information Technology Revolution," *Disasters* **21**, pp. 305–334.

Strom, S. (2006) "Poor Nations Complain Not All Charity Reaches Victims," *The New York Times, International*, 29 January, Section 1, Column 1, p. 8.

Thomas, A. S., and Kopczak, L. R. (2005) "From Logistics to Supply Chain Management: the Path Forward in the Humanitarian Sector," www.fritzinstitute.org/PDFs/WhitePaper/FromLogisticsto.pdf.

Thompson, S., Altay, N., Green, W. G., and Lapetina, J. (2007) "Improving Disaster Response Efforts with Decision Support Systems," *International Journal of Emergency Management* **3**, pp. 250–263.

Trunick, P. A. (2005) "Delivering Relief to Tsunami Victims," *Logistics Today* **46**, p. 8.

United Nations Economic and Social Council (ECOSOC). (2004) "Strengthening the Coordination of Emergency Humanitarian Assistance of the United Nations," http://daccessdds.un.org/doc/UNDOC/GEN/N04/386/60/PDF/N0438660.pdf.

United Nations Economic and Social Council (ECOSOC). (2005) "Strengthening Emergency Relief, Rehabilitation, Reconstruction, Recovery and Prevention in the Aftermath of the Indian Ocean Tsunami Disaster," www.un.org/docs/ecosoc/documents/2005/reports/ECOSOC%20tsunami%20report.pdf.

Van Wassenhove, L. N. (2006) "Humanitarian Aid Logistics: Supply Chain Management in High Gear," *Journal of the Operational Research Society* **57**, pp. 475–489.

Verjee, F. (2005) "The Application of Geomatics in Complex Humanitarian Emergencies," *Journal of Humanitarian Assistance*, www.jha.ac/articles/a179.pdf.

Volz, C. (2005) "Humanitarian Coordination in Indonesia: An NGO Viewpoint," *Forced Migration Review, Special Tsunami Issue*, pp. 26–27.

Walker, P. (2005) "Is Corruption an Issue in the Tsunami Response?," *Humanitarian Exchange* **30**, pp. 37–39.

Whittmore, A. S., and Saunders, S. (1977) "Optimal Inventory under Stochastic Demand with Two Supply Options," *SIAM Journal of Applied Mathematics* **32**, pp. 293–305.

Wolfson, S., and Wright, N. (1995) *A UNHCR Handbook for the Military on Humanitarian Operations*, Geneva, Switzerland.

Wright, G. P. (1969) "Optimal Ordering Policies for Inventory with Emergency Ordering," *Operational Research Quarterly* **20**, pp. 111–123.

Wybo, J. L., and Kowalski, K. M. (1998) "Command Centers and Emergency Management Support," *Safety Science* **30**, pp. 131–138.

6
Large-scale disasters: perspectives on medical response

Jehan Elkholy and Mostafa Gad-el-Hak

The primary aim of disaster response is to restore order, support damaged or nonfunctioning societal functions, and, shortly afterward, reconstruct and rehabilitate the affected society to, at minimum, its predisaster situation. There are three vital phases of the disaster response: preparedness, response, and recovery. Preparedness includes the assessment of risks in relation to material and personnel resources, efficient planning, medical response tactics, and continual training of the staff, including a sufficient amount of field exercise involving various rescue teams. Medical response includes standardization of approach, terminology, communication, and command. Medical response also includes triage of patients and treatment as appropriate for their safe transport for more complete treatment in hospitals. The recovery phase is the prolonged period of adjustment or return to equilibrium that the community and individuals must go through. Evaluation of a disaster begins by reviewing the response plan and its measurable objectives. In the process of conducting an evaluation, assessments are directed for five domains of activity: structure, process, outcomes, response adequacy, and costs. There is a role for each specialist in the management and response to disasters, including surgeons, anesthesiologists, nurses, pediatricians, and psychiatrics. This chapter reviews the general characteristics and phases of large-scale disasters, medical response, and the role of some specialists in that response.

6.1 Introduction

Disasters are predictable, not in time or place, but in their inevitability. In general, a disaster is defined as a situation resulting from an environmental phenomenon or armed conflict that produces stress, personal injury, physical damage, and economic disruption of great magnitude. A more concise definition was provided in Section 1.1 of this book. Disasters result when the extent of damage produced by the force of a natural or manmade hazard exceeds human capacity to cope with its consequences—or when it destroys or places additional burdens on fundamental societal functions such as law and order, communication, transportation, water and food supply, sanitation, and health services.

As a result, order is replaced by chaos. Chaos may be compounded by a disproportionate, inadequate, or disorganized response—the so-called "second disaster" (Pearn, 1998). Aside from the general definition of disaster given previously, others have contributed numerous definitions from their individual perspectives. For example, Cuny (1993) defined a disaster as a situation resulting from an environmental phenomenon or armed conflict that produces stress, personal injury, physical damage, and economic disruption of great magnitude. The definition of a disaster adopted by the World Health Organization and the United Nations is, the result of a vast ecological breakdown in the relations between man and his environment, a serious and sudden (or slow as in drought) disruption on such a scale that the stricken community needs extraordinary efforts to cope with it, often with outside help or international aid (Perez and Thompson, 1994).

Disasters can be viewed solely on the basis of the health impact. In this context, a health disaster occurs when it causes widespread injury or loss of life, or when the social and medical infrastructure of a community is disrupted or so damaged by the event that it significantly reduces or impairs the community's access to the health system (Aghababian et al., 1994).

"Large-scale disasters" refers to events that will produce casualties beyond that which can be planned for by a single hospital. By definition, the event will not impact a single facility, but rather whole communities and regions. These events will require planning beyond the individual hospital, and include all segments of supply and points of care in a region.

This chapter reviews the following:

- General characteristics and phases of large-scale disasters
- Medical response to large-scale disasters
- Role of some specialists in the response to large-scale disasters

6.2 Characteristics of disasters

In contrast to the types of incidents that emergency responders normally face, major disasters share a number of characteristics that create unique difficulties for response organizations: the large number of people affected, injured, or killed; the large geographic scale extending over very large areas; the prolonged duration lasting for days, weeks, or even months; and the large extent of damage to infrastructures. In addition, the hazards that may be multiple and variable, and the need for a wide range of capabilities (Kunkel et al. 1999).

6.3 Classification of disasters

Disasters are classified into simple or compound and/or natural or manmade. Natural disasters include meteorological disasters (cyclones, typhoons, hurricanes, tornadoes, hailstorms, snowstorms, and droughts), topological disasters (landslides, avalanches, mudflows, and floods), disasters that originate underground (earthquakes, volcanic eruptions, and tsunamis), and biological disasters (communicable disease epidemics and insect swarms). Although, manmade disasters include warfare (conventional and nonconventional warfare), civil disasters (riots and strikes), criminal/terrorist action (bomb threat/incident; nuclear, chemical, or biological attack; hostage incident), and accidents (transportation, structural collapse [as buildings, dams, bridges, mines, etc.], explosions, fires, chemical, and biological).

Disasters may be also classified according to the resultant anticipated necessary response:

- Level I disaster is one in which local emergency response personnel and organizations are able to contain and deal effectively with the disaster and its aftermath.
- Level II disaster requires regional efforts and mutual aid from surrounding communities.
- Level III disaster is of such a magnitude that local and regional assets are overwhelmed, requiring statewide or federal assistance (Gunn, 1990).

From a medical point of view, mass disasters are classified into small, medium, and large. Small disasters occur when up to 25 persons are injured, of whom at least 10 need hospitalization; medium disasters occur when at least 100 persons injured, of whom more than 50 need hospitalization; and large disasters occur when at least 1,000 persons are injured (Hadjiiski, 1999).

What makes a disaster different from other medical emergencies? Disaster medicine is perhaps the most interdisciplinary of all fields of medicine. Disaster medical response is complex, and ranges from response to natural disasters to the ravages of warfare and to the medical response to terrorist attacks. In its broadest perspective, it can be defined as the delivery of medical and surgical care under extreme and/or hazardous conditions to the injured or ill victims of disaster (Dara et al., 2005).

The key distinction between a medical disaster and a mass casualty incident (MCI) is that professional search and rescue and the prehospital response is occasionally hampered because of the destruction of the health care infrastructure and the large number of casualties requiring immediate care simultaneously. The survival rate for these individuals declines rapidly unless prompt emergency medical assistance is provided within minutes to a few hours of their injuries. Although the pattern and extent of injuries and illnesses will vary depending on the type and magnitude of the disaster, most disaster response experts agree that the main window of opportunity for saving lives of the critically injured but treatable victims is the first few hours. Response usually occurs at the lowest possible (local) level because there are less logistical hurdles to overcome, and also because immediate, lower-level care through the application of initial life supporting first aid (LSFA) by uninjured bystanders and advanced trauma life support (ATLS) by professional rescuers is more effective in saving lives (Pretto, 2002, 2003).

6.4 Disaster management

An important factor that regulates emergency management system (EMS) is delays between when a decision is made and the time it is carried out. The delay in availability of information, decision making, and final action can take a number of forms. It can be past time delayed (when decisions are made based on late information that is no longer accurate about what is going on in the field), future time delayed (the time between the making of the decision and the action being taken is delayed), and a policy lag (a decision is made based on assumptions about a system of relationships that is no longer applicable). These delays in turn affect the evacuation, supply, transportation, and so on.

EMS and other state and local government response agencies are required to use the standardized emergency management system (SEMS) when responding to emergencies involving multiple organizations. SEMS seeks to address interagency coordination between responding agencies, and to facilitate the flow of emergency information and resources within and between involved agencies and the rapid mobilization, deployment, use, and tracking of resources. SEMS provides a five-level hierarchical emergency response organization (field, local government, operational area, region, and state). This structure uses the incident command system (ICS), multiagency coordination system, the state's master mutual aid agreement, designation of operational areas, and the operational area satellite information system to integrate the disaster response. Each agency organizes itself around a set of common ICS-defined functions (command/management, operations, planning/intelligence, logistics, and finance/administration); specifies objectives that it seeks to accomplish (management by objective is required); and uses preestablished systems to fill positions, conduct briefings obtain resources, manage personnel, and communicate with field operations.

A number of state-related management options could be evaluated to determine their response utility for ICS and SEMS. First, it is crucial to evaluate the ability of the disaster managers to adapt to changing situations. SEMS and ICS help organize the various components of a disaster response. It creates an adaptive priority setting and resource management structure but does not define which tactics might be most useful for organizing individual components into an overall response system for a particular disaster. Second, the various interorganizational interfaces that regulate the response infrastructure self-organizing process need to be evaluated (e.g., how various agencies link together, their social times, and their ability to self-organize and mobilize resources are all directly related to how quickly the response emerges). Third to be evaluated is the possibility to reduce the number of layers of government required to approve the allocation of resources and personnel, and allow existing and emergent organizations in the field to request and approve distribution of state and other resources locally. Last, it is needed to develop a new group of field responders who would be information brokers (e.g., amateur radio operators and others experienced with citizen-level radio communications could be recruited; Hadjiiski, 1999).

6.5 Phases of a disaster

When disaster strikes, the primary aim of disaster response is to restore order, support damaged or nonfunctioning societal functions, and shortly afterward, reconstruct and rehabilitate the affected society to, at minimum, its predisaster situation. At that point, the primary aim of health and medical disaster response is to reduce morbidity and mortality.

There are three vital phases of the disaster response: preparedness, response, and recovery. These are addressed in turn in the following three subsections.

6.5.1 Phase I: disaster preparedness

The severity of disaster outcome is multifactorial, directly proportional to hazard magnitude and intensity, population density, and vulnerability, and inversely proportional to the sum of preparedness and socioeconomic level of a region. Preparedness will include any activities that prevent an emergency, reduce the chance of an emergency occurring, or reduce the damaging affects of unavoidable emergencies (mitigation). Preparedness incorporates plans or preparations to save lives and help response-and-rescue operations. Preparedness activities take place before and after an emergency occurs (UNEP, 2005).

Moralejo et al. (1997) assumed that the medical response to disasters must be planned and trained long before the event occurs. It should include plans for both immediate response and long-term salvage and recovery efforts. The plan should be based on first—local resources immediately available, and second—local and regional reinforcement when needed. The plan should be approved by all agencies that provide authority endorsement. Finally, the plan should be regularly revised and tested.

A prudent first step in disaster planning is the assessment of risks in relation to material and personnel resources; efficient planning medical response tactics; and continual training of the staff, including a sufficient amount of field exercise involving various rescue teams. Risk assessment is necessary to identify various types of accidents that may occur within the area where the medical services are provided. There must be a prioritization to focus on the greatest risk and greatest benefit with allocation of resources based on the likelihood of the threat balanced with potential loss. The first priority in any disaster is human safety. Saving collections is never worth endangering the lives of staff or patrons. In a major event, the fire department, civil defense authorities, or other professionals may restrict access to the building until it can be fully evaluated. Once safety concerns are met, the next consideration will be records and equipment crucial to the operation of the institution, such as registrar's records, inventories, and administrative files. Collections salvage and building rehabilitation will be the next priority.

Disaster mitigation is any medical or nonmedical intervention aimed at reducing injury or damage once the event has occurred. Prevention or mitigation of the human impact of trauma or illness in a disaster can be done through prudent land use location of human settlements in safe, low-risk areas; limited land or building occupancy density low population/building density; good quality infrastructure building stock and materials for shelters; early warning systems and predesignated evacuation routes for populations at risk such as in coastal areas at risk for hurricanes; epidemiological or disease surveillance systems, particularly in preparing for bioterrorism; well-equipped and trained first responders (EMS, police, fire) and other emergency health workers and services capable of rapid deployment and field operations in the local community; a significant proportion of the civilian population (>30%) knowledgeable in what to do in disaster and trained in LFSA; hospitals with structural and nonstructural integrity capable of operating under disaster conditions; a disaster-"resistant" telecommunications infrastructure (with backup systems such as amateur radio network) designed to allow multijurisdictional coordination (police, fire, EMS, hospital, etc.); and enhanced by robust information management and decision support systems (Micheels, 1994; Pretto et al., 1992).

The goal of preevent training is to ensure an adequate, competent, and flexible response that will satisfy the acute needs resulting from the disaster while still meeting baseline demands of the affected community. Management by objective skills training could ensure that the disaster response communication and resource infrastructure is in place and functioning at the highest efficiency. This training should use an interdisciplinary approach and incorporate both medical planners and the numerous agencies functioning as first responders. Disaster scenarios could be scripted to illustrate the progress of the disaster. Disaster managers and responders should be trained to visualize the whole of the response, do process mapping, form "messy" groups, and engage in group problem solving. Each responding unit should have identified roles and responsibilities to prevent excessive overlapping and parallel efforts. In addition, there should be one clearly defined chain of command, usually to ensure proper command and control to optimize the functioning of the responding organizations and proper execution of the response plan.

Thus, to manage disasters quickly and effectively, every major disaster is analyzed and evaluated carefully, and the results of such studies be distributed both nationally and internationally. In addition, persons who are expected to handle a major disaster must have gained experience from their daily work with minor accidents. Although major disasters are different than minor ones, certain patterns are the same, and therefore, persons with experience and training will be better equipped to cope with the next disaster. Training and exercise are crucial and should be performed both as minor exercises around a table, in classrooms, as alarm calls, and in the field in realistic surroundings. In all cases, the exercises should be carried out in cooperation with the partners who participate in the real situation (e.g., fire brigade, police, ambulance personnel; Kunkel et al., 1999).

Integration of disaster preparedness must be brought to the local level. The best plan is ineffective if it is not properly executed or able to adapt and continue to function despite losses. The disaster medicine community must adopt the concept of the business continuity plan used by industry to minimize disruption of normal operations, despite the disorder following disasters. A disaster can cripple the delivery of health care through decreased infrastructure in the face of increased demands that now exceed capabilities. Hospitals must maintain operational effectiveness with minimal interruptions, despite the potential loss of infrastructure and personnel. This is a true marker of resiliency.

The basic unit of medical response in a disaster is the local prehospital EMS/trauma and hospital system. In large-scale disasters where the local community is overwhelmed, health services are backed up by regional and national resources, depending on the magnitude of the needs. In general terms, the phases and concomitant activities and the timing of the delivery of emergency medical services will vary according to disaster type. In sudden impact disasters such as earthquakes or in terror attacks using conventional weapons such as in the 9/11 event, the following phases and medical services are commonly observed. The initial phase (the first 24 hours) is the stage of greatest life-saving potential, and its medical services include alertness, notification, search for victims, rescue, and EMS response. During the next phase (intermediate phase that lasts from 1–12 days), rescue continues, and there are certain public health measures that should be undertaken as epidemic surveillance and vector control. During the final phase (after 12 days), public health measures are carried out, and there is rehabilitation of the injured and reconstruction of the community.

6.5.2 Phase II: medical response

The purpose of MCI plans and guidelines is to ensure that sound medical practice is instituted and followed in the treatment of multiple casualties and to provide means and standardization of approach, terminology, communication, and command. In most scenarios, there will be multiple agencies with varied levels of training that must function under an extraordinary command structure, working in a chaotic and stressful atmosphere. A truly standardized approach is absolutely necessary to ensure that order emerges promptly from this chaos (Magliacani and Masellis, 1999; Norberg, 1998).

During the initial response, organizations involved in disaster response and the potentially affected populations are notified. In the event that the disaster is anticipated, this phase takes place even before the disaster. The METHANE report is a simple way to remember the items of the message where **M** stands for major incident or disaster, either standby or declared; **E** stands for exact location of the incident; **T** stands for type of incident, fire, explosion, traffic, etc.; **H** stands for hazards, actual and potential; **A** stands for access to the incident, for how to reach the emergency services; **N** stands for number of casualties expected; and **E** stands for required type and number of emergency services expected. The first METHANE report may be inaccurate, but updating the report makes it increasingly accurate.

Once the activation phase has begun, the prearranged command and staff structure for responding to the disaster should be arranged and initial communication nets established. During this phase, disaster responders become informed of the health and medical needs generated by a disaster, together with the amount of material and manpower resources required. This must be accomplished before large numbers of responding units and personnel arrive; otherwise, there is the creation of mass confusion and loss of control. The ICS is the management structure through which EMS response to a disaster is organized and carried out in an orderly manner because early organization of on-scene operations is one of the highest priorities. This is one of the most crucial steps to take once the disaster occurs. It is the responsibility of the first arriving ambulances to establish command areas, these will include command post, staging and equipment area, treatment area, and loading area.

The command post will be established by the first arriving unit from each agency, police, fire, and EMS. This provides for a timely passage of information and orderly coordination among all agencies.

The command function must be provided by people with appropriate education and training. They should be well prepared to use modern manual and technological systems for collecting information, calculation, decision making, information/communication, documentation, and follow-up. At any MCI, there are three commanders (police command, fire command, and EMS command) falling under one incident control, who is the highest ranking. For the whole scene, there is one control for the incident that may be the police or the fire commander. EMS responders are assigned the following roles by EMS command: safety officer, forward officer, triage officer, treatment officer, loading officer, parking officer, communication officer, and equipment officer.

It may become necessary to combine roles if resources do not provide EMS personnel for each role. It is vitally important to the success of an MCI that all responding personnel be educated on the role and function of each member of the command structure.

The EMS commander is a manager who does not perform medical tasks but delegates them. The senior emengency medical technician (EMT) or paramedic on the first arriving EMS unit will assume the responsibility of EMS command. At a large or ongoing incident, this person may be relieved by a supervisor or more experienced person. The EMS commander reports to the incident commander. The EMS commander will supervise the placement of the parking, loading, and staging and treatment areas. He or She will then designate and supervise a safety officer, triage officer, treatment officer, parking officer, loading officer, staging officer, and communications officer. While the command structure is being delegated and carried out, safety should be ensured. This includes safety of rescuers, scene, and survivors. The process through which victims of a disaster are located, identified, and evacuated from the disaster area is called search and rescue. Search and rescue may fall under the direction of fire, EMS, or police forces. The act of search and rescue must be highly organized to ensure adequate and complete coverage of all areas.

The ability of the managers during disasters to regulate the medical disaster response depends on availability and quality of communications. However, following a disaster, there is communications failure and information overload. The communications system is damaged, used for other purposes, or is overloaded with information. Communications are not always quickly available where they are needed. The volume of information is so high that it cannot be reduced in a timely and efficient manner into usable information on which to base decisions. Communication can be done face to face, by runners, or by radio. The commander may delegate a communication officer. The wireless internet information system for medical response in disasters (WIISARD) explores the use of scalable wireless networks to facilitate medical care at the site of a disaster.

Casualty triage

Triage is a system used for categorizing and sorting patients according to the severity of their injuries. The concept of triage involves providing the most help for as many as possible (do the most for the most). During disasters, initial treatment is provided to casualties in order to prevent further or rapid deterioration in the patient's status, and in the case of weapons of mass destruction, perform decontamination in predesignated areas to prepare the victim for transportation to a treatment facility and to protect rescuers and other individuals from toxic or infectious substances. Triage principles should be used whenever the number of casualties exceeds the number of skilled rescuers available. Triage is a dynamic process; it is done repeatedly to see if the casualty is improving or deteriorating. The triage officer reviews and classifies all victims for life-threatening injuries. The triage officer reports to the EMS commander and is usually the junior EMT or paramedic on the first arriving ambulance. At a large or ongoing incident, this person may be relived by a more experienced person.

The triage officer will quickly view and classify each patient (15 seconds per patient). During review of each patient, the S-T-A-R-T triage will be used. It is a universal standard of simple-triage-and-rapid-tagging of patients that relies on a three-part assessment: ventilation status, perfusion status, and walking ability. The major role is assessment with actual care limited to airway support, by ensuring an open airway and perfusion support, by control bleeding with a tourniquet, and by placing the patient in shock position.

Once triage and tagging are performed, the triage officer will supervise the evacuation of all victims to the proper patient collection areas. All data on actual numbers of victims is communicated to the EMS commander. Documentation on number of victims, injuries, and care should be provided (Benson et al., 1996; Kossman and Vitling, 1991; Pruitt et al., 2002).

The process of gathering patients from a hazardous location or scene of injury to a staging area near the disaster zone where further harm is avoided is based on the concept of casualty collection points (casualty clearing station and treatment area). In this area, secondary triage takes place, and patients are prepared for transport and loaded onto transport vehicles to hospitals. A loading area should be set up directly next to the treatment area. There should be easy access from the staging area and easy access out of the incident site. Incoming ambulance crews will await patient assignment by the loading officer. The treatment officer has overall responsibility for the treatment of all victims. At a large, ongoing incident, this person may be relived by a more experienced EMT or paramedic. The treatment officer reports to the EMS commander.

The treatment officer (if a paramedic) should contact medical control if ALS orders are needed. Initial treatment of all patients should be at least to the B-A-S-I-C level (B: bleeding control; A: airway support; S: shock therapy; I: immobilization; C: classification). The loading officer supervises the loading of all victims who have been stabilized and treated. The loading officer, in cooperation with the treatment officer, decides the order of transportation. The loading officer reports to the EMS commander. Once the loading officer has supervised the patient loading, the crew will continue care until arrival at the designated hospital. The loading officer must keep the communications officer well informed because the communications officer has to notify the hospital of the impending arrival of victims. The equipment officer coordinated all incoming ambulances, EMTs, and supplies. This person is designated by the EMS commander and does not have to be an EMS provider. If resources are limited, a police officer can be assigned this responsibility. The communications officer is responsible for coordinating the communication between the scene and outside agencies and hospitals concerning the number and severity of injury each can accept. The communications officer reports to the EMS commander. Emergency medical or surgical care is provided to patients at collection points whose severity of illness or injury preclude safe transport to a treatment facility. It involves the delivery of basic first aid by uninjured covictims and is aimed at initiating the trauma life support chain. LSFA entails the following basic maneuvers: calling for help, maintaining a patent airway, controlling external bleeding, positioning for shock, rescue pull, and CPR (if indicated). The second step in the life support chain involves the administration of advanced first aid in the form of airway control, hemorrhage control, immobilization of unstable limb fractures, wound dressing, and burn treatment, among other life-saving maneuvers by trained first responders. The administration of life-saving emergency medical or surgical care to critically injured patients by physicians usually includes intravenous fluid resuscitation, pleural drainage, endotracheal intubation, mechanical ventilation, wound suturing, and other life-saving medical/surgical interventions.

The availability of transport is a key variable regulating how quickly the response system can be organized. Casualties will have to be moved from the site of their injury to emergency care field stations, hospitals, and evacuation areas. A significant percentage of the surface transportation in the disaster area, possibly more than half, will be carried out by the public. The remainder will require public and private agency assistance. During some disasters, the majority of critically injured individuals were taken to only one or two receiving facilities, which were almost overwhelmed. This occurred at a time when other facilities sat dormant awaiting patients (Birnbaum and Sundnes, 2001).

Hospital response

In the event of a disaster, the goal of the hospital's performance is to deliver specialized in-hospital medical or surgical care by physicians, nurses, and other health care professionals to casualties to correct or reverse life-threatening illness or injury and/or to prevent long-term disability. Another goal is to establish a clear chain of command or incident management system/hospital emergency incident command system (HEICS). Confusion and chaos are commonly experienced by the hospital at the onset of a medical disaster. However, these negative effects can be minimized if management responds quickly with structure and a focused direction of activities. The HEICS is an emergency management system that employs a logical management structure, defined responsibilities, clear reporting channels, and a common nomenclature to help unify hospitals with other emergency responders. There are clear advantages to all hospitals using this particular emergency management system.

Based on public safety's ICS, HEICS has already proved valuable in helping hospitals serve the community during a crisis and resume normal operations as soon as possible. HEICS is fast becoming the standard for health care disaster response; it offers predictable chain of management, flexible organizational chart allowing flexible response to specific emergencies, prioritized response checklists, accountability of position function, improved documentation for improved accountability and cost recovery, common language to promote communication and facilitate outside assistance, and cost-effective emergency planning within health care organizations.

The hospital response should provide for personnel (medical and nonmedical) necessary medical equipment, supplies, and pharmaceuticals; establish mutual aid agreements with other hospitals; once alerted/activated, rapidly reorganize resources to convert to disaster operations and reassign personnel to meet the special needs of disaster patients (internal and external disaster plans); and coordinate and maintain open channels of communication with prehospital EMS systems, hospitals, local and state health authorities, families of casualties, and the media. It provides a plan for the transfer of stable inpatients to other health care institutions and establishes predesignated treatment areas within the hospital with prioritization of care, as follows: casualty reception and triage area (emergency room [ER] and adjacent areas), and decontamination area (in the event of casualties of weapons of mass destruction). An area for the diagnosis, treatment, or stabilization of patients who have life-threatening conditions (acute care areas, such as operating room [OR], intensive care unit [ICU]), an area for the diagnosis, treatment, or stabilization of patients having urgent conditions who will be referred for follow-up care; an area for the treatment of patients with nonurgent conditions, with referral for later definitive diagnosis, treatment, and comprehensive care; an area for the reception of hopeless moribund or dead casualties; and an alternative treatment area (outside the hospital) in the event the hospital cannot provide services. Hospital response also includes evaluation of performance after the event and revision of disaster plan, and it provides training based on knowledge gained through experience, after action reports and evaluation (Farmer and Carlton, 2005; Perry, 2003).

Haz-mat response is necessary when approaching an accident or incident site that may involve hazardous materials; all personnel should approach from the upwind side of the incident. They should stop at a safe distance to observe and wait for the fire department to tell them on where to set up a treatment area. They should NOT approach the scene until instructed to do so. The presence of victims at the site and their specific location should be noted. This information should be delivered immediately to the haz-mat response team on their arrival. Personnel not wearing the proper personal protective equipment, unless properly trained, should not enter the contaminated area If any personnel become contaminated, report to the decontamination area established by the haz-mat response team and follow the decontamination procedure prescribed for the material involved. Receiving hospitals must be notified at the earliest possible moment and provided with the nature of injuries sustained by the contaminated patients, hazardous substance involved, and decontamination procedures used on the patient at the site prior to transport.

An orderly and timely termination will be announced after the EMS commander is notified by the loading officer that the last patient has been transported. The EMS commander will notify the incident commander when no further resources or EMS support is needed. The EMS commander, in consultation with the incident commander, will decide which units will remain on scene. Following the incident, it is the responsibility of the EMS commander to submit a list with the number of patients treated, the names of those transported, and the name of the receiving hospital. A list will be submitted to the incident commander and forwarded to the director of operations.

6.5.3 Phase III: recovery

A critical part of handling any serious emergency situation is in the management of the disaster recovery phase. The recovery phase is the prolonged period of adjustment or return to equilibrium that the community and as individuals must go through. It commences as rescue is completed and individuals and communities face the task of bringing their lives and activities back to normal. Much will depend on the extent of devastation and destruction that has occurred as well as injuries and lives lost. The priority during this phase is the safety and well-being of the employees and other involved persons, the minimization of the emergency itself, and the removal of the threat of further injury.

Exposure to chemicals disasters causes damage to lung tissue and respiratory functions. Victims suffer from breathlessness, cough, nausea, vomiting, chest pains, dry eyes, poor sight, photophobia, and loss of appetite. They also manifest psychological trauma symptoms, including anxiety, depression, phobias, and nightmares. Women in the affected areas appear to suffer a variety of medical problems (abortion, irregular menstruation, and failure of lactation). Women also bear a disproportionately large burden of economic and social losses: they lose earning members of their family, with no alternative source of income.

6.6 Role of specialists

In the following context, we discuss the role of specialists in the management of and response to disasters, with emphasis on the role of the anesthesiologist.

The psychological definition of a disaster is a sudden event that has the potential to terrify, horrify, or engender substantial losses for many people simultaneously. Survivors of such large-scale tragedies, whether injured or uninjured survivors and rescuers, invariably experience significant emotional and psychological distress. Even within subjects who have experienced the same disaster, individuals vary greatly in their outcome. Psychological disorders may be classified according to the time of occurrence into immediate disorders (anxiety, upset, and panic), early disorders (feeling guilty), and late disaorders (posttraumatic stress disorder, phobias, and obsession compulsions). Psychiatrists participate in providing a new framework for preparedness and response to disasters and trauma. They also share in cultural, religious, and ethnic differences related to the prevention and treatment of psychological sequelae. Immediate psychotherapy should be available to disaster victims in an effort to alleviate suffering and to prevent chronic distress. Critical incident stress debriefing refers to stress management through psychological support of disaster-affected response personnel (Deahl et al., 2000; Gray et al., 2004; Litz et al., 2002; van Emmerik et al., 2002).

The pediatrician has a multitude of roles in disaster preparedness, including personal preparedness through anticipatory guidance to families and roles in the community. This multitude of roles in emergency preparedness is not limited to general pediatricians; it also applies to pediatric medical subspecialists and pediatric surgical specialists, whose involvement may range from giving advice to families and children to being subject-matter experts for preparedness and critical resources in their communities (Leonard, 1998; Quinn et al., 1994).

Anesthesiologists are acute care physicians with special expertise in airway management, physiological monitoring, patient stabilization and life support, fluid resuscitation, and crisis management. These are the most important aspects of emergency medical care of the disaster patient. As such, anesthesiologists can serve crucial roles not only in the anesthetic management of civilian and combat casualties in the familiar setting of the OR, but also, if called on, can function adequately as team members in field medical teams, the ER, or in the management of ICU patients. Anesthesiologists are also well qualified to sort, triage, stabilize, and resuscitate casualties; to provide external hemorrhage control; to diagnose and treat life-threatening conditions; to establish IV access; and to manage acute pain and ICU patients. The ability of anesthesiologists to more effectively participate in emergency medical care in disasters is enhanced by the acquisition of competency in ATLS, basic principles of mass casualty and disaster management; and knowledge of the anesthetic methods, techniques, and equipment commonly used outside the OR environment.

Patients who are successfully transported to acute care hospitals are triaged according to the nature and extent of their individual injuries, the number and severity of injuries to the entire group, and the magnitude of the ongoing disaster. Casualties who require anesthetic care or intubation for their injuries simply may need routine anesthetic evaluation and care, just like any patient who presents with traumatic injuries. However, patients who have sustained blast injuries may present special and unfamiliar anesthetic considerations concerning protection of the caregivers and the rendering of optimal patient care. Furthermore, in mass disaster situations, usual medical resources may rapidly become less available, and anesthesiologists may be called on to render care in situations resembling "field anesthesia" as practiced by the military in combat situations. The mainstay of techniques under these conditions include total intravenous anaesthesia (TIVA), inhalational anesthesia, and regional anesthesia (Pretto, 2003). TIVA requires only drugs, an airway and the means to ventilate, a syringe pump, and monitors; the pulse oximeter likely being the one of choice in a disaster. The technique is therefore highly portable for use in difficult circumstances. There are various combinations, sequentially and/or simultaneously administered in a mixed infusion, of a relatively short list of familiar drugs: thiopental, midazolam, propofol, narcotics (morphine and fentanyl), ketamine, and muscle relaxants (usually rocuronium or vecuronium). Inhalational anesthesia requires specialized equipment, rudimentary versions of which are discussed here. It is generally not suitable for inductions in trauma patients because they must be presumed to be "full stomachs," but it is commonly used for maintenance of anesthesia in trauma patients after stabilization. Isoflurane and sevoflurane seem to have become the agents of choice. There seems to be a difference of opinion on the suitability of regional and peripheral blocks for use in mass casualty disasters, some contending that sterility can be problematic, and that the blocks are difficult to place and are sometimes ineffective. Others suggest that verbal contact can be maintained, the equipment is simple and portable, and trained personnel can place the blocks quickly. Continuous catheter regional techniques may be used much more commonly in the future because experience with using these techniques on military casualties demonstrates their efficacy for both the immediate situation and for postoperative analgesia.

The main role of the pharmacist during disasters is to develop a pharmacy cache; however, with a minimum of training and effort, the pharmacist can also perform other functions such as cardiopulmonary resuscitation, trauma management, and triage.

Nurses can be optimally prepared for a disaster of any type by being aware of community hazards and vulnerabilities, as well as being familiar with the community health care system and its level of preparedness. They participate a lot in emergency treatment and triage and in securing community resources for victims. Emergency nurses play an important role in the "all hazards" approach to hospital disaster planning. Ideally, emergency nurses act administratively with physicians to coordinate, develop, and support protocols for patient and emergency department management. Emergency nurses are actively involved in decision making with all aspects of disaster planning and implementation. In addition, they continue to refine and improve disaster planning based on performance drills and actual disasters. They amend policies and procedures as needed to ensure safety and standards of care.

6.7 Disaster evaluation

Evaluation has several purposes, the most fundamental of which is to determine the extent to which an organization, program, or unit achieves its clearly stated and measurable objectives in responding to a disaster. Evaluations are used to adjust disaster plans, to focus practice drills and preparedness, to improve planning for rapid assessment and management of daily response operations, and to provide input for the refinement of measures of effectiveness. Disaster evaluation research seeks to obtain information that can be used in preparation for future disasters by developing profiles of victims and types of injuries; assessing whether program adjustments can reduce disability and save lives; determining if better methods to organize and manage a response exist, including the use of resources in a relief effort; identifying measures that can be implemented to reduce damage to community; and assessing the long-term physical and emotional effects of a disaster on individuals and communities.

Evaluations begin by reviewing the disaster response plan and its measurable objectives. Without measurable objectives, a disaster response plan cannot be evaluated (Silverman et al., 1990). In the process of conducting an evaluation, assessments are directed for five domains of activity: structure, process, outcomes, response adequacy, and costs. Evaluation of structure examines how the medical and public health response was organized, what resources were needed, and what resources were available. Process assesses how the system (both medical and public health components) functioned during the impact and post-impact, how well individuals were prepared, and what problems occurred. Outcomes assessments identify what was and was not achieved as a result of the medical and public health response. This assessment focuses on the impact of care provided to patients during the disaster. Assessing the adequacy of the disaster response examines the extent to which the response systems were able to meet the needs of the community during the disaster. The analysis of the adequacy of the response is valuable for planning for future disasters. The main concern is, overall, how much death and disability occurred that could have been prevented? Disaster response costs can be measured in several ways: the total cost of the relief effort, the cost per every life saved, the cost for various subsystems that operated during the response phase, and the costs of preparedness.

6.8 Failure of disaster response and problems encountered during disaster management

There are many common failures that are evident at many MCIs that could be due to poor planning, inadequate on-scene needs assessment, inadequate or inappropriate communications, inadequate on-scene command structure, and the absence of sound medical approaches to the sorting of patients and prioritizing of their care. There are also common problems of disaster/MCI responses. These include problems of communication due to the need to contact a large number of persons and to provide information to the public; extreme crowdedness, which imposes a problem to the hospital staff working in the nearby hospitals; security problems; and infection control. Other problems are associated with the postdisaster period are fatigue and posttraumatic stress disorder of the hospital staff. These problems cannot always be alleviated by calling in more resources or personnel; they require a totally different kind of response.

6.9 Conclusions

Disasters are predictable, not in time or place, but in their inevitability. Large-scale disasters refers to events that will produce casualties beyond that which can be planned for by a single hospital. There are three vital phases of the disaster response: preparedness, response, and recovery. Preparedness includes preplanning, preparation of the proper equipment, and training. Response includes the command structure and coordination between emergency services, triaging casualties, and treating and transporting them to hospitals. The aim of the recovery phase is to return the community and the medical service to the predisaster level. There is a role for each specialist in the management and response to disasters, including surgeons, anesthesiologists, nurses, pediatricians, and psychiatrics.

References

Aghababian, R., Lewis, C. P., Gans, L., and Curley, J. (1994) "Disasters Within Hospitals," *Ann. Emerg. Med.* **23**, pp. 771–777.
Benson, M., Koenig, K. L., and Schultz, C. H. (1996) "Disaster Triage: STAR SAVE—A New Method of Dynamic Triage for Victims of a Catastrophic Earthquake," *Prehospital and Disaster Medicine* **11**, pp. 117–124.
Birnbaum, M. L., and Sundnes, K. O. (2001) "Disaster Evaluation: Guidelines for Evaluation of Medical Response in Disasters," *Prehospital and Disaster Medicine* **16**, pp. S119–S120.
Cuny, F. C. (1993) "Introduction to Disaster Management: Lesson 2—Concepts and Terms in Disaster Management," *Prehospital and Disaster Medicine* **8**, pp. 89–94.
Dara, S. I., Ashton, R. W., Farmer, J. C., and Carlton, P. K., Jr. (2005) "Worldwide Disaster Medical Response: A Historical Perspective," *Crit. Care Med.* **33**, pp. S2–S6.

Deahl, M., Srinivasan, M., Jones, N., Thomas, J., Neblett, C., and Jolly, A. (2000) "Preventing Psychological Trauma in Soldiers: The Role of Operational Stress Training and Psychological Debriefing," *Brit. J. Med. Psychol.* **73**, pp. 77–85.

Farmer, J. C., and Carlton, P. K., Jr. (2005) "Hospital Disaster Medical Response: Aligning Everyday Requirements with Emergency Casualty Care," *World Hosp. Health Serv.* **41**, pp. 21–43.

Gray, M. J., Litz, B., and Maguen, S. (2004) "The Acute Psychological Impact of Disaster and Large-Scale Trauma: Limitations of Traditional Interventions and Future Practice Recommendations," *Prehospital and Disaster Medicine* **19**, pp. 64–72.

Gunn, S. W. A. (1990) *Multilingual Dictionary of Disaster Medicine and International Relief*, Kluwer, Dordrecht, the Netherlands.

Hadjiiski, O. (1999) "Mass Disasters Bulgarian Complex Programme for Medical Care for Patients with Burns after Fire Disasters," *Annals of Burns and Fire Disasters* **XII**, pp. 224–228.

Kossman T., and Vitling I. (1991) "Triage and First Aid During Mass Accidents (with Burns) and Transportation to Trauma Centers," *Acta. Chir. Plast.* **33**, pp. 100–105.

Kunkel, K. E., Pielke, R. A., and Chagnon, S. A. (1999) "Temporal Fluctuations in Weather and Climate Extremes That Cause Economic and Human Health Impacts: A Review, " *Bull. Am. Meteorol. Soc.* **80**, pp. 1077–1098.

Leonard, R. B. (1998) "Role of Pediatricians in Disasters and Mass Casualty Incidents," *Pediatr. Emerg. Care.* **4**, pp. 41–44.

Litz, B., Gray, M., Bryant, R., and Adler A. (2002) "Early Intervention for Trauma: Current Status and Future Directions," *Clinical Psychology: Science and Practice* **9**, pp. 112–134.

Magliacani, G., and Masellis, M. (1999) "Guidelines for Fire Disaster Medical Management in the Mediterranean Countries," *Annals of Burns and Fire Disasters* **XII**, pp. 44–47.

Micheels, J. (1994) "The General Practitioner and Collective Disaster Plans," *Rev. Med. Liege* **49**, pp. 481–491.

Moralejo, D. G., Russell, M. L., and Porat, B. L. (1997) "Outbreaks Can be Disasters: A Guide to Developing Your Plan," *J. Nurs. Adm.* **27**, pp. 56–60.

Norberg, K. -A. (1998) "Disaster Medicine and Disaster Planning: Swedish Perspective," *Internet J. Disaster Medicine* **1**, no. 2.

Pearn, J. A. M. (1998) "Medical Response to Disasters," *The Medical Journal of Australia* **169**, p. 601.

Perez, E., and Thompson, P. (1994) "Natural Disasters: Causes and Effects. Lesson 1— Introduction to Natural Disasters," *Prehospital and Disaster Medicine* **9**, pp. 101–109.

Perry, R. W. (2003) "Incident Management Systems in Disaster Management," *Disaster Prevention and Management* **12**, pp. 405–412.

Pretto, E. A. (2002) "Anesthesia and Disaster Medicine," *Anesthesiology News* **1**, no. 1, www.anes.upmc.edu/anesnews/volume/2002summer/articles/special_article.html.

Pretto, E. A. (2003) "Anesthesia and Disaster Medicine. Part II. Framework for Mass Casualty Management and the Role of the Anesthesiologist," *Anesthesiology News* **1**, no. 2, www.anes.upmc.edu/anesnews/volume/2003winter_spring/articles/special_article.html.

Pretto, E., Ricci, E., and Safar P. L. (1992) "Disaster Reanimatology Potentials: A Structured Interview Study in Armenia III: Results, Conclusions, and Recommendations," *Prehospital and Disaster Medicine* **7**, pp. 327–338.

Pruitt, B. A., Jr., Goodwin, C. W., and Mason, A. D., Jr. (2002) "Epidemiological, Demographic, and Outcome Characteristics of Burn Injury," in *Total Burn Care*, second edition, ed. D. N. Hernon, pp. 16–30, WB Saunders Company, Philadelphia, Pennsylvania.

Quinn, B., Baker, R., and Pratt, J. (1994) "Hurricane Andrew and a Pediatric Emergency Department," *Ann. Emerg. Med.* **23**, pp. 737–741.

Silverman, M., Ricci, E., and Gunter, M. (1990) "Strategies for Improving the Rigor of Qualitative Methods in Evaluation of Health Care Programs," *Evaluation Review* **14**, pp. 57–74.

United Nations Environment Programme (UNEP). (2005) "Environmental Management and Disaster Reduction: Building a Multi-Stakeholder Partnership," program and abstracts of the *World Conference on Disaster Reduction*, session 4.5, 18–20 January, Kobe, Hyogo, Japan. Available at www.unisdr.org/wcdr/thematic-sessions/cluster4.htm.

van Emmerik, A., Kamphuis, J., Hulsbosch, A., and Emmelkamp, P. (2002) "Single-Session Debriefing after Psychological Trauma: A Meta-Analysis," *Lancet* **360**, pp. 766–770.

7

Augmentation of health care capacity in large-scale disasters

Atef M. Radwan

Many health care facilities have their own disaster plans for mitigation and management of major incidents. Based on their determined capacities, these facilities can cope with the expected risks. Large-scale disasters such as wars and weather-related disasters may hit the health care facilities any time. For that, these facilities should plan to cope with a devastating volume of health burden. Hospitals cannot face such large-scale disasters without putting a plan to increase their capacity. As stated by Hick et al. (2004), surge planning should allow activation of multiple levels of capacity from the health care facility level to the national level. Plans should be scalable and flexible to cope with the many types and varied time lines of disasters. Incident management systems and cooperative planning processes will facilitate maximal use of available resources. Plans should involve techniques of how to estimate the number of casualties and the severity of the incident, how to increase the facility capacity, how to adapt it, and how to call the adequate staff in the appropriate time. Besides these issues, this chapter reviews the importance of cooperative regional capacity augmentation, off-site patient care, and the role of the government and how it will collaborate with the health care facilities. This chapter also stresses the important role of the trained community being the first responders in all disasters.

7.1 Introduction

Health care facilities are occasionally at great risk of disasters that generate large numbers of casualties. Few surplus resources to accommodate these casualties exist in our current health care system. The U.S. National Disaster Medical System resource status 2005 report on Hurricane Katrina stated initially that the total patients treated were 16,477. Plans for "surge capacity" must thus be made to accommodate a large number of patients. Surge planning should allow activation of multiple levels of capacity from the health care facility level to the national level. Plans should be scalable and flexible to cope with the many types and varied timelines of disasters. Incident management systems and cooperative planning processes will facilitate maximal use of available resources. However, resource limitations may require implementation of triage strategies. Facility-based or "surge in place" solutions maximize health care capacity for patients during a disaster (Hick et al., 2004).

When these resources are exceeded, community-based solutions, including the establishment of off-site hospital facilities, may be implemented. Selection criteria, logistics, and staffing of off-site care facilities are complex. The World Trade Center attacks of 2001 and the weather-related disasters that followed have revealed potentially large gaps in the ability of the health care system to find either the capacity or the special capabilities to cope with disasters that severely injure a large number of victims. There is an urgent need for communities and regions, not just individual health care facilities,

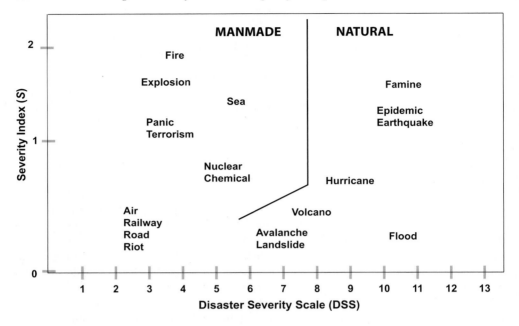

Figure 7.1 Disaster severity scale (DSS) versus severity index (S) for manmade and natural disasters. Adapted from de Boer & van Remmen (2003).

to develop tiered, and generic surge capacity plans to provide care for large-scale disasters.

7.2 Definitions

There are different scales to define the severity or intensity of disasters; one of these is the disaster severity scale (DSS; de Boer & van Remmen, 2003). As shown in Figure 7.1 and Table 7.1, natural disasters mostly have the largest scales. Through defining the disaster scale, health care facilities can estimate what should be prepared to cope with these disasters in advance.

There are three broad areas in which an augmented or "surge" response may be required during a disaster. "Public health surge capacity" refers to the overall ability of the public health system to manage a large incident by increased capacity for patient care and for multiple other patient and population-based activities. "Health care facility–based" and "community-based patient care surge capacity" relate to making available adequate resources for the delivery of acute medical care to large numbers of patients. "Surge capability," in contrast, refers to more specialized resources for specific patient groups (e.g., burns; Hick et al., 2004). From the local health care facility to the state and national level, surge capacity decision making must take place within an incident management system that involves the following key stakeholders (Barbera & Macintyre, 2002):

- Emergency medical service (EMS)
- Emergency management
- Public health
- Public safety/law enforcement
- Health care systems
- Red Crescent and Red Cross

Table 7.1. *Modified disaster severity scale (DSS)*

Classification	Grade	Score
Effect on infrastructure (impact area + filter area)	Simple	1
	Compound	2
Impact time	<1 h	0
	1–24 h	1
	>24 h	2
Radius of impact site	<1 km	0
	1–10 km	1
	>10 km	2
Number of dead	<100	0
	>100	1
Number of injured (N)	<100	0
	100–1000	1
	>1000	2
Average severity of injuries sustained (S) $(T1+T2)/T3$	<1	0
	1–2	1
	>2	2
Rescue time/ clearance time (C) (rescue + first aid + transport)	<6 h	0
	6–24 h	1
	>24 h	2
Total	DSS	1–13

Adapted from de Boer & van Remmen (2003).

- Mental health
- Jurisdictional legal authorities
- Professional associations, including pharmacy, medical, nursing, and mental health
- Health professional training institutions

7.3 Capacity augmentation of health care facility

7.3.1 Variables of health care capacity

The main content of disaster medicine is based on empiricism. During the past few years, a mathematical approach to some aspects has been added. This may result in the creation of some order in chaos. This modeling of medical disaster management is important not only in the preparedness phase, but also during the disaster itself and its evaluation. This may in turn result in a decrease in mortality, morbidity, and disability among disaster casualties (de Boer, 1999).

An accident involving one or more casualties (N), with varying severity of injuries sustained (S), will be met by assistance of a specific capacity (C). Medical assistance consists of that available at the site of the accident, the transportation of the casualty, and that available in the hospital. In an organized context, there are relevant services, such as ambulances and hospitals. All these services within the entire medical assistance chain (MAC) have a certain capacity (C), which is sufficient for the normal, everyday occurrence. If, however, the number of victims (N) with a specific average severity of injuries sustained (S) exceeds the existing capacity (C), a discrepancy then arises (the turning point) between the injured and their treatment, with the result that either additional services have to be called in from outside or existing services have to be intensified.

The turning point will be reached more quickly depending on the number of casualties involved (N) and the seriousness of injuries sustained (S). Conversely, the greater the capacity (C) of the medical assistance services, the later the turning point is reached. In short, it is directly proportional to N and S and inversely proportional to C. This can be illustrated in the following simple formula (de Boer & van Remmen, 2003):

$$\text{Medical severity index} = \frac{N \times S}{C} \qquad (7.1)$$

and if the ratio is more than 1, the facility is considered to be in a disaster situation.

The number of casualties can be determined through the incident reports from the disaster scene. There are also different ways to guess that number. If we have a database of the population number in mobile transport vehicles or immobile buildings, we can easily guess the estimated number of casualties. Every locality can make a map for that and put the database in computer software to be used as an aiding tool during disaster assessment. An example of such empiricism is depicted in Table 7.2, which can be fine-tuned through such a computer database and even satellite photos of the accident scene.

7.3.2 Triage priorities

The medical severity factor (S) can be determined by the medical teams when they triage patients into the four priorities: $T1$, $T2$, $T3$, and $T4$.

Immediate priority ($T1$): casualties who require immediate life-saving procedures (airway obstruction, respiratory, and cardiorespiratory distress and shock).

Urgent priority ($T2$): causalities who require surgical or medical intervention within 2 to 4 hours (open chest wound without respiratory distress, abdominal wounds without shock and fractures).

Delayed priority ($T3$): less serious cases whose treatment can be safely delayed beyond 4 hours (walking wounded victims, such as those suffering from sprains, strains, and contusions).

Expectant priority ($T4$): casualties whose condition is so severe that they cannot survive despite the best available care and whose treatment would divert medical resources from salvageable patients who may then be compromised (massive head injury with signs of impending death).

The ratio of casualty groups $T1$ and $T2$ to the $T3$ casualty group, or that between those requiring and those not requiring hospitalization, is referred to as the medical severity factor S (de Boer & van Remmen, 2003). Thus,

$$S = \frac{T1 + T2}{T3} \qquad (7.2)$$

7.3.3 Capacities in the medical assistance chain

The chain along which the victim receives medical and nursing assistance between the initial site and the hospital is the MAC. It can be divided into three, more or less separate, organizational systems or phases. The first is the site of the accident or disaster, the second is the transporting of casualties and their distribution in the various hospitals in the vicinity, and the third is the hospital. During each phase, personnel are working with specific materials,

Table 7.2. *Basic parameters for contingency planning and estimating N*

Immobile objects			Range for N^a
Residential area[b]	Per 100,000 m²	Low-rise buildings	20–50
		High-rise buildings	50–200
Business area	Per 100,000 m²		0–800
Industrial area	Per 100,000 m²		0–200
Leisure area	Per type	Stadium	h
		Discotheque	h
		Camping site	h
Shops	Per type	Department store	h
		Arcade	h
Mobile objects			
Road transport	Per 100 m[c]	Multiple collision	5–50
	Per type[d]	Coach	10–100
Rail transport[e]		Single deck	5–400
		Double deck	10–800
Air transport[f]	Per type	Small	10–30
		Large	150–500
Inland shipping[g]	Per type	Ferry	10–1000
		Cruise ship	200–300

[a] The range depends on date, time, and other local circumstances.
[b] A residential area is a combination of number of residents per house (1.8–2.8) and number of houses per 100,000 m² (30–70).
[c] Per car; length 5 m and 1.5–3 passengers.
[d] Road transport per type (articulated) local bus or (articulated) double-decker bus.
[e] Rail transport are carriages of three or four wagons.
[f] Air transport with a seat occupancy of 70%.
[g] Inland shipping transport with a seat occupancy of 80%.
[h] Awaiting further research.
From de Boer & van Remmen (2003).

employing specific techniques and with a single aim, viz. to provide the victim with medical and nursing assistance. Therefore, during each phase, personnel, materials, and techniques are providing a certain capacity: the so-called medical rescue capacity (MRC) at the site of the disaster, the medical transport capacity (MTC) during transport to medical facilities, and the hospital treatment capacity (HTC) in the hospital. Those steps are schematically depicted in Figure 7.2.

MRC: An experienced team composed of a doctor/specialist and a nurse, assisted by one or two first aid support staff, would need approximately 1 hour to perform life- and limb-saving procedures for one *T1* and three *T2* casualties.

MTC: The number of ambulances (*X*) required at a disaster is directly proportional to the number of (to be hospitalized) casualties (*N*) and the average time of the return journey between the site of the disaster and the surrounding hospital (t), and inversely proportional to the number of casualties that can be conveyed per journey and per ambulance (*n*) and the total fixed length of time (*T*) during which *N* casualties have to be moved (de Boer & van Remmen, 2003). Thus,

$$X = \frac{N \times t}{T \times n} \quad (7.3)$$

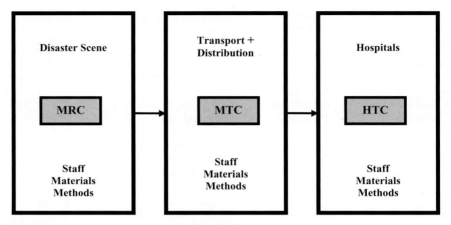

Figure 7.2 The medical assistance chain. Adapted from de Boer & van Remmen (2003).

where T can be fixed at between 4 and 6 hours, or vary according to the needed time to clear the disaster scene from casualties. An example of calculation of number of ambulances required follows:

$N = 96$ victims requiring hospital treatment
$n = 1$ victim per ambulance per journey
$T = 6$ h to transport the 96 victims
$t = 1$ h average traveling time
$X = (96 \times 1)/(6 \times 1) = 16$

HTC: The number of patients who can be treated per hour and per 100 beds. For the day-to-day surgery situation, the HTC for patients with mechanical injuries amounts to 0.5 to 1 patient an hour per 100 beds (de Boer and van Remmen, 2003). Within the framework of a practiced disaster relief plan, this number can be increased to between 2 and 3 patients per hour per 100 beds. These figures, derived from many exercises for mechanical injuries, are primarily determined by the number of available surgeons, anesthesiology specialist nursing staff, and medical equipment available.

Hospital capacity is affected by many bottlenecks: emergency room capacity, operating room capacity, intensive care unit (ICU) capacity, special care capacity, medical team availability, and the time of disaster (day time, night time, holiday time, etc.). This can also be fed to a computer program to determine the capacity at different disaster situations and at different times.

7.3.4 Mass trauma casualty predictor

Past mass trauma events show patterns of hospital use. It is possible to estimate initial casualty volume and pattern after a mass trauma event. Public health professionals and hospital administrators can use this information to handle resource and staffing issues during a mass trauma event. Within 90 minutes of an event, 50% to 80% of the acute casualties will likely arrive at the closest medical facilities. Other hospitals outside the area usually receive few or no casualties. The 1995 Sarin attack in the Tokyo subway produced nearly 6,000 injured (Asai & Arnold, 2003). Explicit planning with sound prediction of patient numbers is required to quickly expand patient capacity through restructuring the existing bed space and

7.3 Capacity augmentation of health care facility

Table 7.3. *Number of major incidents and corresponding mortalities occurring in Zagazig City, Egypt, in 2005*

Date of Major Incident	Total	Alive		Dead	
March 2005	10	10	(100%)	0	
May 2005	15	15	(100%)	0	
July 2005	10	4	(40%)	6	(60%)
7 December 2005	46	45	(97.8%)	1	(2.2%)
8 December 2005	29	29	(100%)	0	
9 December 2005	10	9	(90%)	1	(10%)
Total	120	112	(93.3%)	8	(6.7%)

From Ismail (2006).

Table 7.4. *Number of patients admitted to Zagazig University Hospital, according to their triage priorities*

Triage Sort	Total	Alive		Dead		Chi Square Test of Significance (χ^2)	Probability P
$T1$ (immediate)	11	4	(36%)	7	(64%)		
$T2$ (urgent)	7	5	(85.7%)	2	(14.3%)	50.42	0.001
$T3$ (delayed)	102	102	(100%)	0	(0%)		

Medical severity index is 18 102 (17.6%). From Ismail (2006).

the incorporation of "hasty treatment space" onsite (hallways, sheltered parking) and offsite (adjacent buildings, public shelters) (Shapira & Shemer, 2002). Furthermore, biological and chemical agents often require patient isolation and observation, which demands still more capacity (Keim & Kaufmann, 1999). The less injured casualties often leave the scene under their own power and go to the nearest hospital. As a result, they are not triaged at the scene by EMS, and they may arrive to the hospital before the most injured.

On average, it takes 3 to 6 hours for casualties to be treated in the emergency department before they are admitted to the hospital or released. When trying to determine how many casualties a hospital can expect after a mass trauma event, it is important to remember that casualties present quickly and that approximately half of all casualties will arrive at the hospital within a 1-hour window. This window begins when the first casualty arrives at the hospital. To predict the total number of casualties the hospital can expect, simply double the number of casualties the hospital receives in the first hour.

The total expected number of casualties will be an estimate. There are many factors that may affect the accuracy of this prediction, such as transportation difficulties and delays, security issues that may hinder access to victims, and multiple explosions or secondary effects of explosion. Tables 7.3 and 7.4 summarize a portion of the exhaustive study conducted by Ismail (2006). In Zagazig University Hospital, the number of hospitalized patients ($T1 + T2$) in relation to nonhospitalized ($T3$) patients in six major incidents that occurred in 2005 in Zagazig City, Egypt, was about 0.18. This means that if the estimated number of casualties is 100, the expected patients who need hospital admission will be 18. By simple mathematics, the hospital can prepare itself to the coming patients and manage its resources accordingly. Moreover, hospitals can determine their ability to cope with this number of patients through comparing their HTC to the number of $T1 + T2$. In the case that there is a mismatch between capacity and load, hospitals can ask for help from other facilities. This should be preplanned with the other facilities.

7.3.5 Increasing hospital bed capacity

Most victims of chemical and explosion events will be present within the first 6 hours to hospital emergency departments (Greenberg & Hendrickson, 2003). Expediting the disposition of patients and clearing the emergency department of ambulatory patients accomplishes rapid clearance of the emergency department. Elective procedures and admissions should be held or cancelled. The facility may elect a graded response, which provides different actions to be taken according to the number of casualties expected.

According to the type of incident, additional emergency patient screening areas may be required if the volume of patients overwhelms the resources of the emergency department. Each inpatient unit may be assigned by the disaster plan to automatically accept a predetermined number of patients from the emergency department who do not yet have beds, e.g., two per nursing unit, or 110% unit census (Schultz, 2003). Health care facilities should be prepared to discharge inpatients to make beds available for incoming casualties. The discharge function should receive immediate attention when the health care facility disaster plan is activated. Identification of patients eligible for early discharge is an inexact science, but patients who are clinically stable and whose inpatient requirements are limited to a few parenteral medications may be appropriate candidates. Preplanning with ancillary health care services (e.g., nursing homes) is essential for these early discharges. A discharge holding area (e.g., lounge, cafeteria area) may allow these patients to be moved from their rooms while awaiting appropriate transportation, home care, and pharmacy arrangements (Schultz et al., 2002).

7.3.6 Adaptation of existing capacity

Private rooms may be converted to double rooms, and patients may be placed in halls. Closed areas may be reopened if staffing and supplies allow. Health care facilities may find that patient care areas typically reserved for postanesthesia care, chest pain observation, gastroenterology procedures, pulmonary and cardiac catheterization suites, and outpatient surgery areas provide the best opportunities for increasing critical care capacity (Viccellio, 2003).

Generally, approximately 10% to 20% of a hospital's operating bed capacity can be mobilized within a few hours using these strategies, with an additional 10% from conversion of "flat space" areas such as lobbies, waiting rooms, classrooms, conference facilities, physical therapy areas, and hallways. Inova Health System's four hospitals in northern Virginia, for example, made 343 additional beds (of 1,500 total) and forty-three operating rooms available within 3 hours of the 2001 Pentagon attack. The District of Columbia made available 200 beds from their existing 2,904 staffed beds. Thus, for any given facility, surge capacity planning should incorporate the availability of about 20% to 30% of operating beds for immediate or near-term patient use. Some institutions may have significantly different numbers according to their elective admission rates, critical case resources, and availability of additional staffed beds. ICU bed availability is likely to be significantly lower (SoRelle, 2003). In our disaster plan at the Zagazig University Hospital (Zagazig City, Egypt), we assigned 50% of the thirty-two operating rooms to be an alternative place for critical care in case we have a medical disaster necessitating mechanical ventilation as in chemical or biological disasters; operating room nurses should be trained to cover this situation if or when it occurs.

7.3.7 Staff calling in and staff augmentation plan

In hospital disaster plans, staff augmentation is regularly addressed in a variety of ways, including extending hours of present staff and calling in supplemental staff. A number of basic steps can expand hospital staffing for mass casualty incidents (JACHO, 2003 and Rubinson et al., 2005). These steps are itemized by them as follows:

- Development of a communitywide concept of "reserve staff" identifying physicians, nurses, and hospital workers who (1) are retired, (2) have changed careers to work outside health care services, or (3) now work in areas other than direct patient care (e.g., risk management, utilization review). Although developing the list of candidates for a communitywide "reserve staff" will require limited resources, the reserve staff concept will only be viable if adequate funds are available to (1) regularly train and update the reserves so that they can immediately step into roles in the hospital that allow regular hospital staff to focus on incident casualties, and (2) develop protocols for where and how to use such staff efficiently and safely.
- It may be possible to expand the "reserve staff" concept to include medical, nursing, and allied health students training in programs affiliated with the hospital(s). This potential is untested and may not be feasible if the students actually reduce available staff time because of their need for supervision.
- Medical staff privileges are generally granted on a hospital-by-hospital basis. In a mass casualty incident, physicians may be unable to reach the hospital where they usually admit patients. Hospital preparedness can be increased if medical staff credentials committees develop a policy on the recognition of temporary privileges in emergency or disaster situations and if hospitals in a community regularly share lists of the medical staffs and their privileges.
- In the initial hours of a mass casualty incident, "first responders" (i.e., EMS, police, fire personnel) may be fully occupied in the on-scene care and potential decontamination of casualties. However, as the duration of the incident progresses, first responders may be potential sources to help augment hospital staff.
- In many disaster drills, the incident places a short-term but intense demand on the hospital. As a result, the clinical personnel experience a substantial increase in demands, but the support staff (e.g., food service workers, housekeepers) may have only a limited change in demands. In a mass casualty incident, the demand for both care and support services may be more sustained. Hospital preparedness will be facilitated by providing for augmentation of both clinical and nonclinical support staff.
- Some biological incidents are different because of their risk of infecting hospital staff. Biological terrorism will pose additional challenges of both uncertainty and fear. Staff concerns can be reduced through appropriate education and the use of universal precautions until the nature of the disease agent is understood. However, hospital preparedness plans need to include contingency plans in case medical professionals and/or volunteers do not show up.
- Communities have a long history of helping hospitals in times of crisis. A frequent demonstration of this community support is the willingness of individuals with four-wheel drive vehicles to provide transportation assistance to hospital patients and staff during snowstorms. The potential for untrained volunteers to assist with mass casualty incidents is very limited. Hospital staff will be under enormous demands and stress. There will be only limited time to identify, train on-site, and supervise volunteers. In some cases, volunteers may add to the problems of staff identification and crowd control. In mass casualty incidents, there is unlikely to be time to conduct intensive staff training between identification of the incident and its onset. Weather-related natural disasters—floods and hurricanes—may provide a little lead-time for training, but most of the available time will be consumed by implementing the disaster plan and by staff needs to arrange care for members of their families and for pets. We are developing a system in Zagazig Governorate, Egypt, in which medical students are trained in first aid and disaster management with joint drills with hospital staff. A plan of tasking student groups to be responsible for allocated areas within the city is in process and still under evaluation.

- Although many hospitals rely on information resources originally developed by the military services for addressing chemical and biological exposure, some of the information and supplies assumed in a combat situation are not representative of civilian hospital environments. Standardized materials oriented to training hospital personnel need to be developed and made widely available. Such materials would facilitate training, allow for a more standardized body of information across the hospital field, reduce the loss of expertise that accompanies employee turnover, and be cost efficient. National grants are seen as the only realistic source for developing the necessary training resources.
- Faced with the demands of a mass casualty incident, physicians and hospital staff will be called on to provide extraordinary service to their communities. Pressure and stress will be high. Casualties will be numerous, and may include friends and neighbors. To allow staff to function at their highest potential, the following recommendations of JACHO (2003) should be offered:
    (a) Facing long hours and the likelihood of limited communications, hospital staff does not need the distraction of worrying and arranging for the needs of family members. In some communities, the network of an extended family or established group of friends may provide "coverage" during the incident. In many communities, however, population mobility, nuclear family structures, and single-parent families may mean that many staff members do not have existing arrangements to care for their families. Mass casualty preparedness will be facilitated if hospitals work with community resources—school systems, mosques and churches, and employers—to include in their disaster plan prearranged supervision, shelter, and feeding for the families of those working in the hospital. These prearranged community support systems can be activated using public service announcements on radio and television stations.
    (b) Those who have studied or experienced mass casualty incidents have reported the enormous stress and pressure faced by health workers. Effective response by these workers to the crisis requires that they have the necessary supportive services for themselves. Such services include access to vaccines, infection control advice, adequate rest and relief, and mental health counseling. In a sustained, mass casualty incident, the inclusion of these resources in the disaster plan will assist staff in meeting the other demands the plan places on them.
- At the onset of the mass casualty incident, there is likely to be confusion and conflicting information about the incident. This lack of certainty may distract hospital workers wanting to understand the risks they personally face in caring for incident victims. The use of universal precautions and a system for sharing information with staff prior to any incident are likely to facilitate implementation of the disaster plan. Protocols must be in place for revision of staff work hours (e.g., 12-hour shift standard in disasters), callback of off-duty personnel, use of nonclinical staff (e.g., nurse administration, continuous quality improvement coordinator) in clinical roles as appropriate, reallocation of outpatient staff resources (e.g., cancel subspecialty clinics and reallocate staff), and use of solicited (former employees, retirees) and unsolicited volunteers. Credentialing of nonemployees should be consonant with institutional bylaws and legislative and accrediting body regulations (Joint Commission on Accreditation of Healthcare Organizations [JCAHO], 2003). Army Medical Command should have an emergency management program through which it can deploy teams or resources under the National Response Plan to provide short-duration medical augmentation to civil authorities. Resources that may be available include medical personnel (i.e., special medical augmentation response teams) and mobile support hospitals (Cecchine et al., 2004).

7.3.8 Modification of the standards of care

The critical care mass casualty response plan will not be the same for every ICU, every hospital, or every community. The answer is not to develop a standardized response, but rather to standardize the process that each community uses to prepare a plan. The same questions need to be asked in every community and every hospital. How many patients can be cared for? How can the care of critically ill patients be provided outside the ICU, and where is that care best provided? What drugs and equipment will be necessary, and how can they be provided? What level of care can be provided? What are the priorities of the community and the hospital? The standard of care will have to change to make significant increases in the capacity to care for critically ill patients. Communitywide plans will be much more effective and easier to implement than having a different plan for each hospital.

Decisions on how to decide who will receive mechanical ventilation and other life-saving treatments need to be made before a disaster occurs. Response preparation processes need to include health care worker education and training, with a plan to provide the necessary training rapidly. Each facility needs to assess its ICU capacity and prepare to expand it rapidly. In addition to institutional and local protocols, regional and national protocols need to be developed for rapid expansion of in-hospital and ICU care. Specialty patient care surge capability (e.g., burns, pediatrics) may be created to address patient needs temporarily, but ideally only pending transfer to a specialty center with adequate resources. Consulting with experts from a hospital's usual referral centers may be helpful in planning for these situations (Rubinson et al., 2005).

7.3.9 Triage of patients in mass critical care

Triage is commonly employed during everyday ICU operations (Sinuff et al., 2004). Both the Society of Critical Care Medicine (1994) and the American Thoracic Society (1997) have published consensus statements to guide this process. Normally, patients who are judged to meet a subjective threshold of "likely to benefit" from critical care are admitted to and remain in ICUs on a first-come, first-served basis. It is recommended that in the aftermath of bioterrorist attacks, if there are limited hospital resources and many critically ill patients in need, triage decisions regarding the provision of critical care should be guided by the principle of seeking to help the greatest number of people survive the crisis. This would include patients already receiving ICU care who are not casualties of an attack. Even after a hospital has shifted to delivering only essential elements of critical care to its critically ill patients, it may become the case that there may be insufficient resources to treat all seriously ill patients. The most ethical way to help the greatest number of critically ill people survive in such dire conditions is to give such interventions first to the people deemed most likely to survive.

7.4 Cooperative regional capacity augmentation

A community planning process that integrates regional health care facilities is critical to develop a systematic process to increase health care capacity. The individual health care facility represents the "first tier" of response. The mission for the facility is to take steps to increase facility capacity. Should these responses prove inadequate, the "second tier" of resources—other local health care facilities—is activated. This can be named as tiered system and carried out as follows (Barbera and Macintyre, 2002 and Hick et al., 2004):

- **Tier 1:** individual health care facility (facility-based "surge in place" response)
- **Tier 2:** health care coalitions (multiple facility response using facility and coalition agreements/plans)
- **Tier 3:** jurisdictional incident management (community response, coordination at emergency operations center)
- **Tier 4:** regional incident management (cooperation between jurisdictions/coalitions)
- **Tier 5:** state response (support to jurisdictions)
- **Tier 6:** national response (support to state)

The achievement of weapons of mass destruction (WMD) preparedness depends on the recognition that such incidents are simultaneously complex and large. Internationally, political authorities have emphasized the development of community emergency plans that address not just natural and technological disasters, but also terrorist attacks (Perry & Lindell, 2000). Although historically, hospitals have only marginally participated in such planning, the effective address of WMD incidents demands much fuller participation. The prospect of mass casualties accompanied by significant destruction of community resources requires cooperation among hospitals, as well as among hospitals and organizations that deliver local emergency management, prehospital care, law enforcement, public health services, and handling of the deceased.

The long-standing pattern of mutual aid agreements among local fire departments and law enforcement departments forms an important model for hospitals. Such agreements address sharing of personnel, resources, and equipment designed to forestall situations that may overwhelm the resources of a single institution. Bradley (2004) indicates that alhough patient load can be managed to a certain extent when movement to hospitals is controlled by ambulance services, victim self-referral is more common in terrorist incidents, leading patients to hospitals nearest the incident site. This phenomenon creates the conditions for overload at some institutions, whereas others may have no or few patients. Mutual aid agreements and preplanning allow for systematic redistribution of patients, pharmaceuticals, equipment, medical staff, or support staff to relieve overtaxed hospitals (Bradley, 2004).

The state of medicine relative to many biological, radiological, and chemical threats is such that death tolls are likely to be quite high. Most hospital morgues are small holding areas designed to facilitate transfer of the deceased to private funeral services or government medical examiners. Because hospital efforts must focus on patient care, it is important to engage in planning with both medical examiners and public health authorities to arrange effective handling of remains. In some cases, postmortem studies will be needed to confirm agents and inform treatments, and infectious disease risks from bodies form a public health threat (Morgan, 2004).

The important principle is that WMD terrorist incidents require an awareness of the interdependence of institutions. The demands imposed by a WMD incident are immense and affect virtually all community institutions. No hospital can stand alone in such events; demands quickly outstrip capabilities. Even if hospital facilities are undamaged in an attack, there remains the chance that off-duty personnel can be incapacitated (Frezza, 2005). Collective planning, particularly the integration of hospital plans with those created by local emergency management authorities, is the weakest, and yet the most important, factor in successful response.

Management of the hospital—sometimes called command and control—during any large scale disaster is critical for efficient and effective operations. Following the 2001 earthquake in Gujarat, India, it was found that hospitals with preplanned incident management systems were far better able to sustain delivery of medical care both during the crisis and through the aftermath (Bremer, 2003). Most of the hospital community partners in response—fire departments, EMS, police—routinely use incident command or incident management systems (Perry, 2003). Hospital adoption of a similar system enables the medical staff to fully integrate their activities with those of supporting agencies, as well as to better manage the onslaught of self-referred patients, family members, and other observers who tend to converge on medical facilities (Macintyre et al., 2000). The hospital emergency incident command system has long been available to hospitals and was cited as a "best practice" by the U.S. Occupational Safety and Health Administration (2004). This system provides a predictable chain of management, flexible organization chart, prioritized response checklists, and common language for communication with external agencies. Because WMD incidents may require continuous hospital operations for long periods, it is particularly important that a command center with communication facilities be preplanned.

7.5 Off-site patient care

If a large number of ambulatory "walking wounded" or "potentially exposed" victims are generated by an event, triage and initial treatment sites may be immediately needed to relieve pressure on the emergency transportation and care system. For every casualty injured or infected, hundreds more may seek evaluation (Okumura et al., 1998; Petterson, 1988). Such sites may also be required when the local health care infrastructure is severely damaged. For example, auxiliary care sites have been proposed to cope with health care facility damage after earthquakes (Schultz et al., 1996). Many localities rely on EMS to organize triage and initial treatment sites. A process to provide health care workers from other agencies, special teams, or less affected facilities to the scene should be in place to reduce convergent volunteerism (Cone et al., 2003).

Nonambulatory patients should be accommodated within the existing health care infrastructure as much as possible, with patient transfer as a later step. In selected situations, usually in the setting of a contagious disease epidemic, but sometimes in the setting of a hospital evacuation or other circumstance, these mechanisms will be inadequate, and an off-site hospital facility may be required. Each hospital accredited by JCAHO is required to plan for such facilities (JCAHO, 2003). The need for such a facility should be anticipated as early as possible according to patient load, hospital capacity, and event data. The authority to initiate an off-site facility and the administrative, staffing, logistic, and legal issues should be detailed and drilled in advance.

7.6 Role of government

The primary preparation and response to a disaster is local, but the national authority can and should play an important role in assisting with standardization of the process. Public health policies developed by the government have an important impact on disaster planning (Farmer et al., 2006). The emergency response of the government must be integrated with that of local disaster response teams. The government can provide equipment and drugs from the national stockpile, which can be delivered to different area within 12 hours. The government must then distribute these supplies locally. Input from critical care experts as to what supplies would be needed to provide basic critical care for large numbers of patients will help determine what should be included in the national stockpile. National leadership is important in developing national standards of care, but governments will also need to develop disaster management plans, including broad input from community political leaders, health care workers, and others. Only by having a strong local and regional plan will we have an effective national response to a widespread disaster. In the event of a disaster with many critically ill victims, it may be necessary to mobilize health care workers from other regions to provide sufficient critical care. The involvement of the government is essential in ensuring rapid credentialing of health care professionals across state, and potentially national, borders. The process must be easy to initiate and quick in response.

Governments will also be needed to provide legislative remedies to allow hospitals and ICUs to function under a different standard of care in the event of a disaster that overwhelms our current system. Furthermore, governments have a role in directing and standardizing broad-scale education for health care professionals. For example, governments should consider requiring and funding standardized training in disaster management (Farmer et al., 2006).

7.7 Community involvement

Although some researchers have undertaken the study of volunteerism as it occurs after disasters, few have studied emergency citizen groups as a means to increase community preparedness (Angus et al., 1993). Concerns include issues of training capability and integration with the municipal emergency management agencies. Although many programs have been run in many areas (Simpson, 2000) that have been affected by numerous devastating disasters over the past 10 years, there have been only limited efforts between governmental and voluntary organizations to meet the multiple disaster health needs during disasters (Bowenkamp, 2000). The factual performance of the community response is still not tested accurately. Moreover, there is a potential for nonprofessional personnel to make a scenario more hazardous by becoming involved. Added to that, we need to assess what material can fit training people of different educational and cultural levels, and how this material can be delivered to them.

The events that occurred on September, 11, 2001, put into perspective much more than one could imagine; medical students in New York felt that they were helpless, and they had the worst feeling of helplessness imaginable (Disaster Reserve Partner Group of New York College of Osteopathic Medicine [NYCOM]); this was quoted from Vohra & Meyler (2003). Such a scenario was encountered in Cairo and Alexandria, Egypt, twice in the past 5 years when two large buildings caught fire and collapsed while citizens were trying to help save people. This is a high price to pay, and it is preventable through training. A group of Zagazig University students were trained on structured approach to major incidents, cooperation with emergency services, rescue safety, sorting victims, and other key skills included in other related courses (e.g., CERT) with context tailoring. Contents were chosen after reviewing the COCHRANE library, courses from the Federal Emergengy Management Agency, American Red Cross emergency courses and Disaster Reserve Partner Group of NYCOM (Vohra & Meyler, 2003), and the Ph.D. thesis on spaced teaching versus mass teaching (Bullock, 2002).

The latter study made the following conclusions (Radwan, 2004):

- Continue involving these students in pertinent teaching activities to reinforce their training and to prevent skill attrition; this can be done through videotapes and booster training sessions.
- Teaching should be as simple as possible to avoid decay in the knowledge gained and in the skill acquired.
- Full-scale exercise is an important role-play tool for skill acquisition; it allows implementing the fact that teaching of the skills should be in the order they will be used.
- Building bodies of trained citizens could be a good way to respond in times of disasters. They can offer qualified help in simple first aid measures, transporting less seriously injured patients, and psychological support to victims and their families.

7.8 Summary

The roles and responsibilities of a hospital during a disaster have been assumed by disaster medical planners for a long time. Their focus has been on prehospital rescue and initial medical care. Furthermore, most planners have not considered what happens when or if a hospital is rendered overwhelmed by large-scale disasters. Disaster medical training of hospital personnel should include how to augment hospital capacity in such cases. Recent natural disasters have underscored these issues. To move forward, hospitals and response agencies must do the following:

1. Improve the interface between public/governmental agencies and hospitals.

2. Develop rapidly responsive, "portable," robust critical care capabilities.
3. Better incorporate the hospital into the overall medical disaster planning process, including inter-facility cooperation and alternative plans if a hospital becomes unusable.
4. Improve disaster medical and critical care education of all hospital personnel.
5. Seek synergy between the support requirements for hospital disaster medical infrastructure and other existing hospital programs.

References

American Thoracic Society. (1997) "Fair Allocation of Intensive Care Unit Resources," *Am. J. Respir. Crit. Care Med.* **156**, pp. 1282–1301.

Angus, P. C., Pretto, E. A., Abram, J. I., and Safar, P. (1993) "Recommendations for Life-Supporting First Aid Training of the Lay Public for Disaster Preparedness," *Prehospital and Disaster Medicine* **8**, pp. 157–160.

Asai, Y., and Arnold, J. L. (2003) "Terrorism in Japan," *Prehospital and Disaster Medicine* **18**, pp. 106–114.

Barbera, J., and Macintyre, A. (2002) "Medical and Health Incident Management System: A Comprehensive Functional System Description for Mass Casualty Medical and Health Incident Management," George Washington University Institute for Crisis, Disaster, and Risk Management, Washington, DC, www.gwu.edu/~icdrm.

Bowenkamp, C. D. (2000) "Community Collaboration in Disaster," *Prehospital and Disaster Medicine* **15**, pp. 206/81–208/82.

Bradley, R. N. (2004) "Health Care Facility Preparation for Weapons of Mass Destruction," *Prehosp. Emerg. Care* **4**, pp. 261–269.

Bremer, R. (2003) "Policy Developments in Disaster Preparedness and Management," *Prehospital and Disaster Medicine* **18**, pp. 372–384.

Bullock, I. (2002) "A Combined Paradigm Perspective Evaluating Two Teaching Formats of the Advanced Life Support Resuscitation Course and Their Impact on Knowledge and Skill Acquisition and Retention," Ph.D. Thesis, University of Reading, Reading, United Kingdom.

Cecchine, G., Wermuth, M. A., Molander, R. C., McMahon, K. S., Malkin, J., Brower, J., Woodward, J. D., and Barbisch, D. F. (2004) "Triage for Civil Support Using Military Medical Assets to Respond to Terrorist Attacks," Rand Corporation, Santa Monica, California.

Cone, D. C., Weir, S. D., and Bogucki, S. (2003) "Convergent Volunteerism," *Ann. Emerg. Med.* **41**, pp. 457–462.

de Boer, J. (1999) "Order in Chaos: Modelling Medical Management in Disasters," *European Journal of Emergency Medicine* **6**, pp. 141–148.

de Boer, J., and van Remmen, J. (2003) *Order in Chaos: Modelling Medical Disaster Management Using Emergo Metrics*, LiberChem Publication Solutions, Culemborg, The Netherlands.

Farmer, J. C., Paul, K., and Carlton, J. (2006) "Providing Critical Care During a Disaster: The Interface Between Disaster Response Agencies and Hospitals," *Crit. Care Med.* **34**, supplement, pp. S56–S59.

Frezza, E. E. (2005) "The Challenge of the Hospitals and Health-Care Systems in Preparation for Biological and Chemical Terrorism Attack," *J. Soc. Sci.* **1**, pp. 19–24.

Greenberg, M. I., and Hendrickson, R. G. (2003) "Drexel University Emergency Department Terrorism Preparedness Consensus Panel: Report of the CIMERC/Drexel University. Emergency Department Terrorism Preparedness Consensus Panel," *Acad. Emerg. Med.* **10**, pp. 783–788.

Hick, J. L., Hanfling, D., Burstein, J. L., DeAtley, C., Barbisch, D., Bogdan, G. M. and Cantrill, S. (2004) "Health Care Facility and Community Strategies for Patient Care Surge Capacity," *Ann. Emerg. Med.* **44**, pp. 253–261.

Ismail, A. M. (2006) "Medical Severity Index of Mechanical Major Incidents Admitted to Zagazig University Hospital," M.Sc. Thesis, School of Medicine, Zagazig University, Zagazig City, Egypt.

Joint Commission on Accreditation of Healthcare Organizations (JCAHO). (2003) "Comprehensive Accreditation Manual for Hospitals: Medical Staff Section MS.5.14.4.1," *Disaster Privileging Standard*, Oakbrook Terrace, Illinois.

Keim, M., and Kaufmann, A. F. (1999) "Principles for Emergency Response to Bioterrorism," *Ann. Emerg. Med.* **34**, pp. 177–182.

Macintyre, A. G., Christopher, G. W., Eitzen, E. J., Gum, R., and Weir, S., DeAtley, C. (2000) "Weapons of Mass Destruction Events with Contaminated Casualties: Effective Planning for Healthcare Facilities," *JAMA* **283**, pp. 242–249.

Morgan, O. (2004) "Infectious Disease Risks from Dead Bodies Following Natural Disasters," *Pan Am. J. Public Health* **15**, pp. 307–312.

Okumura, T. Suzuki, K., and Fukada, A. (1998) "The Tokyo Subway Sarin Attack: Disaster Management. Part 2: Hospital Response," *Acad. Emerg. Med.* **5**, pp. 618–624.

Perry, R. W. (2003) "Incident Management Systems in Disaster Management," *Dis. Prevent. Manage.* **12**, pp. 405–412.

Perry, R. W., and Lindell M. K. (2000) "Understanding Human Behavior in Disasters with Implications for Terrorism," *J. Contingen. Cri. Manag.* **11**, pp. 49–61.

Petterson, J. S. (1988) "Perception Versus Reality of Radiological Impact," *The Goiana Model. Nuclear News* **31**, p. 84.

Radwan, A. M. (2004) "Evaluation of Focused Disaster Relief Training for Citizenry (FDRTC)," Thesis in European Master of Disaster Medicine, Eastern Piedmonte University, Piedmont, Italy.

Rubinson, L., Jennifer, B., Nuzzo, S. M., Daniel S. T., O'Toole, T., Bradley, R., Kramer, B. S., and Thomas, V. I. (2005) "Augmentation of Hospital Critical Care Capacity after Bioterrorist Attacks or Epidemics: Recommendations of the Working Group on Emergency Mass Critical Care," *Crit. Care Med.* **33**, pp. 2393–2403.

Schultz, C. H., Koenig, K. I., and Noji, E. K. (1996) "A Medical Disaster Response to Reduce Immediate Mortality after an Earthquake," *N. Engl. J. Med.* **334**, pp. 438–444.

Schultz, C. H., Mothershead, J. L., and Field, M. (2002) "Bioterrorism Preparedness. I: The Emergency Department and Hospital," *Emerg. Med. Clin. North Am.* **20**, pp. 437–455.

Shapira, S. C., and Shemer, J. (2002) "Medical Management of Terrorist Attacks," *Israeli Med. Assn. J.* **4**, pp. 489–492.

Simpson, D. M. (2000) "Non-Institutional Sources of Assistance Following a Disaster: Potential Triage and Treatment Capabilities of Neighborhood-Based Preparedness Organizations," *Prehospital and Disaster Medicine* **15**, pp. 199/73–206/80.

Sinuff, T., Kahnamoui, K., and Cook, D. J. (2004) "Rationing Critical Care Beds: A Systematic Review," *Crit. Care Med.* **32**, pp. 1588–1597.

Society of Critical Care Medicine. (1994) "Consensus Statement on the Triage of Critically Ill Patients. Medical Ethics Committee," *J. Am. Med. Assoc.* **271**, pp. 1200–1203.

SoRelle, R. (2003) "A Real Solution to Overcrowding," *Emergency News* **25**, pp. 1–10.

U.S. Occupational Safety and Health Administration (OSHA). (2004) "Hospital Emergency Incident Command System Update Project," OSHA, Washington, DC.

Viccellio, P. (2003) "Emergency Department Overcrowding: Right Diagnosis, Wrong Etiology, No Treatment," *Emergency News* **25**, p. 25.

Vohra, A., and Meyler, Z. (2003) "The Disaster Reserve Partner Group at NYCOM," *J. Am. Osteopathic Assoc.* **103**, pp. 505–506.

8
Energy, climate change, and how to avoid a manmade disaster

Ahmed F. Ghoniem

Energy powers our life, and energy consumption correlates well with our standards of living. The developed world has become accustomed to cheap and plentiful supplies. Recently, more of the developing world populations are striving for the same, as well as taking steps toward securing their future needs. Competition over limited supplies of conventional resources is intensifying, and more challenging environmental problems are springing up, especially related to carbon dioxide (CO_2) emissions. Strong correlations have been demonstrated between CO_2 concentration in the atmosphere and the average global temperature, and predictions indicate that the observed trends over the past century will continue. Given the potential danger of such a scenario, steps should be taken to curb energy-related CO_2 emissions, enabled technologically and applied in a timely fashion. These include substantial improvements in energy conversion and utilization efficiencies, carbon capture and sequestration, and expanding the use of nuclear energy and renewable sources.

8.1 Introduction

Energy has been regarded as the most challenging need for humanity, and has been ranked highest on the list of priorities and requisites for human welfare. Through rapid industrialization and the implementation of modern economic systems, food production has expanded, at the expense of using more energy in agriculture and food transportation and processing. Similarly, providing clean potable water is an energy-intensive enterprise and can only be secured when energy resources are available. Mobility, lighting, and communications are all indispensables energy-intensive needs of modern life, as are heating and air conditioning. Industrial production continues to be a large consumer of energy, in many forms, and industrialization comes at the cost of accelerated energy use.

Energy, in its raw forms, is a natural resource that exists in abundance but in forms that are not necessarily most useful for many of the functions listed above. Converting natural sources of energy into useful forms, and in the quantities and at the rates needed for industrial, transportation, and domestic use, is a relatively recent development. Consumption rates have been growing steadily as standards of living improve and populations grow. Different sources have been discovered and harnessed, and many different uses have been invented. Conversion technologies have largely kept pace with demands and have adjusted to expanding use. These trends are causes for concern, and recently, even more troubling trends have emerged. Some of the traditional sources of energy are being depleted at faster rates, while the technologies required to harness alternative sources have not kept pace with the rising demands of developed and more significantly developing countries.

Moreover, although environmental concerns related to the use of some sources have been addressed and largely mitigated, others are only beginning to be understood, and some are yet to be explored carefully. For the former, CO_2, and its role in global warming stand out, and currently pose the most vexing problem.

To appreciate the scale of the challenges, we start by reviewing current energy consumption rates and the raw/crude sources of this energy. This is summarized in Section 8.2. We look at the relation between the per capita energy consumption and the gross national product, and how this correlation changed over time for a number of countries. Given that currently more than 85% of our energy is supplied from fossil resources, it is also important to consider how much might be left and the time scales of exhausting these resources.

The current concerns over global climate change and its correlation with the accumulation of greenhouse gases in the atmosphere pose a more urgent set of questions. These are related to predicting the impact of CO_2 on the global temperature, the reliability of the predictions, and anticipating the impact of global warming on life and the health of the planet. This is reviewed in Section 8.3. Because technology has led to the enormous expansion of energy, at the discovery, conversion, and utilization fronts, as well as the vast and concomitant improvement of the quality of life, technology is also expected to propose solutions to the CO_2/climate change predicament. This is discussed in Section 8.4. A one-solution-fits-all scenario cannot work given the wide range of utilization patterns, the distribution of raw energy resources, and implementation flexibility; rather, multiple solutions, to be pursued in parallel, are needed. This is the subject of Section 8.5.

Given the current infrastructure and end product utilization pattern, it is unlikely that the dependence on fossil fuel for electricity generation, fuel production, ground transportation, and domestic and industrial use will change significantly in the coming decades. Thus, unless CO_2 can be captured at large scales and stored safely, the current alarming trends in global temperature are likely to continue. Section 8.5 describes recent progress in developing approaches to decarbonize power generation plants burning heavy hydrocarbons. Efficiency improvements must be at the forefront of the effort to save resources and reduce the impact of energy consumption on the environment. Meanwhile, the use of zero-carbon energy sources, including nuclear energy and renewable sources such as hydraulic, geothermal, wind, solar, and biomass energy, should expand. The potential and recent progress in these areas is described briefly in Section 8.6.

Transportation consumes a significant fraction of our total energy use (close to 27% in the United States) and contributes a proportional fraction of CO_2 emissions. Transportation is largely dependent on fossil fuels—primarily oil, apart from $\sim$1.2% biomass—whereas coal, natural gas, nuclear, and renewable energy contribute to electricity generation. Recent trends in transportation technologies are reviewed in Section 8.7, and relative well-to-wheels efficiencies of different options are discussed.

After traveling the journey of energy from the cradle, that is, from its raw forms to the grave, where it ultimately dissipates into low-temperature heat, or as it is converted from a quasistable chemical, nuclear, thermal, potential, or other " high grade sources" until it is dissipated into low grade forms, and reviewing its growing impact on our environment, especially with regard to CO_2, we conclude by emphasizing the need to pursue the prudent approach of conserving the available resources, harnessing a more diverse portfolio, and controlling the emission of greenhouse gases. Although technology offers the requisite set of solutions to achieve these objectives, economics, policy, and public awareness are necessary for the timely and successful implementation of the technical solutions.

8.2 Energy consumption—now and then

8.2.1 How much we use

The world consumes massive amounts of energy, and the consumption rate is rising steadily with a positive second derivative starting at the onset of the twenty-first century. According to the International Energy Agency (IEA, 2003) the world power capacity in 2003 was close to 14 TW; 3.3 TW were in the United States. Of the current total consumption, close to 82% is produced from fossil fuels (petroleum, natural gas, and oil, with a very small fraction of nonconventional sources such as tar sands). Moreover, 10% comes from biomass (combined agricultural and animal products, mostly converted to thermal energy through combustion). Nuclear fission, hydroelectric, and other renewable energy (geothermal, wind, and solar) supply the rest of the current energy mix. According to the same report, the total world capacity is expected to reach beyond 50 TW by the end of the twenty-first century because of the population growth and the rise in living standards in developing countries, despite the anticipated improvement in energy intensity (defined as the gross domestic product per unit energy used, or GDP/J) and the reduction in the carbon intensity of the fuel mix (defined as the energy produced per unit mass of carbon used, or J/C) as the consumption of lower carbon sources expands beyond current levels. Consumption in developed countries is likely to level off as their energy efficiency improves. It is interesting to note that the massive expansion in energy consumption began with the Industrial Revolution, nearly 150 years ago. Since then, technology has enabled the discovery and harnessing of more raw sources, and more uses, direct and indirect, of energy, such as transportation, lighting, and air conditioning.

Figure 8.1 shows the breakdown of the world primary energy consumption in 2004, as compiled by the IEA. The total is 11,059 Mtoe (million ton oil equivalent). The total consumption of 11,059 Mtoe/year is equivalent to 531 EJ/year for that year. Fossil fuel sources, measured by the total thermal energy equivalent, are currently dominated by oil, followed by coal and natural gas, but the last is catching up fast. Oil is the fuel favored mostly in the transportation sector, while coal is burned mostly in electricity generation, where the use of natural gas has been rising. Of the total consumption of more than 530 EJ/year in 2004, the IEA estimates that oil contributed 34%, followed by natural gas and coal, at 20.9% and 25.1%, respectively. Most of the nonfossil energy is generated from hydroelectric (2.2% at 100% efficiency, or 6.6% of the primary according to the old convention), and nuclear-electric energy (6.5% with 33% efficiency). Biomass, geothermal, solar, and wind generate the rest. Biomass sources contribute the absolute majority of this part of the renewable[1] energy, used mostly in rural communities, where it constitutes a significant source of energy for heating and cooking. As shown in Figure 8.1, the total contributions of nonhydraulic nonbiomass renewable sources is 0.4%. Wind and solar utilization, however, have been growing rapidly, at more than 25% per year. Biomass conversion to liquid fuels is also gaining some momentum in developing countries.

8.2.2 Energy and how we live

Energy is strongly correlated with the quality of life, industrial productivity, abundance of agricultural harvest and clean water, convenience in transportation, and human comfort and

---

[1] Renewable sources have also been called nonexhaustable sources, which is a more technically sound but less frequently used label.

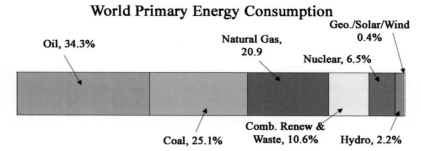

Figure 8.1 The breakdown of the world primary energy consumption in 2004. The total is 11,059 Mtoe (million ton oil equivalent). Except for hydropower, primary energy is meant to measure the thermal energy in the original fuel that was used to produce a useful form of energy (e.g., thermal energy [heat], mechanical energy, electrical energy). When energy is obtained directly in the form of electricity, an efficiency is used to convert it to equivalent thermal energy. *Source:* International Energy Agency (2004).

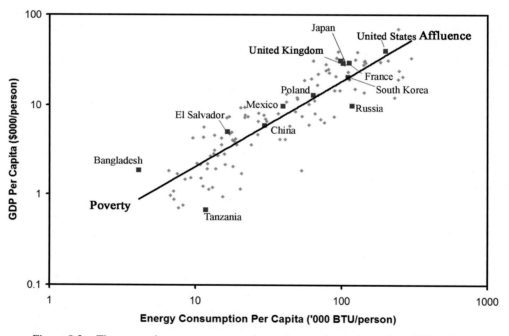

Figure 8.2 The per capita energy consumption and gross domestic product (GDP). GDP per capita is given for the year 2000, expressed in 1995 dollars. *Source:* United Nations Development Program (2003).

health. Our welfare depends on continuous and guaranteed supplies of different forms of energy, on demand and at different scales, at affordable rates and all the time. It has been shown that the per capita gross domestic product (GDP) correlates well with the per capita energy consumption, with developed countries consuming energy at orders of magnitude higher than those of developing poorer nations (Figure 8.2). Even among developed nations, some consume at multiple the rates of others (compare the per capita energy consumption of the United States and that of Japan). Although exact values vary, and the overall energy efficiency of developed countries has constantly improved as the productivity of their economies grew, there is still a significant gap between energy consumption in developed and developing countries. As is shown next, one can easily define an affluence index based on the per capita energy consumption. On average, energy consumption worldwide has grown by close to 1.55% per year from the mid-1980s to the mid-1990s, with U.S. consumption growing at 1.7%, China at 5.3%, and India at 6.6%.

Increasing the per capita GDP of a particular country goes hand in hand with its per capita energy consumption, primarily during the early stages of development. This trend slows down as the economy matures and becomes more energy efficient, as in the case of the United States, the European Union (EU), and Japan. Rising energy prices often promote conservation and slow down consumption, and the impact often persists even after energy prices fall back to more affordable levels. Transition economies are still in the fast consumption rise phase and have not shown moderating trends yet. Some developing countries have begun to take significant steps toward improving their economic conditions through industrialization, agricultural mechanization, and large-scale infrastructure improvement, and with that their energy use has been growing faster in the past few years. In particular, China and India, two of the largest countries in the world, have experienced fast rise in economic activities lately and a concomitant rise in energy production and consumption. Neither country is expected to reach a steady state in its energy consumption per capita soon because of the large population fraction that has yet to participate in the economic improvement.

Consumption patterns vary widely and depend on the economy, weather, and population density, among other factors. The United States, which consumes close to 25% of the energy used worldwide (with less than 5% of the population), consumed more than 100 EJ in 2003 (twice that of China and four times that of India, with both countries planning to double their consumption over the next 15–20 years). In the United States, of the total consumption, 26.86% went into transportation, 32.52% was used in industrial production, 17.55% was used commercially, and 21.23% for residential use (IEA, 2003). Sourcewise, 85% of the total energy in the United States, was generated from fossil fuels; 8% from nuclear; and the rest from renewable sources, including biomass, hydroelectric, and GWS (geothermal, wind, and solar, in that order). Currently, consumption is projected to rise over the next 25 years, with the fossil fuel share reaching 89%. At the other extreme, we note that nearly 25% of the world population does not have access to electricity, and nearly 40% relay on biomass as their primary source of energy.

It should, however, be noted that matching the energy use patterns and measures of developed countries is not necessary for developing nations to lift their standards of living and qualities of life beyond their current states. Attempting such a match would be an impossible goal, given the available resources and the associated costs. Higher qualities of life in developing countries could be achieved at lower energy intensity than currently seen in developed nations. For instance, it has been shown that the United Nations (UN) human development index, which includes data that reflect the physical, social, and economic health of a population such as the per capita GDP, education, longevity, use of technology, and gender development, rises steeply with the early stages of growth in the per capita electricity consumption, before it levels off at much higher electricity consumption rates (United Nations Development Program, 2003). Thus, even small but critical increases of energy availability where it is needed most can have enormous impacts and positive payoffs on people's standards of living without a parallel massive increase in energy consumption.

8.2.3 How much we will use

According to the 2005 *World Energy Outlook* published by the IEA (www.worldenergyoutlook.org/), the total worldwide energy use is expected to rise by more than

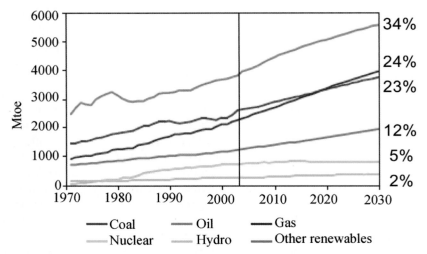

Figure 8.3 World consumption of energy since 1970 and projections to 2030. *Source:* Energy Information Administration, U.S. Department of Energy (2005).

50% over the next 25 years, while the share of the different raw sources is not expected to change significantly. The share of fossil fuels is predicted to grow slightly, as shown in Figure 8.3. Given the large matrix of parameters affecting energy supply and demand, economic conditions, and population growth, predicting the growth in energy consumption and the availability of energy sources is risky and prone to wide uncertainty. However, historical trends show that changes in consumption pattern occur rather slowly, given the large infrastructure that supports resource extraction, conversion, and supply, and the current patterns of energy utilization. Changes require massive investment and a population willing to support such investment. Changes also follow the discovery of more convenient forms, such as the rise in petroleum and fall of coal in mid-twentieth century, and technologies that enable the large-scale introduction of alternatives, such as nuclear energy. Economic factors largely rule energy source utilization and its penetration.

Fossil fuels reserves (those known to exist; that is, those that have been discovered and can be extracted under existing economic and operating conditions) and resources (those believed to exist, but their extraction requires different economic and operating conditions) have a finite lifetime, perhaps 100 to 300 years, depending on the fuel type, recovery rate, search and production technologies, exploration, and consumption rates.[2] Current predictions indicate that the lifetime of oil ranges from 50 to 75 years for the reserve, whereas resources are predicted to last for 150 years. Natural gas is expected to last longer, nearly twice as long as oil. Coal, however, is plentiful and is expected to last for several hundreds to thousands of years. These estimates are approximate at best because the recoverable amount of the resource is strongly dependent on the available technology, cost, and consumption pattern. With the current projections of reserves and resources, it is coal that will last the longest, with oil running out the fastest. Coal is available worldwide and in many of the fast-developing economies, such as China and India. Taking into account other heavy sources of hydrocarbon, such as oil shale and tar sands, recoverable liquid fuel estimates increase substantially. Other hydrocarbon resources include deep ocean methane hydrates, which are believed to be a viable vast source if the technology is developed for

---

[2] For more on the subject, see Goodstein (2004) and the website for the Association for the Study of Peak Oil & Gas at www.peakoil.net.

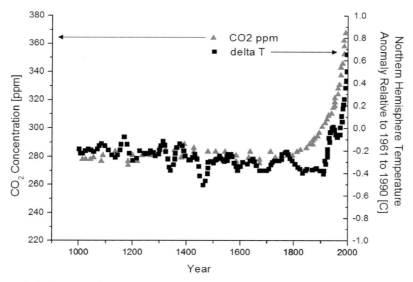

Figure 8.4 The rise in the atmospheric concentration of CO_2 and the global average temperature over the past 1,000 years, IPCC Third Assessment Report, 2001.

bringing them up without disturbing their original state or the health of the ocean. A case for the existence of abiogenic (nonorganic) methane in deep underground formation has been made, and if proven, would be another vast resource (Scott et al., 2004).

The growing evidence of the correlation between the global temperature and the CO_2, concentration in the atmosphere has prompted calls for increasing the use of low-carbon or zero-carbon energy sources, or preventing CO_2 produced in fossil fuel combustion from entering the atmosphere. Since the beginning of large-scale industrialization and the rapid rise in hydrocarbon consumption, atmospheric concentration of CO_2 has grown from 280 to 360 ppm. Electric power generation has been and will remain the major source of these gases, followed by transportation, with industrial and residential contribution following at smaller rates. Electricity generation plants use coal extensively, although the use of natural gas has been rising, and nuclear and hydraulic sources comprise a reasonable share. Because they are stationary, electricity generation plants should be considered as an easier target for reducing CO_2 emission per unit useful energy produced, through efficiency improvement or CO_2 capture and storage, as is explained later.

8.3 Carbon dioxide

The prospect of the rise in fossil fuel consumption, especially those with high-carbon content such as coal, oil, and other heavy hydrocarbons, has led some to warn against irreversible global warming and associated climate change, and prompted others to call for national and international intensive efforts to develop technologies that can generate the extra energy needed by the mid-twenty-first century, that is up to 10 TW in power capacity using carbonfree sources (Smalley). These concerns arose from demonstrated evidence that the rise of atmospheric concentration of CO_2 and the global average temperature are correlated, and that the rate of increase of CO_2 in the atmosphere may accelerate if the projected growth in carbon-based fuels is materialized (Hoffert et al., 1998). This correlation is shown in Figure 8.4. One of the striking features of this correlation is the simultaneous rise of CO_2 and temperature starting around the time of the onset of the Industrial Revolution,

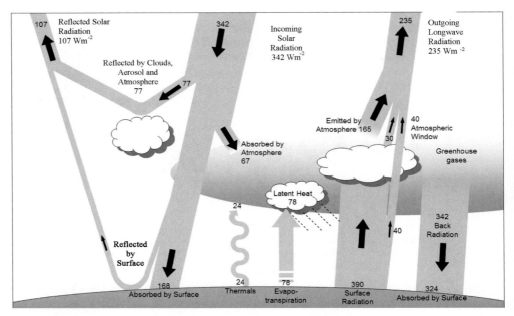

Figure 8.5 Solar energy flux: how much of it reaches the Earth's surface, the radiation emitted by the ground, and the balance that is reradiated back to the Earth's surface, from IPCC Report (2007).

when consumption of fossil fuel experienced a sudden acceleration that continues today.

8.3.1 Greenhouse gases

Global warming—that is, the observed rise in the Earth's surface and near-surface temperature by slightly more than 1°C over the past 150 years, essentially since the beginning of the Industrial Age—is believed to be related to the rise in the concentration of greenhouse gases in the atmosphere during the same period. Greenhouse gases are defined as H_2O, N_2O, CO_2, methane, CH_4, nitrous oxide, CFCs (chlorofluorocarbons), and aerosols. The greenhouse potential of CO_2, CH_4, N_2O, and CFC (taken as averages among different estimates) is 1:11:270:1,300–7,000 (with the latter depending on the particular CFC).[3] Most CO_2 emissions result from fossil fuel combustion, with a small fraction from cement production. It is predicted that continued emissions of greenhouse gases at the anticipated rates would lead to a rise of two to three times in their concentration in the atmosphere by the end of the twenty-first century, in proportion to the rise in energy consumption. Figure 8.4 shows the continuing rise of the global temperature over the past 150 years, following a long plateau since the year 1000, although a slowing trend was observed in the 1950s.

8.3.2 Energy balance

The energy fluxes to and from the Earth, and their change within its atmosphere are shown in Figure 8.5. Solar radiation is concentrated at short wavelengths, within the visible range of 0.4 to 0.7 micron because of the high temperature of the surface of the

---

[3] The global warming potential is a relative measure for the warming potential of different greenhouse gases, accounting for their lifetime in the atmosphere and relative radiative forcing strengths, all normalized with respect to CO_2, both having the same mass.

8.3 Carbon dioxide

sun, estimated to be approximately 6,000°C. Only a small fraction of solar radiation lies in the ultraviolet range, down to 0.1 micron, and in the infrared range, up to 3 micron. On average, 30% of the incoming solar radiation is reflected back by the Earth's atmosphere and surface (the albedo), 25% is scattered and absorbed by the Earth's atmosphere at different levels, and the remaining 45% reaches the surface and is absorbed by the ground and water. The fraction of the incoming radiation that is either absorbed or scattered while penetrating the Earth's atmosphere does so in a spectrally selective way, with ultraviolet radiation absorbed by stratospheric ozone and oxygen, and infrared radiation absorbed by water, carbon dioxide, ozone, O_3, nitrous oxide, and methane in the troposphere (lower atmosphere). Much of the radiation that reaches the ground goes into evaporating water. Outgoing radiation from the cooler Earth's surface is concentrated at the longer wavelengths, in the range of 4 to 100 microns. Greenhouse gases in the atmosphere absorb part of this outgoing radiation, with water molecules absorbing in the 4 to 7 microns range, and CO_2 absorbing in the range of 13 to 19 micron. A fraction of this energy is radiated back to the Earth's surface, and the remaining is radiated to outer space. The change of the energy balance due to this greenhouse gas radiation is known as the radiation forcing of these gases, and its contribution to the Earth's energy balance depends on the concentration of the greenhouse gases in the atmosphere. The net effect of absorption, radiation, and reabsorption is to keep the Earth's surface warm, at average temperature close to 15°C. In essence, the Earth's atmosphere acts as a blanket; without it, the surface temperature can fall to values as low as $-19°C$. Because of its concentration, CO_2 has the strongest radiation forcing among known greenhouse gases, except for that of water, which is least controlled by human activities. Increasing the concentration of greenhouse gases is expected to enhance the radiation forcing effect. Moreover, a number of feedback mechanisms, such as the melting of the polar ice (which reflects more of the incident radiation back to space) and the increase of water vapor in the atmosphere (due to enhanced evaporation) are expected to accelerate the greenhouse impact on the rise of the mean atmospheric temperature.

Current estimates indicate that fossil fuel combustion produces almost 6 GtC/y (this unit is used to account for all forms of carbon injected into the atmosphere, with carbon accounting for 12/44 of carbon dioxide; that is, 1 GtC is equivalent to $44/12 = 3.667$ $GtCO_2$). This should be compared with other sources/sinks that contribute to CO_2 in the atmosphere. As shown in Figure 8.6, CO_2 is injected into the environment through respiration and the decomposition of waste and dead biomatter, and is removed by absorption during photosynthesis and by the phytoplankton living in the oceans. Respiration produces nearly 60 GtC/y, while photosynthesis removes nearly 61.7 GtC/y, with a balance of a sink of 1.7 GtC/y. The surface of the ocean acts as a sink, contributing a net uptake of 2.2 GtC/y, a source/sink balance between production of 90 GtC/y and consumption of 92.2 GtC/y. Changing land use (deforestation) and ecosystem exchange adds/removes 1.4/1.7 GtC/y, for a net balance of a sink of 0.3 GtC/y. The overall net gain of CO_2 in the atmosphere is estimated to be around 3.5 GtC/y. It is within these balances that the contribution of fossil fuel combustion (and a small amount from cement production) appears significant. However, it must be stated that these numbers are somewhat uncertain, and there is 1 to 2 GtC/y unaccounted for in the overall balance, when all the uncertainties are traced. Nevertheless, the clear evidence is that CO_2 concentration in the atmosphere has risen, showing its most visible sign since the start of the Industrial Revolution, that is, with the significant rise in fossil fuel consumption.[4]

---

[4] It is estimated that for each 2.1 GtC introduced into the atmosphere, CO_2 concentration rises by 1 ppm, and the average lifetime of CO_2 in the atmosphere is 100 to 200 years.

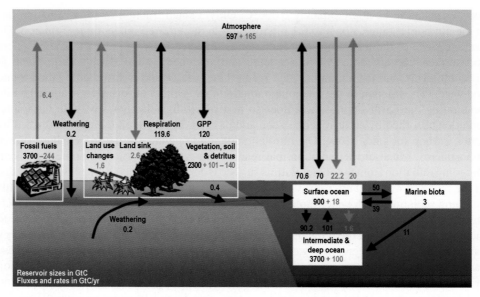

Figure 8.6 The Earth's carbon cycle and the contribution of fossil fuel utilization to it. Adapted from TPCC Report (2007).

8.3.3 Climate modeling

Global climate models, or global circulation models (GCMs), are used to estimate the change in the Earth's temperature and CO_2 concentration in the atmosphere, among other state variables such as pressure and wind velocity, given different scenarios for the introduction of CO_2 into the environment, solar radiation, and other parameters that could affect the atmosphere. These models integrate the time-dependent conservation equations, that is, total mass, momentum, energy, and chemical species, equations over a global grid that covers the entire surface of the Earth and extends vertically from the ground (including the ocean's surface) to some distance in the upper atmosphere (stratosphere) where boundary conditions are imposed (Figure 8.7). These conservation equations are tightly coupled. In particular, although the starting point is the Navier–Stokes equations, which govern the wind speed and pressure distribution, the energy equation, which is used to predict the local temperature, and a number of other transport equations that balance the change of the different chemical species that undergo mixing and reaction in the atmosphere, must be solved simultaneously to supply the source terms in that equation. The energy equation models the response of the atmosphere to the incoming and outgoing radiation across the computational domain, as well as the interior radiation forcing due to the greenhouse gases. It couples the impact of radiation to that generated by mixing, evaporation and condensation, and local reactions. The number of extra transport equations is determined by the number of chemical components, including gases and aerosols, which must be used to define the local chemical state of the atmosphere. The equations describing the GCM may be coupled to those describing ocean circulation models, which are used to predict the change of the water temperature, evaporation rate, and evolution of concentration of different gases within that vast body of water. This coupling adds not only to the accuracy of the overall prediction, but also to the numerical complexity and the computational requirements. Boundary conditions at the ground and on the water surface (or ice surface) must be supplied, depending on the nature of the terrain, the ground cover, and the season. Input regarding land use change is also necessary.

8.3 Carbon dioxide

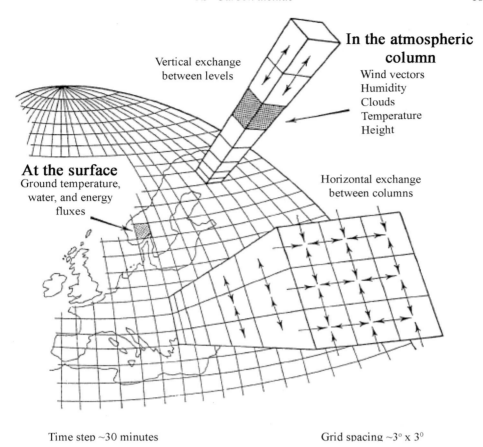

Figure 8.7 Spatial grid used to define a discrete representation of the governing transport equations (mass, momentum, energy, and chemical species) on the Earth's surface and atmosphere, the state variables, and the exchange fluxes at each grid volume. This representation is used in global climate (circulation) models. From Tester et al. (2005).

The solution of these coupled equations predicts the state of the atmosphere at any moment and location, that is, the wind velocity, pressure, temperature, and concentration of relevant gases, over many years. Atmospheric flows are turbulent, driven by local instabilities, and experience chaotic dynamics, and solving these equations on coarse grids, on the order of many kilometers in each coordinate direction, sacrifices resolution for affordability. Unresolved dynamics within the computational cells or across time steps are replaced by local mixing and transport models, and many details might be lost or simply averaged over to reduce the computational complexity. Many constitutive relations describing the transport relations and chemical kinetics are necessary to close the system of equations. The problem is compounded by uncertainty in the emission scenario, the model structure, and the modeling parameters. Because solutions are required for long integration time, modeling and numerical errors, and uncertainty in input parameters, may propagate and contaminate the results. Furthermore, the convective nonlinearities of the governing equations, even at the coarse-grid level, may lead to critical phenomena that depend sensitively on the initial conditions and the model parameters. Solutions could bifurcate to other regimes if some of these initial conditions change or the parameters deviate from their

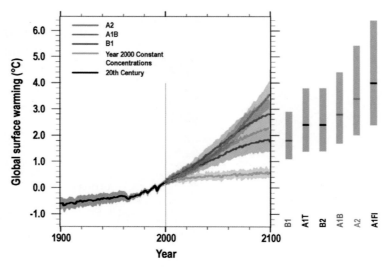

Figure 8.8 Intergovernmental Panel on Climate Change (IPCC) prediction of the temperature rise during the twenty-first century, according to different models that account for scenarios for the introduction of CO_2 into the atmosphere and its response to higher greenhouse gas concentration, IPPC (2007).

average values. Many "local" phenomena can also be unstable and can, if energized, trigger large-scale change, such as the disintegration of the large mass ice sheets. The Intergovernmental Panel on Climate Change (IPCC) predictions of the temperature trajectory during the twenty-first century are shown in Figure 8.8 for different CO_2 emission scenarios and model construction.

Given these complexities and uncertainties, one would like to study the sensitivities of the solution to many input parameters and to bound the response of the model to the known range of each input parameter. However, even with a coarse numerical grid, the computational load is enormous, and relatively few cases can be predicted at reasonable resolution, even on the fastest available supercomputers. Some further simplifications are often made to reduce the model complexity, such as eliminating variation in one of the dimensions, hence allowing more cases to be run and statistical analysis to be applied to the results. Such solutions are then used to construct ensemble probabilities for the different outcomes. An example of the results of such modeling is shown in Figure 8.9 in the form of the probability distribution function (PDF) of the predicted temperature change for these uncertainties. As shown there, the predicted rise of 2°C to 3°C is most probable, and higher and lower values are less probable, given the limitations of the model and input parameters.

Climate sensitivity, or the incremental change in the global mean climatological temperature resulting from the doubling of atmospheric CO_2 concentration, is still being debated, and model estimates indicate a range of 1.5°C to 4.5°C (Caldeira et al., 2003). Cloud feedback is the largest source of uncertainty in these model predictions, with aerosols, non-CO_2 greenhouse gases, internal variability in the climate system, and land use change being significant sources of variability. Uncertainty in aerosols radiative forcing remains large. Another source of uncertainty is the rate of heat diffusion into the deep oceans, given the sensitivity of the predictions to how much energy will be absorbed by this massive heat sink. Most predictions focus on CO_2-induced climate change because CO_2 is the dominant

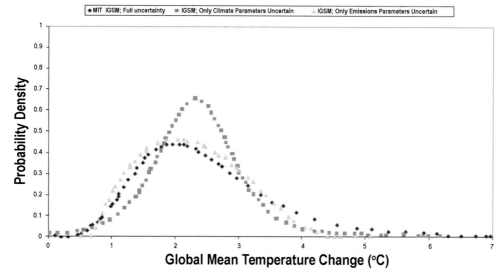

Figure 8.9 Probability distribution of global mean surface temperature change from 1990 to 2100 from all uncertain parameters (*solid line*), only climate model parameters uncertain and emissions fixed (*dotted line*), and only emissions uncertain with climate model parameters fixed (*dashed line*). From Webster et al. (2003). The integrated global system model (IGSM) is a model with intermediate complexity for modeling global circulation coupled with an ocean circulation model and supplemented with necessary emissions models.

source of change in the Earth's radiative forcing in all IPCC scenarios (Nekicenovic and Swart, 2000).

8.3.4 Global warming and climate change

More evidence is being cited for global warming (e.g., 19 of the warmest 20 years since 1860 have all occurred since 1980; 2005 was the warmest year since a record of actual measurements of the Earth's temperature was kept, and probably the warmest in the past 1,000 years (temperatures of back years were inferred from tree rings and ice core)). Data suggest that current temperatures are close to the highest values estimated to have been reached during the past 400,000 years, and that CO_2 concentration is even higher than the highest value estimated during this period. These records are shown in Figure 8.10. As shown in the plot, although wide variation of all three quantities, namely, the temperature (in this case, the temperature over the Antarctic), the atmospheric concentration of CO_2, and that of CH_4, have occurred cyclically during this period, they are all very well correlated. Prior to the 1800s, natural causes were responsible for CO_2 concentration variation. It is interesting to note that the time scales of rise and fall of the quantities of interest are very different; on the scale of the plot, the rise seemed to have occurred very rapidly, whereas the fall occurred slowly. Although cyclic variation over the geological time scale seems to be the norm, current CO_2 levels are higher than the peaks reached previously, and model prediction of future rise suggest more than doubling these value under business as usual scenarios (for CO_2 emission).

It is not easy to predict the course and precise impact of global warming on life on Earth, as defined by the rise of the surface and near-surface temperatures. For instance, although the average temperature will rise, thus extending the growth seasons of plants,

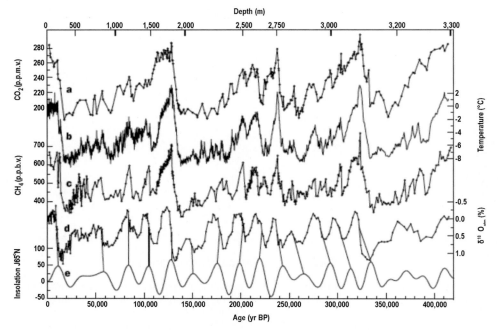

Figure 8.10   Time series of (a) CO_2; (b) isotopic temperature of the atmosphere; (c) CH_4; (d) $\delta^{18}O_{atm}$ and (e) mid-June isolation, Petit et. al. (1991).

it may also support the multiplication of pests that can destroy crops. Also, dry seasons may become longer, and droughts may become more frequent in areas already known for their hot climate and desert topography. Some animal habitats may become endangered, especially in the colder climates. The impact on energy consumption is unclear because more cooling may be necessary during warmer days, but less heat would also be required during cooler days. Melting of glaciers and ice caps would make more land available for agriculture, but water runoffs would damage the soil. Several of these effects would impact poor countries more severely, where adaptability and adjustment to substantial changes are less likely to be successful. Several major trends that could make a strong impact on life on Earth have been suggested with reasonable confidence, including sea-level rise, change of ocean acidity, and increase in violent weather phenomena.

Sea level rise

The sea level rise because of the melting of the polar ice caps, the receding glaciers, and the thermal expansion of the ocean surface waters is an important result of global warming. Records of different geological periods indeed confirm that the rise and fall of the Earth's near-surface temperature is associated with the same trend of the sea level. Estimates of the sea level variation during the twentieth century indicate a rise of nearly 20 cm, but actual values may be different because of the uncertainty associated with the techniques used in these estimates and associated measurements. Interestingly, the melting of the glaciers and ice caps may contribute least to the rise of sea level as the global temperature rises because of the balancing effect of increased evaporation. Most of the impact results from the warming of the surface layer of the ocean waters and its volumetric expansion. Combined, it has been estimated that by the end of the twenty-first century, with 1°C to 2°C rise in temperature, a 30 to 50 cm sea level should be expected (Raper and Braithwaite,

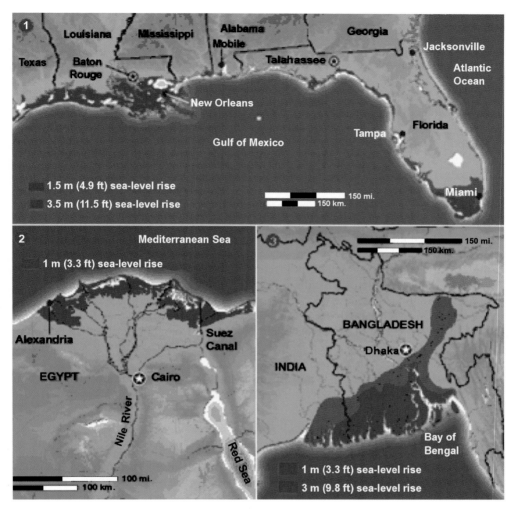

Figure 8.11 Estimates for flooded areas due to predicted sea level rise in the southern United States, northern Egypt, and Bangladesh. *Source:* www.cresis.ku.edu/research/data/sea_level_rise/index.html.

2006). More careful calculations show a total rise of 0.387 m, attributed to 0.288 m of thermal expansion, 0.106 m from the melting of glaciers and ice caps, and 0.024 m from Greenland and 0.074 m from Antarctica (Houghton et al., 2001). Estimates vary rather widely, depending on the melt models and the geometric ice volume models, and how volume shrinkage is treated.[5] These models feed into a global mass balance to account for melting, evaporation, and precipitation, and are coupled with radiation forcing models for the atmospheric temperature variation with regional adjustment, to predict the sea level rise on a reasonably resolved global grid. Sea level rise will have a devastating impact on coastal areas, especially agricultural land in the southern United States, India, Bangladesh, and Egypt, as shown in Figure 8.11.

---

[5] It is estimated that if all glaciers and ice caps melt, the sea level will rise by 50 cm, but the melting of the Greenland and the Antarctic ice sheet, whose ice is mostly above sea level and would require millennia to melt, could lead to a 68-m rise.

Change of ocean acidity

CO_2 absorption in the ocean lowers its pH levels, making it more acidic and impacting near-surface organisms as well as those living at deeper elevations. The average ocean water pH is 8.2, and it is estimated that the rise in atmospheric CO_2 has already lowered the pH by 0.1 from preindustrial levels. Ocean circulation models used in these studies include weathering of carbonate and silicate minerals on land; production of shallow water carbonate minerals; production and oxidation of biogenic organic carbon; production and dissolution of biogenic carbonate minerals on the ocean; air-sea gas exchange of carbon; and transport of all species by advection, mixing, and biological processes. These models predict a pH reduction of 0.7 units over the coming centuries if the current rise in CO_2 continues according to business-as-usual scenarios, and until fossil fuels are exhausted (leading to more than 1,900 ppm in the atmosphere by 2300). Meanwhile, there is no record that ocean pH level ever dropped below 0.6 units lower than their levels today. Because CO_2 solubility in water increases at lower temperatures and higher pressures, CO_2-caused acidity rise might increase at deeper levels, affecting acidity-sensitive corals, including strong reduction in calcification rates. The negative impact of higher acidity would compound the negative impact resulting from rising water temperature alone (which also lowers CO_2 solubility) because that further changes the ocean chemistry and the response of the bioorganisms. Although the full impact of these changes is still under investigation, and it will be centuries until these effects are fully observed, coral reef, calcareous plankton, and other organisms whose skeleton or shells contain calcium carbonate may be endangered sooner (Caldeira and Wickett, 2003). Higher water temperature has been shown to lead to bleaching of coral, killing the living organisms and leaving behind only their calcium carbonate skeleton.

Changes in weather phenomena

With warmer temperatures, on average, a more temperate climate will extend to higher latitudes, and extended periods of rain may occur, due in part to the higher water concentrations in the warmer atmosphere. Hurricanes and typhoons, spawned by waters warmer than 27°C within a band from 5 to 20 degrees north and south latitude, may occur more frequently. Ocean currents, such as the Gulf Stream and the equatorial currents, which are driven by surface winds and density differences in the water, could also become more frequent and violent. Some of these currents can be accompanied by phenomena that cause strong weather perturbations. For instance, El Niño, which arises because of wind-driven surface water currents westward from the South American coast and sets up ocean circulation in which upwelling of colder water replaces the surface warmer waters, is known to increase the frequency of hurricanes and heavy storms. Figure 8.12 shows the change in the total power dissipated annually by tropical cyclones in the north Atlantic (the power dissipation index [PDI]) against the September sea surface temperature (SST). A substantial and dangerous rise in the PDI is observed since the early 1990s.

Regional impact

Studies of the local/regional impact of global warming on climate change demonstrate the urgency for immediate action and the difficulties of determining with certainty the likely consequences. For instance, Hayhoe at al. (2004) conducted a modeling study of the impact of the rise of CO_2 levels to 550 ppm (likely to be achieved with aggressive intervention) or 970 ppm (likely to occur in the absence of mitigation policies) by the end

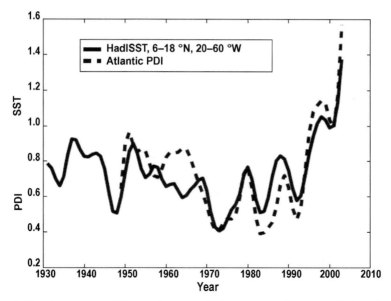

Figure 8.12 A measure of the total power dissipated annually by tropical cyclones in the north Atlantic, the power dissipation index (PDI), compared to September sea surface temperature (SST), measured over the past 70 years. The PDI has been multiplied by 2.1×10^{-12}, and the SST is averaged over 6–18°N latitude and 20–60°W longitude. North Atlantic hurricane power dissipation has more than doubled in the past 30 years. From Emanuel (2005).

of the twenty-first century on the state of California. Focusing on a small region not only allows the models to use finer grids in solving the governing transport equations, and hence achieve higher predictive accuracy, but it also requires the application of downscaling and rescaling methods to relate data and predictions at different resolutions. Statistical methods are often used for this purpose, given the nature of weather phenomena and the probabilistic approaches used in their description. The California study shows that, by the end of the century,

- A statewide temperature rise of 2.3°C to 5.8°C (from current averages of 15°C), with higher values predicted for the summers and under the higher emissions scenario, would be expected.
- This rise was associated with more heat wave days (rise of 50%–600% in extreme cases) and longer heat wave seasons.
- Heat wave mortality in Los Angeles was predicted to rise by a factor of 2 to 7.
- Although one extreme case (low emissions) showed a rise in annual precipitations, others showed up to 30% decrease, and with a drop of 30% to 90% in the Sierra Nevada Mountains snowpack.
- Accordingly, annual reservoir inflow would also decrease.
- Substantial loss in alpine and subalpine forests were also predicted, ranging from 50% to 90%.

The study concludes by stating that "[d]eclining Sierra Nevada snowpack, earlier runoff and reduced spring and summer streamflows will likely affect surface water supply and shift reliance to ground water resources, already overdrafted in many agricultural areas in California. Significant impact on agriculture and the diary industry follow" (Hayhoe et al., 2004, p. 12426).

UN and kyoto

In response to these threats, several actions have been suggested and some have been taken. Worthwhile mentioning here is the UN Framework Convention on Climate Change, signed in 1992. Its ultimate objective was to achieve stabilization of greenhouse gas concentration in the atmosphere at a level that would prevent dangerous anthropogenic interference with the climate system. Such a level should be achieved in a time frame sufficient to allow ecosystems to adapt naturally to climate change, to ensure that food production is not threatened, and to enable economic development to proceed in a sustainable manner. Several years hence, the Kyoto agreement, following intensive debates and deliberation in the UN conference on climate change, was proposed in 1997. Kyoto agreement, which called for reduction of CO_2 emissions to levels 5.2% below 1990 level by 2008 to 2012, was supposed to be enforced by 2005, but that did not occur. The agreement would have impacted the developed countries primarily, with 12.5% CO_2 reduction in the United Kingdom, 8% reduction in the EU, 6% reduction in Japan, and 7% reduction in the United States. Energy conservation efforts were accounted for in the calculations of the mandatory reduction. Although some efforts have been taken toward limiting CO_2 emissions, including considering the establishment of some form of tax on CO_2 emissions, as well as a trading system for CO_2 in some countries, the agreement was never enforced, and the target reductions are very unlikely to be achieved voluntarily in the near future. A combination of economic concerns and technology hurdles must be overcome before steps can be taken in that direction. Although it seems daunting, several proposals have been made to enable capping the emissions, some of which are reviewed in the next few sections.

8.4 CO_2 emission mitigation

The previous discussions show that continuing the process of releasing more energy-related CO_2 into the atmosphere may pose a serious and irreversible risk. Meanwhile, the current energy infrastructure is predominantly dependent on fossil fuels, and changing this infrastructure can only occur gradually, over many decades, and at substantial investment. Depending on the alternatives, other environmental costs should be considered. Thus, CO_2 reduction will have to be achieved while we continue to use fossil fuels as a primary energy source for several decades, although alternatives are being introduced and integrated into our energy infrastructure. Four major approaches to accomplish this task have been proposed and are listed as follows:

1. Improving the efficiency on the supply side (i.e., improving conversion efficiency from the raw sources to the useful form or end product, such as that of electricity generation power plants, vehicle engines and transmission, light bulbs, and all other devices that convert energy from one form to another).
2. Improving efficiency on the demand side, that is, on the energy utilization side through improved building insulation, using natural heating and cooling, use of public transportation, higher efficiency appliance, and so on. This includes city planning, agricultural practices, and water use.
3. Reduced dependence on high-carbon fuels by switching from coal to natural gas or other low-C hydrocarbon fuels, expanding the use of nuclear energy, and increasing reliance on renewable sources, including solar and geothermal sources for heat and electricity, some forms of biomass for fuel and electricity production, and wind and wave energy for electricity.
4. CO_2 capture and sequestration from power plants burning heavy hydrocarbons, directly by injecting CO_2 produced in such plants in deep reservoirs or reacting it into stable disposable chemicals, and indirectly by using biological approaches, such as growing trees and algae.

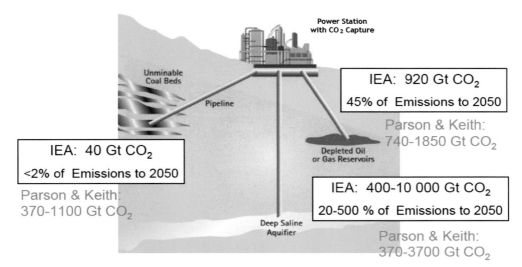

Figure 8.13 CO_2 sequestration potential in coal beds, depleted oil and gas reservoirs, and deep saline aquifers, as estimated by the International Energy Agency and a study by Parson and Kieth (1998). Figure originally prepared by C. Edwards (personal communication).

The four approaches presented here are discussed in some detail next, along with estimates of their impact on CO_2 emission and system efficiency. Perhaps the most unusual approach here is the last, that is, CO_2 capture from the combustion gases and its sequestration (CCS) in underground geological formations. Multiple sequestration strategies in large-capacity reservoirs have been identified, and studies and small-scale experiments are underway to examine their long-term potentials. The major reservoirs and their estimated capacity are shown in Figure 8.13, and several projects/experiments are currently underway to test this concept (with total capacity of 30 $MtCO_2/y$). For comparison, the IPCC estimates that the total cumulative 1990 to 2100 emissions of CO_2 from fossil fuel burning, using business-as-usual global energy scenario (IS92a), is 1,500 GtC. Moreover, the carbon content of "all" remaining exploitable fossil fuels, excluding methane hydrates, is estimated to be 5,000 to 7,000 GtC. Sequestration in the form of solid carbonate minerals have also been proposed (e.g., reacting forsterite [Mg_2SiO_4] with CO_2 via exothermic reactions favored under ambient conditions to form $MgCO_3$ [serpentine ($Mg_3Si_2O_5(OH)_4$)]. Although the resources are abundant to store all the expected-to-be-emitted CO_2 in the form of carbonate carbon and the method is safe, it requires large amounts of material to be transported and processed, making it rather expensive.

8.4.1 Implementing multiple solutions

CO_2 emissions reduction scenarios, using multiple approaches that would be implemented in parallel, collectively, to achieve the overall goal, have been suggested, domestically and at the global scale. Given the amount of CO_2 currently being produced, the anticipated rise in the emission rate and the inertia of the system against rapid change, it is highly unlikely that a single solution can be scaled up to be sufficient. Moreover, contrary to other regulated pollutants, such as CO, SO_2, and NO_x, which impact the local and regional environment, CO_2 footprint is global, and global solutions must be suggested, agreed to, and pursued on that scale.

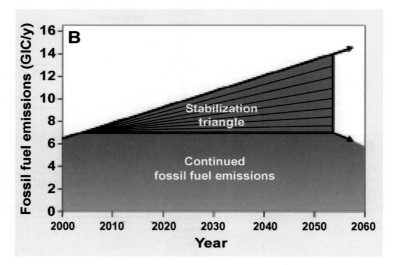

Figure 8.14   The top curve shows the rise in the yearly CO_2 emission, following a business-as-usual scenario in energy sources and cement manufacturing. The bottom curve shows an emission path that leads to stabilizing emissions at 7 GtC/y. The difference between the two curves is divided into seven "wedges," each enabled by one of the options described in Table 8.1. From Pacala and Socolow (2004).

8.4.2 The "wedges"

Pacala and Socolow (2004) describe a scenario for achieving the goal of stabilizing atmospheric CO_2 at the 550 ppm level by 2050 using existing technologies, but with some radical changes on how extensively certain technologies are deployed. Atmospheric modeling shows that the objective of reaching this level of CO_2 concentration could be achieved by holding CO_2 emission at 7 GtC/y over the next 50 years. Alternatively, the BAU rate of increase of 1.5% per year would double the rate of emissions to 14 GtC/y in the year 2050. The authors discuss a number of solutions; each one would prevent the emission of 1 GtC/y by mid-century. Note, for reference, that 1 GtC/y is produced by a 750 to 800 GW coal power plant at the current average efficiency of 34%, or a 1,500 to 1,600 GW NG power plant at the current average efficiency of 46%. Each different solution is expected to be deployed gradually, reaching full maturity in 50 years, but must start immediately to have the expected effect. They divide the different solutions among three categories:

- Improved conversion and utilization efficiency
- Shifting the fuel to lower carbon content
- CCS
- Deployment of renewable resources

Figure 8.14 shows the overall strategy, represented by seven "wedges," each leading to the reduction of CO_2 emission by 1 GtC/y by the year 2050. Deploying seven solutions should lead to the desirable goal of stabilizing the CO_2 emission rate at 7 GtC/y.

Table 8.1 summarizes some of the different solutions, and the necessary strategy for implementation of each solution. Several messages can be gleaned from Table 8.1. All options are somewhat available for deployment, if the will, economic incentive, and engineering is scaled up to the level indicated. All options require large-scale effort, starting with the best available technology that defines the current state of the art, but moving on to

Table 8.1. *CO_2 reduction through efficiency improvement, fuel shift, CO_2 capture, and renewable source*

Option	Technological solution	Needs
Improved conversion and utilization efficiency		
1. Efficient vehicles	Raise fuel economy for 2B cars from 30 to 60 mpg	Novel engine options; reduced vehicle size, weight, and power
2. Less use of vehicles	2B cars @ 30 mpg travel 5,000 instead of 10,000 m/y	Expand public transit options
3. Efficient buildings	One-fourth less emissions: efficient lighting, appliances, etc.	Higher fuel price, insulations, environmentally guided design
4. Efficient coal plants	Raise thermal efficiency from 32% to 60%	Technical improvement in gas separation, high-temperature gas turbines, etc.
Fuel shift		
5. NG instead of coal for electricity	Replace 1.4 TW coal (@ 50%) with gas at 4X current NG plant capacity	Price of NG
CO_2 capture (CCS) and nuclear energy		
6. In power plants	CCS in 0.8 TW coal or 1.6 TW gas (>3,000 time Sleipner capacity)	Improved technology
7. In H_2 production for transportation	CCS in coal producing 250 MtH_2/y or NG plants producing 500 MtH2/y (10X current H_2 production from NG)	Technology and H_2 issues
8. In coal to "syngas" plants	CCS in plants producing 30 Mbarrel/day (200X current Sasol capacity) from coal	Technology and price
9. Nuclear instead of coal for electricity	700 GW fission plants (2X current capacity)	Security and waste disposal
Renewable sources		
10. Wind instead of coal for electricity	Add 2M 1-MW peak turbines (50X current capacity) (need 30×10^6 ha, sparse and offshore)	Land use, material, offshore technology
11. PV instead of coal for electricity	Add 2 TW peak PV (700X current capacity) (2×10^6 ha)	Cost and material
12. Wind for H_2 (for high-efficiency vehicles)	Add 4M 1-MW peak turbines (100X current capacity)	H_2 infrastructure
13. Biomass for fuel	Add 100X current Brazil sugar cane or U.S. corn ethanol (need 250×10^6 ha; one-sixth of total world cropland)	Land use

Adapted from Pacala and Socolow (2004).

make this technology commonplace, economically viable across the world, and adaptable to different conditions. Some options still need proof of concept, such as carbon capture and sequestration from power plants, whereas others pose particular technology challenges in implementation such as using hydrogen extensively in the transportation fleet. Some solutions are interdependent, such as hydrogen produced from fossil fuels with CCS and used for transportation. Current rates of CO_2 underground injection, used primarily for enhanced oil recovery, will have to be scaled up by two orders of magnitudes to satisfy the needs for one wedge. For synfuel production, that is, production of liquid fuel from coal and other heavy hydrocarbons, the maximum capacity of the largest of such plants is that of Sasol of South Africa, which produces 165,000 bpd (barrels per day) from coal. Thus, a wedge requires 200 Sasol scale plants with CCS, and this accounts for the CO_2 production in the synfuel operation alone.

The challenges are not less daunting when we move to the use of carbonfree sources, such as nuclear energy, biofuel, solar, and wind sources. Note, for instance, that the production of biofuels currently consumes substantial amounts of fossil fuels, especially in mechanized agriculture, in the production of fertilizers and transportation of feedstock. If fossil fuels were to be replaced by biofuel, the need for land area, water, fertilizers, and so on would rise. Accelerated progress in improving the efficiency of biofuel production would be required to get closer to achieving these goals. Similarly, energy is used in the fabrication of wind turbines, their installation and maintenance, as well as in photovoltaics (PVs). To reach the capacity required for a single wedge, we need 700 times the current installed capacity of PV, and that does not consider the need for storage and the associated loss of efficiency during storage and recovery of energy, which depends strongly on the storage mode. The production of hydrogen using renewable sources would be even more energy intensive if the energy required for transportation and storage of this light fuel is taken into consideration. Other technologies are not mentioned but could have a similar contribution, such as solar thermal electric, which can be hybridized with fossil sources for dispatchability and higher overall efficiency. Some are even more futuristic, such as space-based solar power (Glaser et al., 1997).

The study of Pacala and Socolow (2004) also considers CO_2 sinks that can be expanded by reducing the deforestation rate and promoting the reforestation of clear-cut forests, especially in tropical areas. These solutions are not shown in Table 8.1 but are discussed in Pacala and Socolow. They estimate that one-half of a wedge would be created by reforesting near 250 million hectares in the tropics or 400 million hectares in temperate zones (while current areas of tropical and temperate forests are 1,500 and 700 million hectares, respectively). Better agricultural practices that reduce the decomposition rate of organic matter could also contribute to reducing the loss of soil carbon. The impacts should be considered temporal because decomposition is inevitable.

8.5 Low-carbon fossil conversion technologies

An effort to address two of the major concerns raised in the review of the recent trends in energy utilization and its impact on the environment, that is, the depletion of fossil fuel resources and the rise in CO_2 concentration in the atmosphere and its alarming consequences, must consider a number of external factors. These include the massive infrastructure employed for recovery, refinement, delivery, conversion, and utilization of this fossil-based energy, as well as the economic, social, political, and security concerns. Realistic strategies are likely to be based on gradual transitions toward more efficient and less carbon-intensive options. As shown in the discussion of the "wedges," multiple solutions that can be implemented in parallel are necessary, keeping in mind that solutions that fit into developed and developing economies might be different. Solutions that address the needs of remote and sparsely populated areas are different from those needed in heavily populated or industrialized areas. Viable CO_2 reduction solutions depend strongly on the availability of different forms of fossil fuels (e.g., NG vs. coal), the local sources of renewal energy, the public perception of nuclear energy safety, and the development of solutions to some of the outstanding problems associated with long-term waste storage and proliferation. Solutions will also be driven by policies and economic incentives, which could change the balance between centralized power and distributed power, or move it from fossil-based to renewable or hybrid-based systems.

Given these factors, a high-priority solution is improving the efficiency of energy and utilization, with the first referring to the production of useful forms of energy (e.g., thermal, mechanical, or electrical energy) from its original form (e.g., chemical or solar). The second refers to how efficiently the final product is being used (e.g., insulation of heated spaces and reduction of aerodynamic and other form of drag resistance in vehicles). Efficiency-related issues can be posed in the following question: "do we have an energy crisis or an entropy crisis?" In other words, are we using the source availability as well as we should, or are we wasting a good fraction of it during conversion or while it is being used to perform certain functions? Improving efficiency on both fronts prolongs the lifetime of available fuels and reduces their environmental impact. Improvements in efficiency are likely to come at the cost of improved systems that convert or use more of the available energy in a given source, at least at the early stages, and can only be offset be charging higher fuel prices, providing other monetary incentives, or relying on cultural and social attitudes.

8.5.1 Chemical energy

The conversion of chemical energy to mechanical or electrical energy is a rich field that offers significant opportunities for efficiency improvement and, with sufficient modification over the current practice, for carbon capture and sequestration (Hoffert et al., 2002). Although the efficiency of coal power plants has been rising because of the implementation of supercritical cycles, and regulated emissions such as NO_x and SO_x have been reduced significantly, CO_2 emissions per unit energy production from coal plants is highest among all fuels. Capture of CO_2 from coal and other fossil-powered plants for storage/sequestration, as well as use in enhanced oil and gas recovery and other industrial processes, is an attractive option for reducing CO_2 if geological storage is successful (Aresta, 2003). The use of natural gas has also expanded significantly over the past two decades. Natural gas (NG) plants have significant efficiency advantages because NG can be easily used in combined cycles and it produces less CO_2 per unit primary energy due to its lower carbon content. NG is also a clean burning fuel, and smaller plants can be built for distributed power, as well as and for combined heat and power, a much more efficient alternatives to centralized, often remote, plants.

Power plants and components that can reach 60% to 70% overall efficiency, measured as the electric energy output as a percentage of the fuel lower heating value, have been proposed. These plants incorporate tightly integrated, high-efficiency thermochemical (reforming, gasification, and combustion), thermomechanical (gas and steam turbines), electrochemical (high-temperature fuel cells) and possibly thermoelectric energy conversion elements (Figure 8.15). The large-scale deployment of these plants poses several challenges, including the development of high-efficiency components, the overall integration of the components, and the environmental control technology needed to remove pollutants and CO_2 from the exhaust stream and store the latter. It is currently possible to reach 55% efficiency in natural gas combined cycle plants, without the need for fuel cells. Advanced power plants employ advanced high-temperature gas turbines with inlet temperature close to 1400°C, integrated with supercritical steam cycle steam pressure exceeding 250 bar and 550°C. Practical efficiencies using natural gas are getting close to their thermodynamic limits.

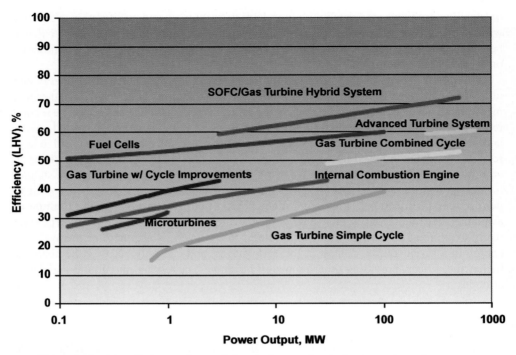

Figure 8.15 The efficiency of several chemical-to-mechanical energy conversion systems and its scaling with the power. The focus of the diagram is electric power generation. From *Fuel Cell Handbook*, Office of Fossil Energy, U.S. Department of Energy, Washington, DC, www.netl.doe.gov/technologies/coalpower/fiuelcells/seca/pubs/FCHandbook7.pdf.

Figure 8.16 shows the component layout of a plant that uses gasification to enable the utilization of a range of solid and liquid fuels, while incorporating high-temperature fuel cells, gas turbines, and steam turbines to maximize the overall conversion efficiency (total electric energy output as a fraction of the input fuel chemical energy). Using a gasifier, a mixture of coal (or other liquid and solid fuels), water, and oxygen is converted into a mixture of carbon monoxide and hydrogen, and other gases (and solids if the fuel is contaminated with uncombustible residues). The "syngas" is cleaned up to remove acidic and other undesirable gaseous compounds and solid residues, and is then used in the fuel cell to generate electricity directly at high efficiency. To avoid poisoning the fuel cell or damaging the gas turbine, the syngas must be free of sulfuric compounds, ashes, and other metallic components. The high-temperature fuel cell exhaust is used directly, or after the combustion of residual fuels, in a gas turbine. The hot exhaust of the gas turbine raises steam for the steam cycle. While using coal as a fuel, and because of the expected high conversion efficiency of the fuel cell, predicted efficiencies for these cycles are in the range of 50%.

8.5.2 CO_2 capture

Reduction of CO_2 emissions from power plants burning hydrocarbons by separating CO_2 has been the subject of extensive research recently, and several schemes have been proposed. Maximizing the plant efficiency counters the efficiency penalty associated with CO_2 separation (or other gas separation processes incorporated for the same goal, such as air separation

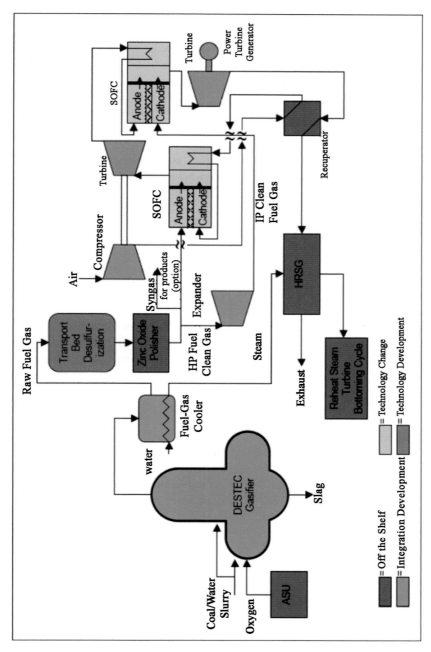

Figure 8.16 Layout of an integrated gasification combined cycle power plant, in which the conventional gas turbine-steam turbine combined cycle is equipped with "topping" high-temperature fuel cells to maximize the overall conversion efficiency.

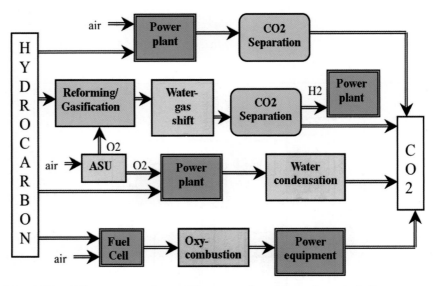

Figure 8.17 Different approaches to CO_2 capture from power plants, including postcombustion capture, oxyfuel combustion, and precombustion capture.

in some designs). Meanwhile, it is important to maximize CO_2 concentration in the stream before separation to minimize separation energy penalty. As shown in Figure 8.17, low carbon energy conversion schemes for power production include postcombustion capture, precombustion capture, oxyfuel combustion, and electrochemical separation. In terms of implementation, the first is the simplest, and requires the least modification of the power cycle, whereas the last is the least technologically developed. The first approach is most suitable for retrofit of existing plants because it requires least modification of the power plant itself. However, it is not necessarily the most efficient low-C plant layout, and that motivates the other options. The second and third options may require some special equipment, such as CO_2 gas turbine for oxyfuel combustion and H_2 turbine for precombustion separation. In the case of coal, precombustion separation requires gasification in an integrated gasification combined cycle plant. The last option relies on the development of robust and efficient fuel cells for high-temperature operation that are also affordable. In general, the efficiency penalty of CO_2 capture depends on the fuel, and the optimal design may not be the same for coal or NG. Note that in coal power plants, in the case of gasification, H_2S is captured from the flue gases before it can be used in a gas turbine or a fuel cell (or before it is emitted since sulfur compounds are heavily regulated). It is possible that the total acid gas (CO_2 + H_2S) can be removed in the same step (instead of removing the two components separately), thus by improving the economics of the capture strategy.

In postcombustion capture the decarbonization process, or the separation of CO_2 from flue gases, uses chemical absorption (Hendricks, 1994). Compressed NG, coal or coal-produced syngas, is used as a fuel. The estimated reduction in efficiency for coal (syngas) and NG are 6 to 10 percentage points and 5 to 10 percentage points, respectively. Given current efficiencies of coal and NG plants, this amounts to increasing the fuel use by 24% to 40% and 10% to 22%, respectively. In the oxyfuel combustion process, an air separation unit is used to deliver oxygen to the gas turbine combustion chamber (Gottlicher, 2004). The fuel here is either NG or syngas. The estimated reduction in efficiency for coal (syngas) and NG are 5 to 12 percentage points and 6 to 9 percentage points, respectively. Given current efficiencies of coal and NG plants, this amounts to increasing the fuel use by 24% to 27% and 22% to 28%, respectively. In precombustion power generation process, syngas is produced either in the natural gas reformer (NGR) or in a coal gasifier. Next, the syngas is cooled, and the cooled gas is introduced into the water-gas shift reactor to convert CO to CO_2 using steam. Following this step, CO_2 is separated from hydrogen, and the latter is burned in air. The estimated reduction in efficiency for coal (syngas) and NG are 7 to 13 percentage points and 4 to 11 percentage points, respectively. Given current efficiencies of coal and NG plants, this amounts to increasing the fuel use by 14% to 25% and 16% to 28%, respectively (Gottlicher, 2004).

8.5.3 Electrochemical separation

Finally, it is possible to use a high-temperature fuel cell, instead of combustion, to convert the chemical energy directly to electricity, especially when NG or coal-produced syngas are used as fuels. High-temperature fuel cells can achieve high efficiency, especially when used at low-current/low-power density mode. Moreover, they produce a product stream of $CO_2 + H_2O$, without nitrogen on the fuel exit side. The fuel gas is introduced on the anode side, and is electrochemically oxidized by the oxygen ions that migrate across the electrolyte from the cathode side to the anode side. On the anode side, CO_2 and H_2O form, without being contaminated with nitrogen, which stays on the cathode side. Thus, using a high-temperature fuel cell substantially reduces the need for an air seperation unit, although more pure oxygen might be necessary to burn the residual fuels leaving the fuel cell. The estimated efficiency penalty due to CO_2 capture is 6 percentage points. SOFC technology is under development (Hoffert et al., 2002).

8.5.4 Synfuel production

Similar decarbonization concepts can be applied to synfuel production plants, including those designed to produce hydrogen from coal or other heavy hydrocarbon sources, and for plants that might be used to generate electricity and synfuel from the same feedstock, simultaneously or on demand. Precombustion capture lends itself well to this application because many synfuel production processes (e.g., so-called "indirect approaches") start with the production of syngas ($CO + H_2$) using traditional gasification of heavier hydrocarbons in pure oxygen. If the plant is used for H_2 production, all CO_2 can be separated. Otherwise, only some would be separated following gasification, and CO would be used in the Fisher–Tropsch process. With CO_2 capture, and depending on the level of plant integration (heat and mass integration), the efficiency of hydrogen or other hydrocarbons production drops. Hydrogen can also be produced by water catalysis without CO_2 emission if the source of electricity is carbon free (e.g., nuclear energy), or directly from high-temperature heat (850°C), using thermochemical cycles.

8.6 Zero-carbon conversion technologies: nuclear and renewable sources

Zero-carbon energy sources are nuclear energy and renewable sources, such as hydraulic, geothermal, wind, solar, and biomass.[6] Nuclear energy is a scalable source that can supply a reasonable fraction of our future energy needs, and that can be integrated well into the

---

[6] Other forms include ocean tidal waves, waves, and ocean thermal energy, although none of them has yet made much impact on the energy resources. All forms of renewable energy originate in solar energy, except

exiting electricity generation and distribution infrastructure. Concerns of waste management and weapon proliferation, and over the public perception of safety, should be addressed before substantial expansion can be expected. As shown next, hydraulic power, which contributes a substantial fraction of renewable electricity, is near its peak and has its own share of environmental problems. Other forms have much lower energy and power density than fossil and nuclear energy, and are characterized by high but varying degrees of intermittency.[7] Biomass is used extensively in rural communities in developing countries to provide thermal energy. More recently, efforts to produce liquid transportation fuels from biomass have intensified, but the potential of bioenergy is limited by land and water resources. Most significant renewable sources are wind and solar, but many technical and economic challenges remain.

8.6.1 Nuclear energy

Nuclear energy currently provides 20% of the electricity needs of the United States, and more that 85% of that of France. Worldwide, it is estimated that nuclear energy supplies 6.4% of the primary energy (2.1% in the form of electricity), which amounts to nearly 17% of the electricity supplies. Nuclear energy has grown slowly because of the concerns over large-scale accidents, and the problems of waste disposal and weapons proliferation. Nuclear electric power plants, totaling about 500 worldwide, use uranium 235 (U-235), which is produced by enriching natural uranium. Light water reactors, both the pressurized and boiling types, represent the majority of current nuclear reactors, but some plants use gas-cooled graphite reactors. Progress has been made in designing passively safe reactors that reduce the chances of accidents, but current systems have yet to incorporate these designs on a large scale. The ultimate limitation on fission energy, besides the waste and security, is believed to be the fuel supply. Current estimates for the ground-based reserves and ultimately recoverable resources of U-235 translate to 60 to 300 TW-year of primary power.[8] Uranium can also be recovered from seawater, and large resources are known to exit thereon, but large-scale extraction has not been attempted. Plutonium 239 is produced during the uranium reactions in power reactors and can be separated from the spent fuel rods for use in sustained nuclear reaction for power generation, or for nuclear weapons. For this reason, reprocessing for spend fuels is currently banned in the United States and most other countries. Fast breeder reactors, such as liquid metal–cooled reactors can be used to produce plutonium 239 and other fissile isotopes, such as thorium 233.

8.6.2 Hydraulic power

Currently, an important source of renewable energy is hydraulic power plants built at natural waterfalls or river dams. There is close to 0.7 TW capacity installed worldwide. Expansion possibilities are limited, with the 18 GW Three Gorges Dam under construction in China

---

for geothermal energy (original hot gases that formed the Earth) and ocean tidal waves (gravitational). The notion of zero carbon power is relative, and for some forms such as biomass, fossil fuels are still used in their production.

[7] Typically, fossil fuel power flow within the components is in the order of 100 kW/m^2 or larger for high-speed propulsion. Renewable sources are three to four orders of magnitude lower, depending on the energy form. For instance, the average (total) solar power reaching the Earth's surface is 300 W/m^2.

[8] If all current energy needs were to be met using nuclear fission energy using available uranium, these estimates would translate to 5 to 25-year supply.

8.6 Zero-carbon conversion technologies: nuclear and renewable sources

being one of the last large-scale projects. Overall, this source, when near full utilization, is not expected to exceed 0.9 TW. The capacity might decrease if climate conditions change and lead to different rainfall patterns. Hydropower is seasonable, but large dams reduce the oscillation in power production between seasons by creating high-capacity reservoirs that regulate the flow of water into the power plants. Moreover, contrary to other renewable sources, hydropower is not intermittent on a day-to-day basis. However, hydropower is not without negative ecological impact, and large reservoirs of water created behind manmade dams can affect the local ecosystems. Downstream of a dam, soil can become less fertile because silt that used to replenish its nutrients is no longer able to flow. River fish populations can also be negatively impacted, and it has been recommended that some dams be removed to revive fish habitats.

8.6.3 Geothermal energy

A scalable renewable energy source is geothermal energy, which relies on drilling deep wells and building thermal conversion power plants that can take advantage of the relatively small temperature difference between the source and the environment. The efficiency of these plants is relatively low since the ideal (Carnot) efficiency itself is low because of the small temperature gradient. However, the potential capacity of this source is large, potentially exceeding 10 TW worldwide, if limited to the use of ground wells. The current installed capacity is less than 10 GW of electricity, and is limited by available and affordable well drilling technology. Most wells have a relatively small lifetime, 5 years on average, and new wells must be drilled to continue the plant operation. To reach its full potential, deeper wells, reaching down 5 to 10 km, will have to be used, and novel drilling technologies are under development for this purpose. Shallow sources of geothermal energy have also been used for heating and cooling. Concepts for hybridizing geothermal energy with fossil fuels are being considered to improve plants overall efficiency and extend their lifetimes.

8.6.4 Wind energy

Although wind and solar contribution to total energy production currently represent a very small fraction of the total supply, both have grown steadily over the past decade, in the range of 25% to 30% per year, and indications are that this trend will continue for some time. Part of the overall improvement in wind energy economics is associated with the design and installation of larger turbines, a trend that is expected to continue. Doubling the per turbine capacity is expected during the next decades, with further innovations such as actively controlled blade pitch for variable wind speed, and the installation of arrays of sensors and actuators to protect against wind gust and violent storms. Wind turbines with 5 MW capacity, at heights exceeding 120 m, have been proposed to harness wind speeds over a wider range of wind velocities. Larger-size turbines are favored in offshore installations, where the wind is stronger and less intermittent, and the impact on the local environment is minimized. Current effort to develop floating turbines, if successful, will be able to exploit the higher more sustained wind conditions deeper offshore, while taking advantage of the experience in building and maintaining offshore oil drilling platforms. Depending on the turbine size and the extent of the wind farm, wind energy technology offers solution for remote, off-grid applications; distributed power applications; and grid-connected central generation facilities. If located away from highly populated areas, wing turbine noise and visual impact can be minimized. Total potential wind capacity that can be used practically is believed to exceed 10 TW, including offshore locations.

8.6.5 Solar energy

Expanding solar energy utilization is important in the effort to meet rising energy demand while limiting CO_2 emissions. Solar thermal and solar thermal electric conversion over a wide range of scales for heat and power applications are important for distributed and centralized energy production (Mancini, 2003). Work is underway to scale up current trough-based technologies to higher capacity using the power tower concept, as well as scaling it down using the solar dish concept. Both of these three-dimensional concentration technologies achieve higher temperatures concentrated heat, and hence higher efficiency thermal conversion cycles. The power tower is intended for large-scale applications, in the O[100 MW] range and above, whereas the solar dish is intended for smaller more modular applications, O[20 kW]. In both cases, spherically shaped concentrators are used to concentrate solar radiation and raise the heat transfer fluid temperature further beyond what the trough can achieve, thus raising the thermodynamic efficiency of the cycle and simplifying the heat storage potential. Hybrid solar-fossil operation can be beneficial in making the plant operable under all conditions without the need for large-scale storage, and in improving the overall efficiency even when the solar energy is sufficient for normal operation. In hybrid operation, the temperature of the working fluid is raised using a combination of solar and fossil energy, making it possible to use higher-efficiency combined cycles.

Photovoltaics are convenient but expensive direct conversion devices that produce electricity from sunlight. Solar PV cells are solid-state devices, and hence require little maintenance. The efficiency of silicon type PV is 10% to 20%. Silicon-based PVs have been used for small, distributed power applications, but the relatively expensive price of their electricity (in excess of 5X fossil based) has hampered their widespread application for large-scale generation. Tax incentives are beginning to encourage adopting distributed solar power, and central generation is planned. Recent development in nanosctructured organic PV cells promises to lower their price and provide more flexibility in installation. Although they have lower efficiencies, they promise to be easier to fabricate, lighter in weight, and more adaptable. Grid-connected distributed PV applications could also become an attractive application for reducing the cost of installation, and tracking would be necessary to maximize power production. As in the case of wind, large-scale storage is necessary to overcome intermittency, especially in decentralized applications. Lack of storage limits the effort to take advantage of the vast solar potential.

More recently, approaches for the direct hydrogen generation from sunlight have been proposed (combined photovoltaic/electrolytic cells), and combined thermo- and photoelectric conversion in the same hardware.

8.6.6 Biomass energy

Biomass is the second largest source of renewable energy worldwide, following hydropower, and the oldest. Plants store energy through photosynthesis, converting radiation into chemical energy by combining CO_2 and H_2O into carbohydrates, such as sugar, starch, and cellulose, in photon-energized reactions. This energy can be converted back to other forms through combustion (as most biomass is currently used), gasification, fermentation, or anaerobic digestion. During this process, or in follow-up conversion processes, CO_2 is released back, making biomass conversion carbon neutral as long as no fossil fuels are used in its production. Although this may be the case in rural and developing economies, it is hardly the case in developed countries where fossil fuels are usually used in agriculture, transportation, and conversion of biomass into transportation fuels. In the event that biomass is used to produce liquid fuels such as ethanol for transportation, it is important that the heating value of the fuel produced is larger than that of the fossil fuel consumed in the production process, that is, the yield (output–input) is positive, or the chemical efficiency is larger than unity. This is the case for high-energy crops, such as sugar cane, and when most of the crop residues are used in the fuel synthesis process. With corn starch, estimates vary rather widely, and depend on the location of fermentation relative to production and on the processes involved in the conversion of starch to ethanol, as well as on the how the byproducts are valued (Patzek (2006)). More recently, efforts have gone into producing organisms for the efficient conversion of cellulose, hemicellulose, and lignin into ethanol, hence increasing the yield beyond that attained from the grain alone.

Similar to hydropower, biomass is not without negative environmental impact, such as the use of large quantities of water and fertilizers, insecticides, and herbicides; soil erosion; and impact on the ecosystems, such as deforestation. In general, the scalability of biomass is limited, given the photosynthesis efficiency, which is less than 1 W/m^2 power density (in thermal energy), more than an order of magnitude lower than that of wind and solar power density (in electrical energy), with the latter technologies producing electricity directly.

8.6.7 Renewable sources and storage

Expanding the use of renewable energy sources, which are characterized by large intermittency or interruptability on scales spanning hours to days (for solar and wind sources), to seasonal (solar, wind, biomass, hydro- and some geothermal), or to longer (some forms of fossil), to a measurable level would require substantial expansion in the use of high-capacity storage technologies. The challenge here is to either expand the use of high-capacity batteries in the case of wind and PV which generate electricity, or develop and use high-efficiency conversion technologies, such as electrolysis, to convert electricity into chemical forms, such as hydrogen. The latter also requires the development of efficient hydrogen storage technologies, another process that is likely to dissipate some of the available energy. Hydrogen can be consumed later to generate mechanical energy, or electrical energy directly in proton exchange membrane (PEM) fuel cells. "Reversible" or two-way PEM fuel cells/electrolysis can be designed for hydrogen production and utilization. Considering the efficiency in each step, the "roundtrip" efficiency, that is, from electricity back to electricity, is rather low. The use of solar thermal electric power simplifies short time storage of renewable resources because these systems can store thermal energy in the form of, e.g., molten salt, to be used later in running the power plant.

Higher-capacity storage options include pumped-hyro and compressed air. Compressed air storage are made compatible with wind turbines, by building a gas turbine power plant close to the wind farm and using the compressors to store some of the extra available electric power in the form of high-pressure air, to be used later to run the gas turbine and reproduce the electricity. Both systems can be hybridized with fossil fuels to overcome intermittency. The potential for chemical storage is discussed briefly later in this chapter, in the context of transportation. It should be mentioned that large-scale storage technologies have environmental footprints that should not be ignored in evaluating their performance, such as the toxicity of battery chemicals and the land use in hydro and air storage projects. Small scale, high energy, and power density, such as batteries, supercapacitors, and flywheels, are important for hybrid power trains for transportation.

8.7 Transportation

The most promising near-term solution for reduction of CO_2 emission in transportation is efficiency improvement, followed by the use of low-carbon fuels such as NG, the use of nuclear or renewable hydrogen to fuel IC engines or fuel cells when they become available, and electric cars if the source of electricity is carbon free. Fossil-based hydrogen is unlikely to reduce CO_2 emission because of the inefficiencies encountered in hydrogen fuel production process (near 60% or 80% when coal or NG are used, respectively), in hydrogen transportation and its onboard storage. All-electric vehicles, charged from high-efficiency fossil electricity (grid electricity near 60%), nuclear electricity, or renewable electricity, would be preferred as higher-energy density batteries become available. Besides engine efficiency, reduction of vehicle weight through the use of light weight material can have a high payoff. Further gain in CO_2 reduction can be attained if these improvements are done in parallel with the expansion of carpooling and public transportation, which themselves can be powered more by low-carbon fuels or electricity.

The efficiency of internal combustion engines has been improving steadily over the years, and diesel engines have reached rather impressive efficiencies. Although the trend is likely to continue, much higher efficiencies are not expected, and transition to different power trains, along with improvements in aerodynamics and reduction in vehicle weight, are necessary to achieve substantial improvement in fuel economy. Higher efficiencies can be achieved using hybrid power trains, in which the internal combustion engine operates near or at its point of maximum efficiency most of the time while the excess energy is stored when not needed and then drawn from the storage device as needed. The elimination of idling loss by allowing the engine to shut off when not needed reduces a significant source of losses. Higher overall efficiency in the hybrid power train is also enabled when regenerative breaking, which recovers and stores some of the breaking energy, is incorporated, but this requires a high-power storage device such as supercapacitors. Plug-in hybrids will have larger batteries with higher energy density that can power the vehicle for longer distances on electricity drawn from the grid (which received its energy from high-efficiency power plants, nuclear plants, or renewable sources) are the next step for these hybrid systems.

Transition to low-temperature PEM or similar fuel cells for transportation, which promises further increase in conversion efficiency, idle elimination, and zero pollution, will require the development of efficient, large-scale hydrogen production and mobile storage technologies. Production of hydrogen as a transportation fuel without CO_2 emission will require the transition from the current practice of producing hydrogen via methane steam reforming to methods that incorporate CCS, using technologies described previously. Renewable electricity can be used to produce hydrogen by electrolysis as well. Another potentially more difficult challenge is the mobile storage of hydrogen. For mobile storage, high-pressure tanks can only be used for limited driving distances, and the requisite high-strength material may add significantly to the weight of the vehicle. Cryogenic storage is energy intensive, using 40% to 100% of the energy that can be recovered from the stored hydrogen itself. Chemical storage of hydrogen in solid hydrides is a promising concept that is still under development. Another "chemical storage" format is regular liquid fuels, such as gasoline, which are essentially hydrogen carriers that can produce hydrogen using onboard reforming. Although reforming efficiency degrades the overall systems efficiency, many studies show better overall efficiency than direct storage. Figure 8.18 shows a comparison between the overall, well-to-wheels efficiency of different transportation options, including the fuel production and the power plant efficiency. Many assumptions go into these calculations, but the overall trends have been demonstrated by other studies.

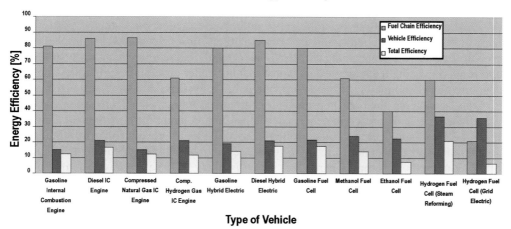

Figure 8.18 Well-to-wheels efficiency of different power trains using different fuels. The total efficiency includes both vehicle operation and the energy required to produce the fuel. Extracting oil, refining gasoline, and trucking the fuel to filling stations for internal combustion engines is more efficient that creating H_2 for fuel cells. From Wald (2004).

Wheel-to-wheel efficiency, obtained using techniques similar to life-cycle analysis, is the overall efficiency of using the fuel, and it is the product of the fuel chain efficiency, so-called "well-to-tank efficiency," and the vehicle efficiency, or tank-to-wheels efficiency. The two values vary widely depending on the fuel source and the drive train. Picking a clear winner is difficult given the assumptions used to obtain these numbers, but fuel cells using centralized production of hydrogen and diesel hybrid engines have some advantages, the former in the vehicle efficiency and the latter in the fuel chain efficiency.

A number of recent developments may help shape the transition strategy for personal transportation, including the introduction of ultra low sulfur diesel fuel, as well as advanced diesel exhaust clean-up technologies. Diesel engines are fully compatible with the existing fuel production and distribution infrastructure, vehicle design, and vehicle production, and are fully scalable in size and power. In larger vehicles, such as buses and trucks, compressed NG is being used more extensively to replace traditional diesel fuels. NG is easier to store because of its higher volumetric energy efficiency and is a cleaner burning fuel than many other liquid fuels. NG also produces the lowest amount of CO_2 per unit energy. Higher compression ratios can be used in NG spark ignition engines because of their higher octane number, and in compression ignition engines because of their higher ignition temperatures, further improving the engine efficiency. In some cases, mixing small amounts of hydrogen (hythane) could further improve the combustion properties of methane and reduce CO_2 emissions further. NG can also be used with onboard reforming to produce hydrogen for PEM fuel cells.

8.8 Conclusions

Current concerns over energy stem from the growing needs of the expanding populations that strive for better standards of living and compete for limited resources, and the strong evidence that continuing use of high-carbon fuels, without measures to reduce CO_2 emissions, may lead to irreversible damage as a result of global warming. The problem is particularly daunting because of the amounts of energy used and its time derivative, and the global nature of the problem and the solution. Besides the growing competition over limited resources, there is a need for global cooperation to find and implement solutions. A single approach will not be sufficient to address these concerns; rather, a portfolio of approaches must be applied simultaneously. Solutions must be endorsed and implemented globally because the scope of the problem is also global. Solutions are technically grounded and rely on existing technologies and some that are under development, but they must be supported by economic incentives and public policies to succeed in large-scale change.

Conservation has the highest priority. It is grounded in improving conversion and end-use efficiencies, both preserving energy resources and reducing the environmental impact. In almost in all applications, conversion efficiency can be improved, either by eliminating sources of losses or by taking advantage of waste energy. Improving conversion efficiency often requires more complex hardware, which is likely to raise the cost. Saving on fuel prices could make up for the extra upfront investment.

Improvements in the accuracy and comprehensiveness of global climate modeling, supported by improved physical submodels; the implementation of more efficient simulation techniques capable of incorporating the impact of uncertainty; and faster computational hardware are necessary to refine predictions related to global warming and its impact on our environment. Although historical records and current reliable predictions agree on the strong correlation among CO_2 emission, its concentration in the atmosphere, and the near Earth temperature, it is important to increase the reliability of predictions under different scenarios, both on a global scale and locally.

Carbon capture and sequestration from power plants, fuel production facilities, and other energy-intensive industries offers an opportunity to continue to use fossil fuels, while mitigating its contribution toward global warming. Given the plentiful supplies of coal and other heavy hydrocarbons, their cheap prices, and the massive infrastructure built around using these fuels or their derivatives, it is unlikely that a shift to alternatives will be sufficiently fast to avoid the predicted trends. Research, development, and demonstration are necessary to enable the widespread adoption of CCS.

Nuclear energy and renewable resources are necessary components. Nuclear power is a scalable resource that can satisfy a larger fraction of electricity generation, but concerns about waste, safety, and security need to be addressed. Only some forms of renewable energy are currently economically competitive, and others require large-scale storage that adds to the complexity of the system and cost of operation. Intensive effort is required on both fronts to increase the contribution of these sources to our energy demands.

References

Aresta, M. (2003) *Carbon Dioxide Recovery and Utilization*, Kluwer, Dordrecht, The Netherlands.

Caldeira, K., Jain, A. K., and Hoffert, M. I. (2003) "Climate Sensitivity Uncertainty and the Need for Energy without CO_2 Emissions," *Science* **299**, pp. 2052–2054.

Caldeira, K., and Wickett, M. (2003) "Anthropogenic Carbon and Ocean pH," *Nature* **425**, pp. 365–365.

Emanuel, K. (2005) "Increasing Destructiveness of Tropical Cyclones over the Past 30 Years," *Nature Letters* **436**, pp. 686–688. See also wind.mit.edu/~emanuel/.

Fuel Cell Handbook, Office of Fossil Energy, U.S. Department of Energy, Washington, D.C., 2004, 7th Edition, www.netl.doe.gov/technologies/coalpower/fiuelcells/seca/pubs/FCHandbooks7.pdf.

Goodstein, D. (2004) *Out of Gas: The End of the Age Of Oil*, W. W. Norton, New York, New York.

Gottlicher, G. (2004) *The Energetics of Carbon Dioxide Capture in Power Plants*, National Energy Technology Laboratory (NETL), Office of Fossil Energy, U.S. Department of Energy, Washington, DC.

Intergovernmental Panel on Climate Change, Working Group I: The Physical Basis of Climate Change (2007), ipcc-wg1.ucar.edu/wg1/report.

Hayhoe, K., Cayan, D., Field, C., Frumhoff, P., Maurer, E., Miller, N., Moser, S., Schneider, S., Cahill, K., Cleland, N., Dale, L., Drapek, R., Hanemann, R., Kalkstein, L., Lenihan, J., Lunch, C., Neilson, R., Sheridan, S., and Verville, J. (2004) "Emissions Pathways, Climate Change and Impact on California," *Proc. National Acad. Sci.* USA 1011, pp. 12422–12427.

Hendricks, C. (1994) *Carbon Dioxide Removal from Coal Fired Power Plants*, Kluwer, Dordrecht, The Netherlans.

Hoffert, M. I., Caldeira, K., Benford, G., Criswell, D. R., Green, C., Herzog, H., Jain, A. K., Kheshgi, H. S., Lackner, K. S., Lewis, J. S., Lightfoot, H. D., Manheimer, W., Mankins, J. C., Mauel, M. E., Perkins, L. J., Schlesinger, M. E., Volk, T., and Wigley, T. M. L. (2002) "Advanced Technology Paths to Global Climate Stability: Energy for a Greenhouse Planet," *Science* **298**, pp. 981–987.

Hoffert, M. I., Caldeira, K., Jain, A. K., Haites, E. F., Danny Harvey, L. D., Potter, S. D., Schesinger, M. E., Schneider, S. H., Watts, R. G.,Wigley, T. M. L., and Wuebbles, D. J. (1998) "Energy Implication of Future Stabilization of Amospheric CO_2 Contents," *Nature* **395**, pp. 881–884.

IPCC Third Assessment Report: Climate Change, 2001. www.ipcc.ch/pnb/reports.htm

International Energy Agency (IEA). (2005) *World Energy Outlook*, IEA, Paris, France, www.worldenergyoutlook.org.

Mancini, T.R. (2003) "An Overview of Concentrating Solar Power," Sandia National Laboratories, Albuquerque, New Mexico. www.eia.doe.gov/cneaf/solar.renewables/page/solarreport/solarsov.pdf.

Nakicenovic, N., and Swart, R. (2000) *Special Report on Emissions Scenarios*, UN's Intergovernmental Panel on Climate Change, Cambridge University Press, Cambridge, United Kingdom.

Pacala, S., and Socolow, R. (2004) "Stabilizing Wedges: Solving the Climate Problem for the Next 50 Years with Current Technologies," *Science* **305**, pp. 968–972.

Parson, E. A., and Keith, D. W. (1998) "Fossil fuels without CO_2 emissions: Program, prospects and policy implication." *Science* **282**, pp. 1053–1054.

Patzek, T. W. (2006) "Thermodynamics of Corn–Ethanol Biofuel Cycle," www.petroleum.berkeley.edu/papaers/patzek/CRPS416-Patzek-Web.pdf.

Petit, J. R., Jonzel, J., Raynand, D., Barkov, N. J., Barnola, J. M., Basile, J., Bender, M., Chappellaz, J., Davis, M., Delaygue, G., Dollmolte, M., Kotlyakov, V. M., Legrand, M., Lipenkov, V. Y., Lorius, C., Pepin, L., Ritz, C., Saltzman, E., Stievenard, M. "Climate and Atmospheric History over the Past 420,000 Years from the Vostok Ice Core," *Nature*, **399**, June 1991, 429–436.

Raper, S. C., and Braithwaite, R. J. (2006) "Low Sea Level Rise Projections from Mountain Glaciers and Icecaps Under Global Warming," *Nature* **439**, pp. 311–313.

Scott, H. P., Hemley, R. J., Mao, H., Herschbach, D. R., Fried, L. E., Howard, W. M., and Bastea, S. (2004) "Generation of Methane in the Earth's Mantle: In Situ Pressure–Temperature Measurements of Carbonate Reduction," *Proc. National Acad Sci.* USA **101**, pp. 14023–14026.

Smalley, R. E. (2003) "Our Energy Challenge," http://smalley.rice.edu.

Tester, J. W., Drake, E. M., Driscoll, M. J., Golay, M. W., and Peters, W. A. (2005) *Sustainable Energy: Choosing Among Options*, MIT Press, Cambridge, Massachusetts.

United Nations Development Program. (2003) *Human Development Report*, Oxford University Press, London, United Kingdom.

Wald, M.L. (2004) "Questions about a Hydrogen Economy," *Scientific American* **290**, pp. 66–73.

Webster, M., Forets, C., Reilly, J., Rabiker, M., Kicjlighter, D., Mayer, M., Prinn, R., Sarofim, M., Solokov, A., and Wang, C. (2003) "Uncertainty Analysis of Climate Change and Policy Response," Climate Change, 61: 295–320.

9

Seawater agriculture for energy, warming, food, land, and water

Dennis M. Bushnell

The combination of the incipient demise of cheap oil and increasing evidence of global warming due to anthropogenic fossil carbon release has reinvigorated the need for and efforts toward renewable energy sources, especially for transportation applications. Biomass/biodiesel appears to have many benefits compared to hydrogen, the only other major renewable transportation fuel candidate. Biomass production is currently limited by available arable land and fresh water. Halophyte plants and direct seawater/saline water irrigation proffer a wholly different biomass production mantra—using wastelands and very plentiful seawater. Such an approach addresses many to most of the major emerging societal problems, including land, water, food, global warming, and energy. For many reasons, including seawater agriculture, portions of the Sahara appear to be viable candidates for future biomass production. The apparent nonlinearity between vegetation cover and atmospheric conditions over North Africa necessitates serious coupled boundary layer meteorology and global circulation modeling to ensure that this form of terra forming is favorable and to avoid adverse unintended consequences.

9.1 Introduction

Beginning with the technological development of fire in the human hunter–gatherer period, biomass was, until the 1800s, the dominant energy source. The subsequent development and utilization of fossil fuels, including coal, oil, and natural gas, powered/enabled tremendous technological progress and major increases in societal population and wealth. The consequent emission into the atmosphere of the products of combustion of these fossil fuels has now altered the atmospheric composition sufficiently to affect the planetary radiation budget, causing increased global warming. Ice cores indicate that atmospheric CO_2 concentration is greater today than at any time in the past 650,000 years. Influences of this warming are already apparent according to some analysts, including some 5 million cases of illness and 150,000 deaths/year. Potential impacts of global warming include Arctic/glacier melting, alteration of species and disease patterns, altered ocean pH, heat waves, floods, increased incidence of and more severe storms, rising ocean levels, droughts, enhanced pollution, extinction of millions of terrestrial species, and resultant tremendous economic impacts. The Arctic region tundra melting is releasing immense amounts of fossil methane, which is a global warming gas with twenty-two times the impact of CO_2, thus greatly accelerating the warming process. Nominally, some 75% of the world's energy is currently generated using fossil fuels (Adams, 2002). Because plants take up CO_2 during growth and then redeposit the CO_2 back into the atmosphere, they are a renewable/green energy source. The use of fossil (vs. renewable) carbon is the warming issue. There are serious calls for and work on CO_2 sequestration, potentially a major engineering and economic endeavor. Potential global warming solutions include green energy (one approach to which is discussed herein), conservation, and genomic biologicals with greatly increased CO_2 update/storage. There is also another set of solutions of the megaengineering genre, for example, trigger calderas (nascent volcanoes) to put massive amounts of particulate matter into the atmosphere, spread nanoparticles on the monolayer of surfactants on the ocean surface to alter ocean albedo, gigantic reflective films/membranes in orbit, and seeding the oceans with iron to provoke/enable phytoplankton blooms.

Global warming is one of the major current concerns regarding use of fossil fuel; the other is the incipient demise of cheap (high-quality, sweet) oil. There are immense deposits of fossil methane and coal still extant, but oil production has peaked in many areas and is expected to peak within 10 years or less in most others, at a time when demand for oil is rising rapidly—driven in a major way by the phenomenal growth of the Asian economies. This combination of reducing production and increasing demand will inexorably drive up petroleum costs. Oil is currently the fuel of choice for transportation. Alternative transportation fuels, at higher costs and with a warming penalty, could be extracted from coal. Other transportation fuel alternatives include hydrogen and biomass/biodiesel. Hydrogen has major production and storage problems and, for global warming aversion, H_2 production should be via renewables. Biomass/biodiesel has production problems associated with increasing shortages of suitable arable land and sweet water. There are suggestions that fresh water scarcity is now the single greatest threat to human health, the environment, and the global food supply.

9.2 Biomass and the Sahara

Of the renewables solar, wind, hydro, and geothermal, only solar and geothermal have the potential to provide the requisite capacity to replace fossil energy sources and solve the global warming problem (Touryan, 1999). Solar energy can be used in many ways, including direct heating, photocatalytic disassociation of water, hydrogen production from genomic and artificial photosynthesis, electricity production from photovoltaics (PV; including the emerging nano/plastic PV), and biomass (Smith et al., 2003). As a potential transportation fuel, biomass/biodiesel requires no new storage- or transportation-specific infrastructure as does H_2, has minimal sulfur, and is relatively inexpensive, competitive with oil at the current nominal oil price of $90+/barrel. Biomass availability is currently limited by a combination of shortages in arable land/sweet water and suitable processing plants. Due to the cost of transporting biomass, the requisite processing plants should be distributed/located near the biomass production sites. Work on bioreactors at the Pacific Northwest National Laboratory in Richland, Washington, and elsewhere indicates bioreactors scale well down to individual holding size utilization/cogeneration. Biodiesel is obviously transportable via the current oil transportation infrastructures (e.g., pipelines, tankers). Biorefineries currently under development include biochemical/fermentation, thermochemical/pyrolysis/gasification, and chemical (Klass, 2004; Morris, 1999).

The potential biomass utilization spectrum encompasses transportation fuels (biodiesel, H_2), heat generation, electricity generation via biofuel cells or conventional steam plants, food, and petrochemical feed stock for plastics. Biomass grown on less than the land mass of the Sahara could supply/replace the world's fossil energy requirements.

As stated previously, shortages of water and arable land currently limit biomass production. This is especially true for the Sahara, which constitutes a portion of the 44% of the planet's land mass considered to be wasteland due primarily to shortages of sweet water. Other such areas include the Middle East, western Australia, and the southwest United States. All of these areas are adjacent to seawater resources. The Sahara's fresh water availability has decreased by a factor of three since the 1950s. The region is increasingly resorting to inherently expensive long-distance water transfer and desalinization, and therefore would appear to be an unlikely candidate for major biomass production. The Sahara does have vast distributed underground aquifers, largely transnational and currently underused (Shahin, 2002). These aquifers are often saline and are becoming more so. Seawater intrusion into coastal aquifers is an increasing issue worldwide. Aquifer utilization is causing land salinization with some 20% of irrigated land in the region affected by salinity, and the percentage growing rapidly. The water age in these aquifers is some 20,000 years, and there is little replenishment. As an example of the sufficiency of these aquifers for biomass production, the Nubian sandstone aquifer in the eastern Sahara would provide only about 60 years of biomass production, altogether not a long-term solution.

A review of the resources of the Sahara region indicates coastlines and sunlight as major advantages. The sunlight could be used either for direct electricity production via the emerging nanoplastic inexpensive and potentially highly efficient photovoltaics or, given suitable water, biomass production. The electricity could be used to produce H_2 directly via electrolysis using saline or salt water. Direct photocatalytic electrolysis could also be used on the Sahara for H_2 production.

9.3 Saline/salt water agriculture

Conventional wisdom throughout most of the world is that saline water/soil is detrimental to disastrous for agriculture. However, there are indications, both historical and recent, that saline/salt water agriculture is a viable-to-desirable alternative to conventional agriculture in at least portions of the Sahara and other arid areas. In several areas around the globe, people have used a class of plants termed halophytes (salt plants) and brackish/saline water for both food and fodder and to reclaim/desalinate land (Ahmed and Malik, 2002; Glenn et al., 1998; Khan and Duke, 2001; National Research Council, 1990; Yensen, 1988). The primary advantage of seawater agriculture is that 97% of water on the planet is seawater, which is unlikely to run out. Also, seawater contains a wide variety of important minerals and 80% of the nutrients needed for plant growth. Nitrogen, phosphorus, and iron are the required additives. Nitrogen fixation from the atmosphere would be the approach of choice for this important growth requirement. There are actually four methods of using seawater for agriculture. Of these, desalinization is, in general, too expensive for agriculture. Deeper/colder ocean water can be used to precipitate moisture from the local atmosphere, and seawater greenhouses use sunlight to vaporize/reprecipitate the seawater. The fourth is direct seawater irrigation using halophyte plant stock.

There are about 10,000 natural halophyte plants, of which about 250 are potential staple crops. Seawater agriculture possibilities are far richer than the conventional approaches using mangrove trees and salicornia. Genomic/bioresearch is ongoing worldwide to enhance the overall productivity of halophytes with the goal of halophilics (salt loving), the more salt the better. Huge land areas worldwide are already salt affected, and major regions overlie saline aquifers. Also, rising sea levels due to global warming are/will increasingly inundate coastal aquifers with seawater. More than 100 halophyte plants are now in trials for commercial applications. Nearly twenty countries are involved with saline farming experiments for food production. Much of this effort is assessable via the website for the International Center for Biosaline Agriculture in Dubai, United Arab

Emirates (www.biosaline.org/html/index.htm). In particular, the Chinese have reported genomic versions of tomato, eggplant, pepper, wheat, rice, and rapeseed grown on beaches using seawater. The outlook for genomic-derived halophyte enhancements appears to be quite favorable, with considerable improvements thus far and research still in the early stages regarding enhanced growth rates, reduced water/nutrient requirements, and plant optimization for specific biorefining processes.

Rough estimates regarding the feasibility of using seawater to irrigate the Sahara for food and primarily biomass are, with oil at \$70/barrel, not ridiculous, and for certain areas appear to be favorable. The higher oil prices predicted for the future improve the economics. A 1.2 acre-meters of water per year are nominally required to produce biomass. An acre might produce 10 to 30 tons of biomass/year, and a ton of biomass is equivalent, energy-wise, to nearly three barrels of oil. Therefore the estimated value of an acre of biomass, after refining, could be as high as \$3,300 or more. It would appear to be worthwhile to pursue further definitization of Sahara biomass production using seawater irrigation near dry and flattish coastal areas, as well as inland wherever the seawater pumping economics are reasonable. Saline irrigation could also be used wherever saline aquifers are available at reasonable pumping depths. It should be emphasized that what makes such an approach interesting now is the increasing cost of oil and the unique characteristics of biodiesel versus hydrogen as a transportation fuel (utilization of existing infrastructures and lack of storage problems). Given the innate desert advantage of plentiful and intense sunlight, the emerging inexpensive plastic/nano photovoltaic could perhaps be used to generate inexpensive pumping power for more inland and higher-elevation seawater irrigation activities. The average inland Sahara elevation is about 450 m, with a vertical lift pumping cost the order of \$1,200 for the requisite 1.2 acre-meters, in the absence of less costly solar power.

An initial very related experiment along the lines of seawater agriculture for global warming mitigation is underway in Eritrea on the horn of Africa—the Manzanar Project (Sato et al., 1998). The approach combines seawater aquaculture (algae growth/conversion to fish and shrimp) and direct seawater irrigation of mangrove trees, with the addition of iron and ammonium phosphate for timber and animal fodder. The project is an obvious existence proof for the overall approach advocated herein. In addition, there is the Eritrea's seawater farms project that incorporates the cultivation of salicornia using seawater directly for human and animal consumption. The credo of this project envisions seawater agriculture as a solution to the global warming problem.

Probably the study most closely related to the present one is that of Glenn et al. (1992), which suggests seawater/saline agriculture for carbon sequestration, energy, food, water, and land. The work provides detailed suggestions for siteing of such farming activities including northeast India, the southeast and lower Indus river regions of Pakistan, the salt flats on the sea of Oman and the Persian Gulf, and the sizable percentage of currently irrigated lands that have become saline. Suggested inland deserts include the Caspian Sea and Aral Sea environs and the Lake Eyre basin in Australia. It is noted therein that 22.5% of the world's arid regions have no inherent soil constraints for growing crops except lack of water.

9.4 Additional impacts/benefits of saline/seawater agriculture

seawater/saline agriculture using improved genomic versions of halophytes offers the potential to significantly contribute to the solution of many to most of the current/emerging societal/global problems. The obvious first-order contribution(s) concern/address global warming, energy, and food. Additional issues addressed by saline/seawater agriculture include land, water, and minerals. Seawater irrigation can convert major areas of the planet's wastelands into productive land, thus addressing the increasing shortage of arable land. By substituting, in many cases, seawater for fresh/sweet water in agriculture, seawater agriculture returns some of the 66%+ of the available fresh water we now use for conventional agriculture crop irrigation back to/for other human uses, creating a major favorable potential impact on the increasing water scarcity problem. In the minerals arena, conventional mining is one of the most environmentally damaging activities conducted by humans. seawater contains a large number of minerals, which could both put missing/trace minerals back into the food supply and be recovered after evaporation. Current seawater mineral extraction activities involve magnesium, bromide, salts, phosphorites, and metallic sulfides. There are also nascent low-cost bio/algae extraction technologies. Although it is projected that the nature of the soils in coastal deserts, such as the Sahara, Peru, Australia, the Middle East, and southwest United States, should allow much of the salts to leach back into the ocean, technological processes could probably be developed to capture/use such if the economics are favorable.

seawater irrigation on coastal deserts could also have impacts on local-to-regional atmospheric water and heat balances (Boucher et al., 2004; Glenn et al., 1992; Pielke et al., 1993). Such irrigation produces cool and wet surfaces that increase low-level atmospheric instabilities, leading to increasing incidence of storms. Various studies indicate that irrigation and associated vegetation changes have a direct influence on atmospheric moisture content and produce increased rainfall. Therefore, irrigation could be considered a mild form of terra forming. For North Africa, there are some indications that greatly increased vegetation cover could affect continent-scale atmospheric motions because the interaction between the atmosphere and land cover appears to be nonlinear (Raddatz and Knorr, 2004). Therefore, serious predictive computations and model studies should probably be undertaken previous to any large irrigation efforts in the region to ensure that any unintended consequences are favorable. Such studies should include the atmospheric particulate loading, which can allow cloud formation at 0.1% saturation (Mason & Ludlam, 1951).

9.5 Summary

The combination of oil price increases and global warming drives efforts toward renewable energy sources, especially for transportation applications. Biofuels appear to be preferable to hydrogen, the only other major candidate for renewable transportation fuel. Biomass production is currently limited by available arable land and fresh water. Conventional wisdom indicates that saline soil and water are highly detrimental to devastating to sweet water agriculture. There is, however, an alternative approach to agriculture using bioengineered versions of the more than 2,000 natural halophytes (salt plants). Experiments indicate that saline/seawater agriculture using this plant stock can provide conventional agriculture productivity using what are regarded as deserts or wastelands (some 44% of the planet's land mass). Many of these deserts are adjacent to salt water oceans and seas. Fuels produced from the enabled/resulting biomass are CO_2 neutral (plants take up the CO_2) and viable alternatives to increasingly expensive petroleum. To the extent that such saline agriculture replaces conventional agriculture for food production, fresh water would be released back for direct human use. Thus far, saline agriculture is essentially an experimental food producing activity worldwide, with considerable promise for major productivity increases. Halophyte plants and seawater irrigation proffer a wholly new biomass production mantra—using wastelands and very plentiful seawater. Such an approach addresses many to most of the major societal problems, such as land, water, food, global warming, and energy. For many reasons, including seawater agriculture, portions of the Sahara appear to be viable candidates for future biomass production. The apparent nonlinearity between vegetation cover and atmospheric conditions over North Africa necessitates serious coupled boundary layer meteorology and global circulation modeling to ensure that this form of terra forming is favorable, thus avoiding any adverse and unintended consequences.

References

Adams, M. L. (2002) "Sustainable Energy from Fission Nuclear Power," *The Bridge* **32**, no. 4, pp. 20–26.
Ahmad, R., and Malik, K. A. (2002) *Prospects for Saline Agriculture*, Kluwer, Dordrecht, The Netherlands.
Boucher, O., Myhre, G., and Myhre, A. (2004) "Direct Human Influence of Irrigation on Atmospheric Water Vapor and Climate," *Climate Dynamics* **22**, pp. 597–603.
Glenn, E. P., Brown, J. J., and O'Leary, J. W. (1998) "Irrigating Crops with Seawater," *Scientific American*, August, pp. 76–81.
Glenn, E. P., Hodges, C. N., Lieth, H., Pielke, R., and Pitelka, L. (1992) "Growing Halophytes to Remove Carbon from the Atmosphere," *Environment* **34**, pp. 40–43.
Khan, M. A., and Duke, N. C. (2001) "Halophytes—A Resource for the Future," *Wetlands Ecology and Management* **6**, pp. 455–456.
Klass, D. L. (2004) "Biomass for Renewable Energy and Fuels," *Encyclopedia of Energy*, vol. 1, pp. 193–212, Elsevier, New York, New York.
Mason, B. J., and Ludlam, F. H. (1957) "The Microphysics of Clouds," *Rep. Prog. Phys.* **14**, pp. 147–195.
Morris, G. (1999) "The Value of the Benefits of U.S. Biomass Power," National Renewable Energy Laboratory Report No. NREL/SR-570-27541, U.S. Department of Energy.
National Research Council. (1990) *Saline Agriculture*, National Academy Press, Washington, DC.
Pielke, R. A., Lee, T. J., Glenn, E. P., and Avissar, R. (1993) "Influence of Halophyte Plantings in Arid Regions on Local Atmospheric Structure," *Int. J. Biometeorology* **37**, pp. 96–100.
Raddatz, T. J., and Knorr, W. (2004) "The Influence of Surface Emissivity on North African Climate," *Geophysical Res. Abstracts* **6**, no. 02880, European Geosciences Union.
Sato, G. H., Ghezae, T., and Negassi, S. (1998) "The Manazar Project: Towards a Solution to Poverty, Hunger, Environmental Pollution and Global Warming Through Seawater Aquaculture and Silviculture in Desert," *In Vitro Cell. Dev. Biol.-Animal* **7**, pp. 509–511.
Shahin, M. (2002) *Hydrology and Water Resources of Africa*, Kluwer, Dordrecht, The Netherlands.
Smith, H. O., Friedman, R., and Venter, J. C. (2003) "Biological Solutions to Renewable Energy," *The Bridge* **33**, no. 2, pp. 36–40.
Touryan, K. J. (1999) "Renewable Energy: Rapidly Maturing Technology for the 21st Century," *J. of Propulsion & Power* **15**, pp. 163–174.
Yensen, N. P. (1988) "Plants for Salty Soils," *Arid Lands Newsletter* **27**, pp. 3–10.

10

Natural and anthropogenic aerosol-related hazards affecting megacities

Hesham El-Askary and Menas Kafatos

The continuous increase of major urban centers in the world presents several challenges for their populations and the world's environment. Specifically, aerosol-related hazards such as sand and dust storms, wildfires, and urban pollution all affect megacities and are tied to the growth of cities as well. In this chapter, we examine natural and anthropogenic aerosol-related hazards affecting the world's megacities. We concentrate on three geographic areas, Egypt, China, and the Indo–Gangetic Basin, all of which contain large population centers and face acute environmental issues. We examine with specific examples the great potential of remote sensing technology to track, model, and analyze such hazards for the benefit of all concerned.

10.1 Introduction

The world's population has more than tripled in the past 70 years. Between 1830 and 1930, the world's population doubled from 1 to 2 billion people. By 1970, it had nearly doubled again, and by the year 2000, there were about 6 billion people on Earth. This rapid increase in population is sometimes called the population explosion because the exponential growth of the human population results in an explosive increase in numbers, illustrated in Figure 10.1. The continuous increase in world population is expected to reach 9 billion by the year 2050 (Figure 10.2).

Megacities are large urban centers characterized by a population of multimillion, ranging from several million to ~20 million or more. The continuous increase of the world's population, accompanied by people moving to the major cities in their countries, leads to an increase in the number of megacities in the world. In many countries, one large urban center dominates over next-size urban centers, ~20% to 40% of the total population (e.g., Athens, Cairo, Istanbul, Mexico City, Seoul, Tokyo). In the largest countries, several megacities exist (e.g., India: Kolkata, Delhi, Mumbai; China: Shanghai, Beijing, Guangzhou, Hong Kong; the United States: Chicago, Los Angeles, New York). Having large numbers of people in one city exposes megacities to a high risk of being adversely affected or even devastated by natural or anthropogenic hazards. The hazards (natural and anthropogenic) affecting megacities are aerosols and pollution, sand and dust storms (SDS), severe weather (in particular, hurricanes and typhoons), fires, heat waves and cold waves, earthquakes, tsunamis, and volcanic eruptions. Hazards become disasters when rapid, large-scale devastation occurs (i.e., depending on space, time, and magnitude). One reason is that megacities have high population density (~100,000/km^2, instead of ~300/km^2, even for densely populated India); hence, a short time scale disaster (minutes to days) can have a large impact. However, over longer time scales (years to decades), cumulative effects (e.g., pollution) may produce larger impacts (health, climate change). We tend to concentrate on short events

10.1 Introduction

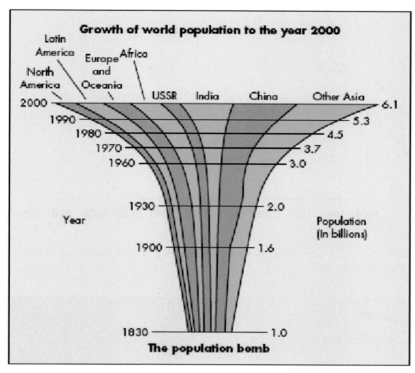

Figure 10.1 Population explosion. Copyright © 2006 Pearson Prentice Hall, Inc. *Source:* Edward A. Keller and Robert H. Blodgett, Natural Hazards, Earth's Processes as Hazards, Disasters and Catastrophes (2006) Pearson Prentice Hall.

(disasters) because they are more dramatic and can cause large loss of life and property in a short time. However, over years to decades, phenomena such as desertification, pollution, and, particularly, climate change may have more profound effects. Feedback between anthropogenic and natural hazards can produce exponentially grown impacts (e.g., SDS and urban pollution: Beijing, Cairo, Indo–Gangetic Basin in India; fires and urban pollution: as recently occurred in Kuala Lumpur). A large cause of the impact of both hazards and disasters (the main one?) is the population and urbanization growth of megacities.

Wars, considered as calamitous events and not natural hazards, have large adverse effects as they target cities, with catastrophic consequences for megacities. Studies by the Army show that large-scale military operations in the desert increase the likelihood of dust storms at least five fold, as illustrated in Figure 10.3.

In this chapter, we only consider aerosol-related hazards (natural, such as SDS, and anthropogenic, such as fires and pollution) because megacities have direct effects on their production and vice versa. Hazards affect megacities and also have long-term climate implications. Moreover, the cumulative effects of different types of aerosols result in nonlinear effects over the cities. They cause substantial air quality, health, and economic problems that are particularly acute for megacities, and they may even inhibit further growth of megacities.

As such, we review some of the manmade and aerosol-related natural hazards affecting megacities. We cover dust storms affecting the North Africa region and highlight some of these impacts over Cairo, Egypt, as well as over India, with dust storms being transported

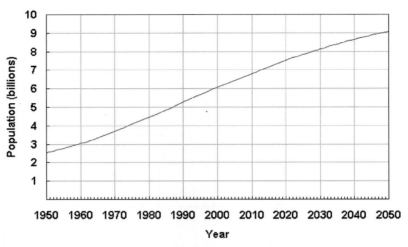

Figure 10.2 World population, 1950–2050. *Source:* U.S. Census Bureau, International Data Base, October 2002.

Figure 10.3 Army operations in Iraq facing severe dust storms.

over the Indo–Gangetic Basin. This work has been introduced and discussed in many research articles (El-Askary et al., 2003, 2004, 2006; Kafatos et al., 2004; Qu et al., 2002). Even though we concentrate on three urban and regional locations, we emphasize the increasingly global effects of these hazards and, ultimately, their major role in climate change.

10.2 Aerosol properties

Aerosols are naturally or manmade tiny particles suspended in the air. They originate from natural hazards such as volcanoes, dust storms, forest and grassland fires, living vegetation, and sea spray. However, they can result from human activities that currently account for 10% of the total aerosol amount (loading) in the atmosphere, such as produced by the burning of fossil fuels and the alteration of natural surface cover. Aerosol particles and gases in the atmosphere affect the incoming solar radiation, causing scattering and absorption of light. How much scattering occurs depends on several factors, including wavelength of the radiation, the abundance of particles or gases, and the distance the radiation travels through the atmosphere.

There are three types of scattering: Rayleigh scattering, Mie scattering, and nonselective scattering. Rayleigh scattering occurs when particles are small compared to the wavelength of radiation. These could be particles such as small specks of dust or molecules. Rayleigh scattering causes shorter wavelengths to be scattered much more than longer wavelengths and is dominant in the upper atmosphere. It gives rise to the blue color of the sky and red sunsets. Mie scattering occurs when the particles are about the same size as the wavelength of radiation. Dust, pollen, smoke, and water vapor are common causes of Mie scattering, which tends to affect longer wavelengths than those affected by Rayleigh scattering. It occurs mostly in the lower portions of the atmosphere where larger particles are more abundant. Nonselective scattering occurs when the particles are much larger than the wavelength of the radiation. Water droplets and large dust particles can cause this type of scattering, which derives its name from the fact that wavelengths are scattered equally. This type of scattering causes fog and clouds to appear white to our eyes because blue, green, and red lights are all scattered in approximately equal quantities. Absorption is the other main mechanism at work when electromagnetic radiation interacts with the atmosphere. In contrast to scattering, it causes molecules in the atmosphere to absorb energy at various wavelengths. Ozone, carbon dioxide, and water vapor are the three main atmospheric constituents absorbing radiation.

10.3 Sand and dust storms

Dust particles comprise a large fraction of natural aerosols. Understanding them on both global and regional scales is important if we are to study climate effects. They are generated by strong windstorms with velocities exceeding 48 km (30 mi) per hour, and dust carried by wind can reduce visibility to less than 0.8 km (0.5 mi) for a significant period of time. Although intense dust storms reduce visibility to near zero in and near source regions, visibility improves away from the source. A typical dust storm can be several hundred kilometers in diameter and may carry more than 100 million tons of dust particles that are comprised of mineral particles less than 0.05 mm in diameter. Natural dust also contains small amounts of biological particles, such as spores and pollen. Hence, dust storms can be a significant health hazard for people with respiratory problems and may adversely impact urban areas, particularly megacities.

Sand and dust storms are climatic, environmental events that are presently worst in the Mediterranean and East Asia regions and have steadily worsened over the past decade due to massive deforestation and increased droughts. They are dynamic events, moving as a wall of debris. Large sand grains wind-lifted into the air fall back to the ground in a few hours, while smaller particles (dust) less than 0.01 mm remain suspended in the air for longer time periods, being swept thousands of kilometers downwind. Unlike dust storms, sand storms are almost exclusively a desert phenomenon, occuring over deserts and arid regions, where sand is transported about 30 m (100 ft) at the surface of the land. Blowing sand is very abrasive. Severe or prolonged dust and sand storms result in major disasters. For example, a 5-hour dust storm near Jingchang, China, caused 640 million yuan in economic damage over a wide area and injured or killed in excess of 300 people. In mid-March 1998, the Middle East was hit by choking sandstorms that claimed four lives and injured twenty-nine people, forcing the Suez Canal and airports to close, resulting in severe economic losses. On May 7, 2002, blowing sand from the Saharan interior caused an Egyptian aircraft to crash into a hillside when attempting an emergency landing near Tunisia, killing eighteen people (Miller, 2003).

Dust storms affect the atmosphere and regional climate as they modify the energy budget by cooling the atmosphere from reflection of solar radiation back into space, as well as by causing atmospheric heating due to the absorption of IR radiation. Dust affects rain forming, acting as cloud condensation nuclei (CCN), which may enhance or suppress rain. In combination with anthropogenic pollutants, they can work toward the formation of heavy rainfall. Dust deposition can affect marine ecosystems (e.g., algal blooms) and transport microbes. Dust storms are connected to a variety of phenomena related to precipitation, soil moisture, and human activities, such as land use/land cover practices.

10.3.1 Remote sensing of sand and dust storms

In arid regions where dust storms form, specifically in the Middle East Gulf countries around the Arabian Sea, Iran, Afghanistan, and Pakistan experience a high frequency of dust storms, as many as thirty events per year (Kutiel and Furman, 2003; Middleton, 1986; Moulin et al., 1998). In Egypt, sandstorms called *khamasin* can last for a number of days, resulting in seasonal hazards (Egyptian Meteorological Authority, 1996). On 2 May 1997, a severe sandstorm, the worst in 30 years, hit much of Egypt, resulting in eighteen deaths. Sherif Hamad, head of the Egyptian Meteorological Service, stated, "we've never experienced such a powerful one."

Dust storms occur at different spatial scales, ranging from mesoscale/regional to continental. Dust storms have increasingly become a serious environmental issue, with effects reaching far beyond the places of origin or the regions adjacent to the deserts and extending for hundreds or even thousands of kilometers, crossing the Mediterranean, the Atlantic Ocean, and even the Pacific Ocean (Prospero, 1996). Dust storms are very dynamic, with particle size, distribution, and direction varying significantly (MacKinnon and Chavez, 1993). Dust storms are generated by strong winds lifting particles of dust or sand into the air from regions that are mainly deserts, dry lakebeds, semiarid desert fringes, and drier regions where vegetation has been reduced or soil surfaces have been disturbed (Williams, 2001). However, Tegen and Fung (1995) discovered that disturbed soil surfaces can contribute up to 50% of the atmospheric dust load. Therefore, dust events may also be considered anthropogenic in nature. Human influences such as deforestation, as well as climate, play an important role (Rosenfeld, 2001). Airborne particles from dust storms can alter the local climatic conditions by intercepting sunlight. They participate in modifying the energy budget by cooling and heating the atmosphere (Liepert, 2002). Specifically, the presence of an absorbing dust layer results in a substantial decrease in the incoming short-wave solar radiation, resulting in a major change to the surface energy balance. Moreover, atmospheric stabilization is affected when dust differentially warms a layer of the atmosphere at the expense of near-surface cooling.

Remote sensing technologies afford human societies the ability to track and follow sand and dust storms over large regions and over time. Ground-based observations, although important, are inadequate as sand storms develop and propagate over inaccessible regions of the Earth. Many remote sensing systems can be used for detecting dust storms. Geostationary operational environmental satellites (GOES) are considered to be most suitable to track the time evolution of active and short-lived dust storms because of their high temporal resolution (15 minutes). Suspended dust can be highlighted as brightness changes. However, due to poor spatial (1 km and 4 km) and spectral (1-visible) resolutions, they are used only to detect and monitor very large dust storms (Chavez et al., 2002). In contrast, Landsat TM has high spatial resolution. Therefore, Landsat can map the dust source location accurately if the image is cloud free. However, it has poor temporal resolution (2 weeks), and just like GOES, has inherent difficulty in penetrating clouds (Chavez et al., 2002). The sea-viewing wide field of view sensor (SeaWiFS) is a useful sensor in detecting dust plumes lasting for a long period of time and having a dark (low-radiance) background (i.e., the ocean) (Chavez et al., 2002). It has been used to detect the very large dust plumes generated by winds in Africa and Asia (Husar et al., 1998; Prospero, 1999). However, SeaWiFS has difficulty detecting small and short-lived dust events over desert areas due to the high radiance of deserts (Chavez et al., 2002).

Different studies of the characteristics of dust events have been carried out on the ultraviolet (UV) region of the electromagnetic spectrum using total ozone mapping spectrometer (TOMS) measurements (Cakmur et al., 2001; Herman et al., 1997; Prospero et al., 2002). The TOMS instrument detects aerosol particles by measuring the amount of backscattered ultraviolet radiation. However, TOMS data may also produce misleading results by confusing dust with other aerosol particles. This is because TOMS responds better to backscattered radiation caused by Rayleigh scattering when particles are smaller than the incident wavelength, while dust particles are coarser and more likely to produce Mie scattering. Moderate resolution imaging spectro radiometer (MODIS) data were found to be suitable for monitoring environmental changes in sand and dust storms as shown in this chapter (El-Askary et al., 2002). In previous work, tropical rainfall measuring mission (TRMM) data indicated that dust storms may amplify desertification effects (Rosenfeld, 2001).

10.3.2 Egypt case study

Egypt is located in the northeastern arid region of the African Sahara. More than 90% of Egypt is desert, which includes the southwestern part of the country. The majority of the population lives in or around the Nile Delta, which makes it one of the most densely populated areas of the world. Dust storms forming in the vast desert areas of Egypt greatly affect life in its urban regions and particularly Cairo. Air pollution from dust storms is a significant health hazard for people with respiratory problems and can adversely impact the environment. It is obvious that timely warnings of dust storms must be initiated in populated regions for health concerns and traffic control. Dust storm detection and tracking could be difficult because dust storms share some similar characteristics with clouds. This section shows, based on our analysis, the different capabilities of several remote sensing instruments with respect to monitoring sand and dust storms. Here, we consider the Nile Delta as a case study. Multisensor data analysis has been carried out in light of different behavior of dust particles at various wavelengths. A combination of optical sensing uses MODIS and microwave sensing of dust storms uses the TRMM microwave imager (TMI). Our approach proves to be superior to currently used methods because it can clearly distinguish aerosol particles from dust. This work is of great significance to monitoring dust storms not only in Egypt but also in the entire Arab region for control measures, and hence for environmental management.

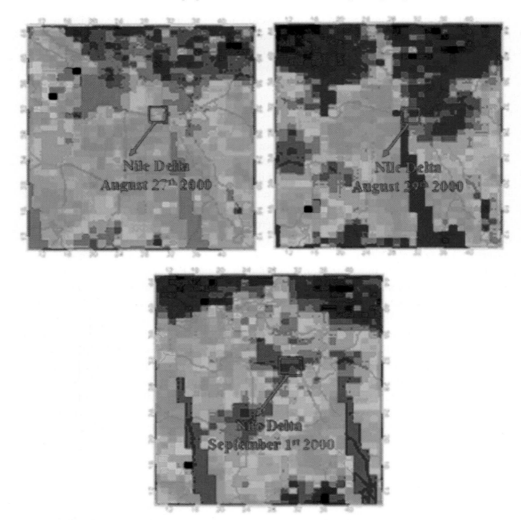

Figure 10.4   TOMS-derived aerosol index over the Nile Delta and adjoining areas.

Dust screening using TOMS aerosol index

TOMS data are widely used for detection of dust storms in the ultraviolet region. TOMS detects aerosol particles by measuring the amount of backscattered ultraviolet radiation. TOMS-derived aerosol index over the Nile Delta and adjacent areas has been calculated (El-Askary et al., 2003b). As is apparent from Figure 10.4, the poor spatial resolution of TOMS is unsuitable for detecting dust storms over a small area, such as the Nile Delta. Furthermore, the analysis is based on qualitative and quantitative examinations, resulting in a general technology for dust storm monitoring and detection. TOMS responds to backscattered radiation for Rayleigh scattering, most prominent at relatively large wavelengths compared to particle size. Dust particles, in contrast, are predominantly coarser and respond to Mie scattering.

Visible sensing of sand and dust storms

The MODIS level 1B radiance dataset contains the radiance counts from thirty-six spectral bands. To extract meaningful information and to reduce the relevant dimensions, a spectral

10.3 Sand and dust storms

principal component analysis (PCA) has been applied (El-Askary et al., 2003a, 2003b). PCA is a coordinate transformation typically associated with multiband imagery. PCA reduces data redundancy by creating a new series of images (components) in which the axes of the new coordinate system point to the direction of decreasing variance. For more details on PCA, interested readers can consult Joliffe (1986). In a specific case, the MODIS radiance bands from 1–3 September 2000 were transformed in the previous manner. The first four principal components (PCs) for each day are found to contain about 90% of the total variance. From the first PC for both days, it was clear that the western part of the Nile Delta was indicating the presence of certain features on 1 September. We suspected such features to be dust because they totally disappeared on 3 September. To study these features, three locations were selected over the Western Desert and the Nile Delta. Spectral patterns are observed and averaged over a 5 × 5 window in these locations. The spectral patterns of these three regions on 1 September totally differ from each other. This is because they represent vegetation mixed with dust, pure vegetation, and pure dust in regions "A," "B," and "C," respectively. The difference in the spectral signatures of regions "A" and "B" is due to light scattered from the suspended dust and does not reach the ground. Vegetation does not absorb all the radiation from the sun because some radiation is scattered back to the sensor without reaching the ground. On 3 September, the spectral patterns of regions "A" and "B" agree with each other because both regions represent pure vegetation. The signature of region "C" is the same on both days, and it represents dust in the Western Desert. By calculating the spectral difference between the three patterns for 1 and 3 September, regions "B" and "C" yielded a zero difference, because they did not change over the 2 days. However, region "A" reveals a difference matching very well the pattern obtained from region "C" in both days, which corresponds to pure dust. Our analysis suggests the signature of a dust storm, over the Nile Delta, transported from the Western Desert (Figure 10.5).

In Figure 10.5, band 7 falls in the short-wave infrared (SWIR) region of the spectrum, whereas band 11 falls in the green portion of the spectrum. Saharan dust is mainly composed of silica particles that produce a high peak in the SWIR region, and a well-known reflectance albedo over the green region of the visible spectrum. This explains the presence of the high peaks over bands 7 and 11. However, the pure vegetation signature taken from region "B" shows a minimum over the SWIR region because the relevant plants do not have enough water supplies, which in turn leads to less reflection in the near infrared and significantly less in the short infrared compared with healthy vegetation.

For further verification of the presence of the Sahara dust over the Nile Delta and the Mediterranean, the first four PCs were subjected to a K-means clustering analysis. The applied K-means algorithm in this study uses the spectral properties of the multispectral image for clustering. The algorithm was implemented with four classes, for a total of twenty iterations. The classified principal component image is shown in Figure 10.6, which shows that the dust particles over the Nile Delta belong to the same class as the Western Desert, as one would expect.

Dust plumes are three-dimensional features characterized by vertical and horizontal motion. The specific direction and trend of dust storms can be clearly delineated and identified with the use of directional filters such as the Sobel filter (Richards and Jia, 1999). We applied a 3 × 3-kernel Sobel filter to distinguish the abnormal concentrations of dust clouds and specify their direction over the Nile Delta. The Sobel filter operates by estimating the magnitude of the directional derivatives along any two given directions. The advantage of this filter is that it is able to capture the changes in the image in two dimensions, as shown in Figure 10.7.

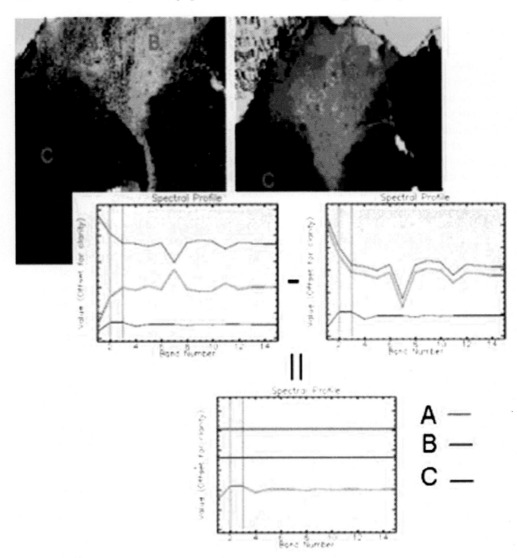

Figure 10.5 MODIS gray-scale first principal component for the dust storms of 1 and 3 September 2000, and the associated spectral patterns obtained from the original bands.

Our analysis showed that the dust storm originated from the Western Desert, toward the Nile Delta on 31 August. On 1 September, it started drifting away from the western part of the Nile Delta. Finally, this direction of the dust cloud matches the known direction of winds prevailing during September, namely, from the south to the southwest direction (Egyptian Meteorological Authority, 1996).

Microwave sensing of sand and dust storms

El-Askary et al. (2003b) emphasized the presence of dust storms over the Nile Delta through the use of passive microwave data. The single scattering albedo over optical wavelengths is different from the microwave. In the microwave range, the single scattering albedo plays a more important role than the emissivity of the medium through which the radiation passes.

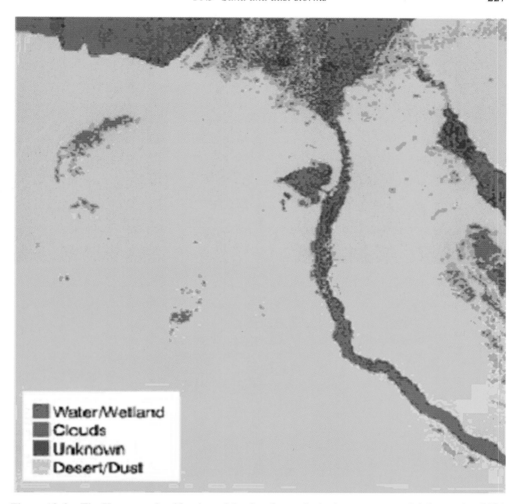

Figure 10.6 The K-means classification of the first four principal components of 1 September 2000.

The single scattering albedo is defined as the probability that given an interaction between a photon and a particle, the photon will be scattered rather than absorbed. For a single particle size, this probability may easily be expressed in terms of the optical efficiencies as shown in Equation (10.1), where (τ_e) is the extinction coefficient and (τ_a) is the absorption coefficient (Zender, 2000).

$$\bar{\omega} = \frac{1 - \tau_a(r, \lambda)}{\tau_e(r, \lambda)} \qquad (10.1)$$

Dust particles are larger than normal aerosol particulates, but are of similar size to the incident microwave radiation. Therefore, microwave radiation responds to dust particles with Mie scattering. The shorter the wavelength of the incident radiation in the microwave range, the greater is the scattering, and hence, the lower the brightness temperature. For example, for quantitative estimation of the amount of scattering produced by dust particles over the Nile Delta, in the 85-GHz vertically polarized channel, the brightness temperature into 19-GHz and 22-GHz vertical channels can be integrated, and the 85-GHz response

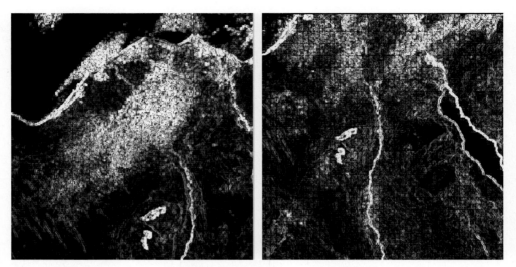

Figure 10.7 Two Sobel-filtered images from 31 August (*left*) and 1 September (*right*) 2000, showing the general direction of the dust storm over the Nile Delta.

subtracted from them. This approach uses the single scattering albedo in Equation (10.1) and has been described in Ferraro and Grody (1994) and shown in Equation (10.2).

$$SI_{85} = 451.9 - 0.44 \times TB_{19v} - 1.775 \times TB_{22v}$$
$$+ 0.00574 \times TB_{22v} \times TB_{22v} - TB_{85v} \quad (10.2)$$

where TB refers to the brightness temperature over the specific channel, and SI_{85} is the scattering index derived for the 85-GHz channel. The SI increases with greater scattering and leads to a decrease in brightness temperature. Brightness temperature drops as less radiation reaches the sensor in the shorter wavelengths due to greater scattering.

The SI was calculated for a number of days from 27 August to 3 September. Brightness temperature was averaged for the Nile Delta over a 20×20-km window, for all the days for which the SI was estimated, as shown in Figure 10.8. The black square in each of them represents the area over which the 20×20 window has been averaged. The TOMS-derived aerosol index (AI) scenes (National Aeronautics and Space Administration, Goddard Space Flight Center, n.d.) is also shown side by side with the daily TMI figures for comparison.

Here, low brightness temperature days that correspond to greater scattering produced by dust particles are shown in brighter color within the blue to yellow range. A bright color (yellow) indicates the increasing presence of dust. A clear correspondence between scattering and microwave brightness temperature is obvious, as the dust storm is seen to oscillate and migrate over the study area. Furthermore, a significant negative correlation ($r = -0.56$, p value = 0.0175 from the two-tailed t-test) between the ultraviolet-derived TOMS AI and the microwave-derived TMI brightness temperature is observed (Figure 10.9). The days with higher values of AI obtained from TOMS show lower brightness temperature and are indicated by yellow color in Figure 10.8. The TOMS product at ultraviolet wavelengths captures the response of smaller particles, namely, aerosols to Rayleigh scattering, which is important at a relatively longer wavelength, compared to particle size. In the microwave range, namely for TMI, the wavelength is significantly larger. Therefore, Mie scattering is applicable when particles are about the same size as the radiation wavelength. In this

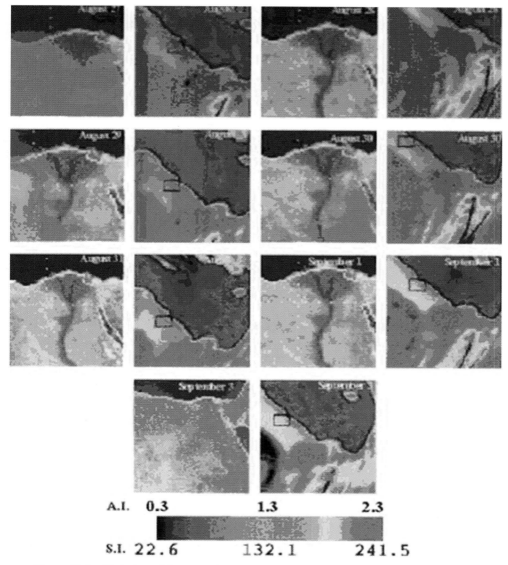

Figure 10.8 Variation of the scattering index and aerosol index derived from TMI brightness temperatures and TOMS shown side by side. TOMS to the left and TMI to the right in each image pair.

respect, the microwave is far more suitable than the ultraviolet spectral region because it may not differentiate dust particles from aerosols. Note that in Wentworth (1922), it was estimated that the sand grain size (diameter) ranges from 2 mm to 1/16 mm, whereas the silt size is 1/16 to 1/32, and the clay size can be from 1/32 to 1/264. Therefore, the expected grain size of the dust settled at the higher levels of the atmosphere ranges from fine sand to clay.

Dust storms can be monitored as well by looking at their effect on water vapor. The dust particles act as small cloud condensation nuclei, around which water vapor droplets can accrete and give rise to smog and fog. This can be verified in Figure 10.10. The

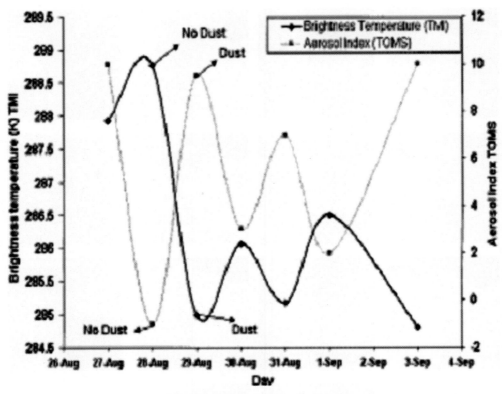

Figure 10.9   Daily variations of the brightness temperature and aerosol index for the selected areas in Figure 10.8.

thing to note from figure 10.10 is the way the water vapor shifts with the drift in dust storm for the 1 and 3 September 2000, and hence, the location of water vapor or smog matches quite well with the position of the haze and dust storm as inferred from Figure 10.7 (El-Askary et al., 2003a, 2003b; see also El-Askary [2006] and El-Askary and Kafatos [2006]).

10.3.3   India case study

The Indo–Gangetic (IG) Basin is one of the world's largest and most productive river basins. Growing population and increasing urbanization has resulted in overexploitation of the natural resources and has caused serious concerns for the sustainability of irrigated agriculture and other water uses. The Indo–Gangetic Basin consists of numerous rivers where the water table goes down in the basin during the summer months. The population living in the basin uses drinking water from rivers, lakes, and ponds. Dust storms contaminate the water resources, especially during the premonsoon and monsoon periods. As such, people living in the basin suffer with waterborne diseases. Dust transportation in the Indo–Gangetic Basin often initiates from the western parts of India and Pakistan. Occasionally, Saharan dust storms travel all the way to Asia with the summer monsoons (Dey et al., 2004; El-Askary et al., 2004). Here, we are concentrating on the IG basin, which experiences dust storms originating in the west during the premonsoon period that are a major threat to agricultural

10.3 Sand and dust storms 231

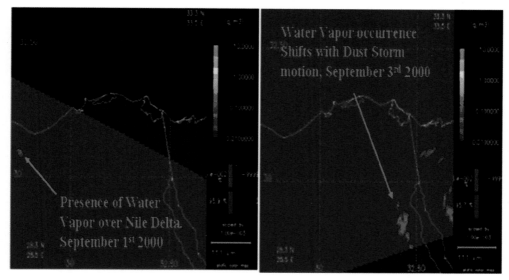

Figure 10.10 Anomalous water vapor concentration over and adjacent to the Nile Delta.

resources and people living in this region. Figure 10.11 shows three main city locations that have been affected by dust events, namely, New Delhi, Kanpur, and Varanasi.

Dust screening using TOMS aerosol index

Different studies of the characteristics of dust events have been carried out in the UV region of the electromagnetic spectrum using TOMS measurements (Cakmur et al., 2001; Herman et al., 1997). We have used TOMS aerosol index data with resolution 1 degree x 1 degree to delineate dust and haze over the Indo–Gangetic Basin and, in particular, the area over the major cities in that basin. Figure 10.12 shows the aerosol index variability over New Delhi and Varanasi during June 2003. The aerosol index shows maximum values due to the dust outbreak in the Indo–Gangetic Basin on 10 and 16 June 2003 over Delhi where two major dust events were recorded. The aerosol index data show some variability as the dust events enter different phases of deposition, origin, and transportation. The corresponding values of the aerosol index over Varanasi show a general major drop in the aerosol index values compared to those over New Delhi. The small values over Varanasi are due to the fact that the dust took several days before reaching it from Delhi. Therefore, on 10 and 16 June, there are small aerosol index values, whereas a few days later, on 14 and 21 June, values of the aerosol index increased. Moreover, the aerosol index values observed on 14 and 21 June over Varanasi are lower, compared to the observed aerosol index values over Delhi on 10 and 16 June. Such observations are due to the fact that the intensity of dust events weakens as the storm travels from one location to another, with very fine aerosols dominating at the last location. Aerosol index values also reflect some local dust contribution from the Thar Desert, in addition to the major events originating from the Saharan Desert. The general trend as seen over both cities verifies the migration of the dust from Delhi to Varanasi, as well as the reduction in its strength as dust deposition occurs during the event propagation.

Aerosol index values can be very helpful in determining the occurrences of aerosol events and anomalies above the normal thresholds; however, further investigation and detailed data

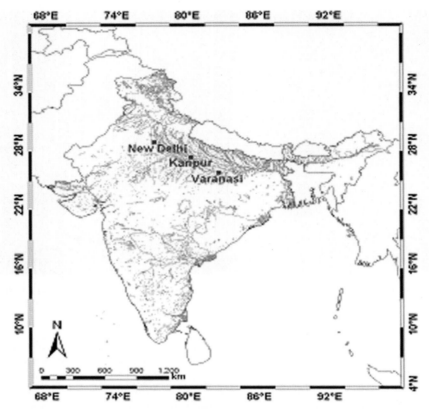

Figure 10.11 Base map showing the Indo–Gangetic Basin drainage network and the three major cities lying within the basin.

analysis are still required to verify the dust occurrence (El-Askary et al., 2003a). This is attributed to the fact that with TOMS data, it is difficult to distinguish between different types of aerosol particles (dust tends to have larger particles size than smoke) and absorbing properties in the UV.

After the observation of the aerosol anomaly over New Delhi, further analysis was carried out to study the relation between the aerosol index and the ground-based respiratory suspended particulate matter (RSPM).[1] The RSPM data show a significant increase of particulate matter concentration in the atmosphere during the month of June, thus confirming the presence of the dust particles. Figure 10.13 shows the relationship between aerosol index and RSPM over New Delhi. It is clear that on the days of the dust outbreak, high aerosol index values were observed, and an obvious increase in the values of the RSPM are observed that matches well with the high aerosol index values of 10 June. Lower values of the RSPM were recorded when the aerosol index values were minimal. The trend lines of both the aerosol index and the RSPM show a general decline in the values after 10 June, as the dust event moved away from New Delhi and toward Kanpur and Varanasi. The decreasing trend of both parameters indicates the moving of the dust storm in the eastern direction across the Indo–Gangetic Basin.

---

[1] http://envfor.nic.in/cpcb/.

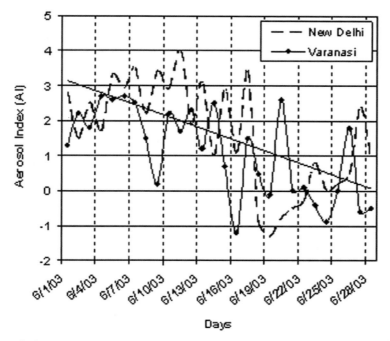

Figure 10.12 Aerosol index variability over New Delhi and Varanasi during June 2003.

Visible sensing of sand and dust storms

The dust storm over India was detected from MODIS measurements on 9 June 2003, where a large dust plume can be clearly seen. Moreover, MODIS aerosol products provide good global spatial and temporal coverage with regional dust properties. Figure 10.14 shows the occurrence of a strong desert dust outbreak (which took some days to reach India from its origin in the Saharan region) detected by MODIS. The large extension of the dust plume is clearly observed. Although the extraction of physical aerosol parameters from the images is more difficult (Stowe et al., 1997; Tanré et al., 1997; Torres et al., 1998), methods are being developed as one of the tasks for aerosol studies, which must be validated and complemented by ground-based measurements (Singh et al., 2004).

Another potential of visible sensing of dust outbreaks uses the multiangle imaging spectroradiometer (MISR), which can be used to detect large dust storms (Figure 10.15) (El-Askary et al., 2004). MISR is characterized by its ability to observe at different viewing angles, and hence, identification of dust storms can be greatly improved (El-Askary et al., 2003a; Kafatos et al., 2004). This is attributed to the fact that dust storm events that are difficult to be detected in the nadir-viewing angle, may be easily detected by the off-nadir, angle views due to the thicker depth of the atmosphere. MISR has the potential to enhance the detection of small dust storms, thus it might be helpful in early detection of dust storms (El-Askary et al., 2003a; Kafatos et al., 2004). In addition, combining the information from different angle views could be useful in discriminating between dust clouds and regular clouds. Different angle views of MISR make it possible to set limits on particle shape, size, composition, and aerosol amount. They also help distinguish between various types of aerosols, such as dust, soot, and sulfates. As a result, using MISR multi-angle viewing could be beneficial in decreasing the background effects for desert regions.

234 *Natural and anthropogenic aerosol-related hazards affecting megacities*

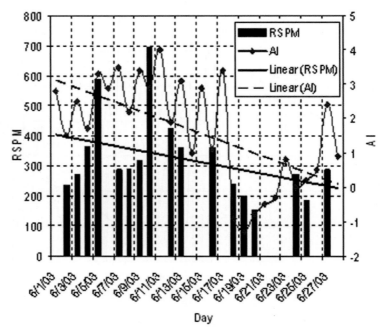

Figure 10.13 Aerosol index variability versus RSPM over New Delhi during June 2003.

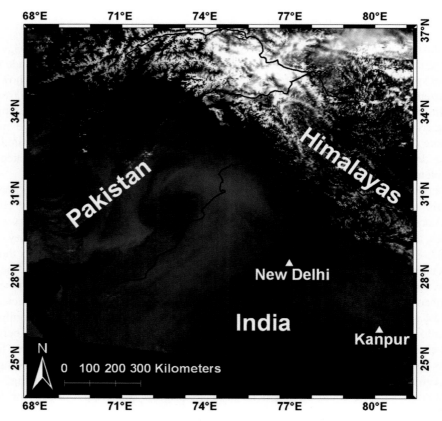

Figure 10.14 MODIS image of a dust outbreak in northern India on 9 June 2003.

10.3 Sand and dust storms

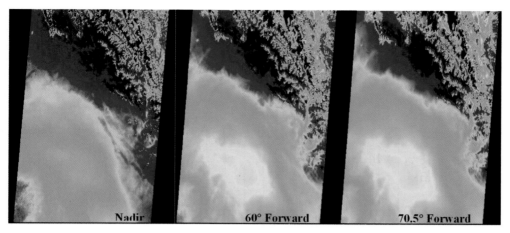

Figure 10.15 Different angle views of a large dust plume on 9 June 2003 over the northwestern part of India.

Microwave sensing of sand and dust storms

As discussed previously, dust particles vary in size distribution and contribute a major portion of the aerosol concentration in any dust outbreak event. Therefore, because the size of dust particles is substantially larger than other aerosol particulates, they are not amenable to proper monitoring by optical sensors. Over the Indo–Gangetic Basin, Dey et al. (2004) found that the origin of dust may be far away from the Indian region and that these dust storms further transport sand from the Thar Desert.

The presence of dust storms over the Indo–Gangetic Basin has been corroborated through the use of passive microwave data (El-Askary et al., 2003b). The advanced microwave sounding unit (AMSU) onboard the Aqua satellite is primarily a temperature sounder operating in fifteen frequency channels ranging from 23.8 to 89 GHz and has a 40-km horizontal spatial resolution at Nadir. The first two channels, 23.8 and 31.4 GHz, provide surface and moisture information. We used the 23.8-GHz frequency channel (vertically polarized), which measures brightness temperature at near surface level. Mie scattering takes place in the lower atmosphere (less than 4.5 km), where larger particles are more abundant, and dominates when cloud conditions are overcast and can affect longer wavelength radiation. Smoke, dust particles, pollen, and water vapor are the dominant sources of Mie scattering in the lower atmosphere (El-Askary et al., 2003a, 2003b; Sabins, 1978). Thus, microwave sensing provides another potential way to distinguish and detect dust in the atmosphere due to the similarity in size of dust particles and wavelength. This means that as the incident radiation is of shorter wavelength in the microwave range, the scattering due to particles is greater, and hence, the brightness temperature is lower. Dust storms produce large scattering of the incoming solar radiation.

Using microwave data, we have found that T_b is negatively correlated with aerosol index due to high scattering in the atmosphere in the presence of a dust storm (9 June). While in dustfree atmosphere, T_b is found to be positively correlated with aerosol index. The daily variations of T_b obtained from the AMSU were plotted against the aerosol index from TOMS, over New Delhi. Low brightness temperature days correspond to greater scattering produced by dust particles indicating the presence of dust. A clear negative correspondence between T_b and the aerosol index is observed to occur during the different phases of the dust storm as it moves over the study area (Figure 10.16).

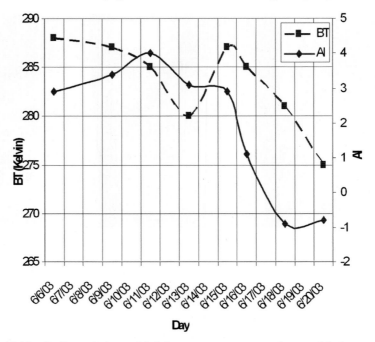

Figure 10.16 Daily variations of brightness temperature and aerosol index over New Delhi during June 2003.

10.3.4 Modeling of dust storms (dust cycle model)

The dust cycle model has been developed by the Atmospheric Modeling and Weather Forecasting Group at the University of Athens (UOA) (Kallos et al., 1997c, 1997b, 2004; Nickovic et al., 1997, 2001). The dust module solves the Euler-type prognostic equation given in Equation (10.3) for the dust concentration:

$$\frac{\partial C}{\partial t} = -u\frac{\partial C}{\partial x} - v\frac{\partial C}{\partial y} - w\frac{\partial C}{\partial z} - (\Delta_H K_h)(\Delta_H C) - \frac{\partial \left(K_z \frac{\partial C}{\partial z}\right)}{\partial z} + S \qquad (10.3)$$

Here, C is the dust concentration; u, v, and w are the wind vector components; K_h is the lateral diffusion coefficient; K_z is the turbulence exchange coefficient; S is the source/sink term; and Δ_H is the horizontal nabla operator. The dust module incorporates the state-of-the-art parameterizations of all major phases of the dust life cycle such as production, diffusion, advection, and removal. The module also includes effects of the particle size distribution. The current dust model uses the Olson World Ecosystem data to specify the desert source area. The dust concentration is set to zero in the initialization and is subsequently generated by the model. During the model integration, the prognostic atmospheric and hydrological conditions are used to calculate the effective rate of the injected dust concentration based on the viscous-turbulent mixing, shear free diffusion, and soil moisture. The dust module has been tested for a number of cases and has shown the capability to simulate realistic dust cycle features (Alpert et al., 2002; Kallos et al., 1997a, 2002; Nickovic et al., 2001; Papadopoulos et al., 1997, 2001).

Nickovic et al. (1997, 2001) showed that the SKIRON/Eta system was able to simulate the long-range transport of dust reasonably well in various places around the Earth. The system

10.3 Sand and dust storms

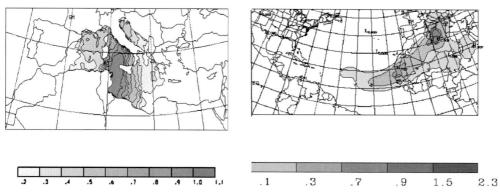

Figure 10.17 Optical thickness of the dust plume over (a) the Mediterranean Sea, and (b) Atlantic Ocean, for 18 June 1993.

is able to describe cross-Atlantic transport of dust quite satisfactorily. Such a typical case has been analyzed at UOA for the period of 10 June to 10 July 1993 (Figure 10.17). The sea level pressure map shows a large anticyclone covering the north Atlantic and a relatively strong pressure gradient in northwestern Africa in mid-June. The atmospheric circulation over the Sahara during the period provided favorable conditions for dust production. As a result, a dust plume was generated and driven by subtropical easterlies toward the Atlantic Ocean. During 18–19 June, a strong low-level flow divergence caused splitting of the dust cloud. One of the flow branches transported the dust toward the Atlantic Ocean (Figure 10.17b), and the other branch of dust moved toward the Mediterranean Basin (Figure 10.17a). The SKIRON/Eta system was also used to simulate a Mediterranean episode, as shown in Figure 10.17a. The model forecast provides a good agreement with the METEOSAT-visible analyzed image of aerosol optical depth (Figure 10.17b).

Operational model boundary and initial conditions

Satellite remote sensing is an effective and economic means of monitoring sand and dust storms, particularly, because many dust storms occur in remote regions of the Earth. As we know, not only can it provide initial parameters for model simulations, but it can also be used for model simulation verification and validation. At the present time, remote sensing of dust storms can only provide near global horizontal coverage with limited vertical resolutions (Qu et al., 2002, 2004). Predictive models for regional applications need initial and boundary conditions from a global model. This occurs in a similar way as in mesoscale weather models (e.g., COAMPS), which also depend on a global weather model (e.g., the Navy Operational Global Atmospheric Prediction System [NOGAPS]; Staudenmaier, 1997). To develop *RS* data requirements for the boundary and initial conditions for one of these models, detailed information about the currently used set of parameters for these models must be obtained.

Meteorological parameters (three-dimensional [3D] fields), such as geopotential height, temperature, humidity, and horizontal wind components from the National Center for Environmental Prediction (NCEP) analysis (or reanalysis), can be used to initialize models such as the SKIRON/Eta system and to provide lateral boundary conditions for model simulation (forecasting or hindcasting mode). For land surface boundary conditions, soil moisture, land surface temperature, snow cover, etc., ground observations and satellite observations from platforms such as Terra/Aqua MODIS and AMSR–E can be used for optimal land initialization.

To improve the model performance, one can use a specific parameterization scheme to interpolate the MODIS aerosol optical depth and particle size products into the 3D dust concentration using climatic aerosol profiles to initialize the dust module. This scheme can establish a lookup library that relates the Terra/Aqua aerosol measurements with the 3D dust concentration. MODIS quantitative data are necessary for the identification of spatiotemporal distribution of dust plumes over the Atlantic and for vertical profiling at preselected locations in order to intercompare models using remote sensing data on vertical distribution. The ability to provide information of 3D dust distributions would add immensely to the information content and value of dust storm monitoring and prediction. The vertical distribution of the dust varies with geolocations, surface type, and atmospheric conditions. Using the LIDAR measurement in Greece showed that transported dust is found to be multilayered with dust heights between 1.5 and 5 km (Hammonou et al., 1999).

We believe that model boundary and initial conditions requirements should include some of the meteorological parameters such as cloud type, atmospheric dust, and land cover in order to augment product scenes. Remotely sensed images, field surveys, and long-term, ground-based instrumentation should be used in order to monitor sand saltation, dust emissions, vegetation cover and change, short-term meteorological parameters, and soil conditions. These parameters, once known, could be passed to the model as initial and boundary conditions. Such models always require global aerosol distributions of natural and anthropogenic aerosols because of their potential climatic implications. As a result, aerosols are considered to be very important in understanding the physics of the storms. Therefore, there has been more recent emphasis on the development of aerosol retrieval algorithms for existing satellites and the deployment of several new satellite sensors designed specifically for aerosol detection. Aerosols in the atmosphere form a mixture of different components. The detection and identification of a specific range of aerosol particles (dust in our context) requires knowledge of the microphysical and optical properties in the visible and infrared spectral range. Once characterized, the ability of dust aerosols to modify the incoming solar radiation can be used to effectively identify the onset of dust storms and monitoring of the same at different times of the year.

Techniques should be developed to convert synoptic reports, such as visibility to aerosol amounts and methods for characterizing source regions of dust and source strength using aerosol retrievals (Tegen et al., 1996). In context, work is needed on adapting data assimilation techniques to aerosol data streams, and generalizing the global aerosol model to use multiple components and sizes as background fields for data assimilation. Parameters that are available from remotely sensed data include the aerosol optical thickness, concentration, and dry mass column for total aerosol or for individual aerosol components of sulfates, dust, black carbon, organic carbon, and sea salt. These parameters could be obtained from the TOMS system in the past and are of great use in the GOCART model. Furthermore, several meteorological forecast products (i.e., winds, precipitation, cloud fractions) should be considered as well. Different conditions characterize each of the previous models, and each condition would need some parameters to be fulfilled. These parameters include (1) aerosol index (AI), (2) snow cover area index (SCAI), (3) desertified area index (DAI), (4) land use/cover change (LU/CC), (5) water deficit index (WDI), (6) vegetation index (VI), (7) land surface temperature, (8) land surface reflectance and albedo, (9) sea surface temperature (SST), (10) leaf area index (LAI/FPAR), (11) vegetation indices with surface flux applications, (12) terrestrial carbon cycle, and (13) net primary productivity (NPP). In summary, future assimilation of RS data into operational models can greatly increase the utility and predictive power of these models to mitigate the effects on megacities.

10.4 Air pollution

The world's megacities are the sites of production of a variety of aerosols and are themselves affected by natural and human-induced aerosols. Sources of aerosols impacting cities include industrial and automobile emissions, sand and dust storms, fire-induced aerosols, and, occasionally, volcanoes. Air pollution problems over megacities differ greatly and are influenced by a number of factors, including topography, demography, meteorology, level and rate of industrialization, and socioeconomic development. Here, we present a specific case study for Cairo because this megacity is affected by all different types of aerosols, and a case study for regional effects in India with large-scale vegetation and even climate impacts.

10.4.1 Cairo air pollution case study

Cairo is the capital of Egypt and is considered to be a key city for economy, education, politics, industry, and technology in the Middle East. The population of the greater Cairo region has exceeded 12 million. Increasing business and industrial activities in the city, accompanied by shortage of the institutional capabilities for monitoring and control, in addition to environmental impact negligence that prevails over many of the production sectors, have resulted in excessive air pollution problems that have reached the level of crisis during the year. The World Health Organization reports that the air pollution in downtown Cairo has reached 10 to 100 times what is considered to be a safe limit. Dust storms and the so-called "black cloud" phenomenon have significantly increased within and around Cairo over the past 5 years. The major source of natural and anthropogenic mineral dust aerosols over Cairo and the greater Delta region is wind-laden dust transported from arid or desert regions of the Middle East and from Sahara during the summer season. Both phenomena (dust storms and black cloud) cover the city at particular times of the year (April and May for the dust, September and October for the black cloud) (Figure 10.18). Cairo is considered to be among the worst cities in the world in terms of air pollution. Although ambient levels of air pollution are still high in other cities, there are also successes. In Bangkok, authorities were able to significantly reduce concentrations of carbon monoxide (by 50%) and lead (by 80%) since 1993. Other success stories include the United Nations Environment Program (UNEP) and its brokering of the Montreal Protocol, as well as the work that is continuing to be done by the International Panel on Climate Change (IPCC) in its efforts to understand and combat global warming.

Aerosol optical depth

Aerosol optical depth (AOD) is a measure of the opaqueness of air, where high values of AOD indicate more absorption of the radiation, and hence, poor visibility. We have used MODIS data for obtaining the AOD, which becomes an indicator of the dust storm and black cloud impacts on the optical properties of the aerosols forming in the atmosphere. AOD data from the MODIS sensor are used to measure the relative amount of aerosols suspended in the atmosphere and how much light is blocked by airborne particles. Among these aerosols that can be detected using the AOD are the dust, sea salts, volcanic ash, and smoke, being either solid or liquid particles. Dust and smoke are the main constituents of the vertical column in our case, where they reflect the visible and near infrared radiation, preventing them from passing through the atmospheric column. The presence of the dust and smoke particles during certain time periods throughout the year results in optical depth increase over Cairo, and the Delta region is shown in Figure 10.19. The monthly variations

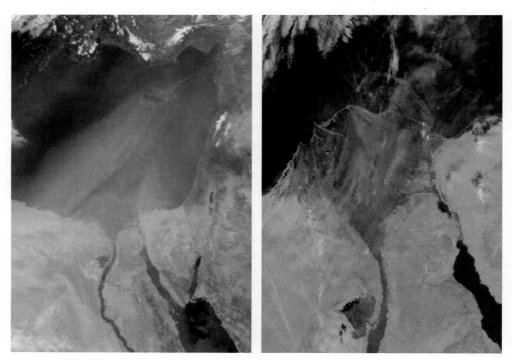

Figure 10.18 MODIS satellite images showing the dust storms blowing during April 2005 (*left panel*) and very dense smog during October 2005 over the greater Delta region (*right panel*).

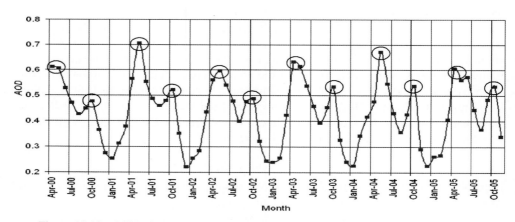

Figure 10.19 MODIS monthly aerosol optical depth (AOD) over Cairo, latitude [29.66 N, 30.43 N] and longitude [30.64 E, 31.84 E], from April 2000 to November 2005. Peak values are circled.

of aerosol optical depths indicate seasonal variations, with relativity higher aerosol optical depth or atmosphere turbidity in the summer and very low aerosol optical depths during the winter. The MODIS AOD product correctly indicates high AOD (>0.5) during dusty and highly polluted time periods, where high AOD is expected, and low values during well-known clear periods. AOD peaks are observed over the months of April, May, and

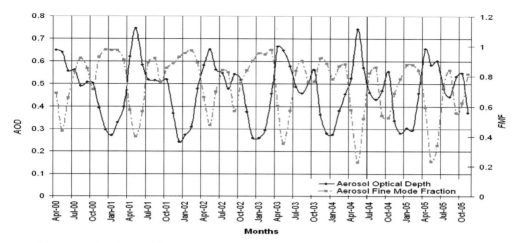

Figure 10.20 Aerosol fine mode fraction (FMF) versus aerosol optical depth (AOD) over Cairo, latitude [29.66 N, 30.43 N] and longitude [30.64 E, 31.84 E], from April 2000 to November 2005.

October, where April and May represent the most frequent time of dust storm occurrences and October represents the time of black cloud occurrences over the past few years.

Aerosol fine mode fraction

Dust and different anthropogenic pollutants forming the black cloud vary widely in grain size distribution and hence have varying optical characteristics. Using the fine mode fraction (FMF) derived from radiance measurements by MODIS on the Terra satellite, aerosols can be partitioned between fine and coarse modes. This partitioning results in the reduction of the average aerosol optical depth to values below those retrieved from MODIS. This is described by the fact that the MODIS-retrieved column AOT at 550 nm ($\tau 550$) is the sum of maritime (τm), dust (τd), and anthropogenic (τ_a) components. The fraction of $\tau 550$ contributed by fine mode aerosol ($f 550$) is also reported from the MODIS retrievals. AOD values showed the intensity of either the dust or black cloud events and revealed higher dust presence during the months of April and May than the black cloud events during the months of October of each year. This is attributed to the fact that dust storms originate from a wider pool and are characterized by higher grain sizes compared to pollutants contributing the black cloud formation. Fine particles with effective radii 0.10, 0.15, 0.20, and 0.25 μm, together with the dust particles with effective radii 1.5 and 2.5 μm, contribute to the value of the AOD intensity. However, FMF shows only the fine particles contribution in the AOD intensity, which is low during the dust storm season because coarser grains dominate these dust events. In contrast, FMF shows high values during the stable times of the year and a matching value with the AOD value during the months of October during the years 2000 to 2003. This is due to the nature of the atmosphere during the stable months of the year and the months of the black cloud events. The latter is evidenced by the fine particles dominancy due to the emissions during this month of the year as proposed with the matching season of the black cloud events. However, a lower matching is observed in the years 2004 and 2005, which might be due to the lower emission rates forced by the governments for a cleaner environment. The dominant presence of the FMF particles during the most stable months of the year is validated by the high negative correlation between AOD and FMF of value -0.75, as shown in Figure 10.20.

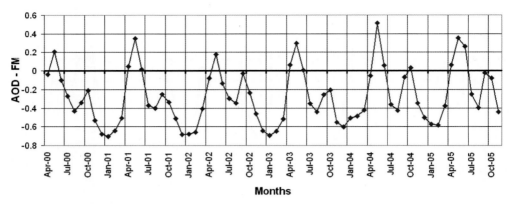

Figure 10.21 Difference between aerosol optical depth (AOD) and aerosol fine mode fraction (FMF) over Cairo, latitude [29.66 N, 30.43 N] and longitude [30.64 E, 31.84 E], from April 2000 to November 2005.

Calculating the difference between the AOD and the FMF revealed the coarse grains dominancy during the months of April and May of each year. Such high differences correspond to the dust season during these months, as shown in Figure 10.21. In contrast, finer grains dominate the rest of the months. Low difference values are observed during the months of January of each year because of the clear atmospheric column, and hence, the dominancy of the fine particles comprising the vertical column of the atmosphere.

Fire aerosols: thermal anomalies and fire pixel count

Fires are a common occurrence in Egypt due to trash burning and other industrial activities. Owing to the high pollution during the month of October, we detected that the active fires occurred over the same region as the block cloud events, contributing to the overall pollution. MODIS thermal anomalies/fire products detect fire locations (i.e., hotspots) using 4- and 11-micrometer brightness temperatures assuming the emissivity are equal to 1. The detection strategy is based on absolute detection (i.e., if the fire is strong enough relative to the background to account for variability of the surface temperature and reflection by sunlight). A corrected fire pixel layer count is produced over Egypt during the month of October of years 2001 to 2004, as shown in Figure 10.22.

High anomalous fire counts of different strengths are detected over the Delta region over three years, where the years 2002 and 2004 showed the highest counts, on the order of 50 to 100 events, as shown in Figure 10.23.

10.4.2 Pollution effects forcing on large-scale vegetation in India

Climate variability can have a large impact on the vegetation pattern of different ecosystems because climate and terrestrial ecosystems are closely coupled, and therefore, affect the population of the Earth. The effect of climate change on vegetation has been studied by numerous researchers (Anyamba and Eastman, 1996; Eastman and Fulk, 1993; Gray and Tapley, 1985; Li and Kafatos, 2000; Lu et al., 2001). Sarkar and Kafatos (2004), studied the impacts of climate factors and pollution on large-scale vegetation in India. A number of studies have shown that the normalized difference vegetation index (NDVI) provides an effective measure of photosynthetically active biomass (Justice et al., 1985; Tucker and Sellers, 1986). NDVI is also found to be well correlated with physical climate geophysical

10.4 Air pollution

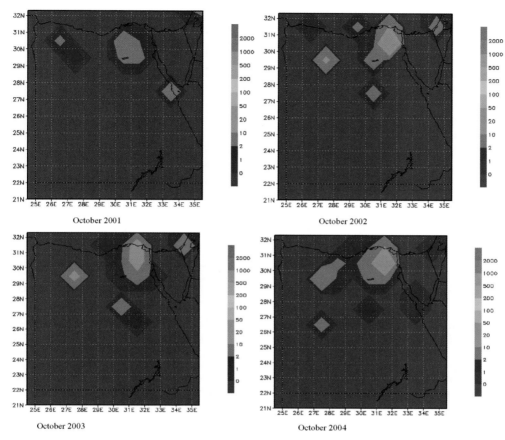

Figure 10.22 MODIS corrected fire pixel count over Egypt during the month of October from 2001 to 2004.

variables, including rainfall, temperature, and evapotranspiration in a wide range of environmental conditions (Gray and Tapley, 1985). The interannual and interseasonal variability of NDVI has been studied for different regions in the past (Anyamba and Eastman, 1996; Eastman and Fulk, 1993), and other work has shown global associations of NDVI anomaly patterns in different continents with sea surface temperature anomaly (SSTA) over the tropical Pacific (Myneni et al., 1996). Li and Kafatos (2000) studied the spatial patterns of NDVI and its relation to ENSO over North America, while Li and Kafatos (2002) performed a statistical spatial analysis of NDVI for different ecoregions. Sarkar and Kafatos (2004) extended the previous work in order to understand the effect of meteorological parameters on NDVI, which is very important in contributing to the agricultural productivity of the Indian subcontinent. They used EOF/PCA (principal component analysis) for analyzing the variability of a single field (i.e., a field of only one scalar variable, (such as NDVI and AOD). The method (Preisendorfer, 1988) finds the spatial patterns of variability (EOF), their time variation, or (PCs), and gives a measure of the importance of each pattern.

They found that the distribution of EOF1 of NDVI shows a good relation, with potential anthropogenic effects as seen from the EOF analysis of the AI. Figure 10.24 shows good correlation of the EOF1 and EOF4 of NDVI from 1982 to 2000 with the EOF1 of AI, despite the limited data span of the AI. The PC1 and PC4 of NDVI show substantial

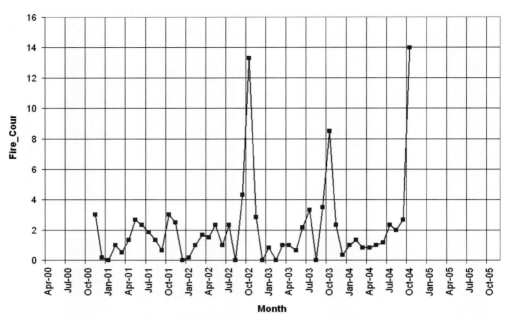

Figure 10.23 MODIS monthly fire count over Cairo, latitude [29.66 N, 30.43 N] and longitude [30.64 E, 31.84 E], from November 2000 to October 2004.

negative correlation coefficients of -0.53 and -0.7, with p values of 0.0175 and 0.043, carried out through a two-tailed t-test with a test size of 0.05, with the PC1 of AI.

Sarkar and Kafatos (2004) also found substantial negative correlation of PC1 of AI with NDVI, apparent over the entire southern edge of the Himalayas and the majority of the IG plains and parts of southern peninsular India. The aerosol over this region is rich in sulfates, nitrates, organic and black carbon, and fly ash. The IG plain receives most of the rainfall because the monsoon winds are obstructed from flowing beyond the Himalayan range; however, even with this effect, the vegetation growth is not high in IG plains principally because of high aerosol, which is due to large population growth and industrialization. The conclusion is that climatic factors being the same over the Indian subcontinent and Southeast Asia, the differences observed in vegetation density between these two areas can only be attributed to anthropogenic influences. The mechanism by which such aerosol buildup can affect vegetation growth can be considered from cloud microphysics. Specifically, the aerosol clouds are comprised of smaller water droplets that are less likely to fall and thus result in reduced rainfall (Jones et al., 1994). Moreover, such toxic buildup of clouds because of aerosols can slash the incoming solar radiation by 10% to 15%, thus cutting off the vital energy source for plant growth. These are coupled to the direct effect of the buildup of tropospheric pollutants on plant growth and development. The increase in pollutants like SO_2, NO_2, and tropospheric ozone can harm the plants and crops in more than one way and can have an impact on the yield, pest infestations, nutritional quality, and disease of different crops and on forests around megacities (Emberson et al., 2001; Kuylenstierna and Hicks, 2003). In India, a number of field studies (Emberson et al., 2001) have been carried out on wheat around industrial sources of pollution, which show large reductions in yield close to the source of pollution. The yield reductions are the result not only of SO_2, but are also associated with NO_2 and particulates that accompany SO_2 concentrations in the field.

10.5 Forcing component

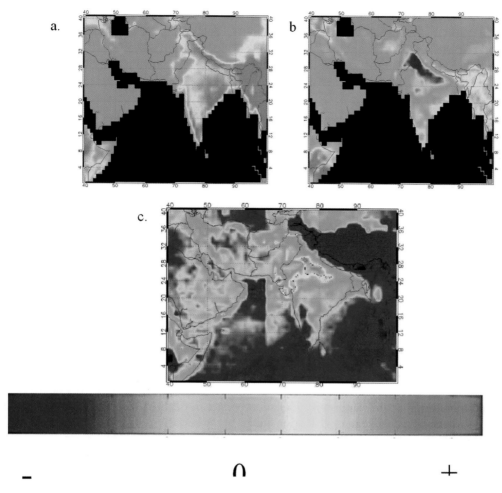

Figure 10.24 The EOFs of NDVI from 1982 to 2000 and EOF1 of aerosol optical depth (AOD). (a) EOF1 of NDVI; (b) EOF4 of NDVI; (c) EOF1 of AOD. Note the bright red streak in (c) signifying high accumulation of suspended particulates and its correspondence with areas of low vegetation in EOF1 and EOF4 shown in (a) and (b).

The strong effect of tropospheric aerosols revealed in remote sensing NDVI data suggests the serious consequences of pollution on vegetation. The effect of aerosol found by this study is clearly local, restricted to areas of industrial belts, and a result of growing urbanization and industrialization. Such effects of pollutants on large-scale vegetation and agricultural productivity need to be taken into account in countries such as India. In particular, the rapid increase in urban pollution predicted over the coming decades in many developing countries may have large impacts on agriculture and thus provide a negative feedback for the further development of large urban centers.

10.5 Forcing component

The implications on climatic effects due to aerosols with a large variability such as mineral dust are important. Airborne mineral dust can influence the climate by altering the radiative properties of the atmosphere. For instance, aerosols in the form of dust particles reflect the incoming solar radiation to space, thereby reducing the amount of radiation available to the ground. This is known as "direct" radiative forcing of aerosols. Aerosols also serve as CCN and change the cloud albedo and microphysical properties of clouds, known as "indirect" radiative forcing of aerosols. Direct and indirect radiative forcing by mineral dust are observed over a desert case study in China, and a highly vegetated case study over the Nile Delta in Egypt, using boundary layer dispersion (BLD), albedo, sensible heat flux (SHF), latent heat flux (LHF), and outgoing long-wave radiation (OLR) parameters. These regional studies show climate effects at regional levels, which, however, may also have consequences for long-term climate effects. The connection of dust-related forcing will not only affect megacities, but may also be related to climate variability. During the presence of a dust event, short-wave fluxes largely decrease, accompanied by an abrupt increase in the down-welling long-wave fluxes, resulting in surface forcing. This leads to absorption of the short- and long-wave radiations, resulting in a positive forcing in the top of the atmosphere.

In this section, we are focusing on the radiative impacts of dust over some meteorological parameters. Aerosol in the atmosphere is a mixture of different components. The detection and identification of a specific range of aerosol particles (dust in our context) requires knowledge of the microphysical and optical properties in the solar and terrestrial spectral range. Once characterized, the ability of dust aerosols to modify the incoming solar radiation can be used to effectively identify the onset of dust storms and monitoring of the same at different times of the year. We have made use of satellite datasets such as the radiation budget data from the AOD from MODIS and AI from TOMS to calculate the radiation at the top of the atmosphere (TOA). Moreover, numerical models can be used to predict the radiation at the surface. The aerosol forcing at TOA, middle atmosphere, and surface can be calculated based on a combination of satellite observations and numerical calculations. Such monitoring of the variation in radiation budget throughout the year will trigger an alarm whenever a change in net radiation budget at the surface is observed, or a significant increase in the atmospheric backscattering is noticed.

Satellite datasets such as the radiation budget data from clouds and earth's radiant energy system (CERES) and the AOD from MODIS and TOMS can be used to calculate the radiation at the TOA. Numerical models can be used to predict the radiation at the surface. The aerosol forcing at TOA, middle atmosphere, and surface can then be calculated based on a combination of these satellite observations and numerical calculations. Such monitoring of the variation in radiation budget throughout the year will trigger an alarm whenever a change in net radiation budget at the surface is observed or a significant increase in the atmospheric backscattering, for instance, over China is noticed as shown in Figure 10.25. Sudipta Sarkar (personal communication) outlined this approach for China, where dust storms originate from the western dust bowl of the country and travel eastward toward the eastern coast during the spring.

Figure 10.25 shows that there occurs a drastic increase in surface forcing over the eastern and northeastern part of China during spring months (MAM) compared to winter and summer. The values increase from about ~ -20 W/m^2 during the winter months (DJF) to ~ -40 during the spring months (MAM) in the eastern part of China, while a larger increase is seen in the northeast (~ -10 to ~ -60 W/m^2). As the phase of dust transport from the western side of the country subsides with the coming of summer (JJA), the surface forcing again drops back to lower values of around -30 W/m^2 in the eastern part, while higher values persist in the northeast, indicating the persistence of suspended dust particles. Two dust events are presented here to monitor their forcing impact on several atmospheric parameters.

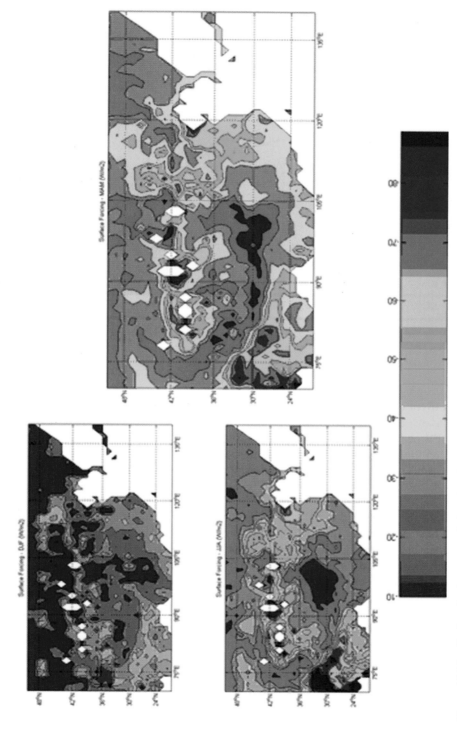

Figure 10.25 Surface forcing pattern during the winter (DJF), summer (JJA), and spring (MAM) seasons over China.

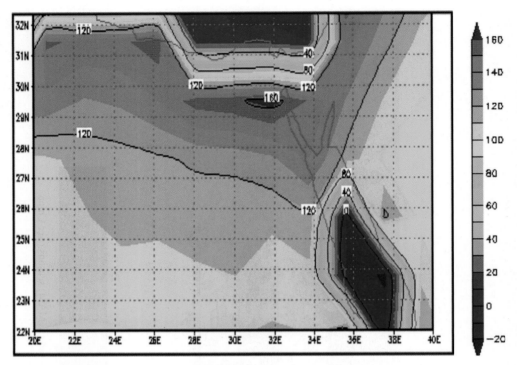

Figure 10.26 Sensible heat flux over the Nile Delta region of Egypt, 31 August 2000.

10.5.1 Egypt case study

This dust event occurred over the Nile Delta region, a highly vegetated area that suffered from a massive dust event prevailing from the Western Desert. A detailed analysis of this dust event is performed by El-Askary et al. (2003b; 2006), where the dust characteristics over different regions of the electromagnetic spectrum have been studied. Over the Delta region, sensible and highly vegetated latent heat fluxes are crucial for analysis.

Sensible heat flux

The positive surface radiation balance is partitioned into heat flux into the ground, and turbulent heat fluxes back into the atmosphere through sensible and latent heat fluxes. Sensible heat is the energy flow due to temperature gradients. SHF had apparent change with the strong sand dust storm's passing. The daytime surface air temperature over a desert is regulated mainly by sensible heat flux from the ground into the surface air. The peak heat flux at the center occurs during the passage of a large dust event, as shown in Figure 10.26, representing the Delta event reaching a high value of 160 W/m^2. However, we note the gradual decrease of the SHF over the Delta when approaching the Mediterranean because the dust event is not covering the northern part of the Delta, as shown in Figure 10.27 depicting MODIS image.

Latent heat flux

SHF and LHF represent energy flowing from the Earth's surface into the atmospheric boundary layer. Latent heat is energy flow due to evapotranspiration (the sum total of water evaporated and transpired by plants into the atmosphere). Hence, heat flow from bare fields

Figure 10.27 MODIS image showing dust invading the Delta region from the Western Desert, 31 August 2000.

is dominated by sensible heat, while heat flow from vegetated areas and water bodies is dominated by latent heat. This is validated from the low SHF values over the Mediterranean Sea, reaching below zero levels shown in Figure 10.26, and the very high LHF over the sea 350 W/m², as shown in Figure 10.28. The LHF is expected to show a high value over the Delta region, owing to the high vegetation characterizing this region. However, a smaller value ranging from 0 to 200 W/m² is observed over different parts of the Delta. This is because the dust occurrence absorbs most of the moisture going into the atmosphere and hence leads to a lower LHF value. However, the northern part of the Delta did not suffer from as much dust as the rest of the area, which leads to a higher LHF value to be observed.

10.5.2 China case study

The second forcing study is related to a very strong sand dust storm that occurred on 5 May 1993 in several remote regions of northwest China, namely, Xinjiang, Gansu, Ningxia, and Inner Mongolia. A detailed analysis of this dust event is found in Zhang et al. (2003), where improved physical processes and a radiation parameterization scheme are introduced, resulting in a better simulation of the dust event, as shown in Figure 10.29. This figure shows the sand-dust wall also proceeding from northwestern Jinchang to the city and further intensified.

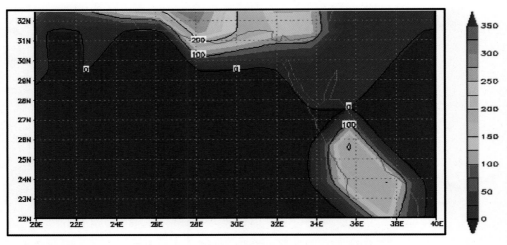

Figure 10.28 Latent heat flux over the Nile Delta region of Egypt, 31 August 2000.

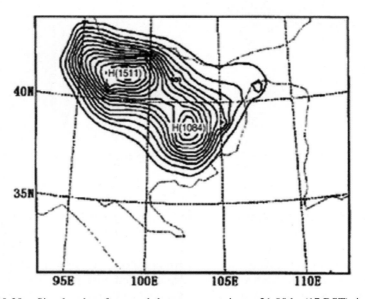

Figure 10.29 Simulated surface sand-dust concentration at 21:00 hr (17 BST), in mg/m^3.

The radiative forcing effect plays an important role during strong dust and sand events; thus, radiation fluxes can be studied in this research through different parameters. Over China, the boundary layer dispersion, albedo, sensible heat flux, and outgoing long-wave radiation can be studied. The extent of the changes is related to the strength of the dust event.

Boundary layer dispersion

Dispersion in the atmospheric boundary layer is a random process driven by the stochastic nature of turbulence (Weil et al., 2006). Uplifting vertical wind currents are needed for dust uptakes and lofting, and such vertical motions are more common when having an unstable boundary layer. This is attributed to the fact that a stable boundary layer suppresses the

10.5 Forcing component

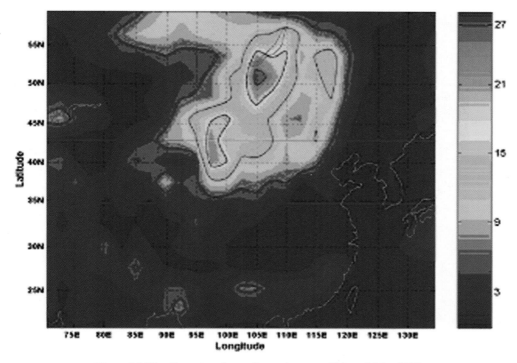

Figure 10.30 Boundary layer dispersion over China, 5 May 1993.

vertical motions and inhibits dust lofting. In the China case study, where there is extensive lack of vegetation over the region under study, extreme daytime heating of the ground was observed, hence contributing to the destabilization of the boundary layer and leading to a high dispersion value. Therefore, as the amount of heating increases, one expects this unstable layer to deepen. Thus, midlatitude deserts, with their high daytime temperatures, are particularly prone to an unstable boundary layer. The contribution to a high dispersion value as observed over China, where the boundary layer dispersion over the dust event reaches a maximum value of ∼26 W/m^2, as shown in Figure 10.30, matches well the simulated dust concentration shown in Figure 10.29. The highest boundary layer heights are associated with regions where the sensible heat flux is greatest and the latent heat flux is smallest due to a lack of vegetation. Boundary layer heights in the deserts may be systematically higher than the slightly wetter regions at the edges of deserts.

Albedo

Large amounts of aerosol particles are emitted through natural and anthropogenic activities into the atmosphere (Kafatos et al., 2005). Dust storms are a major contributor to the aerosols' occurrence in the atmospheric column. Hence, such aerosol concentrations reflected by high AOD values lead to a dual effect. The cooling effect occurs through the scattering of most of the solar radiation, enhancing the local planetary albedo of the atmosphere. The heating effect occurs through strong heat absorption by the aerosol particles, heating the Earth-atmosphere system. In addition, as indicated previously, aerosols act in part as CCNs, which leads to increases in the number of cloud droplets having smaller cloud droplet radii. Such increased number of cloud droplets together with a small droplet radius leads to an increase in the cloud albedo (Twomey, 1977). This has been observed in

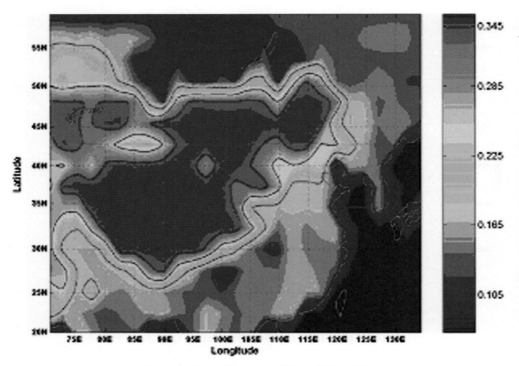

Figure 10.31   Albedo over China, 5 May 1993.

Figure 10.31 over the northeastern region of China, where high albedo values are observed in association with the dust concentration simulated from the model shown in Figure 10.29. The albedo value ranges from 0.1 to 0.345, having the highest value over the region affected by the dust cloud. Compared with the clear-sky background, the albedo values for the area under study during the dust event increased by a great percentage, where the direct solar radiation flux at the surface decreased and the scattered radiation flux at the surface increased as a result of the excess aerosols in the atmosphere.

Outgoing long-wave radiation

The Earth's radiation budget (ERB) is the balance between the incoming energy from the sun and the outgoing reflected and scattered solar radiation plus the thermal infrared emission to space. Low [outgoing long-wave radiation (OLR) is generally due to low thermal emission from the cold tops of extensive cloud cover. However, bright regions correspond to high OLR relating to cloudfree scenes of high surface temperature. This is because direct surface solar heating is considerable over these regions of high surface temperatures, and hence, OLR is also large. During the China dust event, although the surface temperature was high, the dust cloud created a blanket that helped in cool down the temperature and thus resulted in having a low OLR associated with the dust cloud over the region under study, reaching a value of 142.5 W/m^2, as seen in Figure 10.32.

10.6 Conclusions

The world's megacities are increasingly affected by a number of natural and anthropogenic hazards. In particular, natural and anthropogenic aerosols associated with sand and dust

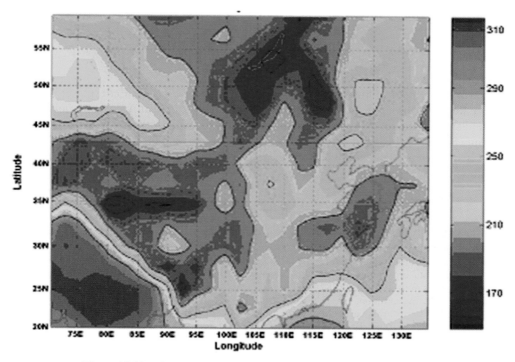

Figure 10.32   Outgoing long-wave radiation over China, 5 May 1993.

storms, urban pollution and fires, cause major economic and health problems, affect local climate, and may even be seriously affecting global climate in the long term. It is, therefore, imperative to use advanced technologies to study the onset of dust storms and the local and regional impacts of urban pollution, and to monitor fires. Remote sensing provides the requisite global and regional coverage as well as temporal coverage; and, when combined with modeling and ground observations, can yield sustained support to mitigate the adverse effects. In this chapter, we demonstrate the potential applied to one large megacity, Cairo, which is affected by several different types of aerosols, and to larger, regional areas, containing several megacities, in India. Lately, dust storms have been receiving more attention due to their increasing frequency and intensity. They are often caused or exacerbated by poor environmental management resulting from a range of factors, including deforestation, degraded rangelands, exhausted cultivated fields, salinized irrigated land, depleted groundwater resources, and the shrinking of water bodies. Dust events would greatly affect tourism, industry, food supply, and health through impacting our monuments and coral reefs, factories, destroying the vegetation cover, and imposing a hazardous health situation, respectively. Environmental management of such occurring events would initially require, to a great extent, their monitoring and tracking. Hence, dust emission, suspension, and deposition should be a major preoccupation of environmental management plans of most industries. For instance, health effects of dust and other pollutants are adequately covered by environmental indicators such as PM10, AOD, and haze. Monitoring and tracking such aerosols would help as a separate environmental indicator, because dust deposition is more a measure of whether, and how well, good practice is being followed. This would deliver some control measures for reducing the manmade part of the crises. This is achieved through increasing the awareness of the dust management in several industries, particularly in mining and in rural areas where episodic wind erosion of soil continues to be a problem.

Acknowledgments

The work presented in this chapter is supported by the Center for Earth Observing and Space Research through grants and projects, among it the National Aeronautics and Space Administration-funded VAccess. The authors want to express their appreciation to Dr. Sudipta Sarkar for his input on the China seasonal forcing and the pollution effects on the biosphere in India. The authors also want to extend their appreciation to Mr. Emmanuel Smith for his support in the final editing of the book chapter.

References

Anyamba, A., and Eastman, J. R. (1996) "Interannual Variability of NDVI over Africa and Its Relation to El Niño/Southern Oscillation," *Int. J. Remote Sens.* **17**(13), pp. 2533–2548.

Cakmur, R., Miller, R., and Tegen, I. (2001) "A Comparison of Seasonal and Interannual Variability of Soil Dust Aerosol over the Atlantic Ocean as Inferred by TOMS and AVHRR AOT Retrieval," *J. Geophys. Res.* **106**, pp. 18287–18304.

Chavez, P. S., Mackinnon, D. J., Reynolds, R. L., and Velasco, M. G. (2002) "Monitoring Dust Storms and Mapping Landscape Vulnerability to Wind Erosion Using Satellite and Ground-Based Digital Images," *Aridlands News Letter* **51**, May/June. see http://ag.arizona.edu/OALS/ALN/aln51/chavez.html

Dey, S., Tripathi, S. N., Singh, R. P., and Holben, B. (2004) "Influence of Dust Storms on the Aerosol Optical Properties over the Indo–Gangetic Basin," *J. Geophys. Res.* **109**, D20211, doi: 10.1029/2004JD004924.

Egyptian Meteorological Authority. (1996) *Climate Atlas of Egypt*, vol. 157, Egyptian Meteorological Authority, Cairo, Egypt.

El-Askary, H. (2006) "Air Pollution Impact on Aerosol Variability over Mega Cities using Remote Sensing Technology: Case Study, Cairo," *Egyptian Journal of Remote Sensing and Space Sciences*. pp. 31–40.

El-Askary, H., Gautam, R., and Kafatos, M. (2004) "Monitoring of Dust Storms over Indo–Gangetic Basin," *Indian J. Rem. Sens.* **32**(2), pp. 121–124.

El-Askary, H., Gautam, R., Singh, R. P., and Kafatos, M. (2006) "Dust Storms Detection over the Indo–Gangetic Basin Using Multi Sensor Data," *Advances in Space Research Journal* **37**, pp. 728–733.

El-Askary, H., and Kafatos, M. (2006) "Potential for Dust Storm Detection Through Aerosol Radiative Forcing Related to Atmospheric Parameters," *IEEE Int. Geosc. and Remote Sensing Symposium IGARSS2006*, Denver, Colorado, July 31–August 4.

El-Askary, H., Kafatos, M., and Hegazy, M. (2002) "Environmental Monitoring of Dust Storms over the Nile Delta, Egypt, Using MODIS Satellite Data," *Proceedings of Third International Symposium Remote Sensing of Urban Areas*, Istanbul, Turkey, June, pp. 452–458.

El-Askary, H., Kafatos, M., Xue, L., and El-Ghazawi, T. (2003a) "Introducing New Techniques for Dust Storm Monitoring," *IEEE Int. Geosc. Remote Sensing Symposium IGARSS2003*, vol. IV, Toulouse, France, July 21–25, pp. 2439–2441.

El-Askary, H., Sarkar, S., Kafatos, M., and El-Ghazawi, T. (2003b) "A Multi-Sensor Approach to Dust Storm Monitoring over the Nile Delta," *IEEE Trans. Geosc. Remote Sensing* **41**(10), pp. 2386–2391.

Emberson, L. D., Ashmore, M. R., Murray, F., Kuylenstierna, J. C. I., Percy, K. E., Izuta, T., Zheng, Y., Shimizu, H., Sheu, B. H., Liu, C. P., Agrawal, M., Wahid, A., Abdel-Latif, N. M., van Tienhoven, M., de Bauer, L. I., and Domingos, M. (2001) "Impacts of Air Pollutants on Vegetation in Developing Countries," *Water Air and Soil Pollution* **130**(1–4), pp. 107–118.

Ferraro, R. R., and Grody, N. C. (1994) "Effects of Surface Conditions on Rain Identification Using the DMSP-SSM/I," *Remote Sensing Reviews* **11**, pp. 195–209.

Gray, T. I. and Tapley, D. B. (1985) "Vegetation Health: Nature's Climate Monitor," *Adv. Space Res.* **5**, pp. 371–377.

Hammonou, E., Chazette, P., Balis, D., Dulax, F., Schneider, X., Ancellet, F., and Papyannis, A. (1999) "Characterization of the Vertical Structure of Saharan Dust Export to the Mediterranean Basin," *J. Geophys. Res.* **104**, pp. 22257–22270.

Herman, J. R., Bhartia, P. K., Torres, O., Hsu, C., Setfor, C., and Celarier, E. (1997) "Global Distribution of UV-Absorbing Aerosols from Nimbus 7/TOMS Data," *J. Geophys. Res.* **102**, pp. 16889–16909.

Husar, R. B., Tratt, D. M., Schichtel, B. A., Falke, S. R., Li, F., Jaffe, D., Gasso, S., Gill, T., Laulainen, N. S., Lu, F., Reheis, M. C., Chun, Y., Westphal, D., Holben, B. N., Gueymard, C., McKendry, I., Kuring, N., Feldman, G. C., McClain, C., Frouin, R. J., Merrill, J., DuBois, D., Vignola, F., Murayama, T., Nickovic, S., Wilson, W. E., Sassen, K., Sugimoto, N., and Malm, W. C. (2001) "Asian Dust Events of April 1998," *J. Geophy. Res.* **106**, pp. 18317–18330.

Joliffe, I. (1986) *Principal Component Analysis*, Springer, Berlin, Germany.

Jones, A. D., Roberts, L., and Slingo, A. (1994) "A Climate Model Study of the Indirect Radiative Forcing by Anthropogenic Sulphate Aerosol," *Nature* **370**, pp. 450–453.

Justice, C. O., Townshend, J. R. G., Holben, B. N., and Tucker, C. J. (1985) "Analysis of the Phenology of Global Vegetation Using Meteorological Satellite Data," *Int. J. Remote Sens.* **6**, pp. 1271–1318.

Kafatos, M., El-Askary, H., and Kallos, G. (2004) "Potential for Early Warning of Dust Storms," *35th COSPAR Science Assembly*, Paris, France, July 18–25.

Kafatos, M., Singh, R. P., El-Askary, H. M., and Qu, J. (2005) "Natural and Anthropogenic Aerosols in the World's Megacities and Climate Impacts," *Eos Trans. AGU*, **85**(17), Joint Assem. Suppl., San Francisco, California, December 5–9.

Kallos, G., Nickovic, S., Jovic, D., Kakaliagou, O., Papadopoulos, A., Misirlis, N., Boukas, L., and Mimikou, N. (1997a) "The ETA Model Operational Forecasting System and Its Parallel Implementation," *Proceedings of the 1st Workshop on Large-Scale Scientific Computations*, Varna, Bulgaria, June 7–11, p. 15.

Kallos, G., Nickovic, S., Jovic, D., Kakaliagou, O., Papadopoulos, A., Misirlis, N., Boukas, L., Mimikou, N., Sakellaridis, G., Papageorgiou, J., Anadranistakis, E., and Manousakis, M. (1997b) "The Regional Weather Forecasting System SKIRON and Its Capability for Forecasting Dust Uptake and Transport," *Proceedings of WMO Conference on Dust Storms*, Damascus, Syria, November 1–6, p. 9.

Kallos, G., Nickovic, S., Papadopoulos, A., Jovic, D., Kakaliagou, O., Misirlis. N., Boukas, L., Mimikou, N., Sakellaridis, G., Papgeorgiou, J., Anadranistakis, E., and Manousakis, M. (1997c) "The Regional Weather Forecasting System SKIRON: An Overview," *Proceedings of the International Symposium on Regional Weather Prediction on Parallel Computer Environments*, Athens, Greece, October 15–17, pp. 109–122.

Kallos, G., Papadopoulos, A., Katsafados, P., and Nickovic, S. (2002) "Recent developments towards the prediction of desert dust cycle in the Atmosphere," Presented at the Bilateral Israeli-Turkish workshop on Atmospheric Deposition of Aerosols and Goses in the Eastern Mediterranean, Deal Sea, Isreal, 6–8 Janvary.

Kallos, G., Papdopoulos, A., Katsafados, P., and Nickovic, S. (2006) "Transatlantic Saharan Dust Transport: Model Simulation and Results," *J. Geophys. Res.* **111**, D09204, doi: 10.1029/2005JD006207.

Kutiel, H., and Furman, H. (2003) "Dust Storms in the Middle East: Sources of Origin and Their Temporal Characteristics," *Indoor Built Environ.* **12**, pp. 419–426.

Kuylenstierna, J., and Hicks, K. (2003) "Air Pollution in Asia and Africa: The Approach of the RAPIDC Programme," *Proc. 1st Open Seminar Regional Air Pollution in Developing Countries*, Stockholm, Sweden, June 4, p. 34.

Li, Z., and Kafatos, M. (2000) "Interannual Variability of Vegetation in the United States and Its Relation to El Niño Southern Oscillation," *Remote Sens. Environ.* **71**(3), pp. 239–247.

Liepert, B. G. (2002) "Observed Reductions of Surface Solar Radiation at Sites in the United States and Worldwide from 1961 to 1990," *Geophysical Research Letters* **29**(10), 1421 pp. 61-1–61-4.

Lim, c., and Kafatos, M. (2002) "Frequency analysis of natural vegetation distribution using NDVI/AVHRR data from 1981 to 2000 for North America: correlations with SOI." Int. J. Remote Sens., 23(17), pp. 3347–3383.

Lu, L., Pielke, R. A., Sr., Liston, G. E., Parton, W. J., Ojima, D., and Hartman, M. (2001) "Implementation of a Two-Way Interactive Atmospheric and Ecological Model and Its Application to the Central United States," *J. Climate* **14**, pp. 900–919.

MacKinnon, D. J., and Chavez, P. S. (1993) "Dust Storms," *Earth Magazine* pp. 60–64.
Middleton, N. J. (1986) "Dust Storms in the Middle East," *J. Arid Environ.* **10**, 83–96.
Miller, S. D. (2003) "Satellite Surveillance of Desert Dust Storms," *NRL Reviews* pp. 69–77.
Moulin, C., Lambert, C. E., Dayan, U., Masson, V., Ramonet, M., Bosquet, P., Legrand, M., Balkanski, Y.J., Guelle, W., Marticonera, B., Bergametti, G., and Dulac, G. (1998) "Satellite Climatology of African Dust Transport in the Mediterranean Atmosphere," *J. Geophys. Res.* **103**, pp. 13137–13144.
Myneni, R. B., Los, S. O., and Tucker, C. J. (1996) "Satellite-Based Identification of Linked Vegetation Index and Sea Surface Temperature Anomaly Areas from 1982–1990 for Africa, Australia and South America," *Geophys. Res. Lett.* **23**, pp. 729–732.
National Aeronautics and Space Administration, Goddard Space Flight Center. (n.d.) Total Ozone Mapping Spectrometer website. http://toms.gsfc.nasa.gov/.
Nickovic, S., Jovic, D., Kakaliagou, O., and Kallos, G. (1997) "Production and long-range transport of desert dust in the Mediterranean Region: Eta model simulations, "Proceedings of the 22nd NATO/CCMS Int. Techn. Meeting on Air Pollution Modeling and Its Application, Clermont Ferrand, France, 2–6 June, P. 8. Edited by Sven-Erik Gryning and Nandine Chaumerliac, plenum Press, New York.
Nickovic, S., Kallos, G., Papadoupoulos, A., and Kakaliagou, O. (2001) "A Model for Prediction of Desert Dust Cycle in the Atmosphere," *J. Geophys. Lett.* **106**(D16), pp. 18113–18129.
Papadopoulos, A., Kallos, G., Mihailovic, D., and Jovic, D. (1997) "Surface Parameterization of the SKIRON Regional Weather Forecasting System," *Proceedings of the Symposium on Regional Weather Prediction on Parallel Computer Environments*, Athens, Greece, October 15–17, p. 6.
Papadopoulos, A., Katsafados, P., and Kallos, G. (2001) "Regional Weather Forecasting for Marine Application," *GAOS* **8**(2–3), pp. 219–237.
Preisendorfer, R. W. (1988) *Principal Component Analyses in Meteorology and Oceanography*, Elsevier, New York, pp. 425 New York.
Prospero, J. M. (1996) "Saharan Dust Transport over the North Atlantic Ocean and the Mediterranean," in *The Impact of Desert Dust across the Mediterranean*, edited by S. Guerzoni and R. Chester, pp. 133–151, Kluwer Acadmic, Norwell, Massachnsetts.
Prospero, J. M. (1999) "Long-Range Transport of Mineral Dust in the Global Atmosphere: Impact of African Dust on the Environment of the Southeastern United States," *Proceedings of the National Academy of Science USA* **96**, pp. 3396–3403.
Prospero, J. M., Ginoux, P., Torres, O., and Nicholson, S. (2002) "Environmental Characterization of Global Source of Atmospheric Soil Dust Derived from the NIMBUS-7 TOMS Absorbing Aerosol Products," *Rev. Geophys.* **40**, pp. 2–31.
Qu, J., Kafatos, M., Yang, R., Chiu, L., and Riebau, A. (2002) "Global Pollution Aerosol Monitoring in the Atmospheric Boundary Layer Using Future Earth Observing Satellite Remote Sensing," *Proceedings SPIE (RS-4882)*, pp. 100–105.
Richards, J. A., and Jia, X. (1999) *Remote Sensing Digital Image Analysis: An Introduction*, Springer-Verlag, Berlin, Germany.
Rosenfeld, D. (2001) "Smoke and Desert Dust Stifle Rainfall, Contribute to Drought and Desertification," *Aridlands News Letter* **49**, May/June. See http://ag.arizona.ed./OALS/ALN/aln49/rosenfeld.html.
Sabins, F. F. (1978) Remote Sensing: Principles and Interpretation, Freeman Hall, san Francisco.
Sarkar, S., and Kafatos, M. (2004) "Interannual Variability of Vegetation over the Indian Sub-Continent and Its Relation to the Different Meteorological Parameters," *Remote Sensing Environment* **90**, 268–280.
Singh, R. P., Dey, S., Triapthi, S. N., Tare, V., and Holben, B. (2004) "Variability of Aerosol Parameters over Kanpur City, Northern India," *J. Geophys. Res.* **109**, D23206.
Staudenmaier, M. (1997) "The Navy Operational Global Atmospheric Prediction System (NOGAPS)," *Western Region Technical Attachment*, pp. 97–109.
Stowe, L. L., Ignatov, A. M., and Singh, R. R. (1997) "Development, validation, and Potential Enhancements to the Second-Generation Operational Product at the National Environment Satellite, Data, and Information Service of the National Oceanic and Atmospheric Administration," *J. Geophys. Res.* **102**, pp. 16923–16934.
Tanré, D., Kaufman, Y. J., Herman, M., and Mattoo, S. (1997) "Remote Sensing of Aerosol Properties over Oceans Using the MODIS/EOS Spectral Radiances," *J. Geophys. Res.* **102**, pp. 16971–16988. 22.

Tegen, I. A., and Fung, I. (1995) "Contribution to the Atmospheric Mineral Aerosol Load from Land Surface Modification," *J. Geophys. Res.* **100**, pp. 18707–18726.

Tegen, I. A., Lacis, A. A., and Fung, I. (1996) "The Influence on Climate Forcing of Mineral Aerosols from Distributed Soils," *Nature* **380**, pp. 419–4

Torres, O., Bhartia, P. K., Herman, J. R., Ahmad, Z., and Gleason, J. (1998) "Derivation of Aerosol Properties from Satellite Measurements of Backscattered Ultraviolet Radiation. Theoretical Basis," *J. Geophys. Res.* **103**, pp. 17099–17110.

Tucker, C. J., and Sellers, P. J. (1986) "Setellite remote sensing of primary production," Int. J. Remote Sens., 7, pp 1395–1416.

Twomey, S. (1997) "The Influence of Pollution on the Shortwave Albedo of Clouds," *Journal of the Atmospheric Sciences* **34**, pp. 1149–1152.

Weil, J. C., Sullivan, P. P., Moeng, C. H., and Patton, E. G. (2006) "Statistical Variability of Dispersion in the Atmospheric Boundary Layer," *14th Joint Conference on the Applications of Air Pollution Meteorology with the Air and Waste Management Assoc.*, January, Atlanta, Georgia.

Wentworth, C. K. (1922) "A Scale of Grade and Class Terms for Classic Sediments," *Journal of Geology* **30**, pp. 377–392.

Williams, M. A. J. (2001) "Interactions of Desertification and Climate: Present Understanding and Future Research Imperatives," *Aridlands News Letter* **49**, May/June. See http://ag.arizona.edu/Oals/ALN/aln49/williams.html

Zender, C. (2000) *Radiative Transfer in the Earth System*, GNU Free Documentation License.

Zhang, X. L., Cheng, L. S., and Chung, Y. S. (2003) "Development of a Severe Sand-Dust Storm Model and Its Application to Northwest China," *Water, Air and Soil Pollution: Focus* **3**(2), pp. 173–190.

11

Tsunamis: manifestation and aftermath

Harindra J. S. Fernando, Alexander Braun, Ranjit Galappatti,
Janaka Ruwanpura, and S. Chan Wirasinghe

Tsunamis are giant waves that form when large sections of seafloor undergo abrupt and violent vertical movement due to fault rupture, landslides, or volcanic activity. A discussion on their formation, propagation in deep and coastal oceans, landfall, and ensuing deadly devastation are described in this chapter, which pays particular attention to the Sumatra Tsunami of the Indian Ocean that occurred on 26 December 2004. Much of the discussion is centered on the observations made in Sri Lanka, where close to 27,000 people lost their lives and another 4,000 remain unaccounted for. The tale of the Sumatra Tsunami in Sri Lanka typifies the mighty destructive forces of nature that control large-scale disasters, the unpredictability of natural phenomena driving such disasters, and the uncontrollability of their manifestation. However, the destructive aftermath could have been mitigated through better alertness and preparedness; education; preservation and reinforcement of natural defenses; sound design of coastal infrastructure; coordinated relief efforts; unselfish corporation across ethnic, social, and political fabrics; and scientifically based reconstruction policies. Inadequate scientific knowledge has been a bane of responding to tsunami disasters, and this chapter highlights some of the key issues for future research.

11.1 Introduction

In Japanese, "tsunami" is synonymous with the term "harbor wave" because it creates large wave oscillations in harbors and enclosed water bodies. Today, "tsunami" is commonly used to describe a series of long waves traveling across the ocean that have much longer wavelengths (wave crest-to-crest distance) compared to the ocean depth. These waves are also referred to as seismic sea waves, considering their genesis as a result of a sudden rise or fall of a section of the Earth's crust at the ocean bed. A seismic event can displace the Earth's crust and the overlaying water column, creating a sudden rise or fall of the sea level directly above. This provides the initial perturbation and energy for the formation of a tsunami. In addition, onshore and submarine volcanic activity and landslides can generate tsunami waves, but the energy associated with such benthic activities is considerably less than those produced by submarine faulting (Silver et al., 2005). Often, a tsunami is incorrectly referred to as a tidal wave. Tidal waves are simply the periodic movement of water associated with the rise and fall of the tides produced by the variation of gravitational attraction forces of celestial bodies. Similarly, tsunamis differ from the usual shorter wavelength waves seen in coastal oceans, which are usually generated by winds and storm systems prevailing over deep water. The latter propagate toward the coast, amplify, and break, thus dissipating energy.

The spread of a tsunami depends on the configuration of the seismic fault zone, and the most energetic rays emanate perpendicular to the fault line. The number of waves generated can vary depending on the nature of the earthquake, and as they propagate, waves can morph into two or more larger waves. Typically, these multiple waves can be separated by a few tens of minutes to an hour. In the deep ocean, the amplitudes of the tsunami waves are on the order of a meter and are imperceptible to ships. However, modern satellite-based technology such as radar altimetry is able to detect variations in the sea level of as little as a centimeter. Because these satellites fly over a given point on the Earth in repeat cycles of 10 to 35 days, the observation of a tsunami wave is mostly serendipitous; thus, satellites cannot be reliably used to monitor the ocean continuously at one location. The speed of the tsunami waves is given by the "long wave speed," $(gh)^{1/2}$, where g is the gravity (gravitational plus centrifugal acceleration) and h is the local ocean depth. For example, if the ocean depth is 5 km, the tsunami speed is about 220 m/s (800 km/h, 450 mi/h), on par with a jet plane. As h decreases toward the shoreline, the wave also slows down, converting a part of its energy to potential energy in the form of increasing wave height. As the wave height increases, water from the shoreline recedes to provide volume flux for the growing wave, thus producing the first visible signs of a tsunami for coastal communities (Figures 11.1 and 11.2). In some instances, however, a small rise in the water level can be observed just before the recession.

Regardless, the incoming wave approaches the coast much like an incoming tide, although at a much faster rate. The wave slumps in the shoaling zone, generating a highly turbulent and entraining bubbly flow with forward momentum (Figure 11.3), thus causing rapid flooding of low-lying coastal areas and mass destruction of property and life. Large bottom shear stresses cause sediment to mobilize, and hence, the arrival waters can be muddy and dark in some cases. Another event that follows a tsunami is a standing wave or seiche (a wave that continually sloshes back and forth in a partially open estuary). When a seiche is generated by a tsunami, subsequent large waves may arrive in phase with seiching, thus resulting in a dramatic increase of wave height. Seiching may even continue over several days after a tsunami.

The maximum vertical height to which the water is observed with reference to the actual sea level is referred to as *runup* (Liu et al., 1991). The maximum horizontal distance that is reached by a tsunami is referred to as *inundation*. Both quantities depend on the local topography and bathymetry, as well as the orientation with respect to the incoming wave, tidal level, and magnitude of the tsunami (Heller et al., 2005; Kaistrenko et al., 1991). For example, a funnel-shaped embayment focuses wave energy and amplifies the height of the waves (Figure 11.1b), leading to a greater level of destruction. The presence of an offshore coral reef may dissipate the energy of a tsunami due to increase in bottom drag on waves (Madsen et al., 1990), thus decreasing the impact on the shoreline. Normal wind swell may ride atop of a tsunami wave, thereby increasing the wave height. *Runup* heights are measured by looking at the distance and extent of salt-killed vegetation and the debris left once the wave has receded. This distance is referenced to a datum level, usually the mean sea level or mean lower low tide level. The reference to mean lower low water is more significant in areas with greater tidal ranges, such as in Alaska, where a smaller tsunami wave can be more devastating during a high tide than a larger wave at low tide.

As large tsunami waves approach islands, they may refract or bend around them and diffract through the channels between the islands. The ability of a tsunami to bend around islands is called "the wraparound effect." During this phenomenon, the energy of the tsunami often decreases, resulting in smaller wave heights. Sometimes, tsunami waves are reflected back by the land mass, instead of bending around, thereby increasing the wave height of the next approaching wave.

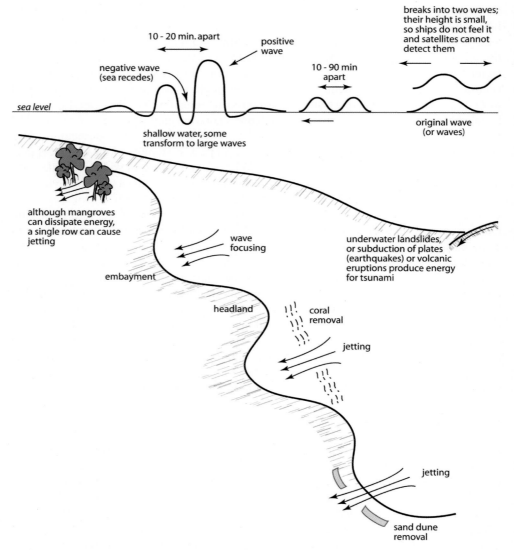

Figure 11.1 Schematic view of various aspects of long waves as they approach the shoreline (*upper*). The offshore wave interaction with the shoreline (*lower*).

When a tsunami finally reaches the shore, it may appear as a rapidly rising or falling tide, a series of breaking waves, or even a bore (Kirkgöz, 1983). Reefs, bays, entrances to rivers, undersea features, and the slope of the beach all contribute to the modification of a tsunami as it approaches the shore. Tsunamis rarely become great, towering breaking waves. Sometimes, the tsunami may break far offshore, or it may form into a bore, a steplike wave with a steep breaking front. A bore can occur if the tsunami moves from deep water into a shallow bay or river. The water level on shore can rise many meters. In extreme cases, the water level can rise to more than 15 m (50 ft) for tsunamis of distant origin and more than 30 m (100 ft) for a tsunami near its earthquake's epicenter. The first wave arriving may not be the largest in the series of waves. One coastal area may see no damaging wave activity, whereas in another area destructive waves can be large and violent. The flooding

Figure 11.2 The receding of ocean tens of minutes prior to the arrival of the Sumatra Tsunami, as observed in Wellawatta, Sri Lanka. (a) Receding water; (b) same location on a typical day (photos courtesy of Mr. M.M.H. Ibrahim); (c) and (d) same as in (a) and (b), respectively, but in Hikkaduwa, Sri Lanka (courtesy of Mr. Mambo Thushara); (e) arrival of a storm surge (5.5–6 m high) during Hurricane Katrina close to a location where the Mississippi River Gulf Outlet enters the Gulf Intercoastal Waterway, the appearance of which has similarities to a tsunami front (courtesy of Dr. Shahrdad Sajjadi); (f) arrival of the Sumatra Tsunami over receded seabeds in areas shown in (a) and (c).

of an area can extend inland by 500 m or frequently more, covering large expanses of land with water and debris. The speed of the waves flooding the land is about 30 km/h. Flooding tsunami waves tend to carry loose objects and people out to sea when they retreat. Tsunamis may reach a maximum vertical runup height of 30 m. A notable exception is the landslide-generated tsunami in Lituya Bay, Alaska, in 1958, which produced a 525-m runup.

Figure 11.3 (a) The arrival of the Sumatra Tsunami in a coastal town of Thailand (CBS News). (b) Same as (a), but in the coast of Galle, Sri Lanka (taken by an amateur photographer and broadcasted by the Rupavahini TV Network of Sri Lanka).

After the runup, part of the tsunami energy is reflected back to the open ocean. In addition, a tsunami can generate a particular type of wave called an edge wave that travels back and forth and parallel to the shore. These effects result in many arrivals of tsunamis and their remnants at a particular point on the coast rather than a single wave. Because of the complicated behavior of tsunami waves near the coast, the first runup of a tsunami is often not the largest, emphasizing the importance of not returning to a beach until several hours after a tsunami hits. The rule of thumb is to leave the beach immediately after symptoms of a major quake and not to return for about 6 hours.

There is one commonalty of all tsunamis: they are capable of unleashing mammoth devastation to human life and property, as was evident from the aftermath of the great tsunami event of 26 December 2004 in the Indian Ocean, which is known as the Sumatra Tsunami. This chapter gives a general description of processes that collude to form a tsunami event, the evolution of a tsunami following its generation, and its devastating aftermath. Special attention is given to the manifestation of the Sumatra Tsunami in Sri Lanka, which has been studied by the authors in detail through several visits to the tsunami-ravaged areas.

11.2 Causes of tsunamis: a general overview

The displacement of water that becomes a tsunami can be caused by dynamic processes of the Earth such as plate tectonics, earthquakes, volcanic eruptions, submarine mudslides, and landslides, but tsunamis can also be triggered by anthropogenic activities such as dam breaks, avalanches, glacier calving, or explosions. Earth is a dynamic planet leading to motions of so-called lithospheric plates moving relative to the underlying asthenosphere. Lithospheric plates are on average 100 km thick, and comprise both oceanic and continental areas. The relative motion between those plates is driven by thermodynamic processes of the Earth's core and mantle. The energy preserved in the core and mantle at the formation of planet Earth is constantly released toward the surface. This leads to convection processes with up- and down-welling patterns driving the plate motion through friction and mantle flow. The rigid lithospheric plates form three different types of boundaries: convergent, divergent, and strike-slip. Depending on the type of boundary, two lithospheric plates can override each other, slip along each other, or separate from each other. Naturally, the convergent plate boundary has the largest potential of accumulating stress, which is eventually released through viscoelastic deformation and earthquakes. These plate boundaries are also known as subduction zones. Two plates that override each other with a convergence rate of a few centimeters per year do not show continuous movement, but do exhibit periodic relative movement. For long time periods, the motion is locked, and the relative motion is transformed into stress and strain. After a certain yield stress has been exceeded, the plates overcome the friction and move again. This causes a sudden displacement of lithospheric material, which also displaces the water column above, since water is incompressible.

Other important causes of tsunamis are landslides. The mechanism is quite straightforward: a large volume of rock, mud, or volcanic material collapses into a water body and displaces the water instantaneously. A prominent possible example is the Cumbra Viejo volcano on the island of La Palma, Canary Islands. It has been proposed as a potential mechanism of generating a tsunami that would affect the East Coast of North America (Ward and Day, 2001).

The causes of tsunamis are multifold, but large tsunamis require a significant amount of displaced water and are thus mostly limited to large earthquakes and large landslides. A sudden release of gas hydrates has also been recently proposed as a possible cause. A meteor falling into the ocean can cause massive waves much higher than a tsunami, but they dissipate in height with distance at a much faster rate than typical tsunamis.

In principle, tsunami waves can be regarded as caused by multiple point sources. Each point source or explosion causes waves to propagate radially outward. In reality, tsunamis are a combination of multiple explosion sources and propagate with the highest energy in a direction perpendicular to the line of sources (e.g., a ruptured fault zone).

11.3 Hydrodynamics of tsunamis

The first mathematical treatment of shallow water waves was published by Joseph-Luis Lagrange (1736–1813) at the end of the eighteenth century, wherein the celerity (propagation speed) of a sinusoidal long wave was shown to be $c = (gh)^{1/2}$, where g is the gravity and h the water depth. As the waves approach the beach, the amplitude increases and the waves break, causing a highly turbulent flow. These waves have periodic (sinusoidal) behavior, and they travel as a wave train. In the 1830s, however, John Scott Russell (1808–1882) observed single-crested lonely waves propagating along a navigational channel, which were termed *solitary waves*. A striking feature of these waves was the absence of the usual wave *trough* following a wave elevation (*crest*). Until Joseph Boussinesq (1842–1929) presented a mathematical theory to explain these nonlinear waves by accounting for streamline curvature effects, wave theorists tried, without much success, to describe solitary waves using linear approaches. The profile derived by Boussinesq took the form

$$\eta/a = sech^2(3a/4h)^{1/2}[(x - ct)/h] \quad (11.1)$$

where $\eta(x)$ is the free-surface elevation above still water depth of h, a the wave amplitude, x the streamwise coordinates, and t the time. The symmetrical wave profile about $x = 0$ tends asymptotically to $\eta(\pm\infty)$, for $\eta/a = 0.05$ the length of the wave is $L/h = 2.12(a/h)^{-1/2}$, and its wave celerity is

$$c = [g(h + a)]^{1/2} \quad (11.2)$$

The wave energy per unit width is $E = (8/3)\rho_w gah^2[a/(3h)]^{1/2}$, where ρ_w is the water density. For all waves in still water of depth h and amplitude a, the solitary wave has the maximum energy because of the absence of wave troughs. Consequently, tsunamis propagating in the form of solitary waves will cause maximum destruction.

The propagation of tsunamis in ocean environments can be analyzed using ray-tracing models (Satake, 1988) or hydrodynamic long-wave models (Choi et al., 2003). The ray-tracing equations for the case of long waves become

$$\frac{d\theta}{dt} = \frac{\cos \xi}{nR} \tag{11.3}$$

$$\frac{d\varphi}{dt} = \frac{\sin \xi}{R \sin \theta} \tag{11.4}$$

$$\frac{d\xi}{dt} = \frac{\sin \xi}{n^2 R} \frac{\partial \eta}{\partial \theta} + \frac{\cos \xi}{n^2 R \sin \theta} \frac{\partial \eta}{\partial \varphi} - \frac{\sin \xi \cot \theta}{nR} \tag{11.5}$$

where θ and φ are the latitude and longitude of the ray, $n = (gh)^{-1/2}$ is the slowness, g the gravitational acceleration, $h(\theta, \varphi)$ the water depth, R the radius of the Earth, and ξ the ray direction measured counterclockwise from the south. The ray tracing technique is usually used as a short wavelength approximation in the theory of long-wave propagation, and it can yield in principal the travel time and wave amplitude of a tsunami at a particular location; however, in practice, the ray theory is effective only in travel time calculations but not the wave amplitude.

In contrast, the shallow water theory on a spherical Earth can be used to describe the spatiotemporal evolution of long-wave profiles, thus allowing calculations of both the arrival time and the amplitude. The relevant equations describing shallow waves on a spherical Earth, based on Choi et al. (2003), can be written as

$$\frac{\partial \eta}{\partial t} + \frac{1}{R \cos \theta} \left\{ \frac{\partial M}{\partial \varphi} + \frac{\partial}{\partial \theta}(N \cos \theta) \right\} = 0 \tag{11.6}$$

$$\frac{\partial M}{\partial t} + \frac{gh}{R \cos \theta} \frac{\partial \eta}{\partial \varphi} = fN \tag{11.7}$$

$$\frac{\partial N}{\partial t} + \frac{gh}{R \cos \theta} \frac{\partial \eta}{\partial \theta} = -fM \tag{11.8}$$

where $M(\theta, \varphi, t)$ and $N(\theta, \varphi, t)$ are the flow discharge in meridional and zonal direction, $n(\theta, \varphi, t)$ is the water displacement, and f the Coriolis parameter.

As waves break and slump on costal areas, existing forward momentum causes the wave to runup along the shoreline, which indeed is the most destructive consequence of tsunamis (Hunt, 2005). Inundations so caused depend on the nature of the wave, and considerable attention has been paid to understand and parameterize wave-induced inundations. The runup of solitary waves in surf zone has been studied by Synolakis (1987a, 1987b), among others, and the turbulent energy dissipation and dispersion pertinent to breaking waves have been reported by Wu (1987) and Chang and Liu (1999).

One of the least understood aspects of nonlinear shoaling waves is the wave runup and its interaction with the coastal environment (Walder et al., 2006). Early laboratory experiments (Camfield and Street, 1969; Hall and Watts, 1953) and theoretical analyses (Carrier and Greenspan, 1958) have provided much insight and impetus for subsequent work. Hall and Watts (1953) proposed an empirical formula for runup height H_R on a 45-degree impermeable slope:

$$\frac{H_R}{h} = 3.1 \left(\frac{a}{h}\right)^{1.15} \quad (11.9)$$

whereas the theoretical and experimental work of Synolakis (1987a, 1987b) on the runup of nonbreaking long waves results in the following equation for smooth and planar beaches:

$$\frac{H_R}{h} = 2.831\sqrt{\cos\beta}\left(\frac{a}{h}\right)^{5/4} \quad (11.10)$$

where β is the beach inclination angle.

The formula of Müller (1995) suggests

$$\frac{H_R}{h} = 1.25 \left(\frac{\pi}{2\beta}\right)^{0.2} \left(\frac{a}{h}\right)^{1.25} \left(\frac{a}{\lambda}\right)^{-0.5} \quad (11.11)$$

where λ is the wavelength. More recently, Gedik et al. (2004) proposed empirical relationships for runup in impermeable and permeable beaches, with the latter focusing on both armored and nonarmored beaches.

Obviously, coastlines are complex, convoluted, inhomogeneous, and replete with natural and manmade features that have severe implications on the runup and inundation, as illustrated in Figure 11.1. To account for such complexities, earlier work with planar beaches has been extended to a variety of situations. For example, Kanoglu and Synolakis (1998) have discussed long-wave evolution and wave runup on piecewise linear bathymetries in one and two dimensions. Pelinovsky et al. (1999) studied tsunami runup on a vertical wall in a bay of different cross sections. In all cases, the effects of the variability of the beach morphology were clear, as were the effects of natural and built features.

Figure 11.1 shows the effects of headlands and embayment, beach defenses such as corals and mangroves (green belts), and sand dunes that generally protect beaches. Wave focusing on beaches can cause jetting of water through an embayment, while protecting the bounding headlands. Mangroves and corals enhance the resistance on the flow and reduce the impact momentum (Danielsen et al., 2005; Fernando et al., 2005; Kunkel et al., 2006). A recent report by the World Conservation Monitoring Center, Environmental Program (United Nations Environment Program [UNEP], 2006) estimates the economic value of beach defenses to be between $100,000 and $600,000 per square kilometer per year for coral reefs and $900,000 per square kilometer per year for mangroves. If healthy, both ecosystems help to absorb 70% to 90% of storm surge energy from hurricanes and tropical storms (UNEP, 2006). In 2004, the Louisiana Wetlands Conservation and Restotation Authority suggested that a storm surge is reduced by 1/4 ft/mi of wetland marsh along the central Louisiana coast. In addition, computer modeling conducted by the Army Corps of Engineers using the ADCIRC model indicate that the storm surge height in New Orleans during Hurricane Katrina would have been 3 to 6 ft higher in the absence of wetlands east of the Missisipi River Gulf Outlet.

11.4 Ecological impacts of tsunamis—a general overview

The physical impact of a tsunami wave is uniquely different from all other natural disasters that impact coastal ecosystems. A tsunami wave causes a flooding event of extremely short duration when compared with river flooding, storm surges, and wind and wave attack during a cyclone or hurricane. Usually, the tsunami wave that does most of the damage lasts for around 10 minutes, and the time taken for the water to recede varies, depending on the topography of the terrain and the availability of drainage paths. Very high water velocities can be present during both flooding and recession. The receding water is usually more destructive in the areas where the return flow has concentrated. The depths and velocities of flow, while they last, can be an order of magnitude higher than those that occur during a severe cyclone (typhoon or hurricane).

High water velocities can cause much physical damage to property, vegetation, and even to the landscape itself, while drowning all nonaquatic life that could not reach high ground or stay afloat for a sufficiently long period. A different type of damage is caused by extended inundation under saline water in some low-lying, poorly drained areas and in others where drainage paths have been blocked by debris. The seriousness of damage done by tsunami inundation is also influenced by whether the fauna and flora in that area are adapted to inundation by high tides and storm surges. In microtidal coastlines, such as those found in Western Sumatra and Sri Lanka, any prolonged flooding due to a tsunami would have more serious impact, for example, than in the northern part of the Bay of Bengal, which has higher tidal ranges and a more frequent incidence of storm surges due to cyclones.

Direct damage due to high flow velocities can be found on coral reefs, beach formations, vegetation, and the built environment (Pennisi, 2005). Debris created by the destruction of buildings (particularly those made of brick and mortar), as well as sand and earth scoured and displaced by the flow of water, cause arguably the greatest damage to the environment by smothering coral reefs and seagrass beds in the sea and lagoons and by blocking drainage paths on land. Coral reefs can also be damaged by moving debris. The land is not only been polluted by debris being strewn about, but also by mud and sludge deposited in large areas. Brick, mortar, and concrete are not biodegradable and will thus not be removed by natural processes, except over a very long time period. Heavy debris found in the sea in zones of active wave action would tend to gradually sink under the sand and disappear from view. Any toxic substance that might leach into the water from such debris will continue to leach into the environment. It is even more serious if chemical storage depots are damaged by a tsunami wave. Chemical and bacterial pollution would have a greater direct human impact in urban areas than in the relatively undeveloped rural areas.

Shallow wells that supply drinking water to rural communities can be adversely affected by being submerged under a tsunami wave. Many such wells are dug in sand aquifers, where a lens of fresh water is perched above the much saltier underlayer. Excessive pumping from such wells can easily expose the underlayer. It has been more effective sometimes to stock these wells with truckloads of fresh water brought in from elsewhere. The hazard of contaminated drinking water is universally recognized. The fact that a tsunami will affect a very narrow coastal strip and that uncontaminated water might be available 1 or 2 km inland, however, is usually forgotten in the immediate relief effort.

Water flowing deep and fast can have a range of impacts on plant life. Such extreme velocities can strip all foliage from plants and shrubs but not necessarily uproot them. Given that the extreme conditions usually persist only for 10 minutes or less, some species of plants can recover if not impeded by persistent salinity. Even among palm trees, there are differences between species with regard to how they survive tsunami flows. Palm trees can resist high-velocity flow due to their slim and flexible trunks that pose low resistance to the flow. They return their initial state immediately after the flow has ceased. Coconut palms have a very high survival rate, even after their crowns have been dragged under the water. In contrast, a brief dousing of the crown of a palmyrah palm in salt water seems to be sufficient to kill the tree. Thus, the regeneration of plant life (and, consequently, animal life) in the affected area will follow a different trajectory if left entirely to the forces of nature. However, the coastal zone is usually one of the most intensively settled areas in most countries (note that ∼50% of the world's population live in coastal areas). The ecosystems in such areas, except where they are inaccessible or where conservation has been aggressively practiced, are not left entirely to the forces of nature but are exposed to anthropogenic influence. Under these circumstances, there is also the danger of the spread of alien invasive species with potential consequences for the endemic life forms.

The salination of soil and surface water is a major problem in poorly drained areas, including areas where drainage is blocked by debris. Such areas, however, will recover naturally as precipitation dilutes and leaches away the salt. The period of recovery, nevertheless, can be very long in poorly drained and/or arid areas. A complete recovery of damaged vegetation must also await the restoration of the original salinity regime. These considerations apply equally to wetlands such as mangroves, marshes, and lagoons (Arnold, 2005; Ebert, 2005). Many livelihoods, particularly in the more impoverished sector of the population, are based on the goods and services provided by wetlands. In such cases, it is not possible to leave the recovery of the wetlands to nature alone, but restoration methods must be carefully considered for reasons of effectiveness, with least intrusiveness to nature, and economic feasibility.

11.5 The Sumatra Earthquake and Tsunami

11.5.1 The Sumatra–Andaman Island Earthquake, 26 December 2004

The Sumatra–Andaman Island Earthquake occurred on December 26, 2004, at 0:58:53 UTC or 6:58:53 local time (Lay et al., 2005). The hypocenter of the earthquake was at 3.316°N, 95.854°E, approximately 250 km west of Banda Aceh, Sumatra, Indonesia (Figure 11.4). The depth of the earthquake was initially detected at 10 km, but later corrected to 30 km. The earthquake had a magnitude of 9.0 based on the teleseismic moment magnitude scale. The seismic moment is a product of the rigidity of the Earth, the average slip, and the fault plane area. It is based on the energy released by the earthquake and is not limited to certain wavelengths of seismic waves, as in the "body wave magnitude" (which uses the magnitude of P-waves) or "surface wave magnitude" (which is based on surfaces waves of period 20 seconds). Although the U.S. Geological Survey lists this earthquake as the fourth largest ever recorded, after the Chile (1960), Alaska (1964), and Andreanof Islands (1957) earthquakes, Stein and Okal (2005) reported that it was the second largest with a moment magnitude of 9.3. They argue that the longest period waves (normal modes) have not been incorporated in the previous analyses (but only periods below 300 seconds), and thus, it underestimated the energy released by the earthquake. Owing to the observed increase of seismic moment with increasing wave period, this leads to an estimated seismic moment of 1×10^{30} dyn.cm compared to the previous estimate of 4×10^{29} dyn.cm. Even the Earth's rotation was said to be affected by the enormity of the energy release (Chao, 2005).

The earthquake followed almost two centuries of plate convergence between the India plate and the Burma microplate, which comprises the tip of Sumatra as well as the Andaman and Nicobar Islands. The last large earthquake along this fault was recorded in 1833, and the stress accumulation ever since contributed to the large magnitude. The Burma microplate is a sliver between the Indian plate and the Sunda plate, both of which move northward. The Indian plate dips to the east under the Burma microplate at a convergence rate of approximately 20 mm/y. However, the subduction process between the India and Burma plates is oblique, meaning that the direction of motion is not normal to the fault line, but

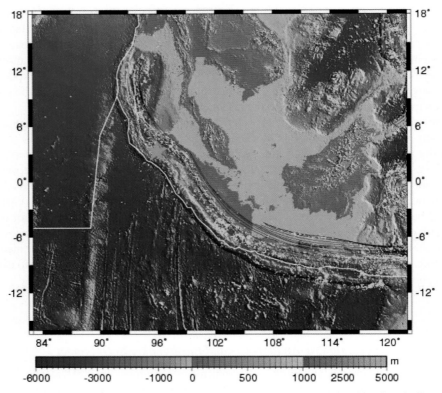

Figure 11.4 Epicenter of the Sumatra–Andaman Island Earthquake. White lines indicate plate boundaries, black lines represent subduction slab depth contours, and yellow dots represent past earthquake locations.

oblique with different angles. Although the southern part has a significant normal component, the northern part becomes a strike-slip motion, which may have been responsible for the cease of the rupture. The fault plane of the Sumatra–Andaman Island Earthquake has been split into two parts: a southern part offshore of Sumatra, where the megathrust event of 26 December took place, and a northern part, where most of the aftershocks indicated a significant amount of rupture as well. The length of the entire fault zone was about 1,200 km, along which the sea floor displaced by an average slip of 11 m, with a maximum of 20 m. The displacement of the sea floor caused the displacement of about a trillion tons of water. Based on the analysis of body waves originating in the southern part, a fast slip had been predicted. However, the series of aftershocks observed in the following months demonstrated that the fault zone reached much farther north to about 15°N. The slip along the northern part must have been slow because it has not been identified in the body wave analysis, but triggered long period normal modes (Stein and Okal, 2005). A fault model with a length of 1,200 km and a width of 200 km with an average slip of 11 m explains the observed seismic moment well. Nevertheless, the southern and northern parts must be distinguished based on the fault mechanism. Tsunami propagation models that considered only the southern part do not show significant wave height reaching Sri Lanka; the northern part must be included to explain the observed wave heights in Sri Lanka.

The earthquake unleashed a series of tsunami waves that crashed into coastal towns, fishing villages, and tourist resorts from Sumatra to Sri Lanka, India, Thailand, Malaysia,

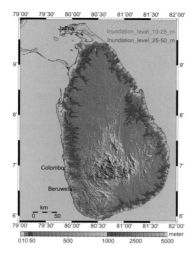

Figure 11.5   Inundation areas for a rising sea level of 10, 25, and 50 m, respectively. Topography taken from the Shuttle Radar Topography Mission 90 m; unfinished product.

and some countries along the East coast of Africa, killing more than 230,000 people in at least nine countries and leaving thousands missing (Schiermeier, 2005). The return period of megathrust earthquakes in this region can be estimated based on the convergence rate to be roughly 500 to 800 years; the accumulation of stress and strain, however, is nonlinear and not homogeneous along the fault zone, as demonstrated by this event.

11.5.2 The Sumatra Tsunami in Sri Lanka

After the earthquake occurred and the water had been displaced, the tsunami propagated through the Indian Ocean. Within about 2 hours, it arrived at the eastern coasts of Sri Lanka, plunging onto beaches and sometimes inundating several kilometers inland (Figure 11.5). Although more than 790 tsunamis have been recorded in the Pacific Ocean since 1900, in recent history, this is the second tsunami reported to have struck Sri Lanka, the first being in 1883. A tsunami of great magnitude, referred to as the *flooding of the ocean*, however, has been reported in the Great Buddhist Chronicle *Mahawamsa*, one of the most revered annals of Sri Lanka's 2,500-year civilization. The story deals with the era of the Great King Kelenitissa (circa 150 BC), who was punished by the gods via flooding his kingdom with ocean waters and waves. This was a punishment for the king's execution of a monk, who was suspected to have helped in an extramarital affair of the queen, by drowning in a cauldron of boiling oil. To appease the gods, the king sacrificed his daughter, Princess Devi (who would later become the mother of the Sinhalese nation as Vihara Maha Devi), by floating her away in a golden boat. She ultimately landed in the eastern fishing town of Kirinda and became the queen of the rival King Kavantissa of the eastern kingdom. Ironically, Kirinda was among the towns badly devastated by the Sumatra Tsunami.

Here, we focus on the impact of the Sumatra Tsunami on various parts of Sri Lanka and try to understand why the tsunami that arrived from the southeast was so strong that it even curved around the southern part of the island and attacked the west coast with a power similar to that of the east coast. This can be attributed to a combination of factors, including the wave energy; bathymetry; natural coastal defenses and their modifications;

wave effects such as reflection, refraction, seiching, and diffraction; and energy dissipation in the flow.

Although the main earthquake shock occurred in the southern part of the fault, the northern part ruptured like a zipper, and it took several minutes to complete the rupture toward the northern end of the fault plane (Ammon et al., 2005). Owing to these unusual fault mechanisms, the tsunami had a very distinct shape. As mentioned previously, it has been found that the southern fault would likely not have led to a tsunami reaching Sri Lanka, but the northern part of the fault was responsible. Tsunami simulations of Song et al. (2005), which included ocean dynamics and different fault mechanisms, have concluded that 30% of the wave energy was created by the vertical displacement of the fault plane, while 70% was associated with the horizontal displacement.

The tsunami propagated with wave heights of 20 to 60 centimeter over the deep ocean as confirmed by satellite altimetry of Jason-1, ENVISAT, and GFO-1 (Smith et al., 2005). These altimetry observations have been used to improve hydrodynamic models of tsunami wave propagation. Although the agreement of purely hydrodynamic models (e.g., MOST; Titov et al., 2005) is quite good, there is a slightly better agreement in amplitude and phase if ocean dynamics are included (Song et al., 2005). All models greatly underestimate the wave amplitudes when compared with tide gauge records. This may be due to the fact that tide gauges are in shallow water wherein the tsunami increases its wave height, but also where numerical models break down. The Indian Ocean Tsunami has been identified in most parts of the world's oceans at numerous tide gauge stations, as far away as Brazil; it was a truly global event.

The fault line that generated the Sumatra Tsunami occurred along a part of the Sunda Trench that ran approximately north–south as described previously. This generated a linear wave front that was able to travel long distances in a westerly direction without significant spreading of energy. The island of Sri Lanka happened to be directly in the path of this wave front.

11.5.3 Wave observations and impacts on Sri Lanka

Several research groups (e.g., Liu et al., 2005) carried out simulation models of the Sumatra Tsunami propagation after the event. Figure 11.6 shows results of simulations done by DHI Water & Environment in 2005. It shows the wave height and wave travel time to various parts of the Indian Ocean. Such models are very useful in determining the optimal placement of sensors and for understanding the effects of tsunamis. Because of the complex nature of the processes involved, however, these models represent approximations and their efficacy needs to be evaluated by comparing with observations. Such observations must be done soon after the disaster because the data are ephemeral. As such, a group of scientists sponsored by the U.S. National Science Foundation, the U.S. Geological Survey, and the Earthquake Research Institute visited the island of Sri Lanka during the period 9 to 16 January (Goff et al., 2006; Liu et al., 2005). The purpose of their visit was to measure or estimate the maximum wave heights and the inundation area in the most affected regions, selected in consultation with local experts, considering the damage, safety, and accessibility. They interviewed eyewitnesses to obtain direct information on the tsunami, including the arrival time of the leading wave, and examined geological evidences such as scour and sediment deposit and structural damages. The goal was to obtain information to further the scientific understanding of these waves and their aftermath, with the hope of improving predictive capabilities and helping the development of effective tsunami warning systems.

11.5 The Sumatra Earthquake and Tsunami

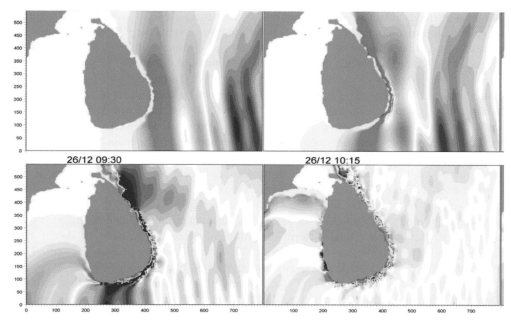

Figure 11.6 Simulation of the Sumatra Tsunami by DHI. Elevated sea level is shown in red, and depressed levels in blue.

As evident from Figure 11.6, Sri Lanka happened to be directly in the path of the tsunami wave front. As a result, it was possible to observe all the possible wave phenomena described previously as the tsunami struck the east coast frontally, refracted around the south coast of the island, and diffracted around the sharp northeastern corner of the Jaffna Peninsula. The waves observed on the southern and western coast are more complex, and influenced by refraction and reflection off the southeast Indian coast and Adam's Bridge, and possibly also due to edge wave phenomena. Wave diffraction from the southernmost point, Dondra Head, was also clearly evident.

The variety of ways in which the waves struck different parts of the coast is a result of this complex interaction of the wave train with complex offshore bathymetry, which has been modified over the years by dredging. The largest waves arrived at 8:50 am (local Sri Lanka time) on the east coast. In some areas of the west coast, such as Beruwela, the highest wave arrived at 11:30 am, although some waves of lesser yet significant amplitude have been noted before that time.

11.5.4 The impact on Sri Lanka

Rapid green assessment

Several "rapid assessments" were carried out in preselected areas in Sri Lanka in the weeks following the tsunami. Each assessment had an objective of obtaining data urgently required by individual agencies and groups for developing policies and for estimating reconstruction costs, as well as for scientific and other purposes. The Ministry of Environment with assistance from UNEP undertook two national studies that became known as the rapid green assessment (RGA) and the rapid brown assessment (RBA), with the objective of covering the entire tsunami-affected area with a uniform methodology to provide a consistent background against which to evaluate all other studies and assessments. The RGA was intended to focus on damage to the biological and physical environment, whereas the RBA concentrated on pollution and related subjects. The RGA comprised a survey of 700 coastal transects spaced at 1-kilometer intervals along the entire affected coastline from the northern coast of the Jaffna Peninsula clockwise along the coast to Negombo on the west coast (north of the capital Colombo). The surveys were carried out by teams of senior students and teaching staff of four coastal universities (University of Jaffna, Eastern University, Ruhuna University, and the University of Colombo), who were provided with intensive field training focused on developing and adopting a uniform methodology. The actual field work was closely supervised by both regional and national technical advisors. These field groups were able to walk the entire length of the coastline, except for some small areas of the northeast coast where they were not permitted access by a terrorist group.

The built environment, vegetation, and the landscape (before and after the tsunami event) were recorded along each transect, together with the inundation level and position, as well as the inundation distance where there was sufficient evidence available. A sketch map was prepared for each site covering at least 500 m on either side of each transect. This information has been gathered in a database and resulted in a "Damage Atlas." A sample from this atlas is shown in Figure 11.7. The layout of transects are shown together with the profile and sketch map of the area around each transect. Some socioeconomic values are also shown graphically. The findings of the RGA are summarized in the rest of this section.

Physical impacts

The physical impacts of the tsunami wave observed in Sri Lanka are consistent with the mechanisms illustrated in Figure 11.1. The number of lives lost, however, not only depended on the physical impacts, but also even more on whether there was any warning, a shelter, or a high elevation point that people could run to, and even the fact that very few Sri Lankans have learned how to swim. The tsunami wave appeared to have arrived without much warning on the east coast, but on the southern and western coasts the destructive wave was preceded by one or more (smaller) waves and clear recession of ocean water, which was not recognized as a warning sign by a coastal population with no previous experience or knowledge of tsunami phenomena. Where there were cognoscenti, thousands were saved, as was in the case of the town of Galbokke, where a former merchant marine officer, Mr. Victor DeSoyza, escorted an entire town to safety well in advance of the tsunami arrival (Figure 11.8). Many in the coastal towns of Sri Lanka lament the inadequacy of knowledge they had on tsunamis, which could have saved scores of lives. This stresses the importance of educating the public on natural disasters, especially how to be prescient and respond to them wisely (Synolakis, 2005). Much has been written about the human cost of this catastrophe. The barest statistics are shown in Figure 11.9.

Although the wave height was more or less uniform along the east coast, much increased spatial variability was observed as the wave traveled to southern and western coasts. The central part of the eastern coast (in the Batticaloa and Amparai districts) is described by Swan (1983) as having a coastal formation known as *barrier and spit*. The towns and villages have been built on a narrow coastal berm between the sea and wetlands (swamps and lagoons). The population density of these habitations was very high. The very high loss of life and infrastructure in these two districts is evident in Figure 11.9.

The high sand dunes found along long stretches of the southeast and the far northeast coasts were very effective in withstanding the tsunami wave. Nevertheless, the wave penetrated all natural gaps and some low points in the dunes, as well as all points where the dunes had been breached by human activities for various purposes. Figure 11.10 shows the

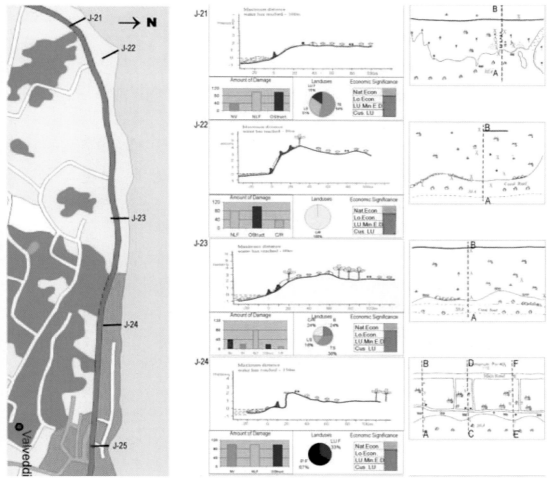

Figure 11.7 An extract from the "Damage Atlas" (MENR/UNEP, 2005).

devastation to the well-known Safari Lodge in Yala, Sri Lanka. During the construction, the investors decided to remove a natural sand dune (Figure 11.11) that protected the beach to provide better visual access of the ocean for the lodge guests, but the wave took advantage of this gap and rushed through the opening with such a force that the entire lodge was strewn into rubble with some 175 human casualties.

Nature also creates gaps in the dune system to allow rivers, tidal creeks, and lagoons to drain into the sea; these are unavoidable. The tsunami wave entered through every one of these gaps and often caused lateral flooding of lands further inland that would otherwise have been protected by the dune system.

Small houses and poorly constructed buildings were often unable to withstand the force of the water and debris, whereas large and strong buildings with concrete rebar were able to withstand the force of the water, with damage only to the ground floors (Figure 11.12). The surprisingly large amount of debris left by the destruction of buildings was strewn on the ground, washed into lagoons, and/or washed back into the sea by the returning water. In the northeastern town of Mullaitivu, which was completely destroyed, the returning

Figure 11.8   Mr. Victor DeSoyza explains to Mr. William Hermann, a reporter from the *Arizona Republic* newspaper, how he recognized the signs of a tsunami. Photograph is by Mr. Jeffrey Topping, *Arizona Republic*, 14 January 2005.

water washed everything into the sea, including large quantities of sand from the beach, even exposing culverts that had remained buried since the Cyclone of 1964.

The force of the water was absorbed by large stands of dense vegetation, reducing the destruction further inland. Coconut trees did not provide such shelter, but unlike other vegetation, the flowing water did not destroy them, even when fully submerged for a brief period. Palmyrah trees, however, appear to have been destroyed easily by immersion in salt water (Figure 11.13). Although the presence of mangroves had partially protected some estuaries, it must be pointed out that mangroves do not occur naturally on exposed coastlines in Sri Lanka; neither can they be grown in the face of such an energetic wave climate. As mentioned, areas of sand dunes (Figure 11.14) have been effective in retarding the waves, but large sand dunes cover only limited stretches of the shoreline. Posttsunami Sri Lanka is placing an increased emphasis on building artificial beach defenses, such as reinforced and armored beaches, against waves (Figure 11.15).

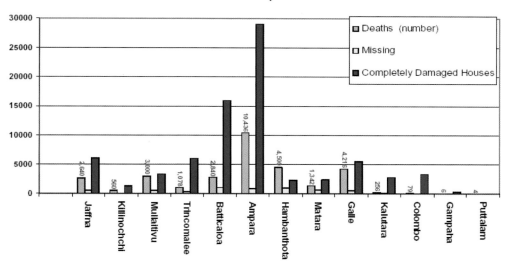

Figure 11.9 The human cost by district (Tittawela, 2005).

Figure 11.10 (a) The rubble left from the Yala Safari Lodge (photographed on 10 January 2005). (b) Among the devastation were some healthy trees, which have adapted to salt water environments and experienced little adverse effects of the tsunami.

Another important issue of anthropogenic-caused damage is the removal of coral reefs. As pointed out by Fernando et al. (2005), the bottom resistance offered by the reefs plays an important role in scattering and dissipating coastal waves (Figure 11.16), and there are indications that areas with reported coral poaching have suffered a great loss of property and life (Figure 11.17). For example, the town of Peraliya, where some 1,800 people on-board the express train Samudra Devi (*Ocean Queen*) perished, drowning in rushing water, was known to be a hotspot of coral poaching. Explosive blasts are used, harvesting both the corals (for making white wash and other paints) and fish casualties (*blast fishing*). As the wave approached the coastline, the water jetted through the path of least resistance created by the removal of corals, creating an effect equivalent to the removal of sand dunes. After the tsunami, the Sri Lanka government erected signs informing that coral mining is illegal, while requesting people to help preserve natural beach defenses such as corals and sand dunes (Marris, 2006).

Figure 11.11 The remnants of the sand dune (*arrow*) that once protected the land. During the construction of the Yala Safari Lodge, part of the dune was removed to provide a better view of the ocean. The opening so generated was the primary cause of the vast devastation in the area, where the collapsing wave jetted through the opening. Remnant ponds generated within the opening are seen in the picture, taken on 10 January 2005.

Figure 11.12 (a) A survived well-built concrete house (to the right) among the collapsed houses (already bulldozed) in the city of Hambantota. The house to the left, which is also shown in (b), survived merely due to its location in the wake of the stronger house. More than 4,000 residents perished in this town. This points to the necessity of developing sound building codes for coastal dwellings.

11.5 The Sumatra Earthquake and Tsunami

Figure 11.13 Dead palmyrah trees.

Figure 11.14 Sand dunes along the coast.

Figure 11.15 Countries such as Sri Lanka are increasing their reliance on artificial beach defenses (March 2006 in the town of Akuralla).

Ecosystem damage

The tsunami caused widespread and severe damage to houses and public infrastructure. The posttsunami landscape was found to be scarred by the removal of material or by the deposition of sand and debris (Figure 11.18). Flooding had caused salt water intrusion into agricultural land, wells, and fresh water aquifers. Further, the visible environment was devastated.

However, the RGA (MENR/UNEP, 2005) that examined damage to the environment and to large ecosystems in the affected coastal zone came to the conclusion that the ecosystem damage was *slight* to *moderate*, considering the impacts on the structure and functioning of the large ecosystems that comprise the coastal zones of Sri Lanka. The implication is that despite the physical damage to their component parts, the basic structure of all the major ecosystems surveyed was still intact, as were their internal and external functions (a case of resilience; Allenby and Fink, 2005). They are expected to regain the former vigor within a short time period. Where livelihood depends on an optimum functioning of an ecosystem, however, it must be considered necessary to intervene to accelerate the recovery.

Both debris left behind by the tsunami and its disposal during cleanup up operations caused environmental damage. Although there was no severe damage to many coral reefs, the smothering of some areas by debris washed from the land required human intervention. Some of the damage had been exacerbated by haphazard disposal of debris during relief operations. Drainage paths of some low-lying lands, including paddy lands, were blocked by debris carried by the tsunami wave and by unplanned disposal of cleared debris.

Figure 11.16 A wave field over a coastal coral bed. The waves start to grow as they arrive at the coastline, but tend to dampen over the coral bed and degenerate into small-scale motions, possibly due to the increased energy dissipation due to enhanced bottom drag (Galle, Sri Lanka).

Long-term impacts on the coastline

It is feared that major changes have occurred in the offshore bathymetry in some areas because of the movement and transfer of sand onshore and offshore. The resulting changes in beach slopes will adjust to a new equilibrium over a period of many years or even decades, depending on the characteristics of the beach processes. Changes in the bathymetry can change wave propagation phenomena and alter the effective wave climate in some coastal areas. Nearshore bathymetric surveys are urgently needed to determine whether this is a serious problem. The deepening of the nearshore area can reduce the natural protection offered to the coastline by a gradually sloping beach. It has not been demonstrated whether this increased wave attack is significant from the point of view of the safety of coastal structures and an increased incidence of coastal flooding. In the case of small fishing harbors, the oncoming wave has damaged the inner slopes of the breakwaters. In contrast, the outgoing water, which tended to concentrate more at the harbor entrance, is expected to have scoured the approach channel. Whether this has been serious enough to affect the stability of the breakwater head is a matter still under study.

The breach of the sand spit in Bentota on the west coast (Figure 11.19) created a second mouth for the Bentara Ganga (river). Although the new river mouth might eventually have migrated toward the old mouth, it would have to migrate through valuable tourist infrastructure as part of this process. Thus, the only alternative was to close the breach as quickly as possible.

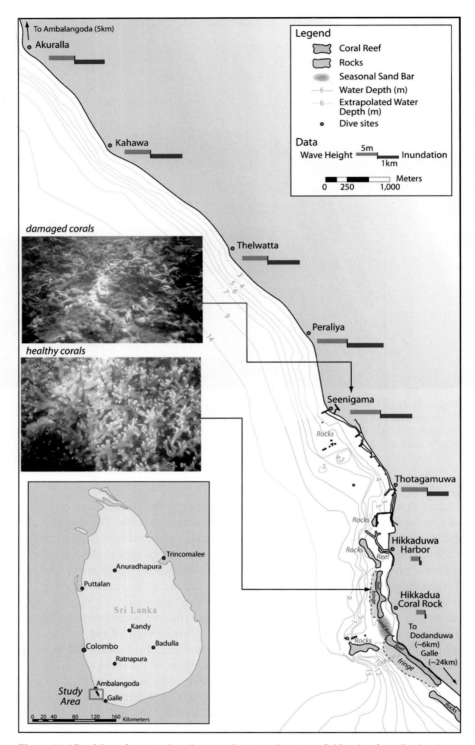

Figure 11.17 Map of a coastal study zone along southwestern Sri Lanka, from Dodanduwa (6 km south of Hikkaduwa) to Akuralla, reported by Fernando et al. (2005). The red square of the inset shows the location of the study area on the island coastline. The wave heights are in green bars (m scale) and the inundation in blue (km scale). Shown in blue are the bathymetric contours. During this study, dives were made, and the inundations were positively correlated with areas of coral mining. Dive sites, coral reefs, sand bars, and rock reefs are also shown. The fringe of coral reefs indicates area where corals are thin and distributed. Typical diver photographs of the areas are also shown in the inset.

11.5 The Sumatra Earthquake and Tsunami

Figure 11.18 Posttsunami debris fields. The picture shows the power of the tsunami wave, which was sufficient to slam a reinforced concrete bar into a coconut tree and break it into pieces (Kahawa, Sri Lanka).

Figure 11.19 Breached spit, Bentara Ganga.

The time it will take for the coastline to recover will greatly depend on the availability of beach material in the nearshore active zone. The sand that was carried onshore by the tsunami wave is no longer available to replenish the beach. Neither is the sand (if any) that was carried into the deeper waters (>8 m of water depth), where normal wave action is rarely felt by the sea bed. Sand carried into estuaries and river mouths will soon reach the active coastal zone again. The natural replenishment of beaches by river sand has, unfortunately, been greatly reduced in recent years by widespread river sand mining.

11.6 Tsunami warning systems

After the experience of the Sumatra Tsunami, the devastating consequences of the size of the tsunami and the lack of preparedness of the affected countries, a potential Indian Ocean tsunami warning system comes to mind (Lomnitz and Nilson-Hofsth, 2005; Marris, 2005). Obviously, there is no technology currently available that would prevent tsunamis from occuring, nor can we predict the occurrence time and location of earthquakes and other tsunami triggering processes with any accuracy. This results in changing the hazard mitigation approach from avoiding tsunamis to monitoring tsunamis.

The objective of a warning system is to continuously monitor the state of the sea surface in near–real time in order to allow for a fast and efficient warning to be issued through the correct channels. It consists of three principal elements: (1) the scientific instrumentation, (2) the communication system, and (3) the warning system. The sensors target the tsunami wave over different ocean depths. Hence, their sensitivity must be sufficient to detect wave heights of a few decimeters over the deep ocean. In addition, all other environmental processes changing the sea level must be understood and monitored in order to filter out these signals from the tsunami wave. As pointed out previously, the wavelength of tsunami waves is different than wind-driven and tidal waves, which allows identification of a tsunami using spectral and time series analysis. Once the observations of water pressure sensors, global navigation satellite systems (GNSS), tide gauges, or satellites have been collected, fast transmission of data to a processing center is required. For this purpose, radio transmission from a buoy to a land station can be used, if the distance between the two antennas is less than about 50 km, or else a satellite link can be established. The processing centers analyze the incoming data in unison and detect potential sources of tsunamis. The results are the basis for the activation of a warning system. The last step is the most critical and onerous because it involves political and social ramifications that go beyond the reliability of scientific data. The principle components of an operational tsunami warning system are the ground sensors, a GNSS buoy, and an ocean bottom pressure gauge (Figure 11.20). The space sensor is a radar altimeter. Communication systems are radio and satellite links from the sensors to the warning system.

At present, there are two tsunami warning systems in place, the deep-ocean assessment and reporting of tsunamis (DART) system of the National Oceanic and Atmospheric Administration and the German and Indonesian tsunami early warning system (GITEWS), the latter still being in the testing stage. Athough the DART system is primarily established in the Pacific Ocean (although some sensors have been recently added to the Atlantic), the GITEWS is focusing on the Indian Ocean. The DART system consists of pressure sensors that are capable of measuring sea level, fluctuation of 1 cm from a water depth of up to 6,000 m. The observations are then transmitted to a control system for processing and status assessment. The GITEWS consists of multiple sensors for monitoring of the seismic activity, water pressure, sea level and meteorological data. A U.S. Indian Ocean tsunami

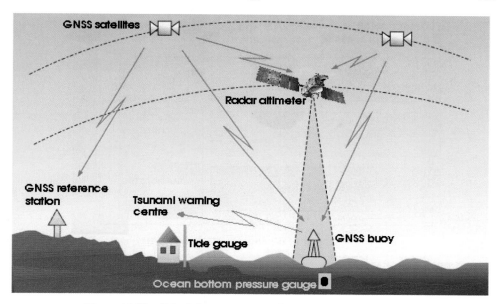

Figure 11.20 Principle components of a tsunami warning system.

warning system is also under consideration by a group of countries led by the United States.

In the case of Sri Lanka, the efforts are aimed at detecting a tsunami approaching from the east. Sri Lanka, however, also faces threats from the west, based on the folklore alluded to before and tectonic plate locations, and possibly from the vast southern oceans where a diffusive tectonic fault is developing. It is important to study these alternative possibilities even though the last tsunami came from the east.

11.7 Planning for tsunamis

The devastating impacts of the Sumatra Tsunami highlighted the need for better planning and preparation for all major natural disasters. Nevertheless, in comparison to major disasters such as floods and hurricanes, for example, the warning time for earthquakes and tsunamis is very short. Various techniques are available to analyze a system of information sources and develop a network to mitigate the impact of a tsunami. As mentioned previously, such a system essentially consists of earthquake and/or tsunami detection, communication of observations, analysis of the available data, cross validation with alternate sources of information where appropriate, determination of the tsunami potential, and issuance of an evacuation order.

The critical path method (CPM) is a widely used scheduling method to analyze such a system. With the CPM it is possible to identify the early and late start/finish dates for each activity and to calculate their "floats." The float time represents a safety factor relative to completing a project or a task on time. Activities with no total floats are called "critical activities," and the path on which they fall is the "critical path." CPM, however, assumes certainty in the durations and relationships among activities. Thus, CPM is not suitable for risk assessment or decision analysis. Monte Carlo simulation-based scheduling using CPM algorithms can overcome some of the early CPM limitations. The "criticality index" (the probability that an activity falls on the critical path) can be calculated using simulation.

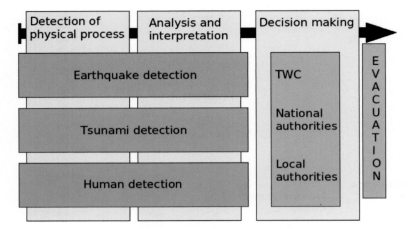

Figure 11.21   Scheme of the critical path method (CPM) network.

It then calculates the criticality of parallel paths of the activity network. Furthermore, the concept of stochastic networks is built on Monte Carlo simulation but incorporates schedule-logic uncertainties. Moussa et al. (2006) developed a simulation-based scheduling networking tool that could stochastically model activity durations and relationships among activities. This tool is called the decision support simulation system (DSSS) and is built on a simulation platform called Simphony (Hajjar and Abourizk, 2002).

The network to be discussed next, developed using DSSS, considers a local area earthquake analysis system, the global earthquake detection system, a deep ocean tsunami warning system with direct radio transmission as well as satellite transmission, local tide gauges, the possibility of immediate human detection, a tsunami warning center for the ocean under consideration, intervention of national and regional authorities, and, finally, the ordering of an evacuation. It is based on a number of assumptions that are necessary to obtain a feasible model of the critical path. The network is designed to be as generic as possible and should apply to many different tsunami events and areas. Where it is required to make specific assumptions, however, Sri Lanka was assumed as the country of interest. The network is just one of the multiple scenarios that could be employed in order to allow for a better planning of tsunami hazard mitigation.

11.7.1 Components of the stochastic scheduling network

The network consists of three principle components: (1) the detection of physical processes, (2) the analysis of observations and interpretation, and (3) decision making (Figure 11.21). These three components describe the horizontal flow of information through the network. In addition, there exist three branches of physical-process detection: (1) scientific detection of earthquakes, (2) scientific detection of tsunamis, and (3) human detection of earthquake/tsunami by direct experience/observation. All observations of these three branches are analyzed and communicated to the decision-making authorities. There are three levels of authority in the system: (1) an oceanic tsunami warning center, (2) national center/authorities, and (3) local authorities. The durations (in minutes) associated with the individual activities have been estimated conservatively to be within a certain range, wherein the minimum time is the fastest possible response. The maximum time represents the longest time anticipated to complete an activity. The range represents the assumed uncertainties for a particular situation/scenario.

The network is simulated in a large number of iterations (say, 1,000 times) to generate the statistical results. The final duration is calculated through forward and backward path calculations using the CPM algorithm to establish the total float, earliest possible start, latest possible start, earliest possible finish, and the latest possible finish of each activity. This process continues during all simulation iterations. At the end, the statistical results that include the average, standard deviation, and minimum and maximum of the duration of each activity are collected. A cumulative probability distribution function is generated, which can be used to analyze the probability of achieving target duration for evacuation.

Earthquake detection

Global or local seismic networks are able to monitor earthquakes that occur anywhere on the Earth. Large earthquakes with a high tsunami potential can be recorded at seismic stations within about 10 minutes; body waves generated by an earthquake within a radius of 50 degrees (about 5,500 km) in the spherical Earth arrive in less than 10 minutes at the stations (Kennett et al., 1995). Seismic waves travel about forty times faster than the tsunami waves and have the potential to alert people before tsunami waves arrive. The detection of earthquakes can be done in local or global networks. The observed earthquake signals are automatically processed and communicated to a seismologist for further analysis and assessment of the tsunami risk. The validation of recorded signals is done by using independent observations from the global network. An alternative path is the detection of the earthquake by humans, which results in an immediate evacuation if the person involved knows about the relationship between earthquakes and tsunamis. This path is thus the most direct one leading to evacuation, and it supersedes all other activities in the network. The seismological analysis results in an alert, indicating the risk of a tsunami based on the location of the earthquake (ocean or land) and the source mechanism (strike-slip, thrust fault, normal fault). Results are then communicated to the appropriate tsunami warning center (TWC).

Tsunami detection

As discussed previously, tsunamis can be detected by a warning system, tidal gauges, or humans. The observations are transmitted via radio communication, satellite, or telephone to the analysis center for further processing. A tsunami triggered by an earthquake should not take longer than about 20 minutes to be observed at the nearest coast or warning system.

Human detection

Humans may sense earthquakes and tsunamis; however, it depends on the knowledge of the individuals and on whether the appropriate actions are taken, (e.g., calling the local authorities and/or initiating an evacuation of the area) (e.g., Figure 11.8). The information is highly valuable in order to establish a fast response to the tsunami threat. This branch in the network passes several steps because human detection is more comprehensive than automated detections systems and thus more efficient and faster. It is important to note, however, that the education about tsunamis and earthquakes is a prerequisite for an efficient response by humans; thus, education must be considered a top priority in any tsunami warning system. Conversely, such a system is somewhat vulnerable for abuse and overreaction.

Authorities' tsunami warning center

Observations of the physical processes are only useful if the appropriate authorities exist to process the data and issue an alert in time. The network proposed here includes the local authorities (e.g., police, fire brigade). The local authorities must have a protocol in place on how to contact and communicate with the national authorities and the responsible TWC. It is important that the links between local and national, as well as national and international, levels are fully established and cross fertilized. An operational center such as the TWC is responsible for an entire ocean, and not only for specific countries. It reports findings to the national authorities who will make a decision for their country and citizens. After the decision is made, local authorities are advised on how to proceed with an evacuation.

Following is a brief explanation of the stochastic scheduling network shown in Figure 11.22 using the data under Scenario 1 in Table 11.1.

- The activities are modeled using uniform distributions to represent the uncertainty involved in the nature of the activities. For example, the activity "Process Info—Human" is estimated using a uniform distribution ranging from 0 to 20 minutes after detecting an earthquake by a human. The variability as indicated through a range depends on various factors. A human may notice an earthquake, but he or she may or may not have access to a communication source to inform local authorities immediately. The scheduling network requires the generation of random numbers for each of the activities estimated as density functions. The simulation starts by generating these random numbers to reflect the continuous distribution function of each activity.
- Links are created among the activities to show their relationships. The most common relationship is called "Finish-to-Start" (FS) indicating the successor activity will only commence once the current activity is finished. There is no lag value considered between the activities. The lag values represent any slack time from the finish of the current activity to the successor activity (e.g., detect EQ—human to process info—human that has an FS relationship).
- This network also considers the "Finish-to-Finish" (FF) relationship among activities. An FF relationship with a lag value of zero indicates that at least two activities involving the FF relationship must be finished before proceeding to a successor activity. For example, activity "Performing Seismic Analysis" can commence when activity "Info. Received by analyst through local network" is finished (FS relationship) or activity "Process Seismic Info. (Global)" is finished. The activity "validate with global network" can commence when activity "Info. Received by analyst through local network" is finished (FS relationship). However, the completion of the activity "Perform Seismic Analysis" depends on the finish of activity "Info. Received by analyst through local network" (FF relationship). This condition restricts the start of the successor activity "Issue a Decision on EQ."

This network also considers either the "And" or "Or" type of relationship between the current activity and the successor activity. An "And" activity relationship is realized when all predecessors' relations are realized. An "Or" activity is realized when any of the predecessors' relations are realized. This network has a combination of both types. The default type of any CPM network is the "And" type. There are a few "Or" type relationships in this network (e.g., activity "Inform National Centre" can commence when one of the two activities "Issue an Alert" or "Issue a Detailed Analysis" are finished). All the "Or" type activities in Figure 11.22 are shown using double lines.

The following are the typical outputs generated from simulation analysis of the network represented previously:

- Early start, late start, early finish, and late finish of each activity and the total duration
- Average, standard deviation, and minimum and maximum duration of each activity and the final duration of the network

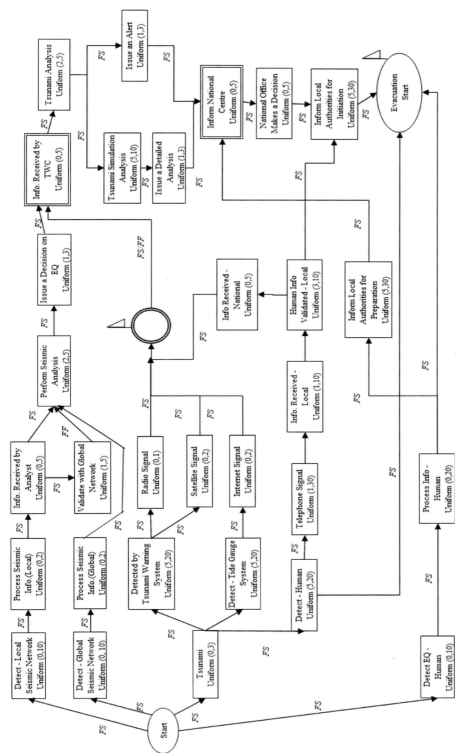

Figure 11.22 Stochastic scheduling network for tsunami planning and hazard mitigation.

Table 11.1. *Input data for stochastic scheduling network*

Code	Activity	Duration Scenario 1	Scenario 2
A1	Detect—Local Seismic Network	Uniform (0, 10)	Triangular (0, 4, 10)
A2	Process Seismic Info. (Local)	Uniform (0, 2)	Uniform (0, 2)
A3	Info. Received by Analyst	Uniform (0, 5)	Uniform (0, 5)
B1	Detect—Global Seismic Network	Uniform (0, 10)	Triangular (0, 4, 10)
B2	Process Seismic Info. (Global)	Uniform (0, 2)	Uniform (0, 2)
B3	Validate with Global Network	Uniform (1, 5)	Uniform (1, 5)
AB1	Perform Seismic Analysis	Uniform (2, 5)	Uniform (2, 5)
AB2	Issue a Decision on EQ	Uniform (1, 3)	Uniform (1, 3)
C1	Tsunami	Uniform (0, 3)	Triangular (0, 2, 3)
C2	Detected by Tsunami Warning System	Uniform (5, 20)	Triangular (5, 10, 20)
C3	Radio Signal	Uniform (0, 1)	Uniform (0, 1)
C4	Satellite Signal	Uniform (0, 2)	Uniform (0, 2)
C5	Detect—Tide Gauge System	Uniform (5, 20)	Triangular (5, 10, 20)
C6	Internet Signal	Uniform (0, 2)	Uniform (0, 2)
C7	Detect—Human	Uniform (5, 20)	Triangular (5, 10, 20)
C8	Telephone Signal	Uniform (1, 30)	Uniform (1, 30)
C9	Info. Received—Local	Uniform (1, 10)	Uniform (1, 10)
C10	Human Info Validated—Local	Uniform (3, 10)	Uniform (3, 10)
C11	Info Received—National	Uniform (0, 5)	Uniform (0, 5)
D1	Detect EQ—Human	Uniform (0, 10)	Triangular (0, 3, 10)
D2	Process Info—Human	Uniform (0, 20)	Triangular (0, 5, 20)
D3	Inform Local Authorities for Preparation	Uniform (5, 30)	Uniform (5, 30)
ABC1	Info. Received by TWC	Uniform (0, 5)	Uniform (0, 5)
ABC2	Tsunami Analysis	Uniform (2, 5)	Uniform (2, 5)
ABC3	Tsunami Simulation Analysis	Uniform (5, 10)	Uniform (5, 10)
ABC4	Issue a Detailed Analysis	Uniform (1, 3)	Uniform (1, 3)
ABC5	Issue an Alert	Uniform (1, 3)	Uniform (1, 3)
ABCD1	Inform National Centre	Uniform (0, 5)	Uniform (0, 5)
ABCD2	National Office Makes a Decision	Uniform (0, 5)	Uniform (0, 5)
ABCD3	Inform Local Authorities for Initiation	Uniform (5, 30)	Uniform (5, 30)

- Cumulative density function of the network duration
- Criticality index (as explained previously) of each activity

In this case, two scenarios were simulated over 1,000 iterations. Scenario 1 uses the uniform distribution for all activities. Because some of the activities are controlled by humans, Scenario 2 uses triangular distributions for eight activities. Those activities use a mode value (most likely value), in addition to the low and high time durations. Figure 11.23 shows the cumulative density curve for both scenarios. For example, there is a 34% probability of starting evacuation within 60 minutes, according to Scenario 1 data. In Scenario 2, the probability to start evacuation within 60 minutes is about 38%. The simulation analysis also identifies the most critical activities in the network. This enables the planners to focus on those activities in developing risk mitigation actions and to avoid any bottlenecks.

The stochastic scheduling network based on simulation and the modified CPM algorithm allows the planners to develop various strategies to plan the evacuation process. This includes a simulation based on the probability distribution of the time for various activities to obtain the cumulative probability of the time between an under ocean earthquake

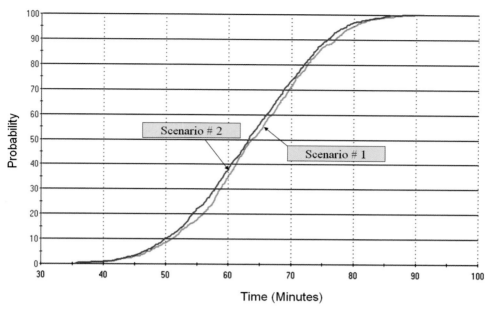

Figure 11.23   Cumulative density curves of scheduling network.

and the issuance of an evacuation order. It also allows us to identify activities that are critical.

11.8 Conclusions

This chapter provides a brief (undoubtedly incomplete) overview of large-scale disasters that follow tsunamis. To this end, discussions are given on their generation mechanisms, propagation, amplification (of amplitude) in coastal waters, and slumping to form an inrush during which the flow inundates for kilometers while leaving a wide swath of death and destruction. The discussion journeyed through various events akin to tsunami disasters, paying particular attention to the Sumatra Tsunami in Sri Lanka, where the authors participated in posttsunami scientific and reconstruction missions. Experiences show that, although the predictability and control of such disasters are untenable in the foreseeable future, mitigation can be accomplished. This requires monitoring of processes that come together to form tsunamis using state-of-the-science technologies, preparedness to respond to warnings, education, unselfish and coordinated response across wide variety of constituencies, and swift and unbiased reactivation of communities after tsunami events. This is true not only for tsunamis, but also for any large-scale natural disaster that threatens human communities and ecosystems.

Acknowledgments

The authors want to acknowledge the support of the University of Calgary, Arizona State University, and the sponsorship of the International Institute for Infrastructure Renewal and Reconstruction. Also, the authors acknowledge the support of the U.S. National Science Foundation, the U.S. Earthquake Research Institute, the British Broadcasting Corporation, the University of Moratuwa (Sri Lanka), and the University of Peradeniya (Sri Lanka), during the early phases of this work.

References

Allenby, B., and Fink, J. (2005) "Toward Inherently Secure and Resilient Societies," *Science* **309**, pp. 1034–1036.

Ammon, C. J., Ji, C., Thio, H. K., Robinson, D., Ni, S., Hjorleifsdottir, V., Kanamori, L. T., Das, S., Helmberger, D., Ichinose, G., Polet, J., and Wal, D. (2005) "Rupture Process of the 2004 Sumatra–Andaman Earthquake," *Science* **308**, pp. 1133–1139.

Arnold, E. (2005) "Nature Bounces Back on Sri Lanka's Coast, National Public Radio: Morning Edition," 24 February 2005.

Camfield, F. E., and Street, R. L. (1969) "Shoaling of Solitary Waves on Small Slopes," *Journal of the Waterways and Harbor Division* **1**, pp. 1–22.

Carrier, G. F., and Greenspan, H. P. (1958) "Water Waves of Finite Amplitude on a Sloping Beach," *J. Fluid Mech.* **4**, pp. 97–109.

Chang, K. A., and Liu, P. L. F. (1999) "Experimental Investigations of Turbulence Generated by Breaking Waves in Water of Intermediate Depth," *Phys. Fluids* **11**, pp. 3390–3400.

Chao, B. R. (2005) "Did the 26 December 2004 Sumatra, Indonesia, Earthquake Disrupt the Earth's Rotation as the Mass Media Have Said?" *EOS* **86**(1), pp. 1–2.

Choi, B. H., Pelinovsky, K. O., Kim, K. O., and Lee, J. S. (2003) "Simulation of the Trans-Oceanic Tsunami Propagation Due to the 1883 Krakatau Volcanic Eruption," *Natural Hazard., and Earth Systems Science* **3**, pp. 321–332.

Danielsen, F., Sørensen, M., Olwig, M., Selvam, V., Parish, F., Burgess, D., Hiraishi, T., Karunagaran, V., Rasmussen, M., Hansen, L., Quarto, A., and Suryadiptra, N. (2005) "The Asian Tsunami: A Protective Role for Coastal Vegetation," *Science* **310**, p. 643.

Ebert, J. (2005) "Coral Survived Tsunami Battering," *Nature News Online*, 12 April 2005.

Fernando, H. J. S., Mendis, G., McCulley, J., and Perera, K. (2005) "Coral Poaching Worsens Tsunami Destruction," *EOS* **86**, pp. 301–304.

Gedik, N., Irtem, E., and Kabdahli, S. (2004) "Laboratory Investigation on Tsunami Runup," *Ocean Engineering* **32**, pp. 513–528.

Goff, J., Liu, P. L. F., Higman, B., Morton, R., Jaffe, B. E., Fernando, H., Lynett, P., Fritz, H., Synolakis, C., and Fernando, S. (2006) "The December 26th 2004 Indian Ocean Tsunami in Sri Lanka," *Earthquake Spectra* **22**, S3, pp. S155–S172.

Hajjar, D., and Abourizk, S. M. (2002) "Unified Modeling Methodology for Construction Simulation," *Journal of Construction Engineering and Management* **128**(2), pp. 174–185.

Hall, J. V., and Watts, J. W. (1953) "Laboratory Investigation of the Vertical Rise of Solitary Waves on Impermeable Slopes," Technical Memorandum No. 33, Beach Erosion Board, Office of the Chief of Engineers, U.S. Army Corps of Engineers, Army Coastal Engineering Research Center, Washington D.C.

Heller, V., Unger, J., and Hagerm, W. (2005) "Tsunami Runup—A Hydraulic Perspective," *Journal of Hydraulic Engineering*, September, pp. 743–747.

Hunt, J. (2005) "Tsunami Waves and Coastal Flooding," *Mathematics Today*, October, pp. 144–146.

Kaistrenko, V. M., Mazova, R. K., Pelinovsky, E. N., and Simonov, K. V. (1991) "Analytical Theory for Tsunami Runup on a Smooth Slope," *Science of Tsunami Hazards* **9**(2), pp. 115–127.

Kanoglu, U., and Synolakis, C. E. (1998) "Long-Wave Runup on Piecewise Linear Topographies," *J. Fluid Mech.* **374**, pp. 1–28.

Kennett, B. L. N., Engdahl, E. R., and Buland, R. (1995) "Constraints on Seismic Velocities in the Earth from Travel Times," *Geophys. J. Int.* **122**, pp. 108–124.

Kirkgöz, M. S. (1983) "Breaking and Runup of Long Wave," *Tsunamis, Their Science and Engineering*, pp. 467–478.

Kunkel, C., Hallberg, R., and Oppenheimer, M. (2006) "Coral Reefs Reduce Tsunami Impact in Model Simulations," *J. Geophys. Res. Lett.*, vol. 33, L23612 (4 pages), doi: 10.1029/ 2006GL027892.

Lay, T., Kanamori, H., Ammon, C. J., Nettles, M., Ward, S. N., Aster, R. C., Beck, S. L., Bilek, S. L., Brudzinski, M. R., Butler, R., DeShon, H. R., Ekstrâm, G., Satake, K., and Sipkin, S. (2005) "The Great Sumatra–Andaman Earthquake of 26 December 2004," *Science* **308**(5725), pp. 1127–1133.

Liu, P. L., Synolakis, C. E., and Yeh, H. H. (1991) "Report on the International Workshop on Long Wave Runup," *J. Fluid Mech.* **229**, pp. 675–688.

Liu, P. L. F., Lynett, P., Fernando, J., Jaffe, B., Fritz, H., Higman, B., Morton, R., Goff, J., and Synolakis, C. (2005) "Observations by the International Tsunami Team in Sri Lanka," *Science* **308**, p. 1595.

Lomnitz, C., and Nilsen-Hofsth, S. (2005) "The Indian Ocean Disaster: Tsunami Physics and Early Warning Dilemmas," *EOS* **86**(7), pp. 65–70.

Madsen, O. S., Mathisen, P. P., and Rosengaus, M. M. (1990) "Movable Bed Friction Factors for Spectral Waves," *Proceedings 22nd International Conference on Coastal Engineering*, ASCE, July 2–6, 1990 in Dalft, The Netherlands. pp. 420–429.

Marris, E. (2005) "Inadequate Warning System Left Asia at the Mercy of Tsunami," *Nature* **433**, p. 3.

Marris, E. (2006) "Sri Lankan Signs Warn Coral Thieves of Nature's Wrath," *Nature* **440**, p. 981.

Ministry of Environment and Natural Resources, Sri Lanka (MENR) and United Nations Environment Programme (UNEP). (2005) "Assessment of Damage to Natural Ecosystems in Coastal and Associated Terrestrial Environments," MENR and UNEP, Colombo, Sri Lanka, April.

Moussa, M., Ruwanpura, J. Y., and Jergeas, G. (2006) "An Integrated Simulation Using Multi-Level Stochastic Networks for Cost and Time Risk Assessment," *Journal of Construction Engineering and Management*, April, volume **132** (12), 1254–1266.

Müller, D. (1995) "Auflaufen und Berschwappen von Impulswellen an Talsperren," in *Versuchsanstalt für Wasserbau, Hydrologie und Glaziologie*, ed. D. Vischer, VAW Mitteilung 137, ETH Zürich, Switzerland.

Pelinovsky, E., Troshina, E., Golinko, V., Osipenko, N., and Petrukhin, N. (1999) "Runup of Tsunami Waves on a Vertical Wall in a Basin of Complex Topography," *Physics and Chemistry of the Earth (B)* **24**(5), pp. 431–436.

Pennisi, E. (2005) "Powerful Tsunami's Impact on Coral Reefs Was Hit and Miss," *Science* **307**, p. 657.

Satake, K. (1988) "Effects of Bathymetry on Tsunami Propagation: Application of Ray Tracing to Tsunamis," *PAGEOPH* **126**, pp. 26–36.

Schiermeier, Q. (2005) "On the Trail of Destruction," *Nature* **433**, pp. 350–353.

Silver, E., Day, S., Ward, S., Hoffman, G., Llanes, P., Lyons, A., Driscoll, N., Perembo, R., John, S., Saunders, S., Taranu, F., Anton, L., Abiari, I., Applegate, B., Engels, J., Smith, J., and Tagliodes, J. (2005) "Island Arc Debris Avalanches and Tsunami Generation," *EOS* **86**(47), pp. 485–489.

Smith, W. H. F., Scharroo, R., Titov, V., Arcas, D., and Arbic, B. (2005) "Satellite Altimeters Measure Tsunami," *Oceanography* **18**, p. 10.

Song, Y. T., Ji, C., Fu, L. L., Zlotnicki, V., Shum, C. K., Yi, Y., and Hjorleifsdottir, V. (2005) "The 26 December 2004 Tsunami Source Estimated from Satellite Radar Altimetry and Seismic Waves," *Geophys. Res. Lett.* **23**, doi: 10.1029/2005GL023683.

Stein, S., and Okal, E. A. (2005) "Speed and Size of the Sumatra Earthquake," *Nature* **434**, pp. 581–582.

Swan, B. (1983) "An Introduction to the Coastal Geomorphology of Sri Lanka," National Museums of Sri Lanka, Colombo, Sri Lanka.

Synolakis, C. E. (1987a) "The Runup of Solitary Waves," *J. Fluid Mech.* **185**, pp. 523–545.

Synolakis, C. E. (1987b) "The Runup and Reflection of Solitary Waves," *Proceedings ASCE Conference on Costal Hydrodynamics*, ed. R.A. Dalrymple, pp. 533–547, American Society of Civil Engineers, New York, New York.

Synolakis, C. E. (2005) "Self-Centered West's Narrow Focus Puts Lives at Risk," *The Times Higher Educational Supplement*, 14 January page 1–4.

Titov, V., Rabinovich, A. B., Mofjeld, H. O., Thomson, R. E., and Gonzalez, F. I. (2005) "The Global Reach of the 26 December 2004 Sumatra Tsunami," *Science* **309**, pp. 2045–2048.

Tittawela, M. (2005) "Rebuilding Sri Lanka Post-Tsunami," Presentation to the ADB, Manila, The Phillipines.

United Nations Environment Program (UNEP) (2006) "In the Front Line—Shoreline Protection and Other Ecosystem Services from Mangroves and Coral Reefs," UNEP-WCMC, Biodiversity Series No. 24, ISBN 92-807-2681-1.

Walder, J. S., Watts, P., and Waythomas, C. F. (2006) "Case Study: Mapping Tsunami Hazards Associated with Debris Flow into a Reservoir," *Journal of Hydraulic Engineering* **132**(1), pp. 1–11.

Ward, S. N., and Day, S. (2001) "Cumbre Vieja Volcano—Potential Collapse and Tsunami at La Palma, Canary Islands," *Geophys. Res. Lett.* **28**(17), 10.1029/2001 GL013110, pp. 3397–3400.

Wu, C. S. (1987) "The Energy Dissipation of Breaking Waves," *Proceedings ASCE Conference on Costal Hydrodynammics*, ed. R. A. Dalrymple, pp. 740–759, American Society of Civil Engineers, New York, New York.

12

Intermediate-scale dynamics of the upper troposphere and stratosphere

James J. Riley

Atmospheric dynamics often play an important role in certain large-scale disasters and, for example, are sometimes the cause, as in tornadoes and hurricanes or typhoons. Therefore, the understanding of, and the ability to predict, atmospheric motions can be crucial in the prediction, prevention, control, and mitigation of atmospheric-related disasters. In this chapter, the dynamics of winds in the upper troposphere and the stratosphere are discussed. Understanding and predicting these winds can be critical in dealing with natural disasters such as the fates of large-scale forest fire plumes and volcanic plumes, and manmade disasters such as global warming and ozone depletion.

12.1 Background

The dynamics of the upper troposphere and the stratosphere at the mesoscale (ranging from about 100 m to several hundred kilometers) play an important role in our weather and climate, acting as a bridge between the larger, synoptic-scale motions and the microscale. Motions in this regime are the cause of the lateral, quasi-horizontal spreading of plumes (e.g., volcanic plumes) that have been injected into this upper troposphere-stratosphere region. These motions are also the cause of the ultimate intermittent, smaller-scale turbulence that results in the mixing of natural and manmade chemical species. The air motions in these regions are important in the exchange of chemicals (e.g., various pollutants such as chlorofluorocarbons and carbon dioxide) between the troposphere and the stratosphere. This, in turn, affects the levels of ozone in the stratosphere and carbon dioxide in the troposphere, and hence, such phenomena as global warming and the depletion of ozone in the Arctic and Antarctic stratosphere. The air motions in this regime act as forcing on the larger-scale atmospheric flows and must be accurately parameterized in any attempt to predict these larger-scale flows. Motions on these scales have a close analogy to mesoscale currents in the open ocean, as they are also affected by stable density stratification and the rotation of the Earth. Therefore, their understanding can lead to a better understanding of such currents in the ocean and vice versa.

A major advance in understanding the wind field in this regime was made in the mid-1980s by Nastrom and Gage (1985; Nastrom et al., 1984), who analyzed wind data taken from the global atmospheric sampling program (GASP). These data, collected from more than 7,000 flights from 1975 to 1979, were automatically recorded with instruments aboard Boeing 747 airliners in routine commercial operation between 9 and 14 km altitude. Figure 12.1 and gives the variance power spectra of the zonal wind and meridional wind, and the potential temperature. The latter two spectral are shifted one and two decades to the right, respectively. It is seen that, for the range of scales of interest (from about 2 km to about 200 km), $k^{-5/3}$ horizontal wave number spectra are observed in all three quantities.

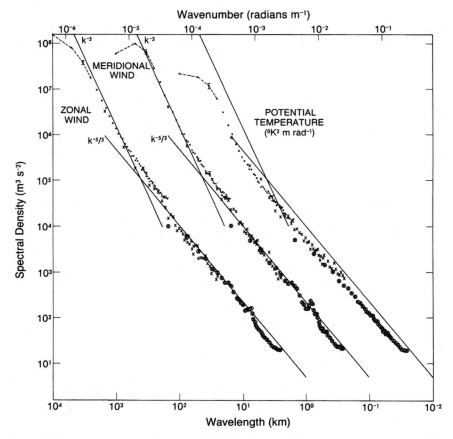

Figure 12.1 Power spectra of wind and potential temperature near the tropopause from the global atmospheric sampling program (GASP) aircraft data. The spectra of the meridional wind and potential temperature are shifted one and two decades to the right, respectively. Lines with slops −3 and −5/3 are entered at the same relative coordinates for each variable for comparison. From Nastrom and Gage (1985).

These results have suggested an inertial subrange as theorized by Kolmogorov (1941, 1962) for smaller-scale turbulence. Early on, however, it was realized that Kolmogorov's theory did not directly apply, and other hypotheses are required.[1] This, in turn, led to several hypotheses in regard to the underlying dynamics. In one hypothesis, it is speculated that the spectra are due to strong nonlinear, upscale transfer of energy (Gage, 1979; Lilly, 1983), similar to that in two-dimensional turbulence (Kraichnan, 1970). The source of energy could be, for example, in the outflow from convective clouds and thunderstorms. In another hypothesis, the results are due to the downscale transfer of energy due to weak internal wave interactions (Dewan, 1979; van Zandt, 1982), much the same as has been speculated for the ocean (Garrett and Munk, 1979). In this case, the internal wave energy could come from the larger-scale, synoptic motions; larger-scale internal waves could be generated that feed the smaller-scale waves, providing a forward cascade of energy.

---

[1] It is argued below that ideas analogous to Kolmogorov's do offer an explanation for the results.

12.2 More recent interpretation of data

Some of the most profound and useful theoretical predictions of turbulence come Kolmogorov (1941, 1962), who made a series of predictions regarding the behavior of the smaller-scale turbulent motions. One of these predictions relates to the third-order structure function $D_{LLL}(r)$, defined as the average of the cube of the difference between the fluid velocity in a particular direction, say the $\mathbf{x}_1$-direction, at a point $\mathbf{x}$, and the same component of velocity at a location displaced a distance r in the same direction; that is,

$$D_{LLL}(r) = \langle [u_1(\mathbf{x} + \mathbf{e}_1 r) - u_1(\mathbf{x})]^3 \rangle \tag{12.1}$$

where $\langle \cdot \rangle$ denotes the averaging operator (usually a space or a time average), $\mathbf{e}_1$ is a unit vector in the $\mathbf{x}_1$-direction, and statistical isotropy has been assumed, the latter one of the hypotheses of Kolmogorov for the smaller-scale motions. For length scales in the so-called "inertial subrange," defined as scales much larger than those affected by viscosity (characterized by the Kolmogorov length scale $(v^3/\epsilon)^{1/4}$, where v is the fluid kinematic viscosity, and ϵ is the average kinetic energy dissipation rate of the turbulence), but much smaller than the energy-containing scales, his hypotheses led to the prediction that

$$D_{LLL}(r) = -\frac{4}{5}\epsilon r \tag{12.2}$$

In particular, a cascade of energy from larger to smaller scales would imply that this structure function is negative.

Lindborg (1999) derived an analogous expression for mesoscale atmospheric motions and suggested that, depending on whether the corresponding third-order structure function was positive or negative, one could determine whether the energy flux was upscale or downscale, thus distinguishing between the two main competing hypotheses proposed for these mesoscale motions. Lindborg and Cho (2001; Cho and Lindborg, 2001) subsequently slightly extended this analysis, and used atmospheric data obtained from the MOZAIC (measurement of ozone by airbus in-service aircraft) program, with more than 7,000 commercial airline flights from 1994 to 1997, to obtain third-order structure function estimates. In this case, the inertial subrange is in the mesoscale, and is highly nonisotropic with regard to the horizontal and vertical directions; the prediction is

$$\langle [u_1(\mathbf{x} + \mathbf{e}_1 r) - u_1(\mathbf{x})]^3 \rangle + \\ \langle [u_1(\mathbf{x} + \mathbf{e}_1 r) - u_1(\mathbf{x})][u_2(\mathbf{x} + \mathbf{e}_1 r) - u_2(\mathbf{x})]^2 \rangle = 2 P_s r \tag{12.3}$$

where now the velocity difference of the transverse velocity u_2 comes into play. If P_s is determined to be positive, then an inverse cascade would be expected and vice versa.

Figure 12.2 is a plot giving this structure function as a function of r. Of interest is the mesoscale range from about 1 km to about 200 km. The data show a growth proportional to r, with a negative coefficient, indicating a forward flux of energy to smaller scales. This result tends to support the internal wave interpretation of these mesoscale motions and is inconsistent with the upscale energy transfer interpretation. As we see in the next section, however, other possible motions can come into play that are not only consistent with this interpretation, but that are also consistent with numerical simulations of such flows. Thus, a third hypothesis for the mesoscale regime is relevant.

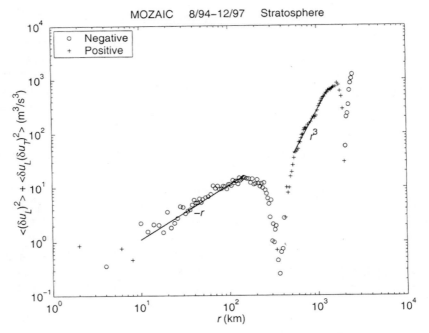

Figure 12.2 Plot showing a fit of $-r$ for $10\,\text{km} < r < 150\,\text{km}$, and r^3 for $540\,\text{km} < r < 1{,}400\,\text{km}$ to the sum of the measured stratospheric third-order structure functions. Straight lines are fits to the $-r$ and r^3 in the respective subranges. Circles and crosses indicate negative and positive values, respectively. From Cho and Lindborg (2001).

12.3 Results from numerical simulations

Assume that atmospheric mesoscale motions represented by the spectra in Figure 12.1 are characterized by a length scale L_H, the scale where most of the energy lies (about 100 km in this case), and a velocity scale u_H characterizing the mean square velocity in this range. The important parameters governing this flow regime are a Froude number, $F_H = u_H/NL_H$, a Reynolds number $Re = u_H L_H/\nu$, and a Rossby number $Ro = u_H/(\Omega L_H)$, where ν is a representative kinematic viscosity of the air, N a representative buoyancy frequency, and Ω the vertical component of the local rotation rate of the Earth. For the mesoscale motions, the Rossby number is very high and can be shown to be not important. In this regime, the Reynolds number is very high, and the Froude number is very low. The former indicates that the effect of the Reynolds number should be weak, whereas the latter indicates that density stratification plays a dominate role.

Riley and deBruynKops (2003) performed direct numerical simulations of flows at small enough Froude number that the effect of density stratification was dominate, but at high enough Reynolds number that the effect of viscosity was weak. Solving the Navier–Stokes equations subject to the Boussinesq approximation, but without the influence of system rotation, their flow was initiated to model the later stages of decay of some laboratory experiments, where the effects of stratification were becoming dominate. They found that, consistent with the analysis of Lindborg and Cho, there was a strong forward cascade of energy from large to small scales. Figure 12.3 depicts the temporal development of the horizontal spectra of the horizontal kinetic energy. Initially, the spectrum is fairly narrow banded and at large scales (low wave numbers), due to the initialization selected. As time develops,

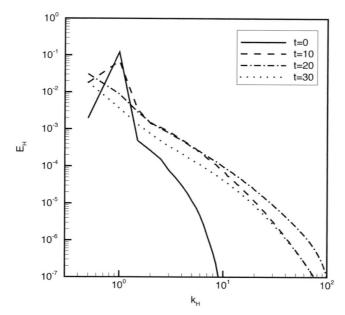

Figure 12.3 Plot showing the time evolution of the horizontal wavenumber spectra of the horizontal velocity from the simulations of Riley and deBruynKops (2003).

however, considerable energy is transferred to small scales where it is ultimately dissipated. Other simulations indicated that these results were only weakly dependent on the Reynolds number, as expected. Furthermore, about one decade of a $k_H^{-5/3}$ spectrum developed, as can be seen in Figure 12.3 at $t = 20$ and 30, again consistent with the atmospheric measurement. (Here, time has been nondimensionalized by the initial value of L_H/u_H, and the buoyancy period is $2\pi/N = 4$.) Riley and deBruynKops speculated that Kolmogorov's ideas might apply in this case, but for energy transfer in the horizontal, and not the isotropic transfer envisioned by Kolmogorov.

Working independently, Lindborg (2005, 2006) performed numerical simulations of strongly stratified flow using the Navier–Stokes equations subject to the Boussinesq approximation. In his simulations, he used forcing at the scale L_H and a hyperviscosity at small scales in order to reduce viscous effects (and artificially increase the Reynolds number). The forcing allowed his flows to reach an approximately statistically steady state. Lindborg again found a strong energy transfer to smaller scales, and he also found an inertial subrange consisting of a $k_H^{-5/3}$ spectra for both the kinetic $E_K(k_H)$ and the potential $E_P(k_H)$ energies. He found that, analogous to Kolmogorov's original theory, as well as those of Corrsin (1951) and Oboukov (1949), the spectra were, approximately,

$$E_K(k_H) = \mathcal{C}\epsilon^{2/3}k_H^{-5/3} \qquad (12.4)$$
$$E_P(k_H) = \mathcal{C}\epsilon^{-1/3}\chi k_H^{-5/3} \qquad (12.5)$$

where χ is the potential energy dissipation rate and $\mathcal{C}$ a universal constant, with the value $\mathcal{C} \sim 0.51$. Figure 12.4 is a plot of these spectra taken from Lindborg (2006), where the spectra have been plotted in a compensated form to determine the constancy of $\mathcal{C}$ and its value. When the data of Riley and deBruynKops were plotted in the same manner, using

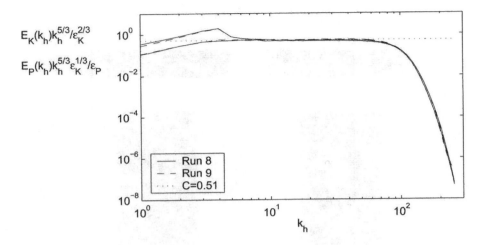

Figure 12.4 Compensated horizontal kinetic and potential energy spectra from Lindborg (2006).

the same scaling with ϵ and χ and the same value of C as suggested by Lindborg, their data also approximately collapsed in an inertial subrange.

As originally suggested by Lilly (1983) and the scaling analysis of Billant and Chomaz (2001), these flows become unstable at vertical scales of order $L_v = u_H/N$ (implying a Richardson number of order 1 and a Froude number based on this scale of order 1). For low Froude number flows, the ratio of the energy-containing scale to this instability scale is

$$\mathcal{R} = \frac{L_H}{L_v} = \frac{L_H N}{u_H} = \frac{1}{F_H} \quad (12.6)$$

a very large number. Therefore, there is a large separation between the energy-containing scale and the instability scale. Furthermore, the transfer of energy between such scales is inviscid if the Reynolds number is large enough. This indicates that, from dimensional reasoning, the parameters governing the horizontal energy spectra are therefore k_H, ϵ, and χ, the latter two being the energy transfer rates. Dimensional analysis then leads to the expressions for E_K and E_P given by Equations (12.4) and (12.5).

It must be remembered, however, that this possible inertial regime in the mesoscale is fundamentally different from that envisioned by Kolmogorov, Corrsin, and Oboukov. In this stratified flow regime, the flow is highly nonisotropic due to the strong effect of density stratification. Furthermore, the energy dynamics are dominated by the density stratification.

As argued by Lindborg (2005), it is suggested here that the atmosphere dynamics in the mesoscale region of the upper troposphere and the stratosphere, being at very low Froude number but very high Reynolds number, develop an inertial subrange similar to what is seen in the numerical simulations, and that this inertial subrange is what is observed in the original data of Nastrom and Gage.

12.4 Implications

The results of the data analysis by Lindborg and Cho, and of the simulations of Riley and deBruynKops and of Lindborg all indicate that the energy transfer in the mesoscale regime is not an upscale transfer analogous to that in two-dimensional turbulence, as suggested by Gage and by Lilly. The energy cascade in the simulated flows are also not consistent with an

12.4 Implications

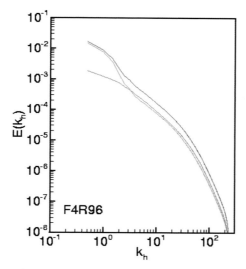

Figure 12.5 Horizontal energy spectra of the kinetic energy in the wave field, the vortex field, and the total velocity field, from the simulations of Riley and deBruynKops (2003).

internal wave cascade, as postulated by Dewan and van Zandt. This latter fact is understood in the following manner. Assuming that internal waves exist in the flow field, it is appropriate to do a kinematic wave/vortex decomposition of the flow field (Riley and Lelong, 2000), splitting the flow field into its internal wave and nonwave (vortex) components. Figure 12.5 is a plot of the horizontal energy spectra of the velocity fields for the wave and vortex components, taken from the simulated results of Riley and deBruynKops (2003) using this kinematic decomposition. It can be seen that about equal amounts of energy are included in each component. Therefore, these flows are clearly more than internal waves; the internal wave hypothesis is not adequate to describe these motions.

We speculate that the energy in the mesoscale comes from a weak forward cascade from the synoptic, quasigeostrophic motions. These motions have little wave component, but essentially consist of the vortex component. This forward cascade of energy feeds mainly vortex energy into the larger-scale motions in the mesoscale, and these motions begin to cascade energy down to smaller scales, as in the numerical simulations. At the same time as this is occurring, however, internal waves are being continually generated as the vortex motions continually go out of adjustment (cyclostrophic adjustment). In addition, there is probably some weak wave and wave/vortex interactions, causing some downscale transfer of energy as well, as observed by Waite and Bartello (2004).

In future research, a better understanding and determination of the energy flux ϵ_f from the synoptic scales into the mesoscale is needed, a topic that is starting to receive considerable interest (see, e.g., Koshyk and Hamilton [2001], Tung and Orlando [2003], Waite and Bartello [2006], and Kitamura and Matsuda [2006]). In addition, it is important to examine simulations of the type presented here to determine the types of motions and nonlinear interactions that are controlling the energy cascade in these flows. For example, can internal waves really be distinguished in these flows (i.e., do parts of the flow field satisfy the internal wave dispersion and polarization relationships)? Finally, the ramifications of these different types of motions can be explored. For example, internal waves can only weakly disperse a chemical contaminant, whereas the vortical mode can result in very strong dispersion.

With the understanding being gained from these studies, it is becoming possible to better estimate horizontal plume dispersion in the upper troposphere and stratosphere, and the mixing of chemical species across the tropopause. For example, the same scaling arguments suggest that, given the local energy flux ϵ_f from a larger-scale atmospheric simulation, the horizontal growth of a plume can be estimated as

$$d^2(t) = \mathcal{C}_1 \epsilon_f t^3 \qquad (12.7)$$

where d is a measure of the width of the plume, and $\mathcal{C}_1$ is a universal constant to be determined. The local mixing rate of chemical species can be expressed in terms of a diffusivity κ_m, and should depend mainly on ϵ_f and N, the local buoyancy frequency. This suggests estimating the local mixing rate as

$$\kappa_m = \mathcal{C}_2 \frac{\epsilon_f}{N^2} \qquad (12.8)$$

where $\mathcal{C}_2$ is another universal constant to be determined.

12.5 Summary

The understanding of and the ability to predict the intermediate-scale dynamics of the upper troposphere and the stratophere can play an important role in the prediction, prevention, control, and mitigation of atmospheric-related disasters. In this chapter, some recent results pertaining to the dynamics in this regime from numerical simulations, from theoretical arguments, and from the analysis of field data are presented and discussed. In addition, the implications of these new results regarding processes that may affect disasters (e.g., horizontal plume dispersion, turbulent mixing) are presented.

References

Billant, P., and Chomaz, J. M. (2001) "Self-Similarity of Strongly Stratified Inviscid Flows," *Phys. Fluids* **13**, pp. 1645–1651.

Cho, J. Y. N., and Lindborg, E. (2001) "Horizontal Velocity Structure Functions in the Upper Troposphere and Lower Stratosphere. Part 1. Observation," *J. Geophys. Res.* **106**, pp. 10223–10232.

Corrsin, S. (1951) "On the Spectrum of Isotropic Temperature Fluctuations in Isotropic Turbulence," *J. Applied Physics* **22**, p. 469.

Dewan, E. M. (1979) "Stratospheric Spectra Resembling Turbulence," *Science* **204**, pp. 832–835.

Gage, K. S. (1979) "Evidence for a $k^{-5/3}$ Law Inertial Range in Mesoscale Two-Dimensional Turbulence," *J. Atmos. Sci.* **36**, pp. 1950–1954.

Garrett, C., and Munk, W. (1979) "Internal Waves in the Ocean," *Ann. Rev. Fluid Mech.* **11**, pp. 339–369.

Kitamura, Y., and Matsuda, Y. (2006) "The k_h^{-3} and $k_h^{-5/3}$ Energy Spectra in Stratified Turbulence," *Geophys. Res. Lett.* **33**, L05809.

Kolmogorov, A. N. (1941) "The Local Structure of Turbulence in Incompressible Viscous Fluid for Very Large Reynolds Numbers," *Kok. Akad. Nauk SSSR* **30**, pp. 299–303 [in Russian].

Kolmogorov, A. N. (1962) "A Refinement of Previous Hypotheses Concerning the Local Structure of Turbulence in a Viscous Incompressible Fluid at High Reynolds Number," *J. Fluid Mech.* **13**, pp. 82–85.

Koshyk, J. N., and Hamilton, K. (2001) "The Horizontal Kinetic Energy Spectrum and Spectral Budget Simulated by a High-Resolution Troposphere–Stratosphere–Mesosphere GCM," *J. Atmos. Sci.* **58**, pp. 329–348.

Kraichnan, R. H. (1970) "Inertial-Range Transfer in Two- and Three-Dimensional Turbulence," *J. Fluid Mech.* **47**, pp. 535–535.

Lilly, D. K. (1983) "Stratified Turbulence and the Mesoscale Variability of the Atmosphere," *J. Atmos. Sci.* **40**, pp. 749–761.

Lindborg, E. (1999) "Can the Atmospheric Kinetic Energy Spectrum Be Explained by Two-Dimensional Turbulence?," *J. Fluid Mech.* **388**, pp. 259–288.

Lindborg, E. (2005) "The Effect of Rotation on the Mesoscale Energy Cascade in the Free Atmosphere," *Geophys. Res. Lett.* **32**, L01809.

Lindborg, E. (2006) "The Energy Cascade in a Strongly Stratified Fluid," *J. Fluid Mech.* **550**, pp. 207–242.

Lindborg, E., and Cho, J. Y. N. (2001) "Horizontal Velocity Structure Functions in the Upper Troposphere and Lower Stratosphere. Part 2. Theoretical Considerations," *J. Geophys. Res.* **106**, pp. 10233–10241.

Nastrom, G. D., and Gage, K. S. (1985) "A Climatology of Atmospheric Wavenumber Spectra of Wind and Temperature Observed by Commercial Aircraft," *J. Atmos. Sci.* **42**, pp. 950–960.

Nastrom, G. D., Gage, K. S., and Jasperson, W. H. (1984) "Kinetic Energy Spectrum of Large- and Mesoscale Atmospheric Processes," *Nature* **310**(5972), pp. 36–38.

Oboukov, A. M. (1949) "Structure of the Temperature Field in Turbulent Flows," *Izvestiya Akademii Nauk SSSR, Geogr. & Geophys.* **13**, p. 58.

Riley, J. J., and deBruynKops, S. M. (2003) "Dynamics of Turbulence Strongly Influenced by Buoyancy," *Phys. Fluids* **15**, pp. 2047–2059.

Riley, J. J., and Lelong, M.-P. (2000) "Fluid Motions in the Presence of Strong Stable Stratification," *Ann. Rev. Fluid Mech.* **32**, pp. 613–657.

Tung, K. K., and Orlando, W. W. (2003) "The k^{-3} and $k^{-5/3}$ Energy Spectrum of Atmospheric Turbulence: Quasi-Geostrophic Two-Level Simulation," *J. Atmos. Sci.* **60**, pp. 824–835.

van Zandt, T. E. (1982) "A Universal Spectrum of Buoyancy Waves in the Atmosphere," *Geophys. Res. Lett.* **9**, pp. 575–578.

Waite, M. L., and Bartello, P. (2004) "Stratified Turbulence Dominated by Vortical Motion," *J. Fluid Mech.* **517**, pp. 281–308.

Waite, M. L., and Bartello, P. (2006) "Stratified Turbulence Generated by Internal Gravity Waves," *J. Fluid Mech.* **546**, pp. 313–339.

13

Coupled weather–chemistry modeling

Georg A. Grell

In the past, much of the development for the simulation of different Earth system components such as weather and air chemistry has occurred independently. As a result, most atmospheric predictive models treat chemistry and meteorology decoupled from each other. Yet accurate forecasting of air quality and aerial transport of hazardous materials may depend strongly on interactions of chemistry and meteorology. This chapter describes state-of-the-art approaches that couple these two and allow feedback at each model time-step, both from meteorology to chemistry and from chemistry to meteorology. Dispersion forecasts and global climate change are described as two archetypal examples of coupled weather–chemistry systems.

13.1 Introduction

Many of the current environmental challenges in weather, climate, and air quality involve strongly coupled systems. It is well accepted that weather is of decisive importance for air quality or for the aerial transport of hazardous materials. It is also recognized that chemical species will influence the weather by changing the atmospheric radiation budget and through cloud formation. However, a fundamental assumption in traditional air quality modeling procedures is that it is possible to make accurate air quality forecasts (and simulations) even while ignoring much of the coupling between meteorological and chemical processes. This commonly used approach is termed "offline." Here, we describe a modeling system— and some relevant applications—that represents an opportunity to include these coupled interactions. The resulting advanced research capabilities will lead to an improvement of the understanding of complex interactive processes that are of great importance to regional and urban air quality, global climate change, and weather prediction. The resulting improved predictive capabilities will lead to more accurate health alerts, to a larger confidence when using the modeling system for regulatory purposes, and to better capabilities in predicting the consequences of an accidental or intentional release of hazardous materials. In this chapter, some aspects of this state-of-the-art coupled modeling system are described, and examples of how important online coupled modeling may be for applications such as dispersion forecasts or global climate change studies are given.

13.2 Fully coupled online modeling

In online modeling systems, the chemistry is integrated simultaneously with the meteorology, allowing feedback at each model time-step both from meteorology to chemistry and from chemistry to meteorology. This technique more accurately reflects the strong coupling of meteorological and chemical processes in the atmosphere. While the "online" coupled modeling approach was pioneered by Jacobson (1994, 1997a, 1997b, 1997c) and Jacobson et al. (1996), here we focus on a new state-of-the-art fully compressible community modeling system, based on the Weather Research and Forecast (WRF) model. It is designed to be modular, and a single source code is maintained that can be configured for both research and operations. It offers numerous physics options, thus tapping into the experience of the broad modeling community. Advanced data assimilation systems are being developed and tested in tandem with the model. Although the model is designed to improve forecast accuracy across scales ranging from cloud to synoptic, the priority emphasis on horizontal grid resolutions of 1 to 10 km makes WRF particularly well suited for newly emerging numerical weather prediction applications in the nonhydrostatic regime. Meteorological details of this modeling system can be found in Skamarock et al. (2005); the details of the chemical aspects are covered in Grell et al. (2005) and Fast et al. (2006).

13.2.1 Grid scale transport of species

Although WRF has several choices for dynamic cores, the mass coordinate version of the model, called advanced research WRF (ARW) is described here. The prognostic equations integrated into the ARW model are cast in conservative (flux) form for conserved variables; nonconserved variables such as pressure and temperature are diagnosed from the prognostic conserved variables. In the conserved variable approach, the ARW model integrates a mass conservation equation and a scalar conservation equation of the form

$$\mu_t + \nabla \cdot (\mathbf{V} \mu) = 0 \qquad (13.1)$$
$$(\mu\phi)_t + \nabla \cdot (\mathbf{V} \mu\phi) = 0 \qquad (13.2)$$

where μ is the column mass of dry air, $\mathbf{V}$ is the velocity vector (u, v, w), and ϕ is a scalar mixing ratio. These equations are discretized in a finite volume formulation, and as a result, the model exactly (to machine roundoff) conserves mass and scalar mass. The discrete model transport is also consistent (the discrete scalar conservation equation collapses to the mass conservation equation when $\phi = 1$) and preserves tracer correlations (c.f. Lin and Rood, 1996). The ARW model uses a spatially fifth-order evaluation of the horizontal flux divergence (advection) in the scalar conservation equation and a third-order evaluation of the vertical flux divergence coupled with the third-order Runge–Kutta time integration scheme. Both the time integration scheme and the advection scheme are described in Wicker and Skamarock (2002). Skamarock et al. (2005) also modified the advection to allow for positive definite transport.

13.2.2 Subgrid scale transport

Typical options for turbulent transport in the boundary layer include a level 2.5 Mellor–Yamada closure parameterization (Mellor and Yamada, 1982), or a nonlocal approach implemented by scientists from the Yong-Sei University (the YSU scheme; Hong and Pan, 1996). Transport in unresolved convection is handled by an ensemble scheme developed by Grell and Devenyi (2002). This scheme takes time-averaged rainfall rates from any of the convective parameterizations from the meteorological model to derive the convective fluxes of tracers. This scheme also parameterizes the wet deposition of the chemical constituents.

13.2.3 Dry deposition

The flux of trace gases and particles from the atmosphere to the surface is calculated by multiplying concentrations in the lowest model layer by the spatially and temporally varying deposition velocity, which is proportional to the sum of three characteristic resistances (aerodynamic resistance, sublayer resistance, and surface resistance). The surface resistance parameterization developed by Weseley (1989) is used. In this parameterization, the surface resistance is derived from the resistances of the surfaces of the soil and the plants. The properties of the plants are determined using land use data and the season. The surface resistance also depends on the diffusion coefficient, reactivity, and water solubility of the reactive trace gas.

The dry deposition of sulfate is described differently. In the case of simulations without calculating aerosols explicitly, sulfate is assumed to be present in the form of aerosol particles, and its deposition is described according to Erisman et al. (1994).

When employing the aerosol parameterization, the deposition velocity, $\hat{v}_{dk}$, for the kth moment of a polydisperse aerosol is given by

$$\hat{v}_{dk} = (r_a + \hat{r}_{dk} + r_a \hat{r}_{dk} \hat{v}_{Gk})^{-1} + \hat{v}_{Gk} \tag{13.3}$$

where r_a is the surface resistance, $\hat{v}_{Gk}$ is the polydisperse settling velocity, and $r_d k$ is the Brownian diffusivity (Pleim et al., 1984; Slinn and Slinn, 1980).

13.2.4 Gas-phase chemistry

Coupled state of the art meteorology–chemistry models such as WRF/Chem (Grell et al., 2005) typically include hundreds of reactions and dozens of chemical species. Solving the corresponding huge systems of ordinary differential equations requires highly efficient numerical integrators. In the case of hard-coded manually "tuned" solvers, even minor changes to the chemical mechanism, such as updating the mechanism by additional equations, often require recasting the equation system and, consequently, major revisions of the code. This procedure is both extremely time consuming and error prone. In recent years, automatic code generation has become an appreciated and widely used tool to overcome these problems. The Kinetic PreProcessor (KPP; Damian et al., 2002; Sandu et al., 2003; Sandu and Sander, 2006) is a computer program that reads chemical equations and reaction rates from an ASCII input file provided by the user and writes the program code necessary to perform the numerical integration. Efficiency is obtained by automatically reordering the equations in order to exploit the sparsity of the Jacobian. Salzmann and Lawrence (2006) adapted KPP for WRF/Chem, slightly modifying the latest KPP version (V2.1) to produce modules that can be used in WRF/Chem without further modifications. Furthermore, a preprocessor for WRF/Chem has been developed (Salzmann and Lawrence, 2006) that automatically generates the interface routines between the KPP-generated modules and WRF/Chem. This WRF/Chem/KPP coupler considerably reduces the effort necessary to add chemical compounds and/or reactions to existing mechanisms, as well as the effort necessary to add new mechanisms using KPP in WRF/Chem.

In addition to KPP, two hard-coded mechanisms are also part of WRF/Chem. These include the well-known and widely used mechanism for the regional acid deposition model, version 2 (RADM2; Chang et al., 1989), originally developed by Stockwell et al. (1990), and CBM-Z, which is an extension of the widely used carbon bond IV mechanism, as described by Zaveri and Peters (1999).

The RADM2 mechanism is a compromise between chemical detail, accurate chemical predictions, and available computer resources. The inorganic species included in the RADM2 mechanism are fourteen stable species, four reactive intermediates, and three abundant stable species (oxygen, nitrogen, and water). Atmospheric organic chemistry is represented by twenty-six stable species and sixteen peroxy radicals. The RADM2 mechanism represents organic chemistry through a reactivity aggregated molecular approach (Middleton et al., 1990). Similar organic compounds are grouped together in a limited number of model groups through the use of reactivity weighting. The aggregation factors for the most emitted volatile organic compounds are given in Middleton et al. (1990).

The CBM-Z photochemical mechanism (Zaveri and Peters, 1999) contains 55 prognostic species and 134 reactions, and has been incorporated into WRF/Chem by Fast et al. (2006). CBM-Z uses the lumped structure approach for condensing organic species and reactions. CBM-Z extends the original CBM-IV to include reactive long-lived species and their intermediates; revised inorganic chemistry; explicit treatment of lesser reactive paraffins such as methane and ethane; revised treatments of reactive paraffin, olefin, and aromatic reactions; inclusion of alkyl and acyl peroxy radical interactions and their reactions with NO_3; inclusion of longer-lived organic nitrates and hydroperoxides; revised isoprene chemistry; and chemistry associated with dimethyl sulfide emissions from oceans. CBM-Z is implemented using a regime-dependent approach in which the kinetics are partitioned into background, anthropogenic, and biogenic submechanisms to reduce the overall computational time.

13.2.5 Parameterization of aerosols

Several choices of aerosol modules are available in WRF/Chem. A modal approach is based on the modal aerosol dynamics model for Europe (MADE; Ackermann et al., 1998), which itself is a modification of the regional particulate model (Binkowski and Shankar, 1995). Secondary organic aerosols (SOAs) have been incorporated into MADE by Schell et al. (2001), by means of the secondary organic aerosol model (SORGAM). The size distribution of the submicrometer aerosol is represented by two overlapping intervals, called modes, assuming a log-normal distribution within each mode. The most important process for the formation of secondary aerosol particles is the homogeneous nucleation in the sulfuric acid-water system. MADE/SORGAM uses the method given by Kulmala et al. (1998). The inorganic chemistry in MADE/SORGAM is based on MARS (Saxena et al., 1986), and its modifications by Binkowski and Shankar (1995). The organic aerosol chemistry is based on SORGAM (Schell et al., 2001), which assumes that SOA compounds interact and form a quasi-ideal solution.

Another choice for aerosols is based on the model for simulating aerosol interactions and chemistry (MOSAIC), implemented and described in detail by Fast et al. (2006). In contrast to the modal approach for the aerosol size distribution employed by MADE/SORGAM, MOSAIC employs a sectional approach, which is not restricted by as many physical and numerical assumptions inherent to the modal approach. In contrast to the modal approach, the sectional approach divides the aerosol size distribution into discrete size bins. A large number of size bins can theoretically better represent the size distribution of various aerosol constituents, gas-aerosol partitioning, solid-liquid partitioning, and scattering and absorption of radiation that is a function of particle diameter.

Another major advantage of this approach is its sophisticated coupling to an atmospheric radiation scheme (Fast et al., 2006). The extinction, single-scattering albedo, and the asymmetry factor for scattering are computed as a function of wavelength and three-dimensional position. In MOSAIC, each chemical constituent of the aerosol is associated with a complex index of refraction. For each size bin, the refractive index is found by volume averaging, and Mie theory is used to find the extinction efficiency, the scattering efficiency, and the intermediate asymmetry factor, as functions of the size parameter. Optical properties are then determined by summation over all size bins. Once the aerosol radiative properties are found, they are fed to the Goddard shortwave radiative transfer model (Chou and Suarez, 1999) to calculate the direct aerosol forcing (Fast et al., 2006). This implementation allows the use of this approach to study processes of great importance to global change.

13.2.6 Photolysis frequencies

Photolysis frequencies for the photochemical reactions of the gas-phase chemistry model are calculated online at each grid point according to Madronich (1987) and Wild et al. (2000). The profiles of the actinic flux are computed at each grid point of the model domain. For the determination of the absorption and scattering cross sections needed by the radiative transfer model, predicted values of temperature, ozone, and cloud liquid water content are used below the upper boundary of WRF. Above the upper boundary of WRF, fixed typical temperature and ozone profiles are used to determine the absorption and scattering cross sections. These ozone profiles are scaled with total ozone mapping spectrometer satellite observational data for the area and date under consideration.

The radiative transfer model in Madronich (1987) permits the proper treatment of several cloud layers, each with height-dependent liquid water contents. The extinction coefficient of cloud water β_c is parameterized as a function of the cloud water computed by the three-dimensional model based on a parameterization given by Slingo (1989). For the Madronich scheme used in WRF/Chem, the effective radius of the cloud droplets follows Jones et al. (1994). For aerosol particles, a constant extinction profile with an optical depth of 0.2 is applied.

13.3 Online versus offline modeling

Both online and offline modeling can be used to predict the consequences of a release of hazardous materials. Offline models require initially running a meteorological model independently of a chemical transport model. The output from the meteorological model, typically available once or twice an hour, is subsequently used to drive the transport in the chemistry simulation. Although this methodology has computational advantages, the separation of meteorology and chemistry can also lead to a loss of potentially important information about atmospheric processes that often have a time scale much smaller than the meteorological model output frequency (e.g., wind speed and directional changes, cloud formation, rainfall). This may be especially important in future air quality prediction systems because horizontal grid sizes on the order of 1 km may be required to match the operational models. This could also be of great importance when considering the dispersion of hazardous materials. At these cloud-resolving scales (horizontal grid spacing less than 5 km), it is important to note that almost all (with the exception of small-scale turbulence) vertical transport comes from resolved, explicit vertical motion fields and not from convective parameterizations. These explicit vertical motion fields usually exhibit very large variability on short time and space scales. It is therefore important to ask the question of how much information—in particular, for the vertical mass transport—is lost during transition from an online to an offline approach.

Grell et al. (2005, 2007) tested the sensitivity to this transition with two different modeling systems: MM5/Chem (Grell et al., 2000, 2005) and WRF/Chem (Grell et al.,

13.3 Online versus offline modeling

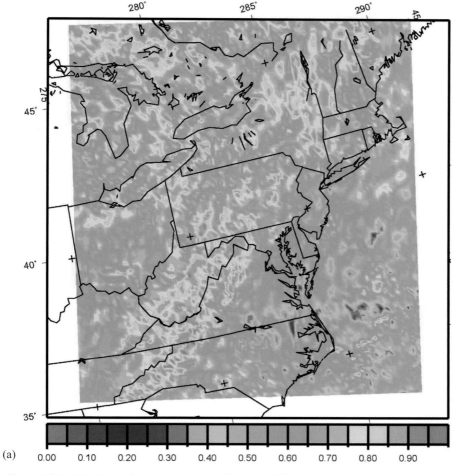

Figure 13.1 Fraction of the vertical mass flux variability that is captured (compared to an online run) when using (a) 60-minute, (b) 30-minute, and (c) 10-minute meteorological time coupling intervals.

2007). Here, we summarize the most important findings from Grell et al. (2007), which used the previously described modeling system.

In a first experiment, WRF/Chem was run for a single-day forecast. The setup of the model was chosen to represent currently used operational applications (horizontal resolution of 12 km). During this simulation, meteorological wind fields were saved at every time-step (60 s) for frequency analysis. Power spectra were then calculated for every horizontal grid point and for the vertical mass flux at levels that may be of importance for transporting constituents out of the planetary boundary layer (PBL). This analysis is limited to a 4-hour period centered around midmorning (1400–1800 UTC). Results are then compared in terms of the total energy that would be captured for various meteorological coupling intervals.

One way to look at the dependence on the meteorological output interval is seen in Figure 13.1, which displays the fraction of the variability of the vertical mass flux that is captured as a function of output frequency. Here, 100% is defined as the sum of the spectral

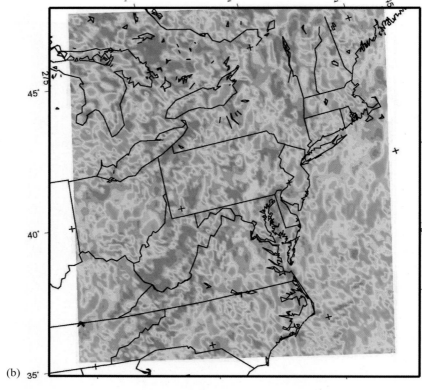

Figure 13.1 (*continued*)

power for all frequencies (because the energy contribution is extremely small, we neglect all wave numbers larger than 50). Figure 13.1a shows that on average approximately 50% of the total energy is captured, although areas in the model domain exist with only 20% to 40% of the total energy captured. As we go to a 30-minute coupling interval (Figure 13.1b), the amount of energy that is captured goes up to 70% to 90%, while for 10-minute coupling interval (Figure 13.1c), almost all energy is captured. Of course, the amount of energy that is captured clearly depends on the model's "effective" resolution. This effective resolution can usually be found by analyzing the power spectrum. It may depend not only on the spatial resolution, but also on factors such as accuracy of numerics, numerical diffusion, and choice of physics parameterizations. For our case, using a horizontal resolution of 12 km with the WRF model, it has been shown that the effective resolution is somewhere near 100 km (or about 10 dt in time). This corresponds well with Figure 13.1 because a coupling interval of 10 minutes equals 10 dt. It is clear that the higher the effective resolution, the more information will be lost at constant coupling intervals.

One might ask whether it might be enough to capture 50% of the variability, in particular, with respect to air quality forecasting or simple dispersion calculations, when trying to reproduce the online results. To test this, in a second set of experiments, WRF/Chem is first run in an online mode, twice daily, from 5 to 29 July.

In addition, three offline simulations over the same time period are performed with 10-, 30-, and 60-minute coupling intervals, initialized on 14 July using the online run from 13 July. Average biases were then compared for the different model runs and observations.

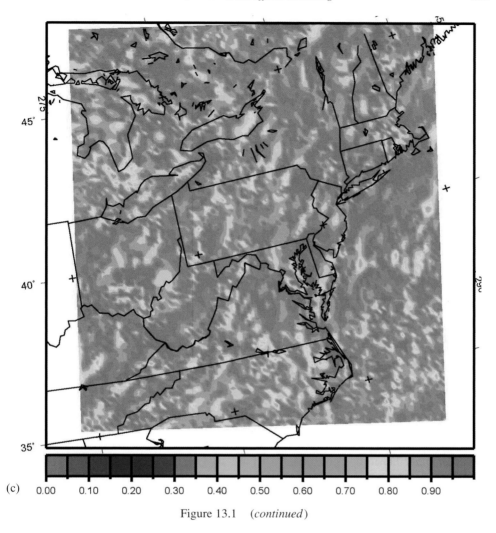

Figure 13.1 (*continued*)

Figure 13.2 shows an example of the difference in mean bias when predicting 8-hour peak ozone values, calculated from a model run with a 60-minute coupling interval and compared to observations. When compared to observations, and averaged over even longer time periods, the bias can be drastically increased in some areas (up to 8 ppb), while other areas show small changes. The same behavior was shown with respect to root-mean-square (RMS) errors and correlation coefficients. RMS errors increased, and correlation coefficients decreased, when going to larger coupling intervals. The largest differences in the model simulations are usually seen near major emission sources, probably also coinciding with strong horizontal and vertical gradients of tracer concentrations.

Grell et al. (2007) also compare a time series of carbon monoxide (CO) averaged over the lower 600 m of the atmosphere and in a region that showed large biases (a box with its southwest corner in New York City and its northeast corner on the New Hampshire coastline). The resulting time series for the various coupling intervals are shown in Figure 13.3 as a function of the output time interval. It should be expected that particulate

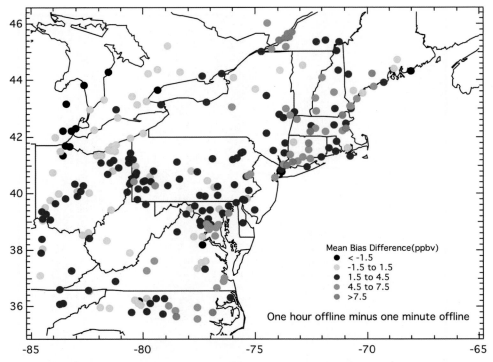

Figure 13.2 Difference in bias for predicted 8-hour ozone peak values. Produced by running WRF/Chem offline with a 1-hour coupling interval. Both models were compared to observations over a 2-week period. Mean bias to observations is determined for each model simulation, and then the displayed differences are calculated.

matter or other hazardous constituents behave in a way similar to CO. Differences for CO—an almost inert tracer—are dramatic, even for smaller coupling intervals. It becomes evident that in order to capture the vertical transport of mass properly for CO or similarly distributed tracers (strong source in the boundary layer), high-frequency coupling intervals are required. It is also clear that some days are particularly sensitive to the coupling interval. One such example is 15 July 2004, 0000UTC, where significant differences still exist with a 10-minute coupling interval. To examine this further, Figure 13.4 shows a vertical profile of CO in the region that was used to calculate the time series shown in Figure 13.3, and for a time (15 July, 0000UTC), that corresponds to large differences seen in Figure 13.3.

Note that even for a 30-minute output interval and at this resolution, differences are still as high as 50 ppb. Although the differences in the low levels are apparent, the simulations for the upper levels are even more striking. Note that the upper levels—because of the much higher wind speeds—are much more likely to also experience the effects of upstream convection. It should also be noted that Figure 13.4 represents not a single grid point, but an average over part of the domain. The differences at single grid points or smaller domains can be even larger.

In general, the online simulation is much more effective in cleansing the air in the PBL (through transporting polluted air upward to the free troposphere and cleaner air downward into the PBL). This is most dramatic during—but not restricted to—occurrences of deep convection. In addition, in Figure 13.5, we show the CO mixing ratio near the tropopause and the 1-hour rainfall rate, indicating the importance of the explicitly resolved transport

13.3 Online versus offline modeling

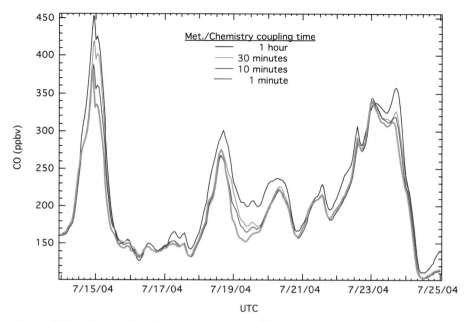

Figure 13.3 Time series of carbon monoxide (CO) mixing ratios for online (green) as well as offline WRF/Chem simulations using 60- (gray), 30- (red), and over 10-minute (blue) coupling intervals. Results were averaged over the lowest 600 m of the atmosphere and over a horizontal area about 250 × 250 km.

of tracers. Lower ozone ratios can usually be found in the lower levels where convection is active. The CO concentrations show the strongest increases near and somewhat downstream of convection, with maximum difference values reaching more than 250 ppb (note that Figure 13.5 does not show differences smaller than 30 ppb).

Explicitly resolved convective systems may have a particularly severe effect on the vertical redistribution of mass because no convective parameterization can be used to accommodate for the amount of mass flux that occurs in the meteorological simulation; all mass transport has to come from the model-predicted vertical velocity fields. In this case, it is not only important to capture most of the variability, but also most of the peaks (positive and negative) of the vertical velocity fields, especially in low levels. For a constituent that has its sources and highest concentration in the PBL, both upward motion out of the PBL and downward motion into the PBL will decrease its concentration in the boundary layer. Although the accuracy of the simulated vertical winds cannot be guaranteed, capturing most of the variability in the vertical velocity field is still essential in order to properly simulate the vertical redistribution of mass. It may also be critical to use very high time resolution meteorological output when performing dispersion simulations in the case of the release of toxic substances. The introduced errors will become larger with a further increase in resolution. It was expected that the effective time resolution of our modeling system was somewhere around 10 dt. For our simulations, this would mean that a 10-minute coupling interval would capture most of the energy in the smaller scales. However, averaging the wind fields over 10 minutes may still create flow inconsistencies in the presence of fast-evolving circulation systems such as convection. This is of particular importance for unbalanced flow regimes, where the divergent component becomes important. The vertical redistribution of the constituents may well be determined more by the nature of these circulation systems

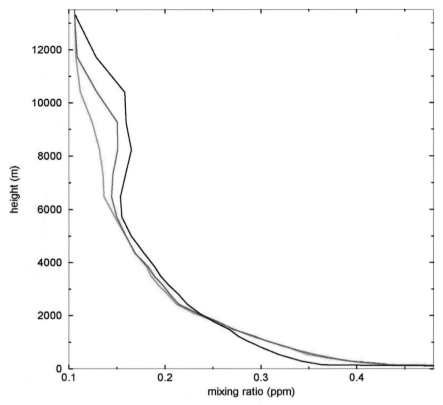

Figure 13.4 Vertical profile of horizontally averaged carbon monoxide (CO) mixing ration. Displayed are results for a run with 60- (green) and 30- (red) minute coupling interval, as well as the online (black) simulation.

than by the larger scale flow. As we go to even smaller scales, this effect will be increased because more and more of these types of circulation systems will be resolved.

13.4 Application in global change research

In this section, we discuss fully compressible, coupled weather–chemistry models as tools for global climate change research. Global climate model predictions contain major uncertainties associated with the direct and indirect effects of aerosols. Our understanding of the life cycle of aerosols, including the distribution of particulate mass, composition, size distribution, physical characteristics, and connection between the physical and optical properties of suspended particulate matter, needs improvement to more accurately simulate aerosol radiative forcing. The coarse spatial resolution and the necessity of using somewhat simplistic parameterizations within global climate models may be a significant source of uncertainty in estimating direct and indirect forcing that may lead to erroneous conclusions regarding spatial variations of future climate change. Simple cloud modeling studies such as Haywood and Shine (1997) have estimated the uncertainties in clear sky and cloudy sky direct radiative forcing. However, there is a large need for process studies related to the direct and indirect forcing using sophisticated, state-of-the-art coupled modeling systems. A multiscale model, such as WRF, that can resolve local and regional atmospheric processes that affect the life cycle of aerosols, can be used to improve our understanding of those processes in which aerosols play an important role.

13.4 Application in global change research 313

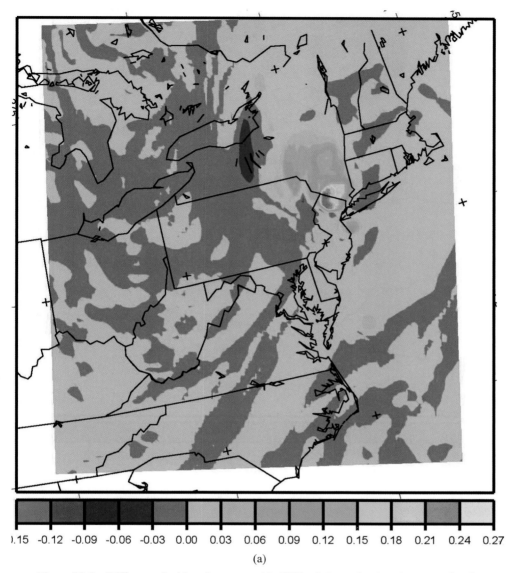

(a)

Figure 13.5 Difference in (a) carbon monoxide (CO) mixing ratios (ppm) at upper levels (above 10 km) between an online and a 60-minute offline simulation on 15 July 2004, 0000UTC, and (b) hourly precipitation rate in millimeters valid from 14 July, 23000UTC, to 15 July, 0000UTC, as predicted by the model.

Organic carbon constitutes a large fraction of the total particulate mass exported from urban areas, yet our understanding of the processes associated with secondary organic aerosol formation and the properties of organic aerosols is limited. The complex hydrocarbon chemistry involved with gas–aerosol exchange requires a large number of trace gases and organic carbon aerosols to adequately represent the wide range of processes that may also vary from region to region. Most models underestimate the amount of organic carbon mass (e.g., Tsigaridis and Kanakidou, 2003; Zhang et al., 2004) that subsequently affects the magnitude of the direct radiative forcing (e.g., Fast et al., 2006). Recent research has also shown that some types of organic aerosols are hydrophilic and may affect the development of clouds (e.g., Chung and Seinfeld, 2005; Novakov and Penner, 1993). Cloud–aerosol

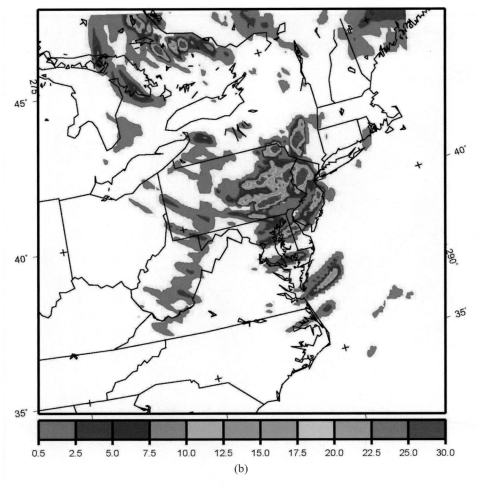

Figure 13.5 (*continued*)

interactions effectively double the number of transported variables in order to differentiate interstitial aerosols from those that are activated in cloud drops. Again, a modeling system such as WRF/Chem can be an extremely useful tool to study these processes with much more realistic physics and chemistry than global climate models. Recent examples of studies with this type of modeling system can be found in Fast et al. (2006), Gustafson et al. (2007), Longo and Freitas (2007), and Longo et al. (2006). Other examples of modeling studies with global climate models that consider interactive effects are found in Jacobson (2002, 2004), Roberts and Jones (2004), Wang (2004), Chung and Seinfeld (2005), and Ramanathan et al. (2005).

13.5 Concluding remarks

Chemical transport models have played a critical role in air quality science and environmental management. These models were designed to describe the interaction and transport of atmospheric chemical constituents associated with the gas and aerosol phases, and have become an essential element in atmospheric chemistry studies. At the same time, with the dramatic increase in computer power in recent years, weather prediction models now use higher and higher resolution to explicitly resolve fronts, convective systems, local wind systems, and even clouds. In the past, much of the development for the simulation of different earth systems components (e.g., weather, air chemistry) has occurred independently. With the realization of the importance of the coupling between the various components, this aspect is now changing. Increasingly more earth systems models are being developed. Although the coupling of these processes is clearly most important for climate models that run on very long time scales, it should not be neglected on shorter and smaller scales.

A modeling system such as WRF/Chem represents an opportunity to include these coupled interactions in future research. Advanced research capabilities will lead to an improvement of the understanding of complex interactive processes that are of great importance to regional and urban air quality, global climate change, and weather prediction. The resulting improved predictive capabilities will lead to more accurate health alerts, to a larger confidence when using the modeling system for regulatory purposes, and to better capabilities in predicting the consequences of an accidental or intentional release of hazardous materials. Finally, coupling weather and air chemistry is only a beginning in the development of a regional earth systems model. Biological, fire weather, ocean, lake, and wave models are already being incorporated. Ultimately, this type of modeling system will be a powerful tool for climate change research, weather prediction, dispersion predictions of intentional or hazardous release of dangerous substances, spread and effects of large wild fires, prediction of wave heights, and many other applications.

References

Ackermann, I. J., Hass, H., Memmesheimer, M., Ebel, A., Binkowski, F. S., and Shankar, U. (1998) "Modal Aerosol Dynamics Model for Europe: Development and First Applications," *Atmospheric Environment* **32**, no. 17, pp. 2981–2999.

Binkowski, F. S., and Shankar, U. (1995) "The Regional Particulate Matter Model, 1. Mode Description and Preliminary Results," *Journal of Geophysical Research* **100**, pp. 26191–26209.

Chang, J. S., Binkowski, F. S., Seaman, N. L., McHenry, J. N., Samson, P. J., Stockwell, W. R., Walcek, C. J., Madronich, S., Middleton, P. B., Pleim., J. E., and Lansford, H. H. (1989) "The Regional Acid Deposition Model and Engineering Model," State-of-Science/Technology, Report 4, National Acid Precipitation Assessment Program, Washington, DC.

Chou, M. D., and Suarez, M. J. (1999) "A Solar Radiation Parameterization for Atmospheric Studies, in Technical Report Series on Global Modeling and Data A1 Simulation," NASA/TM-1999-104606, vol. 15, ed. M. J. Suarez, National Aeronautics and Space Administration, Greenbelt, MD.

Chung, S. H., and Seinfeld, J. H. (2005) "Climate Impact of Direct Radiative Forcing of Anthropogenic Black Carbon," *J. Geophys. Res.* **110**, D11102, doi: 10.1029/2004JD005441.

Damian, V., Sandu, A., Damian, M., Potra, F., and Carmichael, G. R. (2002) "The Kinetic Preprocessor KPP—A Software Environment for Solving Chemical Kinetics," *Computers and Chemical Engineering* **26**, pp. 1567–1579.

Erisman, J. W., van Pul, A., and Wyers, P. (1994) "Parameterization of Surface Resistance for the Quantification of Atmospheric Deposition of Acidifying Pollutants and Ozone," *Atmospheric Environment* **28**, pp. 2595–2607.

Fast, J. D., Gustafson, W. I., Jr., Easter, R. C., Zaveri, R. A., Barnard, J. C., Chapman, E. G., Grell, G. A., and Peckham, S. E. (2006) "Evolution of Ozone, Particulates, and Aerosol Direct Radiative Forcing in the Vicinity of Houston Using a Fully Coupled Meteorology–Chemistry–Aerosol Model," *J. Geophys. Res.*, vol. III, D21305, doi: 10.1029/2005JD006721.

Grell, G. A., and Devenyi, D. (2002) "A Generalized Approach to Parameterizing Convection Combining Ensemble and Data Assimilation Techniques," *Geophys. Res. Let.* **29**, p. 1693, doi: 10.1029/2002GL015311.

Grell, G. A., Emeis, S., Stockwell, W. R., Schoenemeyer, T., Forkel, R., Michalakes, J., Knoche, R., and Seidl, W. (2000) "Application of a Multiscale, Coupled MM5/Chemistry Model to the Complex Terrain of the VOTALP Valley Campaign," *Atmospheric Environment* **34**, pp. 1435–1453.

Grell, G. A., Peckham, S. E., McKeen, S. A., and Skamarock, W. C. (2007) "A Comparison of Online versus Offline WRF/Chem Real-time Air Quality Predictions," Report to NOAA/OAR.

Grell, G. A., Peckham, S. E., Schmitz, R. M., McKeen, S. A., Frost, G., Skamarock, W. C., and Eder, B. (2005) "Fully Coupled "Online" Chemistry within the WRF Model," *Atmos. Environ.* **39**, pp. 6957–6975, doi: 10.1016/j.atmosenv.2005.04.27.

Gustafson, W. I., Jr., Chapman, E. G., Ghan, S. J., Easter, R. C., and Fast J. D. (2007), Impact on modeled cloud characteristics due to simplified treatment of cloud condensation nuclei during NEAQS 2004, *Geophys. Res Lett.*, **34**, L19809, doi: 10.1029/2007GL030021.

Haywood, J. M., and Shine, K. P. (1997) "Multi-spectral Calculations of the Direct Radiative Forcing Tropospheric Sulphate and Soot Aerosols Using a Column Model," *Q. J. R. Meteorol. Soc.* **123**, pp. 1907–1930.

Hong, S. Y., and Pan, H. L. (1996) "Nonlocal Boundary Layer Vertical Diffusion in a Medium-Range Forecast Model," *Mon. Weather Rev.* **124**, pp. 2322–2339.

Jacobson, M. Z. (2002) "Control of Fossil-Fuel Particulate Black Carbon and Organic Matter, Possibly the Most Effective Method of Slowing Global Warming," *J. Geophys. Res.* **107**, p. 4410, doi: 10.1029/2001JD001376.

Jacobson, M. Z. (1994) Developing, coupling, and applying a gas, aerosol, transport, and radiation model to study urban and regional air pollution. PhD. Dissertation, Dept. of Atmospheric Sciences, UCLA, 436 pp.

Jacobson, M. Z., Lu, R., Turco, R. P., and Toon, O. B. (1996) Development and application of a new air pollution modelsystem—Part I: Gas-phase simulations. *Atmos. Environ.*, **30**, 1939–1963.

Jacobson, M. Z. (1997a) Development and application of a new air pollution modeling system. Part II: Aerosol module structure and design, *Atmos. Environ.*, **31A**, 131–144.

Jacobson, M. Z. (1997b) Development and application of a new air pollution modeling system. Part III: Aerosol-phase simulations, *Atmos. Environ.*, **31A**, 587–608.

Jacobson, M. Z. (1997c) Testing the impact of interactively coupling a meteorological model to an air quality model. In *Measurements and Modeling in Enviromental pollution*, pp. 241–249, R. San Jose, C. A. Brebbia, eds., Computational Mechanics Publications, Southampton.

Jacobson, M. Z. (2004) "Climate Response of Fossil Fuel and Biofuel Soot, Accounting for Soot's Feedback to Snow and Sea Ice Albedo and Emissivity," *J. Geophys. Res.* **109**, doi: 10.1029/2004JD004945.

Jones, A., Roberts, D. L., and Slingo, A. (1994) "A Climate Model Study of Indirect Radiative Forcing by Anthropogenic Sulphate Aerosols," *Nature* **370**, pp. 450–453.

Kulmala, M., Laaksonen, A., and Pirjola, L. (1998) "Parameterization for Sulphuric Acid/Water Nucleation Rates," *Journal of Geophysical Research* **103**, pp. 8301–8307.

Lin, S. J., and Rood, R. B. (1996) "Multidimensional Flux-Form Semi-Lagrangian Transport Schemes," *Monthly Weather Review* **124**, pp. 2046–2070.

Longo, K., Freitas, S. R., Silva Dias, M. A., and Silva Dias, P. (2006) "Numerical Modelling of the Biomass-Burning Aerosol Direct Radiative Effects on the Thermodynamics Structure of the Atmosphere and Convective Precipitation," in *8th International Conference on Southern Hemisphere Meteorology and Oceanography*, INPE, Foz do Iguaçu, PR/São José dos Campos.

Longo, K. M., and Freitas, S. R. (2007) "The Coupled Aerosol and Tracer Transport Model to the Brazilian Developments on the Regional Atmospheric Modeling System. 3: Modeling the Biomass-Burning Aerosol Direct Radiative Effects on the Thermodynamics Structure of the Atmosphere and Precipitation," *Atmos. Chem. Phys.*

Madronich, S. (1987) "Photodissociation in the Atmosphere, 1, Actinic Flux and the Effects of Ground Reflections and Clouds," *Journal of Geophysical Research* **92**, pp. 9740–9752.

Mellor, G. L., and Yamada, T. (1982) "Development of a Turbulent Closure Model for Geophysical Fluid Problems," *Reviews of Geophysics and Spacephysics* **20**, pp. 851–887.

Middleton, P., Stockwell, W. R., and Carter, W. P. L. (1990) "Aggregation and Analysis of Volatile Organic Compound Emissions for Regional Modeling," *Atmospheric Environment* **24**, pp. 1107–1133.

Novakov, T., and Penner, J. E. (1993) "Large Contribution of Organic Aerosols to Cloud-Condensation Nuclei Concentrations," *Nature* **365**, pp. 823–826.

Pleim, J. E., Venkatram, A., and Yamartino, R. (1984) "ADOM/TADAP Model Development Program," in *The Dry Deposition Module*, vol. 4, Ontario Ministry of the Environment, Ottowa, Ontario, Canada.

Ramanathan, V., Chung, C., Kim, D., Bettge, T., Buja, L., Kiehl, J. T., Washington, W. M., Fu, Q., and Sikka, D. R. (2005) "Atmospheric Brown Clouds: Impacts on South Asian Climate and Hydrological Cycle," *Proc. Natl. Acad. Sci. USA* **102**, pp. 5326–5333, doi: 10.1073/pnas.0500656102.

Roberts, D. L., and Jones, A. (2004) "Climate Sensitivity to Black Carbon Aerosol from Fossil Fuel Combustion," *J. Geophys. Res.* **109**, D16202, doi: 10.1029/2004JD004676.

Salzmann, M., and Lawrence, M. (2006) "Automatic Coding of Chemistry Solvers in WRF-Chem Using KPP," Preprints of the 8th Annual WRF Users Workshop, National Center for Atmospheric Research, Boulder, Colorado.

Sandu, A., Daescu, D., and Carmichael, G. R. (2003) "Direct and Adjoint Sensitivity Analysis of Chemical Kinetic Systems with KPPI—Theory and Software Tools," *Atmos. Environ.* **37**, pp. 5083–5096.

Sandu, A., and Sander, R. (2006) "Technical Note: Simulating Chemical Systems in FORTRAN90 and MATLAB with the Kinetic Preprocessor KPP-2.1," *Atmos. Chem. Phys.* **6**, pp. 187–195.

Saxena, P., Hudischewskyj, A. B., Seigneur, C., and Seinfeld, J. H. (1986) "A Comparative Study of Equilibrium Approaches to the Chemical Characterization of Secondary Aerosols," *Atmospheric Environment* **20**, pp. 1471–1483.

Schell, B., Ackermann, I. J., Hass, H., Binkowski, F. S., and Ebel, A. (2001) "Modeling the Formation of Secondary Organic Aerosol within a Comprehensive Air Quality Model System," *Journal of Geophysical Research* **106**, pp. 28275–28293.

Skamarock, W. C., Klemp, J. B., Dudhia, J., Gill, D. O., Barker, D. M., Wang, W., and Powers, J. G. (2005) "A Description of the Advanced Research wrf Version 2," Tech. Rep. NCAR/TN-468+STR, National Center for Atmospheric Research, Boulder, Colorado.

Slingo, A. (1989) "A GCM Parameterization for the Shortwave Radiative Properties of Water Clouds," *Journal of the Atmospheric Sciences* **46**, pp. 1419–1427.

Slinn, S. A., and Slinn, W. G. N. (1980) "Prediction for Particle Deposition on Natural Waters," *Atmopsheric Environment* **14**, pp. 1013–1016.

Stockwell, W. R., Middleton, P., Chang, J. S., and Tang, X. (1990) "The Second Generation Regional Acid Deposition Model Chemical Mechanism for Regional Air Quality Modeling," *Journal of Geophysical Research* **95**, pp. 16343–16367.

Tsigaridis, K., and Kanakidou, M. (2003) "Global Modeling of Secondary Organic Aerosol in the Troposphere: A Sensitivity Analysis," *Atmos. Chem. Phys. Discuss.* **3**, pp. 2879–2929.

Wang, C. (2004) "A Modeling Study on the Climate Impact of Black Carbon Aerosols," *J. Geophys. Res.* **109**, D03106, doi: 10.1029/2003JD004084.

Weseley, M. L. (1989) "Parameterization of Surface Resistance to Gaseous Dry Deposition in Regional Numerical Models," *Atmospheric Environment* **16**, pp. 1293–1304.

Wicker, L. J., and Skamarock, W. C. (2002) "Time Splitting Methods for Elastic Models Using Forward Time Schemes," *Mon. Weather Rev.* **130**, pp. 2088–2097.

Wild, O., Zhu, X., and Prather, M. J. (2000) "Fast-J: Accurate Simulation of In- and Below-Cloud Photolysis in Tropospheric Chemical Models," *J. Atmos. Chem.* **37**, pp. 245–282, doi: 10.1023/A: 1006415919030.

Zaveri, R. A., and Peters, L. K. (1999) "A New Lumped Structure Photochemical Mechanism for Large-Scale Applications," *J. Geophys. Res.* vol. **104**, pp. 30, 387–303, 415.

Zhang, Y., Pun, B., Vijayaraghavan, K., Wu, S. Y., Seigneur, C., Pandis, S. N., Jacobson, M. Z., Nenes, A., and Seinfeld, J. H. (2004) "Development and Application of the Model of Aerosol Dynamics, Reaction, Ionization, and Dissolution (MADRID)," *J. Geophys. Res.* **109**, doi: 10.1029/2003JD003501.

14

Seasonal-to-decadal prediction using climate models: successes and challenges

Ramalingam Saravanan

If climatic disasters such as severe droughts can be predicted a season or two in advance, their impacts can be considerably mitigated. With advances in numerical modeling capabilities, it has become possible to make such long-term forecasts, although their skill is quite modest when compared to short-term weather forecasts. This chapter reports on the current status of "dynamical climate prediction" (i.e., prediction of climate variations on seasonal-to-decadal time scales using comprehensive computer models of the climate system). Dynamical climate predictions are now competitive with empirical predictions made using statistical models based on historical data. Dynamical prediction skill is currently limited by errors in the formulation of numerical climate models, as well as by errors in the initial conditions. Increases in computational power, better model formulations, and new ocean observing systems are expected to lead to improved prediction skill in the future.

14.1 Introduction

Predictions, prognostications, and prophecies have always fascinated the human mind over the ages. In the early days of civilization, the ability to predict often connoted power and religious authority. Regardless of whether the predictions were right or wrong, they still affected people's lives. Predictions have lost much of their magical aura in the modern age, but they still play an important role in our lives. In the realm of weather and climate, we use weekly forecasts to schedule our picnics and outings; government agencies and private corporations use seasonal forecasts for planning and resource allocation.

Climate is what you expect, and weather is what you get–so goes the old adage. Climate prediction tells us what to expect the weather to be like in the future. A variety of techniques have been used to predict climate. The simplest technique, of course, is to use the annual cycle. In the Northern Hemisphere, we can predict that the coming January will be cold and July will be warm. This simple, yet highly successful technique exploits an underlying cyclical property of climate. In this case, the cyclical behavior is a response to the annual solar cycle. The challenge of climate prediction is then to go beyond just predicting the annual cycle. To this end, many scientific (and nonscientific) studies have tried to identify other types of cyclical behavior in the climate system. If we can identify such cyclical behavior, and we know which phase of the cycle we are currently in, we can predict what will occur during the next phase. In the context of disasters, the occurrence of extreme events such as droughts and hurricanes is believed to exhibit cyclical, or periodic, behavior on decadal and longer time scales. Perhaps the best-known cyclical phenomenon related to climate prediction is the El Niño phenomenon, which is quasiperiodic, with a period ranging from 4 to 7 years (Philander, 1990).

14.1 Introduction

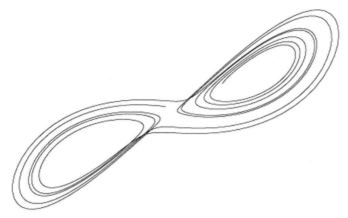

Figure 14.1 A trajectory in the attractor of the Lorenz model, illustrating the two-lobed structure.

Although it is fascinating to study climate cycles such as El Niño, we need to recognize another important property of weather and climate—that apart from the diurnal and annual cycles, they do not evolve in a purely cyclical fashion. Even El Niño exhibits rather irregular behavior. Indeed, it is the study of the erratic behavior of weather that led to the foundation of chaos theory. Ed Lorenz (1963), a professor of meteorology at the Massachusetts Institute of Technology, was experimenting with a simple computer model of the atmosphere when he made a profound discovery. The weather simulated by this computer model was deterministic, meaning that it was completely determined by the initial conditions, but it was chaotic, not cyclical. Lorenz showed that even with near-perfect knowledge of the initial condition, one could not predict weather far into the future. Figure 14.1 shows the trajectory, or time evolution, of weather as simulated by this computer model, now known as the Lorenz model. The trajectory exhibits the famous two-lobed shape that defines the "strange attractor" of the Lorenz model. If the state of the Lorenz model lies initially in the left lobe of the strange attractor, it can evolve into states that stay within the left lobe or switch to the right lobe. It is difficult to predict if and when this switch will occur.

To highlight the chaotic nature of the atmosphere, Ed Lorenz came up with the following argument: consider the flapping of a butterfly's wings in Brazil that perturbs the local winds ever so slightly. According to chaos theory, even this small perturbation could trigger the evolution of a powerful tornado in Texas a week or so later. This so-called "butterfly effect" is a severe limiting factor in our ability to predict weather events accurately (see Chapter 2 for a related discussion). In fact, meteorologists do not anticipate that we will ever be able to predict individual weather events more than about 2 weeks in advance (Chapter 19). The reason being that the atmosphere is a highly nonlinear system, and such systems tend to be quite chaotic in their behavior.

If we cannot predict weather beyond a few weeks because it is chaotic, how can we hope to predict the climate for the next year, or even for the next season? To resolve this apparent paradox, recall that we can already predict climate years in advance because of the annual cycle. This predictability arises from knowledge of the boundary conditions that affect the atmosphere, not from any knowledge of the initial state of the atmosphere. The "butterfly effect" only limits predictability of individual weather events based on accurate knowledge of the initial conditions. Although the atmosphere is chaotic, we can predict the *average* behavior of the atmosphere as it responds to the boundary conditions, provided we can predict the boundary conditions themselves (Palmer, 1993; Shukla, 1982). Lorenz (1975) formalized the distinction between these two kinds of predictability. Predictions based on the knowledge of the initial condition are referred to as *predictability of the first kind*. There is also *predictability of the second kind*, which relies on the knowledge of boundary conditions. It is the latter that allows us to make climate predictions. Whereas the first kind of predictability yields deterministic forecasts of atmospheric conditions, the second kind only yields probabilistic forecasts. A statement about the probability of weather conditions is essentially a statement about the climate.

The atmosphere is not a closed system; it has boundaries above and below. The solar forcing at the top is essentially described by the diurnal and annual cycles, and shows little year-to-year variation. At the surface, the temperature of the land and the ocean control the heating of the atmosphere, and hence affect the evolution of atmospheric flows. To exploit predictability of the second kind for climate prediction, we need to make forecasts of the atmospheric boundary conditions at least a season in advance. It turns out that it is quite difficult to forecast land temperatures because they are essentially determined by atmospheric flow conditions, due to the rather small heat capacity of land. The ocean, in contrast, has a much larger heat capacity. In fact, the heat capacity of just the top 2.5 m of the ocean equals the heat capacity of the entire atmosphere. Therefore, oceanic temperatures vary rather slowly and exhibit predictability on seasonal-to-decadal time scales.

Using the terminology of chaos theory, one can crudely explain the scientific basis for climate prediction in the ocean-atmosphere system as follows: the atmosphere, being a highly nonlinear system, has a Lyapunov exponent that limits its deterministic predictability to about 1 or 2 weeks. The ocean, also a highly nonlinear system, has a smaller Lyapunov exponent, and its deterministic predictability limit is much longer than a season. Climate prediction, in a sense, involves making forecasts of ocean temperatures and using that information to make probabilistic forecasts of atmospheric conditions. (It is worth cautioning that this is a rather simplistic view because the atmosphere and the ocean are intimately coupled, and one cannot make ocean forecasts without simultaneously making an atmospheric forecast as well.)

Because climate prediction is inherently probabilistic, it is natural to consider purely statistical, or empirical, approaches. By this, we mean a statistical model that is "trained" to predict past climate variations using historical climate data. Once trained, the model can be used to forecast future climate variations, given knowledge of the current state of the climate system. Statistical models have been and continue to be used for climate prediction with some success. However, as noted previously, the equations governing atmospheric and oceanic flow are highly nonlinear. Traditional statistical models do not handle nonlinearity very well. Furthermore, only a limited amount of historical climate data is available to train the statistical models. This means that statistical models will never be able to fully exploit the potential for climate predictability.

The alternative approach to climate prediction relies on our knowledge of the equations governing the climate system components, such as the atmosphere, ocean, and land. Although these equations are quite complex, we can solve them numerically using powerful supercomputers. This approach is referred to as dynamical (or numerical) climate prediction, to distinguish it from the statistical (or empirical) approach to climate prediction. The dynamical approach uses complex numerical models of the climate system, often referred to as general circulation models (GCMs). Typically, the atmosphere and the ocean would each be represented by their own GCM, with the two GCMs exchanging fluxes of momentum, heat, and water to simulate the coupling between the atmosphere and the ocean.

There is a fairly standard approach to using climate models to make climate predictions. First, numerical models are constructed for each component of the climate system, such as the atmosphere, ocean, land, and sea ice. Next, the component models are coupled together to construct a "climate system model." Then, the best estimate for the initial condition for the climate system is obtained from observational data. This would include estimating the state of the atmosphere and the state of the ocean. There are significant errors in this state estimation process, primarily in the ocean, because of the limited availability of subsurface observations. Finally, the coupled climate system model is initialized with the best estimate of the initial condition and is numerically integrated forward in time to produce a climate forecast.

In practice, an ensemble of initial conditions is used to make climate forecasts, each slightly different from the others, to represent the uncertainties in the initial state estimates. The time evolution of this ensemble of integrations provides a probabilistic forecast of the atmospheric conditions. *If* we can solve the equations of motion accurately enough, and *if* the ensemble size is large enough, the dynamical approach should yield a better climate prediction than the statistical approach. However, these are two big *"if"s*. It is difficult to solve the equations of motion accurately because it is not possible to simulate all the fine scales of atmospheric motion, even on the most powerful computer. Computational power also limits the size of the ensemble. For these reasons, dynamical predictions are not markedly superior to statistical models at the present time, although they continue to improve.

As one may surmise, making dynamical climate predictions is an expensive proposition. It requires the use of powerful supercomputers, as well as a fairly large scientific staff to develop and maintain state-of-the-art climate models. Often, dynamical climate prediction is carried out as an offshoot of dynamical weather prediction, by national and international weather prediction centers (see Chapters 18 and 19 for a discussion of weather prediction). The quality of the dynamical climate predictions is affected by both the accuracy of the initial state estimate and the fidelity of the climate system model. Various sophisticated mathematical and computational techniques, collectively referred to as "data assimilation," are used to refine the estimates of the initial state.

In this chapter, we present a brief report of the current status and important challenges in dynamical climate prediction (for a comprehensive review of this topic, see Goddard et al., 2001). We focus on predictions that can be made from a season to a decade into the future. On these time scales, we can ignore the human influences on climate ("anthropogenic climate change"). Beyond the decadal time scale, such changes become quite important. Making climate predictions for the next century requires more than just the knowledge of atmospheric and oceanic conditions. We also need to be able to predict the emission levels of greenhouse gases, which are determined by the interplay between political, economic, and technological developments in the future. Indeed, the uncertainties involved in making climate predictions on a centennial time scale are so large that such predictions are more appropriately referred to as *climate projections* (for more on this topic, see Chapters 15 and 16). In the next section of this chapter, we discuss climate phenomena around the globe that could potentially be predicted. In Section 14.3, we assess the actual skill of numerical models in predicting some of these phenomena. In Section 14.4, we discuss the challenges that remain in improving the skill of dynamical climate prediction, followed by concluding remarks in Section 14.5.

14.2 Potentially predictable phenomena

Phenomena that are potentially predictable on time scales of seasons to decades are usually identified through statistical analysis of historical climate data. As described previously, this predictability arises from knowledge of conditions at the surface boundary of the atmosphere, typically the sea surface temperatures. Anomalously warm or cold sea surface temperatures trigger anomalous atmospheric flows that affect weather patterns. Anomalies in tropical sea surface temperatures have a much stronger effect on the atmosphere as compared with other oceanic regions. Therefore, the tropical oceans serve as the dominant source of predictability in the climate system. Of the three tropical ocean basins, the Pacific is by far the most important because of the El Niño phenomenon (Philander, 1990).

During an El Niño event, the tropical Pacific Ocean warms up by a degree or two, and generates atmospheric waves that propagate around the globe, affecting the weather and climate over many continents. El Niño affects the rainfall over equatorial South America, the West Coast of the United States, the southeastern United States, and the Sahel region of Africa. For example, extreme rainfall events over California can be triggered by El Niño events. The occurrence of droughts in the United States is related to El Niño as well. El Niño also affects the genesis of Atlantic hurricanes, with fewer hurricanes typically observed during El Niño years.

The El Niño phenomenon is the "800lb gorilla" of dynamical climate prediction, usually accounting for much of the predictive skill. However, there are also a few, lesser-known, phenomena in the Atlantic Ocean that make more modest contributions to the overall dynamical climate prediction skill (see Xie and Carton, 2004, for a review). These include the following:

1. Atlantic Niño: This is the Atlantic analogue to the Pacific El Niño. It is characterized by anomalous warming of the eastern equatorial Atlantic Ocean.
2. Interhemispheric gradient mode: This mode of climate variability, sometimes called the "meridional mode" or the "Atlantic dipole mode," is characterized by the north–south gradient of sea surface temperature in the tropical Atlantic Ocean. Thermodynamic air–sea interaction is believed to be responsible for the long time scales associated with this mode.
3. Atlantic multidecadal oscillation: This is a multidecadal basin-scale variation in North Atlantic sea surface temperatures. It is believed to be the surface manifestation of variations in the deep oceanic circulation.

Seasonal climate prediction, especially prediction of rainfall, is particularly useful in semiarid regions of the world. Fairly small variations in the seasonal rainfall can make the difference between a normal year and a drought year, which can have disastrous socioeconomic consequences. In the vicinity of the Atlantic Ocean, two semiarid regions exhibit significant climate predictability: the Northern Nordeste region of Brazil, and the Sahel region of Africa. Both regions are affected by the Pacific El Niño, which explains some of the predictability. The Northern Nordeste region of Brazil is also significantly affected by the Atlantic interhemispheric gradient mode. Other potentially predictable phenomena in the Atlantic, such as the Atlantic Niño and the multidecadal oscillation, have an influence on the Sahel rainfall.

14.3 Successes in dynamical climate prediction

The benchmark of any dynamical climate prediction system is its ability to predict the El Niño phenomenon. The area-average of sea surface temperature over a region of the tropical Pacific known as NINO3 is used as a standard measure to represent El Niño. The statistical correlation between the observed temperature and the predicted temperature in the NINO3 region is a common measure of El Niño prediction skill. Table 14.1 shows the NINO3 correlation skill as function of forecast lead-time for dynamical predictions

Table 14.1. *Comparing El Niño prediction skill of dynamical and empirical approaches: correlation of sea surface temperature in the NINO3 region (150°W–90°W; 5°S–5°N) between model predictions and observations, for 1-, 3-, and 5-month forecast lead-times, for the verification period 1987–2001*

Prediction Model	+1 mon corr.	+3 mon corr.	+5 mon corr.
ECMWF (dynamical)	0.94	0.87	0.76
Constructed analogue (empirical)	0.83	0.76	0.70

Data from van Oldenborgh et al. (2005).

carried out using a state-of-the-art climate model developed at the European Center for Medium-range Weather Forecasting (ECMWF) (van Oldenborgh et al., 2005). One-month lead-time forecasts of the anomalous sea surface temperatures associated with the El Niño phenomenon have a correlation of 0.94, but the correlation drops to 0.76 for a 5-month forecast. Although El Niño has an underlying periodicity of 4 to 7 years, it is quite difficult to make predictions more than about 6 months in advance, with the current state of dynamical climate prediction. Also shown in Table 14.1 is the correlation skill for forecasts using one of best empirical models for El Niño prediction, known as the constructed analogue model (van den Dool and Barnston, 1994). This model has a correlation of 0.83 at 1-month lead-time and 0.70 at 5-month lead-time. Although the empirical model does not perform as well as the dynamical model, it is remarkably close, given that the dynamical prediction is orders of magnitude more costly to make than the empirical one.

The dynamical prediction skill of a single climate model is often degraded by systematic errors, or biases, in its climate simulations. In recent years, a multimodel approach has been used to mitigate this problem. Ensembles of dynamical climate predictions obtained from a number of models are averaged together, usually with equal weights, to produce a multimodel forecast (e.g., Palmer at al., 2004). The multimodel forecasts are often superior to forecasts using any individual model, presumably because there is some cancellation of the systematic errors associated with individual models.

In addition to Pacific El Niño prediction, there have also been some notable successes in predicting the Atlantic climate. Figure 14.2 shows the correlation skill for predicting sea surface temperature in a region of the North Tropical Atlantic, as a function of forecast lead-time (Saravanan and Chang, 2004). The solid curve shows the reference benchmark for forecast skill (i.e., the correlation skill of a persistence forecast), where it is assumed that the sea surface temperature simply persists indefinitely. The dashed curve shows the correlation skill obtained for forecasts initialized with global sea surface temperature initial conditions. Note that there is significant skill in climate prediction up to 5 to 7 months ahead, with correlation values in the 0.8 to 0.6 range. The third (dotted) curve in Figure 14.2 shows the correlation skill when only knowledge of Atlantic sea surface temperatures is used in the initial conditions. The lower skill for this case indicates that even for predicting Atlantic sea surface temperatures, knowledge of the global initial conditions is essential. It turns out that the seasonal predictive skill in the Atlantic region is associated both with the local off-equatorial sea surface temperatures (i.e., the interhemispheric gradient mode) and with the remote influence of El Niño. The skill in predicting tropical Atlantic sea surface temperatures also allows us to predict more economically useful quantities such as the seasonal rainfall over the Northern Nordeste region of Brazil, with correlation skill as high as 0.6 for 3-month forecasts (Chang et al., 2003).

324   *Seasonal-to-decadal prediction using climate models: successes and challenges*

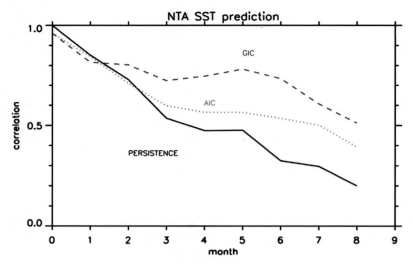

Figure 14.2   Prediction skill for North Tropical Atlantic region (76°W–26°W; 9°N–21°N): correlation between observed and predicted sea surface temperatures: persistence forecast (*solid*), global initial condition (*dashed*), and Atlantic initial condition (*dotted*). Month 0 represents December. Reproduced from Saravanan and Chang (2004).

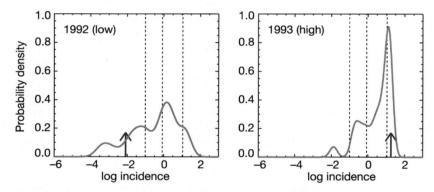

Figure 14.3   Forecast probability distribution function of standardized log malaria annual incidence for Botswana. The probability distribution functions of predicted standardized log malaria annual incidence for the years 1992 (anomalously low incidence, *left*) and 1993 (anomalously high incidence, *right*) computed with the DEMETER multimodel ensemble forecast system are depicted by the solid line. The vertical dashed lines denote the quartiles of the distribution of the standardized log malaria incidence index, and the vertical arrows indicate the values recorded by the Botswana Ministry of Health. Reprinted by permission from Macmillan Publishers Ltd: *Nature* (Thomson et al., 2006), copyright 2006.

Although skill measures such as NINO3 correlation serve as convenient benchmarks of climate prediction, they are not directly useful for practical applications. However, the multivariate information obtained from the ensemble of climate forecasts can be used to provide probabilistic forecasts of more practical value. For example, Thomson et al. (2006) use a multimodel dynamical climate prediction, computed as part of a European project called DEMETER, to provide probabilistic early warnings for incidence of malaria for the nation of Botswana in southern Africa. Figure 14.3 illustrates their predictions as a probability density for two different years, one with low incidence of malaria (1992) and another with high incidence (1993). Clearly, seasonal prediction is capable of distinguishing between low and high incidence. In this case, dynamical predictions can add up to 4 months of lead-time over malaria warnings issued using contemporaneous observed precipitation.

14.4 Challenges that remain

Perhaps the greatest challenge to improving the skill of dynamical climate prediction is to reduce the systematic errors in the climate simulations of computer models (e.g., Davey et al., 2002). Even if we had perfect knowledge of the initial conditions, the systematic model errors would degrade the skill of climate forecasts rapidly. This is the main reason dynamical climate prediction is only marginally superior to empirical climate prediction at present (Goddard and DeWitt, 2005; van Oldenborgh et al., 2005). To address this issue, many climate modeling centers around the world are working hard on improving their models, not only for the purpose of climate prediction on seasonal-to-decadal time scales, but also to improve our confidence in centennial projections of anthropogenic climate change.

Another major challenge is to improve the estimate of the initial conditions used for climate forecasts. As noted by Lorenz, small errors in the initial conditions can grow exponentially in time, causing the simulated trajectory of a dynamical system to diverge from the real trajectory, even with a perfect model. The details of the initial state of the atmosphere are not very important for dynamical climate prediction. Therefore, the initial condition errors arise primarily due to the limited availability of in situ oceanic observations. Combining the limited in situ ocean observations with global satellite observations of the ocean surface to produce the best estimate of the ocean state (referred to as "ocean data assimilation") is an active area of research.

Predicting the frequency and intensity of hurricanes and typhoons is another active area of research. However, seasonal prediction of these phenomena relies primarily on empirical approaches. Climate models currently do not have the numerical resolution needed to simulate the fine spatial scales associated with hurricanes/typhoons. However, climate models can simulate aspects of the global climate, such as El Niño and the Atlantic multidecadal oscillation, that affect the genesis of hurricanes. It is well known that El Niño increases the vertical wind shear over the main development region for Atlantic hurricanes, which inhibits their development (Gray, 1984). Thus, a good dynamical El Niño forecast can be incorporated into an empirical hurricane forecast.

Unlike in the Pacific, the equatorial sea surface temperatures in the Atlantic are notoriously difficult to predict because of a "destructive interference" between the remote influence of the Pacific El Niño and local air–sea interaction (Chang et al., 2006). Improved observations of the subsurface oceanic conditions would help reduce the initial condition uncertainty and lead to better forecasts of equatorial sea surface temperatures in the Atlantic. This is a major challenge for improving predictions of the West African monsoon. Another region that is a prime candidate for dynamical climate prediction is the Sahel region of Africa. Sahel rainfall is known to be affected by the sea surface temperatures in the tropical oceanic regions (Folland et al., 1986; Lamb, 1978). Figure 14.4 compares the observed rainfall over Sahel from 1930 to 2000 with atmospheric model simulations, which use observed sea surface temperatures as the boundary conditions (Giannini et al., 2003). It is clear from Figure 14.4 that if we can predict the global sea surface temperatures, we will also be able to predict a significant portion of Sahel rainfall. The challenge of translating this knowledge into actual climate predictions still remains. Part of this challenge is to improve the often poor simulation of Sahel rainfall in coupled climate models. Another aspect of climate models that needs to be improved is the interaction of the atmospheric

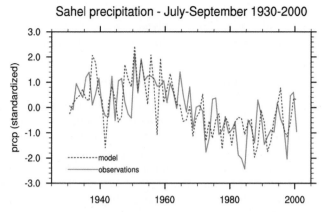

Figure 14.4 Indices of Sahel rainfall variability. Observations used the average of stations between 10°N and 20°N, 20°W and 40°E. Model numbers were based on the ensemble-mean average of gridboxes between 10°N and 20°N, 20°W and 35°E. The correlation between observed and modeled indices of (July–September) rainfall from 1930 to 2000 is 0.60. (Time series are standardized to allow for an immediate comparison because variability in the ensemble mean is muted in comparison to the single observed realization. The ratio of observed to ensemble-mean standard deviations in the Sahel is 4.) Reproduced from Giannini et al. (2003).

circulation with land moisture and vegetation, which could play a significant role in the Sahel region.

Although our discussion thus far has focused on predicting climate on seasonal-to-interannual time scales, climate models suggest that there is potential for predicting climate even on decadal time scales (e.g., Grötzner et al., 1999; Saravanan et al., 2000). Decadal climate predictions rely on temperature and salinity anomalies that exist in the deep ocean, not just in the upper ocean. Estimating the initial state of the ocean for decadal predictions remains a great challenge, given the paucity of observations. New research programs such as Argo (www.argo.ucsd.edu), which features a global array of 3,000 free-drifting floats to measure ocean properties, are actively addressing this challenge.

14.5 Summary

Skill in climate prediction on seasonal-to-decadal time scales arises primarily from the predictability of tropical sea surface temperatures. The primary source of predictability is the El Niño phenomenon in the tropical Pacific, but there are also other tropical phenomena that contribute significantly, such as the interhemispheric gradient mode in the tropical Atlantic. Although climate predictions are most skillful in the tropics, atmospheric wave propagation spreads the tropical influence globally, permitting predictions in other regions as well.

Climate prediction using fluid dynamical models is still an evolving field of research. It has only recently caught up with the prediction skill of empirical approaches to climate prediction. The skill of the empirical approach is limited by the availability of historical data for training the statistical model, and we would need to wait another 30 to 50 years to double the amount of good quality data available. Only the dynamical approach has the potential to substantially improve climate prediction skill without having to wait that long. To achieve this potential, systematic errors in climate models need to be reduced, their numerical resolution needs to be increased substantially, and estimates of initial conditions need to be improved, especially in the ocean. Anticipated increases in computing power over the next decade, improvements in ocean observing systems, and sustained research in climate model development can help us achieve this potential.

With regard to climate prediction, the dissemination of the climate forecasts to the public is also a major challenge. Deterministic weather forecasts have rather high skill over the short term. The longer-term climate forecasts have much lower skill and are inherently probabilistic. Even a "successful" climate prediction would never tell us whether a particular disaster, such as a flood or a hurricane, will *actually* occur. It will only provide us with a probability for its occurrence. Such information is rarely of much use to individuals. However, government agencies and private corporations may be able to incorporate this information into their planning strategies for resource allocation.

References

Chang, P., Fang, Y., Saravanan, R., Ji, L., and Seidel, H. (2006) "The Cause of the Fragile Relationship Between the Pacific El Niño and the Atlantic Niño," *Nature* **443**, pp. 324–328.

Chang, P., Saravanan, R., and Ji, L. (2003) "Tropical Atlantic Seasonal Predictability: The Roles of El Niño Remote Influence and Thermodynamic Air–Sea Feedback," *Geophysical Research Letters* **30**, 10.1029/2002GL016119.

Davey, M., Huddleston, M., Sperber, K. R., Braconnot, P., Bryan, F., Chen, D., Colman, R. A., Cooper, C., Cubasch, U., Delecluse, P., DeWitt, D., Fairhead, L., Flato, G., Gordon, C., Hogan, T., Ji, M., Kimoto, M., Kitoh, A., Knutson, T. R., Latif, M., Le Treut, H., Li, T., Manabe, S., Mechoso, C. R., Meehl, G. A., Power, S. B., Roeckner, E., Terray, L., Vintzileos, A., Voss, R., Wang, B., Washington, W. M., Yoshikawa, I., Yu, J. Y., Yukimoto, S., and Zebiak, S. E. (2002) "STOIC: A Study of Coupled Model Climatology and Variability in Tropical Ocean Regions," *Climate Dynamics* **18**, pp. 403–420.

Folland, C. K., Palmer, T. N., and Parker, D. E. (1986) "Sahel Rainfall and Worldwide Sea Temperatures, 1901–85," *Nature* **320**, pp. 602–607.

Giannini, A., Saravanan, R., and Chang, P. (2003) "Oceanic Forcing of Sahel Rainfall on Interannual to Interdecadal Time Scales," *Science* **302**, pp. 1027–1030.

Goddard, L., and DeWitt, D. G. (2005) "Seeking Progress in El Niño Prediction," U.S. CLIVAR Variations **3**, no. 1, pp. 1–4, www.usclivar.org/.

Goddard, L., Mason, S. J., Zebiak, S. E., Ropelewski, C. F., Basher, R., and Cane, M. A. (2001) "Current Approaches to Seasonal-to-Interannual Climate Predictions," *International Journal of Climatology* **21**, pp. 1111–1152.

Gray, W. M. (1984) "Atlantic Seasonal Hurricane Frequency. Part I: El Niño and 30 mb Quasi-Biennial Oscillation Influences," *Monthly Weather Review* **112**, pp. 1649–1668.

Grötzner, A., Latif, M., Timmermann, A., and Voss, R. (1999) "Interannual to Decadal Predictability in a Coupled Ocean–Atmosphere General Circulation Model," *Journal of Climate* **12**, pp. 2607–2624.

Lamb, P. J. (1978) "Case Studies of Tropical Atlantic Surface Circulation Patterns During Recent Sub-Saharan Weather Anomalies: 1967 and 1968," *Monthly Weather Review* **106**, pp. 482–491.

Lorenz, E. N. (1963) "Deterministic Nonperiodic Flow," *Journal of the Atmopheric Sciences* **20**, pp. 130–141.

Lorenz, E. N. (1975) "Climatic Predictability," in *The Physical Basis of Climate and Climate Modelling*, GARP Publication Series, vol. 16, ed. B. Bolin, World Meteorological Organization, pp. 132–136.

Palmer, T. N. (1993) "Extended-Range Atmospheric Prediction and the Lorenz Model," *Bulletin of the American Meteorological Society* **74**, pp. 49–65.

Palmer, T. N., Alessandri, A., Andersen, U., Canteloube, P., Davey, M., Délécluse P., Dequé, M., Díez E., Doblas-Reyes, F. J., Feddersen, H., Graham, R., Gualdi, S., Guérémy, J.-F., Hagedorn, R., Hoshen, M., Keenlyside, N., Latif, M., Lazar, A., Maisonnave, E., Marletto, V., Morse, A. P., Orfila, B., Rogel, P., Terres, J.-M., and Thomson, M. C. (2004) "Development of a European Multimodel Ensemble System for Seasonal-to-Interannual Prediction (DEMETER)," *Bulletin of the American Meteorological Society* **85**, pp. 853–872.

Philander, S. G. H. (1990) *El Niño, La Niña, and the Southern Oscillation*, Academic Press. San Diego, California.

Saravanan, R., and Chang, P. (2004) "Thermodynamic Coupling and Predictability of Tropical Sea Surface Temperature," in *Earth's Climate: The Ocean–Atmosphere Interaction*, Geophysical Monograph 147, eds. C. Wang, S.-P. Xie, and J. A. Carton, pp. 171–180, American Geophysical Union, Washington, DC.

Saravanan, R., Danabasoglu, G., Doney, S. C., and McWilliams, J. C. (2000) "Decadal Variability and Predictability in the Midlatitude Ocean–Atmosphere System," *Journal of Climate* **13**, pp. 1073–1097.

Shukla, J. (1982) "Dynamical Predictability of Monthly Means," *Journal of the Atmospheric Sciences* **38**, pp. 2547–2572.

Thomson, M. C., Doblas-Reyes, F. J., Mason, S. J., Hagedorn, R., Connor, S. J., Phindela, T., Morse, A. P., and Palmer, T. N. (2006) "Malaria Early Warnings Based on Seasonal Climate Forecasts from Multi-Model Ensembles," *Nature* **439**, pp. 576–579.

van den Dool, H. M., and Barnston, A. G. (1994) "Forecasts of Global Sea Surface Temperature Out to a Year Using the Constructed Analogue Method," in *Proceedings of the 19th Annual Climate Diagnostics Workshop*, pp. 416–419, 14–18 November, College Park, Maryland.

van Oldenborgh, G. J., Balmaseda, M. A., Ferranti, L., Stockdale T. N., and Anderson, D. L. T. (2005) "Did the ECMWF Seasonal Forecast Model Outperform Statistical ENSO Forecast Models Over the Last 15 Years?" *Journal of Climate* **18**, pp. 2960–2969.

Xie, S.-P., and Carton, J. A. (2004) "Tropical Atlantic Variability: Patterns, Mechanisms, and Impacts," in *Earth's Climate: The Ocean–Atmosphere Interaction*, Geophysical Monograph 147, eds. C. Wang, S.-P. Xie, and J. A. Carton, pp. 121–142, American Geophysical Union, Washington, DC.

15

Climate change and related disasters

Ashraf S. Zakey, Filippo Giorgi, and Jeremy Pal

This chapter presents a discussion of climate extremes within the context of the global change debate. This is preceded by a brief review of regional climate modeling, which is a particularly useful tool to simulate extreme climatic events.

15.1 Introduction

Regional climate models became essential tools for studying climate changes on a regional basis, taking into account the highest accuracy of simulation due to high resolution as well as better understanding in both physical and dynamical parameterizations. These models can be coupled to other components of the climate system, such as hydrology, ocean, sea ice, chemistry/aerosol, and land-biosphere models.

Extreme events such as droughts, floods and associated landslides, storms, cyclones and tornadoes, ocean and coastal surges, and heat waves and cold snaps have been detected using the regional climate models. A warmer world should in theory be wetter as well because the rate of evaporation is increased and the atmosphere will contain more moisture for precipitation. Changes in precipitation, however, will not be the same all over the world. Wet areas are likely to become wetter, with more frequent episodes of flooding, while dry areas may become drier, with longer periods of drought leading to an increased threat of desertification. In general, as more heat and moisture is put into the atmosphere, the likelihood of storms, hurricanes, and tornadoes will increase. Any shift in average climate will almost inevitably result in a change in the frequency of extreme events. In general, more heat waves and fewer frosts could be expected as the average temperature rises, while the return period of severe flooding will be reduced substantially if precipitation increases. A 1 in 100-year event, for example, may become a 1 in 10-year event, while a 1 in 10-year event may become a 1 in 3-year event. For less adaptable societies in the developing world, a shorter return period of extreme weather events may not allow them to fully recover from the effects of one event before the next event strikes.

Every region of the world experiences record-breaking climate extremes from time to time. Droughts are another devastating type of climate extreme. Early in the twentieth century, a trend toward increased drought in the North American Midwest culminated in the "Dust Bowl" decade of the "dirty thirties," after which conditions eased. During the 1970s and 1980s, annual rainfall over the Sahel zone of northern Africa dropped 25% below the average, leading to severe desiccation and famine.

Frequent reports of record-breaking events suggest that climate extremes are becoming more common. There is no scientific evidence, however, that this is the case at the global level. Throughout the twentieth century, there does not appear to be any discernible trend in extreme weather events. In addition, given the large natural variability in climate and the general rarity of climatic extremes, it is hard to determine whether they are now occurring as a result of global warming. A more likely explanation is that increased human vulnerability to climate extremes, particularly in developing countries, is transforming extreme events into climatic disasters. The communications revolution has also made people much more aware of the occurrence of extreme events and of their impact.

In its Third Assessment Report (TAR), the Intergovernmental Panel on Climate Change (IPCC) concluded that the globally averaged surface temperature increased $0.6 \pm 0.2°C$ in the twentieth century. This trend is expected to persist, with a $1.4°C$ to $5.8°C$ increase predicted for the twenty-first century (Houghton et al., 2001). Warming will vary by region (Giorgi, 2006) and will be accompanied by significant changes in precipitation, sea level rise, and frequency and intensity of some extreme events. These changes will affect natural and human systems independently or in combination with other determinants to alter the productivity, diversity, and functions of many ecosystems and livelihoods around the world (Houghton et al., 2001).

The main tools used to produce future climate scenarios are coupled atmosphere–ocean general circulation models (GCMs), which provide three-dimensional representations of the global climate system. The horizontal resolution of GCMs, however, is still relatively coarse (order of 100–300 km) and so is of limited use for providing information on climate change–related disasters related to atmospheric extreme events. For this reason, a new generation of research models, called regional climate models (RCMs) has been produced in order to regionally enhance the information of GCMs (Giorgi and Mearns, 1991). These models can be especially useful in the simulation of extreme events, which often occur at small spatial and temporal scales.

15.1.1 Definitions of climate parameters

Extreme weather event: An extreme weather event is an event that is rare within its statistical reference distribution at a particular place. Definitions of "rare" vary, but an extreme weather event would normally be as rare as or rarer than the tenth or ninetieth percentile. By definition, the characteristics of what is called extreme weather may vary from place to place. An extreme climate event is an average of a number of weather events over a certain period of time, an average that is itself extreme (e.g., rainfall over a season).

Climate: Climate in a narrow sense is usually defined as the "average weather," or more rigorously, as the statistical description in terms of the mean and variability of relevant quantities over a period of time ranging from months to thousands or millions of years. The classical period is 30 years, as defined by the World Meteorological Organization. These quantities are most often surface variables such as temperature, precipitation, and wind. Climate in a wider sense is the state, including a statistical description, of the climate system.

Climate system: The climate system is the highly complex system consisting of five major components: the atmosphere, the hydrosphere, the cryosphere, the land surface and the biosphere, and the interactions between them. The climate system evolves in time under the influence of its own internal dynamics and because of external forcing such as volcanic eruptions, solar variations, and human-induced forcing, such as the changing composition of the atmosphere and land use change.

Climate model (hierarchy): A numerical representation of the climate system based on the physical, chemical, and biological properties of its components; their interactions and feedback processes; and accounting for all or some of its known properties. The climate system can be represented by models of varying complexity (i.e., for any one component or combination of components, a hierarchy of models can be identified, differing in such aspects as the number of spatial dimensions, the extent to which physical, chemical, or biological processes are explicitly represented, or the level at which empirical parameterization are involved. There is an evolution toward more complex models with active chemistry and biology. Climate

15.1 Introduction

models are applied, as a research tool, not only to study and simulate the climate, but also for operational purposes, including monthly, seasonal, and interannual climate predication.

Climate prediction: A climate prediction or climate forecast is the result of an attempt to produce a most likely description or estimate of the actual evolution of the climate in the future (e.g., at seasonal, interannual, or long-term time scales).

Climate projection: A climate projection is a projection of the response of the climate system to emission or concentration scenarios of greenhouse gases and aerosols, or radiative forcing scenarios, often based on simulations by climate models. Climate projections are distinguished from climate predications in order to emphasize that climate projections depend on the emission/concentration/radiative forcing scenario used, which are based on assumptions, concerning, for example, future socioeconomic and technological developments that may or may not be realized and are therefore subject to substantial uncertainty.

Climate scenario: A climate scenario is a plausible and often simplified representation of the future climate, based on an internally consistent set of climatological relationships, that has been constructed for explicit use in investigating the potential consequences of anthropogenic climate change, often serving as input to impact models. Climate projections often serve as the raw material for constructing climate scenarios, but climate scenarios usually require additional information such as about the observed current climate. A climate change scenario is the difference between a climate scenario and the current climate.

Climate change: Climate change refers to a statistically significant variation in either the mean state of the climate or in its variability, persisting for an extended period (typically decades or longer). Climate change may be due to natural internal processes or external forcing, or to persistent anthropogenic in the composition of the atmosphere or in land use. Note that the United Nations Framework Convention on Climate Change (UNFCCC), in its Article 1, defines "climate change" as "a change of climate which is attributed directly or indirectly to human activity that alters the composition of the global atmosphere and which is in addition to natural climate variability observed over comparable time periods." The UNFCCC thus makes a distinction between "climate change," attributable to human activities altering the atmospheric composition, and "climate variability," attributable to natural causes.

Climate feedback: An interaction mechanism between processes in the climate system is called a climate feedback, when the result of an initial process triggers changes in a second process that in turn influences the initial one. A positive feedback intensifies the original process, and a negative feedback reduces it.

Climate outlook: A climate outlook gives probabilities that conditions, averaged over a specified period, will be below normal, normal, or above normal.

Climate variability: Climate variability refers to variations in the mean state and other statistics (e.g., standard deviations, the occurrence of extremes) of the climate on all temporal and spatial scales beyond that of individual weather events. Variability may be due to natural internal processes within the climate system (internal variability), or to variations in natural or anthropogenic external forcing (external variability).

Climate sensitivity: In IPCC reports, equilibrium climate sensitivity refers to the equilibrium change in global mean surface temperature following a doubling of the atmospheric (equivalent) CO_2 concentration. More generally, equilibrium climate sensitivity refers to the equilibrium change in surface air temperature following a unit change in radiative forcing ($°C/W\ m_{-2}$). In practice, the evaluation of the equilibrium climate sensitivity requires very long simulations with coupled GCMs (climate model). The effective climate sensitivity is a related measure that circumvents this requirement. It is evaluated from model output for evolving nonequilibrium conditions. It is a measure of the strengths of the feedback at a particular time and may vary with forcing history and climate state.

Climatology: (1) The description and scientific study of climate. (2) A quantitative description of climate showing the characteristic values of climate variables over a region.

Climatological outlook: An outlook based on climatological statistics for a region, abbreviated as CL on seasonal outlook maps. CL indicates that the climate outlook has an equal chance of being above normal, normal, or below normal.

15.2 A brief review of regional climate modeling

Regional climate change is influenced greatly by regional features, such as topography, soil texture, vegetation cover, and aerosols that are not well represented in GCMs because of their coarse resolution. To overcome this problem, limited area RCMs with resolution of up to 50 km or less can be used over given regions of interest with meteorological lateral boundary conditions provided from global model simulations. The regional modeling technique essentially originated from numerical weather prediction, and the use of RCMs for climate application was pioneered by Dickinson et al. (1989) and Giorgi (1990). RCMs are now used in a wide range of climate applications, from palaeoclimate (Hostetler et al., 1994) to anthropogenic climate change studies. They can provide high-resolution and multidecadal simulations, and are capable of describing climate feedback mechanisms acting at the regional scale. A number of widely used limited area modeling systems have been adapted to, or developed for, climate application. RCMs are used for understanding physical and dynamic processes underpinning climatic conditions at regional scales. These models are based on fundamental laws of physics and, as such, have applicability to all regions of the Earth. Some physical and dynamic processes essential to accurate simulation of weather and climate occur at scales too small to be resolved by regional models or too complex to be rendered into forms for exact numerical computation. These processes are represented by parameterizations that approximate the underlying physical behavior. Regional climate models cover domains typically on the size of continents or large fractions of continents. As an example, Figure 15.1 shows the horizontal domain of a regional model for the European and Mediterranean areas.

Such models typically have horizontal resolution of about 50 km and have fifteen to thirty levels in the vertical. Because they employ subglobal domains, regional models can be used to focus on the regional processes of interest (e.g., topographical forcing) and to represent them with high horizontal resolution. The regional climate modeling technique consists of using initial conditions, time-dependent lateral meteorological conditions, and surface boundary conditions from GCMs (or from global analyses of observations) to drive high-resolution RCMs. A variation of this technique is to also force the large-scale component of the RCM solution throughout the entire domain (e.g., Cocke and LaRow, 2000; Kida et al., 1991; von Storch et al., 2000). This technique has been mostly used in a one-way mode (i.e., with no feedback from the RCM simulation to the driving GCM). The basic strategy is, thus, to use the global model to simulate the response of the global circulation to large-scale forcings and the RCM to (1) account for sub-GCM grid scale forcings (e.g., complex topographical features, land cover inhomogeneity) in a physically based way, and (2) enhance the simulation of atmospheric circulations and climatic variables at fine spatial scales. RCMs can be coupled to other components of the climate system, such as hydrology, ocean, sea ice, chemistry/aerosol, and land-biosphere models. In addition, to reduce systematic errors, careful consideration needs to be given to the choice of physics parameterizations, model domain size and resolution, technique for assimilation of large-scale meteorological conditions, and internal variability due to nonlinear dynamics not associated with the boundary forcing (e.g., Giorgi and Mearns, 1991, 1999; Ji and Vernekar, 1997). Depending on the domain size and resolution, RCM simulations can be computationally demanding, which has limited the length of many experiments. GCM fields are not routinely stored at high temporal frequency (6-hourly or higher), as required for RCM boundary conditions, and thus careful coordination between global and regional modelers is needed in order to perform RCM experiments. Because regional climate models consist of different components, as shown in Figure 15.2, and the interactions between these components are based on fundamental laws of physics, they are applicable anywhere on Earth. However,

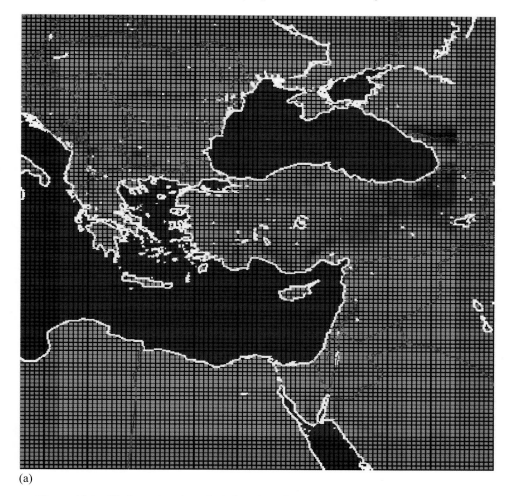

(a)

Figure 15.1 Horizontal and vertical domains of a regional climate model (RCM) for Europe and Mediterranean areas. (a) Horizontal domain of RCM; (b) Vertical levels of RCM.

the importance of physical processes represented by use of parameterizations may differ from region to region.

For example, a simulation for the tropics would not require fine detail about physical processes in frozen soils, whereas an Artic simulation might not be as dependent on a wide range of vegetation classes or soil conditions. Development of regional climate models for different regions by different research groups has tended to focus on accurate simulation for the range of climatic processes indigenous to that particular region, with less attention to processes not playing a dominant role. The choice of an appropriate domain is not trivial. The influence of the boundary forcing can reduce as region size increases (Jacob and Podzun, 1997; Jones et al., 1995) and may be dominated by the internal model physics for certain variables and seasons (Noguer et al., 1998). This can lead to the RCM solution significantly departing from the driving data, which can make the interpretation of downscaled regional climate changes more difficult (Jones et al., 1997). The domain size has to be large enough so that relevant local forcings and effects of enhanced resolution

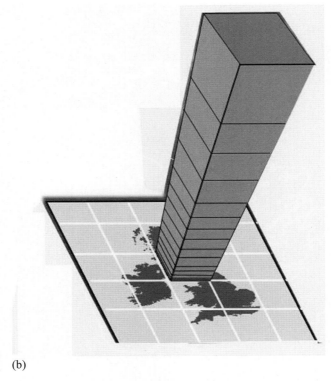

(b)

Figure 15.1 (*continued*)

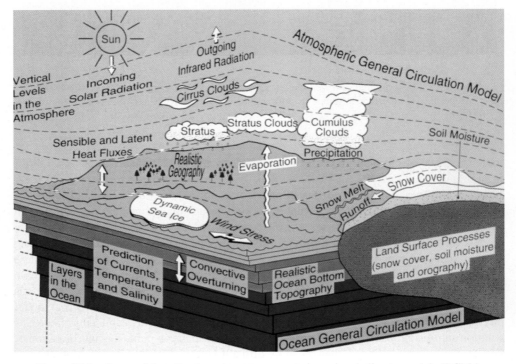

Figure 15.2 Some of the climate system components in regional climate models (RCMs).

are not damped or contaminated by the application of the boundary conditions (Warner et al., 1997). The exact location of the lateral boundaries can influence the sensitivity to internal parameters (Seth and Giorgi, 1998) or may have no significant impact (Bhaskaran et al., 1996). The location of boundaries over areas with significant topography may lead to inconsistencies and noise generation (e.g., Hong and Juang, 1998). The choice of RCM domain and resolution can modulate the effects of physical forcings and parameterizations (Giorgi and Marinucci, 1996; Laprise et al., 1998).

Resolving more of the spectrum of atmospheric motions at high resolution improves the representation of cyclonic systems and vertical velocities, but can sometimes worsen aspects of the model climatology (Kato et al., 1999; Machenhauer et al., 1998). Different resolutions may be required to capture relevant forcings in different subregions, which can be achieved via multiple one-way nesting (Christensen et al., 1998; McGregor et al., 1999), two-way nesting (Liston et al., 1999), or smoothly varying horizontal grids (Qian and Giorgi, 1999). Only limited studies of the effects of changing vertical resolution have been published (Kato et al., 1999).

RCM physics configurations are derived either from a preexisting (and well-tested) limited area model systems with modifications suitable for climate application (Copeland et al., 1996; Giorgi et al., 1993b, 1993c; Leung and Ghan, 1995, 1998; Liston and Pielke, 2000; Miller and Kim, 1997; Pielke et al., 1992; Rummukainen et al., 2000) or are implemented directly from a GCM (Christensen et al., 1996; Jones et al., 1995; Laprise et al., 1998; McGregor and Walsh, 1993). In the first approach, each set of parameterizations is developed and optimized for the respective model resolutions. However, this makes interpreting differences between nested model and driving GCM more difficult because these will not result only from changes in resolution. Also, the different model physics schemes may result in inconsistencies near the boundaries (Machenhauer et al., 1998; Rummukainen et al., 2000). The second approach maximizes compatibility between the models. However, physics schemes developed for coarse resolution GCMs may not be adequate for the high resolutions used in nested regional models and may require recalibration (Giorgi and Marinucci, 1996; Laprise et al., 1998). Overall, both strategies have shown performance of similar quality, and either one may be preferable (Giorgi and Mearns, 1999). In the context of climate change simulations, if there is no resolution dependence, the second approach may be preferable to maximize consistency between RCM and GCM responses to the radiative forcing. In view of the technical impossibility of routinely integrating high-resolution global models on climate time scales, Dickinson et al. (1989), Giorgi (1990), and Giorgi and Marinucci (1996) have shown that it is possible to nest a high-resolution regional model within a coarse-resolution global GCM. The strategy consists in interpolating coarse-resolution GCM atmospheric fields on the regional grid in order to provide time-dependent lateral atmospheric boundary conditions required by the regional model.

15.3 ICTP regional climate model

At the Abdus Salam International Centre for Theoretical Physics (ICTP), Italy, we develop one of the most widely used regional modeling systems, called the regional climate model (RegCM). Regional climate network (RegCNET) has been established at ICTP to promote climate research and ameliorate problems of scientific isolation in economically developing nations by fostering research partnerships, developing collaborative projects, and providing an internationally recognized scientific and educational research forum (www.ictp.trieste.it/RegCNET/). The RegCNET was established in June 2003 during the First ICTP Workshop on the Theory and Use of Regional Climate Models. The workshop was part of the ICTP educational program and was attended by more than seventy participants from nearly thirty countries, mostly economically developing nations. The primary research topics of the RegCNET are in the areas of regional climate change, prediction, and variability, as well as their associated societal and ecological impacts. Currently, the ICTP regional climate model (RegCM3) is the central research tool of the network, although researchers employ other models and methods. RegCM3 is used by scientists in Africa, Asia, Europe, North America, and South America.

The most recent version of the RegCNET model has been released in May 2006 for both parallel and serial computation. The first version of the RegCM series was completed by Dickinson et al. (1989), Giorgi and Bates (1989), and Giorgi (1990) at the National Center for Atmospheric Research (NCAR). It was built on the NCAR-Pennsylvania State University (PSU) Mesoscale Model, version 4 (MM4) (Anthes et al., 1987). However, to adapt the MM4 to long-term climate simulations, the radiative transfer package of Kiehl et al. (1987) was added along with Biosphere-Atmosphere Transfer Scheme (BATS), version 1a (Dickinson et al., 1986). In addition, the existing convective precipitation (Anthes, 1977) and planetary boundary layer (PBL) (Deardorff, 1972) parameterizations were improved on. The second version of the RegCM series was developed by Giorgi et al. (1993a, and 1993c). The dynamical core was upgraded to the hydrostatic version of the NCAR-PSU Mesoscale Model, version 5 (MM5) (Grell et al., 1994). The radiative transfer package was also upgraded according to that of Community Climate Model, version 2 (CCM2) (described by Briegleb, 1992). The Grell (1993) convective parameterization was included as an option, and the explicit cloud and precipitation scheme of Hsie et al. (1984) was used. BATS was updated from version 1a to 1e (Dickinson et al., 1993) and the nonlocal PBL parameterization of Holtslag et al. (1990) was implemented. An intermediate version, REGional Climate Model version 2.5 (RegCM2.5), was developed as described in Giorgi and Mearns (1999). It included an option for the Zhang and McFarlane (1995) convection scheme, the Community Climate Model, version 3 (CCM3) radiative transfer package (Kiehl et al., 1996), a simplified version of the Hsie et al. (1984) explicit cloud and precipitation scheme (SIMEX) (Giorgi and Shields, 1999), and a simple interactive aerosol model (Qian and Giorgi, 1999).

RegCM3 is an integration of the main improvements that have been made to RegCM2.5 since the description in Giorgi and Mearns (1999). These improvements are in the representation of precipitation physics, surface physics, atmospheric chemistry and aerosols, model input fields, and the user interface. In addition, the dynamic code has been modified for parallel computing. An important aspect of the RegCM3 is that it is user friendly and operates on a variety of computer platforms. To that end, substantial changes have been made to the preprocessing, running, and postprocessing of the model. Furthermore, the RegCM3 has options to interface with a variety of reanalysis and GCM boundary conditions. The other primary changes include a large-scale cloud and precipitation scheme that accounts for the subgrid-scale variability of clouds (Pal et al., 2000), new parameterizations for ocean surface fluxes (Zeng et al., 1998), and a new cumulus convection scheme (Emanuel, 1991; Emanuel and Zivkovic-Rothman, 1999). Also new in the model is a mosaic-type parameterization of subgrid-scale heterogeneity in topography and land use (Giorgi et al., 2003). An important new component of the model is the incorporation of interactive aerosols (Qian et al., 2003; Solmon et al., 2006; Zakey et al., 2006). Different aerosols are included, such as sulfate, organic carbon, black carbon, and desert dust. Other improvements in RegCM3 involve the input data. As an example of RegCM application we can refer to the high-resolution (20-km) climate change simulation over the Mediterranean region by Gao et al. (2006). Figure 15.3a–d shows mean precipitation changes in DJF,

MAM, JJA, and SON. In DJF, maximum precipitation increase over the westward side of the mountain chains of the western and central Europe, down to the mid to northern areas of the Iberian, Italian, and Balkan peninsulas, while the minimum precipitation increase is found over the eastward sides. In SON, the main circulation change is in the easterly and southeasterly direction, which explains a positive precipitation change south of the Alps, east of the Pyrenees and Jura mountains. In JJA, most of the region undergoes a large drying due to greater broad scale anticyclones circulation.

15.4 Climate change and extreme events

15.4.1 Defining changes of extremes

To understand how changes in weather and climate extremes could influence society and ecosystems, it is useful first to conceptually address how such extremes could change in a statistical sense. Figure 15.4 presents a typical distribution of a climate variable that is normally distributed, such as temperature. The solid curve represents the present-day frequency distribution of a weather phenomenon (e.g., the daily maximum temperature). Shading indicates the extreme parts of the distribution, representing events in the tails of the distribution that occur infrequently (i.e., values that are far from the mean or median value of the distribution). If there is a simple shift of the distribution in a future climate, there will be an increase in extreme events on one end and a decrease at the other (Figure 15.4a).

This can occur through a change of the mean, where, for example, if the temperature at a location warms by a certain amount, this will almost certainly produce an increase in the number of extreme hot days and a decrease in the number of extreme cold days. It is important to note that the frequency of extremes changes nonlinearly with the change in the mean of a distribution; that is, a small change in the mean can result in a large change in the frequency of extremes (Mearns et al., 1984). Other aspects of the distribution may also change. For example, the standard deviation in a future climate may increase, producing changes in extreme events at both ends of the frequency distribution (Figure 15.4b). A change in the variance of a distribution will have a larger effect on the frequency of extremes than a change in the mean (Katz and Brown, 1992), although these events must be "extreme" enough (i.e., more than one standard deviation from the mean) for this result to hold. Relatively speaking, a 1°C change in the standard deviation of the distribution will have a greater impact on the frequency of an extreme temperature than a 1°C change in the mean of the distribution. To complicate matters, the mean, standard deviation, and even the symmetry of the distribution may change at the same time, consequently altering the occurrence of extremes in several different ways (Figure 15.4c, showing changes in both mean and variance). Not only can the parameters of the distribution change as noted previously (mean variance, etc.), but for variables such as precipitation, which is not normally distributed but better represented by a gamma distribution, a change in the mean also causes a change in the variance (Groisman et al., 1999). This helps explain why increases in total precipitation are disproportionately expressed in the extremes. Fortunately, it is often possible to estimate changes in infrequent extremes, such as those that might occur once every 10 to 100 years, without detailed knowledge of the parent distribution. This is because statistical science provides a well-developed asymptotic theory for extreme values (Leadbetter et al., 1983), which predicts that the largest observation in a large sample, such as the annual maximum temperature or 24-hour precipitation amount, will tend to have one of only three extreme value distributions, depending only on the shape of the upper tail of the parent distribution.

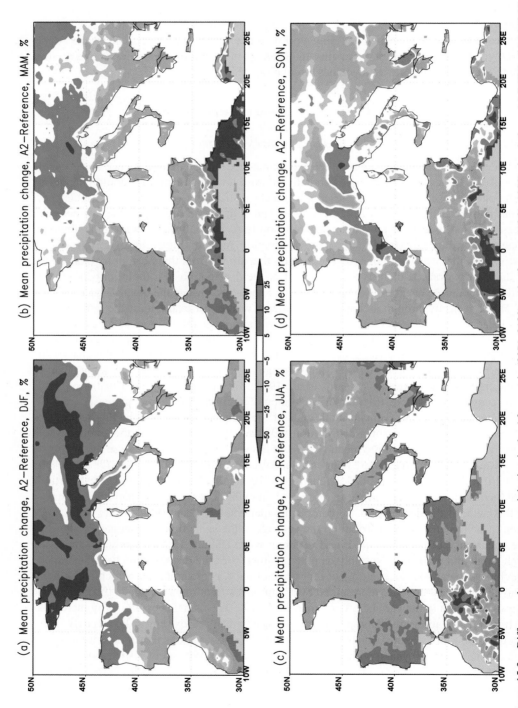

Figure 15.3 Difference between mean precipitation in the A2 scenario (2071–2100) and reference (1961–1990) simulations. (a) DJF%; (b) MAM%; (c) JJA%; (d) SON%.

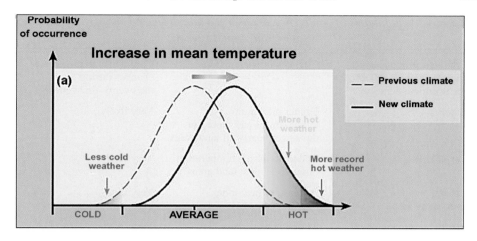
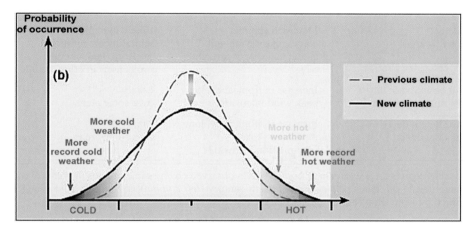
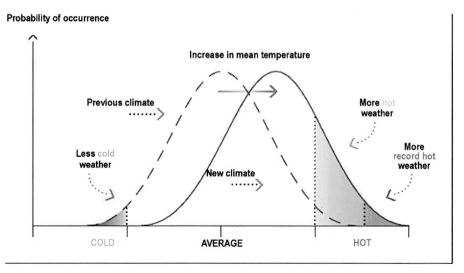

Figure 15.4 Typical distribution of a climate variable. Probabilities of occurrence: (a) Increase in mean temperature; (b) Increase in temperature variance; (c) Change in both mean and variance temperature.

Table 15.1. *Estimates of confidence in observed and projected changes in extreme weather and climate events*

Confidence in observed changes, latter half of the twentieth century	Changes in phenomenon	Confidence in projected changes during the twenty-first century
Likely	Higher maximum temperatures and more hot days over nearly all land areas	Very likely
Very likely	Reduced diurnal temperature range over most land areas	Very likely
Likely over many areas	Increase of heat index over land areas	Very likely over most areas
Likely over many Northern Hemisphere mid- to high latitude land areas	More intense precipitation events	Very likely over many areas
Likely in a few areas	Increased summer continental drying and associated risk of drought	Likely over most midlatitude continental interiors (lack of consistent projections in other areas)
Not observed in the few analyses available	Increase in tropical cyclone peak wind intensities	Likely over some areas
Insufficient data for assessment	Increase in tropical cyclone mean and peak precipitation intensities	Likely over some areas

The table depicts an assessment of confidence in observed changes in extremes of weather and climate during the latter half of the twentieth century (*left column*) and in projected changes during the twenty-first century (*right column*).

The frequency and/or intensity of extremes can change, and both can cause major problems. When we discuss extremes, we must consider them from the point of view of the statistical characteristics described previously and from the socioeconomic or ecological effects of the event. The latter can be thought of in terms of thresholds of the physical systems beyond which serious impacts occur. For example, there is an effect on human mortality and morbidity if there is an occurrence of a series of days in summer when minimum temperatures exceed 30°C (Kalkstein and Smoyer, 1993). The impact of climate on society and ecosystems could change due to changes in the physical climate system (including both natural and anthropogenic causes) or due to changes in the vulnerability of society and ecosystems (even if the climate does not change; e.g., Kunkel et al., 1999). For example, if the level of hurricane activity of the more active hurricane seasons in the 1940s and 1950s were to occur today, societal impacts would be substantially greater than in earlier decades. The impact of a higher tropical storm frequency could be made worse by population increases and density; more people living near the coast, greater wealth, and other factors (Diaz and Pulwarty, 1997; Pielke and Pielke, 1997), although human choices have produced this situation.

Table 15.1 shows the most recently changes in extremes of weather and climate observed. More hot days and heat waves are likely over nearly all land areas. These increases are projected to be largest mainly in areas where soil moisture decreases occur. Increases in daily minimum temperature are projected to occur over nearly all land areas and are generally larger where snow and ice retreat. Frost days and cold waves are likely to become fewer. The changes in surface air temperature and surface absolute humidity are projected to result in increases in the heat index, which is a measure of the combined effects of temperature and moisture. The increases in surface air temperature are also projected to result in an increase in the "cooling degree days," which is a measure of the amount of cooling required on a given day once the temperature exceeds a given threshold, and a decrease in "heating degree days." Precipitation extremes are projected to increase more than the mean, and the intensity of precipitation events are projected to increase. The frequency of extreme precipitation events is projected to increase almost everywhere. There is projected to be a general drying of the midcontinental areas during summer. This is ascribed to a combination of increased temperature and potential evaporation that is not balanced by increases of precipitation. There is little agreement yet among models concerning future changes in midlatitude storm intensity, frequency, and variability. There is little consistent evidence that shows changes in the projected frequency of tropical cyclones and areas of formation. However, some measures of intensities show projected increases, and some theoretical and modeling studies suggest that the upper limit of these intensities could increase. Mean and peak precipitation intensities from tropical cyclones are likely to increase appreciably. For some other extreme phenomena, many of which may have important impacts on the environment and society, there is currently insufficient information to assess recent trends, and confidence in models and understanding is inadequate to make firm projections. In particular, very small-scale phenomena such as thunderstorms, tornadoes, hail, and lightning are not simulated in global models. Insufficient analysis has occurred of how extratropical cyclones may change.

During 2005 across the United States as shown in Figure 15.5, lingering areas of moderate to severe drought persisted over the center. More significant drought plagued northern Illinois, while exceptional drought classification was noted over parts of the Arklatex region. Grassfires affected areas of Texas and Oklahoma due to the tinder-dry conditions. Over Africa, the lack of any significant rainfall across the eastern half of Kenya and southern Somalia has left the region in both a hydrological and an agricultural drought. A heat wave affected eastern Australia during the last 10 days of December, extending to the major coastal cities by the 31 December. The northeast monsoon produced extremely heavy rainfall in parts of the Malay Peninsula during mid-December. In Thailand, fifty-two deaths were attributed to flooding in what was described as the region's worst flooding in nearly 30 years, according to Thailand's Interior Ministry. The floods (associated with the heavy seasonal tropical rains) damaged fourteen bridges, cut 463 roads, and inundated 15,000 hectares (37,000 acres) of agricultural land, including many rubber plantations. In Vietnam, flooding claimed at least sixty-nine lives. Farther south in Malaysia, at least nine people were killed by flooding, and more than 17,000 people were driven into relief shelters.

15.5 Extremes and climate variability

Since Frich et al. (2002) completed a near-global study of trends in indices of extreme temperature and precipitation, several studies have looked in more detail at Europe. Klein Tank and Konnen (2003) examined daily temperature and rainfall series for more than 100 European stations from 1946 to 1999. Averaged over all stations, indices of temperature extremes indicate symmetrical warming of the cold and warm tails of the distributions of daily minimum and maximum temperature in this period. They found a reduction in temperature variability for the earlier part of the period and an increase in variability for the much warmer later part. For precipitation, all Europe-average indices of wet extremes

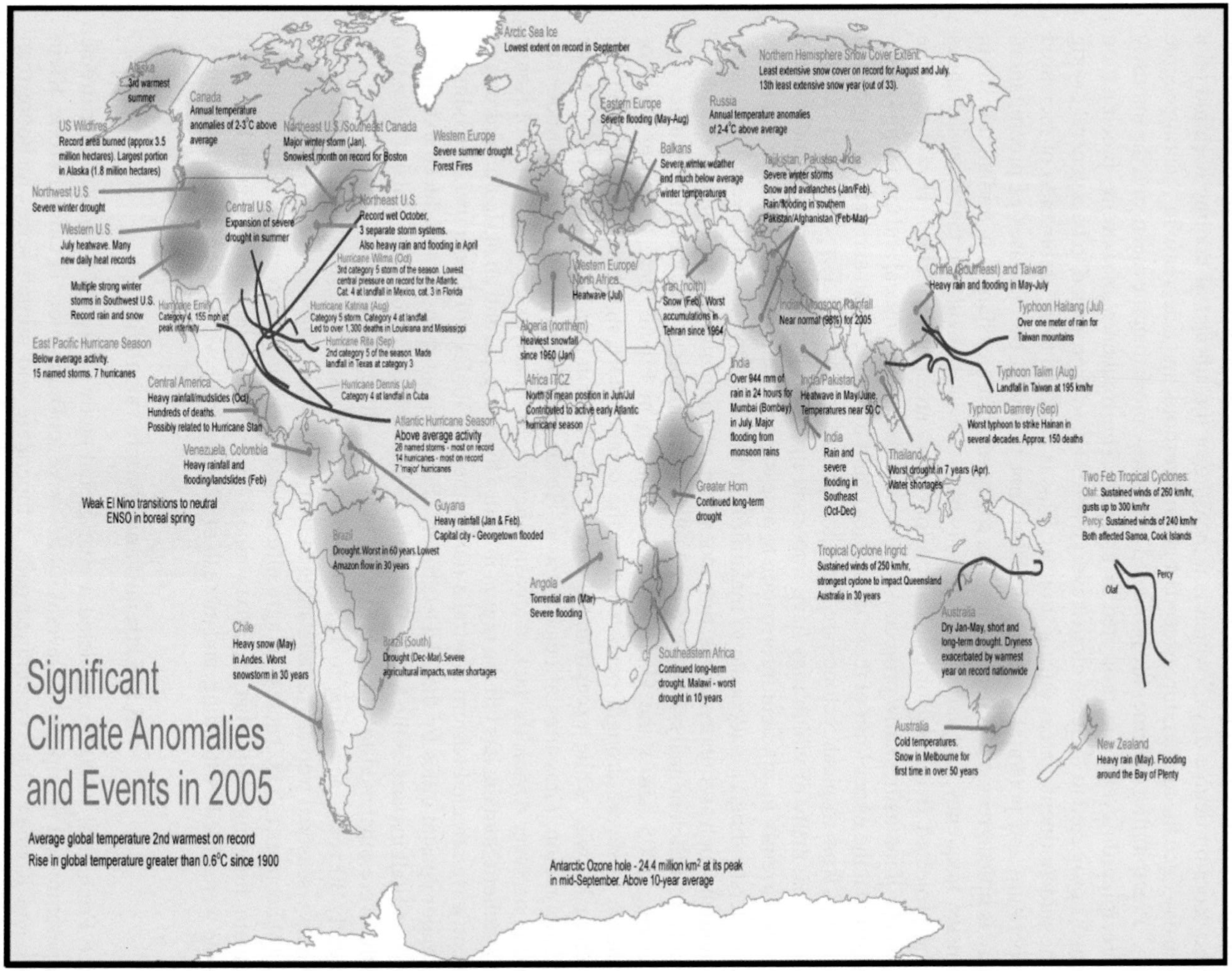

Figure 15.5   Worldwide significant extreme events in 2005.

increase in the 1946 to 1999 period, although the spatial coherence of the trends is low. Wijngaard et al. (2003) assessed the homogeneity of the European Climate Assessment dataset (Klein Tank et al., 2002), a subset of which is used in Klein Tank and Konnen (2003). They found that in the subperiod 1946 to 1999, 61% of the temperature series and 13% of the precipitation series are considered "doubtful" or "suspect." About 65% of the statistically detected inhomogeneities in the temperature series labeled "doubtful" or "suspect" in the period 1946 to 1999 can be attributed to observational changes that are documented in the metadata. For precipitation, this percentage is 90%. Haylock and Goodess (2004) examined daily winter rainfall at 347 European stations from 1958 to 2000. They found that a large part of the observed trends and interannual variability in the maximum number of consecutive dry days and the number of days above the 1960 to 1990, ninetieth percentile of wet-day amounts could be explained by changes in large-scale mean atmospheric circulation. The NAO was important in explaining large-scale observed trends in these indices. Moberg and Jones (2004) compared mean and extreme temperatures in a recent run of the Hadley Centre regional climate model with observations from 1961 to 1990. They found good agreement for latitudes 50–55°N, but seasonal biases up to 5K over other regions. Even larger errors (up to 15K) exist in extreme temperatures. A number of studies have focused on changes in extremes for smaller regions within Europe.

Brunetti et al. (2004) examined trends in temperature and rainfall over Italy during the twentieth century. They found trends toward warmer and drier conditions, but with an increase in rainfall intensity and longer dry spells. The 2003 European heat wave was examined by Beniston (2004), Schar et al. (2004), and Luterbacher et al. (2004). Luterbacher et al. showed that summer 2003 was by far the hottest since 1500. Beniston (2004) showed that the 2003 heat wave bears a close resemblance to what many regional climate models are projecting for summers in the latter part of the twenty-first century. Schar et al. (2004) showed that this heat wave is statistically unlikely given a shift in the mean temperature. An increase in variability is needed to adequately explain the probability of such an event occurring. Domonkos et al. (2003) examined the variability of winter extreme low-temperature events and summer extreme high-temperature events using daily temperature series (1901–1998) from eleven sites in central and southern Europe. They found a slight warming tendency, but only a few of the changes, mostly in the northernmost sites, are statistically significant. Strong connections are present between the frequencies of extreme temperature events and the large-scale circulation on various time scales. Extreme temperatures in central Europe are examined by Kysely (2002a, 2002b). Kysely (2002a) compared extreme temperatures in GCMs, observations, and results downscaled from GCMs and observations. Kysely (2002b) examined daily temperature at Prague-Klementinum to determine changes in heat waves and their relationship with circulation. Garcia et al. (2002) analyzed daily maximum temperature at Madrid and found circulation patterns associated with extremely high temperatures. A circulation index was derived to characterize and forecast a hot day occurrence. Jungo and Beniston (2001) highlighted the change in the seasonal temperature limits at different latitudes and altitudes in Switzerland. Winter minimum temperatures at high-altitude sites and summer maximum temperatures at low-altitude sites in the north, in particular, changed considerably during the 1990s. Brabson and Palutikof (2002) carried out an extremely valuable analysis of the central England temperature record from 1772 to the present to show that both cold summer and hot winter extremes have evolved differently from their means. The reasons for the trends in extremes occurrence are related to changes in the underlying atmospheric circulation. Xoplaki et al. (2003b) examined the interannual and decadal variability of summer (June–September) air temperature over the Mediterranean area from 1950 to 1999. They showed that more than 50% of the total summer temperature variability can be explained by three large-scale predictor fields (300 hPa geopotential height, 700–1,000 hPa thickness, and SSTs). Xoplaki et al. (2003a) found similar results when examining twenty-four stations in the northeastern Mediterranean.

Frei and Schar (2001) presented a statistical framework for the assessment of climatological trends in the frequency of rare and extreme weather events based on the stochastic concept of binomial distributed counts. The results demonstrate the difficulty of determining trends of very rare events. The statistical method is applied to examine seasonal trends of heavy daily precipitation at 113 rain gauge stations in the Alpine region of Switzerland (1901–1994). For intense events (return period: 30 days), a statistically significant frequency increase was found in winter and autumn for a high number of stations. Hand et al. (2004) carried out a study of extreme rainfall events over the UK during the twentieth century. They identified fifty flood-producing events with durations up to 60 hours. The rainfall events were classified by meteorological situation, location, and season, allowing the identification of conditions under which extreme rainfall occurred. Short-term rainfall rates between 5 minutes and 24 hours for Barcelona were examined by Casas et al. (2004). Events were clustered to give four main classes of extreme rainfall events in the area. Skaugen et al. (2004) downscaled daily precipitation using the Max Planck Institute GCM to generate scenarios of extreme rainfall under enhanced greenhouse conditions. The analysis of changes in extreme value patterns shows tendencies toward increased extreme values and seasonal shifts for the scenario period. Fowler and Kilsby (2003) examined 1-, 2-, 5-, and 10-day annual maximum rainfall for 1961 to 2000 from 204 sites across the UK Little change is observed at 1- and 2-day duration, but significant decadal-level changes are seen in 5- and 10-day events in many regions. European flood frequency was examined by Mudelsee et al. (2003). They presented longer-term records of winter and summer floods in two of the largest rivers in central Europe, the Elbe and Oder rivers. For the past 80 to 150 years, they found a decrease in winter flood occurrence in both rivers, while summer floods show no trend, consistent with trends in extreme precipitation occurrence.

Frei et al. (2003) examined mean and extreme precipitation for five RCMs. They found that, despite considerable biases, the models reproduce prominent mesoscale features of heavy precipitation. Benestad and Melsom (2002) examined the relationship between Atlantic SST and extreme autumn precipitation in southeast Norway. They found that the SSTs could explain as much of the rainfall during unusually wet November months in 2000 and 1970 as the SLP. Alpert et al. (2002) analyzed daily rainfall over the Mediterranean to determine the changes in rainfall intensity categories for 1951 to 1995. They found increases in extreme rainfall in Italy and Spain, but no change in Israel and Cyprus. Brunetti et al. (2002) studied daily rainfall over Italy to determine changes in the longest dry period, the proportion of dry days, and the greatest 5-day rainfall totals. There has been a large increase in summer droughts over all of Italy, but there is no significant trend in the extreme rainfall intensity. Crisci et al. (2002) examined daily rainfall in Tuscany to determine 30-year return period intensities. Booij (2002) compared extreme precipitation at different spatial scales by comparing results from using stations, reanalysis projects, global climate models, and regional climate models. A 100-year daily rainfall record for Uccle (Belgium) was examined by Vaes et al. (2002). They found no significant trend in extreme rainfall. Fowler et al. (2005) used two methods to assess the performance of the HadRM3H model in the simulation of UK extreme rainfall: regional frequency analysis and individual grid box analysis. Both methods use L-moments to derive extreme value distributions of rainfall for 1-, 2-, 5-, and 10-day events for both observed data from 204 sites across the

UK (1961–1990) and gridded 50 km by 50 km data from the control climate integration of HadRM3H. Although there are some problems with the representation of extreme rainfall by the HadRM3H model, almost all are related to the orographic enhancement of mean rainfall. The same methods were used by Ekström et al. (2004) to examine results from HadRM3H for a future scenario ensemble of enhanced greenhouse conditions. The authors suggested that by the end of the twenty-first century, the return period magnitude for a 1-day event would have increased by approximately 10% across the UK, with values for 10-day events increasing more in Scotland (up to +30%) than England (−20% to +10%). Osborn and Hulme (2002) examined daily precipitation in the UK from 1961 to 2000. They showed that it has become generally more intense in winter and less intense in summer. Recent increases in total winter precipitation are shown to be mainly due to an increase in the amount of precipitation on wet days, with a smaller contribution in the western UK from a trend toward more wet days. Palmer and Ralsanen (2002) presented a probabilistic analysis of nineteen GCM simulations with a generic binary decision model. They estimated that the probability of total boreal winter precipitation exceeding two standard deviations above normal will increase by a factor of 5 over parts of the UK over the next 100 years.

Analysis of global climate model performance in reproducing observed regional climate variability has given widely varying results depending on model and region. Interannual variability in temperature was assessed regionally, as well as globally, in a long control simulation with the HadCM2 model (Tett et al., 1997). Many aspects of model variability compared well against observations, although there was a tendency for temperature variability to be too high over land. In the multiregional study of Giorgi and Francisco (2000), both regional temperature and precipitation interannual variability of HadCM2 were found to be generally overestimated. Similar results were obtained in the European study of Machenhauer et al. (1998) using the ECHAM/OPYC3 model. However, in a 200-year control simulation with the CGCM1 model, Flato et al. (2000) noted that simulated interannual variability in seasonal temperature and precipitation compared well with observations both globally and in five selected study regions (Sahel, North America, Australia, southern Europe, and Southeast Asia). Comparison against observations of daily precipitation variability as simulated at grid boxes in GCMs is problematic because the corresponding variability in the real world operates at much finer spatial scale.

Results based on the GCM used in the reanalysis were found to be in good agreement with observations over Oregon and Washington. Equilibrium of $2 \times CO_2$ simulation with the Hadley Centre model shows a tendency for daily temperature variability over Europe to increase in JJA and to decrease in DJF. Daily high temperature extremes are likely to increase in frequency as a function of the increase in mean temperature, but this increase is modified by changes in daily variability of temperature. There is a corresponding decrease in the frequency of daily low temperature extremes. The seasonally warm and humid areas such as the southeastern United States, India, Southeast Asia, and northern Australia can experience increases in the heat index substantially greater than that expected due to warming alone. There is a strong correlation between precipitation interannual variability and mean precipitation. Increases in mean precipitation are likely to be associated with increases in variability, and precipitation variability is likely to decrease in areas of reduced mean precipitation. In general, where simulated changes in regional precipitation variability have been examined, increases are more commonly noted. The tendency for regional interannual variability of seasonal mean precipitation has been increased in HadCM2 simulations in many of the regions. Increases in interannual variability also predominated in the CGCM1 simulation, although there were areas of decrease, particularly in areas where mean rainfall decreased. An increase in monthly precipitation variance over southern Europe, northern Europe, and central North America has been found. There is a tendency for increased daily rainfall variability in two models over the Sahel, North America, south Asia, southern Europe, and Australia. It should also be noted that in many regions, interannual climatic variability is strongly related to ENSO, and thus will be affected by changes in ENSO behaviors. The tendency for increased rainfall variability in enhanced GHG simulations is reflected in a tendency for increases in the intensity and frequency of extreme heavy rainfall events. Such increases have been documented in regionally focused studies for Europe, North America, south Asia, the Sahel, southern Africa, Australia, and the South Pacific, as well as in the global. Under $2 \times CO_2$ conditions, the 1-year return period events in Europe, Australia, India, and the United States increased in intensity by 10% to 25% in two models. Changes in the occurrence of dry spells or droughts have been assessed for some regions using recent model results.

15.6 Regional impact studies

15.6.1 Severe summertime flooding in Europe

Using a high-resolution climate model, the influence of greenhouse gas–induced global warming on heavy or extended precipitation episodes that inflict catastrophic flooding has been detected. It was noticed that an increase in the amount of precipitation that exceeds the ninety-fifth percentile is very likely in many areas of Europe, despite a possible reduction in average summer precipitation over a substantial part of the continent. These indicate that episodes of severe flooding may become more frequent, despite a general trend toward drier summer conditions. Figure 15.6a shows the relative change in mean precipitation during July, August, and September according to the high-resolution (50-km grid) regional climate model (HIRHAM4) for the A2 scenario (drawn up by the IPCC; Nakicenovic et al., 2000). The relative change in the mean 5-day precipitation for July to September that exceeds the ninety-ninth percentiles in scenario A2 with respect to the control is shown in Figure 15.6b. Even when a reduction in total mean precipitation is simulated, the amounts of precipitation in the intensive events are much less reduced, and even increase in many places. The higher the percentile considered, the larger are the areas that show a positive change.

Figure 15.7 shows the analysis of observed recent trends in summer (defined as June-July-August) large-scale circulations and precipitation over Europe. The comparison of the present-day RegCM simulations to summer observations shows generally good agreement with a slight tendency to overestimate precipitation. Compared to the B2 simulation, the A2 scenario simulation shows larger changes in magnitude (due to the greater anthropogenic forcing), but nearly the same spatial patterns of change.

The simulated changes in average summer 500-hPa geopotential height and precipitation for the B2 scenario are presented in Figure 15.8. The positive values of geopotential height change, reflecting the higher temperatures and thicker lower atmosphere simulated in the future run.

The change patterns seen in Figure 15.8 imply an increased likelihood for drier summers over most of western and southern Europe and wetter summers over Finland and western Russia. The changes in the mean precipitation, however, do not indicate how the frequency of flood or drought might change. Due to availability limitations of daily Europeanwide precipitation data, we are unable to provide statistics of recent trends in the occurrence of flood and drought, and thus, only analyze the simulated changes.

15.6 Regional impact studies

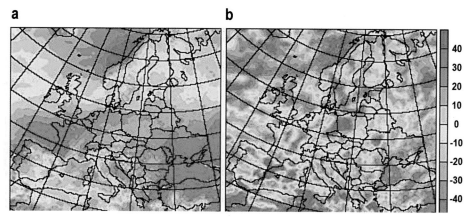

Figure 15.6 Relative percentage change in precipitation for July to September in the Intergovernmental Panel on Climate Change's A2 scenario with respect to the present day. Relative change is shown for (a) the seasonal mean, and (b) the 5-day mean exceeding the ninety-ninth percentile.

15.6.2 Warming and heat wave

A changing climate is expected to increase average summer temperatures and the frequency and intensity of hot days. Heat waves in Europe are associated with significant morbidity and mortality. A record-breaking heat wave affected the European continent in summer 2003. In a large area, mean summer (June, July, and August, referred to as JJA below) temperatures have exceeded the 1961 to 1990 mean by 3.8°C, corresponding to an excess of up to five standard deviations (Figure 15.9). Even away from the center of action, many long-standing temperature records have tumbled.

A preliminary analysis of the 2003 heat wave in Europe estimated that it caused 14,802 excess deaths in France, 2,045 excess deaths in the UK, and 2,099 in Portugal. Ongoing epidemiological studies will better describe and contribute substantial evidence to the understanding of health effects of heat waves in Europe and add significantly to targeting interventions (Table 15.2).

Recent scientific assessments indicate that, as global temperatures continue to increase because of climate change, the number and intensity of extreme events are likely to increase. New record extreme events occur every year somewhere around the globe, but in recent years the numbers of such extremes have been increasing. The impact of extreme summer heat on human health may be exacerbated by increases in humidity. Heat waves usually occur in synoptic situations with pronounced slow air mass development and movement, leading to intensive and prolonged heat stress. However, even short or moderate heat episodes adversely affect human health. The TAR of the IPCC (Houghton et al., 2001) stated that "there is new and stronger evidence that most of the warming observed over the past 50 years is attributable to human activities." Detecting climate change is difficult because any climate change "signal" is superimposed on the background "noise" of natural climate variability. Nevertheless, there is now good evidence that the climate is changing. The global average land and sea surface temperature increased by 0.6 ± 0.2°C over the twentieth century (Houghton et al., 2001). Nearly all of this increase occurred in two periods: 1910 to 1945 and since 1976. At the regional scale, warming has been observed in all continents, with the greatest temperature changes occurring at middle and high latitudes in the Northern Hemisphere. Extreme weather events are, by definition, rare stochastic

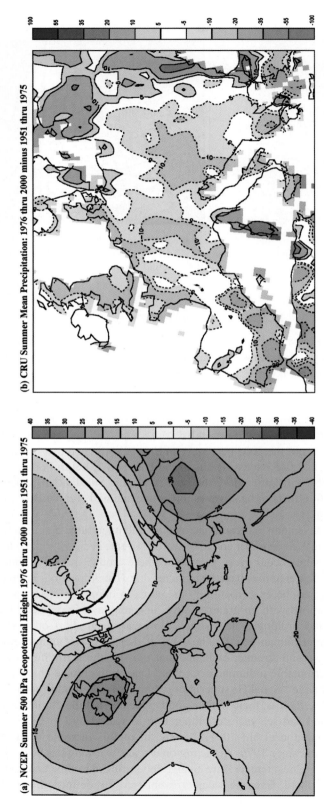

Figure 15.7 Average summer difference in (a) NCEP 500-hPa geopotential height (m), and (b) CRU mean precipitation (percent, land only) for the period 1976 through 2000 minus 1951 through 1975. In (a), the contour interval is 5 m, with positive values shaded red and negative values shaded blue. In (b), the contour intervals are ±5, 10, 20, 35, 55, and 100%, with positive values shaded red and negative values shaded blue.

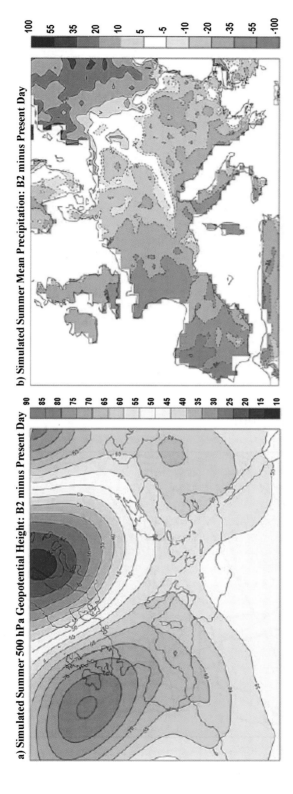

Figure 15.8 Average summer difference in (a) 500-hPa geopotential height (m), and (b) mean precipitation (percent, land only) between the B2 and present-day simulations. In (a), the contour interval is 5 m, with values greater than 50 hPa shaded red and values less than 50 hPa shaded blue. In (b), the contour intervals are ±5, 10, 20, 35, 55, and 100%, with positive values shaded red and negative values shaded blue.

b) Simulated Summer Mean Precipitation : B2 minus Present Day

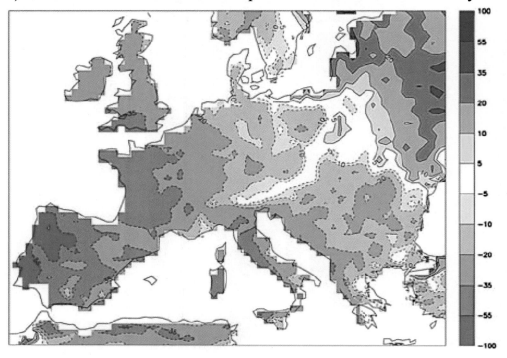

Figure 15.8 (*continued*)

Temperature anomaly (C)

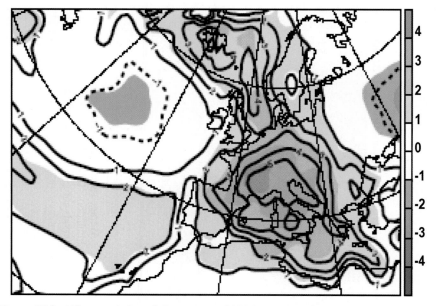

Figure 15.9 Characteristics of the summer 2003 heat wave. JJA temperature anomaly with respect to the 1961 to 1990 mean. Color shading shows temperature anomaly (8°C), bold contours display anomalies normalized by the 30-years standard deviation.

Table 15.2. *Number of deaths during the 2003 heat wave over Europe*

Country	Number of Fatalities	Other Details
France	14,802	Temperatures soared to 40°C in parts of the country. Temperatures in Paris were the highest since record-keeping began in 1873.
Germany	7,000	High temperatures of up to 41°C, the hottest since records began in 1901, raised mortality some 10% above average.
Spain	4,230	High temperatures coupled with elevated ground-level ozone concentrations exceeding the European Union's health risk threshold.
Italy	4,175	Temperatures in parts of the country averaged 9°C higher than previous year.
United Kingdom	2,045	The first triple-digit (Fahrenheit) temperatures were recorded in London.
The Netherlands	1,400	Temperatures ranged some 8°C warmer than normal.
Portugal	1,316	Temperatures were above 40°C throughout much of the country.
Belgium	150	Temperatures exceeded any in the Royal Meteorological Society's records dating back to 1833.

events. With climate change, even if the statistical distribution of such events remains the same, a shift in the mean will entail a nonlinear response in the frequency of extreme events.

15.6.3 Wind storms (hurricanes)

The year 2005 was marked by weather-related natural disaster. Roughly half of all the loss events recorded were windstorms, with costs to be borne by the world's economies exceeding $185 billion. In January, winter storm Erwin crossed Scotland and southern Scandinavia at up to 120 km/h on a path that took it as far as Russia. It was the strongest storm in Norway for over 10 years and in Sweden for over 30 years. For the European insurance industry, it was the fifth most expensive storm of the past 50 years. The hurricanes in the United States, the Caribbean, and Mexico alone destroyed insured values. In the Atlantic, twenty-seven tropical storms and hurricanes broke all meteorological and monetary records. For the first time since its introduction in 1953, the official list of twenty-one names was not long enough and had to be supplemented by the first six letters of the Greek alphabet.

Katrina was the sixth strongest hurricane since recordings began in 1851. Rita, the fourth strongest hurricane ever registered, reached mean wind speeds of up to 280 km/h. Stan progressed at a relatively slow speed but carried enormous amounts of rain into Middle America, causing thousands of landslides, under which more than 800 people were buried. Wilma was the strongest hurricane ever registered in the Atlantic. At the end of November, Delta became the first tropical cyclone ever to be registered in the Canaries.

In 2005, six natural hazard events complied with the definition of "great natural disaster." They accounted for more than 91,000 deaths (out of a total of 100,000). Figure 15.10 shows for each year the number of great natural disaster divided by type of event.

New meteorological records and unusual tracks for hurricanes have been recorded in 2005, many of which have never been observed before, at least not since recordings of the weather in the Atlantic have been made on a systematic basis. Figure 15.11 shows the track of all tropical cyclones in the North Atlantic, the Gulf of Mexico, and the Caribbean during the 2005 hurricane season.

Hurricane Katrina, which developed out of a low-pressure vortex over the Bahamas on 23 August and made landfall on the evening of 25 August as a Category 1 hurricane near Miami, was the eleventh tropical cyclone of the season, as shown in Figure 15.12. In the days that followed, Katrina moved over the eastern part of the Gulf of Mexico with a rapid increase in intensity. On 28 August, it was already a Category 5 hurricane over those areas where the water temperature was currently $1°C$ to $3°C$ above the long-term average, with peak gusts of around 340 km/h. Katrina maintained this strength as it crossed the oilfields off the coast of Louisiana and Mississippi. On 29 August, shortly before hitting the mainland some 50 km east of New Orleans, it weakened to a Category 3 windstorm. The wind and storm surge damage was horrendous: parts of New Orleans were flooded when levees along Lake Pontchartrain and artificial drainage canals failed, many offshore plants in the Gulf of Mexico were destroyed, and more than 1,300 people were killed.

Hardly a month had passed after Katrina when Rita formed in the southern part of the Bahamas as the second Category 5 hurricane of the season. It had a central pressure of 897 hPa, one of the lowest readings ever for a North Atlantic hurricane (Figure 15.13). Some of the prediction models used by the National Hurricane Center in Miami (Florida) indicated temporarily that Rita would make landfall near Galveston/Houston, Texas, with a force of 4 to 5. However, when Rita made landfall on 24 September near Sabine Pass in the border area between Texas and Louisiana, it was a Category 3 hurricane with peak gusts of 25 km/h.

15.7 Summary

Natural variability in the climate often produces extremes in the weather. An important question that scientists are trying to answer is whether mankind's interference with the global climate through the enhancement of the natural greenhouse effect will increase the frequency or magnitude of extreme weather events. Some definitions of climatic extremes choose to separate the nature of the event from its social and economic consequences. A climate extreme then, is a significant departure from the normal state of the climate, irrespective of its actual impact on life or any other aspect of the Earth's ecology. When a climate extreme has an adverse impact on human welfare, it becomes a climatic disaster. Extreme weather can permanently and profoundly affect lives and livelihoods, and place unexpectedly high pressures on social and health support structures. As current climate patterns suggest a trend toward global warming and an increase in these extreme weather events. Recent episodes of extreme weather events in Europe were accompanied by a significant and somehow unexpected toll of deaths and diseases. These events will continue to pose additional challenges to current and future populations, in terms of risk management and of reliability of infrastructure, including health services, power supply, and others. The scientific and decision-making communities should therefore take action to put in place evidence-based interventions, or precautionary measures, to limit the impacts on the environment and to actively reduce the health effects on human populations and ecosystems.

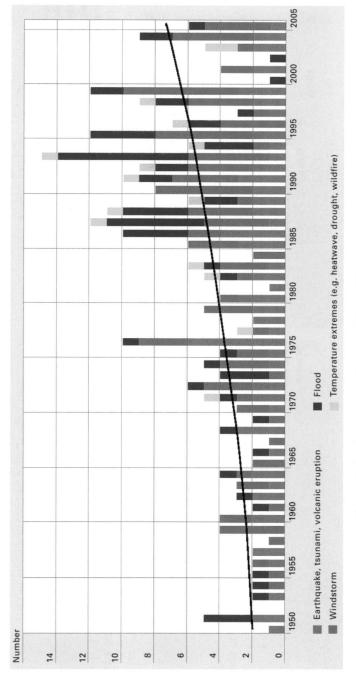

Figure 15.10 Comparison of decades 1950 to 2005 of great natural disaster.

353

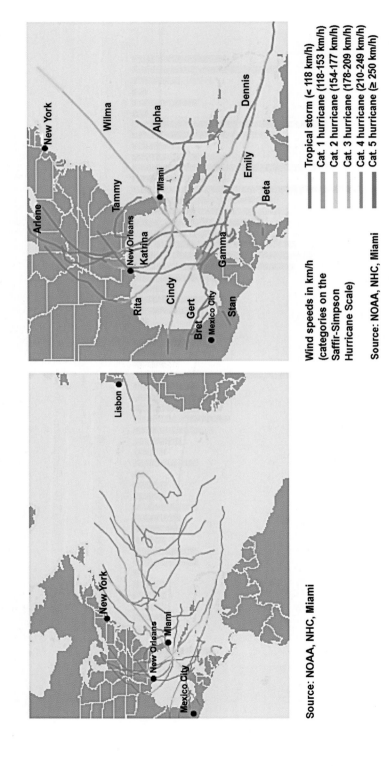

Figure 15.11 Track of all tropical cyclones in the North Atlantic, the Gulf of Mexico, and the Caribbean during the 2005 hurricane season. *Source*: National Oceanic and Atmospheric Administration (NOAA), National Hurricane Center (NHC), Miami, Florida.

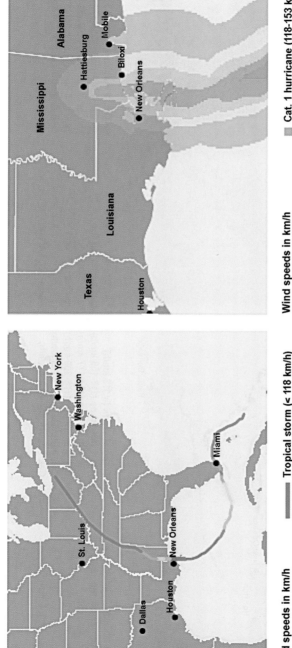

Figure 15.12 Track of Hurricane Katrina and the wind field. *Source*: NOAA, NHC, Miami, Florida.

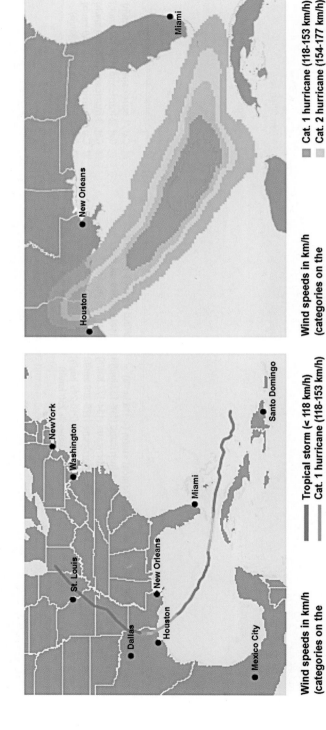

Figure 15.13   Track of Hurricane Rita and the wind field. *Source*: NOAA, NHC, Miami, Florida.

Overall, in the United States, the researchers found a slight downward trend in the number of extreme temperature events, despite an overall warming in the mean temperature. For the 1910 to 1998 periods, there has been a slight decrease in the number of days below freezing over the entire country. Regionally, the start of the frostfree season for the Northeast occurred 11 days earlier in the mid-1990s than in the 1950s. The growth of population, demographic shifts to more storm-prone locations, and the growth of wealth have collectively made the nation more vulnerable to climate extremes.

Acknowledgments

The authors want to thank the regional climate model (RegCM) group at Abdus Salam International Centre for Theoretical Physics (ICTP) for providing the required information for this work.

References

Alpert, P., Ben-Gai, T., Baharad, A., Benjamini, Y., Yekutieli, D., Colacino, M., Diodato. L., Ramis, C., Homar, V., Romero, R., Michaelides, S., and Manes, A. (2002) "The Paradoxical Increase of Mediterranean Extreme Daily Rainfall in Spite of Decrease in Total Values," *Geophysical Research Letters* **29**, article no. 1536.

Anthes, R. A. (1977) "A Cumulus Parameterization Scheme Utilizing a One-Dimensional Cloud Model," *Mon. Wea. Rev.* **105**, pp. 270–286.

Anthes, R. A., Hsie, E.-Y., and Li, Y.-F. (1987) "Description of the Penn State/NCAR Mesoscale Model Version 4 (MM4)," NCAR Tech. Note NCAR/TN-282+STR, National Center for Atmospheric Research, Boulder, Colorado.

Benestad, R. E., and Melsom, A. (2002) "Is There a Link Between the Unusually Wet Autumns in Southeastern Norway and Sea-Surface Temperature Anomalies?," *Climate Research* **23**, pp. 67–79.

Beniston, M. (2004) "The 2003 Heat Wave in Europe: A Shape of Things to Come? An Analysis Based on Swiss Climatological Data and Model Simulations," *Geophysical Research Letters* **31**, article no. L02202.

Bhaskaran, B., Jones, R. G., Murphy, J. M., and Noguer, M. (1996) "Simulations of the Indian Summer Monsoon Using a Nested Regional Climate Model: Domain Size Experiments," *Clim. Dyn.* **12**, pp. 573–587.

Booij, M. J. (2002) "Extreme Daily Precipitation in Western Europe with Climate Change at Appropriate Spatial Scales," *International Journal of Climatology* **22**, pp. 69–85.

Brabson, B. B., and Palutikof, J. P. (2002) "The Evolution of Extreme Temperatures in the Central England Temperature Record," *Geophysical Research Letters* **29**, article no. 2163.

Briegleb, B. P. (1992) "Delta-Eddington Approximation for Solar Radiation in the NCAR Community Climate Model," *J. Geophys. Res.* **97**, pp. 7603–7612.

Brunetti, M., Buffoni, L., Mangianti, F., Maugeri, M., and Nanni, T. (2004) "Temperature, Precipitation and Extreme Events During the Last Century in Italy," *Global and Planetary Change* **40**, pp. 141–149.

Brunetti, M., Maugeri, M., Nanni, T., and Navarra, A. (2002) "Droughts and Extreme Events in Regional Daily Italian Precipitation Series," *International Journal of Climatology* **22**, pp. 543–558.

Casas, M. C., Codina, B., Redano, A., and Lorente, J. (2004) "A Methodology to Classify Extreme Rainfall Events in the Western Mediterranean Area," *Theoretical and Applied Climatology* **77**, pp. 139–150.

Christensen, J. H., Christensen, O. B., Lopez, P., van Meijgaard, E., and Botzet, M. (1996) "The HIRHAM4 Regional Atmospheric Climate Model," *Sci. Rep.* **96-4**, p. 51.

Christensen, O. B., Christensen, J. H., Machenhauer, B., and Botzet, M. (1998) "Very High-Resolution Regional Climate Simulations over Scandinavia—Present Climate," *J. Climate* **11**, pp. 3204–3229.

Cocke, S. D., and LaRow, T. E. (2000) "Seasonal Prediction Using a Regional Spectral Model Embedded within a Coupled Ocean–Atmosphere Model," *Mon. Wea. Rev.* **128**, pp. 689–708.

Copeland, J. H., Pielke, R. A., and Kittel, T. G. F. (1996) "Potential Climatic Impacts of Vegetation Change: A Regional Modeling Study," *J. Geophys. Res.* **101**, pp. 7409–7418.

Crisci, A., Gozzini, B., Meneguzzo, F., Pagliara, S., and Maracchi, G. (2002) "Extreme Rainfall in a Changing Climate: Regional Analysis and Hydrological Implications in Tuscany," *Hydrological Processes* **16**, pp. 1261–1274.

Deardorff, J. W. (1972) "Theoretical Expression for the Counter-Gradient Vertical Heat Flux," *J. Geophys. Res.* **77**, pp. 5900–5904.

Diaz, H. F., and Pulwarty, R. S. (1997) *Hurricanes, Climate and Socioeconomic Impacts*, Springer-Verlag, Berlin, Germany.

Dickinson, R. E., Errico, R. M., Giorgi, F., and Bates, G. T. (1989) "A Regional Climate Model for the Western United States," *Climatic Change* **15**, pp. 383–422.

Dickinson, R. E., Kennedy, P. J., Henderson-Sellers, A., and Wilson, M. (1986) "Biosphere-Atmosphere Transfer Scheme (BATS) for the NCAR Community Climate Model," Tech. Rep. NCAR/TN-275+STR, National Center for Atmospheric Research, Boulder, Colorado.

Dickinson, R. E., Henderson-Sellers, A., and Kennedy, P. J. (1993) "Biosphere-Atmosphere Transfer Scheme (BATS) Version 1e as Coupled to the NCAR Community Climate Model," Tech. Note, NCAR/387+STR, National Center for Atmospheric Research, Boulder, Colorado.

Domonkos, P., Kysely, J., Piotrowicz, K., Petrovic, P., and Likso, T. (2003) "Variability of Extreme Temperature Events in South-Central Europe During the 20th Century and Its Relationship with Large-Scale Circulation," *International Journal of Climatology* **23**, pp. 987–1010.

Ekström, M., Fowler, H. J., Kilsby, C. G., and Jones, P. D. (2004) "New Estimates of Future Changes in Extreme Rainfall Across the UK Using Regional Climate Model Integrations. 2. Future Estimates and Use in Impact Studies," *Journal of Hydrology* **300**, pp. 234–251.

Emanuel, K. A. (1991) "A Scheme for Representing Cumulus Convection in Large-Scale Models," *J. Atmos. Sci.* **48**, pp. 2313–2335.

Emanuel, K. A., and Zivkovic-Rothman, M. (1999) "Development and Evaluation of a Convection Scheme for Use in Climate Models," *J. Atmos. Sci.* **56**, pp. 1766–1782.

Flato, G. M., Boer, G. J., Lee, W. G., McFarlane, N. A., Ramsden, D., Reader, M. C., and Weaver, A. J. (2000) "The Canadian Center for Climate Modelling and Analysis Global Coupled Model and Its Climate," *Clim. Dyn.* **16**, pp. 451–467.

Fowler, H. J., Ekström, M., Kilsby, C. G., and Jones, P. D. (2005) "New Estimates of Future Changes in Extreme Rainfall Across the UK Using Regional Climate Model Integrations. Part 1: Assessment of Control Climate," *Journal of Hydrology*. **300** (1–4), pp. 212–231.

Fowler, H. J., and Kilsby, C. G. (2003) "Implications of Changes in Seasonal and Annual Extreme Rainfall," *Geophysical Research Letters* **30**(13), p. 1720.

Frei, C., Christensen, J. H., Déqué, M., Jacob, D., Jones, R. G., and Vidale, P. L. (2003) "Daily Precipitation Statistics in Regional Climate Models: Evaluation and Intercomparison for the European Alps," *Journal of Geophysical Research—Atmospheres* **108**, no. D 3, p. 4124, doi: 10.1029/2002JD002287.

Frei, C., and Schar, C. (2001) "Detection Probability of Trends in Rare Events: Theory and Application to Heavy Precipitation in the Alpine Region," *Journal of Climate* **14**, pp. 1568–1584.

Frich, P., Alexander, L. V., Della-Marta, P., Gleason, B., Haylock, M., Tank, A. M. G. K., and Peterson T. (2002) "Observed Coherent Changes in Climatic Extremes During the Second Half of the Twentieth Century," *Climate Research* **19**, pp. 193–212.

Gao, X., Pal., J. S., and Giorgi, F. (2006) "Projected Changes in Mean and Extreme Precipitation over the Mediterranean Region from a High Resolution Double Nested RCM Simulation," *Geophysical Research Letters* **33**, no. L03706, doi: 10.1029/2005GL024954.

Garcia, R., Prieto, L., Diaz, J., Hernandez, E., and del Teso, T. (2002) "Synoptic Conditions Leading to Extremely High Temperatures in Madrid," *Annales Geophysicae* **20**, pp. 237–245.

Giorgi, F. (1990) "Simulation of Regional Climate Using a Limited Area Model Nested in a General Circulation Model," *J. Climate* **3**, pp. 941–963.

Giorgi, F. (2006) "Climate Change Hot-Spots," *Geophys. Res. Lett.* **33**, L08707, doi: 10.1029/2006 GL025734.

Giorgi, F., and Bates, G. T. (1989) "The Climatological Skill of a Regional Model over Complex Terrain," *Mon. Wea. Rev.* **117**, pp. 2325–2347.

Giorgi, F., Bates, G. T., and Nieman, S. J. (1993a) "The Multi-Year Surface Climatology of a Regional Atmospheric Model over the Western United States," *J. Climate* **6**, pp. 75–95.

Giorgi, F., and Francisco, R. (2000) "Evaluating Uncertainties in the Prediction of Regional Climate Change," *Geophys. Res. Lett.* **27**, pp. 1295–1298.

Giorgi, F., Francisco, R., and Pal, J. (2003) "Effects of a Subgrid-Scale Topography and Land Use Scheme on the Simulation of Surface Climate and Hydrology. Part 1: Effects of Temperature and Water Vapor Disaggregation," *J. Hydrometeorology* **4**, pp. 317–333.

Giorgi, F., and Marinucci, M. R. (1996a) "An Investigation of the Sensitivity of Simulated Precipitation to Model Resolution and Its Implications for Climate Studies," *Mon. Wea. Rev.* **124**, pp. 148–166.

Giorgi, F., Marinucci, M. R., and Bates, G. T. (1993b) "Development of a Second-Generation Regional Climate Model (RegCM2). Part I: Boundary-Layer and Radiative Transfer Processes," *Mon. Weath. Rev.* **121**, pp. 2794–2813.

Giorgi, F., Marinucci, M. R., Bates, G. T., and DeCanio, G. (1993c) "Development of a Second-Generation Regional Climate Model (RegCM2). Part II: Convective Processes and Assimilation of Lateral Boundary Conditions," *Mon. Wea. Rev.* **121**, pp. 2814–2832.

Giorgi, F., and Mearns, L. O. (1991) "Approaches to the Simulation of Regional Climate Change: A Review," *Rev. Geophys.* **29**, pp. 191–216.

Giorgi, F., and Mearns, L. O. (1999) "Regional Climate Modeling Revisited. An Introduction to the Special Issue," *J. Geophys. Res.* **104**, pp. 6335–6352.

Giorgi, F., and Shields, C. (1999) "Tests of Precipitation Parameterizations Available in Latest Version of NCAR Regional Climate Model (RegCM) over Continental United States," *J. Geophysical Research—Atmosphere* **104**(D6), pp. 6353–6375.

Grell, G. A. (1993) "Prognostic Evaluation of Assumptions Used by Cumulus Parameterizations," *Mon. Wea. Rev.* **121**, pp. 764–787.

Grell, G. A., Dudhia, J., and Stauffer, D. R. (1994) "Description of the Fifth Generation Penn State/NCAR Mesoscale Model (MM5)," Tech. Rep. TN-398+STR, National Center for Atmospheric Research, Boulder, Colorado, p. 121.

Groisman, P. Y., Karl, T. R., Easterling, D. R., Knight, R. W., Jamason, P. F., Hennessy, K. J., Suppiah, R., Page, C. M., Wibig, J., Fortuniak, K., Razuvaev, V. N., Douglas, A., Forland, E., and Zhai, P.-M. (1999) "Changes in the Probability of Heavy Precipitation: Important Indicators of Climatic Change," *Climatic Change* **42**, pp. 243–283.

Hand, W. H., Fox, N. I., and Collier, C. G. (2004) A Study of Twentieth-Century Extreme Rainfall Events in the United Kingdom with Implications for Forecasting," *Meteorological Applications* **11**, pp. 15–31.

Haylock, M., and Goodess, C. (2004) "Interannual Variability of European Extreme Winter Rainfall and Links with Mean Large-Scale Circulation," *International Journal of Climatology* **24**, pp. 759–776.

Holtslag, A. A. M., de Bruijn, E. I. F., and Pan, H. L. (1990) "A High Resolution Air Mass Transformation Model for Short-Range Weather Forecasting," *Mon. Wea. Rev.* **118**, pp. 1561–1575.

Hong, S. Y., and Juang, H. M. H. (1998) "Orography Blending in the Lateral Boundary of a Regional Model," *Mon. Wea. Rev.* **126**, pp. 1714–1718.

Hostetler, S. W., Giorgi, F., Bates, G. T., and Bartlein, P. J. (1994) "Lake-Atmosphere Feedbacks Associated with Paleolakes Bonneville and Lahontan," *Science* **263**, pp. 665–668.

Houghton, J. T., Ding, Y., Griggs, D. J., Noguer, M., van der Linden, P. J., Dai, X., Maskell, K., and Johnson, C. A. (2001) "Climate Change 2001: The Scientific Basis. Contribution of Working Group I to the Third Assessment Report of the Intergovernmental Panel on Climate Change," Cambridge University Press, Cambridge, United Kingdom.

Hsie, E. Y., Anthes, R. A., and Keyser, D. (1984) "Numerical Simulation of Frontogenisis in a Moist Atmosphere," *J. Atmos. Sci.* **41**, pp. 2581–2594.

Jacob, D., and Podzun, R. (1997) "Sensitivity Study with the Regional Climate Model REMO," *Meteorol. Atmos. Phys.* **63**, pp. 119–129.

Ji, Y., and Vernekar, A. D. (1997) "Simulation of the Asian Summer Monsoons of 1987 and 1988 with a Regional Model Nested in a Global GCM," *J. Climate* **10**, pp. 1965–1979.

Jones, R. G., Murphy, J. M., and Noguer, M. (1995) "Simulations of Climate Change over Europe Using a Nested Regional Climate Model. I: Assessment of Control Climate, Including Sensitivity to Location of Lateral Boundaries," *Quart. J. R. Met. Soc.* **121**, pp. 1413–1449.

Jones, R. G., Murphy, J. M., Noguer, M., and Keen, A. B. (1997) "Simulation of Climate Change over Europe Using a Nested Regional Climate Model. II: Comparison of Driving and Regional Model Responses to a Doubling of Carbon Dioxide," *Quart. J. R. Met. Soc.* **123**, pp. 265–292.

Jungo, P., and Beniston, M. (2001) "Changes in the Anomalies of Extreme Temperature Anomalies in the 20th Century at Swiss Climatological Stations Located at Different Latitudes and Altitudes," *Theoretical and Applied Climatology* **69**, pp. 1–12.

Kalkstein, L. S., and Smoyer, K. E. (1993) "The Impact of Climate Change on Human Health: Some International Perspectives," *Experientia* **49**, pp. 969–979.

Kato, H., Hirakuchi, H., Nishizawa, K., and Giorgi, F. (1999) "Performance of the NCAR RegCM in the Simulations of June and January Climates over Eastern Asia and the High-Resolution Effect of the Model," *J. Geophys. Res.* **104**, pp. 6455–6476.

Katz, R. W., and Brown, B. G. (1992) "Extreme Events in a Changing Climate: Variability Is More Important Than Averages," *Climatic Change* **21**, pp. 289–302.

Kida, H., Koide, T., Sasaki, H., and Chiba, M. (1991) "A New Approach to Coupling a Limited Area Model with a GCM for Regional Climate Simulations," *J. Met. Soc. Japan* **69**, pp. 723–728.

Kiehl, J. T., Hack, J. J., Bonan, G. B., Boville, B. A., Breigleb, B. P., Williamson, D., and Rasch, P. (1996) "Description of the NCAR Community Climate Model (ccm3)," Tech. Rep. NCAR/TN-420+STR, National Center for Atmospheric Research, Boulder, Colorado.

Kiehl, J. T., Wolski, R. J., Briegleb, B. P., and Ramanathan, V. (1987) "Documentation of Radiation and Cloud Routines in the NCAR Community Climate Model (CCM1)," Tech. Note NCAR/TN-288+1A, National Center for Atmospheric Research, Boulder, Colorado.

Klein Tank, A. M. G., Wijngaard, J. B., Konnen, G. P., Bohm, R., Demaree, G., Gocheva, A., Mileta, M., Pashiardis, S., Hejkrlik, L., Kern-Hansen, C., Heino, R., Bessemoulin, P., Muller-Westermeier, G., Tzanakou, M., Szalai, S., Palsdottir, T., Fitzgerald, D., Rubin, S., Capaldo, M., Maugeri, M., Leitass, A., Bukantis, A., Aberfeld, R., Van Engelen, A. F. V., Forland, E., Mietus, M., Coelho, F., Mares, C., Razuvaev, V., Nieplova, E., Cegnar, T., Lopez, J. A., Dahlstrom, B., Moberg, A., Kirchhofer, W., Ceylan, A., Pachaliuk, O., Alexander, L. V., and Petrovic, P. (2002) "Daily Dataset of 20th-Century Surface Air Temperature and Precipitation Series for the European Climate Assessment," *International Journal of Climatology* **22**, pp. 1441–1453.

Klein Tank, A. M. G. K., and Konnen, G. P. (2003) "Trends in Indices of Daily Temperature and Precipitation Extremes in Europe, 1946–99," *Journal of Climate* **16**, pp. 3665–3680.

Kunkel, K. E., Pielke, R. A., Jr., and Changnon, S. A. (1999) "Temporal Fluctuations in Weather and Climate Extremes That Cause Economic and Human Health Impacts: A Review," *Bull. Amer. Meteor. Soc.* **80**, pp. 1077–1098.

Kysely, J. (2002a) "Comparison of Extremes in GCM-Simulated, Downscaled and Observed Central-European Temperature Series," *Climate Research* **20**, pp. 211–222.

Kysely, J. (2002b) "Temporal Fluctuations in Heat Waves at Prague-Klementinum, the Czech Republic, from 1901–97, and Their Relationships to Atmospheric Circulation," *International Journal of Climatology* **22**, pp. 33–50.

Laprise, R., Caya, D., Gigure, M., Bergeron, G., Côte, H., Blanchet, J. P., Boer, G. J., and McFarlane, N. A. (1998) "Climate and Climate Change in Western Canada as Simulated by the Canadian Regional Climate Model," *Atmosphere–Ocean* **36**, pp. 119–167.

Leadbetter, M. R., Lindgren, G., and Rootzen, H. (1983) *Extremes and Related Properties of Random Sequences and Process*, Springer-Verlag, Berlin, Germany.

Leung, L. R., and Ghan, S. J. (1995) "A Subgrid Parameterization of Orographic Precipitation," *Theor. Appl. Climatol.* **52**, pp. 95–118.

Leung, L. R., and Ghan, S. J. (1998) "Parameterizing Subgrid Orographic Precipitation and Surface Cover in Climate Models," *Mon. Wea. Rev.* **126**, pp. 3271–3291.

Liston, G. E., and Pielke, R. A. (2000) "A Climate Version of the Regional Atmospheric Modeling System," *Theorl. Appl. Climatol.* **66**, pp. 29–47.

Liston, G. E., Pielkec, R. A., Sr., and Greene, E. M. (1999) "Improving First-Order Snow-Related Deficiencies in a Regional Climate Model," *J. Geophysical Res.* **104**, pp. 19559–19567.

Luterbacher, J., Dietrich, D., Xoplaki, E., Grosjean, M., and Wanner, H. (2004) "European Seasonal and Annual Temperature Variability, Trends, and Extremes Since 1500," *Science* **303**, pp. 1499–1503.

Machenhauer, B., Windelband, M., Botzet, M., Christensen, J. H., Deque, M., Jones, R., Ruti, P. M., and Visconti, G. (1998) "Validation and Analysis of Regional Present-Day Climate and Climate Change Simulations over Europe," MPI Report No. 275, MPI, Hamburg, Germany.

McGregor, J. L., Katzfey, J. J., and Nguyen, K. C. (1999) "Recent Regional Climate Modelling Experiments at CSIRO," in *Research Activities in Atmospheric and Oceanic Modelling*, ed. H. Ritchie, CAS/JSC Working Group on Numerical Experimentation Report, WMO/TD no. 942, Geneva, Switzerland.

McGregor, J. L., and Walsh, K. (1993) "Nested Simulations of Perpetual January Climate over the Australian Region," *J. Geophys. Res.* **98**, pp. 23283–23290.

Mearns, L. O., Katz, R. W., and Schneider, S. H. (1984) "Extreme High Temperature Events: Changes in Their Probabilities with Changes in Mean Temperature," *J. Climate Appl. Meteor.* **23**, pp. 1601–1613.

Miller, N. L., and Kim, J. (1997) "The Regional Climate System Model," in *Mission Earth: Modeling and Simulation for a Sustainable Global System*, eds. M. G. Clymer and C. R. Mechoso, Society for Industrial and Applied Mathematics, Philadelphia, Pennsylvania.

Moberg, A., and Jones, P. D. (2004) "Regional Climate Model Simulations of Daily Maximum and Minimum Near-Surface Temperature Across Europe Compared with Observed Station Data 1961–1990," *Climate Dynamics* **23**, pp. 695–715.

Mudelsee, M., Borngen, M., Tetzlaff, G., and Grunewald, U. (2003) "No Upward Trends in the Occurrence of Extreme Floods in Central Europe," *Nature* **425**, pp. 166–169.

Noguer, M., Jones, R. G., and Murphy, J. M. (1998) "Sources of Systematic Errors in the Climatology of a Nested Regional Climate Model (RCM) over Europe," *Clim. Dyn.* **14**, pp. 691–712.

Osborn, T. J., and Hulme, M. (2002) "Evidence for Trends in Heavy Rainfall Events over the UK," *Philosophical Transactions of the Royal Society of London Series A—Mathematical Physical and Engineering Sciences* **360**, pp. 1313–1325.

Pal, J., Small, E., and Eltahir, E. A. B. (2000) "Simulation of Regional-Scale Water and Energy Budgets: Representation of Subgrid Cloud and Precipitation Processes within RegCM," *J. Geophys. Res. Atmospheres* **105**, pp. 29579–29594.

Palmer, T. N., and Ralsanen, J. (2002) "Quantifying the Risk of Extreme Seasonal Precipitation Events in a Changing Climate," *Nature* **415**, pp. 512–514.

Pielke, R. A., Jr., and Pielke, R. A., Sr. (1997) *Hurricanes: Their Nature and Impacts on Society*, John Wiley & Sons, New York, New York.

Pielke, R. A., Sr., Cotton, W. R., Walko, R. L., Tremback, C. J., Lyons, W. A., Grasso, L. D., Nicholls, M. E., Moran, M. D., Wesley, D. A., Lee, T. J., and Copeland, J. H. (1992) "A Comprehensive Meteorological Modeling System—RAMS," *Meteor. Atmos. Phys.* **49**, pp. 69–91.

Qian, Y., and Giorgi, F. (1999) "Interactive Coupling of Regional Climate and Sulfate Aerosol Models over East Asia," *J. Geophys. Res.* **104**, pp. 6501–6514.

Qian, Y., Leung, R., Ghan, S. J., and Giorgi, F. (2003) "Regional Climate Effects of Aerosols over China: Modeling and Observation," *Tellus B* **55**, pp. 914–934.

Rummukainen, M., RLisLnen, J., Bringfelt, B., Ullerstig, A., Omstedt, A., Willen, U., Hansson, U., and Jones, C. (2000) "RCA1 Regional Climate Model for the Nordic Region—Model Description and Results from the Control Run Downscaling of Two GCMs," *Clim. Dyn.* **17** pp. 339–359.

Schar, C., Vidale, P. L., Luthi, D., Frei, C., Haberli, C., Liniger, M. A., and Appenzeller, C. (2004) "The Role of Increasing Temperature Variability in European Summer Heatwaves," *Nature* **427**, pp. 332–336.

Seth, A., and Giorgi, F. (1998) "The Effects of Domain Choice on Summer Precipitation Simulation and Sensitivity in a Regional Climate Model," *J. Climate* **11**, pp. 2698–2712.

Skaugen, T., Astrup, M., Roald, L. A., and Forland, E. (2004) "Scenarios of Extreme Daily Precipitation for Norway Under Climate Change," *Nordic Hydrology* **35**, pp. 1–13.

Solmon, F., Giorgi, F., and Liousse, C. (2006) "Aerosol Modeling for Regional Climate Studied: Application to Anthropogenic Particles and Evaluation over a European/African Domain," *Tellus B* **58**(1), pp. 51–72.

Tett, S., Johns, T., and Mitchell, J. (1997) "Global and Regional Variability in a Coupled AOGCM," *Climate Dynamics* **13**, pp. 303–323.

Vaes, G., Willems, P., and Berlamont, J. (2002) "100 Years of Belgian Rainfall: Are There Trends?" *Water Science and Technology* **45**, pp. 55–61.

von Storch, H., Langenberg, H., and Feser, F. (2000) "A Spectral Nudging Technique for Dynamical Downscaling Purposes," *Mon. Wea. Rev.* **128**, pp. 3664–3673.

Warner, T. T., Peterson, R. A., and Treadon, R. E. (1997) "A Tutorial on Lateral Conditions as a Basic and Potentially Serious Limitation to Regional Numerical Weather Prediction," *Bull. Am. Met. Soc.* **78**, pp. 2599–2617.

Wijngaard, J. B., Klein Tank, A. M. G., and Können, G. P. (2003) "Homogeneity of 20th Century European Daily Temperature and Precipitation Series," *International Journal of Climatology* **23**, pp. 679–692.

Xoplaki, E., Gonzalez-Rouco, J. F., Gyalistras, D., Luterbacher, J., Rickli, R., and Wanner H. (2003a) "Interannual Summer Air Temperature Variability over Greece and Its Connection to the Large-Scale Atmospheric Circulation and Mediterranean SSTs 1950–1999," *Climate Dynamics* **20**, pp. 537–554.

Xoplaki, E., Gonzalez-Rouco, J. F., Luterbacher, J., and Wanner, H. (2003b) "Mediterranean Summer Air Temperature Variability and Its Connection to the Large-Scale Atmospheric Circulation and SSTs," *Climate Dynamics* **20**, pp. 723–739.

Zakey, A. S., Solmon, F., and Giorgi, F. (2006) "Implementation and Testing of a Desert Dust Module in a Regional Climate Model," *Atmos. Chem. Phys.* **6**, pp. 4687–4704.

Zeng, X., Zhao, M., and Dickinson, R. E. (1998) "Intercomparison of Bulk Aerodynamic Algorithms for the Computation of Sea Surface Fluxes Using TOGA COARE and TAO Data," *J. Climate* **11**, pp. 2628–2644.

Zhang, G. J., and McFarlane, N. A. (1995) "Sensitivity of Climate Simulations to the Parameterization of Cumulus Convection in the CCC-GCM," *Atmos–Ocean* **3**, pp. 407–446.

16

Impact of climate change on precipitation

Roy Rasmussen, Aiguo Dai, and Kevin E. Trenberth

Current climate models suggest that global warming will result in more frequent extreme hydrological events (floods and droughts). These results, however, must be tempered with the fact that current climate models do not realistically represent many of the processes important to the formation of clouds and precipitation at various time and space scales. For instance, the diurnal cycle of precipitation is poorly represented in most climate models. The proper representation of precipitation is a major challenge to global climate models, which typically only resolve processes at 200- to 400-km scales, and is a focus of current scientific research. This chapter addresses the current understanding of the likely climate impact on precipitation, as well as some of the key challenges facing climate modelers with regard to improving future projections of precipitation.

16.1 Introduction

Heated by sunlight and atmospheric radiation, water evaporates from the ocean and land surfaces, moves along with winds in the atmosphere, condenses to form clouds, and falls back to the Earth's surface through rain and snow, some of which flows back to oceans through rivers, thereby completing this global water cycle (Figure 16.1).

Daily newspaper headlines of floods and droughts reflect the critical importance of the water cycle, in particular, precipitation in human affairs. Flood damage estimates are in the billions of dollars annually, with thousands of lives lost, whereas drought costs are of similar magnitude and often lead to devastating wildfires. Floods are often fairly local and develop on short time scales, whereas droughts are extensive and develop over months or years. Droughts and floods are two extreme manifestations of the water cycle, which are controlled by processes at a wide range of both spatial and temporal scales. For instance, whether a thunderstorm occurs may depend on water vapor amounts that vary on small scales and depend on boundary layer and surface processes, such as those associated with heterogeneous vegetation and soil moisture. However, large-scale and systematic patterns, such as the diurnal cycle, are also important. Thus, to accurately predict the water cycle in both weather and climate models, including both extreme and more moderate manifestations, detailed knowledge of processes at all scales is necessary.

To predict the impact of future climate change on the water cycle and on precipitation in particular, climate models must be able to realistically represent these processes across scales (Figure 16.1). This presents a major challenge to global climate models, which typically only resolve processes at 200- to 400-km scales. Because many of the processes important to the water cycle are subgrid scale for climate models, they must be parameterized using the resolved fields. Much of the current research related to improving climate predictions of the water cycle has been focused on improving parameterizations

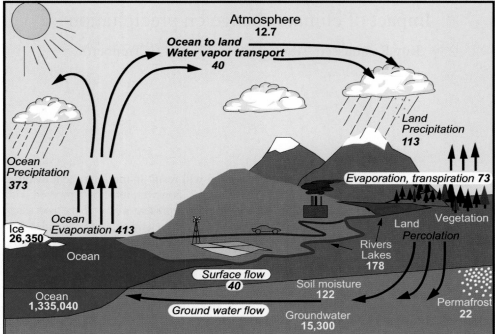

Figure 16.1 Estimated global water cycle. Water storage is given by the nonitalic numbers in cubic kilometer, while fluxes of water are given in thousands of cubic kilometer per year by the italic numbers. From Trenberth et al. (2007).

of the subgrid processes, such as those related to precipitation, evaporation, soil moisture, and runoff. This chapter addresses some of the current challenges facing climate modelers with regard to precipitation. Section 16.2 discusses precipitation processes in both observations and models, and how its character plays a major role in determining the water cycle. Section 16.3 describes how precipitation (including floods and droughts) is likely to be impacted by future climate change. Questions and issues are raised in Section 16.4, and a summary and conclusion is given in Section 16.5.

16.2 Precipitation processes in observations and models

The character of precipitation (i.e., its frequency, intensity, timing, liquid or solid phase, etc.) depends not only on the nature of the storm, but also on the available moisture. Rainfall, or more generally, precipitation, is often discussed together with other atmospheric variables, such as temperature, pressure, and wind. Yet, most of the time it does not rain. And when it does, the rain rate varies. At the very least, we could always consider how frequent and how intense the rain is when it does fall, in addition to the total amounts. Steady moderate rains soak into the soil and benefit plants, while the same rainfall amounts in a short period of time may cause local flooding and runoff, leaving soils much drier at the end of the day.

Take, for example, two hypothetical surface stations shown in Figure 16.2. Each station receives the same amount of total precipitation (75 mm) during the month, but it is distributed with different frequency. Station A receives all its precipitation on 2 days, and

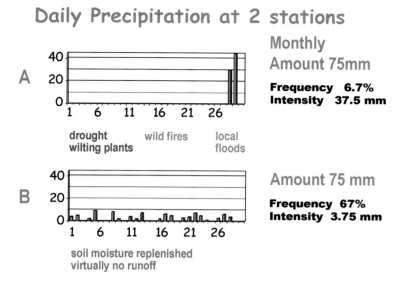

Figure 16.2 Example of two hypothetical surface weather stations receiving the same total monthly amount of precipitation, but with different frequency and intensity. Shown are the amounts each day of the month in millimeters, with consequences noted underneath.

thus, its frequency of precipitation is only 6.7%, but with a high average intensity of 37.5 mm/day. Station B, in contrast, has precipitation falling on 20 days, with a frequency of 67%, but of low average intensity (3.75 mm/day). These two scenarios lead to different climates. The climate of station A will be dry and arid until the heavy precipitation occurs late in the month. This heavy precipitation may result in floods and strong erosion events. Station B, in contrast, has a moister climate with soil moisture close to saturation. Most precipitation in this case results in almost no runoff. At station B, there may be hydrological drought (but no agricultural drought) because of the absence of river flow, whereas at station A, runoff occurs, and thus, soil moisture is much less than at A and agricultural drought may occur. Current climate models produce frequent drizzle of low intensity, and so tend to produce a climate more similar to that at station B.

This example highlights the fact that the characteristics of precipitation are just as vital as the amount. In models, it may be possible to "tune" parameters to improve amounts, but unless the amounts are right for the right reasons—and these include the correct combination of frequency and intensity of precipitation—it is unlikely that useful forecasts or simulations will result. In addition, it is likely that the characteristics of rain will change as climate changes. For example, it is argued based on atmospheric thermodynamics (Trenberth et al., 2003) and largely confirmed by coupled climate models (e.g., Sun et al., 2007) that extreme weather events, such as heavy rainfall (and thus floods), will become much more frequent as the climate warms. These aspects have enormous implications for agriculture, hydrology, and water resources, yet they have not been adequately appreciated or addressed in studies of impacts of climate change, although the climate impact community has begun addressing some of the issues over certain regions (Houghton et al., 2001).

In addition, most climate models do not even agree on the global mean precipitation rate expected as a result of global warming, despite agreement in the mean temperature rise (Figure 16.3). Note that all sixteen climate models shown in Figure 16.3 are consistent in their prediction of global temperature increase of approximately 2°C, but that they differ

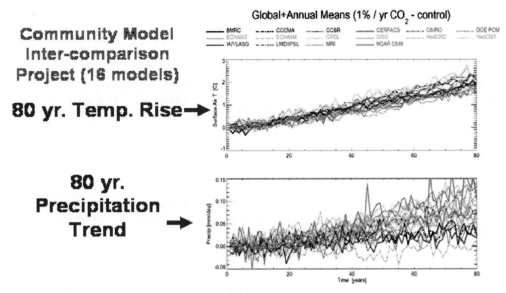

Figure 16.3 Comparison of the predicted 80-year temperature and precipitation trend for sixteen different climate models from Covey et al. (2003). Simulations assumed 1% per year increase of CO_2. No other anthropogenic climate forcing factors, such as anthropogenic aerosols, were included. Adapted from Covey et al. (2003).

widely in their prediction of the global mean precipitation rate, ranging from nearly 0 to 0.2 mm/day. This variability relates to uncertainty in the treatment of the water cycle, especially related to the formation of clouds and precipitation and its appropriate frequency, intensity, timing, and phase.

16.2.1 Evaluation of model simulated changes in precipitation by examination of the diurnal cycle

One way to address the issues concerning frequency and intensity of precipitation is to systematically examine the timing and duration of precipitation events as a function of time of day. This allows the systematic errors in predicting onset time and duration of precipitation in models to be explored. Hence, the diurnal cycle allows us to begin to come to grips with the characteristics of precipitation and how well they are modeled. The diurnal cycle in precipitation is particularly pronounced over the United States in summer (Figure 16.4) and is poorly simulated in most numerical models (Dai, 2006a).

The mean pattern of the diurnal cycle of summer U.S. precipitation is characterized by late afternoon maxima over the Southeast and the Rocky Mountains, and midnight maxima over the region east of the Rockies and the adjacent plains (Figure 16.4). Diurnal variations of precipitation are weaker in other seasons, with early to late morning maxima over most of the United States in winter. The diurnal cycle in precipitation *frequency* accounts for most of the diurnal variations, whereas the diurnal variations in precipitation intensity are small (see Dai et al., 1999).

Over the central plains, diurnal variations in climatologically significant circulations, such as the low-level jet (Higgins et al., 1997) and the mountain-plains circulation, interact to produce a nocturnal environment with little inhibition to convective onset, while maintaining

16.2 Precipitation processes in observations and models

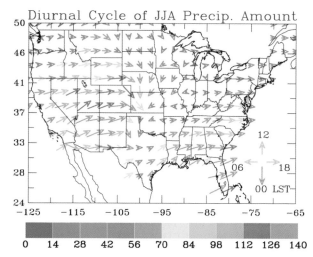

Figure 16.4 Diurnal cycle of total precipitation amount from hourly data for June, July, and August (JJA) for the United States. Vector length and color indicate the amplitude as a percent of the diurnal mean precipitation, and the direction indicates the timing of maximum in local solar time or LST, as given by the key at lower right. The diurnal cycle in amount comes almost entirely from changes in frequency (timing) rather than changes in intensity. Adapted from Dai et al. (1999).

high values of convective instability (Trier and Parsons, 1993). Thunderstorms initiated over the Rocky Mountains in the late afternoon propagate into and through this favorable environment, leading to the observed nocturnal precipitation maximum observed in this region (Carbone et al., 2002).

Over the Southeast and the Rockies, both the static instability and the surface convergence favor afternoon moist convection in summer, resulting in very strong late afternoon maxima of precipitation over these regions.

Models can typically simulate some, but not all, of the diurnal cycle pattern, and some models are wrong everywhere (Dai 2006a). In the National Center for Atmospheric Research (NCAR) Community Climate System Model (CCSM) and the corresponding atmospheric module, the diurnal precipitation occurs about 2 hours before it does in nature, and the complex structure of nocturnal maxima in the U.S. Great Plains is absent (Figure 16.5).

Figure 16.5 also illustrates the distinctive nature of the diurnal cycle in many other parts of the globe. Over many parts of the oceans, there is often a minimum in precipitation from about midnight to early morning (2200-0600 LST), which is out of phase with most continental regions (e.g., Bowman et al., 2005; Dai, 2001; Sorooshian et al., 2002).

Model criteria for the onset of moist convection are often too weak, and so moist convection in the model starts too early and occurs too often. In the models, premature triggering leads to weaker convection and precipitation, as well as weaker downdrafts and gust fronts. Subsequent convection is also expected to be weaker due to the prominent role gust fronts play in initiating convection. Premature convection reduces convective available potential energy and results in convection that does not propagate, in contrast to the observations of the dominance of propagating systems over the central United States (Carbone et al., 2002). Premature triggering of convection and associated cloudiness disrupts the proper heating at the surface of the continent, and thus prevents the continental-scale "sea breeze" and its associated convergence and divergence patterns from developing properly. As a result, the transport of moisture and its role in setting up convective instabilities is also disrupted. It

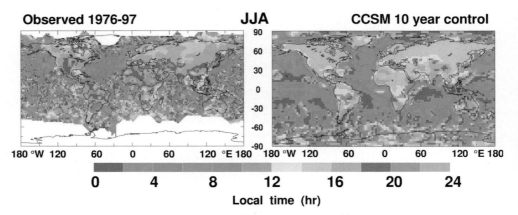

Figure 16.5 The local solar timing (h) of the maximum of the diurnal cycle of precipitation for June, July, and August (JJA) is given for convective precipitation from (left) observations for 1976 to 1997 (Dai, 2001) versus (right) a control run (10-year average) of the National Center for Atmospheric Research (NCAR) Community Climate System Model (CCSM). From Trenberth et al. (2003).

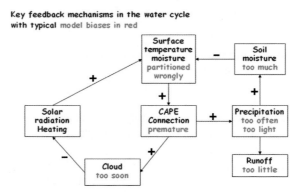

Figure 16.6 Key feedback mechanisms in the diurnal cycle along with model biases in red. Adapted from Trenberth et al. (2003).

further causes the premature release of convective available potential energy, which results in subsequent convection being weaker. Figure 16.6 presents the typical processes, the typical errors in models, and their feedbacks. It is these kinds of interactions and processes that must be better simulated in models in order for the water cycle to be properly predicted in climate projections.

16.2.2 Observed trends in moisture and extreme precipitation events

Atmospheric moisture amounts are generally observed to be increasing in the atmosphere since about 1973, prior to which reliable moisture soundings are mostly unavailable (Ross and Elliott, 2001; Trenberth et al., 2005). In the Western Hemisphere north of the equator, annual mean precipitable water amounts below 500 mb increased over the United States, the Caribbean, and Hawaii by about 5% per decade as a statistically significant trend from 1973 to 1995 (Ross and Elliott, 2001). Most of the increase is related to temperature and hence in atmospheric water holding capacity. In China, analysis by Zhai and Eskridge (1997) also reveals upward trends in precipitable water in all seasons and for the annual mean from 1970 to 1990. Earlier, Hense et al. (1988) revealed increases in moisture over the western Pacific. Precipitable water and relative humidities have not increased over much of Canada and decreases are evident where temperatures declined in northeast Canada (Ross and Elliott, 1996). More comprehensive analysis of precipitable water amounts is possible after 1987 with the SSM/I instrument on satellites. Total water vapor over the oceans (Figure 16.7a) has gone up by 0.4 mm/decade from 1988 to 2004, which is equivalent to 1.4%/decade (Trenberth et al., 2005). The increase is strongly tied to increases in sea surface temperature during this period (Figure 16.7b), and the relationship is strong enough to suggest that water vapor amounts have increased by about 4% since the mid-1970s, consistent with roughly constant relative humidity and the Clausius–Clapeyron relationship that determines the change in water holding capacity as temperatures change. An analysis of surface humidity records over the globe by Dai (2006b) shows significant increasing trends in surface specific humidity from 1976 to 2004 over most land areas and many oceans, whereas trends in surface relative humidity are relatively small. In summary, although uncertainties exist due to errors in humidity measurements (e.g., Guichard and Parsonsk, 2000; Wang et al., 2002), most studies indicate that water vapor increases are present in many regions where air temperature has increased.

Atmospheric moisture amounts over the adjacent oceans around North America are critical for precipitation in the United States, and there have been upward trends in precipitable water in all these regions since 1973 by more than 10% (e.g., Ross and Elliott, 1996; Trenberth et al., 2005). All things being equal, that should lead to 10% stronger rainfall rates when it rains because low-level moisture convergence will be enhanced by that amount. It has been argued that increased moisture content of the atmosphere favors stronger rainfall and snowfall events, thus increasing the risk of flooding. There is clear evidence that rainfall rates have changed in the United States, for instance, for total annual precipitation from 1-day extremes of more than 2 in (50.8 mm) amounts after 1910 (Karl and Knight, 1998). The "much above normal" area, defined as the upper 10% overall, increased steadily throughout the twentieth century from less than 9% to more than 11%, a 20% relative change in total. Karl and Knight, in further analysis of U.S. precipitation increases, showed how it occurs mostly in the upper tenth percentile of the distribution and that the portion of total precipitation derived from extreme and heavy events increased at the expense of more moderate events. Kunkel et al. (1999) showed that extreme precipitation events of 1- to 7-day duration in the United States increased at a rate of about 3% per decade from 1931 to 1996. For the contiguous United States, Kunkel et al. (2003) and Groisman et al. (2004) confirmed earlier results and found statistically significant increases in heavy (upper 5%) and very heavy (upper 1%) precipitation, by 14% and 20%, respectively, for the twentieth century. Much of this increase occurred during the last three decades of the century, and it is most apparent over the eastern parts of the country. Globally, there is also evidence of an increasing contribution of heavy events to total precipitation and increases in heavy and very heavy precipitation, even in places where total amounts are declining (Alexander et al., 2006; Groisman et al., 2005).

Although enhanced rainfall rates increase the risk of flooding, mitigation of flooding by local councils, the Corps of Engineers, and the Bureau of Reclamation in the United States is continually occurring, and flooding records are often confounded by changes in land use and increasing human settlement in flood plains. Nevertheless, great floods have been found to be increasing in the twentieth century (Milly et al., 2002).

370 *Impact of climate change on precipitation*

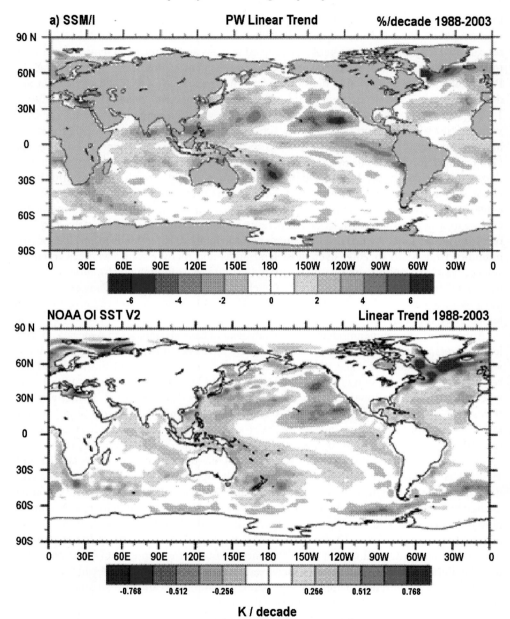

Figure 16.7 (a) Linear trend in vertically integrated precipitable water for 1988 to 2003 in percent per decade over the ocean. (b) Linear trend of sea surface temperature (SST) from 1988 to 2003. The color key is chosen to match the expected change in precipitable water from regression from 30°N to 30°S of 7.8% K in panel (a). Note the good correlation between the precipitable water trends in panel a with SST trends in panel (b). Adapted from Trenberth et al. (2005).

Natural variations in the climate system can also play an important role in the formation of floods and droughts. Research has shown a strong link of floods and droughts to (1) El Niño-Southern Oscillation events (Diaz and Markgraf, 2000; Trenberth and Guillemot, 1996); (2) the North Atlantic oscillation (Hurrell et al., 2003; Malmquist, 1999; Rodwell et al., 1999); (3) the quasibiennial oscillation (Elsner et al., 1999; Gray, 1984; Shapiro, 1989); (4) the northern annular mode (Thompson and Wallace, 1998); and (5) the Madden–Julian oscillation (MJO) (Malone and Hartmann, 2000a; 2000b).

ENSO, for instance, has been shown by Trenberth and Guillemot (1996) to have played a role in the 1988 drought and 1993 flood in North America. They showed that both the drought and the flood developed as a result of a change in the large-scale atmospheric circulation in the Pacific and across North America that altered the storm tracks into and across North America. ENSO altered the tropical Pacific SST and therefore the atmospheric convection, which in turn changes the latent heating pattern and the quasistationary planetary wave pattern, the jet stream location, and associated storm tracks.

The MJO has been shown to impact midlatitude weather through tropical convection and latent heating. In this case, Rossby wave packets are generated that have been linked to extreme flooding events in the United States and Europe. The MJO has also been shown to impact tropical cyclone formation. Tropical cyclone formation is significantly more likely when the MJO is producing westerly winds and enhanced precipitation (Malone and Hartmann, 2000a, 2000b). Tropical cyclones typically have the highest property damage loss of any extreme event, and are therefore of great interest to state and local disaster preparedness organizations, as well as the insurance industry (Murnane, 2004).

Beyond climate impacts on extremes, local and regional weather factors play an important role in the production of extreme weather events. The recent studies of the Buffalo Creek flood (Chen et al., 2001, Warner et al., 2000, Yates et al., 2000) provide evidence that burned regions in a river basin can be preferred regions for flash flood formation. Mesoscale model simulations indicated preferred cloud and precipitation formation over the burn region due to enhanced sensible heat flux. The sensible heat flux was increased due to the darker soil and low water content in the soil after the fire. Runoff was also increased due to the more impervious soil. Under synoptic conditions associated with the Buffalo Creek flood (high/low-level moisture, weak steering level winds), strong precipitation developed over this region in good comparison with radar and rain gauge estimates of precipitation. It was also found that the moisture advection into the region was affected by farm irrigation within 100 km of the site. Thus, local and regional characteristics of the soil and vegetation and its coupling to the boundary layer and clouds may play an important role in extreme weather events, especially flood formation.

16.3 How should precipitation change as the climate changes?

Changes in radiative forcing associated with increasing greenhouse gases in the atmosphere produce increased heating at the surface. The actual amount of heating depends critically on various feedbacks, including water vapor feedback, ice-albedo feedback, and effects of changes in clouds. Uncertainties associated with these feedbacks are largely responsible for the wide range of climate sensitivity of 1.5°C to 4.5°C in global mean surface temperature among climate models to doubling of carbon dioxide (Houghton et al., 2001).

For precipitation, the models predict on average an increase of 1% to 2% per 1°C warming in global mean temperature (see also Allen and Ingram, 2002). However, increasing aerosols prevent solar radiation from reaching the surface and lead to lower heating rates and diminished precipitation. These effects are highly regional and add considerably to complications in projections of precipitation with models.

Generally, evaporation at the surface cools and hence acts to "air condition" the planet (Trenberth and Shea, 2005), and therefore, increased surface evaporation would be expected to moderate temperature increases. In fact, a very robust finding in all climate models (Houghton et al., 2001) with global warming is for an increase in potential evapotranspiration. In the absence of precipitation, this leads to increased risk of drought, as surface drying is enhanced. It also leads to increased risk of heat waves and wildfires in association with such droughts because once the soil moisture is depleted, then all the heating goes into raising temperatures and wilting plants. An analysis by Dai et al. (2004) of the Palmer Drought Severity Index calculated using observed precipitation and temperature suggests global drought areas have more than doubled since the 1970s, partly resulting from enhanced drying associated with recent rapid warming at northern middle and high latitudes.

Although enhanced drying occurs at the surface as a result of higher temperatures, the water holding capacity of the atmosphere is also increased by about 7% K^{-1} as governed by the Clausius–Clapeyron equation (Trenberth et al., 2003). Moreover, models suggest that changes in atmospheric relative humidity are small, presumably because of precipitation and atmospheric mixing processes, and this is borne out by available observations (e.g., Dai, 2006b; Soden et al., 2002; Trenberth et al., 2005). Hence, the actual absolute moisture content of the atmosphere should also increase at something like this rate because relative humidity does not change much, which is again consistent with observations in areas of the world not dominated by major pollution clouds. However, it is unlikely that the moisture changes will be geographically uniform, even though models predict increases in surface temperatures almost everywhere after a few decades with increases in greenhouse gases. Atmospheric dynamics play a role through favored regions of convergence and subsidence. The predicted larger increases in surface temperatures projected at high latitudes means that there will a bigger absolute increase in moisture amount in higher latitudes due to the nonlinear nature of the Clausius–Clapeyron equation.

Heavy rainfall rates typically greatly exceed local surface evaporation rates and thus depend on low-level moisture convergence. In this case, rainfall intensity should also increase at about the same rate as the moisture increase, namely, 7% K^{-1} with warming. In fact, the rate of increase can even exceed this because the additional latent heat released feeds back and invigorates the storm that causes the rain in the first place, further enhancing convergence of moisture. This means that the changes in rain rates, when it rains, are at odds with the 1% to 2% K^{-1} for total rainfall amounts. The implication is that there must be a decrease in light and moderate rains, and/or a decrease in the frequency of rain events, as found by Hennessey et al. (1997). Thus, the prospect may be for fewer but more intense rainfall—or snowfall—events.

Of course, these general arguments must be tempered by regional effects and changes in teleconnections (see also Trenberth, 1998). Typically, neither observational nor model data have been analyzed in ways that can check on these concepts, although some recent model analyses are moving in this direction. Often daily mean amounts are used and may be analyzed in terms of the "shape" and "scale" parameters of a gamma distribution fit to the data. Wilby and Wigley (2002) used this approach to demonstrate increases in extremes of precipitation with anthropogenic forcing in the NCAR climate system model and the Hadley Centre coupled model despite quite different spatial patterns of precipitation change. Expectations outlined here are realized in the ECHAM4/OPYC3 model (Semenov and Bengtsson, 2002). Another recent check of this was performed in the Hadley Centre model (Allen and Ingram, 2002), and it indeed shows at the time of doubling of carbon dioxide in the model simulations that rainfall intensities less than about the 85th percentile decrease in frequency, whereas heavy events increase at close to the Clausius–Clapeyron rate. An analysis of the future climate simulations by the latest generation of coupled climate models (Sun et al., 2007) shows that globally for each 1°C of surface warming, atmospheric precipitable water increases by ~9%, daily precipitation intensity increases by ~2%, and daily precipitation frequency decreases by 0.7%. However, for very heavy precipitation (>50 mm day^{-1}), the percentage increase in frequency is much larger than in intensity (31.2% vs. 2.4%) so that there is a shift toward increased heavy precipitation.

In extratropical mountain areas, the winter snowpack forms a vital resource, not only for skiers, but also as a freshwater resource in the spring and summer as the snow melts. Yet, warming makes for a shorter snow season with more precipitation falling as rain rather than snow, earlier snowmelt of the snow that does exist, and greater evaporation and ablation. These factors all contribute to diminished snowpack. In the summer of 2002 in the western parts of the United States, exceptionally low snowpack and subsequent low soil moisture likely contributed substantially to the widespread intense drought because of the importance of recycling. Could this be a sign of the future?

16.4 Questions and issues

Climate change is certainly likely to locally change the intensity, frequency, duration, and amounts of precipitation. Testing of how well climate models deal with these characteristics of precipitation is an issue of significant societal importance. The foremost need is better documentation and processing of all aspects of precipitation.

There is a need for improved parameterization of convection in large-scale models. Parameterization of convection needs to be improved to appropriately allow convective available potential energy (CAPE) to build up as observed, and likely involves both the improvement of "triggers" and the suppression of convection by the presence of convective inhibition (CIN). Parameterizations are a scale interaction problem in part because the triggers are often subgrid scale (e.g., outflows and other small-scale boundaries, gravity wave motions, convective rolls), whereas larger-scale motions may suppress or enhance the magnitudes of CAPE and CIN. Some processes are difficult to include in global climate models, such as convection, which is typically simulated at grid points in single columns that are not directly related to events at adjacent grid point columns, whereas in the atmosphere, mesoscale convective systems can be long lasting and move from one grid column to another (e.g., Carbone et al., 2002). There is also a need for improved observations and modeling of sources and sinks of moisture for the atmosphere, especially over land. This relates to recycling and the disposition of moisture at the surface in models, and whether the moisture is or is not available for subsequent evapotranspiration. It relates to improved and validated treatment of runoff, soil infiltration, and surface hydrology in models, including vegetation models.

Finally, we believe that improved simulation of the diurnal cycle of precipitation in models is essential. This probably also requires improved simulation of the diurnal cycle of temperature, cloud amount, and atmospheric circulation, and especially the buildup and release of CAPE. We believe the best approach is a hierarchical one using models ranging from single-column models, to cloud-resolving models, mesoscale regional models, global atmospheric models, and coupled climate models. Land surface processes and atmosphere–ocean–land interactions are clearly important. The replication of the diurnal cycle and the associated precipitation is a key framework for testing these different models. Accordingly, at NCAR, we have established a "water cycle across scales" program to address the issues outlined previously, among others. A focus of the program is on the warm season diurnal cycle in North America. Hence, the diurnal cycle is being exploited as a test bed to examine systematic timing and duration of precipitation events, and as a vehicle to improve model performance. More information about the water cycle initiative is available online at www.tiimes.ucar.edu/wcas/index.htm.

16.5 Summary

This chapter highlights some of the key issues that need to be addressed in order to properly simulate the water cycle and precipitation in climate models. These include

1. The proper simulation of the frequency, intensity, duration, timing, amount, and phase of precipitation.
2. The proper treatment of local evapotranspiration versus large-scale moisture advection.
3. The proper treatment of water runoff and soil infiltration in order to properly model soil moisture and its impact on latent and sensible heat fluxes.

The recent Intergovernmental Panel on climate change Fourth Assessment report (Solomon et al. 2007) on the impacts of global climate change on the earth system also emphasizes the need to improve the modeling of these key issues.

It is critical that scientists address these issues in order to provide natural disaster managers and other users of climate information with the proper guidance to make the difficult decisions facing our global society in the near future.

Acknowledgments

A significant part of this chapter is reproduced from an article in the *Bulletin of the American Meteorological Society* entitled "The Changing Character of Precipitation," by Trenberth, Dai, Rasmussen, and Parsons (2003). The material from that article is used by permission of the American Meteorological Society. The authors also acknowledge support from the National Center for Atmospheric Research (NCAR) Water Cycle Program funded by the National Science Foundation.

References

Alexander, L. V., Peterson, T. C., Caesar, J., Gleason, B., Tank Klein, A. M. G., Haylock, M., Collins, D., Trewin, B., Rahimzadeh, F., Tagipour, A., Rupa Kumar, K., Revadekar, J., Griffiths, G., Vincent, L., Stephenson, D. B., Burn, J., Aguilar, E., Brunet, M., Taylor, M., New, M., Zhai, P., Rusticucci, M., and Vazquez-Aguirre, J. L. (2006) "Global Observed Changes in Daily Climate Extremes of Temperature and Precipitation," *J. Geophys. Res.* **111**, D05109, doi: 10.1029/2005JD006290.

Allen, M. R., and Ingram, W. J. (2002) "Constraints on Future Changes in Climate and the Hydrologic Cycle," *Nature* **419**, pp. 224–232.

Bowman, K. P., Collier, J. C., North, G. R., Wu, Q. Y., Ha, E. H., and Hardin, J. (2005) "Diurnal Cycle of Tropical Precipitation in Tropical Rainfall Measuring Mission (TRMM) Satellite and Ocean Buoy Rain Gauge Data," *J. Geophys. Res.* **110**, D21104, doi: 10.1029/2005JD005763.

Carbone, R. E., Tuttle, J. D., Ahijevych, D. A, and Trier, S. B. (2002) "Inferences of Predictability Associated with Warm Season Precipitation Episodes," *J. Atmos. Sci.* **59**, pp. 2033–2056.

Chen, F., Warner, T., and Manning, K. (2001). "Sensitivity of Orographic Moist Convection to Landscape Variability: A Study of the Buffalo Creek, Colorado, Flash Flood Case of 1996," *J. Atmos. Sci.* **58**, pp. 3204–3223.

Covey, C., AchutaRao, K. M., Cubasch, U., Jones, P., Lambert, S. J., Mann, M. E., Phillips, T. J., and Taylor, K. E. (2003) "An Overview of Results from the Coupled Model Intercomparison Project," *Global Planet Change* **37**, pp. 103–133.

Dai, A. G. (1999) "Recent Changes in the Diurnal Cycle of Precipitation over the United States," *Geophys. Res. Lett.* **26**, pp. 341–344.

Dai, A. G. (2001) "Global Precipitation and Thunderstorm Frequencies. Part II: Diurnal Variations," *J. Climate* **14**, pp. 1112–1128.

Dai, A. G. (2006a) "Precipitation Characteristics in Eighteen Coupled Climate Models," *J. Climate* **19**, pp. 4605–4630.

Dai, A. G. (2006b) "Recent Climatology, Variability and Trends in Global Surface Humidity," *J. Climate* **19**, pp. 3589–3606.

Dai, A., Giorgi, F., and Trenberth, K. E. (1999) "Observed and Model-Simulated Diurnal Cycles of Precipitation over the Contiguous United States," *J. Geophysical Research—Atmospheres* **104**, pp. 6377–6402.

Dai, A. G., Trenberth, K. E., and Qian, T. T. (2004) "A Global Dataset of Palmer Drought Severity Index for 1870–2002. Relationship with Soil Moisture and Effects of Surface Warming," *J. Hydrometeorology* **5**, pp. 1117–1130.

Diaz, H. F., and Markgraf, V., eds. (2000) *El Niño and the Southern Oscillation, Multiscale Variability and Global and Regional Impacts*, Cambridge University Press, Cambridge, United Kingdom.

Elsner, J. B., Kara, A. B., and Owens, M. A. (1999) "Fluctuations in North Atlantic Hurricane Frequency," *J. Climate* **12**, pp. 427–437.

Gray, W. M. (1984) "Atlantic Seasonal Hurricane Frequency. Part I. El Niño and the 30 mb Quasi-Biennial Oscillation Influences," *Mon. Wea. Rev.* **112**, pp. 1649–1668.

Groisman, P. Ya., Knight, R. W., Easterling, D. R., Karl, T. R., Hegerl, G. C., and Razuvaev, V. N. (2005) "Trends in Intense Precipitation in the Climate Record," *J. Climate* **18**, pp. 1326–1350.

Groisman, P. Ya., Knight, R. W., Karl, T. R., Easterling, D. R., Sun, B., and Lawrimore, J. H. (2004) "Contemporary Changes of the Hydrological Cycle over the Contiguous United States: Trends Derived from In Situ Observations," *J. Hydrometeorology* **5**, pp. 64–85.

Guichard, F. D., and Parsonsk, E. M. (2000) "Thermodynamic and Radiative Impact of the Correction of Sounding Humidity Bias in the Tropics," *J. Climate* **13**, pp. 3611–3624.

Hennessy, K. J, Gregory, J. M., and Mitchell, J. F. B. (1997) "Changes in Daily Precipitation Under Enhanced Greenhouse Conditions," *Climate Dynamics* **13**, pp. 667–680.

Hense A., Krahe P., and Flohn H. (1988) "Recent Fluctuations of Tropospheric Temperature and Water Vapour Content in the Tropics," *Meteor. Atmos. Phys.* **38**, pp. 215–227.

Higgins, R. W., Yao, Y., Yarosh, E. S., Janowiak, J. E., and Mo, K. C. (1997) "Influence of the Great Plains Low-Level Jet on Summertime Precipitation and Moisture Transport over the Central United States," *J. Climate* **10**, pp. 481–507.

Houghton, J. T., Ding Y., Griggs, D. J., Noguer, M., van der Linden, P. J., Dai, X., Maskell, K., and Johnson, C. A. (2001) *Climate Change 2001: The Scientific Basis*, Third Assessment Report of the Intergovernmental Panel on Climate Change, Cambridge University Press, Cambridge, United Kingdom.

Hurrell, J. W., Kushnir, Y., Ottersen, G., and Visbeck, M. (2003) *The North Atlantic Oscillation: Climate Significance and Environmental Impacts, Geophys. Monogr.* **134**, American Geophysical Union, Washington, DC.

Karl, T. R., and Knight, W. R. (1998) "Secular Trends of Precipitation Amount, Frequency, and Intensity in the United States," *Bull. Am. Met. Soc.* **79**, pp. 231–241.

Kunkel, K. E., Easterling, D. R., Redmond, K., and Hubbard, K. (2003) "Temporal Variations of Extreme Precipitation Events in the United States: 1895–2000," *Geophys. Res. Lett.* **30**, pp. 1900–1910.

Kunkel, K. E., Andsager, K., and Easterling, D. R. (1999) "Long Term Trends in Extreme Precipitation Events over the Conterminous United States and Canada. *J. Climate*, **12**, 2515–2527.

Malmquist, D. L., ed. (1999) "European Windstorms and the North Atlantic Oscillation: Impacts, Characteristics, and Predictability," a position paper based on the *Proceedings of the Risk Prediction Initiative Workshop on European Winter Storms and the North Atlantic Oscillation*, Hamilton, Bermuda, 18–19 January 1999.

Malone, E. D., and Hartmann, D. L. (2000a) "Modulaton of Eastern North Pacific Hurricanes by the Madden–Julian Oscillation," *J. Climate* **13**, pp. 1451–1460.

Malone, E. D., and Hartmann, D. L. (2000b) "Modulation of Hurricane Activity in the Gulf of Mexico by the Madden–Julian Oscillation," *Science* **287**, pp. 2002–2004.

Meehl, G., Covey, C., McAvaney, B., Latif, M., and Stouffer, R. (2005) "Overview of the Coupled Model Intercomparison Project," *Bull. Am. Met. Soc.* **86(1)**, pp. 89–93.

Milly, P. C. D., Wetherald, R. T., Dunne, K. A., and Delworth, T. L. (2002) "Increasing Risk of Great Floods in a Changing Climate," *Nature* **415**, pp. 514–517.

Murnane, R. J. (2004) "Climate Research and Reinsurance," *Bull. Amer. Met. Soc.* **85**, pp. 697–707.

Rodwell, M. J., Rowell, D. P., and Folland, C. K. (1999) "Oceanic Forcing of the Wintertime North Atlantic Oscillation and European Climate," *Nature* **398**, pp. 320–323.

Ross, R. J., and Elliott, W. P. (1996) "Tropospheric Water Vapor Climatology and Trends over North America: 1973–1993," *J. Climate* **9**, pp. 3561–3574.

Ross, R. J., and Elliott, W. P. (2001) "Radiosonde-Based Northern Hemisphere Tropospheric Water Vapor Trends," *J. Climate* **14**, pp. 1602–1612.

Semenov, V. A., and Bengtsson, L. (2002) "Secular Trends in Daily Precipitation Characteristics Greenhouse Gas Simulation with a Coupled AOGCM," *Climate Dynamics* **19**, pp. 123–140.

Shapiro, L. J. (1989) "The Relationship of the Quasi-Biennial Oscillation to Atlantic Tropical Storm Activity," *Mon. Wea. Rev.* **117**, pp. 2598–2614.

Soden, B. J., Wetherald, R. T., Stenchikov, G.L., and Robock, A. (2002) "Global Cooling after the Eruption of Mount Pintubo: A Test of Climate Feedback by Water Vapor," *Science* **296**, pp. 727–730.

Sorooshian, S., Gao, X., Maddox, R. A., Hong, Y., and Imam, B. (2002) "Diurnal Variability of Tropical Rainfall Retrieved from Combined GOES and TRMM Satellite Information," *J. Climate* **15**, pp. 983–1001.

Solomon, S., Quin, D., and Manning, M. (2007) "Climate Change 2007. The Physical Basis Contribution of Working Group I to the Fourth Assessment Report of the Intergovernment Panel on Climate Change, Cambridge University Press, 1996 pp.

Sun, Y., Solomon, S., Dai, A., and Portmann, R. W. (2007) "How Often Will It Rain?" *J. Climate.* **19**, pp. 916–934.

Thompson, D. W., and Wallace, J. M. (1998) "The Artic Oscillation Signature in the Wintertime Geopotential Height and Temperature Fields," *Geophys. Res. Lett.* **25**, pp. 1297–1300.

Trenberth, K. E. (1998) "Atmospheric Moisture Residence Times and Cycling: Implications for Rainfall Rates and Climate Change," *Climatic Change* **39**, pp. 667–694.

Trenberth, K. E., Dai, A., Rasmussen, R. M., and Parsons, D. B. (2003) "The Changing Character of Precipitation," *Bull. Am. Met. Soc.* **84**, pp. 1205–1217.

Trenberth, K. E., Fasullo, J., and Smith, L. (2005) "Trends and Variability in Column-Integrated Water Vapor," *Clim. Dyn.* **24**, pp. 741–758.

Trenberth, K. E., and Guillemot, C. J. (1996) "Physical Processes Involved in the 1988 Drought and 1993 Floods in North America," *J. Climate* **9**, pp. 1288–1298.

Trenberth, K. E., and Shea, D. J. (2005) "Relationships Between Precipitation and Surface Temperature," *Geophys. Res. Lett.* **32**, L14703, doi: 10.1029/2005GL022760.

Trenberth, K. E., Smith, L., Qian, T., Dai, A., and Fasullo, J. (2007) "Estimates of the Global Water Budget and Its Annual Cycle Using Observational and Model Data," *J. Hydrometeorology*, **8**, pp. 758–769.

Trier, S. B., and Parsons, D. B. (1993) "Evolution of Environmental Conditions Preceding the Development of a Nocturnal Mesoscale Convective Complex," *Mon. Wea. Rev.* **121**, pp. 1078–1098.

Wang, J. H., Cole, H. L., Carlson, D. J., Miller, E. R., Beierle, K., Paukkunen, A., and Laine, T. K. (2002) "Corrections of Humidity Measurement Errors from the Vaisala RS80 Radiosonde—Application to TOGA COARE Data," *J. Atmosph. & Ocean. Tech.* **19**, pp. 981–1002.

Warner, T., Brandes, E., Sun, J., Yates, D., and Mueller, C. (2000) "Prediction of a Flash Flood in Complex Terrain. Part I. A Comparison of Rainfall Estimates from Radar, and Very Short Range Rainfall Simulations from a Dynamic Model and an Algorithmic System," *J. Appl. Met.* **39**, pp. 797–814.

Wilby, R. L., and Wigley, T. M. L. (2002) "Future Changes in the Distribution of Daily Precipitation Totals Across North America," *Geophys. Res. Lett.* **29**, 10.1029/2001GL013048.

Yates, D., Warner, T., and Leavesley, G. (2000) "Prediction of a Flash Flood in Complex Terrain. Part II. A Comparison of Flood Discharge Simulations Using Rainfall Input from Radar, a Dynamic Model, and an Automated Algorithmic System," *J. Appl. Met.* **39**, pp. 815–825.

Zhai P., and Eskridge, R. E. (1997) "Atmospheric Water Vapor over China," *J. Climate* **10**, pp. 2643–2652.

17

Weather-related disasters in arid lands

Thomas T. Warner

Arid lands are vulnerable to many of the same kinds of weather-related disasters that are experienced in temperate climates, but there are some that can be much more severe. One type of disaster may be generally referred to as severe weather, which encompasses dust storms and flash floods. Another type of disaster, desertification, is related to the vulnerability of arid land vegetation and substrates to human- and climate-induced perturbations. This chapter describes the physical processes associated with each of these types of disasters and provides examples of each.

17.1 Introduction

Desert weather and climate, and associated disasters, have historically been primarily of academic concern to all but the few local residents, even though some broader long-term interest has existed because of military activities, the exploitation of the desert's great natural resources, agricultural reclamation through irrigation, and the perception of some regional climatic trends toward desertification. However, more recently, population growth in arid areas has often surpassed that in more temperate zones, the attractions being unpolluted air, abundant sunshine, beautiful landscapes, and endless open space. In North America, for example, population growth in urban areas of the Sonoran Desert in the second half of the twentieth century occurred much more rapidly than in New England and in the Midwest. Population growth in the arid states of the United States has been about five times the national average. Worldwide, about 35% of the population is estimated to live in dry lands. This fraction will likely increase because most inhabitants of arid lands are in developing countries where the rate of population growth is greatest. Thus, deserts are becoming less deserted, and weather-related disasters will impact a growing fraction of the world population.

Deserts also influence the weather and people far outside their borders because there are dynamic connections in the atmosphere between weather processes that are geographically remote. The fact that warm, arid climates comprise more than 30% of the land area of Earth implies that their aggregate effect on nondesert areas and global climate in general can be significant. If the state of the soil and vegetation is used, in addition to the climate, to define the area of deserts, the desert fraction of the total land area increases by another 10% because of the human degradation (United Nations, 1978). That is, all things considered, the total is close to 40%. By comparison, only 10% of Earth's land surface is cultivated. Figure 17.1 illustrates the fraction of the land in different arid and nonarid categories. The designated hyperarid, arid, and semiarid areas collectively represent Earth's traditionally defined dry climates. Dry subhumid climates are typically grassland and prairie. The humid land area consists of mostly tropical and midlatitude forest, grassland, and rainfed agricultural land.

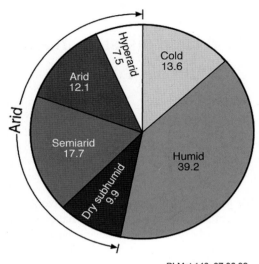

Figure 17.1   Global land area by aridity zone (percentage). Data from United Nations Environment Program (UNEP) (1992).

17.2 Severe weather in arid lands

If one defines severe weather as simply conditions that are threatening to life, even the daily desert sky, barren of clouds, would be included because of its causal association with the temperature extremes, desiccating humidity, and lack of liquid water. Acknowledging that even these undisturbed conditions would be severe for the unprepared resident or traveler of the desert, we will limit the discussion here to more disturbed weather that is especially challenging for survival. This disturbed weather falls into two categories: high winds with associated dust and sand storms, and rains that produce flash flooding. Both can cause disasters when they occur in populated areas and on sufficiently large scales.

17.2.1 Dust storms and sand storms

Dust suspended in the atmosphere, as well as dust storms, are common in arid and semiarid areas because most of the surface is covered by mobile, loose sediments. Even though most of us think we know a dust or sand storm when we see it, there is a formal international meteorological definition. When the visibility is less than 1,000 m and the dust is being entrained into the atmosphere within sight of the observer, then a dust storm is reported. However, the popular usage of the term includes a dense dust cloud that has been produced somewhere upstream. Despite this international standard, many national weather services and investigators have their own particular definitions. For example, Tao et al. (2002) reported on the criteria used in Inner Mongolia:

- Dust storm: at least three stations reporting horizontal visibility of less than 1,000 m and an average wind speed of 10.8 to 20.7 m/s
- Strong dust storm: at least three stations with horizontal visibility of less than 500 m and an average wind speed of 17.2 to 24.4 m/s
- Very strong dust storm: at least one station with horizontal visibility of less than 50 m and an average wind speed of 20.8 m/s or greater

When the horizontal visibility is greater than 1,000 m and moderate winds are causing dust to be elevated into the atmosphere at a location, the condition is defined as rising dust rather than as a dust storm (Abdulaziz, 1994). When dust has been entrained into the atmosphere at some upstream location, it is locally defined as suspended dust.

Dust elevated into the atmosphere by dust storms has numerous environmental consequences. These include contributing to climate change; modifying local weather conditions; producing chemical and biological changes in the oceans; and affecting soil formation, surface water and groundwater quality, and crop growth and survival (Goudie and Middleton, 1992). Societal impacts include disruptions to air, land, and rail traffic; interruption of radio services; the myriad effects of static electricity generation; property damage; and health effects on humans and animals.

The extremely arid desert is almost exclusively an *aeolian* environment, where particles are transported primarily by the wind rather than by a combination of wind and water. The particulate material originates from the chemical and mechanical weathering of rock. Most aeolian particles in deserts are silt and not sand, silt being composed of particles of much smaller size and weight. The finer clay and silt particles, with diameters of up to 0.05 mm, are elevated into the atmosphere as "dust" and suspended for long or short distances, depending on the amount of wind energy. Larger particles (e.g., sand) are entrained into the airstream only at higher speeds, do not remain airborne for long, and sometimes accumulate as dunes. The particles that are too heavy to be entrained remain behind and form an armor on the surface. Thus, despite the popular symbolic connection between sand storms and deserts, the laden winds of a "sand storm" are in fact generally heavy with the smaller silt particles.

Dust storms and sand storms require both a dry, granular surface substrate, as well as winds to elevate the grains. Despite the varied dynamics, it is possible to divide the wind-generating processes into convective scale and large scale. On the large scale, the pressure gradients associated with synoptic-scale weather systems force the winds, and the high winds and dust or sand storm can last for days. On the convective scale, cold air downdrafts are produced by the frictional drag of falling rainfall and the cooling associated with precipitation evaporation in the unsaturated subcloud layer. When these downdrafts near the ground, they spread horizontally outward from the thunderstorms, as shallow fast-moving density currents with strong winds. Durations of such sand or dust storms would be typically an hour or less.

The dangers associated with sand storms and dust storms are numerous. First, the amount of solid material in the air is sufficient to cause suffocation if inhaled and accumulated in the respiratory tract. Second, the high wind exacerbates the normal desiccating effects of the high temperature and low relative humidity, with the clothed human body losing up to a quart of water per hour in such conditions. Bodies recovered soon after such storms have been completely mummified. A third risk is that of live burial if the sand is allowed to drift over the body.

The physical manifestation of dust storms and sand storms depends on whether dust or sand particles are involved, and whether the winds are of large scale or convective origin. If only sand is involved, and not dust, the particles will generally not be elevated through as deep a layer. In addition, the air will clear more rapidly after the high winds cease. As wind speeds increase early in the event, a few grains of sand begin to move within the first centimeter above the surface. With higher winds, individual narrow streams of sand merge into an unbroken thin carpet that moves at great speed and obscures the ground from horizon to horizon. The depth of the current of moving sand depends on the wind speed and the size of the sand grains. George (1977) stated that depths of about 2 m are typical of his experience in the Sahara, with the top of the layer being quite distinct. His account follows, of one long-lived sand storm event in the Sahara:

The spectacle I beheld was unique. I stood under a blazing sky and looked down at the top of a seemingly infinite sea of sand flowing along at high speed. The brow of the hill on which I stood formed an island in the midst of this surging golden yellow sea. At some distance I could make out the brows of other hills, other "islands," but the most curious part of the experience was the sight of the heads and humps of several camels that appeared to be floating on the surface of the sand, like ducks drifting in the current of a river. The camels' heads and humps, all that showed above the upper limit of the sandstorm, were moving in the same direction as the drifting sand, but a great deal more slowly. (p. 19)

With very high winds, the depth of the layer of elevated sand can far exceed the 2 m in this example. The existence of clear sky above the sand in the situation just noted means that no dust was elevated throughout the boundary layer, before the winds reached the speed at which the sand was entrained into the air. Also, because the sky was cloud free and the sand storm was long lasting, these winds were likely forced by a large-scale pressure gradient, and not by a convective event. In contrast, a convectively forced sand/dust storm in the Gobi Desert of Mongolia is described in Man (1999):

Now the storm was coming for us with a sort of focused rage that had not been there when I first noticed it. To our left, that hard blade was the leading edge of a great grey-brown dome of dust that formed a semicircle, perhaps 15 km across and nearly a kilometer high. It was the mouth of the beast, and it was ringed above by a mantle of charcoal clouds, and inside it thunder boomed and lightning flashed. Round us, the desert stirred with the storm's first menacing breath. It was coming at us fast, at a gallop. As the edge of the disc ate up the desert to either side of us, the sky darkened, the distant mountains vanished, and the clouds began to close around us.

Andrews (1932) described another brief but extreme dust storm in the Gobi:

... the air shook with a roar louder than the first, and the gale struck like the burst of a high-explosive shell. Even with my head covered, I heard the crash and rip of falling tents. . . . For 15 minutes we could only lie and take it. The sleeping bag had been torn from under me and the coat of my pajamas stripped off. The sand and gravel lashed my back until it bled. . . . Suddenly the gale ceased, leaving a flat calm. (p. 144)

Because of the shorter lifetime of convective events, the most severe part of the latter storms were over within a half hour.

The immense amount of sand and dust transported in these storms, over short and long distances, is a testimony to their severity and energy content. A dramatic example is related to the great wealth of unprecedentedly well-preserved dinosaur bones that have been found in the Gobi Desert. It is speculated that suffocation in sand storms is an explanation for the fact that many of the deaths seem to have occurred in some cataclysm, even though with apparently none of the trauma associated with floods and other severe events. In the Museum of Natural History in Ulaanbaatar, Mongolia, are the fossilized bones of two dinosaurs that are still locked in combat, the way they died from some other act. The theory is that, in the Late Cretaceous, 80 million years ago, the area was a richly populated mixture of desert and savanna, with lush marshlands, mudflats, and lakes. However, periodically, the placid environment of the mammals and reptiles was abruptly interrupted by a tidal surge of sand from the sky, carried aloft by some severe Cretaceous storm system. The weight of the winds and the sand deluge caused dunes to collapse and sand to accumulate rapidly, repeatedly suffocating and preserving a wealth of faunae. There are sand storms today of this apparent severity.

The physics and characteristics of sand and dust transport

Sediment movement is a function of both the power of the wind and the characteristics of the grains that hold them in place. The effect of the wind is twofold. First, the air flowing over the grains causes a decreased pressure on top. Because the pressure under the grain is not affected, there is an upward force. This causes the grains to be "sucked" into the airflow. There is also a frictional drag effect between the wind and the grain that creates a force in the direction of the wind. In opposition, three effects tend to keep the particles in place: weight, packing, and cohesion. Packing is simply the physical forces of particles on each other when they are compacted into a shared volume. Cohesion represents any process that causes particles to bind to one another. One example would be surface tension effects among particles in damp soil. There are also organic and chemical compounds that can cause grains in desert surface crusts to adhere to each other. The weight simply represents the effect of gravity, where the force is proportional to the mass of a particle. As the effect of the wind overcomes the cohesive forces, particles begin to shake and then lift into the airstream. Chepil (1945) showed that particles with greater diameter and greater density have a higher wind threshold required for entrainment into the airstream. However, particles with diameters smaller than about 0.06 mm (about the size boundary between silt and sand) also have a high threshold because there are greater electrostatic and molecular forces of cohesion, the smaller particles retain moisture more efficiently, and the smaller particles are shielded from the wind in the spaces among the larger particles. Thus, the particles that are most effectively entrained into the airstream have diameters between about 0.04 and 0.40 mm (Wiggs, 1997). There are many factors that complicate this analysis, however, such as substrate moisture, the mixtures of particle sizes, surface slope, vegetation, and surface crusting.

Substrate moisture is an especially poorly understood factor with respect to its effect on the particle entrainment threshold. The general concept is that the surface tension associated with pore moisture binds the grains together, and the entrainment threshold is higher for a wet than for a dry substrate. However, there is considerable disagreement among studies in terms of the particular sensitivity of the entrainment threshold to the moisture content. This is at least partly due to the fact that the moisture effect is sensitive to the grain-size distribution. Wiggs (1997) has a more detailed discussion of this effect.

There are two general mechanisms by which particles can be transported by the wind. For small particles with diameters of less than about 0.06 mm, the settling velocity is low and turbulence can mix the particles upward throughout the boundary layer where they can remain suspended for many days. Such suspensions of dust can be sustained by convective boundary-layer turbulence, even in the absence of significant large-scale winds for elevating more dust. Because this dust can limit visibility to a few meters, and in a gentle breeze it moves like a mist, it has been called a "dry fog" (George, 1977). Of course, the term "fog" is inappropriate in a literal sense because there is no condensate of water, so the sometimes used terms "dust haze," "dry haze," or "harmattan haze" might be preferred. If a traveler is reliant on the sun or landmarks for navigation, the threat can be very great indeed.

Larger particles move through three different modes of contact, or *bedload*, transport: saltation, creep, and reptation. Saltation is the process by which grains are elevated from the grain bed, gain horizontal momentum through exposure to the higher-velocity airstream above the ground, descend back to the grain bed and impact other grains, and bounce upward again into the airstream (Anderson, 1987). Some of the momentum from the original grain is transferred to the other grains by the impact, and they may be ejected from the grain bed into the airstream. When the newly ejected grains from this "splash" effect have sufficient

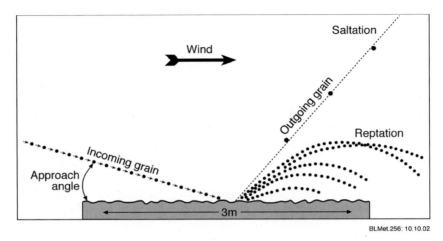

Figure 17.2 The process of reptation where a high-speed saltating grain ejects other grains from the surface. The vertical scale in this schematic is exaggerated for clarity.

momentum to enter the airstream, they may begin saltation themselves. When these splashed grains take only a single hop, the process of grain movement is called reptation. Figure 17.2 shows a schematic of the reptation process. Last, surface creep describes the rolling of larger particles as a result of wind drag and the impact of saltating or reptating grains. Thus, most of the larger-particle transport takes place near the surface, with the particle concentration decreasing exponentially with height. For hard surfaces, Pye and Tsoar (1990) suggested a maximum saltation height of 3 m, with a mean height of 0.2 m. Heathcote (1983) estimated that 90% of sand is transported in the lowest 50 cm of atmosphere, Wiggs (1992) measured up to 35% of sand transport taking place in the lowest 2.5 cm, and Butterfield (1991) stated that 80% of all transport occurs within 2 cm of the surface. Clearly, the wind field and the particle size strongly control the depth over which large particles are transported.

Figure 17.3 summarizes the type of particle transport in terms of the grain diameter and the friction velocity (sometimes called "shear velocity") of the wind. The ratio of the fall velocity of the particles to the friction velocity of the wind is a criterion that can be used to separate the typical degree of particle suspension in the airstream (Chepil & Woodruff, 1957; Gillette et al., 1974). Suspension of particles is clearly favored for smaller particle sizes and higher friction velocities. The short-term suspension category implies time periods of a few hours, whereas long-term suspension is days to weeks. Despite the implication from this figure that sand particles only remain in suspension for very short periods, even for large friction velocities, there have been numerous observed instances of the long-distance transport of large particles. For example, Glaccum and Prospero (1980) found 0.09-mm diameter quartz grains over Sal Island in the Cape Verde Group, more than 500 km off west Africa. Betzer et al. (1988) collected large numbers of quartz grains with diameters greater than 0.075 mm over the Pacific Ocean, more than 10,000 km from where they were entrained in an Asian dust storm. Sarre (1987) summarized threshold friction velocities required for entrainment of larger sand particles, based on a number of literature sources. There is much yet to be understood about the entrainment and transport of large particles. The correlation between the measured rate of sand transport and the friction velocity is shown in Figure 17.3. There is a lot of scatter in the data points because of the factors discussed previously. Further discussion of the relationship between threshold

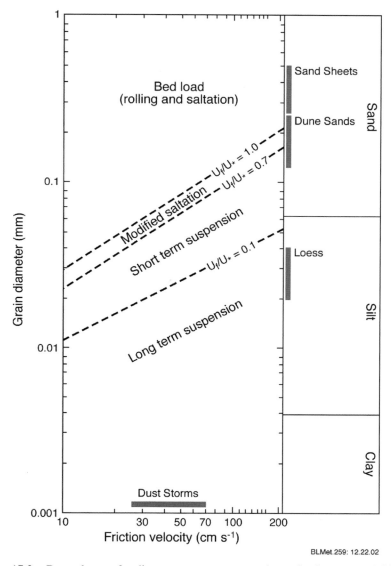

Figure 17.3 Dependence of sediment transport type on the grain diameter and friction velocity. The U_f is the settling velocity, and U^* is the friction velocity. The modified saltation region represents a boundary zone between suspension and saltation processes. Adapted from Tsoar and Pye (1987).

friction velocity and surface characteristics, such as soil particle size and surface roughness, can be found in Marticorena et al. (1997).

Convective- and turbulence-scale variations in the wind speed have a strong effect on sediment transport, as can be anticipated from Figure 17.4. Convective-scale "gust fronts" are caused by downdrafts emanating from the base of precipitating convective clouds, and these wind speed spikes are likely responsible for much of the elevated sand and dust associated with haboobs. Also, turbulence intensity is strong near the surface when wind speeds are high, whether from synoptic-scale or mesoscale forcing, and the turbulent eddies produce much irregularity in the wind speed and direction. So, even though the

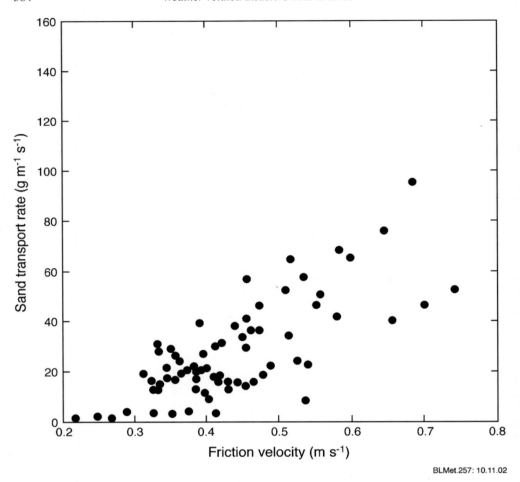

Figure 17.4 Measured relationship between friction velocity and sand transport rate. Adapted from Wiggs (1997).

time-averaged wind speed, as might be reported by an anemometer, is below the threshold for particle entrainment into the airstream, there might still be considerable transport as a result of the wind speed peaks. This is illustrated in Figure 17.5 in terms of the correlation between the measured friction velocity and sediment transport rate during a 2-minute period. During this short time interval, the sediment transport rate varied by a factor of ten.

It should be noted that the near-surface transport of large grains, even though perhaps not as visually dramatic as the kilometers-deep clouds of dust, may represent the greatest hazard of the storm. That is, the airborne sediment load near the ground is what suffocates or buries those who are trying to survive.

Electrical effects within dust and sand storms

It is well known that lightning discharges occur in the dust clouds over volcanic eruptions, and there is evidence of lightning within large dust storms on other planets (Uman, 1987). Within dust storms and sand storms on Earth, there have been numerous scientific observations of strong atmospheric electrical charges (e.g., Kamra, 1972), as well much anecdotal

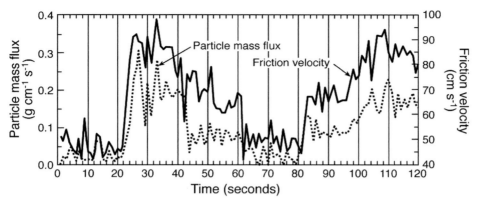

Figure 17.5 Correlation between measured sediment transport rate and friction velocity for a 2-minute period. Adapted from Butterfield (1993).

data on the response of humans and equipment to electrical effects in deserts. Stowe (1969) reported potential gradients of 20,000 to 200,000 volts per meter in the lowest meter above the ground in a Saharan dust storm. Within sand storms, point discharges of up to 1 m in length have been observed (Uman, 1987), and Kamra (1969) reported a "feeble lightning discharge" within a dust storm in India. Note that these electrical phenomena can occur within dust storms that are not associated with cumulus clouds, and thus, any discharges observed are not related to the traditional mechanisms for the production of lightning in thunderstorms.

There are numerous personal accounts of the electrical effects of dust storms and sand storms. Clements et al. (1963) reported that ignition systems of automobiles have failed to work unless the frame was grounded, telephone systems have failed, and workman have been knocked to the ground by electrostatic voltages that have developed. George (1977) reported an experience wherein walking in a sand storm produced an extreme headache that was largely relieved when a metal rod was used as a walking stick (and an electrical ground).

Dust storm hazards

Dust storms have detrimental physiological effects, and can represent hazards to road, rail, and air transportation. A few of the many possible examples, in the latter category, follow. Thirty-two multiple-vehicle traffic accidents resulted from haboob-related dust storms on Interstate Route 10 in Arizona, in North America, between 1968 and 1975 (Brazel and Hsu, 1981). The safety and efficiency of the aviation system is also seriously impaired by dust storms. Severe prefrontal dust storms in 1988 in South Australia caused many airport closures (Crooks and Cowan, 1993). In 1973, a Royal Jordanian Airlines aircraft crashed in Nigeria, with 176 fatalities, as a result of thick, harmatten-related dust (Pye, 1987). A very severe dust storm, called a "black storm" in China, occurred in northwestern China in 1993. The direct economic costs resulting from structural damage to buildings, broken power lines, traffic accidents, and so on were equivalent to US $66 million. In addition, there were 85 fatalities, 31 missing, and 264 injured. Over long periods, blowing sand can cause serious abrasive damage to wood structures such as telephone and electric poles. The greatest abrasion is often found at a distance of about 0.20 to 0.25 m AGL, that is, the height of the optimal combination of wind speed and particle concentration (Oke, 1987).

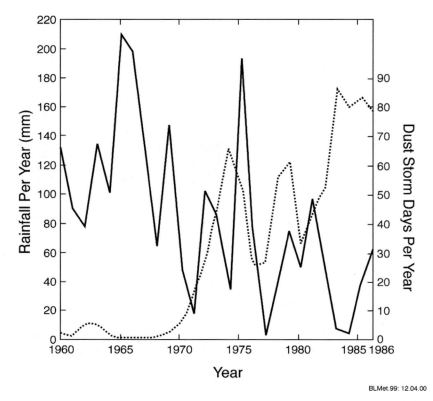

Figure 17.6 The 25-year annual total rainfall trend (*solid line*) plotted versus the number of dust storm days per year (*dotted line*) for Nouakchott, Mauritania. From Middleton et al. (1986).

Surface and meteorological conditions: geographic areas favorable for the occurrence of dust and sand storms

Dust production is most favorable for surfaces that are free of vegetation; do not have crusts; are composed of sand and silt, but little clay; have been disturbed by human or animal activity; are relatively dry; and are free of chemical (e.g., salts) or organic particle cements (Middleton, 1997). Such conditions are often found where recent climate change, water action, or human activity have mobilized or concentrated dust or sand. Significant natural dust sources are floodplains, alluvial fans, wadis, glacial outwash plains, salt pans, other desert depressions, former lake beds, active dunes, devegetated fossil dunes, and loess (Middleton, 1997). Many of these dust-conducive conditions are caused by the action of water.

The relationship between recent or long-term precipitation amounts and entrainment of dust has been extensively studied (e.g., Brazel et al., 1986). Goudie (1983a, 1983b), for example, found that dust storm frequencies are highest when mean annual precipitation is in the 100 to 200 mm range. Greater precipitation would generally cause (1) more vegetation, which would bind the soil and reduce the wind speed near the surface; and (2) a more cohesive moist soil. Where precipitation is below 100 mm, human disruption of the soil might not be as great as it is in the 100 to 200 mm range, where marginal agriculture is attempted at desert edges (Pye, 1987). There are also clear correlations between short- and long-term droughts and dust storm frequency. For example, Figure 17.6 shows the

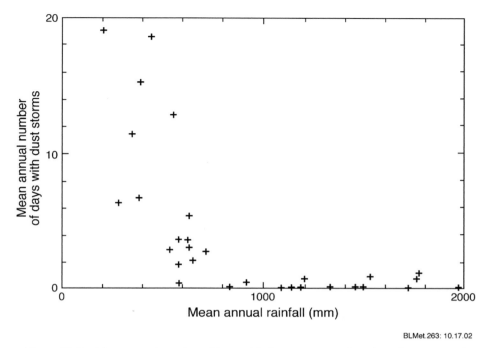

Figure 17.7 Mean annual number of days with dust storms as a function of mean annual precipitation, for thirty stations in China. Adapted from Goudie (1978).

25-year rainfall trend plotted versus the number of dust storm days per year for Nouakchott, Mauritania. Increased anthropogenic disturbance of the surface during the period might have contributed to the rising number of dust storms, but the rainfall deficit was likely the major factor. Similar plots for other locations are found in Goudie and Middleton (1992). Figure 17.7 shows the appearance of a rough threshold effect in the relationship between dust storm frequency and mean annual precipitation for different locations in China. Goudie (1978) and Littmann (1991) showed similar plots for other locations, with all illustrating a negative correlation between mean annual precipitation and dust storm occurrence. Some have a similar threshold in the relationship. Holcombe et al. (1997) also showed the effect of prior precipitation on threshold wind velocities.

In contemporary times, many geographic areas have been under sufficient climate stress and anthropogenic stress (e.g., agriculture) to have experienced at least mild to moderately severe dust storms. Some areas experience such storms with sufficient regularity that special names are associated with them. Middleton (1986) listed seventy such local names, and Middleton (1997) noted the geographic areas where the storms are most frequent. Figure 17.8 shows the major global dust trajectories that result from various atmospheric processes. In the desert Southwest of North America, haboobs are a major source, while farther to the north high-speed downslope winds (katabatic winds) are important. On the larger scale, strong pressure gradients around fronts and surface cyclones entrain dust in the semiarid Great Plains. In South America, strong upper-level westerly winds may mix down to the surface, for example, through the dynamic effects of the Andes Mountains, to elevate dust. Asia and Africa are major dust storm regions, with fronts and surface cyclones being responsible for many of the storms. Monsoonal winds, katabatic winds, and haboobs are locally important. The dust storm day frequencies for some of the source regions indicated

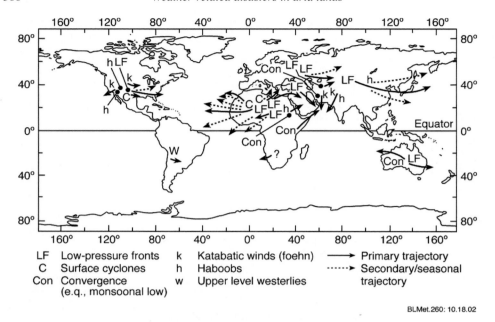

Figure 17.8 Major global dust trajectories and the various atmospheric processes that cause the dust storms. From Middleton et al. (1986).

in Figure 17.8 range from twenty to eighty per year. Further discussion of favored dust storm regions may be found in Middleton (1997), Goudie (1978), and Littmann (1991).

Dust storms often have a strong seasonal variation in their occurrence (Littmann, 1991) that is related to the annual distribution of precipitation (Yu et al., 1993), vegetation cover, freezing of the soil, meteorological systems with high winds, and agricultural activities that disturb the surface. Figure 17.9 shows the number of official meteorological reports of dust at 750 stations for a 6-year period in Asia. For this area, the maximum monthly dust frequency in spring is about five times greater than in other seasons of the year. The minimum in winter is because of frozen ground and snow cover. Duce (1995) showed the monthly distribution of dust storms for selected locations in Australia, illustrating about a fivefold greater frequency in summer. A factor of five to ten greater area of global ultraviolet-absorbing aerosol has been measured for summer relative to winter (Herman et al., 1997). There is also a distinct diurnal maximum in dust storm frequency in many places, when strong afternoon convection mixes higher-speed air to the surface or when thunderstorms cause haboobs (Hinds and Hoidale, 1975). Membery (1985) showed the diurnal occurrence of dust storms in the Tigris–Euphrates flood plain in the summer (Figure 17.10). Kuwait, near this source, commonly experiences low visibilities between about 1200 and 1900 LT. Bahrain experiences low visibilities from Iraq's dust in the early morning, after it has been transported the 500 km to the southeast by the northwesterly shamal (Houseman, 1961; Khalaf and Al-Hashash, 1983).

Description of a haboob, a convection-generated dust storm

In a mature thunderstorm, the latent heat generated by condensation causes buoyancy-driven upward motion in only part of the cloud. In another part of the cloud, the frictional drag of the downrush of rain and hail causes air within some areas of the cloud to subside. This subsidence is enhanced because the rain falling through unsaturated air below the

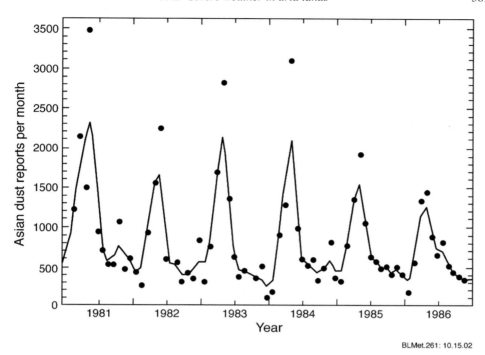

Figure 17.9 Number of monthly official meteorological reports of dust for a 6-year period in Asia. From Prospero et al. (1989).

cloud cools and becomes more dense. When the cool, dense rapidly subsiding air (called a "downburst") reaches the ground, it moves outward away from the storm that created it, flowing rapidly along the ground as a density current. Figure 17.11 shows a cross-section schematic of the cool outflow, with the leading edge called an "outflow boundary," that is propagating ahead of a mature thunderstorm. The strong, gusty winds that prevail at the boundary are defined as a "gust front." The high wind speeds within this cold outflow region entrain dust into the atmosphere, and the turbulence associated with the large vertical shear of the wind mixes the dust upward, sometimes through a layer thousands of meters deep. The "dust front" propagates at up to 15 m/s^{-1}, with an associated advancing, seemingly opaque, near-vertical wall of dust that can merge at its top with the dark water cloud of the thunderstorm. There is sometimes lightning visible through the dust cloud and above it, probably related to the thunderstorm, but possibly also to the electrification of the dust cloud itself. When the dust storm arrives, visibility can be reduced to zero, but it is commonly 500 to 1,000 m. Additional information about the outflow boundaries that are responsible for haboobs can be found in Wilson and Wakimoto (2001), Wakimoto (1982), and Droegemeier and Wilhelmson (1987).

Typical outflow boundaries last less than an hour, and this time scale thus represents the lifetime of most haboobs. This was confirmed in a study by Freeman (1952) of eighty-two haboobs in the Sudan. There are rare haboobs, however, that last much longer. A description of two long-lasting ones on the eastern Arabian Peninsula is provided by Membery (1985). Both occurred within a 2-day period, reduced visibility to less than 500 m for more than 4 hours, had propagation speeds of about 15 to 20 m/s, and produced dust storms that lasted for at least 9 hours as they traversed hundreds of kilometers. That the motion of these dust storms resulted from the propagation of a density current is supported by the observed

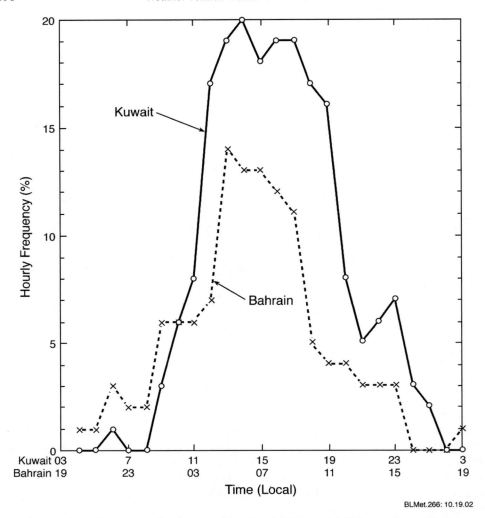

Figure 17.10 Frequency distribution of hourly visibilities of 1,500 m or less at Kuwait (near the dust source in the Tigris–Euphrates Valley), and of visibilities of 3,000 m or less at Bahrain, about 500 km to the southeast, for June, July, and August 1982. These data are representative of the conditions in other years. Note that the Bahrain time axis is shifted from the Kuwait axis to account for the typical advective times scales associated with the dust transport from Iraq to Bahrain in the northwest summer wind. Adapted from Membery (1985).

vertical profile of wind speed (Figure 17.12). The large-scale wind speed at the 850-mb level above the cold density current was only 2 m/s, whereas within the density current the air was moving at 15 m/s.

Mitigation of dust storms

Human impacts have made some places more like deserts, with the effect of increasing dust storm frequency and severity (see the next section). Thus, reversing the various factors that have lead to desertification would contribute to a reduction in dust storms. There are also situations where dust storms from naturally occurring deserts affect populated areas,

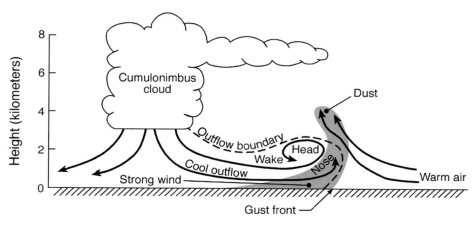

Figure 17.11 Cross-section schematic of a haboob caused by the cool outflow from a thunderstorm, with the leading edge that is propagating ahead of the storm called an outflow boundary. The strong, gusty winds that prevail at the boundary are defined as a gust front. The leading edge of the cool air is called the nose, and the upward-protruding part of the feature is referred to as the head. Behind the roll in the wind field at the leading edge is a turbulent wake. The rapidly moving cool air and the gustiness at the gust front raise dust (*shaded*) high into the atmosphere.

and measures have been taken to reduce their impact. For example, when cold fronts move southward from Siberia, the strong frontal winds entrain dust from the Gobi and Taklamakan deserts, and the dust-laden air causes air quality problems in downwind urban areas such as Beijing. With the intent of reducing the entrainment of dust at its source, and dust transport in the lower boundary layer, a massive forestation project was undertaken wherein 300 million trees were planted throughout northern China in the 1950s. Figure 17.13 illustrates the geographic extent of the forestation project, known as the "Great Green Wall." The map is indicative of only the general area of tree planting because clearly trees will not grow in the central Taklamakan Desert. In any case, the intent of planting the new vegetation was to reduce the wind speed near the ground and increase soil moisture, and thus reduce the entrainment of dust in the source regions. Also, the trees would reduce the low-level wind speed downwind of the dust source, and thus, some of the dust might more quickly settle out of the atmosphere. After the forestation, the dust storm frequency and duration in Beijing decreased dramatically. The frequency decreased from 10 to 15 events per month to almost none, and the duration decreased from longer than 50 hours to less than 10 (Figure 17.14). During this period, the overall rainfall and wind speed climate for the upwind area did not change appreciably. At Bayinmaodao, near the upwind, northern edge of the forested area, there was little change during this period in the dust storms' frequency and duration. There have been many other more modest attempts to modify the land surface in order to reduce the severity of dust storms.

17.2.2 Rainstorms, floods, and debris flows

Although desert weather is considered inherently severe, primarily because of a lack of rain, weather events that mitigate the dryness are ironically also often severe, both meteorologically and hydrologically. When rain does reach the ground, the surface is relatively impervious to water and devoid of vegetation that is sufficient to slow the runoff. The

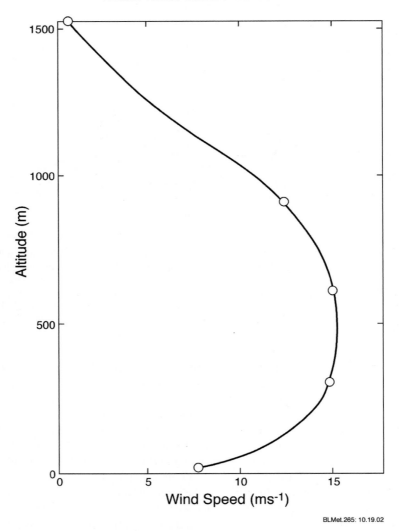

Figure 17.12 Vertical profile of the low-level wind speed within a density current in the eastern Arabian Peninsula. From Membery (1985).

rapid concentration of this runoff from thunderstorms causes flash floods that account for considerable loss of life. It was noted previously that more people currently perish in the desert from an excess of surface water (i.e., drowning in floods) than from a lack of it (i.e., thirst) (Nir, 1974). In general, compared to floods in humid climates, the river hydrographs[1] rise more quickly, have a sharper peak, and recede more quickly. As evidence of the "flashiness"[2] of arid watersheds, in the United States the twelth largest flash floods all occurred in arid or semiarid areas (Costa, 1987), and more than half of railroad track washouts occur in the driest states. Figure 17.15 illustrates hydrographs that are typical of desert flash floods, even though the rate of rise and fall of the curve and the width of the peak

---

[1] A hydrograph shows river discharge (volume per unit time) plotted as a function of time.
[2] This refers to how prone the watershed is to the occurrence of flash floods. One definition of a flash flood is an event in which river discharge rises from normal to flood level within 6 hours.

17.2 Severe weather in arid lands

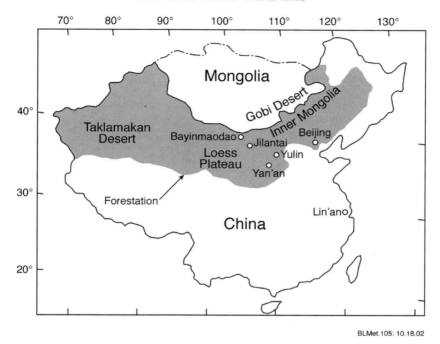

Figure 17.13 Area forested with 300 million trees in the 1950's (*shaded*) in order to reduce dust transport southward into urban areas. The locations shown are dust monitoring stations. From Parungo et al. (1994).

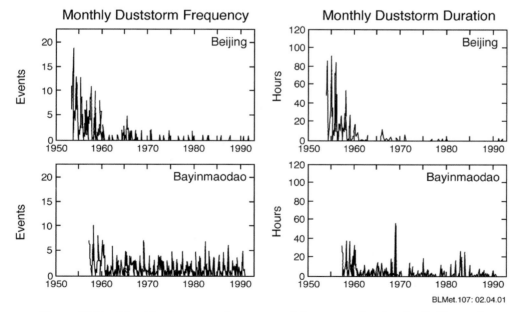

Figure 17.14 Monthly dust storm frequency and dust storm duration for Beijing and Bayinmaodao, China. See Figure 17.13 for locations. From Parungo et al. (1994).

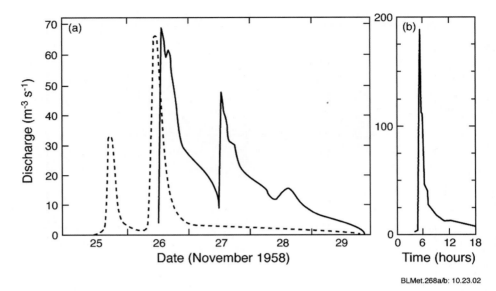

Figure 17.15   Discharge as a function of time (hydrograph) for floods in (a) two central Sahara wadis associated with the same storm (adapted from Goudie and Wilkinson, 1977); and (b) Tanque Verde Creek, Arizona, in the northern Sonoran Desert (adapted from Hjalmarson, 1984).

depend on many factors such as the size of the contributing watershed. Other examples of flood hydrographs in arid lands are found in Reid and Frostick (1997) and Knighton and Nanson (1997). It should not be assumed that most floods and flash floods in desert areas result from thunderstorm rainfall. For example, Kahana et al. (2002) showed that 80% of the most severe floods in the Negev Desert during a 30-year period were associated with synoptic-scale precipitation events.

In desert floods, there is a spectrum of material involved. Some floods can consist almost exclusively of water, with little solid material. Others can be composed of 90% mud, boulders, and vegetation. Fifty-ton boulders have been carried for many kilometers (Childs, 2002). Toward the latter end of the spectrum, the floods are called mud flows or debris flows. A single flood can have both types of flows at different stages—mostly water and mostly debris. In general, for a variety of reasons, the sediment load in the flood water in ephemeral desert rivers is often many times higher than in rivers in humid climates. This is shown in Figure 17.16, where the sediment content in runoff is plotted against mean annual precipitation. The curves are extrapolated to show low sediment yield for extremely arid climates simply because the sediment must go to zero as the rainfall does.

When the flood material consists mostly of debris, it is much more viscous than water and moves much less rapidly, almost with the consistency of wet cement. In large debris flows, the leading wall is almost vertical and can be 10 to 15 m high. After the flow has exited its wadi or canyon, and spent its energy as it spread across a flood plain, it can dry with its near-vertical wall intact, sometimes 5 m in height. Some floods have been known to run 400 to 500 km into the desert plains before coming to rest (Peel, 1975). Personal accounts of desert floods can be found in Childs (2002), Hassan (1990), and Jahns (1949).

The response of the surface-hydrologic system to rainfall in the desert is complex, almost precluding hydrologic modeling (Reid and Frostick, 1997). It first should be recognized that most floods in the desert occur because of the unusual character of the surface rather

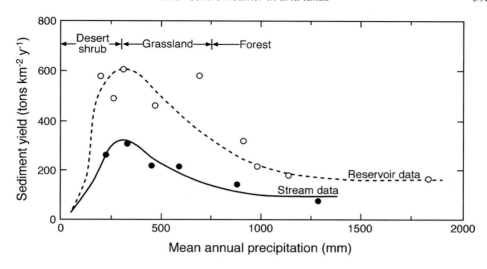

Figure 17.16 Sediment yield in runoff, plotted as a function of mean annual precipitation. The empty circles are data based on rates of sediment accumulation in reservoirs, and the solid circles are based on sediment content of stream water. From Langbein and Schumm (1958).

than of the rainfall. That is, the rain is not likely of much greater intensity than what would be experienced from a similar type of storm in more humid areas. Regarding the surface, however, runoff of rainfall can be greater in deserts because there is less organic matter in the soil to absorb the rainfall; the lack of vegetation means that raindrop impact can seal the soil surface; the surface can be composed of impervious rock; in the open desert there are few animals, insects, and worms that make the soil permeable; and chemical action can produce impervious layers (e.g., caliche). An illustration of the complexity of the flood generation process is how runoff is influenced by slope in the desert. Many studies have shown that slopes of 1 to 2 degrees generate twice the runoff per unit area than do slopes of 20 degrees. One contributing factor is that steeper slopes have had much of the fine material washed away, whereas this accumulated fine material on more horizontal surfaces permits the development of low permeability crusts. Another counterintuitive observation is that rain intensities in excess of 15 mm/h can produce less proportional runoff than does a rainfall of 3 to 10 mm/h. Again, the factor is surface sealing related in that heavy rainfalls can break up certain crusts (Peel, 1975). Even though these and other factors make it perhaps difficult to be quantitative about desert runoff and flood production, the fact remains that flash floods are much more common in the desert, in general, because more water runs off.

Other factors help determine the severity of floods that results from desert rain events. If the storm movement and the channel routing conspire so that the water arrives nearly simultaneously in the main channel from many directions, a major flood will result. In contrast, if the tributaries feed the water into the main channel over a longer period of time, the flood will not be as great. This factor is in addition to the obvious one that storms that remain stationary for significant periods and pour water onto the same watershed are more dangerous than storms that move rapidly and spread the water over a larger area.

Flood frequency and severity in the desert vary from year to year as much as does the rainfall that causes the floods. One of the cyclic patterns of the atmosphere that strongly controls rainfall and floods in some desert areas is the EL Niño Southern Oscillation (ENSO) cycle. For example, eleven of the forty major floods of the past century in the desert Southwest of North America occurred in El Niño years (Childs, 2002). On longer time scales, Ely et al. (1993) showed that, over the past 5,000 years, by far the greatest number of floods in Arizona (northern Sonoran Desert) and southern Utah (Great Basin Desert) occurred in the last 200 years when there have been more frequent strong El Niño events.

17.3 Desertification

If the inability of the land and/or the climate to sustain abundant life is the primary criterion for naming a place desert, there can be no argument that many areas are now more desertlike than they were in past times. Of the effects of natural processes such as changes in climate, it is clear from archaeological evidence that animals that are now found only in humid Africa, and humans that hunted them, once thrived in areas that are now part of the Sahara. Of anthropogenic causes, one need only recall images of the midwestern United States in the 1930s, where plows exposed the underbelly of the natural High Plains grassland to winds, and the land ceased to produce anything for years except meters-deep piles of dust. In the same vein, Bingham (1996) referred to "virtual deserts" of human origin, such as urban areas, acid rain dead lakes, charred rainforests, and rivers and oceans depleted of life by poison and exploitation. He pointed out some of the ramifications of human-caused desertification over the past few millennia.

Even Robert Malthus, the great pessimist who predicted that population would outrun the productivity of land, did not factor in the active and evidently permanent destruction of its underlying fertility, but that destruction has occurred so uniformly in so many otherwise unrelated situations that one might suspect that desertification is the dark companion of all human progress.

Much of the rich history of the Mediterranean was bought at the price of desert, from the vanished cedars of Lebanon to the long-barren fields of Carthage and the naked Illyrian coast, whence came, once upon a time, the masts of Venetian ships. The legendary Timbuktu, the vanished myrrh-rich forests of Arabia and Yemen, the bitter shriveled Aral Sea, the lost gardens of Babylon, the scorced plains of Sind, the scoured fringes of the Gobi, the once-teeming veldt of southern Africa, and, of course, the vanished cattle and sheep empires of the American West all testify to the amazing diligence of humankind.

Desert lands interact with the atmosphere very differently than do those that are more moist and more vegetated. Thus, surfaces that have been temporarily or permanently rendered more dry and barren of vegetation by either natural or anthropogenic forces, that is, desertified, will have different associated microclimates and boundary layer properties. The question of the feedbacks among surface conditions, the regional climate, and the human response to the physical system is thus important to the topic of desert meteorology. Literature abounds on this subject of desertification, but regardless of whether it is caused by climate, humans, or both in concert, the existence of a local effect on the atmosphere is without dispute. It should be noted that more has been written about desertification than about any other environmental degradation process, or disaster, probably because the actual and potential human consequences are so terribly immense.

17.3.1 What is desertification?

The working definition of desertification to be used here is intuitive and simple; it is the development of any property of the climate or land surface that is more characteristic of a desert, whether the change is natural or anthropogenic. More formal and elaborate definitions are plentiful, and include a change in the character of land to a more desertic condition involving the impoverishment of ecosystems as evidenced in reduced biological productivity and accelerated deterioration of soils and in an associated impoverishment of dependent human livelihood systems (Mabbutt, 1978a, 1978b). Glantz and Orlovsky (1983), Odingo (1990), and Rozanov (1990) provided a thorough review of various definitions, where the former source lists more than 100. A popular and encompassing one is offered by Dregne (1977): desertification is the impoverishment of arid, semiarid, and some subhumid ecosystems by the combined impact of human activities and drought. It is the process of change in these ecosystems that can be measured by reduced productivity of desirable plants, alterations in the biomass and the diversity of the micro- and macrofauna and flora, accelerated soil deterioration, and increased hazards for human occupancy. In this definition is included deterioration in the soil or the biomass, by natural and/or human processes, in arid as well as subhumid areas. He later expanded the definition even further, referring to the "impoverishment of terrestrial ecosystem," in general, thus allowing the term "desertification" to be used even for humid climates (Dregne, 1985). Dregne (1977) pointed out that desertification in the more subhumid zones is of greater importance than in truly arid areas because the impact is greater—a greater number of people are affected, there is more soil to be lost, and there is more of a local economy to be impacted. Such liberal extension of the term "desertification" to humid areas is in keeping with the first use of the term by Aubreville (1949), who referred to the creation of desert conditions by deforestation in humid parts of west Africa.

A caveat that is sometimes implicit in the definition of desertification is that the process leads to long-lasting, and possibly irreversible, desertlike conditions (Hellden, 1988). Thus, the reduction in vegetation along a desert margin, associated with normal interannual or interdecadal rainfall variability, would not be referred to as desertification unless, perhaps, resulting wind or water erosion of the soil was so great that the process was not reversible on similar time scales—that is, vegetation did not return with the rains. The term "irreversible" should be used carefully because soils can regenerate on geological time scales, and it would be a rare situation indeed where life did not return in some form if the climate eventually became more favorable or the negative human impact ceased.

Desertification is popularly thought of as a problem of only Third World countries, such as those of the Sahel. Indeed, if the previously mentioned degree of impoverishment of the human condition is a primary criterion, it is mainly a Third World problem. Without the social support infrastructure that is common in many prosperous countries, starvation and malnutrition of millions of people result. Nevertheless, there is significant desertification in First World countries, and it has many immediate implications. It portends very sobering long-range problems in areas such as food and energy supply, the environment, the economy, and the quality of life. Even though Sears (1935) and others make the distinction between desertification, which is caused by humans, and desertization, which is natural, we adopt the former term for both processes.

A list of basic indicators of the desertification process is provided by Grainger (1990): ground conditions, climatic indicators, data on agricultural production, and socioeconomic indicators. A more specific list of major symptoms of desertification includes the following (Sheridan, 1981):

- Declining water tables
- Salinization of top soil and water
- Reduction of surface waters
- Reduction of substrate moisture content

- Unnaturally high soil erosion (by wind and water)
- Desolation of native vegetation

Additional signs are loss of soil organic matter, a decline in soil nutrient levels, and soil acidification (Bullock and Le Houérou, 1996). There are many geographic areas that exhibit all six symptoms, but any one symptom is sufficient to qualify the area as undergoing desertification. The overall process often becomes first apparent in one aspect, but frequently progresses to more. For example, salinization or drying of the substrate may lead to a reduction of the vegetation cover, which permits greater erosion by wind and water. The wind-driven sand can then be responsible for stripping remaining vegetation and for further scouring the surface.

Some erroneously imagine desertification as an advance of the desert into surrounding less desertlike areas. Rather, it seems to be a situation in which marginal areas, near a desert but not necessarily on its edge, degrade as a result of the combined effects of drought and human pressures of various sorts. Such areas will possibly grow and merge with each other, to the point that the patchy zones of destruction will become part of a nearby desert. Except for the special case of drifting desert sand encroaching on surrounding nondesert, this is closer to the real process of desertification (Dregne and Tucker, 1988). The concept is further complicated by the fact that deserts exist in degrees, so that the further impoverishment of the vegetation in an existing vegetation-impoverished desert would also be rightly called desertification. This has been discussed in the context of anthropogenic change, where, for example, the Sonoran and Chihuahuan deserts have become markedly more barren of vegetation and wildlife in the past 100 years.

As noted previously, there have been numerous attempts at quantifying and mapping the status, or degree of, desertification. For example, Dregne (1977) presented a classification system with the following subjective criteria that are generally related to vegetation, erosion, and salinity:

Slight desertification

- Little or no deterioration of the plant cover and soil has occurred.

Moderate desertification

- Plant cover has deteriorated to fair range condition.
- Hummocks, small dunes, or small gullies indicate that accelerated wind or water erosion has occurred.
- Soil salinity has reduced crop yields by 10% to 50%.

Severe desertification

- Undesirable forbs and shrubs have replaced desirable grasses or have spread to such an extent that they dominate the flora.
- Wind and water erosion have largely denuded the land of vegetation, or gullies are present.
- Salinity that is controllable by drainage and leaching has reduced crop yields by more than 50%.

Very severe desertification

- Large, shifting, barren sand dunes have formed.
- Large, deep, and numerous gullies have developed.
- Salt crusts have developed on nearly impermeable irrigated soils.

It is important to recognize that these criteria refer to a change in the state of the system, are evidence of a process, and do not reflect the actual condition. Thus, there must be some

17.3 Desertification

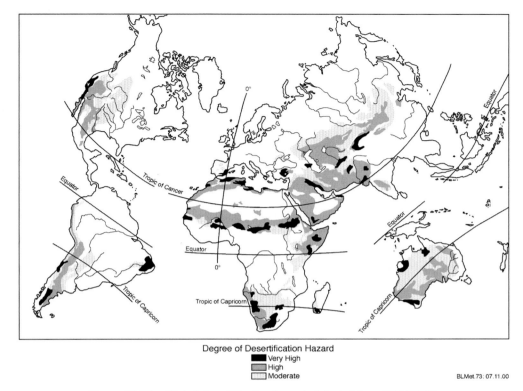

Figure 17.17 Global desertification hazard. Adapted from UN (1977).

period of time implied, during which these changes have occurred to some base or reference state. For the maps of Dregne (1977) that show the worldwide status of desertification, this original pristine state is that which prevailed before human impact, a condition that is more difficult to estimate for the Old World through which so many civilizations have passed. In addition, if the desertification definition is limited to human impacts, the assumption is that the climate has been stable for that period. In Dregne's classification, areas of extreme aridity were automatically placed in the "slight desertification" category because the reference state was already extremely arid, and only slight additional degradation of the vegetation and soil was possible. Indicators of desertification have been discussed by many authors, including Reining (1978) and Warren and Maizels (1977).

Many of the definitions of desertification agree that, most often, the process occurs when human and natural pressures on the land coincide, for example, during periods of drought. Thus, the natural drought cycle can be one of the contributors. In contrast, significant interannual variability in the rainfall apparently can reduce the potential for desertification if it discourages agricultural use of an area. For example, Dregne (1986) pointed out that desertification has been limited to a moderate level in semiarid northeastern Brazil by the large interannual variability in the rainfall there. That is, if droughts had been less prevalent, there would have been more agricultural pressure and land degradation.

17.3.2 Extent of desertification

Some desertification maps depict the areas that are susceptible to desertification, whereas others show the areas that are actually undergoing the process. Figure 17.17 displays

Table 17.1. *Extent of desertification of different degrees of severity, for different continents*

Region	Light Area		Moderate Area		Strong Area		Severe Area	
Africa	1,180	(9%)	1,272	(10%)	707	(5.0%)	35	(0.2%)
Asia	1,567	(9%)	1,701	(10%)	430	(3.0%)	5	(0.1%)
Australia	836	(13%)	24	(4%)	11	(0.2%)	4	(0.1%)
N. America	134	(2%)	588	(8%)	73	(0.1%)	0	(0.0%)
S. America	418	(8%)	311	(6%)	62	(1.2%)	0	(0.0%)
Total	4,273	(8%)	4,703	(9%)	1,301	(2.5%)	75	(0.1%)

The area is given in 1,000 km^2, and the percent area is the fraction of all drylands in that region.
After Bullock and Le Houérou (1996); based on Oldeman et al. (1990), Le Houérou (1992), Le Houérou et al. (1993), and UNEP (1992).

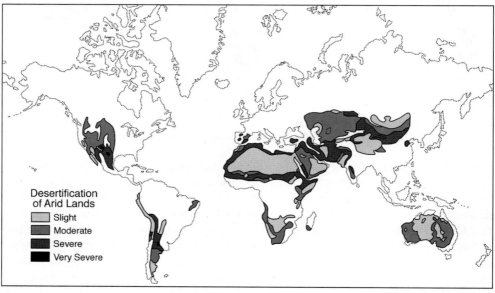

Figure 17.18   Desertification severity. From Dregne (1983).

an estimate of desertification hazard that is based on climate conditions, the inherent vulnerability of the land, and the human pressures. Based on this map, at least 35% of Earth's land surface is now threatened by desertification, an area that represents places inhabited by 20% of the world population. There are also a number of estimates of the degree of actual worldwide desertification (not desertification hazard). Examples of such desertification severity maps can be found in UNEP (1992), Dregne (1977), and Dregne (1983), with the latter depicted in Figure 17.18 . Table 17.1 lists the areas and percents of lands in different regions with light, moderate, strong, and severe desertification. Africa and Asia are comparable in terms of the percent of the land in each severity category that has been desertified. Of the remaining areas, North America has the greatest percent of its arid land with moderate or worse desertification, and this is followed by South America. If all severity categories are included, the order of decreasing percent of the area affected

17.3 Desertification

Table 17.2. *Areas suffering at least moderate desertification (1,000 km²), by region*

Region	Rainfed Cropland Area		Irrigated Cropland Area		Rangeland Area		Total Area	
Africa	396	(23%)	14	(5%)	10,268	(34%)	10,678	(33%)
Asia	912	(53%)	206	(76%)	10,889	(36%)	12,007	(37%)
Australia	15	(1%)	2	(1%)	3,070	(10%)	3,087	(10%)
Europe (Spain)	42	(2%)	9	(3%)	155	(1%)	206	(1%)
N. America and Mexico	247	(14%)	28	(10%)	2,910	(10%)	3,185	(10%)
S. America	119	(7%)	12	(5%)	3,194	(10%)	3,325	(10%)
Total	1,731	(100%)	271	(100%)	30,486	(100%)	32,488	(100%)

The numbers in parentheses are the percentage of total area.
From Dregne (1983).

Table 17.3. *Areas and numbers of people affected by at least moderate desertification, by region*

Region	Affected Area (1,000 km²)	Percent of Total Area	Affected Population (Millions)	Percent of Total Population
Africa	7,409	37%	108.00	38%
Asia	7,480	37%	123.00	44%
Australia	1,123	6%	0.23	0%
Med. Europe	296	1%	16.50	6%
N. America	2,080	10%	4.50	2%
S. America and Mexico	1,620	8%	29.00	10%
Total	20,008	100%	281.23	100%

From Mabbutt (1984).

is Africa, Asia, Australia, South America, and North America. Of the arid, semiarid, and dry subhumid climate zones that were considered susceptible to desertification, 70% of the area, representing 3,592 million hectares, has actually been affected (UNEP, 1992). A similar table can be found in Dregne (1983). Table 17.2 shows that rangeland is most affected by moderate or worse desertification, in comparison with rainfed and irrigated cropland. Seventy percent of the desertified area is in Africa and Asia. An estimate of the number of people affected by at least moderate desertification is provided in Table 17.3, with roughly four-fifths being in Africa and Asia. The physical causes of soil degradation, which is an eventual symptom of desertification, are estimated in Table 17.4. Even though this partitioning sheds no light on the complexity of factors that have led to the wind and water erosion, for example, the numbers are interesting. Both Africa and Asia have roughly equal contributions from wind and water erosion. Chemical degradation is primarily salinization, and physical effects include soil compaction. Note that these data do not represent areas with "light" degradation. The total areas reportedly affected by desertification on each continent depend considerably on the sources of data and the methods used, as indicated in these tables compiled by different investigators.

17.3.3 Anthropogenic contributions to desertification

Based on climatic data, more than one-third of the Earth's surface is desert or semidesert. If we go by data on the nature of soil and vegetation, the total area is some 43% of the Earth's

Table 17.4. *Soil degradation in the susceptible drylands by process and continent, excluding degradation in the light category (million ha)*

	Africa	Asia	Australia	Europe	N. Am.	S. Am.	Total
Water	90.6	107.9	2.1	41.7	28.1	21.9	292.3
Wind	81.8	72.7	0.1	37.3	35.2	8.1	235.2
Chemical	16.3	28.0	0.6	2.6	1.9	6.9	66.3
Physical	12.7	5.2	1.0	4.4	0.8	0.4	23.9
Total	201.4	213.8	3.8	86.0	66.0	37.3	617.7
Area susceptible dryland	1,286.0	1,671.8	663.3	299.7	732.4	516.0	5,169.2
Percent degraded	15.6%	12.8%	0.6%	28.6%	9.0%	7.2%	11.9%

From Thomas and Middleton (1994).

land surface. The difference is accounted for by the estimated extent of the manmade deserts (9.1 million km^2), an area larger than Brazil (UN, 1978). Most human contributions to desertification are related to agricultural exploitation of lands that are semiarid. Movement of agriculture into such areas whose natural condition is already agriculturally marginal may be caused by a number of economic or social factors. These include the need to provide food for a larger population, the decreasing productivity of existing agricultural land, and greed. We see that human attempts to use arid lands in ways that disregard their fragility generally result in the land becoming even more unproductive—perhaps even permanently damaged. Even land in nonarid climates that is sufficiently abused can take on the characteristics of a desert. For example, tillage of steep slopes can cause the topsoil to be lost by erosion during heavy rain events. Furthermore, pollution can poison the soil, leaving it bare for generations.

There is much literature that emphasizes the importance of the human contribution to the problem, for a wide range of geographic areas. The following are examples (Bullock and Le Houérou, 1996):

- China—Chao (1984a, 1984b)
- Australia—Perry (1977), Mabbut (1978a)
- South America—Soriano (1983)
- North America—Dregne (1983); Schlesinger et al. (1990)
- Europe—Lopez-Bermudez et al. (1984); Rubio (1987); Katsoulis and Tsangaris (1994); Puigdefabregas and Aguilera (1996); Quine et al. (1994)
- North and West Africa, and the Sahel—Pabot (1962); Le Houérou (1968, 1976, 1979); Depierre and Gillet (1971); Boudet (1972); Lamprey (1988); Nickling and Wolfe (1994); Westing (1994)
- East Africa—Lusigi (1981); Muturi (1994)
- South Africa—Acocks (1952); Hoffman and Cowling (1990); Bond et al. (1994); Dean and McDonald (1994)

Table 17.5 lists estimates of the various causes of desertification in different areas of the world. The following paragraphs discuss human activities that can contribute to the desertification process.

Exposing soils to erosion through plowing and overgrazing

Any process that enhances the wind or water erosion of topsoils can have an aridifying effect, especially in an area that is already semiarid. As such, plowing and overgrazing of semiarid lands worldwide have been a common cause of humanmade "natural" disasters that have

Table 17.5. *Human causes of desertification, in percent of desertified land*

Region	Over Cultivation	Over Stocking	Fuel and Wood Collection	Salinization	Urbanization	Other
Northwest China	45	16	18	2	3	16
North Africa and Near East	50	26	21	2	1	—
Sahel and East Africa	25	65	10	—	—	—
Middle Asia	10	62	—	9	10	9
United States	22	73	—	5	—	—
Australia	20	75	—	2	1	2

After Le Houérou (1992).

temporarily or permanently rendered the land relatively uninhabitable and desertlike. An example of one of the most rapidly occurring, if not the worst, such disaster is the evolution of the Great Plains of North America from a lush (by semiarid standards) grassland to a "dust bowl." During the 1910s and 1920s, rainfall was extraordinarily high in the area, and millions of acres of already overgrazed shortgrass prairie were plowed up and planted in wheat. This was done despite the knowledge that the 50-cm rainfall isohyet has historically wandered capriciously over a large distance between the Rocky Mountains and the Mississippi River. In fact, the semiarid grasslands of the Great Plains of west-central North America were known as the Great American Desert in the nineteenth century.

Exacerbating the problem was the pronouncements of the land companies and railroads that "rain follows the plow"—that by some miraculous physical mechanism, the rainfall would follow the farmer into the semidesert. In contrast to these optimistic, self-serving statements, a federally appointed Great Plains Committee evaluated the condition of the plains before the dust bowl debacle and made the following assessment: current methods of cultivation were so injuring the land that large areas were decreasingly productive even in good years, whereas in bad years they tended more and more to lapse into desert (Great Plains Committee, 1936). However, predictably, the wet years waned in the early 1930s, and the consequences were disastrous. The wheat stubble had been turned under, and the loose, dry soil lay exposed to the first of a long series of wind storms that began on 11 November 1933. Some farms had lost virtually all their topsoil by nightfall of the first day. During the second day, the dust-laden skies were black. As more storms stripped away the land's productivity, houses and machinery were buried under meters of silt, some of which reached high into the atmosphere to the jet stream, to be transported as far away as Europe. The drought and the damage continued throughout the decade. There are few that claim that the drought alone would have caused a similar calamity in an undisturbed prairie, without the unwitting collaboration of the farmer. Since this experience of the 1930s, droughts in the Great Plains in the 1950s and 1970s caused soil loss of equal or greater magnitude in some areas.

There are myriad other examples of such desertification by agricultural abuse, on both grand and small scales, throughout the world. A small-scale one, to contrast with the large-scale experience of the Great Plains, is the semiarid Rio Puerco Basin of New Mexico in North America. This area was so agriculturally productive 100 years ago that it was referred to as the "bread basket of New Mexico." But by then, the land had already begun to erode and desiccate as a result of long-term overgrazing. By 1950, the agricultural settlements had been completely abandoned, and the process of extraordinarily severe soil loss and arroyo deepening and widening continues today. Sheridan (1981) reported that the Rio Puerco supplies less than 10% of the Rio Grandes' water, but contributes more than 50% of its load of sediment. As with the attribution of the causes of the phenomenally rapid erosion of the Great Plains in the 1930s, there are those who claim that the climate of the Rio Puerco changed a bit in terms of the seasonal distribution of the rainfall, and that this could help explain the severe erosion during the past 100 years. Nevertheless, most analysts suggest that overgrazing was the primary cause, if not the only cause. Schlesinger et al. (1990) discussed a study in the same area in which desertification of grassland is attributed to the effects of overgrazing on an increase in the heterogeneity of soil resources, including water. This leads to an invasion of the native grasslands by shrubs. Another example of regional desertification is the semiarid San Joaquin Valley of California, one of the most productive agricultural areas in the world. Here, 80% of the privately owned rangeland is overgrazed (Sheridan, 1981).

Figure 17.19 illustrates the extent to which disturbing the natural vegetation and soil structure contributes to water and sediment runoff. In a semiarid area of Tanzania, the percent of rainwater lost to runoff and the amount of soil loss were compared for four areas of about the same size and slope. The ungrazed grassland and shrubland had very little loss of soil and water, whereas the area planted with millet and the area that was bare, for example, from overgrazing, had high rates of soil and water loss.

It is ironic that the "rain follows the plow" proclamation is not only untrue, but it also describes the opposite of the actual probable response of the Earth-atmosphere system. That is, plowing under natural vegetation and exposing bare ground produces effects that could diminish the rainfall. First, stripping the ground of vegetation diminishes its water retention capacity, allowing more rainfall to run off or evaporate rapidly. Also, bared ground freezes more quickly and deeply than ground covered by vegetation, and thus water is more likely to run off. These effects of natural vegetation removal will reduce soil moisture, and therefore evaporation from the soil. In addition, when the surface is plowed, there is no transpiration of water vapor by vegetation. This lower water vapor flux to the atmosphere by evaporation and transpiration could inhibit the development of rainfall. Thus, modification of semiarid land by agricultural activities can potentially affect the regional climate and make it less suitable for agriculture.

The topsoil loss that results from erosion of plowed land by wind and water can naturally limit the ability of vegetation to survive, even if the rainfall amounts are not affected. In the extreme, the topsoil loss may be total, and leave exposed only unproductive subsoils, or even bedrock. Nowhere in the world is there a greater abundance of extreme examples of desertification by soil erosion than in the Middle East, where civilization after civilization, and conflicts that have laid waste to civilizations, have taken their toll on the environment over the millennia. Lowdermilk (1953) described a typical example of the degradation that has taken place long before contemporary times:

We crossed the Jordan again (into Syria) into a region famous in Biblical times for its oaks, wheat fields, and well-nourished herds. We found the ruins of Jerash, one of the 10 cities of the Decapolis, and Jerash the second. Archaeologists tell us that Jerash was once the center of some 250,000 people. But today only a village of 3,000 marks this great center of culture, and the country about it is sparsely populated with seminomads. The ruins of this once-powerful city of Greek and Roman culture are buried to a depth of 13 feet with erosional debris washed from eroding slopes.... When we examined the slopes surrounding Jerash we found the soils washed off to bedrock in spite of rock-walled terraces. The soils washed off their slopes had lodged in the valleys.... Still farther north in Syria, we came upon a region where erosion had done its worst in an area of more than a million acres... French archaeologists, Father Mattern, and others found in this manmade desert more than

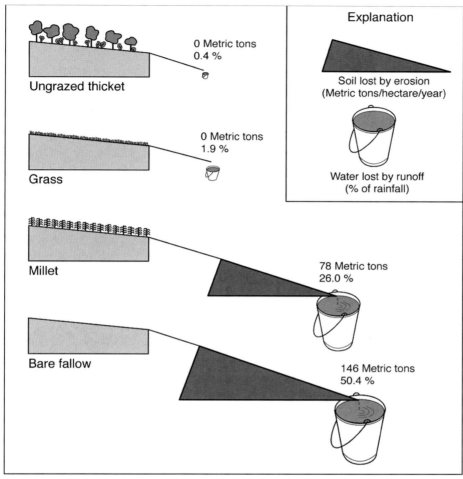

Figure 17.19 Water and soil loss from runoff in four areas of about the same size and slope in semiarid Tanzania. Adapted from Goudie and Wilkinson (1977).

100 dead cities. . . . Here, erosion had done its worst. If the soils had remained . . . the area might be repeopled again and the cities rebuilt. But now that the soils are gone, all is gone. (p. 9)

Salinization of soils through irrigation and excessive well pumping in coastal areas

Rainwater that provides for the needs of native vegetation and agricultural crops does not increase the salinity of the soil. The rainwater is relatively free of minerals because its source is atmospheric water vapor. However, irrigation water, whether it is obtained from groundwater or surface water, has dissolved minerals. Even when this groundwater or river water does not have an especially high original mineral content, the high evaporation rate in semiarid and arid regions causes the minerals to become more concentrated (1) as the water evaporates from the surface after it is applied; (2) while the water is in transit through long irrigation canals; or (3) while waiting for use, in impoundments. We know from previous discussions of the "oasis effect" that water surfaces or moist soils exposed to the ambient extremely dry, hot desert air evaporate very rapidly. The ultimate reservoir for the salts that remain is the soil from which the water evaporates. When these salt concentrations become excessive, agriculture can become prohibitive, and, unfortunately, even the original native vegetation can often no longer survive. An additional problem is that applying water to many arid soils causes salts already in the soil to rise toward the surface and concentrate. Salinization is especially a problem where downward drainage of mineralized irrigation water is blocked by an impermeable layer near the surface. After the land is sufficiently salinized that it is unable to support any vegetation, it thus becomes more susceptible to erosion by wind and water, and unable to provide any water vapor to the atmosphere through transpiration.

Large areas of semiarid land have reverted to a more barren state because of this effect. For example, more than 25% of the land of Iraq has been rendered unsuitable for producing native, or any other kind of, vegetation (Reisner, 1986). Eckholm (1976) pointed out the following:

The first recorded civilization, that of the Sumerians, was thriving in the southern Tigris–Euphretes valley by the fourth millennium B.C. Over the course of two thousand years, Sumerian irrigation practices ruined the soil so completely that it has not yet recovered.... Vast areas of southern Iraq today glisten like fields of freshly fallen snow.

There are many other examples throughout history of a decline in agriculture as a result of salinization. Thomas and Middleton (1994) listed the settlements of the Khorezm oasis in Uzbekistan, many oases in the Taklamakan Desert, and numerous other sites in China, as having been abandoned for this reason. In the Nile Valley, virtually the only agriculturally productive area of Egypt, yield reductions in excess of 25% are attributed to salinization (Dregne, 1986). It is interesting to note that the natural annual flooding of the Nile has historically flushed salts from the soil. This has allowed it to remain one of the world's most agriculturally productive and densely populated areas for thousands of years, in contrast to the situation in Iraq where the Tigris–Euphretes system does not flood the soil in a similar way. However, the Aswan Dam now prevents the natural soil-cleansing Nile flooding. In addition, some areas of the immensely productive San Joaquin Valley in North America are beginning to show lost productivity through salinization.

Salinization of soils also results from excessive pumping of groundwater near coastlines with the sea. When the (fresh) water table drops, hydraulic pressure allows salt water to encroach inland, and this results in soil salinization. Oman and Yemen, among other areas, are experiencing this problem (Grainger, 1990).

Groundwater mining

Groundwater mining is the use of nonrenewable, or only very slowly renewable, reservoirs of underground water for agricultural and other human needs in arid lands. This exploitation can lead to desertification through a few different mechanisms. Perhaps the most damage is done when the water is used for irrigation. When the water is gone, not only will the land lose its greenness and revert back to the desert that the rainfall statistics say that it is, but also the condition of the desert will be the worse for its brief impersonation of farmland. The soil will be more saline, the monoculture style of farming will have made it easier for invader species to replace native ones, the lower water table will make it impossible for some deep-rooted native plants to survive as they once did, streams and marshes and wet salt flats that relied on a high water table will not return, and the loss of soil through wind and water erosion of the bare surface will mean that it is less hospitable for the original vegetation. If the desert surface is not physically disturbed to a great degree, as is the case if the wells are used only to supply water for human consumption in urban

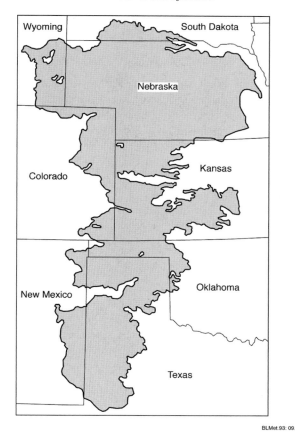

Figure 17.20 Geographic extent of the Ogallala Aquifer in North America. Adapted from Gutentag and Weeks (1980).

areas, the permanent damage may be less severe. However, there are still the impacts of the lowered water table on streams and vegetation. In addition, the ground compacts and settles as water is removed from its interstitial spaces, and this means that the aquifer may never again naturally recharge, even if left alone for many generations. Because the settling is often differential, the surface can develop deep fissures and the slope can change. This severely alters the surface hydrology, and, in turn, the surface's interaction with the atmosphere. Subsidence of more than 7 m has occurred in some places (de Villiers, 2000).

There are many examples, worldwide, of how groundwater mining has changed the complexion of the desert, and how major human, ecological, and meteorological adjustments will be inevitable when the water can no longer be economically retrieved. For example, in the second half of the twentieth century, the Ogallala Aquifer beneath the Great Plains in North America (Figure 17.20) supported the irrigation of more than 60,000 km^2 of agricultural land that was once semiarid. However, the prairie grass sod that once prevented serious erosion no longer exists. In addition, the dryland farming that may possibly ensue after the Ogallala Aquifer is no longer usable in the first half of the twenty-first century will allow wind erosion that could cause the transformation to a desert. Reisner (1986) and others suggested that this could easily lead to another dust bowl. This is not an isolated "doomsday" example. The remaining groundwater in many areas of the world will not be

Deforestation

Removal of trees over large or small areas can cause desertification. The devegetation process is generally motivated by a need for more agricultural land or wood for fuel and construction. A classic example is that hardly a tree survives within a radius of 100 km of Khartoum, Sudan. A variety of mechanisms may contribute to the resulting desertification. Erosion of soils by water may increase because there is less vegetation canopy to intercept rainfall. There is less leaf litter and other decaying biomass to replenish the organic material in soils, which means that rainfall absorption is less and runoff is greater. Removal of vegetation exposes soils to higher wind speeds and more sunlight, causing greater drying and erosion. There is potentially less transpiration to supply the boundary layer with moisture, which may reduce rainfall. Last, the greater demands on soil fertility of some agricultural crops that replace the trees can lead to a soil environment that no longer supports crops or the original vegetation.

There seems to be no necessity for this type of desertification to be limited to large scales, even though we tend to think of such processes in Sahel-size areas. For example, Stebbing (1935, 1938) reported that indigenous residents in northern Nigeria were permitted to fell about 240 ha of forest for new farms because their existing lands were worn out. Much of the disafforested land was not cultivated and has since become a complete waste of drifting sandy soil now incapable of cultivation, and on which a first attempt to reafforest by the Forest Department has failed, owing to the desert conditions becoming so rapidly established (Stebbing, 1935).

There are even numerous examples of the desertification effects of deforestation in humid tropical climates with copious rainfall. Here, removal of the virgin tropical forest exposed the sometimes-thin humus layer that had always been protected by the dense canopy above. Beard (1949) pointed out that deforestation of most of the Lesser Antilles Islands in the Caribbean caused numerous symptoms of desertification. In Anthes (1984), it is claimed that the rainfall over the deforested islands appears to be significantly less than that over the few islands that were mostly spared the process of agricultural exploitation.

Where deforestation and other processes have lead to desertification, it has been suggested that conserving, planting, or replanting trees might be a way to inhibit the further spread of the desert. For example, Stebbing (1935), referring to northern Nigeria, stated:

Plans have been drawn up to create belts of plantations (of trees) across the countryside, selected where possible from existing scrub forest or assisting the latter . . . also to sow seed of the Dom palm along the International Frontier to create a thick belt of the palm. This is a wise step and taken none too soon in the interests of the Province . . . the northern belt is required to counter one of the most silent menaces of the world, if not the most silent one, the imperceptible invasion of sand. (p. 511)

This shelterbelt, of proposed 25 km width and 2,200 km length, was intended to inhibit sand encroachment from the desertified areas. This grand plan was never implemented, but more recently, Anthes (1984) proposed that bands of vegetation in semiarid regions would be useful for a different reason, claiming that convective rainfall might be increased. An interesting description is provided by Lowdermilk (1953) of how deforestation created moving sand dunes over a vast area, and how reforestation was able to reverse the process:

It is recorded that the Vandals in A.D. 407 swept through France and destroyed the settlements of the people who in times past had tapped pine trees of the Les Landes region and supplied resin to Rome.

Vandal hordes razed the villages, dispersed the population, and set fire to the forests, destroying the cover of a vast sandy area. Prevailing winds from the west began the movement of sand. In time, moving sand dunes covered an area of more than 400,000 acres that in turn created 2? million acres of marshland (through blockage of rivers)....

Space will not permit my telling the fascinating details of this remarkable story of how the dunes were conquered by establishing a littoral dune and reforesting the sand behind.... Now this entire region is one vast forest supporting thriving timber and resin industries and numerous health resorts. Fortunately for comparison, one dune on private land was for some reason left uncontrolled. This dune is 2 miles long, ? mile wide, and 300 feet high. It is now moving landward, covering the forest at the rate of about 65 feet a year. As I stood on this dune and saw in all directions an undulating evergreen forest to the horizon, I began to appreciate the magnitude of the achievement of converting the giant sand dune and marshland into profitable forests and health resorts. (p. 19)

Off-road vehicle damage to soil structure and vegetation

Operating vehicles on desert terrain that has fragile vegetation and substrates can cause long-term damage. Not only is vegetation removed, but the soil surface is destabilized through disturbance of desert pavement or other crusts. Both processes lead to increased wind and water erosion. Naturally, military and civilian-recreational off-road vehicles contribute to the problem.

Urbanization and industrial exploitation

Much arid and semiarid land has undergone substantial anthropogenic change through urbanization and exploitation of mineral resources. Arid lands provide 82% of the world's oil, 86% of the iron ore, 79% of the copper, and 67% of the diamonds (Heathcote, 1983; Lines, 1979). In terms of urbanization, it was noted in the introduction that population increases have often been far greater in semiarid areas than in others with harsher winters. One example is that urbanization consumed almost 200,000 ha of the immensely agriculturally productive San Joaquin Valley in California during the last quarter of the twentieth century.

17.3.4 Natural contributions to desertification

Some definitions of desertification include only anthropogenic factors; however, most allow for the fact that the negative impact of agriculture on the biophysical system will be greatest during periods of low rainfall. Thus, drought is the most commonly discussed natural contribution to desertification. In general, any drought that degrades the vegetation cover or causes any of the aforementioned desertification symptoms would be sufficient.

The encroachment of drifting sands into surrounding vegetated areas, and the suffocation of the vegetation there, could also be considered as a natural desertification process. This occurs on local scales, such as associated with mobile sand dunes, as well as over long distances through accumulation of aeolian dust.

17.3.5 Additional selected case studies and examples of desertification

In the previous discussion of the causes of desertification, some examples were provided for each category. This section briefly reviews some additional dramatic instances of desertification in different areas. An immense amount has been written describing various incidences of desertification, and much debate has resulted regarding the relative contribution of drought versus human factors. What is not in doubt is that there are very few areas of Earth that have not experienced some degree of land degradation, so the following discussions cover only a few of the problem areas.

Thar Desert of India and Pakistan

The deforestation and agricultural exploitation of this 1,300,000 km^2 area of western India and eastern Pakistan began millennia ago, and the possible atmospheric consequences are described further in the next section about desertification feedback mechanisms. This Indus Valley area was once the cradle of the Indus civilization, a highly developed, agriculturally based society. About 3,000 years ago, the monsoon rainfall on which it depended began to fail, and the ensuing dry period lasted for almost a millennium. Subsequently, the rains increased somewhat, but they are today only about one-third of what they were estimated to be 3,000 years ago (de Vreede, 1977). This may not be entirely natural variability. It is proposed in Bryson and Baerreis (1967) that the troposphere-deep layers of dense dust caused by the soil dessication have augmented the natural subsidence in these latitudes, which has enhanced the desertification process. Hora (1952) summarized the consensus of the New Delhi Symposium on the Rajputana Desert:

One thing, which was pointedly brought out in the Symposium, was that the Rajputana Desert is largely a manmade desert . . . by the work of man in cutting down and burning forests . . . (and by) the deterioration of the soils. (p. 3)

Punjab, India—north of the Thar Desert

Stebbing (1938) provided the following description of the large-scale conversion of a forest to the bare rock of a desertlike surface:

Perhaps one of the best-known often-quoted examples of this type of damage is the case of the Hoshiarpur Chos in the Punjab, India. This part of the Siwalik range of hills consists of friable rock. The hills were formerly covered with forest. In the latter half of the last century, cattle owners settled in the area, and under the grazing and browsing of buffaloes, cows, sheep, and goats, all vegetative growth disappeared and the trampling of the animals on the slopes loosened the already loose soil. Heat and the annual monsoon rains helped to carry on the process of erosion commenced by the animals. Gradually, ravines and torrents were formed which have cut the hill range into a series of vertical hollows and ridges of the most bizarre shapes; the material thus removed and carried down to the lower level forming fan-shaped accumulations of sand extending for miles out into the plains country, covering up extensive areas of valuable agricultural land. The loss has been enormous. . . . (p. 16)

The Sahel, North Africa

The Sahel is a region on the southern margin of the Sahara that is subject to recurrent drought. Agnew (1990) reviewed the various definitions of its extent in terms of different annual isohyet belts between 200 and 750 mm. Its annual precipitation is associated with the summer monsoon. When the monsoon is weak, the result is a dry year. It is during these drought years that human contributions to desertification can become especially important because of the combined stress of the natural, periodic rainfall deficit and the human pressures. Figure 17.21 shows a close-up view of the large climatological gradient in annual rainfall for one part of the Sahel in Niger. Clearly, modest variation in the location of the large gradient would have significant consequences in terms of annual rainfall.

The works survive of Arab historians who describe the condition of the present Sahara and Sahel a number of centuries ago (Stebbing, 1935). For example, an author recounts a pilgrimage from Mali, through central Niger, and eastward to Mecca, from 1496 to 1497,

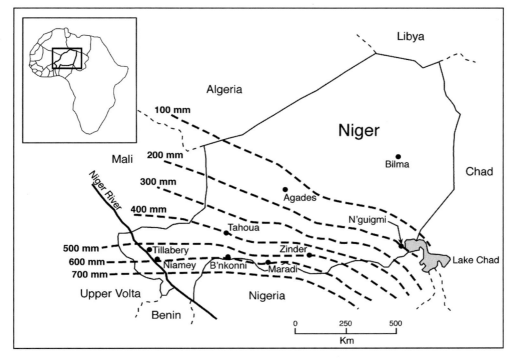

Figure 17.21   Average annual rainfall (millimeters) for Niger, in the Sahel. From Agnew (1990).

where horses and donkeys were used for riding. It is suggested that this would have been possible only if the area had considerably more water and pasture than is the case today. In this area are ruins of permanent villages that had been occupied into the eighteenth century, until they were overwhelmed by sand. More recently, in a journey up the Nile in 1820 and 1821, Linant de Bellefonds reported on the "woodedness" of northern Sudan, which today is complete desert with hardly any vegetation away from the river (Cloudsley-Thompson, 1993).

More recently, there have been substantial variations in the rainfall, with the most well documented being in the last third of the twentieth century. During the wet years of the 1960s, human population and livestock numbers increased. However, when rainfall amounts decreased after that, famine and large-scale loss of livestock resulted, despite massive international aid. An essential question is the degree to which human factors contributed to the decrease in green vegetation, through overgrazing, for example. Even though Sahel green vegetation decreased during the drought, it rebounded during subsequent more wet years that were still drier than the twentieth-century normal. Clearly, the desertification, defined in terms of green vegetation change, was not irreversible in this case. The Sahara "advanced" during dry years, but it "retreate" when rain was more plentiful. This natural ebb and flow of the Sahara's vague edge has been occurring for centuries, without irreversible effects. In this regard, Nicholson (1989a, 1989b) discussed the long-term variability of Sahelian rainfall, noting that there were numerous other centuries with severe drought. Walsh et al. (1988) stated that, in the context of the long-term Sahel climate, the real twentieth-century anomaly was the midcentury wet period, not the late-century dry one.

There are numerous areas in Africa in addition to the Sahel that have experienced desertification in the twentieth century. In west Africa, the annual rate of forest clearing was 4% during some periods. In terms of grasslands, the rich savanna, a vast area that extended from western Africa to the Horn in the east, was only 35% of its original extent in the early 1980s (Xue and Shukla, 1993).

The arid western United States

A comprehensive survey was conducted of the land and water resources in this area by the Council on Environmental Quality of the United States (Sheridan, 1981). The following is an excerpt from the conclusions of this report:

Desertification in the arid United States is flagrant. Groundwater supplies beneath vast stretches of land are dropping precipitously. Whole river systems have dried up; others are choked with sediment washed from denuded land. Hundreds of thousands of acres of previously irrigated cropland have been abandoned to wind or weeds. Salts are building up steadily in some of the nation's most productive irrigated soils. Several million acres of natural grassland are, as a result of cultivation or overgrazing, eroding at unnaturally high rates. Soils from the Great Plains are ending up in the Atlantic Ocean. All total, about 225 million acres of land in the United States are undergoing severe desertification—an area roughly the size of the 13 original states.

...The long-term prospects for increased production from U.S. arid land look unpromising, however. The rich San Joaquin Valley is already losing about 14,000 acres of prime farmland per year to urbanization and could eventually lose 2 million acres to salinization. Increased salinity of the Colorado River could limit crop output in such highly productive areas as the Imperial Valley. Economic projections in Arizona indicate a major shrinkage in cropland acreage over the next 30 years. On the High Plains of Texas, crop production is expected to decline between 1985 and 2000 because of the depletion of the Ogallala Aquifer. And certainly the end is in sight for irrigation-dependent increased grain yields from western Kansas and Nebraska as their water tables continue to drop. (pp. 121–122)

Other assessments of desertification in North America are equally bleak. For example, Dregne (1986) stated that about 90% of the arid lands of North America are moderately or severely desertified.

Hastings and Turner (1965) conducted an interesting study in which they made numerous photographs of the landscape at various places in the northern Sonoran Desert, where they tried to duplicate the viewing location and angle of photographs that were taken earlier in the twentieth century, perhaps 75 years before. Comparisons were revealing in terms of how the vegetation had changed during the period. Sheridan (1981) summarized their evidence that all the plant species had undergone a change during the period that could reflect a warming and drying. The lower elevation desert shrub and cactus communities are now thinner, the higher-level desert grasslands have receded upslope, areas that were originally oak woodlands are now primarily mesquite, and the timberline has moved upward. The pattern is one in which species have moved upslope toward more favorable environmental conditions. Hastings and Turner (1965) stated the following:

Taken as a whole, the changes constitute a shift in the regional vegetation of an order so striking that it might better be associated with the oscillations of Pleistocene time than with the "stable" present.

Even though the impact of long-term overgrazing quite likely contributed to some aspects of the documented change, there is evidence that regional climate change acted in concert. Sheridan (1981) reported that, since near the beginning of the twentieth century, rainfall in Arizona and New Mexico has decreased by about 2.5 cm every 30 years, with most of the reduction in the winter season. Also, during the twentieth century, the mean annual temperature appears to have risen by 1.7°C to 1.9°C (3.0°F–3.5°F). Mechanisms by which the surface environmental change associated with the overgrazing may have contributed to these meteorological changes are discussed in the next section.

Groundwater mining is also creating major changes in the southwestern United States' ecosystems. For example, in the San Pedro River Basin in southern Arizona, grasses, cottonwoods, willows, and sycamores are being replaced by tamarisk and mesquite, and it is reasonable to assume that this is a result of a lowering water table. The taproots of mesquite can extend to depths of greater than 30 m, so it can thrive with deep water tables where other shallower-rooted species such as cottonwood cannot. However, even the deep-rooted mesquite has suffered from the precipitous overdrafts that prevail in this region. South of Tucson, Arizona, in the Santa Cruz Valley, about 800 ha of mesquite have died. Here, overdrafts are greater than in virtually any other area of the North American deserts (Sheridan, 1981).

The mining of the Ogallala Aquifer in the semiarid High Plains of North America is a prime example of how groundwater mining can temporarily green the face of the desert. This area was originally grassland, but the conversion to irrigated crops on an immense scale has desertification consequences. Figure 17.20 shows the great extent of this aquifer, and, by implication, the associated irrigation area. The exposed soils are often susceptible to wind and water erosion. Sheridan (1981) quoted a U.S. Soil Conservation Service representative in Texas:

We are creating a new Great American Desert out there, and eventually the basic resource, soil, will be exhausted. (pp. 94–95)

A specific example—in 1977 high winds in eastern Colorado and New Mexico scoured plowed wheat fields to depths of greater than 1 m, with the silt and sand being visible over the mid-Atlantic Ocean in satellite imagery. This event followed a period of prolonged drought. Another consequence of the conversion to plowed fields that the temporary use of the Ogallala has permitted is that, when the plow abandons the land for lack of water, there will likely not be anyone willing to pay to convert it back to grass, and it will lay open to invasion by non-native plant species and to erosion.

The middle Asian plains

In the 1950s and 1960s, food shortages in the Soviet Union caused the government to open up grasslands in northern Kazakhstan and western Siberia to settlement and farming. Forty million hectares of virgin land were brought under cultivation by hundreds of thousands of settlers. Before farming practices became more conservation conscious, 17 million hectares were damaged by wind erosion, and 4 million hectares were entirely lost to production (Eckholm, 1976). Zonn et al. (1994) further described the environmental consequences of this program.

The loess soil areas of eastern China

Loess is a loamy type of soil consisting of particles that were deposited by the wind and is very susceptible to water erosion. Because it is one of the most productive types of soils in the world, there is a great deal of economic and social pressure to exploit this land, regardless of the environmental consequences. Of the large areas of Asia where desertification borders on being classified as very severe, are parts of China that have large deposits of 100-m deep loess that are cultivated. In the Huang He (Yellow) River watershed, more than 300,000 km^2 experience major soil loss into the river. The sediment load in the river is sometimes as high as 46% (by mass). In some areas, soil erosion has eaten out gullies that are almost 200 m deep. Of the 600,000 km² of loess, it is estimated that 26,000 km² are gullies (Kuo, 1976). However, the land areas around temples, where the natural vegetation is untouched and abundant, has not significantly eroded. This leads Lowdermilk (1953) to conclude that human misuse of the land caused the damage, rather than climate variability.

Caribbean islands

This is an illustration of the possible desertifying effects of deforestation that can occur even in the humid tropics (Anthes, 1984). Beginning in the fifteenth century, many of the Windward and Leeward islands of the Lesser Antilles were colonized. A consequence of the development of the agricultural industry was a deforestation of most of the islands. This caused a marked dessication of the soil, drying up of many ponds, and increased erosion and runoff (Beard, 1949; Bridenbaugh and Bridenbaugh, 1972). However, Dominica retains much of its virgin forest, and Anthes (1984) pointed out that its average annual rainfall is greater by more than a factor of three compared to other islands in the Lesser Antilles that were extensively deforested. It is, however, difficult to separate the deforestation effects from differences in the size and elevation of the islands.

The Tigris–Euphrates River Valley, Iraq

This area was discussed previously in the context of soil salinization. As a result of millennia of irrigation, 20% to 30% of the country's irrigable land is now unsuitable for agriculture, or even native vegetation. Eckholm (1976) summarized the situation well:

that Mesopotamia is possibly the world's oldest irrigated area in not an encouraging observation. The end result of six millennia of human management is no garden spot. The region's fertility was once legendary throughout the Old World, and the American conservationist Walter Lowdermilk has estimated that at its zenith Mesopotamia supported between 17 and 25 million inhabitants (Lowdermilk, 1953). One early visitor, Herodotus, wrote that "to those who have never been in the Babylonian country, what has been said regarding its production will be incredible." Today, Iraq has a population of 10 million, and on the portions of these same lands that have not been abandoned, peasants eke out some of the world's lowest crop yields.

The Mediterranean region

It is hard to estimate the pristine state of this region, which has been disturbed by occupation and exploitation for so many millennia. However, there is plenty of evidence that it at least looks more desertlike now than before, even though there is little indication of any significant climate change. Deforestation is one of the greater changes. More than 4,500 years ago, Egyptians depended on Phoenician (currently, Syria, Lebanon, and Israel) forests to supply them with heavy timber for ships and temples, and for the resins that were used in mummification. Continued demand for eastern Mediterranean cedars, pines, and firs was so great that, by 500 years ago, timber was a scarce material (Mikesell, 1969). The same degradation has occurred throughout the lands that neighbor the Mediterranean Sea, as reported by Tomaselli (1977), Mikesell (1960), Brandt and Thornes (1996), and Beals (1965).

17.3.6 Physical process feedbacks that may affect desertification

It is worth being reminded that the term "feedback" implies that a series of cause-and-effect relationships exist, such that an initial perturbation sets in motion a series of responses wherein the original perturbation is either enhanced or reduced. In the context of desertification, for example, an initial natural or anthropogenic nudge of the system toward

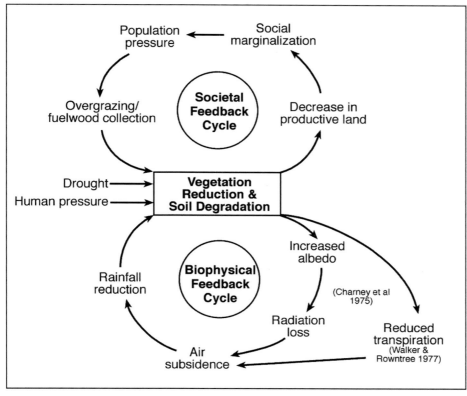

Figure 17.22 An example of combined biophysical and societal feedback mechanisms. The biophysical component is the vegetation-albedo feedback. Based on Scoging (1991), with modification by Thomas and Middleton (1994).

greater aridity would cause a series of responses that might contribute toward greater aridity (positive feedback) or less aridity (negative feedback). No enhancement or mitigation of the original perturbation of the system would be neutral feedback. There are numerous atmosphere-surface feedbacks that might contribute to the maintenance and formation of deserts, and this section reviews some of them in the context of desertification. Societal feedbacks may be intertwined with physical feedbacks, to the same initial perturbation. Figure 17.22 depicts this situation for the vegetation-albedo/transpiration feedback discussed first. On the biophysical side, the vegetation reduction contributes to reduced rainfall, which leads to a further reduction in vegetation, and so on. On the societal side, the initial perturbation toward desertification, the drought, results in overgrazing that contributes to the vegetation reduction. Other feedback diagrams treat different aspects of system interactions. For example, Schlesinger et al. (1990) presented a diagram that links ecosystem properties with global biogeochemistry during desertification. Hulme and Kelly's (1993) feedback diagram includes global warming and the ocean as part of the mechanism. The Walsh et al. (1988) diagram focuses on the hydrologic response.

Vegetation—albedo/transpiration feedback

One of the most discussed desertification feedback mechanisms is the albedo feedback proposed by Otterman (1974), Charney (1975), and Charney et al. (1977). The concept is

Table 17.6. *Ratio of albedoes for anthropogenically impacted desert and nearby protected areas based on satellite data for various parts of the world*

Wavelength (Micron)	0.5–0.6 (Visible)	0.6–0.7 (Visible)	0.7–0.8 IR	0.8–1.1 IR
Sinai/Negev	1.55	1.88	1.87	1.73
Afghanistan/Russia	1.16	1.21	1.21	1.19
Sahel, overgrazed/ranch	1.20	1.37	1.41	1.39

Adapted from Otterman (1981).

that removal of vegetation by overgrazing in arid lands with light-colored sandy soils will cause an increased albedo because the removed vegetation is less reflective than the bare soil that is exposed. This will result in lower net radiation, as well as lower surface and boundary layer temperatures. The more stable vertical profile of temperature will inhibit convection, and less rainfall will result, thus increasing the loss of vegetation. This further baring of the soil will cause the process to continue, which makes it a positive feedback loop. The work by Otterman (1977a, 1977b, 1981) provides convincing documentation of the effects on albedo of grazing and other disturbances of arid lands. It is shown that disruption or removal by humans of the dead vegetation debris between live plants in deserts can have as large an effect on average albedo as the removal of the live vegetation. Table 17.6 shows the ratios of the albedoes of overgrazed areas on one side of a boundary to the albedoes of less damaged areas only a few hundred meters away on the other side of the boundary. The ratios are for bands in the visible and solar infrared. Comparisons are for the Sinai/Negev Desert boundary, where the Sinai is under heavier human pressure; for the Afghanistan-Russia border, where the Afghanistan side is more overgrazed; and for a border between the overgrazed Sahel and a protected ranch. The ratios that range from about 1.2 to almost 1.9 clearly illustrate the albedo response to human pressure, and the potential effect on the surface energy budget and atmospheric processes.

Vegetation removal in the form of deforestation can have similar possible feedback effects, as implied in the previous reference to Anthes (1984) regarding the relation of forests to precipitation. Additional evidence is found in Eltahir (1996), who has shown in a modeling study that deforestation reduces the surface net radiation, which leads to tropospheric subsidence that reduces the rainfall. Also, model experiments by Zheng and Eltahir (1997) showed that deforestation of the coastal areas of the Gulf of Guinea reduces the strength of the monsoon flow that provides most of the rainfall for the Sahel and the southern Sahara.

Dust-radiation feedback

A possible positive feedback that could lead to desertification is related to the radiative effects of desert dust. Bryson and Baerreis (1967) and Bryson et al. (1964) were among the first to discuss how anthropogenic factors such as deforestation and agricultural exploitation can lead to dessication of the soil and the elevation of dust throughout the troposphere, and how this can lead to a positive feedback. In a study near the Thar Desert of India and Pakistan, they showed that the dust has the effect of increasing the midtropospheric subsidence rate by about 50%. The enhanced subsidence decreases the depth of the monsoon layer and reduces the monsoon's penetration into the desert. By this mechanism, a natural drought period or human factors could cause an increase in tropospheric dust, and this could sustain or enhance the initial perturbation toward soil dessication and desertification. Bryson and Baerreis (1967) described archeological evidence of earlier thriving civilizations and agriculture in this region, suggesting that the aridity has increased dramatically over the past few thousand years. They speculate that the widespread documented deforestation and overgrazing that has occurred in this area may have been the initial perturbation that started the feedback that led to a self-sustaining desert.

Water-table deepening-vegetation-runoff feedback

A positive feedback that does not directly involve atmospheric processes was described previously in the discussion of water table deepening as an anthropogenic effect on the desert surface, and ultimately, the atmosphere. Here, partial devegetation of the surface, often a result of agricultural exploitation such as cattle grazing, increases runoff of rainfall because the rain is not intercepted by the foliage and there is less biological litter on the surface to absorb it. This, in turn, causes water courses to be scoured out and deepened. Where water tables are high, as is sometimes the case in deserts, this results in a lowering of the water table, and a reduction in vegetation whose roots can no longer reach the water table. This reduction in deep-rooted vegetation causes increased runoff. Thus, we have a positive feedback in that a reduction in vegetation can lead to a further reduction in vegetation through the increased runoff and water table deepening.

17.3.7 Satellite-based methods for detecting and mapping desertification

Dregne (1987) emphasized the importance of satellite data in shedding new light on some major questions related to desertification:

Claims that the (desert) is expanding at some horrendous rate are still made despite the absence of evidence to support them. It may have been permissible to say such things ten or twenty years ago when remote sensing was in its infancy and errors could easily be made in extrapolating limited observations. It is unacceptable today.

Prior to the availability of satellite data, identification of desertification trends relied on the knowledge of local inhabitants, written historical references, and ground reconnaissance by scientists. Indeed, to this day, there is often no substitute for close firsthand examination, on the ground, of changes in the vegetation or soil characteristics. Nevertheless, the availability of remotely sensed imagery from satellites allows for subjective and objective large-scale analysis of trends in surface conditions. In the simplest approach, spatial and temporal contrasts in the surface brightness (albedo) in visible imagery can be revealing because differences in the amount of vegetation can be easily seen. That is, a vegetation canopy over a sandy substrate surface has a lower albedo than does the sand. Thus, a more vegetated surface appears darker in the visible wavelengths.

Satellite radiances can also detect other disturbances of desert surfaces, in addition to the direct effects on live vegetation. For example, the plant debris and dead plants that often cover the areas between the live vegetation can be removed through livestock grazing and trampling, cultivation, or fuel wood collection. Otterman (1977a, 1977b, 1981) explained that satellite imagery in the reflective (solar) infrared and visible wavelengths dramatically shows these human effects on the dead vegetation, and how they are important to the surface energy budget. Otterman (1981) showed reflectivity for different wavelengths in the visible and reflective infrared based on Landsat satellite imagery for surfaces in the adjacent Sinai and Negev deserts. The Sinai has been much more heavily impacted by overgrazing than has the Negev, and both live vegetation and dead vegetation in the large interstitial spaces are less prevalent in the Sinai. An exception is an area in the Sinai that was enclosed 3 years before the measurements, where natural vegetation had partially regrown. Also shown are the spectral reflectivities, based on a handheld field radiometer, for live plants, dead plant material, and bare, disturbed soil. The crumbled soil is what remains after the plant debris has been removed by the previously mentioned human processes. Having multispectral imagery, rather than just visible imagery, leads to the conclusion that the low reflectivities in the Negev Desert result from the fact that the dead plant material had not been disturbed to the extent that it had been in the Sinai, rather than because the Negev had more live vegetation. It is interesting that none of the satellite imagery shows the low reflectance in the visible red part of the spectrum (0.65 μm) that would be associated with the strong absorption by chlorophyll. This is presumably because vegetation is sparse in both deserts. Thus, visible imagery alone would only have left uncertainty about differences in human impact on the two deserts. However, the satellite-observed reflectivities in the solar infrared are revealing. In this band, live vegetation is very reflective, as is bare soil, but the dead plant material has very low reflectivity. Thus, one of the major differences between the two deserts is the degree to which human activity has denuded the large areas between the live vegetation.

In an effort to use satellite radiances in the visible and solar infrared for detection of changes in desert vegetation and associated desertification, approaches have combined information from different wavelength channels to construct indices such as the normalized difference vegetation index (NDVI) and the green vegetation fraction. Both indices are proportional to the amount of green vegetation. For example, Dregne and Tucker (1988) and Tucker et al. (1991) described the use of the NDVI to trace annual changes in the boundary between the Sahara and the Sahel—that is, the expansion and contraction of the Sahara. It is first demonstrated that mean precipitation and NDVI are approximately linearly related for the area, and the Sahara-Sahel boundary is defined as coinciding with the 200 mm/y isohyet. Thus, the NDVI field can be used to define the boundary location, from which insight can be gained about desertification. To provide a Sahel-average view of vegetation changes, Figure 17.23 shows the average annual NDVI from 1980 to 1990 for the area of the Sahel having a long-term mean rainfall of 200 to 400 mm/y. Also shown is the corresponding NDVI for a zone in the central Sahara. These significant variations in the NDVI correspond well with the Sahel rainfall variability for the period. The year 1984 was one of the driest in the twentieth century, and the vegetation clearly reflects this. To map the advances and retreats of the Sahara during this period, the NDVI was used to estimate the location of the 200 mm/y^{-1} isohyet, defined as the Sahara's southern boundary. The 1984 longitudinal-mean position of the "boundary" was mapped to be more than 200 km farther south than in 1980. By 1988, it had receded northward by most of this distance.

Satellites can also be used to estimate soil moisture in arid areas. There are various possible approaches, where Milford (1987) and Bryant et al. (1990) summarized some applications and limitations. Additional discussions of the use of satellite data to monitor land surface conditions in arid areas can be found in Chen et al. (1998), Robinove et al. (1981), Justice and Hiernaux (1986), and Nicholson and Farrar (1994).

17.4 Summary

This chapter summarizes two classes of disasters that can occur in arid lands: severe weather and desertification. Severe weather phenomena, which include dust storms and flash floods, occur on time scales of a few days to a few hours. In contrast, desertification is a more

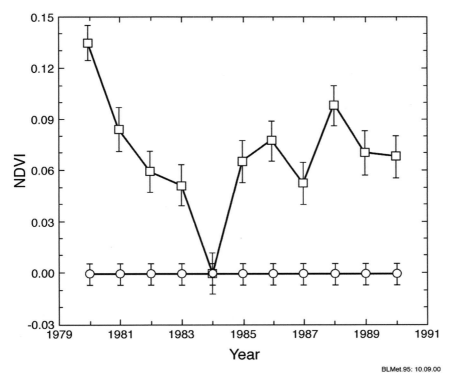

Figure 17.23 Satellite derived normalized difference vegetation index (NDVI) for a zone (the Sahel) with long-term average rainfall between 200 and 400 mm/y along the southern boundary of the Sahara Desert (*squares*), and for the Sahara Desert itself (*circles*) from 1980 to 1990. From Tucker et al. (1991).

subtle and longer-term process that can occur over a period of years to decades and impact areas that represent significant fractions of a continent.

Even though there are various formal definitions of dust storms, they are separable into two types. In one, the high winds that elevate the dust are generated by thunderstorm downdrafts, and the lifetime of the dust storm may be a half hour to a few hours. In the other, the winds are related to weather map scale processes, and the dust storms can last for days to a week. It is shown that there is a strong seasonality to dust storms, for multiple reasons. One is related to the fact that the winds that elevate the dust are from processes that typically occur in certain seasons. Another is that dust can be most readily elevated into the atmosphere when the substrate is dry, during seasons with little rainfall, and when the surface is unfrozen, during nonwinter.

Floods and debris flows are more common in deserts because the sparse vegetation and packed soils cannot absorb rainwater as effectively as can highly vegetated landscapes. The result is that even modest rain amounts falling on desert watersheds can cause flash floods that claim lives and destroy property.

Perhaps the most insidious environmental disaster is desertification, which may occur on local or continental scales. In this process, typically human use of the land and a perturbation from nature, such as a drought, combine to cause a deterioration in the ability of the land to sustain life. Through various positive feedback processes, this can continue until the damage, if not irreversible, persists for a long period of time and causes human starvation and migration on truly massive scales. This chapter reviews various aspects of the desertification process, provides numerous historical examples, and describes positive feedback processes that exacerbate the problem.

References

Abdulaziz, A. (1994) "A Study of Three Types of Wind-Blown Dust in Kuwait: Duststorms, Rising Dust and Suspended Dust," *J. Meteor.* **19**, pp. 19–23.

Acocks, J. P. H. (1952) *Veld Types of South Africa*, Department of Agriculture and Water Supply, Pretoria, South Africa.

Agnew, C. T. (1990) "Spatial Aspects of Drought in the Sahel," *J. Arid Environments* **18**, pp. 279–293.

Anderson, R. S. (1987) "Eolian Sediment Transport as a Stochastic Process: The Effects of a Fluctuating Wind on Particle Trajectories," *J. Geology* **95**, pp. 497–512.

Andrews, R. C. (1932) *The New Conquest of Central Asia*, The American Museum of Natural History, New York, New York.

Anthes, R. A. (1984) "Enhancement of Convective Precipitation by Mesoscale Variations in Vegetative Covering in Semi-Arid Regions," *J. Climate Appl. Meteor.* **23**, pp. 541–554.

Aubreville, A. (1949) "Climats, Forêts et Désertification de l'Afrique Tropicale," Societe de Éditions Géographiques, *Maritime et Coloniales*, Paris, France.

Beals, E. W. (1965) "The Remnant Cedar Forests of Lebanon," *J. Ecology* **53**, pp. 679–694.

Beard, J. S. (1949) *The Natural Vegetation of the Windward and Leeward Islands*, Oxford University Press, London, United Kingdom.

Betzer, P. R., Carder, K. L., Duce, R. A., Merrill, J. T., Tindale, N. W., Uematsu, M., Costello, D. K., Young, R. W., Feely, R. A., Breland, J. A., Berstein, R. E., and Grecom A. M. (1988) "Long-Range Transport of Giant Mineral Aerosol Particles," *Nature* **336**, pp. 568–571.

Bingham, S. (1996) *The Last Ranch: A Colorado Community and the Coming Desert*, Harcourt Brace, San Diego, California.

Bond, W. J., Stock, W. D., and Hoffman, M. T. (1994) "Has the Karoo Spread? A Test for Desertification Using Stable Carbon Isotopes," *South African J. of Science* **90**, pp. 391–397.

Boudet, G. (1972) "Deésertification de l'Afrique Tropicale Séche," *Adansonia* **12**, ser. 2, pp. 505–524.

Brandt, C. J., and Thornes, J. B., eds. (1996) *Mediterranean Desertification and Land Use*, John Wiley & Sons, Chichester, United Kingdom.

Brazel, A. J., and Hsu, S. (1981) "The Climatology of Hazardous Arizona Dust Storms," in *Desert Dust*, T. L. Péwé, ed., Geological Society of America, Special Paper 186, pp. 293–303.

Brazel, A. J., Nickling, W. G., and Lee, J. (1986) "Effect of Antecedent Moisture Conditions on Dust Storm Generation in Arizona," in *Aeolian Geomorphology*, ed. W. G. Nickling, pp. 261–271, Allen and Unwin, Boston, Massachusetts.

Bridenbaugh, C., and Bridenbaugh, R. (1972) *No Peace Beyond the Line—The English in the Caribbean 1624–1690*, Oxford University Press, New York, New York.

Bryant, N. A., Johnson, L. F., Brazel, A. J., Balling, R. C., Hutchinson, C. F., and Beck, L. R. (1990) "Measuring the Effect of Overgrazing in the Sonoran Desert," *Climatic Change* **17**, pp. 243–264.

Bryson, R. A., and Baerreis, D. A. (1967) "Possibilities of Major Climatic Modification and Their Implications: Northwest India, a Case Study," *Bull. Amer. Meteor. Soc.* **48**, pp. 136–142.

Bryson, R. A., Wilson, C. A., III, and Kuhn, P. M. (1964) "Some Preliminary Results of Radiation Sonde Ascents over India," *Proceedings WMO-IUGG Symposium Tropical Meteorology*, November 1963, New Zealand Meteorological Service, Wellington, New Zealand.

Bullock, P., and Le Houérou, L. H. (1996) "Land Degradation and Desertification," in *Climate Change 1995: Impacts, Adaptations and Mitigation of Climate Change*, eds. R. T. Watson, M. C. Zinyowera, R. H. Ross, and D. J. Dokken, pp. 171–189, Cambridge University Press, Cambridge, United Kingdom.

Butterfield, G. R. (1991) "Grain Transport Rates in Steady and Unsteady Turbulent Airflows," *Acta Mechanica* **1**, pp. S97–S122.

Butterfield, G. R. (1993) "Sand Transport Response to Fluctuating Wind Velocity," in *Turbulence, Perspectives on Flow and Sediment Transport*, eds. N. J. Clifford, J. R. French, and J. Hardisty, pp. 305–335, John Wiley & Sons, Chichester, United Kingdom.

Chao, S. C. (1984a) "Analysis of the Desert Terrain in China Using Landsat Imagery," in *Deserts and Arid Lands*, ed. F. El Baz, pp. 115–132, Martinus Nijhoff, The Hague, The Netherlands.

Chao, S. C. (1984b) "The Sand Deserts and the Gobis of China," in *Deserts and Arid Lands*, ed. F. El Baz, pp. 95–113, Martinus Nijhoff, The Hague, The Netherlands.

Charney, J. (1975) "Dynamics of Deserts and Drought in the Sahel," *Quart. J. Roy. Meteor. Soc.* **101**, pp. 193–202.

Charney, J., Quirk, W. J., Chow, S. H., and Kornfield, J. (1977) "A Comparative Study of the Effects of Albedo Change on Drought in Semiarid Regions," *J. Atmos. Sci.* **34**, pp. 1366–1385.

Chen, Z., Elvidge, C. D., and Groeneveld, D. P. (1998) "Monitoring Seasonal Dynamics of Arid Land Vegetation Using AVIRIS Data," *Remote Sensing Environ.* **65**, pp. 255–266.

Chepil, W. S. (1945) "Dynamics of Wind Erosion. I: Nature of Movement of Soil by Wind," *Soil Science* **60**, pp. 305–320.

Chepil, W. S., and Woodruff, N. P. (1957) "Sedimentary Characteristics of Dust Storms. II: Visibility and Dust Concentration," *American J. of Science* **255**, pp. 104–114.

Childs, C. (2002) *The Desert Cries: A Season of Flash Floods in a Dry Land*, Arizona Highways Books, Phoenix, Arizona.

Clements, T., Mann J. F., Jr., Stone, R. O., and Eymann, J. L. (1963) *A Study of Wind-Borne Sand and Dust in Desert Areas*, U.S. Army Natick Laboratories, Earth Sciences Division, Technical Report No. ES-8, Natick, Massachusetts.

Cloudsley-Thompson, J. L. (1993) "The Future of the Sahara," *Environmental Conservation* **20**, pp. 335–338.

Costa, J. E. (1987) "Hydraulics and Basin Morphometry of the Largest Flash Floods in the Conterminous United States," *J. of Hydrology* **93**, pp. 313–338.

Crooks, G. A., and Cowan, G. R. C. (1993) "Duststorm, South Australia, November 7th, 1988," *Bull. Aust. Meteor. & Oceanog. Soc.* **6**, pp. 68–72.

de Villiers, M. (2000) *Water: The Fate of Our Most Precious Resource*, Houghton Mifflin, Boston, Massachusetts.

Dean, W. R. S., and McDonald, I. A. W. (1994) "Historical Changes in Stocking Rates of Domestic Livestock as a Measure of Semi-Arid and Arid Rangeland Degradation in the Cape Province," *J. of Arid Environments* **28**, pp. 281–298.

de Vreede, M. (1977) *Deserts and Men: A Scrapbook*, Government Publishing Office, The Hague, The Netherlands.

Depierre, D., and Gillet, H. (1971) "Désertification de la Zone Sahélienne du Tchad," *Bois et Forêts des Tropiques* **139**, pp. 2–25.

Dregne, H. E. (1977) "Desertification of Arid Lands," *Economic Geography* **3**, pp. 322–331.

Dregne, H. E. (1983) *Desertification of Arid Lands*, Harwood Academic. New York, New York.

Dregne, H. E. (1985) "Aridity and Land Degradation," *Environment* **27**, pp. 16–33.

Dregne, H. E. (1986) "Desertification of Arid Lands," in *Physics of Desertification*, eds. F. El-Baz and M. H. A. Hassan, pp. 4–34, Martinus Nijhoff, Dordrecht, The Netherlands.

Dregne, H. E. (1987) "Reflections on the PACD," in *Desertification Control Bulletin*, United Nations Environment Program, Special Tenth Anniversary of UNCOD Issue (November 15, 1987), pp. 8–11.

Dregne, H. E., and Tucker, C. J. (1988) "Desert Encroachment," *Desertification Control Bulletin*, United Nations Environment Program, No. 16, pp. 16–19.

Droegemeier, K. K., and Wilhelmson, R. B. (1987) "Numerical Simulation of Thunderstorm Outflow Dynamics. Part I: Outflow Sensitivity Experiments and Turbulence Dynamics," *J. Atmos. Sci.* **44**, pp. 1180–1210.

Duce, R. A. (1995) "Sources, Distributions, and Fluxes of Mineral Aerosols and Their Relationship to Climate," in *Aerosol Forcing of Climate*, eds. R. J. Charlson and J. Heintzenberg, pp. 43–72, John Wiley & Sons, Chichester, United Kingdom.

Eckholm, E. P. (1976) *Losing Ground: Environmental Stress and World Food Prospects*, W. W. Norton and Company, New York, New York.

Eltahir, E. A. B. (1996) "Role of Vegetation in Sustaining Large-Scale Atmospheric Circulations in the Tropics," *J. Geophys. Res.* **101**, pp. 4255–4268.

Ely, L. L., Enzel, Y., Baker, V. R., and Cayan, D. R. (1993) "A 5000-Year Record of Extreme Floods and Climate Change in the Southwestern United States," *Science* **262**, pp. 410–412.
Freeman, M. H. (1952) *Duststorms of the Anglo-Egyptian Sudan*, Meteorological Report No. 11, Meteorological Office, London, United Kingdom.
George, U. (1977) *In the Deserts of This Earth*, Harcourt Brace Jovanovich, New York, New York.
Gillette, D. A., Bolivar, D. A., and Fryrear, D. W. (1974) "The Influence of Wind Velocity on the Size Distributions of Aerosols Generated by the Wind Erosion of Soils," *J. Geophys. Res.* **79C**, pp. 4068–4075.
Glaccum, R. A., and Prospero, J. M. (1980) "Saharan Aerosols over the Tropical North Atlantic— Mineralogy," *Marine Geology* **37**, pp. 295–321.
Glantz, M. H., and Orlovsky, N. (1983) "Desertification: A Review of the Concept," *Desertification Control Bulletin*, United Nations Environment Programme, No. 9, pp. 15–22.
Goudie, A. S. (1978) "Dust Storms and Their Geomorphological Implications," *J. Arid Environments* **1**, pp. 291–310.
Goudie, A. S. (1983a) "Dust Storms in Space and Time," *Progress in Physical Geography* **7**, pp. 502–530.
Goudie, A. S. (1983b) *Environmental Change*, Clarendon Press, Oxford, United Kingdom.
Goudie, A. S., and Middleton, N. J. (1992) "The Changing Frequency of Dust Storms Through Time," *Climatic Change* **20**, pp. 197–225.
Goudie, A. S., and Wilkinson, J. (1977) *The Warm Desert Environment*, Cambridge University Press, London, United Kingdom.
Grainger, A. (1990) *The Threatening Desert*, Earthscan, London, United Kingdom.
Great Plains Committee. (1936, December) *The Future of the Great Plains*, Report of the Great Plains Committee, Government Printing Office, Washington, DC.
Gutentag, E. D., and Weeks, J. B. (1980) *Water Table in the High Plains Aquifer in 1978 in Parts of Colorado, Kansas, Nebraska, New Mexico, Oklahoma, South Dakota, Texas and Wyoming*, Hydrologic Investigations Atlas HA-642, U.S. Geological Survey, Reston, Virginia.
Hassan, M. A. (1990) "Observations of Desert Flood Bores," *Earth Surface Processes and Landforms* **15**, pp. 481–485.
Hastings, J. R., and Turner, R. M. (1965) *The Changing Mile*, University of Arizona Press, Tucson, Arizona.
Heathcote, R. L. (1983) *The Arid Lands: Their Use and Abuse*, Longman, London, United Kingdom.
Hellden, U. (1988) "Desertification Monitoring," *Desertification Control Bulletin*, United Nations Environment Programme, No. 17, pp. 8–12.
Herman, J. R., Bhartia, P. K., Torres, O., Hsu, C., Seftor, C., and Celarier, E. (1997) "Global Distribution of UV-Absorbing Aerosols from Nimbus 7/TOMS Data," *J. Geophys. Res.* **102**, pp. 16911–16922.
Hinds, B. D., and Hoidale, G. B. (1975) *Boundary Layer Dust Occurrence. Volume 2: Atmospheric Dust over the Middle East, Near East and North Africa*, Technical Report, U.S. Army Electronics Command, White Sands Missile Range, New Mexico.
Hjalmarson, H. W. (1984) "Flash Flood in Tañque Verde Creek, Tucson, Arizona," *J. Hydraulic Engineering* **110**, pp. 1841–1852.
Hoffman, M. T., and Cowling, R. M. (1990) "Desertification in the Lower Sundays River Valley, South Africa," *J. Arid Environments* **19**, pp. 105–117.
Holcombe, T. L., Ley, T., and Gillette, D. A. (1997) "Effects of Prior Precipitation and Source Area Characteristics on Threshold Wind Velocities for Blowing Dust Episodes, Sonoran Desert 1948–1978," *J. Appl. Meteor.* **36**, pp. 1160–1175.
Hora, S. L. (1952) "The Rajputana Desert: It's Value in India's National Economy," *Proceedings of the Symposium on the Rajputana Desert, Bull. National Institute Sci. India*, **1**, pp 1–11.
Houseman, J. (1961) "Dust Haze at Bahrain," *Meteor. Mag.* **19**, pp. 50–52.
Hulme, M., and Kelly, M. (1993) "Exploring the Links Between Desertification and Climate Change," *Environment* **35**, pp. 4–19.
Jahns, R. H. (1949) "Desert Floods," *Engineering and Science* **12**, pp. 10–14.
Justice, C. O., and Hiernaux, P. H. Y. (1986) "Monitoring the Grasslands of the Sahel Using NOAA AVHRR Data: Niger 1983," *Int. J. Remote Sensing* **7**, pp. 1475–1497.
Kahana, R., Ziv, B., Enzel, Y., and Dayan, U. (2002) "Synoptic Climatology of Major Floods in the Negev Desert, Israel," *Int. J. of Climatology* **22**, pp. 867–882.
Kamra, A. K. (1969) "Electrification in an Indian Dust Storm," *Weather* **24**, pp. 145–146.

Kamra, A. K. (1972) "Measurements of the Electrical Properties of Dust Storms," *J. Atmos. Res.* **77**, pp. 5856–5869.
Katsoulis, B. D., and Tsangaris, J. M. (1994) "The State of the Greek Environment in Recent Years," *Ambio.* **23**, pp. 274–279.
Khalaf, F. I., and Al-Hashash, M. K. (1983) "Aeolian Sedimentation in the Northwest Part of the Arabian Gulf," *J. Arid Environ.* **6**, pp. 319–332.
Knighton, D., and Nanson, G. (1997) "Distinctiveness, Diversity and Uniqueness in Arid Zone River Systems," in *Arid Zone Geomorphology: Process Form and Change in Drylands*, ed. D. S. G. Thomas, pp. 185–203, John Wiley & Sons, Chichester, United Kingdom.
Kuo, L. T. C. (1976) *Agriculture in the People's Republic of China*, Praeger, New York, New York.
Lamprey, H. (1988) "Report on the Desert Encroachment Reconnaissance in Northern Sudan: 21 October to 10 November 1975," *Desertification Control Bulletin*, United Nations Environment Programme, No. 17, pp. 1–7.
Langbein, W. B., and Schumm, S. A. (1958) "Yield of Sediment in Relation to Mean Annual Precipitation," *Trans. of the Amer. Geophys. Union* **39**, pp. 1076–1084.
Le Houérou, H. N. (1968) "La Désertisation du Sahara Septentrional et des Steppes Limitrophes," *Ann. Algér de Géogr.* **6**, pp. 2–27.
Le Houérou, H. N. (1976) "The Nature and Causes of Desertification," *Arid Zone Newsletter* **3**, pp. 1–7, Office of Arid Land Studies, Tucson, Arizona.
Le Houérou, H. N. (1979) "La Désertisation des Régions Arides," *La Recherche* **99**, pp. 336–344.
Le Houérou, H. N. (1992) "An Overview of Vegetation and Land Degradation in the World Arid Lands," in *Degradation and Restoration of Arid Lands*, ed. H. E. Dregne, pp. 127–163, International Center for Arid and Semi-Arid Lands Studies, Texas Tech University Press, Lubbock, Texas.
Le Houérou, H. N., Popov, G. F. and See, L. (1993) *Agrobioclimatic Classification of Africa*, Agrometeorology Series, Working Paper No. 6, Food and Agriculture Organization of the United Nations, Rome, Italy.
Lines, G. C. (1979) "Hydrology and Surface Morphology of the Bonneville Salt Flats and the Pilot Valley Playa, Utah," Geological Survey Water-Supply Paper 2057, U.S. Government Printing Office, Washington, DC.
Littmann, T. (1991) "Dust Storm Frequency in Asia: Climatic Control and Variability," *Int. J. Climatol.* **11**, pp. 393–412.
Lopez-Bermudez, F., Romero-Diaz, A., Fisher, G., Francis, C., and Thornes, J. B. (1984) "Erosion y Ecologia en la Espana Semi-Arida (Cuenca de Mula, Murcia)," *Cuadernos de Investig. Geofra* **10**, pp. 113–126.
Lowdermilk, W. C. (1953) *Conquest of the Land Through 7000 Years*, U.S. Department of Agriculture, Natural Resources Conservation Service, Agriculture Information Bulletin No. 99.
Lusigi, W. (1981) *Combating Desertification and Rehabilitating Degraded Production Systems in Northern Kenya*, IPAL Technical Report A-4, MAB Project No. 3, Impact of Human Activities and Land-Use Practices on Grazing Lands, United Nations Educational, Scientific and Cultural Organization, Nairobi, Kenya.
Mabbutt, J. A. (1978a) *Desertification in Australia*, Water Research Foundation of Australia, Report No. 4, Canberra, Australia.
Mabbutt, J. A. (1978b) "The Impact of Desertification as Revealed by Mapping," *Environmental Conservation* **5**, p. 45.
Mabbutt, J. A. (1984) "A New Global Assessment of the Status and Trends of Desertification," *Environmental Conservation* **11**, pp. 100–113.
Man, J. (1999) *Gobi: Tracking the Desert*, Yale University Press, New Haven, Connecticut.
Marticorena, B., Bergametti, G., Gillette, D., and Belnap, J. (1997) "Factors Controlling Threshold Friction Velocity in Semiarid and Arid Areas of the United States," *J. Geophys. Res.* **102**, pp. 23277–23287.
Membery, D. A. (1985) "A Gravity Wave Haboob," *Weather* **40**, pp. 214–221.
Middleton, N. (1986) *The Geography of Dust Storms*, Dissertation, Oxford University, Oxford, United Kingdom.
Middleton, N. (1997) "Desert Dust," in *Arid Zone Geomorphology: Process Form and Change in Drylands*, ed. D. S. G. Thomas, pp. 413–436, John Wiley & Sons, Chichester, United Kingdom.

Middleton, N. J., Goudie, A. S., and Wells, G. L. (1986) "The Frequency and Source Areas of Dust Storms," in *Aeolian Geomorphology*, ed. W. G. Nickling, pp. 237–259, Allen and Unwin, Boston, Massachusetts.

Mikesell, M. W. (1960) "Deforestation in Northern Morocco," *Science* **132**, pp. 441–454.

Mikesell, M. W. (1969) "The Deforestation of Mount Lebanon," *Geographical Rev.* **59**, pp. 1–28.

Milford, J. R. (1987) "Problems of Deducing the Soil Water Balance in Dryland Regions from Meteosat Data," *Soil Use and Management* **3**, pp. 51–57.

Muturi, H. R. (1994) *Temperature and Rainfall Trends in Kenya During the Last 50 Years*, Department of Research and Development, Ministry of Research, Technology and Training, Nairobi, Kenya.

Nicholson, S. E. (1989a) "African Drought: Characteristics, Causal Theories, and Global Teleconnections," in *Understanding Climate Change*, eds. A. Berger, R. E. Dickinson, and J. W. Kidson, pp. 79–100, American Geophysical Union, Washington, DC.

Nicholson, S. E. (1989b) "Long Term Changes in African Rainfall," *Weather* **44**, pp. 46–56.

Nicholson, S. E., and Farrar, T. J. (1994) "The Influence of Soil Type on the Relationships Between NDVI, Rainfall, and Soil Moisture in Semiarid Botswana. Part I. NDVI Response to Rainfall," *Remote Sens. Environ.* **50**, pp. 107–120.

Nickling, W. G., and Wolfe, S. A. (1994) "The Morphology and Origin of Nabkhas, Region of Mopti, Mali, West Africa," *J. Arid Environments* **28**, pp. 13–30.

Nir, D. (1974) *The Semi Arid World: Man on the Fringe of the Desert*, Longman, London, United Kingdom.

Odingo, R. S. (1990) "Review of UNEP's Definition of Desertification and Its Programmatic Implications," in *Desertification Revisited*, pp. 7–44, United Nations Environment Programme (UNEP), Nairobi, Kenya.

Oke, T. R. (1987) *Boundary Layer Climates*, Methuen, London, United Kingdom.

Oldeman, L. R., Hakkeling, R. T. A., and Sombroek, W. G. (1990) *World Map of Status of Human-Induced Soil Degradation: An Explanatory Note*, Nairobi, Wageningen, and UNEP, ISRIC, Nairobi, Kenya.

Otterman, J. (1974) "Baring High-Albedo Soils by Overgrazing: A Hypothesized Desertification Mechanism," *Science* **186**, pp. 531–533.

Otterman, J. (1977a) "Anthropogenic Impact on Surface Characteristics and Some Possible Climatic Consequences," Israel Meteorological Research Papers, pp. 31–41, Israel Meteor. Service, Bet Dagen, Israel.

Otterman, J. (1977b) "Monitoring Surface Albedo Change with LANDSAT," *Geophys. Res. Ltrs.* **4**, pp. 441–444.

Otterman, J. (1981) "Satellite and Field Studies of Man's Impact on the Surface of Arid Regions," *Tellus* **33**, pp. 68–77.

Pabot, H. (1962) *Comment Briser le Cercle Vicieux de la Désertification dans les Régions Séches de l'Orient*, Food and Agriculture Organization of the United Nations, Rome, Italy.

Parungo, F., Li, Z., Li, X., Yang, D., and Harris, J. (1994) "Gobi Dust Storms and the Great Green Wall," *Geophys. Res. Lett.* **21**, pp. 999–1002.

Peel, R. F. (1975) "Water Action in Desert Landscapes," in *Processes in Physical and Human Geography*, eds. R. F. Peel, M. Chisolm, and P. Haggett, pp. 110–129, Heinemann Educational Books, London, United Kingdom.

Perry, R. A. (1977) "The Evaluation and Exploitation of Semi-Arid Lands: Australian Experience," *Philosophical Transactions of the Royal Society of London, Ser. B* **278**, pp. 493–505.

Prospero, J. M., Uematsu, M., and Savoie, D. L. (1989) "Mineral Aerosol Transport to the Pacific Ocean," in *Chemical Oceanography*, eds. J. P. Riley, R. Chester, and R. A. Duce, vol. 10, pp. 188–218, Academic Press, London, United Kingdom.

Puigdefabregas, J., and Aguilera, C. (1996) "The Rambla Honda Field Site: Interactions of Soil and Vegetation Along a Catena in Semi-Arid Southwest Spain," in *Mediterranean Desertification and Land Use*, eds. C. J. Brandt and J. B. Thornes, pp. 137–168, John Wiley & Sons, Chichester, United Kingdom.

Pye, K. (1987) *Aeolian Dust and Dust Deposits*, Academic Press, London, United Kingdom.

Pye, K., and Tsoar, H. (1990) *Aeolian Sand and Sand Dunes*, Unwin Hyman, London, United Kingdom.

Quine, T. A., Navas, A., Walling, D. E., and Machin, J. (1994) "Soil Erosion and Redistribution on Cultivated and Uncultivated Land Near Las Bardenas in the Central Ebro River Basin, Spain,"

Land Degradation and Rehabilitation **5**, pp. 41–55.

Reid, I., and Frostick, L. E. (1997) "Channel Form, Flows and Sediments in Deserts," in *Arid Zone Geomorphology: Process Form and Change in Drylands*, ed. D. S. G. Thomas, pp. 205–229, John Wiley & Sons, Chichester, United Kingdom.

Reining, P. (1978) *Handbook on Desertification Indicators*, American Association for the Advancement of Science, Washington, DC.

Reisner, M. (1986) *Cadillac Desert: The American West and Its Disappearing Water*, Penguin, New York, New York.

Robinove, C. J., Chavez, P. S., Jr., Gehring, D., and Holmgren, R. (1981) "Arid Land Monitoring Using Landsat Albedo Difference Images," *Remote Sensing Environ.* **11**, pp. 133–156.

Rozanov, B. (1990) "Assessment of Global Desertification," in *Desertification Revisited*, pp. 45–122, United Nations Environment Programme, Nairobi, Kenya.

Rubio, J. L. (1987) "Desertificacion en la Communidad Valenciana Antecedentes Historicos y Situacion Actual de Erosion," *Rev Valenc ed Estud. Autonomicos* **7**, pp. 231–258.

Sarre, R. D. (1987) "Aeolian Sand Transport," *Progress in Physical Geography* **11**, pp. 157–182.

Schlesinger, W. H., Reynolds, J. F., Cunningham, G. L., Huenneke, L. F., Jarrell, W. M., Virginia, R. A., and Whitford, W. G. (1990) "Biological Feedbacks in Global Desertification," *Science* **247**, pp. 1043–1048.

Scoging, H., (1991) Desertification and its management. In *Global Change and Challenge. Geography for the 1990s*. Bennet, R., and Estall R. (Eds.), Routledge, London, 57–79.

Sears, P. (1935) *Deserts on the March*, University of Oklahoma Press, Norman, Oklahoma.

Sheridan, D. (1981) *Desertification of the United States*, Council on Environmental Quality, U.S. Government Printing Office, Washington, DC.

Soriano, A. (1983) "Deserts and Semi-Deserts of Patagonia," in *Temperate Deserts and Semi-Deserts, Ecosystems of the World*, ed. N. E. West, vol. 5, pp. 423–460, Elsevier Scientific, New York, New York.

Stebbing, E. P. (1935) "The Encroaching Sahara: The Threat to the West African Colonies," *Geographical J.* **85**, pp. 506–524.

Stebbing, E. P. (1938) *The Man-Made Desert in Africa: Erosion and Drought*, supplement to *J. Roy. African Soc.* **37**, pp. 1–40.

Stowe, C. D. (1969) "Dust and Sand Storm Electrification," *Weather* **24**, pp. 134–140.

Tao, G., Jingtao, L., Xiao, Y., Ling, K., Yida, F., and Yinghua, H. (2002) "Objective Pattern Discrimination Model for Dust Storm Forecasting," *Meteor. Appl.* **9**, pp. 55–62.

Thomas, D. S. G., and Middleton, N. J. (1994) *Desertification: Exploding the Myth*, John Wiley & Sons, Chichester, United Kingdom.

Tomaselli, R. (1977) "The Degradation of the Mediterranean Maquis," *Ambio* **6**, pp. 356–362.

Tsoar, H., and Pye, K. (1987) "Dust Transport and the Question of Desert Loess Formation," *Sedimentology* **34**, pp. 139–154.

Tucker, C. J., Dregne, H. E., and Newcomb, W. W. (1991) "Expansion and Contraction of the Sahara Desert from 1980 to 1990," *Science* **253**, pp. 299–301.

Uman, M. A. (1987) *The Lightning Discharge*, Academic Press, Orlando, Florida.

United Nations. (1977) *World Map of Desertification*, United Nations Conference on Desertification, Nairobi, Kenya, 29 August–9 September 1977, UN Document A/CONF.74/2, UN, New York, New York.

United Nations. (1978) *United Nations Conference on Desertification: Roundup, Plan of Action and Resolutions*, UN, New York, New York.

United Nations Environment Program (UNEP) (1992) *World Atlas of Desertification*, UNEP, Edward Arnold Press, London, United Kingdom.

Wakimoto, R. M. (1982) "The Life Cycle of Thunderstorm Gust Fronts as Viewed with Doppler Radar and Rawinsonde Data," *Mon. Wea. Rev.* **110**, pp. 1060–1082.

Walsh, R. P. D., Hulme, M., and Campbell, M. D. (1988) "Recent Rainfall Changes and Their Impact on Hydrology and Water Supply in the Semi-Arid Zone of the Sudan," *Geographical J.* **154**, pp. 181–198.

Warren, A., and Maizels, J. K. (1977) "Ecological Change and Desertification," in *Desertification: Its Causes and Consequences*, pp. 169–260, Pergamon Press, New York, New York.

Westing, A. H. (1994) "Population, Desertification and Migration," *Environmental Conservation* **21**, pp. 109–114.

Wiggs, G. F. S. (1992) "Airflow over Barchan Dunes: Field Measurements, Mathematical Modelling and Wind Tunnel Testing," Dissertation, University of London, London, United Kingdom.

Wiggs, G. F. S. (1997) "Sediment Mobilisation by the Wind," in *Arid Zone Geomorphology: Process Form and Change in Drylands*, ed. D. S. G. Thomas, pp. 351–372, John Wiley & Sons, Chichester, United Kingdom.

Wilson, J. W., and Wakimoto, R. M. (2001) "The Discovery of the Downburst: T. T. Fujita's Contribution," *Bull. Amer. Meteor. Soc.* **82**, pp. 55–62.

Xue, Y., and Shukla, J. (1993) "The Influence of Land Surface Properties on Sahel Climate. Part I: Desertification," *J. Climate* **6**, pp. 2232–2245.

Yu, B., Hesse, P. P., and Neil, D. T. (1993) "The Relationship Between Antecedent Rainfall Conditions and the Occurrence of Dust Events in Mildura, Australia," *J. Arid Environments* **24**, pp. 109–124.

Zheng, X., and Eltahir, E. A. B. (1997) "The Response of Deforestation and Desertification in a Model of the West African Monsoon," *Geophys. Res. Lett.* **24**, pp. 155–158.

Zonn, I., Glantz, M. H., and Rubinstein, A. (1994) "The Virgin Lands Scheme in the Former Soviet Union," in *Drought Follows the Plow: Cultivating Marginal Areas*, ed. M. H. Glantz, pp. 135–150, Cambridge University Press, Cambridge, United Kingdom.

18
The first hundred years of numerical weather prediction

Janusz Pudykiewicz and Gilbert Brunet

Modern weather prediction is one of the best tools available to reduce the losses of life and property due to extreme weather events at sea, in the air, and on land. The success rate attained by modern forecasting systems is the result of more than 100 years of complex and often dramatic developments linking the technology with the advancements in theoretical meteorology and physics. A general overview of the process that leads from initial empirical predictions to modern-day forecasting systems is presented. We briefly outline the major conceptual aspects and the profiles of the crucial personalities on the meteorological scene. This chapter concludes with general remarks concerning the future of the predictive meteorological models, including their potential use in geoengineering and climate control.

18.1 Forecasting before equations

Initial attempts to predict droughts, flooding, harvests, plagues, wars, strong wind, ocean currents, and positions of the moon and planets started at the dawn of human civilization with methods that are now either forgotten or discredited for the lack of rational foundation. The examination of the annals of science shows that many philosophers and scientists entertained different aspects of forecasting the state of the environment. The first quantitative basis for the environmental prediction was created within the framework of the reality characteristic of the Cartesian-Galilean science, and that of the empiricist and rationalist philosophies of the seventeenth and eighteenth centuries.

In the mid-nineteenth century, the development of telegraphy first permitted meteorologists to create synoptic weather maps in real time. By superposing a time sequence of such maps reflecting observations taken simultaneously over very large areas, it became possible to observe as storms and other weather phenomena develop and move, as well as to issue warnings to the areas located downwind. The forecast technique at that time was based on a set of empirical rules and was inherently inaccurate; however, the newly acquired ability to provide a spectacular large-scale view of the evolving weather brought significant visibility to meteorology. Governments, military organizations, and agriculture began to consider weather science as a new indispensable element of their activities.

The primary stimulus for the development of modern weather forecasting systems in the 1800s was provided by the maritime disaster of 14 November 1854, during the Crimean war. The flotilla of three warships and thirty-four supply vessels of the British and French coalition was destroyed by a storm of unprecedented strength, which ravaged the Dardanelles and the Black Sea. Upon analysis of the meteorological data, it became evident that the perilous storm had formed on November 12 and swept suddenly across the Europe from the Southeast. It suddenly became clear that with a minimum monitoring system the disaster was fully avoidable. This observation gave Urbain Le Verrier, the famous French astronomer, a

Figure 18.1 Vice-Admiral Robert FitzRoy (1805–1865), the captain of the *HMS Beagle*, Governor of New Zealand (1843–1845), and the creator of one of the first national weather services.

strong motivation to seek the support of the Emperor of France, Napoleon III, to organize a meteorological service based at the Paris observatory. Shortly after, the national storm warning service was established in France and the era of modern meteorology began. Other countries followed the example set by France and organized their own weather services.

The prime example of the meteorology from that era was the service established in Britain in 1854 by Robert FitzRoy (Figure 18.1), the former captain of the *HMS Beagle* (Figure 18.2), the same vessel that hosted Darwin in his seminal voyage (Agnew, 2004). The department organized by FitzRoy was associated with the British Board of Trade and was initially charged with distributing standard instruments to naval and merchant seagoing vessels and collecting back record books. However, the initial activity expanded significantly in 1861; storm warnings were issued to seaports using the information from a telegraphic network of about two dozen coastal weather stations in Britain and five continental observatories. Within 1 year, the operation grew significantly, culminating in a 2-day forecast based on the empirical method, and was produced for five regions of the British Isles. Predictions were based on a daily set of data from the thirty-two stations connected by a telegraphic network. The forecast was sent to the Board of Trade, the Admiralty, Lloyd's, the Horse Guards, and eight daily papers, and attracted significant attention (Anderson, 1999).

The estimates of the forecast accuracy varied; it is certain, however, that even the best predictions never came close to the standards set by astronomical predictions (despite the evident progress, the same situation is true even today in the beginning of the twenty-first century). In the nineteenth century, the standards of accuracy for the prediction of natural systems were set indeed by astronomy. The unreasonable expectations with respect to the program established by FitzRoy exposed science to considerable ridicule with tragic

18.2 The birth of theoretical meteorology

Figure 18.2 The *HMS Beagle* on a surveying voyage to South America and around the world from 1831 to 1836. Robert FitzRoy was promoted to the rank of captain during this trip. He invited the young naturalist Charles Darwin to join him on this expedition and provided him with significant help during the crucial stages of his research.

consequences. In 1865, FitzRoy dramatically abandoned his efforts and committed suicide. President of the Board of Trade Lord Stanley summarized the entire tragic affair in the significant statement: "You all know that Admiral FitzRoy died as literally a victim to the public service as if he had fallen at his station. His zeal led him to labour beyond his strength (from speech to Liverpool Chamber of Commerce. Quoted in "The Times" 3 September 1866). The fate of the program established by FitzRoy was the effect of unreasonable expectations that meteorology can achieve the same accuracy as those typical for astronomical calculations. Despite the tragic and untimely termination of his forecast program, Admiral FitzRoy showed for the first time that modern infrastructure can be employed in order to enable society to predict the weather. According to some estimates, the forecast issued by FitzRoy achieved a success rate of the order of 80% and was often used to avoid losses due to severe maritime weather. The important legacy of FitzRoy was also his struggle to remove meteorology from the realm of the so-called "weather prophets" and to bring it toward standards established by modern science.

18.2 The birth of theoretical meteorology

When the nineteenth century ended, most nations with telegraph networks operated national weather services responsible for collecting meteorological data and preparing manual weather forecasts. This new practical development of meteorology driven by technological developments was not closely connected to the development of the theoretical meteorology that emerged as a new separate scientific discipline in some academic institutions.

The main element on which these development were founded was a field description of media in motion due to Euler; his method was, in fact, the crucial element of the development of physics from the eighteenth century until our time. By using the Eulerian methods, we can describe complex dynamic and microphysical processes occurring in the atmosphere, hydrosphere, and other terrestrial systems. The techniques of modeling natural systems based on the solution of the conservation laws became an element of routine activities in most environmental prediction centers around the world in our time.

The importance of the Eulerian technique was described particularly well by Trusdell (1954) in his seminal paper on the kinematics of vorticity:

D'Alambert and Euler share in the brilliant discovery of the field description of media in motion—the very foundation stone of modern physics, though seldom so much as mentioned in histories of that subject but to Euler alone we owe the idea of formulating a field theory as the theory of the integrals of a set of partial differential equations. (p. 31)

The development of theoretical meteorology in the second half of the twentieth century was based on a consequent application of the field description adopted in physics. Unfortunately, the major developments in this area had a relatively small impact on operational forecasting. The formal mathematical methods of theoretical physics had no role in the practice of meteorologists forecasting the weather. The major prediction method until the 1920s was based on pattern matching on synoptic maps. Forecasters were seeking in the large libraries of past weather maps situations similar to the current one; the meteorological forecast was based on the analogy to how the past had evolved. This situation prevailed until the 1940s, when the theoretical developments finally began to influence forecasting practice.

The initial records indicating the application of scientific methods for weather prediction date to the beginning of the twentieth century. The first paper that set explicitly the formulation of weather prediction based on Eulerian methods was published in the *Monthly Weather Review* by an American meteorologist, Cleveland Abbe (1901). The paper, entitled "The Physical Basis of Long-Range Weather Forecasts," was the culmination of a long carrier aimed toward the application of the methods of physics to predict the weather. After several years of practical experience gained performing astronomical calculations in the 1860s, Abbe (Figure 18.3) entertained the idea of applying equations of dynamics to predict atmospheric motion. The synthesis of his thoughts in this area finally materialized in 1901 in the paper published in the *Monthly Weather Review*. The concept of the application of primitive meteorological equations to predict the weather can be firmly associated with Abbe's contribution. To the present day, the paper of Abbe impresses by its clarity of thought, as well as by its recognition of major problems of meteorology. The second contribution to the formulation of the scientific forecast was published by Vilhelm Bjerknes (1904) in German; the English translation of the title is "The Problem of Weather Prediction, Treated from the Standpoint of Mechanics and Physics."

A comparison of both papers shows that although the paper by Abbe is more detailed and characterized by a surprising modernity of thoughts, that by Bjerknes is more concise and readable. This fact probably explains why Bjerknes' paper was recognized as the foundation of the weather prediction, whereas Abbe's contribution was unjustly overlooked by several generations of meteorologists. The visibility of the Bjerknes' work was additionally enhanced by his reputation in the area of fluid dynamics and the support of a large scientific community that included such great scientists as Hertz, Arrhenius, and Poincaré. Abbe, on the other hand, faced significant hardships, and he did not receive much support. At the critical stage of his scientific carrier, he was even exposed to a harsh and undeserved criticism of his superiors, including a demotion and salary reduction. These unfortunate events coincided with the transfer of weather prediction activities into the U.S. Department

Figure 18.3 Cleveland Abbe (1838–1916) was the first meteorologist who formulated the problem of weather prediction in precise mathematical terms. His contribution to the development of numerical weather prediction was acknowledged only more recently.

of Agriculture in 1891. According to Willis and Hooke (2006), that action created an unfavorable climate for the continuation of Abbe's work with scientifically based weather prediction.

Despite the fact that Bjerknes did not quote the work of Abbe, both men enjoyed good scientific relations. In 1905, Abbe welcomed Bjerknes in the United States. He had arranged for him a lecture in Washington, DC, during which the Norwegian scientist presented the basis for scientific weather prediction. The lecture was well received and resulted in a yearly grant from the Carnegie Institution of Washington. Bjerknes retained this grant until World War II. Historical records show that this assistance enabled him to employ and educate a considerable number of research assistants, all of whom later became well-known geophysicists.

It is quite significant to note that neither Abbe nor Bjerknes believed that the actual forecast based on the solution of the primitive meteorological equations is possible within the foreseeable future. Despite this fact, Abbe (1901) proposed several methods that were applied about 50 years later. In particular, he suggested to use the expansion of the meteorological fields using spherical harmonics:

But the problem is undoubtedly too complex for plane harmonics; we shall need to develop the original functions and integrals in a series of spherical harmonics or equivalent equations in which latitude, longitude, and altitude must play equal parts. A first step in this direction was taken by Oberbeck, who adopted a few terms of the harmonic series, but a second equally important step was taken by Margules, who, assuming a simple distribution of temperature over the earth's surface,

Figure 18.4 Max Margules (1856–1920) made a series of fundamental contributions to theoretical meteorology and physical chemistry.

worked out a system of teasers, dividing the globe up into regions of low pressure and high pressure with the attending winds. (Abbe, 1901, p. 558)

The solution suggested by Abbe was quite timely but, according to all known records, it never led to any actual forecasting attempt. The main reason of this situation was most likely the general lack of interest among American synopticians and the conviction that the traditional methods are the most appropriate for solving the prediction problem.

18.3 Initial attempts of scientifically based weather prediction

The authors of the seminal papers formulating the concept of weather prediction as a problem of physics were not among those who actually performed the first forecast. The initial attempts came from the famous Vienna school of dynamic meteorology. One of the leading personalities of this school was Max Margules (Figure 18.4), whose theoretical study of the Laplace tidal equation applied to the Earth's atmosphere constitutes a fundamental contribution to dynamic meteorology (Margules, 1893). In this study, he identified two types of solutions of the tidal equation as inertia gravity waves and rotational waves. The significance of this discovery was realized only in the 1930s, after some insight was provided by both Rossby and Haurwitz. Margules also laid foundations for the convective theory of cyclones and created the first theory of atmospheric fronts. From the current perspective, it is quite clear that Margules was the first dynamic meteorologist in the modern sense.

In addition to the basic theoretical studies, Max Margules was, according to all records, the first meteorologist who attempted a physically based weather forecast. In his paper published in 1904 in tribute to the achievements of Ludvig Boltzmann, Margules (1904) considered the prediction of a surface pressure directly from the continuity equation. His result was rather pessimistic; Margules concluded that the forecast based solely on the continuity equation is impossible because the initial wind cannot be specified with sufficient accuracy. Today, it is clear that the problem identified by Margules is that of the balance in the initial conditions. An excellent analysis of this problem is presented by Lynch (2001).

Despite the fundamental contribution to dynamic meteorology, Margules' papers never received the attention that they deserved. The author became disillusioned with atmospheric sciences and retired from his post in the Austrian Weather Service after 24 years of work. His attention was later turned to chemistry. He died in 1920 from starvation, being too proud to accept any help that was extended to him. According to Gold (1920):

Margules retired from active participation in the work of the Austrian Meteorological Service during the directorship of the late Prof. Pernter and applied himself to the study of chemistry. He fitted up a small laboratory in his own house, were he lived in comparative retirement. The present writer was saddened to see him there in 1909 entirely divorced from the subject of which he had made himself a master. Meteorology lost him some 15 years ago and is forever the poorer for a loss which one feels might and ought to have been prevented. (pp. 286–287)

It is remarkable that working in a small improvised chemical laboratory, after his retirement from the Austrian Weather Service, Margules managed to make a seminal contribution to thermodynamics and physical chemistry. His theories and equations concerning properties of chemical solutions are still valid and used in chemical laboratories around the world. In particular, the equation describing the vapor pressure of different species in multicomponent mixtures provides an inspiration for research in chemical thermodynamics even today. When assessing the importance of the contribution of Margules to science, we see that his work on seemingly unrelated subjects, such as meteorology and the phase equilibria, was so successful because of his deep understanding of thermodynamics and statistical physics.

The second major contribution to the weather prediction that came from the Vienna school is that by Exner (1908), who applied a technique based on maximum simplification of the primitive equations proposed by Abbe and Bjerknes. Exner assumed geostrophic flow and constant-in-time thermal forcing. The forecast was based on the equation governing the advection of the pressure pattern with a constant speed in the westerly current. Although his method was of limited applicability, it lead to a relatively reasonable prediction. Exner's work did not receive any enthusiastic review; it established, however, the trend to simplify primitive equations before using them for actual prediction. This trend was about to strongly affect the future directions of forecasting. The general philosophy proposed by Exner was dormant for the next 40 years, just to be reborn in a slightly different form in the late 1940s. In the meantime, the weather forecasting experienced turbulent years, mimicking in many ways the dramatic parallel developments of other disciplines.

18.4 Bergen school of meteorology

Vilhelm Bjerknes remained one of the main figures on the meteorological scene at that time. In the years following publication of his seminal paper on weather prediction, he devoted his attention to the study and propagation of the theoretical meteorology, mainly at Stockholm University. In 1907, he moved from Sweden to his native Norway to Christiania, where he received professorship at the university. This move reflected the changing political picture of Scandinavia, where on 7 June 1905 a new Kingdom of Norway was created after a peaceful separation from Sweden. During his first years in Christiania, in cooperation with his assistants Johan Sandström, Olaf Devik, and Theodor Hesselberg, founded by the Carnegie grant, Bjerknes published a substantial work entitled *Dynamic Meteorology and*

Hydrography." The book attracted considerable international attention, which led directly to the offer of directorship of the new geophysical institute at the University of Leipzig (Eliassen, 1995). In 1913, Bjerknes established his scientific activity in Germany, bringing with him two new Carnegie assistants, T. Hesselberg (later director of the Norwegian Meteorological Institute) and H.U. Sverdrup (later director of the Scripps Oceanographic Institution in the United States). After World War I started, many German students of Bjerknes were called to the military service, and Sverdrup and Hasselberg also abandoned their professor. However, Bjerknes remained at his post because Norway was a neutral country at that time. In 1916, in the midst of the war hostilities in Europe, Bjerknes received unexpected help from his son; Jacob Aall Bonnevie Bjerknes (who is known to most people as Jack Bjerknes), not yet 19 years old, left his studies in Norway and joined his family in Leipzig. This sudden move proved to have a critical significance for the development of meteorology. Immediately after his arrival, Jack took over the research of Herbert Petzold, a German doctoral student who perished at Verdun in 1916. He discovered that the convergence lines in the wind field are thousands of kilometers long, have a tendency to move eastward, and are associated with cloud systems. Jack Bjerknes reported these findings in a scientific paper that was published before he was 20 years old. With the progression of war activities, the situation at the Geophysical Institute in Leipzig became quite difficult, leading to both staff and severe food shortages. However, the Bjerknes family received a significant help from Norway, where oceanographers Fridtjöf Nansen and Björn Helland-Hansen, were instrumental in establishing a professorship for Vilhelm Bjerknes in the famous city of Bergen. In 1917, Bjerknes moved to his new post of oceanography professor, which was established at Bergen Museum because there was no university at Bergen at that time. He was accompanied by two assistants: Jack Bjerknes and Halvor Solberg. The activities of Vilhelm and Jack Bjerknes in the following years led both to the formation of the university at Bergen and to the establishment of one of the most influential schools of meteorology (Reed, 1977). The so-called "Bergen school" had created the theoretical models of cyclones and fronts that influenced meteorology between the 1920s and 1950s.

The Bergen school was widely accepted by the meteorologists and saw a number of successes in the practice of forecasting between the 1920s and 1940s. The weather predictions based on the Bjerknes frontal cyclone model contributed to saving many lives and reducing losses caused by extreme weather at sea, in the air, and on land. The main criticism directed at the method of the Bergen school was related to the fact that there was no sufficient theoretical foundation of the core frontal instability concept. The method of the Bergen school was further discredited in the 1940s by the emerging observational and theoretical evidence for baroclinic development involving the entire troposphere. Despite this criticism, the Bjerknes frontal cyclone model is still used as a tool in practical weather prediction.

18.5 First numerical integration of the primitive meteorological equations

The initial attempts at forecasting, along with the papers of Abbe and Bjerknes, set the stage for the first genuine attempt to predict the future state of the atmosphere from the known initial conditions. The author of this first attempt was L.F. Richardson (Figure 18.5), a scholar with an exceptional originality and independence of thought that set him apart from his generation. His biography is discussed, among many other authors, by Richardson (1957). The numerous contributions of Richardson in different areas, described in his biographies, create an impression of one of the most exciting scientific carriers of the twentieth century.

18.5 First numerical integration of the primitive meteorological equations

Figure 18.5 Lewis Fry Richardson (1881–1953) created the basic concepts of numerical weather prediction. He also contributed to other areas of research in fluid mechanics and in the field of the mathematical theories of war.

During World War I, Lewis Fry Richardson (1922) carried out a manual calculation of the pressure change over Central Europe using the initial conditions obtained from a series of synoptic charts prepared in Leipzig by Vilhelm Bjerknes. It is quite remarkable that Richardson, in his initial consideration of the problem of weather forecasting, mentioned astronomy as a source of the methodology to follow, which was the same inspiration as that guiding Abbe in his work.

Richardson, being a skilled mathematician, formulated the numerical scheme for the solution of primitive meteorological equations in a manner that is quite similar to the methods used later in the future three-dimensional atmospheric models. It is remarkable that the general structure of the system proposed by Richardson is essentially the same as that in the models based on primitive equations 50 years later. Richardson also considered the equation for a generic atmospheric tracer that made him the first to propose the prediction of chemical constituents in the atmosphere.

The forecast of Richardson was not successful; in fact, his results were so bad that the future generations of meteorologists were discouraged from repeating the attempt to numerically solve the set of primitive equations. We know now that this was a serious mistake because the method proposed by Richardson was basically sound. His forecast failed only because he selected an erroneous method to estimate the rate of pressure change. To calculate the surface pressure changes, Richardson employed the continuity equation, using exactly the same method that was discredited by Margules more than 10 years earlier (Lynch, 2003). The resulting prediction of the pressure change, not surprisingly, was completely unrealistic.

After completing the work on formulation of the methods for the numerical weather prediction, Richardson realized that very similar techniques could be used for the prediction of an even more complex nonlinear system involving human relations. One of the best known examples of his work in this field was dedicated to the modeling of arm race using a set of coupled differential equations. Richardson verified his calculation by collecting a significant amount of data about the conflicts from the past. In order to organize the data, he classified all human conflicts in terms of the number of casualties represented on a logarithmic scale. This method was borrowed from astronomy and other natural sciences where logarithmic scale is adopted to represent both objects and processes.

Further refinement of Richardson's model involved the analysis of the lengths of common boundaries between countries. His geometrical considerations led him to the conclusion that the measured length of political boundaries and coastlines depends heavily on the length of a sampling interval as the line on a map can be followed more closely with a short measuring yardstick. This observation went unnoticed despite being published, but it received wide publicity only after Benoit Mandelbrot found it by chance many years later. Today it is well known that the work of Richardson in the field of conflict analysis has led to the theories of fractals, scale invariance, and chaos; all of them being an integral part of our present knowledge, while predicting the behaviour of complex nonlinear systems including weather, economy, and society.

18.6 Weather forecasting after Richardson

When World War I ended, meteorologists developed three different scientifically based approaches to predict the weather. The first one was derived from the philosophy of using simplified equations as proposed by Exner, the second used the methods of the Bergen school developed by Vilhelm and Jack Bjerknes, and, finally, the third was founded on the numerical approximation of equations proposed by Abbe and Bjerknes. The three schools of forecast met in Bergen in 1920, during a congress where the Bergen school presented the most recent version of the frontal model. Despite its shortcomings, the conceptual models of Jack Bjerknes contributed significantly to the improvement of the quality of the routine meteorological forecasts. On the other hand, the advancements of the Vienna school of theoretical meteorology created the foundation for further theoretical advancements, leading to a better understanding of the Earth's atmosphere. The prevailing trend in the work performed during the 1930s and 1940s was to create simplified mathematical models explaining atmospheric motions. In mathematical terms, the process could be described as follows: let one start with the dynamic equations describing the full spectrum of atmospheric motion and simplify them maximally while retaining the essential terms governing the large-scale motions.

In the study of the baroclinic instability, Charney (1947) derived equations for unstable waves by neglecting meteorologically unimportant acoustic and shearing gravitational oscillations. By applying the technique of scale analysis to the primitive equations, Charney was able to simplify them in such a manner that gravity wave solutions were completely eliminated, leading to the so-called "quasigeostrophic system" (Charney, 1948). In the special case of the horizontal flow with the constant static stability, the quasigeostrophic potential vorticity equation reduces to a form equivalent to nondivergent barotropic vorticity equation

$$\frac{d(f + \zeta)}{dt} = 0 \qquad (18.1)$$

where ζ is the relative vorticity, $f = 2\Omega \sin(\phi)$, Ω is the angular frequency of the planet rotation, and ϕ is the latitude.

This equation is particularly conducive to further studies of large-scale atmospheric flow. The barotropic equation was used by Rossby (1939) (Figure 18.6) in his study of atmospheric waves, leading to a better understanding of synoptic-scale circulation systems.

18.6 Weather forecasting after Richardson

Figure 18.6 Carl-Gustaf Rossby (1898–1957) was a Swedish–American meteorologist who greatly contributed to the understanding of the large-scale motion of the atmosphere in terms of fluid mechanics. In the early stage of his education, he was a student of Vilhelm Bjerknes at the University of Stockholm (Photo: MIT PAOC Program).

The intensive development of meteorology during the early 1940s coincided with World War II, mainly due to the strong demand for meteorologists urgently needed to support air and naval operations. At that time, it became clear that the reliability of prediction techniques was rather low and that future progress definitely required new methods. The person who made it all happen was John von Neumann (Figure 18.7), a scientist with significant mathematical skills and keen interest in applying them to practical problems. Recognizing the importance of numerical techniques for solution of hydrodynamical problems, von Neumann secured support of the Institute of Advance Studies (IAS), Radio Corporation of America, and Princeton University in order to create the first computer based on his theoretical concept. Furthermore, the significant influence and contacts of von Neumann helped him obtain grants from the Weather Bureau, the Navy, and the Air Force in order to assemble a group of theoretical meteorologists at IAS under the leadership of Jule Charney (Lynch, 2002).

The group was composed of Eliassen, Thompson, and Hunt. Arnt Eliassen arrived at Princeton in the summer of 1948, bringing experience and knowledge based on the tradition of the Bergen school, which was at this time second to none. At the later stage Eliassen accepted a position at the University of California with Jack Bjerknes and was replaced by another Norwegian meteorologist, Ragnar Fjörtoft, September, in 1949.

Despite the best efforts, computers produced at IAS had very limited power, and from the beginning, it became clear that the integration of the equations attacked by Richardson in 1916 was out of question. The reason for this was the presence of gravity waves that imposed

Figure 18.7 John (Janos) von Neumann (1903–1957), a Hungarian-born U.S. mathematician who contributed to quantum physics, functional analysis, set theory, topology, economics, computer science, numerical methods, hydrodynamics, statistics, and several other areas of mathematics. He combined mathematical genius with the keen interest to apply theory to solve practical problems. He was one of the most important contributors to the first successful numerical integrations of equations describing large-scale atmospheric motions. He was also the first scientist who suggested the concept of geoengineering by covering polar ice caps with colorants to enhance the absorption of radiation and consequently raise global temperatures. Today, the ideas of John von Neumann are being echoed, but in terms of reducing global temperatures.

a strong limitation on the time step; furthermore, there was also a problem with the model initialization. The meteorology group at IAS decided to accept a compromise solution, and the experimental forecast was run based on the barotropic vorticity equation. The equation was solved on a grid with 700-km resolution, with a time step of 3 hours. Results were far from perfect, but the entire experiment was described in the paper published in *Tellus* (Charney et al., 1950). Despite the fact that the work reported was far from the original program of Richardson, the sample forecast was forwarded to him; his wife commented that indeed the 24-hours forecast looks closer to the verification map than to the initial conditions. This opinion describes concisely the level of accuracy of early numerical weather prediction models.

Nevertheless, the initial forecast at IAS created significant enthusiasm in some parts of the meteorological community and, subsequently, led to the creation of a significant number of models based on quasigeostrophic equations in many countries around the world throughout the 1950s. These models were essentially restricted to the middle and high latitudes (Bengtsson, 1999), and by nature of the quasigeostrophic theory, they were unable to represent small-scale circulation systems. Furthermore, the incorporation of physical processes, radiation, clouds, and precipitation in the filtered equation models was difficult, and the state of the entire numerical weather prediction was still far from full scientific acceptability. The recognized limitations of models based on filtered equations led to the intensive work on models based on the solution of the set of primitive meteorological equations.

18.7 Richardson's experiment revisited and the birth of forecasting based on primitive equations

One of the crucial problems waiting to be addressed was that concerning the specification of initial conditions. The pioneering study of Hinkelmann (1951) suggested the solution to the problem of "(meteorological noise)" (*meteorologischen Lärmes*). The first successful experiments with the model based on primitive equations followed in a few years time (Hinkelmann, 1959; Phillips, 1956). Abandoning rather inadequate models based on filtered equations was further facilitated by the development of efficient methods of time integration of the primitive equations. The two most notable methods were the semi-implicit algorithm proposed by Robert et al. 1972 and a split-explicit integration technique introduced by Marchuk (1974). In both cases, the same time step could be used as that for the quasigeostrophic equations, and hence, the advantage in using the quasigeostrophic models came to an end.

In 1965, Mintz reported results obtained from the global circulation models. A few years later, Miyakoda et al. (1972) performed a medium-range prediction with the hemispheric model. This work provided significant stimulus to the Global Atmospheric Research Programme and the creation of the European Center for Medium Range Weather Forecasting. First primitive equations models were based on the finite-difference techniques for discretization of the spatial derivatives in a manner similar to the technique used by Richardson. The situation changed, however, in the 1970s, when the spectral technique was proposed by Arnt Eliassen. Based on the work with efficient Fourier transforms, spectral models were developed by Robert (1969). Hemispheric spectral models became operational in Australia and Canada in 1976. These models were naturally suited for solving weather prediction in spherical geometry; furthermore, they enabled simple implementation of the semi-implicit treatment of gravity-wave terms (Robert 1982). The landmark contributions of Robert, combined with the availability of a well-equipped computer center in the Canadian Weather Service in the early 1970s, were significant for the development of forecasting based on primitive equations.

The development of more realistic models also created strong motivation for the improvement of the methodology used to prepare the initial conditions. The old methods based on the analysis of synoptic charts and various forms of interpolation procedures were replaced by data assimilation techniques based on the optimum control theory developed by Lions (1968). The essence of these new methods is the minimization of a functional indicating the discrepancy between the model and the observations. The model input and parameters are considered in this formulation as control parameters (Le-Dimet and Talagrand, 1986). The successful application of the optimum control methods for problems of meteorology was built on a strong foundation created by applied mathematics and engineering sciences.

18.8 Expansion of the scope of traditional meteorological prediction

With the availability of relatively realistic models of the atmospheric circulation that emerged in the late 1970s, the entirely new set of problems entered into the realm of modeling. In general terms, these problems were related to the extension of the set of model variables to include chemical species and aerosols. In many ways, this trend was consistent with the initial idea of Richardson to include some kind of a tracer variable in meteorological models. The broadening of the scope of atmospheric modeling also followed the examples set by Rossby, who was one of the first to recognize the importance of large-scale mixing processes in the atmosphere. The ideas of Rossby also contributed to the investigation of the long-range transport of air pollutants using meteorological models.

The expansion of the scope of weather models set the trend toward unifying climate and atmospheric chemistry studies. In the early 1980s, atmospheric scientists considered two opposing problems associated with aerosols and chemical constituents. The first one was connected to the hypothetical cooling of the atmosphere due to fine particles; the second was related to the issue of global warming caused by emission of greenhouse gases. The dispute set more than 20 years ago is still actual today and the atmospheric models are often used as the tool for environmental and economical studies related to the climate change.

The expansion of the scope of atmospheric forecasts also included nuclear and chemical substances that could be transported by the synoptic and mesoscale circulation systems. In 1986, the major nuclear accident at Chernobyl led to a significant release of radioactive substances. One of the first responses to the disaster was from Canada, with the simple hemispheric two-level model that provided quite accurate indication about the spread of radioactivity around the Northern Hemisphere (Pudykiewicz, 1988). The model consisted of a set of advection–diffusion equations governing the horizontal transport of radionuclides released from the damaged reactor. The meteorological input data in the form of a time sequence of the velocity, geopotential, temperature, and humidity fields was obtained directly from the objective analysis performed at the Canadian Meteorological Centre. This configuration permitted the most effective use of meteorological expertise to solve, in the operational mode, the problem of atmospheric dispersion. The study of the atmospheric transport also benefitted from significant expertise in the semi-Lagrangian methods, which had been developed by A. Robert for efficient integration of the primitive equations.

Chemical tracer transport models predict atmospheric concentrations of contaminants for a given emission–source distribution. However, the inverse problem can be posed, which consists of evaluating unknown sources of atmospheric tracers on the basis of a given set of measured concentrations. The solution of the inverse problem can be used to estimate and verify emission inventories of many toxic and radioactive species, as well as to detect unknown sources of atmospheric tracers. For example, the detection and localization of nuclear testing can be achieved using observations of radionuclides in the atmosphere (Pudykiewicz, 1998). The identification of a nuclear event using atmospheric tracer models is the important part of the international system designed to verify compliance with the Comprehensive Nuclear Test Ban Treaty. It also demonstrates the significant expansion of the scope and importance of meteorological forecasts.

18.9 Development of the modern atmospheric prediction systems

The increase in prediction skill of numerical weather prediction (NWP) in the past few decades was due to close interplay of research and development advancements in numerical methods, subgrid-scale physics parameterizations (cloud, mountain, etc.), data assimilation from surface and space observation networks, and high-performance computer (HPC) systems. The HPC system amelioration drives the NWP science by permitting a time-space resolution increase in the modeling of dynamical and physical processes in the atmosphere.

As an example, let us look at Figure 18.8, depicting the increase in forecast quality at the Canadian Weather Service over a period of almost a half-century of HPC system improvements (from Bendix to IBM machines). The positive linear trend in the forecast quality of an almost 1 day per decade of HPC system amelioration is clearly visible. In particular, the skill of a forecast at 1 day in 1960 is equivalent to that at 5 days in 2000. Note that for the 1990s, the peak computing power went from tens to tens of thousands of gigaFLOPS. Hence, the increase factor of 1,000 in computing power was needed to achieve an improvement of 1 day in forecast skill.

Numerical weather prediction models rely increasingly on HPC systems. The increasing computational potential allows creation of multiscale models and data assimilation systems, with realistic coupling to chemical, hydrological, and surface processes. A good example of such system is the global environmental model developed in Canada (Côté et al., 1998a, 1998b). The extensive calculations performed with very high-resolution models also requires extensive development of new numerical techniques. It is becoming evident that finite volume solvers for conservation laws on unstructured and adaptive meshes will slowly gain more acceptance as the best tool for construction of the global cloud scale resolving meteorological models. Clearly, a modern NWP system closely builds on research and development teams composed of research scientists from different areas such as meteorology, physics, mathematics, chemistry, and computing. An HPC system used by interdisciplinary teams must be capable in the next few years of dealing with many terabytes of data and to attain performance peak of tens of teraFLOPS.

At the dawn of the twenty-first century, significant research and development work is needed before acceptable meteorological and environmental operational forecasts can be produced over a wide range of scales (from urban to planetary). Internationally, the increasing demand for accurate weather and environmental prediction has led to a significant attention being given to investments in the numerical weather prediction and the HPC system. This is particularly true in the European Community, Japan, and the United States. At this moment, only a few countries in the world are equipped with resources to address the issue of scale coupling in meteorological models. A successful numerical experiment to investigate a complex multiscale system was conducted recently by Japanese and Canadian researchers (Figure 18.9). The international team performed high-resolution computer simulation of a full life cycle of Hurricane Earl (1998), which transformed into an extratropical system with heavy precipitation affecting the Canadian Maritime Provinces. The simulation was performed in 2004 on one of the world's most powerful computers, the Earth Simulator, in Japan. The Canadian Mesoscale Community Model (Tanguay et al., 1990) has achieved 10 teraFLOPS on an 11,000 × 8,640-km domain with fifty vertical levels that cover North America and North Atlantic (Desgagné et al., 2006).

18.10 From weather prediction to environmental engineering and climate control

New atmospheric models will ultimately lead to the improved meteorological prediction, including small-scale weather events. In addition to perfecting the purely meteorological forecasts, the trend to include various environmental couplings in atmospheric models, initiated in the 1970s, will continue. It is very likely that the models will be used increasingly to address problems of environmental emergencies and general management of ecosystems. The questions related to the decline, invasion, and adaption of species will be most likely resolved using large models with extensive representation of atmosphere, land surface, and soil system. The future environmental models may also help in reducing the losses to forests related to fires and other factors such as insects. Coupling the atmospheric models with

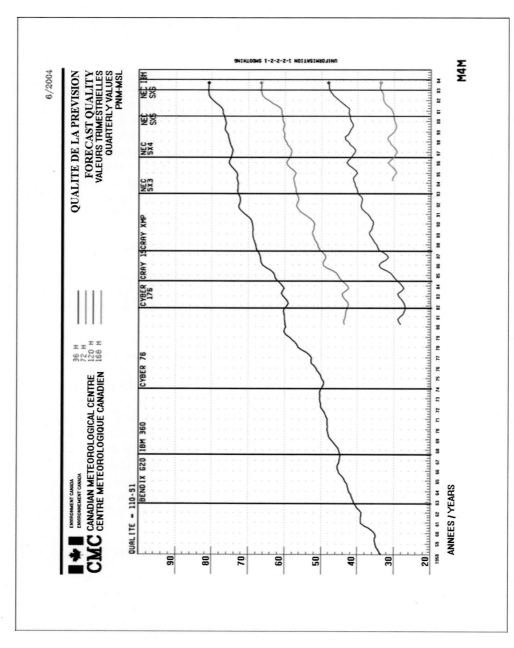

Figure 18.8 The quality of weather forecasts has systematically increased over the past four decades at Canadian Meteorological Centre in conjunction with the power of high-performance computers (HPCs) doubling each year, increasing surface- and space-based observing systems, high-bandwidth telecommunication, and major scientific advances. Note that the HPCs were Bendix G20 and IBM370 in the 1960s; Control Data 7600 and Cyber 176 in the 1970s; Cray 1S, Cray XMP-2/8, Cray XMP-4/16 in the 1980s; NEC SX-3/44, SX-3/44R, SX-4/64M2, and SX-5/32M2 in the 1990s; and, finally, an

18.10 From weather prediction to environmental engineering and climate control

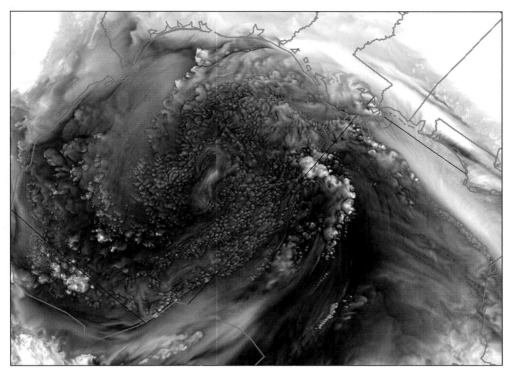

Figure 18.9 This image shows the simulated specific humidity field (altitude 325 m) at 06Z on 1, September 1998, for a 1,000 × 1,000 km subdomain over the Gulf of Mexico. The image shows only ~1% of the total surface of the computing domain. The pixels are at the model 1 km horizontal resolution. The simulation shows realistic instabilities and sheared disturbances evolving around the spinning hurricane.

hydrological processes will permit the prediction of energy production from hydroelectric power plants, evaluation of the potential for bulk water exports, mapping of flood plains, planning of river diversions, and construction of new dams.

The newly acquired prediction potential of environmental models will most likely lead to the discussion of optimum control of the emission of pollutants and energy production. In such applications, the ecological cost calculated by a model added to the cost of control actions will form a rational basis for the evaluation of the solution minimizing the overall cost. The potential for new applications linking the environmental models and the economy is limitless. In the future, it could include, for example, even the optimum control of weather systems (Hoffman, 2002). The control parameters in such a case will be the positions of the orbital reflectors of radiation and the functional to minimize the total cost of operating such reflectors plus the total cost of damages caused to the infrastructure due to wind speeds above a particular threshold. At this moment, such a possibility of loss minimization by diverting cyclones seems to be in the realm of dreams. There are, however, even more drastic suggestions of applying the optimum control theory to the environment. Nobel laureate Dr. P. Crutzen (2006) suggested controlling the global temperature to counteract the global warming trend by selective injection of SO_2 to the stratosphere. The manmade injection of this gas will contribute to the enhancement of a persistent sulphate aerosol layer that always exists in the stratosphere and can, consequently, lead to the compensation of the global warming in the manner similar to that after major volcanic eruptions.

However, the question is "is our understanding of the atmosphere sufficient to allow controlling the weather systems and global temperature trends?" Despite the progress in numerical modeling and the vast increase of data sources, there are still significant gaps in the understanding of the atmosphere that form an obstacle for the application of the formal optimum control methods.

Future meteorological research will be assisted by computer models solving the set of nonhydrostatic equations with a resolution permitting explicit representation of cloud-scale processes. Models of this kind will eliminate the artificial separation of scales that is one of the fundamental inadequacies of the current algorithms used for research in both weather prediction and climate studies. Practical applications of global-scale cloud resolving models will require, however, computers with speeds of the order of tens of petaFLOPS. This estimate most likely will be changed to an even larger number with the increasing realism of the submodels representing chemical processes, the biosphere, and ocean circulation.

The new class of atmospheric models will improve the forecast skills, and it will likely contribute to the increase of the general understanding of the atmosphere by enabling meteorologists to chart the phase space of the atmospheric system. This new potential insight will likely lead to new and significant results in the field of theoretical meteorology. We cannot exclude the possibility that one day, new theoretical results will match the increases of the model resolution, and the dream of predicting the weather will be replaced by the even bigger dream of controlling it.

18.11 Conclusions

In this chapter, we outline the historical background of the development of weather prediction that recently culminated with models of the atmosphere capable of high-resolution, 5-day forecasts. Besides being able to predict the wind, temperature, and precipitation, current models can also warn us about the dangers associated with bad air quality or spread of toxic chemicals and nuclear contaminants. The capability to forecast a wide range of environmental parameters makes current weather prediction centers well suited to predict the consequences of a large spectrum of both natural and manmade disasters. The best known examples include the prediction of tropical cyclones in the southern latitudes and severe winter weather in countries such as Canada. The models can also forecast the consequences of nuclear accidents caused by the atmospheric transport of contaminants, as was demonstrated following the 1986 nuclear disaster in Chernobyl. The increasing scope of applications of meteorology creates the demand for potential optimum control of weather and climate in analogy to the equivalent actions in traditional engineering. The expected effect of these actions may be the diversion of tropical cyclones and mitigation of global warming. However, the current state of debate about these matters shows that our understanding of geophysical systems is still far from complete and that the potential use of geoengineering to avoid disasters will remain in the research arena for the foreseeable future. In contrast, the increase of computing power to the petaFLOPS range combined with new data and expected theoretical advances may increase the role of weather prediction models as a tool helping to eliminate the causes of meteorologically caused disasters.

References

Abbe, C. (1901) "The Physical Basis of Long-Range Weather Forecasts," *Mon. Wea. Rev.* **29**, pp. 551–561, http://docs.lib.noaa.gov/rescue/mwr/029/mwr-029-12-0551c.pdf.

Agnew, D. C. (2004) "Robert FitzRoy and the Myth of the 'Marsden Square': Transatlantic Rivalries in Early Marine Meteorology," *Notes Rec. R. Soc. Lond.* **58**, pp. 21–46.

Anderson, K. (1999) "The Weather Prophets: Science and Reputation in Victorian Meteorology," *Hist. Sci.* **37**, pp. 179–216.

Bengtsson, L. (1999) "From Short-Range Barotropic Modelling to Extended-Range Global Weather Prediction: A 40-Year Perspective," *Tellus* **51 A–B**, pp. 13–32.

Bjerknes, V. (1904) "Das Problem der Wettervorhersage, betrachtet vom Standpunkte der Mechanik und der Physik," *Meteor. Z.* **21**, pp. 1–7, www.history.noaa.gov/stories_tales/bjerknes.html.

Charney, J. G. (1947) "The Dynamics of Long Waves in a Baroclinic Westerly Current," *J. Appl. Met.* **4**, pp. 135–162.

Charney, J. G. (1948) "On the Scale of Atmospheric Motions," *Geofys. Publ.* **17**, pp. 3–17.

Charney, J., Fjörtoft, R., and von Neumann, J. (1950) "Numerical Integration of the Barotropic Vorticity Equation," *Tellus* **2**, pp. 237–254.

Côté, J., Desmarais, J.-G., Gravel, S., Méthot, A., Patoine, A., Roch, M., and Staniforth, A. (1998a) "The Operational CMC-MRB Global Environmental Multiscale (GEM) Model. Part II: Results," *Mon. Wea. Rev.* **126**, pp. 1397–1418.

Côté, J., Gravel, S., Méthot, A., Patoine, A., Roch, M., and Staniforth, A. (1998b) "The Operational CMC-MRB Global Environmental Multiscale (GEM) Model. Part I: Design Considerations and Formulation," *Mon. Wea. Rev.* **126**, pp. 1373–1395.

Crutzen, P. (2006) "Albedo Enhancement by Stratospheric Sulphur Injections: A Contribution to Resolve a Policy Dilemma," *Climatic Change* **77**, pp. 211–220.

Desgagné, M., McTaggart-Cowan, R., Ohfuchi, W., Brunet, G., Yau, P., Gyakum, J., Furukawa, Y., and Valin, M. (2006) "Large Atmospheric Computation on the Earth Simulator: The LACES Project," *Scientific Programming* **14**, pp. 13–25.

Eliassen, A. (1995) "Jacob Aall Bonnevie Bjerknes," *Biographical Memoirs* **68**, pp. 1–21.

Exner, F. M. (1908) "Über eine erste Annäherung zur Vorausberechnung synoptischer Wetterkarten," *Meteor. Zeit.* **25**, pp. 57–67. (English translation published as "A First Approach Towards Calculating Synoptic Forecast Charts," with a biographical note on Exner by Lisa Shields and an introduction by Peter Lynch, Historical Note No. 1, Met Eireann, Dublin, Ireland, 1995.)

Gold, E. (1920) "Dr. Max Margules (Obituary)," *Nature* **106**, pp. 286–287.

Hinkelmann, K. (1951) "Der Mechanism des Meteorologischen Lärmes," *Tellus* **3**, pp. 283–296.

Hinkelmann, K. (1959) "Ein numerisches Experiment mit den primitiven Gleichungen," in *The Atmosphere and the Sea in Motion: Scientific Contributions to the Rossby Memorial Volume*, eds. B. Bolin and E. Eriksson, pp. 486–500, Rockefeller Institute Press, New York, New York.

Hoffman, R. N. (2002) "Controlling the Global Weather," *Bull. Am. Met. Soc.* **82**, pp. 241–248.

Lions, J. L. (1971) "Optimal control of systems governed by partial differential equations," Springer-Verlag, Berlin, 1971.

Le-Dimet, F. X., and Talagrand, O. (1986) "Variational algorithms for analysis and assimilation of meteorological observations: theoretical aspects," *Tellus*, **38A**, 97–110.

Lynch, P. (2001) "Max Margules and His Tendency Equation (includes a translation of Margules' 1904 paper "On the Relationship Between Barometric Variations and the Continuity Equation")," Historical Note No. 5, Met Eireann, Dublin, Ireland.

Lynch, P. (2002) "Weather Forecasting from Woolly Art to Solid Science," in *Meteorology at the Millennium*, ed. R. P. Pearce, Academic Press, London, United Kingdom.

Lynch, P. (2003) "Margules' Tendency Equation and Richardson's Forecast," *Weather* **58**, pp. 186–193.

Marchuk, G. I. (1974) *Numerical Methods in Weather Prediction*, Academic Press, New York, New York.

Margules, M. (1893) "Luftbewegungen in einer rotierenden Sphäroidschale," *Sitzungsberichte der Kaiserliche Akad. Wiss. Wien, IIA* **102**, pp. 11–56.

Margules, M. (1904) "Über die Beziehung zwischen Barometerschwankungen und Kontinuit Ätsgleichung, in *Ludwig Boltzmann Festschrift*, pp. 585–589, J. A. Barth, Leipzig, Germany.

Mintz, Y. (1965) *Very Long-Term Global Integration of the Primitive Equations of Atmospheric Motion*, WMO Technical Note No. 66, pp. 141–155, Geneva, Switzerland.

Miyakoda, K., Hembree, D. G., Strickler, R. F., and Shulman, I. (1972) "Cumulative Results of Extended Forecast Experiment. I: Model Performance for Winter Cases," *Mon. Wea. Rev.* **100**, pp. 836–855.

Phillips, N. A. (1956) "The General Circulation of the Atmosphere: A Numerical Experiment," *Q. J. Roy. Met. Soc.* **82**, pp. 123–164.
Pudykiewicz, J. (1988) "Numerical Simulation of the Transport of Radioactive Cloud from the Chernobyl Nuclear Accident," *Tellus* **40B**, pp. 241–259.
Pudykiewicz, J. (1998) "Application of Adjoint Tracer Transport Equations for Evaluating Source Parameters," *Atmospheric Environment* **32**, pp. 3039–3050.
Reed, R. J. (1977) "Bjerknes Memorial Lecture," *BAMS* **58**, pp. 390–399.
Richardson, L. F. (1922) *Weather Prediction by Numerical Process*, Cambridge University Press, London, United Kingdom. (Reprinted with a new introduction by S. Chapman, Dover Publications, New York, New York.)
Richardson, S. A. (1957) "Lewis Fry Richardson (1881–1953): A Personal Biography," *J. Conflict Resolution* **1**, 300–304.
Robert, A. J. (1969) "The Integration of a Spectral Model of the Atmosphere by the Implicit Method," in *Proceedings of the WMO IUGG Symposium on Numerical Weather Prediction*, pp. VII-9–VII-24, World Meteorological Organization and International Union of Geodesy and Geophysics, Meteorological Society of Japan, Tokyo, Japan.
Robert, A. J. (1982) "A Semi-Lagrangian and Semi-Implicit Numerical Integration Scheme for the Primitive Meteorological Equations," *J. Meteor. Soc. Japan* **60**, pp. 319–324.
Robert, A. J., Henderson, J., and Turnbull, C. (1972) "An Implicit Time Integration Scheme for Baroclinic Models in the Atmosphere," *Mon. Wea. Rev.* **100**, pp. 329–335.
Rossby, C.-G. (1939) "Relation Between Variations in the Intensity of the Zonal Circulation of the Atmosphere and the Displacement of the Semi-Permanent Centers of Action," *J. Marine Res.* **2**, pp. 38–55.
Tanguay, M., Robert, A., and Laprise, R. (1990) "A Semi-Implicit Semi-Lagrangian Fully Compressible Regional Forecast Model," *Mon. Wea. Rev.* **118**, pp. 1970–1980.
Trusdell, C. (1954) *The Kinematics of Vorticity*, Indiana University Publications No. 19, Indiana University Press, Bloomington, Indiana.
Willis, E. P., and Hooke, W. H. (2006) "Cleveland Abbe and American Meteorology, 1871–1901," *Bull. Am. Met. Soc.* **87**, pp. 315–326.

19

Fundamental issues in numerical weather prediction

Jimy Dudhia

Weather-related disasters, unlike many other disasters, are often predictable up to several days in advance. This predictability leads to issues of preparedness that would depend directly on how much the predictions are trusted. This chapter gives insight into predicting the various types of weather disaster and how weather prediction products should be used by emergency managers who want to take into account the state of the science to determine how much they trust the forecasts. The chapter also explains the areas of current uncertainty in weather prediction, which are primarily in the areas of model physics and initialization data.

19.1 Introduction

A key tool in current-day weather forecasting is the numerical weather prediction (NWP) model, and it is usually the primary guidance for forecasts ranging from a few hours to a few days. NWP covers a vast span of temporal and spatial scales, and model domains range from global to continental to national to local. In this chapter, the range of scientific and technical issues that form fundamental limits to model predictions are addressed, with particular attention to the way these problems greatly depend on the particular scales of application required for the forecasts. Therefore, this chapter is organized somewhat according to the type of application, which is largely a function of the required forecast range. Models are used quite differently for short-range local forecasts than for medium-range global forecasts, but each can be a key part of the guidance, not only in day-to-day forecasting, but also in forecasting potential disasters. Also, different limitations of NWP show up in these various applications, so it is important to separate them in order to correctly prioritize the main issues for each type of forecast.

Here, we start with the forecast strategies used for disasters, and then go into uncertainties that limit the accuracy of NWP, some of which are fundamental, and some of which are technical.

19.2 Disaster-related weather

The focus here is on the types of weather associated with disasters. These can be categorized into local events and large-scale events, and, as is seen, it is important to distinguish these because the forecasting and warning strategies would often be different. Local events include hurricanes, wind storms, heavy localized rain or snowfall, and severe storms that may be associated with hail, lightning, or tornadoes. Large-scale events include heat waves, droughts, regional heavy precipitation, and extreme cold spells. Local events may only affect town-size areas, as with severe storms, or slightly larger areas or swaths, as with hurricanes or tornado outbreaks, that may be of order 100 km wide. Large-scale events may affect regions with scales up to thousands of kilometers simultaneously.

19.3 Disaster prediction strategies

Because NWP is not perfect, uncertainties have to be taken into account, and in model forecasts, the character of the uncertainties changes with the length of the forecast and with the scale of the systems to be forecast. Hence, this section is subdivided by forecast range into (1) medium-range prediction (5–10 days), (2) short-range prediction (3–5 days), (3) day-to-day prediction (1–3 days), and (4) very short-range prediction (<1 day). Note that these subdivisions are arbitrary, serving the purposes of the discussion in this chapter, and do not represent any official naming.

19.3.1 Medium-range prediction (5–10 days)

Global models are usually used for this range. For local events, it is very unlikely that a useful forecast can be made in this range. At best, the potential for a hurricane formation, or a severe storm environment, can be seen, but with no certainty in where exactly, when, and even if such an event will occur; thus, a normal strategy is to wait for better shorter-range forecasts before any precautions are taken, given the long lead-time.

Even for large-scale events, such as heat waves, the medium-range prediction is not reliable in a deterministic sense. This has been the reason for several global weather forecast centers to develop an ensemble of forecasts to try to capture the uncertainties, primarily in the initial conditions due to lack of data. Ensembles may have twenty to fifty members typically, and can provide valuable information on potential disasters and their probability, but only if the ensemble is well diversified in the sense that the true outcome is usually captured within their range of solutions. Generating a good ensemble is a major research priority, and relies not only on a good model, but also on an understanding of the characteristics of the important initial analysis uncertainties.

The 5- to 10-day forecast may, for example, give a probability greater than 50% for a heat wave, in which case disaster management may be able to start preparations. Another scenario might be a 10% chance of a major storm in a region, but the use of this would greatly depend on the customer, their needed lead-time, and the possible impact of such a storm. However, this points out that if probabilities are given, they need to be reliable in the sense that 10% of the times that storms are predicted with this probability, they also occur. Too high a predicted probability would result in false alarms, whereas too low a probability would lead to a false sense of security. Given a reliable probability, a proper risk–benefit analysis can be carried out by emergency managers to determine their actions. The ECMWF and NCEP have ensemble forecasts, and evaluation of their usefulness is an ongoing process (e.g., Atger, 1999).

19.3.2 Short-range prediction (3–5 days)

Global and/or regional models may be used in this range. For local events, positioning and timing is still poor in this range when using a deterministic forecast, but now ensembles give more valuable information on probable outcomes than in the medium range. All the criteria related to large-scale events in the previous subsection, now apply to local events. For example, at 3 to 5 days from landfall, disastrous hurricanes are often already formed, and their tracks are routinely forecast using an ensemble of models at the U.S. National

Hurricane Center. The landfall warning region often narrows dramatically in this range, together with time of landfall, and the intensity prediction. In the 2005 season, hurricanes Katrina, Rita, and Wilma, were well forecast even at 3 days from landfall, giving confidence in the state of weather prediction for these events, but at 5 days, there remains a degree of uncertainty that has to be accounted for by emergency management.

Large-scale events are generally deterministically forecastable within 5 days because these usually accompany large-scale air mass motions and jet stream meandering that current-day models capture well up to 5 days. Comparing global forecasts and verifying at a given time for different forecast lengths, it is generally seen that the forecast stabilizes to the correct solution at around the 5 day range, when looking at air masses (as opposed to individual precipitation systems). For example, while the 6- and 7-day forecasts often differ from each other greatly, the 5-day and less forecasts more often agree well on large scales. This determininisticness means that disaster management can take actions with some certainty for large-scale events in the 3- to 5-day range.

19.3.3 Day-to-day prediction (1–3 days)

In this range, the regional models are the main NWP tool. The primary public forecasts focus on this range, which is also critical for disaster management, because evacuation orders, supply lines, preparations of infrastructures, and so on have to be certain within this range. Although local events such as hurricanes have now become quite deterministic within this range, others such as tornadoes clearly cannot be pinpointed a day ahead. Severe storm regions can be predicted quite well a day in advance, and certain events such as frontal passages are also well forecast, allowing for possible timing errors. For severe storms, there is promise as cloud-resolving models are starting to show skill in predicting storm types about a day ahead, but these are still experimental in the United States and new in other countries, such as the United Kingdom and Germany. A typical daily forecast can warn of the possibility of tornadoes in a region that could include several cities, but only a small fraction of that region would actually get a tornado. The conditions for tornadoes are possibly better predicted at present than those required for severe lightning or hail, which appear to rely more on factors that are locally determined by the storms themselves.

19.3.4 Very short-range prediction (<1 day)

Regional or local models may be used, but NWP is accompanied by nowcasting, or observational analyses in real time, in issuing forecasts, particularly as the range falls within a few hours of an event. For example, hurricane forecasts will be compared with observed positions to correct the track forecast. For local storm events, such as tornadoes and flash floods, observations are almost entirely used in the period after the storm forms, with no use for model products. It is an active research area to design models that can be initialized with observed fully developed storms to predict their development in real time, but this task is challenging given the complexity of storm structure, even with good Doppler radar data.

19.4 Fundamental issues: atmospheric predictability

In the remainder of this chapter, we examine what limits the range of forecasts and their accuracy. Atmospheric predictability has inherent limits due to the general nature of atmospheric turbulence. Lorenz and others have defined simple chaotic systems mathematically, and it is obvious that the atmosphere shares this chaotic nature, which means that small perturbations grow to significant scales given sufficient time. This is most clearly seen with the growth of midlatitude cyclones from small frontal disturbances or hurricanes from small convective disturbances or severe thunderstorms from individual cumulus clouds. These processes also require favorable large-scale conditions, but pinpointing the time and place of development is clearly an impossible task in a finite numerical model with limited data. It has been said that the limit for atmospheric deterministic predictability is about 14 days, but currently the best global models cannot give statistically successful forecasts to more than 7 or 8 days on average. This, of course, means that some forecasts may be successful at 10 days, but some may fail at only 5 days, and an important product therefore is some measure of certainty to go along with each forecast, as can be gained from ensembles. Generally, the conditions under which forecasts fail are predictable because the cause is most often the rapid growth of a system that had not been initialized correctly in the model, and certain regions favor such growth, but the model may mislocate the growth, miss it, or overdevelop a neighboring feature. Whether dealing with cyclones in global models or thunderstorms in regional models, these same issues apply.

19.5 The model

The NWP model has some fundamental, some technical, and some scientific limitations. Model results are generally sensitive to changes in the physics packages, the grid size, and the dynamics.

19.5.1 Model physics

The physics includes radiation, cumulus parameterization, cloud physics, land surface, and boundary layer. Radiative properties of clear air are well known, and the error due to model vertical grid resolution is small. The main problems come from clouds that partially fill model grid columns and aerosols that cannot be accounted for accurately in real-time forecasts because of poor information about initial conditions and sources. The cloudiness problem is compounded by the model's cloud physics that itself has uncertainties in the processes that produce ice clouds, and the simplifying assumptions needed to represent microphysical processes with only a few mixing ratio variables (single-moment bulk microphysics schemes). Many microphysics schemes of varying sophistication have been developed, and these can capture gross features of clouds, but the sensitivity of the model to the different schemes shows that there is an inherent uncertainty. The main problem is that the grid volume is not uniformly mixed in reality, and so process rates cannot be accurately computed even if the size distributions assumed by the bulk schemes were correct. At grid sizes greater than 5 to 10 km, whole convective clouds need to be represented within single-grid columns, and this physics is known as cumulus parameterization. The diversity of such schemes shows that no single method has come to dominate in NWP because all schemes show benefits under different circumstances. It seems inherently impossible to have a cumulus scheme that can represent convective development globally. Coarse model results often show great sensitivity to varying the cumulus scheme, particularly in regions such as the tropics or spring/summer continents, where convection drives the weather.

Even leaving the uncertainties of clouds aside, the surface also presents a physics challenge. The land-surface properties depend crucially on its vegetation, soil, and elevation characteristics. The soil temperature and moisture are often predicted in NWP models because of the importance of soil moisture to the boundary layer fluxes. However, no data exist to initialize soil conditions, and these are almost entirely model products themselves.

Snow cover prediction is also part of the land-surface parameterization, and adds to the complexity of accurate surface forecasts, particularly because anthropogenic effects such as snow removal in urban areas cannot be represented in the model. Boundary layer parameterizations also diverge because of the complexity of the problem of representing boundary layer unstable eddies and their transports in the lowest model layers. Different schemes will have different efficiencies of these eddies, and again, no single scheme has come to dominate in NWP applications. Hurricane simulations show sensitivity to how these eddies are treated. The nocturnal boundary layer is also difficult because most of the processes governing surface fluxes occur in a very shallow layer and at fine horizontal scales that are not represented in models.

Given these physics uncertainties, it is perhaps surprising that forecasts are possible at all, but it appears that our knowledge accounts for the primary effects of physics well, and these uncertainties lie in secondary processes, as far as weather prediction is concerned.

19.5.2 Model dynamics

The equations for a compressible gas on thin rotating sphere are well known, and as far as the dry dynamics of the atmosphere are concerned, there are no choices. Chapter 2 introduces the basic Navier-Stokes equations. To these Coriolis and curvature terms are added, and a Reynolds stress treatment of sub-grid turbulance is typically used. Additional equations carry moisture variables. Some models make approximations, such as being hydrostatic, incompressible, or not using the full three-dimensional Coriolis force, but when these approximations are well justified from scaling considerations, they do not interfere with the model accuracy. Therefore, this is an area of certainty and agreement in NWP.

19.5.3 Model numerics

Computer models either deal with grids or spectral modes to reduce the infinite atmospheric degrees of freedom to a computationally manageable size. Global model grid sizes are now coming down to 20 to 30 km, as the ECMWF model just went to T799 on February 1 2006. Regional model grids are now in the 4 to 12 km range, and just starting to resolve individual thunderstorms at the smaller end of the range. The choices of numerical schemes (numerics) in some way equate to the choice of resolution in considering their effects on the results, and generally, with everything else equal, numeric choices of comparable cost have less effect on the results than physics.

19.6 Model data

Initial conditions are vital for forecasts, and, for regional models, boundary conditions become similarly important as the range extends. A perfect model is not going to give a perfect result unless the initial conditions are also perfect, and similarly, with perfect initial data, a perfect result will not be achieved without a perfect model. In reality, of course, neither is ever going to be perfect, but this underscores the equal standing that the data has with the model in producing good forecasts.

As has been mentioned, the main challenge in forecasts is to represent the growth of large features from small initial features, and when those small features are in data voids, such as the ocean regions, this has historically been a major limitation. Now, with the advent of satellite data and the growth of the science of data assimilation to make the best use of it, these data voids are becoming filled. A major illustration of this has been at the

Table 19.1. *Numerical weather prediction*

Forecast	Forecast Range (Days)	Local Events Guidance	Large-Scale Events Guidance
Medium-range	5–10	Little	Probabilistic
Short-range	3–5	Probabilistic	Deterministic
Day-to-day	1–3	Probabilistic/deterministic	Deterministic
Very short-range	<1	Deterministic	Deterministic

ECMWF, where historically the Southern Hemisphere forecasts were always worse than the Northern Hemisphere because of the large ocean areas in the south. However, with the use of satellite data assimilation, the Southern Hemisphere has now almost caught up, showing the dominance of these new data sources in global forecasting (e.g., Rabier, 2005). Chapter 21 describes the use of satellite data in more details, where Fig. 21.11 illustrates the ECMWF improvements.

The story for storm-scale prediction is different because in short-term regional forecasts aimed at predicting storm outbreaks, the scales of the important heterogeneities cannot be resolved by any currently conceivable observing system. Also, the time constraints make data assimilation more challenging as the scales become smaller. As mentioned previously, it is a major challenge to initiate observed thunderstorms in models, and various data assimilation techniques are being applied in the hope that increasing computer power will eventually help overcome the technical limitations. However, even with Doppler radar, there are significant voids at fine scales, and data itself remain limited.

19.7 Conclusions

Forecasts have to be used consistently with their reliability in disaster prediction. This chapter attempts to categorize scenarios in which deterministic and probabilistic forecasts should be used based on the current state of NWP forecasts. It is important for the producers of forecasts to also give a reliable measure of how certain or uncertain their forecast is, and what else could occur instead, and this requires the judicious use of ensembles when deterministic forecasts are not reliable. Table 19.1 provides a summary of the kind of guidance given by current-day NWP.

The models and data are equally important aspects in producing a good forecast. Progress is still being made in both these areas and is largely helped by increasing computer power and increasing satellite data.

Finally, the reduction in computing costs will also mean that more nations will be able to produce their own forecasts in the future, and it is important to disseminate NWP knowledge and models to aid these nations as they start up.

References

Atger, F. (1999) "The Skill of Ensemble Prediction Systems," *Mon. Wea. Rev.* **127**, pp. 1941–1953.
Rabier, F. (2005) "Overview of Global Data Assimilation Developments in Numerical Weather-Prediction Centres," *Quart. J. Roy. Meteor. Soc.* **131**, pp. 3215–3233.

20

Space measurements for disaster response: the International Charter

Ahmed Mahmood and Mohammed Shokr

The history of satellite remote sensing applications is not so recent, and there have been numerous national activities and international initiatives to promote the use of these applications for managing disasters of natural or manmade causes. Space-based remote sensing is carried out by means of both passive and active sensors onboard polar and geostationary orbital platforms. A major program developed to use space technology in disaster management is the International Charter "Space and Major Disasters." It was established to achieve cooperation among space agencies and space system operators to deliver data and information products to help civil protection, rescue, and relief organizations in the wake of disasters: hurricanes, tsunamis, floods, earthquakes, volcanic eruptions, landslides, forest fires, and oil/chemical spills. In this chapter, in addition to remote sensing principles, policies, and programs, some typical cases of disaster coverage by the International Charter are described to demonstrate the growing relationship between space data and service providers and the user communities in need.

20.1 Introduction

One of the immediate casualties in the event of a major disaster is the very system on the ground that is expected to be of help in responding to the disaster. The management of a major disaster is often beyond the scope of the ground-based or airborne systems, hence the use of space technologies, which is not only beneficial but also necessary. These technologies, labeled as Earth observation (EO), have been commonly used for satellite imaging of the disaster-stricken regions to assess damage and brace for emergencies. Satellite systems other than those for EO have been providing data complements and communications. Geopositioning and communications satellites are important assets in establishing logistical support. In the case of the Kashmir Earthquake of October 2005, for instance, in addition to high-resolution pre- and postdisaster satellite imagery, there was an equally pressing demand for very small aperture terminals (VSATs) to support voice and data applications and remote location determination with global positioning systems (GPSs).

Few EO systems and the associated ground segments were developed in the past specifically for managing disasters, but their ad hoc use has increased over time with the result that several new and planned space missions are now focused on the remote sensing data application of disaster management. Likewise, some of the currently implemented major environmental security initiatives have space-relied disaster management as one of their major performance indicators. The RESPOND service element of the European Global Monitoring for Environment and Security (GMES) is based on the provision of maps updated with satellite imagery delivered in situations of crisis caused by disasters. One of the nine societal benefits to be derived from the Global Earth Observation System of Systems that the Group on Earth Observations is leading pertains to the reduction of loss of life and property from natural and human-induced disasters by employing, among others, space-based observations. The Integrated Global Observing Strategy (IGOS) is an international partnership among thirteen founding members that was established in 1998 and that brings together the efforts of a number of international bodies, including the Committee on Earth Observation Satellites (CEOS), concerned with the observational component of global environmental issues, both from research and operational points of view. One of these IGOS issues is natural disaster mitigation processed by means of analysis of known space-based and in situ observing systems. World Weather Watch (WWW) of the World Meteorological Organization (WMO), which preceded IGOS as a global observing effort, also called for an integrated approach for surface- and space-based information. There have been several other initiatives with the direct involvement of space organizations for environmental and disaster management, including the Global Disaster Observation System of the Society of Japanese Aerospace Companies, Earth Watching of the European Space Agency (ESA), the Disaster Watch program of the Canadian Space Agency (CSA), and the Center for Satellite Based Crisis Information (ZKI) of the German Space Agency DLR's remote sensing sector to supply data and the relevant information in the event of emergencies.

A truly operational space-based concept, however, had escaped implementation until the signing by the ESA, French Centre National d'Etudes Spatiales (CNES), and the CSA space agencies in 2000 of what is now known as the International Charter "Space and Major Disasters," pursuant to the declaration of the third UNISPACE conference a year earlier. The charter is the first instance of joint satellite operations among the member agencies whose number has now grown from the initial three to the current eight; it is also the first operational application of satellite data for disaster response. The charter, which is about space data provision, is at the same time an attempt to shift the space data use priorities to the general public and the communities in need. This chapter is dedicated to the charter functions and operations.

Section 20.2 introduces the use of remote sensing for disaster management. Section 20.3 addresses a few theoretical aspects of remote sensing that are related to data acquisition, analysis, and information retrieval. Section 20.4 summarizes the international space-based initiatives for disaster management and the potential benefit of space information in this field. The basic information about the International Charter is introduced in Section 20.5. The sensors and the missions that comprise the data sources for the charter are included in Sections 20.5.2 and 20.5.3, respectively. These two sections should be consulted while reading the entire chapter as they contain explanations of many acronyms used throughout. The structure and operational mechanisms of the charter are explained briefly in Section 20.6. Finally, Section 20.7 introduces case histories that were activated during several disasters that struck various parts of the world. The chapter ends with concluding remarks.

20.2 Space remote sensing and disaster management

The use of satellites during disaster management activities has a history of nearly four decades. It started in the late 1960s, with the development of satellite disaster communication systems (Richards, 1982). In the case of remote sensing satellites, the use was tested soon after the launch in 1972 of the first Landsat satellite, which was found to be a valuable space-based asset for floods, drought, grass and forest fires, glacier movements, seismic mapping, and lava flows (Robinov, 1975). The capabilities of Landsat even for oil slick detection were reported by Deutsch et al. (1977). The geostationary National Oceanic and Atmospheric Administration (NOAA) weather satellites, such as GOES (geostationary operational environmental satellite), also provide valuable remote sensing capability. These geostationary satellites were advocated as data collection systems (DCS) with considerable value in disaster warning and disaster preparedness planning. The GOES DCS transmitted weather pictures through the spacecraft transponder and in addition relayed in situ data obtained from a variety of remotely located land or marine sensors and data collection platforms. The system was identified for use in disaster warning for tsunamis, volcanic eruptions, and other environmental disturbances. Geosynchronous satellites are commonly used to track ocean storms (hurricanes, cyclones, etc.), and techniques such as scatterometry are used for measuring surface wind velocities. The choice of spaceborne instruments varies according to the disaster type. For example, the synthetic aperture radars (SARs), like the ones onboard the European remote-sensing satellites (ERSs) and environment satellite (EN-VISAT) and the Canadian RADARSAT-1, provide a rapid view of water-saturated surfaces, but their relatively coarse resolution precludes their effective use in detecting earthquake damage to buildings and other urban structures. However, the requirements for earthquake prediction are different. Here, the technique of SAR interferometry (Hein, 2004), based on exploiting the phase difference of two SAR data takes with the exact same viewing geometry, can reveal new information on terrain movement. In conjunction with the GPS, it is now possible to make accurate measurements to determine the rate of movement along faults. Where there is no movement, these faults may be locked and may consequently be the locus of earthquake occurrence when the lock ruptures (Walter, 1999).

In view of their usefulness, space systems have been popularly factored in the overall disaster planning by governments. The U.S. State Department has responded favorably to proposals from the U.S. Agency of International Development, specifically its Office of Foreign Disaster Assistance, to employ the various sensors and systems onboard the U.S. civil weather and land remote sensing satellites for natural disaster early warning capabilities (Rose and Krumpe, 1984). India is one of the few other countries that have been using space technology inputs for near real-time monitoring of drought, flood, and cyclone as part of their national disaster management mission (Venkatachary et al., 1999). Under the central sector scheme on Remote Sensing Application Mission for Agriculture, India's Department of Agriculture and Cooperation and Department of Space initiated two national programs, namely, the National Agricultural Drought Assessment and Monitoring System and the Near-Real-Time Flood Monitoring, which were continued in the country's seventh and eighth 5-year plans corresponding, respectively, to the periods 1987 to 1992 and 1992 to 1997. In fact, it is the Department of Space that was assigned the implementation of these programs in cooperation with the concerned states and central government departments. The daily data from NOAA-AVHRR (advanced very high-resolution radiometer) sensor had been analyzed since 1988 under the Drought Assessment program to provide near real-time information on local and national levels for biweekly/monthly and seasonal crop conditions. Likewise, near real-time flood maps are prepared using Indian WiFS (wide field sensor) and LISS (linear imaging self-scanning sensor) data under the Flood Monitoring program.

The Japanese Aerospace Exploration Agency (JAXA) has recently introduced a Disaster Management Support Office within its Office of Space Applications pursuant to JAXA's 2025 vision. The recently launched advanced land observing satellite (ALOS) is part of JAXA's vision for establishing a system for natural disaster management.

Globally, each major disaster type can be covered with a variety of satellite sensors. There are a number of satellite-based programs that render climatic change forecasts for drought early warning. The type of data used for such forecasts has been derived from several sources: AVHRR for sea surface temperature (SST), TIROS (television infrared observing satellite) for atmospheric profiles, operational vertical sounder, ERS scatterometer for wind velocities, special sensor microwave/imager on the satellites of the U.S. Defense Meteorological Satellite Program (OMSP), GOES-East and GOES-West, Meteosat-GMS (geostationary meteorological satellite) of Japan, Indian National SATellite (INSAT), and others. Multichannel, multisensor data from these sources have been used for meteorological parameters, such as precipitation intensity, amount and coverage, atmospheric moisture, winds, and surface wetness. Improvements in spatial distribution of rainfall are achieved by integrating radar, rain gauges, and remote sensing techniques. The vegetation condition, which is another input for drought monitoring, can be assessed by means of NOAA-AVHRR and Indian remote sensing satellite (IRS) data. The normalized difference vegetation index and temperature condition index derived from satellite data are largely accepted. High-resolution data from polar orbiting remote sensing satellites such as Landsat, French SPOT (Systme pour l'Observation de la Terre), and IRS are used for the assessment of the impact of droughts. In the area of drought monitoring and early warning, EO from satellites is highly complimentary to in situ observations.

The earthquake hazard can be covered by space-based systems in its various phases: mitigation, warning, and response. A good knowledge of a region's seismicity and its tectonic setting is essential for earthquake mitigation. Several space-based techniques continue to contribute significantly to our understanding of regional tectonics, including satellite geodesy, which translates into laser ranging, very long baseline interferometry, and the use of GPS. Both optical and radar data are used for geological and structural mapping. Hyperspectral satellite data may also be used for lithological discrimination to detect such destabilizing phenomenon as soil and rock creep. Visible and infrared satellite imagery with moderate resolution (>10 m) is used to map lineaments, which are often surface expressions of deep-seated faults and fractures.

Like other geophysical and intuitive methodologies, space observation is not adequate for earthquake prediction. The earthquake warning requires reliable methods to detect earthquake precursors. Constant monitoring of earthquake-prone regions may be a first step. There have been a few instances where satellite images of the site were acquired immediately preceding a major seismic event, which might have contained the intrinsic record of stress release. One such instance was the 1999 earthquake in the Izmit region of Turkey on the morning of 17 August. Canadian RADARSAT-1 imaged the site in the course of its baseline data acquisitions called "Background Mission" (Mahmood et al., 1998) only hours before the region was struck by an earthquake. Further data acquisitions over the earthquake region were realized in the successive duty cycles of the spacecraft with a view to establishing plausible interferometric pairs. Processing interferometric pairs produced maps of surface elevation or deformation. However, scene coherence problems precluded the extraction of any useful information from the interferograms.

For the non real-time response phase, satellite data contained in a geographic information system (GIS) database of the earthquake site can be used to generate different images for damage assessment. Very high-resolution images, such as those provided by commercial satellite operators, are a necessity for this kind of damage assessment work. A 1-m resolution IKONOS image of Bhuj (Gujrat) was procured for the charter activation over this Indian earthquake of January 2001. Some useful results could be obtained also by pre- and postearthquake IRS data of 5.8-m resolution. Data from the SPOT series of satellites were also used.

The management of forest fire disasters can be carried out in three phases: preparedness, detection and response, and postfire or burnt area assessment. Preparedness involves risk assessment, which in turn requires the knowledge of such variables as land use and land cover, wildfire history, demographic distribution, infrastructures, and urban interfaces. In this regard, space remote sensing is used to derive vegetation stress variables that can subsequently be related to wildfire occurrence. The most frequently used data source for this information is NOAA-AVHRR. Other alternative data sources are ATSR-2 (along track scanning radiometer), VEGETATION onboard SPOT 4, Landsat, and commercial IKONOS. For the identification of wildfire risk area, moderate resolution imaging spectroradiometer (MODIS) is also useful.

The detection of wildfires is made possible by either sensing their thermal or midinfrared signature during day or night, or detecting the light emitted by them at night. The satellite sensors that have been commonly used for fire detection include NOAA-GOES, NOAA-AVHRR, Meteosat-GMS, DMSP-Operational Linescan System, and BIRD, which detects not only hot spots, but also information about the physical properties of the fire (Dech, 2005).

In the postfire assessment phase, the most important consideration is the mapping of burnt areas. On national and international scales, NOAA-AVHRR data are most commonly used for burnt area mapping. The VEGETATION instrument onboard SPOT 4 is another choice. The 1.6-μm data from ASTR-2 provide improved fire scar delineation. At regional scales, within national boundaries, high-resolution data from Landsat TM (thematic mapper) and SPOT VIR (visible infrared) are used to determine the extent of wildfire damage. The medium spatial resolution data from the Indian and the Russian remote sensing satellites in conjunction with the same medium spatial but high spectral resolution EOS-MODIS and ENVISAT-MERIS (medium-resolution image spectrometer) are also recommended for burnt area cartography on regional or global scale.

Flood forecasting requires hydrological modeling and climatological prediction information. The GOES and polar-orbiting operational environmental satellite (POES) are used for such climatological information. On a local scale, land surface data on topography, hydrographics, and the fluvial makeup of the region are needed. Cartographic updates by means of digital elevation models (DEMs) are a critical aspect of remote sensing. The satellite data most suited for these inputs are from both high- to medium-resolution radar and optical sensors: SPOT, Landsat, IKONOS, ERS, and RADARSAT.

The flood preparedness warning is based on the information about weather and watershed conditions. Again, GOES and POES weather satellites provide information on precipitation, moisture, temperature, winds, and soil wetness, whereas active and passive microwave sensors have been used increasingly for estimates of snowpacks and other watershed studies. Medium- to high-resolution space imagery integrated with GIS can help in preparing flood recovery plans. The medium resolution imagery, in particular, can be used to establish new flood extent boundaries. There have been some model examples, namely, the Red River flooding in North America, of flood extent mapping on an operational basis with SAR systems (RADARSAT-1). As in the case of other disasters, high-resolution imagery can be used for locating damaged sites and facilities.

Landslide mitigation primarily involves mapping the zones that are at risk, along with their relevant terrain features. In terms of space remote sensing, experience has shown that high-resolution stereo SAR and optical images, combined with topographic and geological information, are useful for landslide risk mapping. The multi-incidence, stereo, and high-resolution attributes of RADARSAT-1 data can be used for landslide inventory. High-resolution optical systems such as IKONOS, IRS, and the stereo capability of SPOT 4, along with the more recent ENVISAT and soon to be operational or launched systems like ALOS and RADARSAT-2, should be useful for landslide recognition and related land use mapping. Radar interferometry can be used for observing slope stability, which is a critical factor in the generation of landslides. In this regard, the ERS1/2 tandem, RADARSAT-1 interferometric mission, and shuttle radar topography mission (SRTM) data are the potential source of information for terrain analysis.

Landslide warning and prediction to prepare for the disaster has been attempted by establishing the rainfall threshold where a landslide triggers. The tropical rainfall measuring mission has provided useful rainfall information for tropical areas. Multitemporal SPOT images have been applied successfully for very large, slow-moving landslides. Radar interferometry can be a useful technique in detecting soil creep as a precursor to slope failure; however, the limitation on horizontal resolution of these images and the scene coherence, particularly for repeat-pass interferometry, may render the technique ineffective. Nonetheless, interferometric SAR (InSAR) techniques have been applied to monitor slow movements over annual and monthly intervals, with typical displacements in the order of a few millimeters to several centimeters within these time periods. The future 1- to 3-m resolution of TerraSAR and RADARSAT-2, combined with the current capabilities of ENVISAT-ASAR, may successfully address some of the issues related to horizontal resolution and repeat-pass interferometry. For landslide disaster response, predisaster images from the satellite data archives can be readily provided for comparison with postdisaster images for change detection.

The dangers associated with a volcanic eruption are both proximal (lava flows, volcanic cloud, pyroclastics) and distal (volcanic ash plumes). The most serious hazard is nonetheless the worldwide propagation of airborne ash that poses a threat to aviation. Consequently, space remote sensing, including the meteorological satellites, play an important role in tracking and monitoring the ash plumes. Given the global nature of this hazard, nine regional centers of expertise, known as volcanic ash advisory centres (VAACs), provide updated information bulletins to meteorological watch offices (MWOs). These centers are part of the International Airways Volcano Watch program established by the International Civil Aviation Organization.

The monitoring of proximal volcanic hazard depends also on the use of low-orbit EO systems. The available space systems are Landsat 7, NASA's Terra satellite with MODIS, and ASTER (advanced spaceborne thermal emissive radiometer) package of payloads in companionship with AQUA, ENVISAT, ALOS, and the very high-resolution IKONOS.

In the case of technological hazards such as oil and chemical spills, produced either as a result of accidents or deliberate tanker cleaning/bilge pumping, EO data are already used operationally for enforcement and monitoring. An example is the CSA's integrated satellite tracking of polluters project (ISTOP), now conducted by the Canadian Ice Services of Environment Canada, to monitor illegal oil dumping in Canadian waters. Other Canadian federal departments, namely, Transport Canada and Fisheries and Oceans-Canadian Coast Guards, participate in the project. The principal satellite data source is the spaceborne SAR, RADARSAT-1 in this case. Low-resolution SAR images are generated rapidly for an overview of the suspected zone and are passed on to the authorities to optimize the flight plan for surveillance aircraft if the authorities decide to follow up on the satellite observations. In some cases, the suspected spills are zeroed in by finer-resolution SAR imagery, for instance, by employing RADARSAT-1 Fine beam mode instead of the reconnaissance ScanSAR beam mode. The interpretation of SAR imagery in the coastal zones of major accidental oil spillage is problematic in view of wind shadows that have the same dampening effect on radar backscatter as an oil slick. In such situations, the more easily interpretable optical data (SPOT VIR, Landsat TM) may be used. In the postspill phase, the authorities may be interested in knowing where the oil is likely to come ashore. The space EO data-derived spill vector outlines that can be integrated with meteorological satellite data and marine current data may predict the potential beaching zones of the oil.

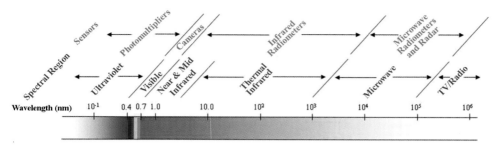

Figure 20.1 Electromagnetic wave spectrum used in remote sensing observations.

20.3 General principles of remote sensing

Remote sensing refers to observing objects by sensors not in direct contact with the object. Based on the platform location, remote sensing is categorized into surface-based, airborne, and spaceborne. This chapter addresses main concepts in the acquisition and analysis of spaceborne remote sensing (also called satellite remote sensing).

Satellite remote sensing refers to observations of emitted or reflected radiation by sensors onboard satellite platforms. The observed radiation is emitted from the Earth's surface and the atmosphere. These radiometric observations can be converted into geophysical parameters of the surface if the atmospheric contribution is accounted for. In contrast, it can be converted into atmospheric parameters if the surface radiation is accounted for. Retrieval of surface or atmospheric parameters from satellite observations involves using statistical, empirical, physical, and/or radiative transfer models. Examples of such parameters include, but are not limited to, soil moisture, vegetation density, SST, wind speed over ocean surface, forestfire extent, flood extent, atmospheric water vapor column, aerosol type, and optical depth. Some of these parameters are indicators of natural hazards.

Remote sensing observations are acquired in predetermined spectral bands (electromagnetic waves). The three main categories are optical, infrared, and microwave bands (Figure 20.1). For the microwave, there are passive sensors that measure the emission from the "footprint" of the sensor (also called "effective field of view") and active sensors that transmit their own signal and measure the backscatter from the surface. Active microwave sensors are known as "radar sensors." The most commonly used sensor of this type is called SAR. The principles of SAR operation, along with examples from the SAR system onboard the Canadian satellite RADARSAT-1, are introduced in the following sections.

Remote sensing observations in different spectral bands are capable of retrieving information on different surface or atmospheric parameters. For example, optical bands are capable of identifying vegetation, snow, and water surface parameters. Infrared bands are best in retrieving temperature of the surface, including land, sea, or ice. Passive microwave bands are used to retrieve parameters such as surface wind speed over ocean, soil moisture, sea ice concentration, integrated water vapor, and cloud liquid content over land or ocean. Radar sensors have a wide range of applications in land and ocean environments, ranging from ocean wave spectrum, forestry clear-cut, oil spill in ocean, terrain elevation, and deformation of the Earth's surface, to name a few. Both types of microwave sensors (passive and active) are not sensitive to the atmosphere. This is a significant advantage over optical and infrared sensors.

Spaceborne remote sensing can be categorized into two main classes, based on the type of the platform: geostationary and polar orbiting. Geostationary satellites orbit the Earth at an altitude of 36,000 km in the equatorial plane, with the same angular speed of the Earth's

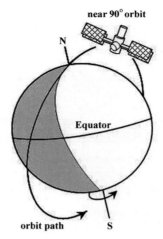

Figure 20.2   Orbit configuration of near-polar orbit satellites.

rotation. The satellite can then be viewed to be suspended in space, looking at the same side of the Earth. Image data from a geostationary satellite are typically available every 6 hours, which is an excellent temporal resolution. However, the spatial resolution is rather coarse (a few tens of kilometers at best). Each image is a snapshot of the entire part of the Earth's surface as viewed from the satellite point. Geostationary satellites are the primary meteorological observation platforms.

Polar orbiting satellites, in contrast, orbit the Earth at lower altitudes (500–1,400 km) in a near-vertical plane. The orbit is typically inclined by 10 to 25 degrees to the vertical (Figure 20.2). The measured radiation is sampled at relatively finer resolution but with a repeat cycle in the range of a few days to a few tens of days. Repeat cycle is the number of days between two identical orbits (i.e., until the satellite passes over the same point on the Earth's surface to within 1 km). Due to their finer resolution, polar orbiting satellites are useful in understanding and monitoring natural hazards. However, their temporal resolution is rather coarse.

As noted previously, sensors measure electromagnetic radiation or reflection/scattering from the Earth's surface as modulated by the atmosphere. The source of the electromagnetic radiation can be in the following forms:

- **The Sun:** The sensor measures the reflected solar radiation in the visible bands as it reaches the top of the atmosphere (TOA).
- **Emitted radiation:** The sensor measures the emission from the surface as modulated by the atmosphere. This is usually performed in the infrared or microwave bands.
- **Transmitted radar pulses from an antenna onboard the satellite:** The sensor measures the backscatter returned from the surface. The atmosphere does not usually contribute to the observed backscatter.

The first two sources encompass what is known as passive remote sensing category, whereas the third identifies the active remote sensing (Figure 20.3). After recording the reflected, emitted, or scattered radiation, the system has to transmit the recorded observations to a ground station, where data are processed into digital images. A digital image is composed of pixels. A pixel represents the area on the ground from which a single observation is acquired. An observation is represented by a digital number. The relationship

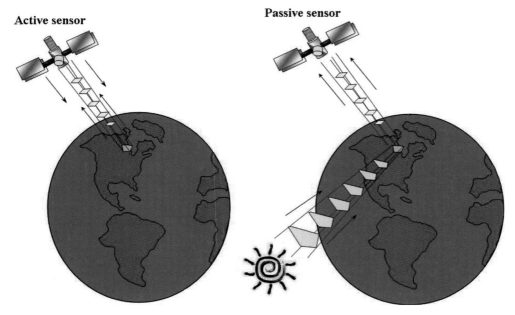

Figure 20.3   Illumination source for active and passive sensors.

between the digital number and the original observation is loosely known as "radiometric calibration."

Radiometric calibration may be relative or absolute. Relative calibration is a simpler process, which aims at normalizing data from multiple satellites using the same sensor. It usually involves the selection of ground targets whose reflectance values are considered constant over time. This process is important for change detection, which is an important technique for hazard monitoring. Absolute calibration accounts for sensor parameters (e.g., gain and offset), illumination and viewing geometry of the sensor, as well as absorption and scattering by the atmosphere. Usually, image data provided by a ground station are not absolutely calibrated. Absolute calibration is necessary if the purpose is to retrieve geophysical parameters using radiometric observations.

Geometric correction is another important process, which transforms a raw satellite image into another image with a different geometry suitable for cartographic applications. This includes methods to register multitemporal imagery to a master image (so that all images have the same geometry) or register an image to a known geographic map format. It also accounts for sensor peculiarity, such as attitude, altitude, earth rotation, velocity, or scan skew. More information on the fundamentals of optical, thermal, and microwave remote sensing are presented in the following section, along with brief information on image processing, interpretation, and classification.

20.3.1 Optical, thermal, and microwave imaging

The satellite imaging systems can be classified into optical, thermal infrared, and microwave sensors. Optical remote sensing makes use of the visible (wavelength 0.4–0.7 μm), near-infrared (wavelength 0.7–1.3 μm), and combined short- and midwave infrared (wavelength 1.3–8 μm) to form images by detecting the solar radiation reflected off the Earth's surface. Thermal infrared sensing (wavelength 8–14 μm) makes use of the infrared radiation emitted

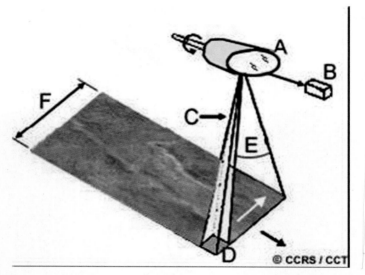

Figure 20.4 Configuration of an across-track scanning radiometer. Reprinted from CCRS training course material. © Canada Centre for Remote Sensing.

from Earth and the atmosphere. Microwave sensing is composed of two categories: passive and active. The wavelength range is between 1 and 1,000 mm. A passive microwave system measures the microwave radiation emitted from the surface. Only in cases of relatively high microwave frequencies (e.g., 85 GHz) does the atmosphere contribute to the observed emitted radiation. An active microwave system (radar system) measures the reflected radar signal, which is scattered from the surface back to the sensor. It can be an imaging sensor such as SAR or a profile sensor such as scatterometer (Elachi, 1988). A brief background on the optical, infrared, and imaging radars systems is included in the following account.

Optical, thermal, and passive microwave remote sensing

Optical, thermal, and passive microwave sensors on satellite platforms use scanning radiometers to measure the received radiation at the top of the atmosphere. Two main scanning modes are used: across track and along track. The across-track system (Figure 20.4) scans the swath width (F) in a series of lines using a rotating mirror (A). The lines are oriented perpendicular to the direction of motion of the satellite platform (i.e., across the swath). As the platform moves forward, successive scans can be combined to produce two-dimensional image of the surface. Observations in different spectral bands can be obtained using a bank of internal detectors (B), each sensitive to a specific range of wavelengths. Hence, these sensors are called multispectral scanners (MSSs).

A single observation acquired by the scanner is the integration of the radiation over a ground area called the integrated field of view (IFOV), which is determined by the angular beamwidth (C). The IFOV can be small (a few tens of meters in case of Landsat and SPOT satellites, see Section 20.5.2) or large (a few kilometers in the case of AVHRR, especially near the end of the swath).

Optical sensors measure the ratio of the reflected to the incident radiation in a given spectral band. This ratio, expressed as a decimal, is called "spectral albedo." The Earth's surface reflects roughly 39% of all visible light impinging on it.

20.3 General principles of remote sensing

Thermal infrared and passive microwave sensors measure the emitted radiation in terms of brightness temperature using a radiometer system. Full description of radiometry is presented in Ulaby et al. (1981). Brightness temperature is defined as the ratio of the radiation from a given body at a given physical temperature to the radiation from a "blackbody" at the same temperature. A blackbody is an ideal object that absorbs all incident radiation (hence reflects nothing) and reradiates energy uniformly in all directions at a different frequency f and temperature T (in K), according to the following expression, which is known as Planck's equation:

$$B_f = \frac{2hf^3}{c^2}\left(\frac{1}{e^{hf/kT}-1}\right) \quad (20.1)$$

where B_f is the radiation flux density, h is Planck's constant, k is Boltzmann's constant, and c is the speed of light. In the microwave region, the term hf/kT is very small; hence, Planck's equation can be rewritten into what is known as Rayleigh–Jean's equation

$$B_f = \frac{2hkTf^2}{c^2} \quad (20.2)$$

The radiation flux density can be determined from the previous equations. Therefore, any measured radiation can then be expressed in terms of the physical temperature of blackbody that gives rise to the measured radiation. This is called brightness temperature (T_B).

The radiation emitted from any surface varies with the wavelength. The wavelength at which the radiation peaks depends on the physical temperature. It decreases as the physical temperature increases. The Earth–atmosphere system emits radiation in the range of 3- to 100-μm wavelength band, which is called "outgoing long-wave radiation." For the Earth surface at an average physical temperature around 300°K, the spectral radiance peaks at a wavelength around 10 μm. For this reason, most of the satellite thermal infrared sensors have a spectral channel detecting radiation of wavelength around 10 μm.

Remote sensing of thermal infrared is often used to measure land surface temperature and SST, and to detect forest fires or other warm/hot objects. For typical fire temperatures from about 500°K to more than 1,000°K, the radiation peaks at around a 3.8-μm wavelength. Sensors such as the NOAA-AVHRR, ERS-ATSR, and Terra-MODIS are equipped with this band that can be used for detection of fire hotspots.

The real challenge in using optical and infrared sensors to retrieve surface information is to account for the atmospheric contribution in the observed reflected or emitted radiation. The radiative processes that contribute to the observed radiation in the optical band are summarized in Figure 20.5. This figure shows the three radiometric components that are received by an optical sensor: the terrestrial L_T, the cloud reflection L_C, and the path radiation L_P. Terrestrial radiation is the desired contribution from the ground target within the IFOV, whereas path radiation is the contribution from the atmosphere and areas adjacent to the IFOV. Further explanation of each component, following the component numbers in the figure, are now discussed.

As solar irradiance penetrates through the atmosphere, some of it reaches the IFOV (1), while the rest is absorbed or scattered by the atmosphere (2). Absorption is the process by which radiant energy is absorbed and converted into other forms of energy. Atmospheric gases absorb radiation only at certain wavelengths. The two gases that absorb most of the solar energy are O_3 and H_2O. In general, the atmosphere absorbs 25% of solar radiation. Scattering is the redirection of energy by particles in the atmosphere. The scattering mechanism depends on the relative size of the scattering particles to the incident wavelength. In

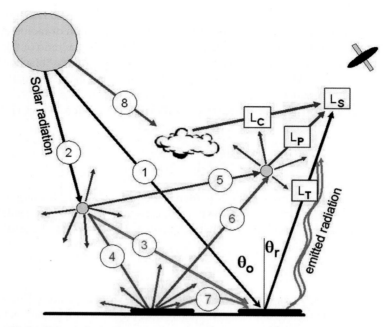

Figure 20.5 Solar radiation interaction with atmosphere and surface, with the net radiation that reaches the satellite point of observation.

general, the atmosphere scatters 6% of the incident solar radiation. Scattering is inversely related to the fourth power of the radiation's wavelength. Blue light (wavelength 0.4 μm) is scattered sixteen times more than near-infrared light (0.8 μm). As seen in Figure 20.5, some of the energy scattered by the atmosphere reaches the IFOV (3), while the rest falls on neighboring areas (4). The scattered and emitted radiation by the atmosphere (5) constitutes part of L_P, whereas the other part is engendered by reflection from neighboring areas (6). The contribution of neighboring areas to the illumination of the IFOV (7) is usually negligible. Reflection by clouds represents, in general, 19% of the incident solar radiation (8).

Figure 20.5 also shows that the reflectance of a target is a function of the illumination geometry of the source and the viewing geometry of the sensor. In addition, it is determined by the structural and optical properties of the surface. As mentioned previously, these factors have to be accounted for in the radiometric calibration.

In the case of the emitted infrared radiation, the contribution of the atmosphere is manifested through the absorption, scattering, and reradiation by atmospheric molecules and particles. Two atmospheric gases, namely, H_2O and CO_2, absorb most of the infrared radiation. These are called "greenhouse gases." These gases also radiate in the infrared band. Absorption by atmospheric gases normally restricts thermal sensing to two specific spectral bands of wavelength 3 to 5 μm and 8 to 14 μm. Because of the relatively long wavelength of thermal radiation (compared to radiation in the optical band), atmospheric scattering is minimal. Description of basic principles of atmospheric sensing and radiative transfer is presented in Elachi (1987).

Given that emitted energy from the Earth's surface decreases as the wavelength increases, thermal sensors generally have large IFOV to ensure that enough energy reaches the sensor in order to make a reliable measurement. Therefore, the spatial resolution of thermal sensors is usually coarser than the resolution of optical sensors. Passive microwave sensors are

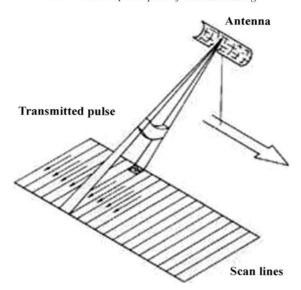

Figure 20.6 A transmitted radar pulse from a right-side–looking airborne (or spaceborne) antenna as it scans a line across the swath.

characterized with much coarser resolution, typically a few kilometers. Thermal imagery can be acquired during daylight or at night and is used for a variety of applications in disaster management, such as forest fire mapping and heat loss monitoring.

Microwave remote sensing

Microwave remote sensing falls into two main categories: passive and active. In principle, passive microwave operates very similarly to the optical and infrared sensing (i.e., through measurement of the emitted radiation from the Earth as modulated by the atmosphere). The source of illumination is still the Sun. In contrast, active microwave systems (radar systems) generate their own source of illumination, namely, radar pulses with predetermined pulse width and repetition frequency. The rest of this section is dedicated to the description of the imaging radar system, as well as estimates of the resolution in range and azimuth directions of the one that is most commonly used, the SAR.

An imaging radar system transmits pulses from its antenna. The pulses travel through the atmosphere, unaffected by atmospheric conditions, and fall onto the Earth's surface. As the satellite moves along its orbit, the pulses illuminate continuous strips of the surface to one side of the orbit direction (Figure 20.6). Interaction with surface roughness and subsurface composition causes a fraction of the incident energy to be scattered off the surface. Part of the scattered energy travels back to a receiver onboard the satellite, (which is usually the same transmitting antenna). The received power (P_r) is related to the transmitted power (P_t) through what is known as "radar equation" (Fitch, 1988; Ulaby et al., 1986). Figure 20.7 is a graphical illustration that shows the geometry and quantities involved in the radar equation. For convenience of interpretation, the transmitted and received points are shown at different locations, but in all operating SAR they are but one point (i.e., the same antenna). The radar equation usually takes the form

$$P_r = \frac{P_t}{4\pi R_t^2} G_t \sigma \frac{1}{4\pi R_r^2} A_r \qquad (20.3)$$

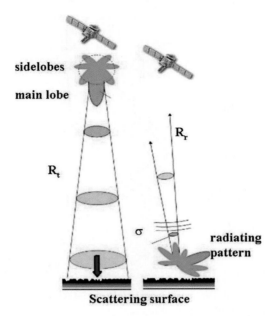

Figure 20.7 Transmission, surface interaction, and reception mechanisms of radar signal. Parameters used in the "radar equation" are also shown.

The transmitted power spreads over the entire space according to the antenna pattern (also called "antenna gain"), which consists of a main lobe and side lobes. The power spreads with reduction in intensity inversely proportional to the square of the distance from the transmitter R_t. The power received at the scattering surface is determined by the value of the antenna gain G_t in the direction of the transmitted pulse. This power is intercepted by the "effective receiving area" of the surface A_{rs}. This is the area of the incident beam from which all power would be removed if one assumed that the power going through the rest of the beam continued uninterrupted. On interception, a fraction of the power p_a is absorbed, whereas the rest excites currents on the scatterer, which then becomes an antenna reradiating with its own antenna pattern G_{rs}. The reradiating power from the scattering surface is measured by what is known as the radar cross section, which is defined as the area in meters squared of an isotropic scatterer that would be required to return the observed scattering energy from the surface. It is a description of how strongly an object reflects an incident radar wave. The portion of the scattering signal that travels back to the antenna and received by its effective aperture A_{rs} is called backscatter cross section σ,

$$\sigma = A_{rs}(1 - p_a)G_{rs} \qquad (20.4)$$

This reradiated power spreads with loss inversely proportional to the square of the distance from the scattering surface, R_r, as indicated in Equation (20.3).

The radar cross section is highly dependent on the radar wavelength, polarization, and incident angle. It is also a function of the surface composition and roughness. However, it is not used in practice because it depends on the resolution of the radar system. When the same surface is imaged by two radar sensors of different resolution, σ will be different. A more useful quantity is obtained by dividing σ by the area of the observed terrain. This is

20.3 General principles of remote sensing

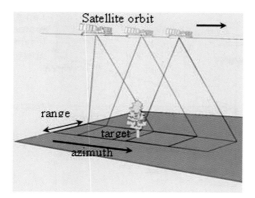

Figure 20.8 The concept of synthetic aperture radar (SAR): a single observation is obtained from several pulses as the satellite moves. This is equivalent to using a very wide antenna, which improves the resolution in the azimuth direction significantly.

called backscatter coefficient or "sigma naught" (σ^0). It is a dimensionless quantity, which can be treated as an intrinsic property of the surface at a given wavelength, polarization, and incidence angle.

As mentioned previously, SAR is the most commonly used imaging radar system. It produces high-resolution images. All radar images that are presented in the following sections are from SAR systems on different platforms. The principle of SAR operation, along with its resolution in the range (across flight path) and azimuth (along flight path) directions, are described here. The image is constructed from the returned echo of a few pulses. Figure 20.8 shows three pulses that capture the shown target (the tree). Location of target in the range direction is achieved by precisely measuring the time from transmission of a pulse to reception of the echo. Range resolution is determined by the transmitted pulse width; narrow pulses yield fine range resolution.

The target's position and resolution in the azimuth direction are obtained using Doppler frequency of the echoed pulse (Franceschetti and Lanari, 1999). Targets ahead of the satellite position produce a positive Doppler offset, whereas targets behind the satellite produce a negative offset. As the satellite platform moves, several echoes are processed simultaneously to determine the location of the target; that is, the Doppler frequency is used to determine the location. This is precisely what is meant by synthetic aperture.

Resolution in the azimuth direction can be formulated using the definitions presented in Figure 20.9. The length of total beam footprint L of the synthetic aperture that results from successively transmitted pulses is defined as

$$L = \alpha R = \frac{\lambda R}{D} \qquad (20.5)$$

where λ is the wavelength of the transmitted signal. The exposure time T_e of a given target within that beam is

$$T_e = \frac{L}{V} = \frac{\lambda R}{DV} \qquad (20.6)$$

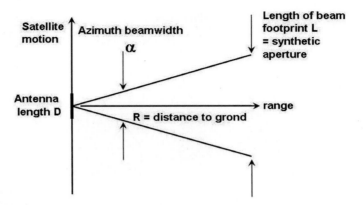

Figure 20.9 Geometric configuration to determine azimuth resolution of synthetic aperture radar (SAR) system.

The Doppler shift of the returned signal of the last pulse that "sees" the ground target with respect to that of the first pulse is denoted ϕ, and the rate of this shift versus the azimuth time is

$$\frac{d\Phi}{dt} = \frac{-2V^2}{\lambda R} T_e = K_a T_e \qquad (20.7)$$

The total Doppler bandwidth (DB) between the first and last pulses results from substituting the expression in Equation (20.5) into Equation (20.6).

$$DB = \frac{2V}{D} \qquad (20.8)$$

The resolution in time is the inverse of the previous expression. When multiplied by the velocity of the antenna (i.e., the satellite), it produces the resolution in space (in the azimuth direction), which is $D/2$. Thus, SAR has the remarkable property that its resolution in the azimuth direction is independent of the distance of the platform from the target and also independent of the wavelength of the transmitted signal.

Most of the radar sensors operate at wavelengths between 2.4 and 30 cm. The most common bands are the C-band (wavelength 5.3 cm) and the L-band (wavelength 20 cm). The capability of SAR to penetrate the surface is increased with longer wavelengths. For wavelengths smaller than 2 cm, radars are not significantly affected by cloud cover. However, at wavelengths greater than 4 cm, rain starts to affect the signal.

An important radiometric property of SAR systems is the polarization of the transmitted and receiving signals. Polarization refers to the alignment of the electric field in electromagnetic wave in a plane perpendicular to the wave propagation direction (Figure 20.10). If the electric field remains in the same plane, the wave is called "linearly polarized." This is the case in all operational SAR systems so far. Polarization of the transmitted and received signals is denoted by a pair of symbols (a combination of H and V for horizontally and vertically polarized signals, respectively). Therefore, HH SAR signal means that the signal is horizontally transmitted and horizontally received, HV means that it is horizontally transmitted and vertically received, and so on. Because the scattering surface can change the polarization of the incident wave, the radar system is often designed to transmit and receive waves at more than one polarization. A full account of the polarization of SAR systems is provided in Ulaby et al. (1990).

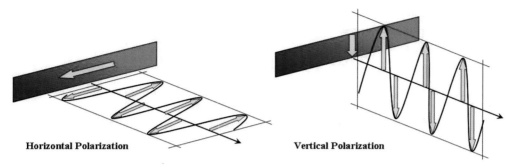

Figure 20.10 Definition of horizontal and vertical polarization from synthetic aperture radar (SAR) antenna. Adapted from CCRS training course.

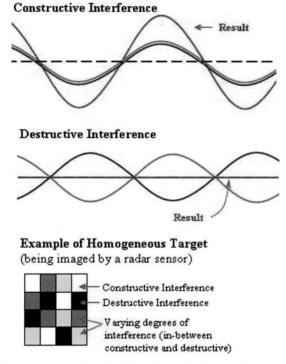

Figure 20.11 Coherent averaging of reflected signals from different scatterers within a single observation cell (pixel) produces random intensity values as a result of constructive and destructive interferences. This is known as speckle in radar images.

Radar images are also characterized by what is known as "fading and speckle," which appear in the image as randomly distributed fine grains. They are noiselike processes generated by the coherent radar signal. When reflected off the surface, the radar waves interfere in a random constructive and destructive manner, and therefore produce random bright and dark spots in radar imagery. Figure 20.11 shows a sample output from the constructive and destructive interference, along with an output of an original image with high speckle level and another output of the same image after reducing the speckle by averaging adjacent pixels. To reduce the speckle level in the imagery product from a SAR processor, a technique called "multilook" is used. Further reduction is usually implemented by users through applications of standard techniques that are mostly included in commercial software for remote sensing data analysis.

20.3.2 Image processing, information contents, and interpretation

The digital or analogue signal received from spaceborne sensors is processed into an image. The fidelity of the image for interpretation and information extraction will depend on the processing techniques that are applied. The various data processing and analysis issues are discussed in this section.

Image processing

Satellite receiving and processing stations typically provide imagery data at three levels of processing called 1A, 1B, and 2. Level 1A, often called the preprocessing level, includes data in raw form intended primarily for mapping, stereo viewing (using stereo image pairs), and special radiometric studies. Minor linear models are usually applied to compensate for instrument effects.

Level 1B is the basic level that incorporates both radiometric and geometric corrections of the image, although not necessarily fully corrected. As mentioned previously, radiometric correction compensates for any undesirable spatial or temporal variations in image brightness, which are not associated with physical variations in the imaged scene. Examples include variation of sun illumination across the swath (or incidence angle in the case of radar data), absorption of radiation by clouds, sensor noise, and viewing geometry of the sensor. This is important for quantitative retrieval of surface parameters or for comparing images from the same spectral bands obtained by different sensors.

For their part, geometric corrections compensate for systematic effects, including panoramic distortion, Earth's rotation and curvature, Earth's surface topography, and variations in the satellite's orbital altitude relative to the reference ellipsoid (Mather, 2004; Verblya, 1995). Each sensor has its own peculiarities of viewing geometry that should be accounted for in a geometric correction scheme. Level 1B product is usually processed according to user-specified parameters such as geographic projection, resampling method, and pixel size. To generate maps, most remote sensing systems expect to convert image data to a distance on the ground. This requires matching up points in the image (typically thirty or more) with points in a precise map, or providing accurate latitudinal and longitudinal estimates of a few points in the image. This process is called "georeferencing." Since the early 1990s, most satellite images are provided fully georeferenced.

Level 2 products are more sophisticated. In addition to the features of level 1B, data are processed into a variety of geophysical parameters, including surface albedo, surface temperature, soil moisture, vegetation indices, cloud thickness, ocean wave heights and spectrum, surface wind, atmospheric water vapor, and aerosol parameters. Data may also be gridded into a standard latitude/longitude grid. Physical, empirical, and electromagnetic wave propagation models are usually used in parameter retrieval.

Image data from a receiving station may be provided fully calibrated, or more likely partially calibrated. In the latter case, the user may further process the data to complete the calibration. This is the case, for example, of the imaging radar data. Calibration of radar data accounts for the nonuniform distribution of transmitted energy. Because the antenna transmits more power in the midrange portion of the illuminated swath than at the near and far ranges (see the antenna pattern shown in Figure 20.6), stronger scattering is expected from the center portion of the swath than the edges. Calibration also accounts for the fact that the energy returned to the radar decreases significantly as the range distance increases.

Thus, for a given surface, the strength of the returned signal becomes smaller as distances increase across the swath.

Image processing at a receiving station outputs images represented by positive digital numbers (intensity levels), which occupy a range from zero to one less than a selected power of 2. Although optical sensors such as Landsat and SPOT typically produce 256 intensity levels, radar systems can differentiate up to around 100,000 intensity levels. Such information cannot be discerned by visual interpretation or computer-assisted techniques alone. Therefore, most radars record and process the original data as 16 bits (65,536 levels of intensity), which are then further scaled down to 8 bits (256 levels) for visual interpretation and/or digital computer analysis.

Processing radar data at a receiving station further aims to reduce the speckle level (see the previous section). For this purpose, a technique known as "multilook imaging" is usually implemented. It uses the division of the radar beam into several narrower subbeams, typically 2 to 5, as shown in Figure 20.8. Each subbeam provides an independent "look" at the illuminated scene. Each "look" is subject to speckle. By summing and averaging them together, the final output image is formed and the amount of speckle is reduced.

The user may further process the imagery data in preparation for visual interpretation or quantitative analysis. For radar images, speckle reduction by spatial filtering (i.e., convolution of each pixel with a small spatial filter) is performed so that the visual appearance of the speckle is reduced. Speckle reduction, however, is achieved at the expense of spatial resolution. Advanced mathematical techniques have been developed to maintain good resolution while achieving the smoothening effect (Richards, 2005).

Further postprocessing of remote sensing images is usually required to minimize geometric distortions in the images. Any remote sensing image, regardless of whether it is acquired by a space- or airborne sensor, will have various geometric distortions. This is an inherent problem in remote sensing because the actual three-dimensional surface of Earth is represented as a two-dimensional image. Geometric distortion of objects in the image is caused by several factors such as the viewing geometry of the sensor, the instability of the platform (speed and orbit track), the curvature of Earth, and the terrain relief. In most instances, geometric distortion can be removed or at least reduced.

For optical sensors, geometric distortion is caused by the scanning mechanism. The sensor scans the surface in subsequent lines using a scanning mirror with constant rotational speed. Because the distance from the sensor to the ground increases further away from the center of the swath, the sensor scans a larger area as it moves closer to the far edge of the swath. This causes compression of the image near the edge, a phenomenon called "tangential scale distortion." Moreover, the eastward rotation of the Earth during a satellite orbit causes the sweep of scanning systems to cover an area slightly to the west of each previous scan. The resultant imagery is thus skewed across the image. This is known as "skew distortion." Corrections of these geometric distortions are usually implemented at the user's end unless the data are provided in gridded format.

Radar imagery suffers from different geometric distortions. The pixel location in the range direction is determined based on the time it takes the radar signal to travel from and return to the transmitting antenna. This time is a function of the distance from the antenna to the ground target, measured in slant range (Figure 20.12).

If the angle of a hillside or a mountain slope is large enough, the top point (B) may appear closer to the base point (A) in the image. In other words, the slope (A–B) will appear compressed, and the length of the slope will be represented incorrectly. This relief displacement is called "radar foreshortening." The severity of foreshortening increases as the tall feature tilts more toward the radar, or roughly speaking, when slope increases.

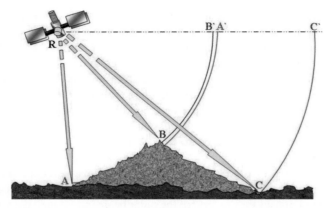

Figure 20.12   Geometric distortion in radar images is caused by the surface elevation. Pixel location in the image is determined by the return time of the signal. Therefore, point B appears in the image before point A.

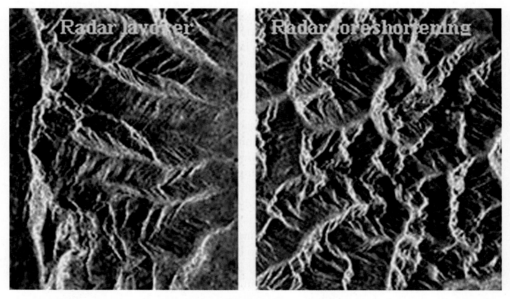

Figure 20.13   Examples of radar layover and foreshortening in radar images. Adapted from CCRS training course.

Maximum foreshortening occurs when the radar beam is perpendicular to the slope, so that the base, the top, and the slope are imaged simultaneously. In this case, the length of the slope will be zero. Foreshortening slopes appear as bright features in the radar image (Figure 20.13).

If the slope increases further, the radar signal reaches the top of the tall target (B) before it reaches the base (A). Hence, the return signal from the top will be received before the signal from the base, resulting in the odd appearance of the top of a mountain preceding its bottom point in the image. This displacement effect is called "layover," and it looks very similar to the foreshortening effect (Figure 20.13). Both foreshortening and layover are most severe for small incidence angles (at the near range of a swath), and in mountainous terrain. For most applications, especially mapping, correction for these forms of geometric distortion is necessary. This can be done using commercial software. However, for the purpose of visual image interpretation, foreshortening and layover are considered useful information.

To take full advantage of this data source, it must be overlayed to existing map data. The geometry and projection information must be equivalent; in other words, there must be an orthogonal projection or correlation between the point on the ground and some reference surface. This process is referred to as "rectification." The output satellite or airphoto images are referred to as orthorectified images or ortho images. Many SAR orthorectification methods have been developed recently (Pierce et al., 1996; Toutin, 2003; Zhou and Jezek, 2004).

Information contents

To interpret remote sensing images, it is important to be familiar with the surface information that produces the measured radiometric parameters. Optical systems (visible reflection and infrared radiation) are predominantly affected by the optical and thermal properties of the surface and the atmosphere through which the electromagnetic wave is generated and transmitted. Information contents in the visible band, for example, are mainly about the surface reflectivity, the color of the observed object, and its directional albedo. In contrast, information contents in the infrared observations are mainly a function of the physical temperature of the object and its emissivity (i.e., the ability of the material to radiate energy in a given wavelength and polarization).

Microwave remote sensing observations are generated by three surface parameters: geometric structure of the observed object, surface roughness, and the dielectric constant of the material, that is, its ability to reflect or scatter the incident energy. The dielectric constant is a function of the physical composition of the surface and subsurface layers through which microwave energy passes and the wetness of these layers. It is worth noting that these factors vary with three sensor's parameters: wavelength, incidence angle, and wave polarization.

For any type of remotely sensed data, image information used in visual interpretation or quantitative retrieval of geophysical parameters can be grouped into the following four items:

- Gray tone intensity information (related to the radiometric observation)
- Spectral information (i.e., combination of measurements from several channels)
- Textural information (spatial arrangement of radiometric observations)
- Geometric and contextual information (structures in imagery data)

A brief description of the use of this information for image interpretation and geophysical parameter extraction is provided in the following paragraphs.

Image interpretation

Image interpretation is the process of identifying objects and their spatial relationship in the image. This has traditionally involved the visual analysis of the image, perhaps with a few measurement tools and a simple table. Modern image analysis has become increasingly sophisticated and automated. Object information in image data includes type, shape, location, and quality. Interpretation can be performed using a single channel image, a pair of stereoscopic images, or multichannel images. Visual image interpretation is common in certain applications, notably geological and meteorological ones. It employs rules of thumb that relate apparent gray tone distribution, texture, and shapes in the image

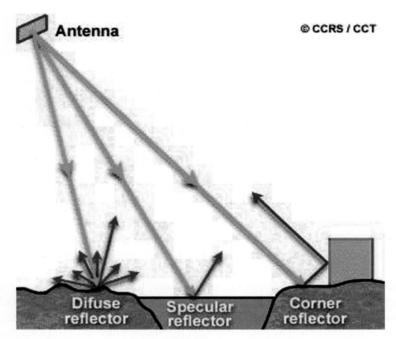

Figure 20.14 Scattering mechanisms of radar signal from rough and smooth surfaces as well as geometric structure. Adapted from CCRS training course. © Canada Centre for Remote Sensing.

to the surface type and properties. This, of course, requires familiarity with the interaction of electromagnetic wave with the surface and peculiarities of the imaging sensor. Interpretation skills required for weather satellite image analysis are explained in Mogil (2001).

In an image acquired with a visible channel, for example, urban areas have a high reflected intensity (or equivalent gray tone), whereas vegetated areas have low intensity. Likewise, river water with sediment appears bright, whereas clear seawater appears dark. In infrared images, bright areas indicate pixels of high physical temperature, and the reverse is also true. Therefore, bright areas may be interpreted as centers of population, high-temperature cloud top and the location of forest fires.

Imaging radar data reveal different information about the surface because the interaction of the microwave signal with the surface is controlled by different surface parameters. Generally speaking, smooth surfaces reflect the signal away from the radar (Figure 20.14), whereas rough surfaces, in contrast, scatter more energy back to the antenna and hence look brighter in a radar image. This is called "specular reflection." A surface appears to be rough to the radar system if its height variation is greater than one-eighth of the incident radar wavelength. Lower radar incidence angles (shallower angles) accentuate surface roughness, whereas higher angles suppress it in the image. Corner reflection is another mechanism that prompts higher brightness in radar images. It occurs when two smooth surfaces form a right angle facing the radar beam. In this case, the beam bounces off the surfaces twice.

Gray tone in a radar image can also be attributed to the dielectric constant of the surface, which is mainly affected by moisture content. When moisture increases, the dielectric constant increases, and so does the ability of the surface to scatter the incident microwave

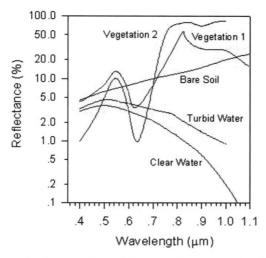

Figure 20.15 Spectral reflectance from different surface types of surface. It can be used to identify the surface in a multispectral data set.

signal. Thus, bright areas in a radar image can possibly be interpreted as being reflection from wet soil.

The spectral reflectance of a material, which is a plot of the fraction of reflected radiation as a function of the incident wavelength, is usually considered a "signature" of the surface that can be used to identify it in multispectral images. This is particularly important if the sensor has sufficient spectral resolution to distinguish spectra from different surfaces. Figure 20.15 shows the reflectance spectrum of five types of land cover.

Image texture is a function of the spatial variation in gray tone values of neighboring pixels. It is useful in a variety of applications. Many methods of texture quantifications have been developed (Rajesh et al., 2001; Tuceryan and Jain, 1998). One immediate application of image texture is the classification of ground targets in radar images (crop types, ice types, rock types, etc.). Texture is also the most important visual cue in image interpretation. The goal of texture analysis is to produce a classification map of the input image where each uniform textured region is identified with ground cover class. Texture boundaries, however, are hard to establish accurately.

Contextual information is also important for image interpretation, especially in the case of single-channel data, which is the case for the majority of imaging radar systems. Ancillary information such as climatic records, recent history of surface changes, and ground measurements of surface parameters is also important.

Quantitative interpretation entails identification of surface type at the pixel level or at the region level in the image. This involves applications of image analysis techniques such as image classification, segmentation, pattern recognition, neural network, and wavelet analysis (Jain et al., 2000; Richards and Jia, 1999; Zhang, 2000). These techniques may require feeding an algorithm with "training" data or generating these data automatically within the algorithm.

20.3.3 Geophysical parameter retrieval and value adding

Radiometric observations obtained by remote sensing are related to the geophysical parameters of the object of interest. The relations can be expressed in the form of empirical equations or mathematical models. The forward problem in such a model involves calculation of the radiation at the point of the satellite observation, using the surface geophysical parameters along with the existing atmospheric conditions. The inverse problem involves retrieval of the geophysical parameters from the radiometric observations. Generally speaking, remote sensing works on the principle of the inverse problem. This is analogous to determining the type of animal from its footprints. An example of an inverse problem in remote sensing is the retrieval of temperature in the upper atmosphere using observations of spectral emissions from different gases in the atmosphere. The required temperature is related to the observed wavelength of the emission through various thermodynamic relations. Another example is the retrieval of soil moisture from radar observations, which are affected by soil moisture, surface roughness, and the canopy structure, if it exists.

Retrieval of geophysical parameters from remote sensing data may employ empirical, physical, electromagnetic, radiative transfer, or thermodynamic models. Empirical models are the simplest relation that can be constructed, for example, from regression of field measurements and the coincident radiometric observations from a spaceborne sensor. Physical models describe interactions among physical parameters of the Earth's surface or the atmosphere, such as the relation between seawater salinity, density, and its temperature. Electromagnetic models describe the interaction of the electromagnetic wave as it propagates through material. They account for the transmission, absorption, and scattering of the wave as a function of the electrical properties of the material and its composition. Radiative transfer models account for the absorption and emission of radiation from successive layers in the atmosphere in order to determine the radiation at the TOA. Thermodynamic models are concerned with heat and energy exchange within the material and with the atmosphere. They are useful in modeling the thermal infrared radiation.

Retrieval of land surface temperature or SST from infrared observations is an example of using empirical relations combined with radiative transfer models. This is a well-established retrieved parameter from using two infrared channels (bandwidth 10.3–11.3 μm and 11.5–12.5 μm). These two spectral bands exist on AVHRR, ATSR, and MODIS sensors and are subject to small atmospheric absorption. The technique that is used for this purpose is known as "split window" (Becker and Li, 1990; Price, 1984). The basic form of the technique is an empirical equation, which can be written for AVHRR measurements from channels 4 (T_4) and 5 (T_5) as follows

$$T_S = a + T_4 + b(T_4 - T_5) \tag{20.9}$$

where a and b are constants that can be estimated from regression of results from a simulation using forward radiative transfer model. SST can also be retrieved using satellite microwave radiometers, which allow retrieval under cloud conditions. Clouds are nearly transparent at microwave frequency less than 10.8 GHz. Furthermore, unlike SST retrieval from infrared data, microwave SST retrievals are not affected by aerosols and are insensitive to atmospheric water vapor. However, the microwave retrievals are sensitive to sea surface roughness, whereas the infrared retrievals are not. SST and land surface temperature anomalies play an important role in predicting weather-related natural disasters.

Geophysical parameter retrieval from remote sensing data has advanced significantly during the past decade. Many parameters have been provided as higher-level products from many sensors. Examples include parameters of surface reflectance, fire and thermal anomalies, land cover, vegetation products, ocean wave height and spectrum, ocean surface wind, and land elevation. Complete lists of such products from each sensor are available on the websites of major space agencies (see, e.g., the NASA website http://landqa2.nascom.nasa.gov for lists of land and ocean surface products from MODIS.)

Value-added products are imagery data that are further processed, combined, or augmented with external data to add value for geoscientific work. A common value-added product is a false color combination of multichannel images. Three channels may be combined such that each channel is represented as one of the primary color (blue, green, or red). Sometimes the combination is done with images of the same ground scene acquired at different times. Some of the examples in Section 20.7 are either of multichannel or multitemporal images. The human eye is more sensitive to variations in color than variations in gray tone.

Multichannel images may also be combined to accentuate a geophysical parameter. A well-known combination is the ratio of the near-infrared band (*NIR*) to the red band (*R*). This is known as the vegetation index. Because vegetation has high near-infrared reflectance but low red reflectance, vegetated areas will have a higher vegetation index values compared to nonvegetated areas. This ratio is usually normalized and known as the normalized difference vegetation index (*NDVI*),

$$NDVI = (NIR - R)/(NIR + R) \qquad (20.10)$$

NDVI is a standard level 2 product from many visible/infrared sensors. It is especially important in delineating areas devastated by drought.

Value adding is particularly required for radar images of high-elevation terrain. This is due to the greater radiometric and geometric sensitivity of imaging radar to local topographic relief. A digital elevation model is usually included as a complementary data set to a radar image scene in order to facilitate radiometric and geometric corrections, and hence, aid in subsequent scene interpretation. In many applications, SAR images are provided with the "added value" of an overlaid geographic map. In this case, the image has to be orthorectified, as explained previously.

In addition to determination of geophysical parameters, SAR is capable of determining the digital elevation (DEM) or the deformation of the surface. The technique, which is used for DEM production, is called "SAR interferometry" (also known as "InSAR"). It entails using two images acquired over the same area at two different times (separated by a few days or a month), but with a slight displacement of the orbit. This displacement causes a phase difference in the received signals. This phase difference is converted into digital elevation value through a sophisticated mathematical manipulation (Massom and Lubin, 2006). Surface deformation requires input from three images plus an exiting DEM or four images (i.e., the production of two DEMs at different times). The technique is called "differential interferometry."

20.3.4 Image classification and change detection

Applications of remote sensing for disaster response require employing image classification and change detection techniques. Image classification in its simplest form entails labeling each pixel or segment in an image to a land cover class based on its radiometric measurements. The output is a thematic map of land cover classes. When followed by application of a change detection technique, the entire approach can be used to assess the damage from a disaster. Thematic maps can also be imported to a GIS. A brief account of image classification and change detection is provided.

Image classification techniques are grouped into two categories: supervised or unsupervised (Verbyla, 1995). Supervised classification requires a set of "training data" to assist in assigning pixels or segments to the appropriate class. Training data are radiometric measurements obtained from areas of homogeneous cover of the relevant class. The areas are usually determined from maps, ground observations, interpreted remote sensing data, or other information. This approach may also require the probability of occurrence of the radiometric measurements for each class. The problem with supervised classification is that the radiometric measurements and their probabilities usually overlap. Maximum likelihood techniques are usually used to resolve these situations (Mesev et al., 2001). They examine the probability function of a pixel value (radiometric measurement) for each class and assign the pixel to the class with maximum probability.

Unsupervised image classification entails examination of radiometric measurements from several classes in a given measurement space (one, two, or higher dimensions). The technique usually employs a cluster-seeking method to examine the data distribution in the given space and identify fairly established clusters. Because no "training" data are provided, it is up to the user to label the identified clusters. This can be achieved using ground validation data, but this is the real challenge in using this approach. The accuracy of classification results, when based on their clustering pattern in the given measurement space, hinges on the degree of separability of those clusters.

Surface cover may change due to natural or manmade effects. The latter is becoming increasingly more important through many activities such as agriculture, building sites, and firewood, to name a few. Natural disasters contribute heavily to surface changes. From a remote sensing point of view, change detection requires the availability of two sets of observations, acquired at different dates. Data from the same sensors facilitate change detection because the two data sets will have the same spatial and radiometric resolutions and same spectral bands. Cloud cover in optical data is certainly an obstacle in achieving accurate change detection, hence, the advantage of using radar images that are not sensitive to cloud cover. If optical data are used for this purpose, then calibration that accounts for solar radiation, solar zenith angle, and solar azimuth should be performed first. Scene-to-scene variation in atmospheric effects (atmospheric absorption, scattering, and emission) should also be accounted for. Moreover, the two image data sets should be coregistered prior to the application of the change detection technique.

Among the first change detection techniques are postclassification comparison methods. Here, two land cover classifications from two images are compared. In this case, there is no need for radiometric normalization between the two images. The accuracy of the change detection hinges on the accuracy of the adapted classifier. Another more involving category of techniques is known as "spectral change identification methods." In this category, data from two images are used to generate a new image, single or multiband, that represents the spectral change between the two images. Most of the techniques under this category are based on spectral difference or ratio between measurements from the same pixel. Accurate image registration is critical for these methods. In case of multispectral data, correlation analysis is performed between components of multispectral vectors. This eliminates the need to accurately account for radiometric influences induced by solar irradiance, sun zenith angle, and atmospheric effects.

Principal component analysis is a popular method to analyze multispectral data. This method can be applied to two or more scenes of the same area, acquired at different dates. The method produces multiband images, called principal components, where significant change in the surface between the different dates provided by the raw image data set appear in the major component images, whereas minor changes appear in the minor component images. A survey of change detection methods is presented in Yuan et al. (1998).

20.4 Space-based initiatives for disaster management

There have been a number of initiatives in the form of programs and commissions under the aegis of international organizations and the United Nations (UN) system, ad hoc groups, and multinational agreements seeking the use of space technologies for disaster management. A brief account of these initiatives follows.

IGOS unites major satellite and ground-based systems for global environmental observation of the atmosphere, oceans, and land. The strategy applies to a process of linking research, long-term monitoring, and operational programs, and identifying gaps and resources needed to fill these gaps in the global environmental monitoring system. In practice, IGOS linkages cover CEOS, specifically its Science and Technology Subcommittee; integrated research programs on global change within the World Climate Research Program and the International Geosphere-Biosphere Program; international group of funding agencies for Global Change Research; and internatioal agencies sponsoring global observations: The Food and Agriculture Organization of the United Nations, Intergovernmental Oceanographic Commission (IOC) of UNESCO, International Council for Science, United Nations Environmental Program (UNEP), and the WMO. The global observing systems, namely, the Global Climate Observing System, the Global Ocean Observing System, and the Global Terrestrial Observing System, are also part of IGOS network. Like the marine-related disaster studies (hurricanes, tsunamis, harmful algal blooms, coral bleaching, invasive species, oil spills, and coastal flooding) that are IOC's focus, these IGOS-linked initiatives employ in situ and space-based data collections. The role of space technologies is especially visible with regard to tsunami warning systems, like the one currently implemented for the Indian Ocean region.

The Disaster Management Support Group replaced the Disaster Management Support project, which began as a short-term task of the Strategic Implementation Team (SIT) created by CEOS in 1997 to develop the concepts of IGOS. The purpose of developing these concepts was to support natural and technological disaster management on a worldwide basis by fostering improved utilization of existing and planned EO satellite data. NOAA provided the lead of this ad hoc group throughout its existence; it held its final meeting in Brussels in June 2001, and the following year submitted its report, "The Use of Earth Observing Satellites for Hazard Support: Assessments and Scenarios." The report remains one of the most comprehensive reviews of space-based disaster remote sensing; it also contained advice on disaster scenarios for the International Charter. The foregoing account on space remote sensing for global disaster management draws significantly on the findings of this report. The group carried out its mandate following hazard theme teams: drought, earthquake, fire, flood, ice, landslide, oil spill, and volcanic hazard. The team membership was drawn from space agencies, satellite operators, and emergency management users, as well as from international organizations. The teams were assigned with compiling user requirements, identifying discrepancies in the provision of satellite data, and developing recommendations for improvement. In addition to the previous hazard theme teams, an information tool team supervised the development of a server. The group formulated its findings with respect to the entire disaster management cycle, from mitigation and warning to response and recovery. The group also assisted the CEOS member agencies and CEOS SIT with actions in response to the recommendations made in the report and worked with relevant international bodies, such as the United Nations Office for Outer Space Affairs (OOSA) and the United Nations International Strategy for Disaster Reduction. With funding support from NOAA, OOSA organized several regional workshops on behalf of CEOS on the use of EO satellite data for disaster support. The closing workshop took place in Munich, Germany, in October 2004.

The Global Disaster Information Network (GDIN) is a nonprofit association for facilitating the provision and use of information to those involved in managing natural or technological disasters. It is an informal international body with members from nongovernmental organizations, government departments, international organizations, industry, academia, and funding/aid agencies. The U.S. government is a key partner and, along with some industrial donors, helps cover the cost of the GDIN secretariat. The individual activities are funded by the host countries or organizations. GDIN has facilitated the development of novel GIS products based on remote sensing for Vietnam, Mozambique, and Turkey, and has assisted others in finding maps, which are a serious requirement of disaster management.

The United Nations Office for the Coordination of Humanitarian Affairs (UN OCHA) has been administering *ReliefWeb* since 1996 for the online display of maps and satellite imagery and other documents on humanitarian emergencies and disasters. Likewise, the UNEP data portal contains information on disaster occurrences. For those interested in the availability of satellite imagery and high-resolution remote sensing data, the Remote Sensing Task Group of the United Nations Geographic Information Working Group acts as facilitator by working in conjunction with specialized UN entities, namely, UNOSAT, which is the United Nations Institute for Training and Research (UNITAR) Operational Satellite Applications Programme implemented through the United Nations Office for Project Services. Another access channel to natural disaster database is the geodata portal of the United Nations System Wide EarthWatch.

WMO's WWW is meant to take advantage of the evolving meteorological satellite technologies for climate prediction and understanding. The Tropical Cyclone Program (TCP) is a subset of WWW and is meant to help with reliable forecasts of tropical cyclone track and intensities, as well as the associated flooding phenomenon, and promote awareness to warnings in an attempt to reduce to a minimum the loss of life and damage to property caused by cyclones. TCP planning and implementation is carried out by means of regional bodies such as the WMO Typhoon Committee under the United Nations Economic and Social Commission for Asia and the Pacific (UNESCAP). The Information, Communication and Space Technology Division of UNESCAP has been active in hazard and disaster management outreach and information dissemination activities.

Within the UN system, the International Telecommunications Union (ITU) represents one of the few non-EO, communication satellite technology organizations that has been recently called on to develop ICT-based solutions in emergency telecommunications directed at improving early warning, disaster preparedness, and mitigation.

The first truly global initiative that would embrace the entire slew of space technologies and the total cycle of disaster management came into being following the recommendations of the third UN space conference UNISPACE III held in Vienna, Austria, in 1999. These recommendations gave rise to several action teams, including Action Team 7, which was tasked to investigate the implementation of an integrated, global disaster management system. This action team, like others, was accountable to CEOS through its STC, attracted membership from the UN member states, and was led by Canada, China, and France (CCF). In the course of delivering on its 3-year-long mandate, the team was supported by the OOSA. A yearly work plan was implemented, and the action team conducted its business through regular plenary meetings and task-oriented working groups. In addition, the three cochairs from CCF held monthly discussions to monitor progress and generate records. The main challenge faced by the action team was to relate information on available space technologies to the needs of the user communities, which had varying degrees of familiarity and knowledge of these technologies. The action team, therefore, surveyed national authorities to find answers to such cross-cutting questions as user needs, national capacity, and space system capabilities before treating the practical issues for the same specific disaster types. The survey questions about user needs focused on the spatial information requirements and the frequency of information generation that would in turn determine the space sensor resolution and revisit. For national capacity, the first question addressed in the survey was whether there existed in the country concerned a designated state authority mandated or entitled to request, receive, and use space-based information for disaster management. Whereas this part of the survey shed light on the ability to access space data records, the map scales at which a country was covered, and the training needs, it was noteworthy that the main obstacle to using space-derived information was systematically identified to be the delay in information dissemination. Individuals dealing with the disaster crisis needed to have faster transmission devices to receive the information in real time and believed that space could not only generate information on the disaster but could also ensure its rapid delivery. Moreover, the action team internally undertook a detailed review of space systems. The information compiled in the document that resulted from this review assisted the action team in evaluating the effectiveness of space technologies to meet the needs of users and the ability of their respective countries to integrate the technologies into their disaster management structures. In addition to describing the programs, initiatives, and space systems and sensors of choice for disaster management, the document examined the type of products offered by space data providers, as well as the policies that governed the use and access to those products.

The action team observed that the potential benefits of space information in disaster management could be grouped into two primary phases: a "hot phase" of warning and crisis management, and a "cold phase" of risk reduction and damage assessment. Risk reduction is the ultimate goal; nevertheless, it is clear that the international community is continuously struck and challenged by "hot" situations of distress caused by floods, forest fires, ice hazards, and such technological disasters as oil spills. Based on the survey results, the action team constituted working groups to determine possible features of an integrated global disaster management system in terms of these various disaster types.

For flood-related disasters, for example, it was suggested that the available resolution of space data systems was appropriate to map flooded areas; however, small targets such as buildings, bridges, and banks were hard to recognize without high-resolution imaging. The frequency of coverage by a single satellite is not adequate, but can be augmented by combining data from different satellites. The most important piece of information needed immediately following a flood is a wide area map with appropriate referential vectors to give an appreciation of the extent of the flood by comparison with the map or imagery predating the event. A second set of products is needed to monitor the evolution of the flood and to plan recovery by merging the image maps with geospatial data, which is done through land use maps, digital elevation models, geological maps, and demographic surveys in a GIS format.

In the case of forest fires, the action team findings pointed to the insufficiency of spectral bands of the existing sensors. Moreover, the temporal frequency was not considered to be adequate. The satellite data products that are compatible with ground-based products and services are generally not available in an appropriate format for those dealing with the hazard. There are only a few specialized institutions dedicated to developing and providing products, technology transfer, and education to deal with forest fires.

In the opinion of the working group on drought, this is an evolving disaster, and as such, does not have an emergency "hot" phase. Drought does not have special requirements for spatial and temporal resolution, although spectral resolution becomes important at various stages of soil and land cover dryness, as well as humidity levels of soil and vegetation.

In the case of earthquakes, very high-resolution sensors are necessary, and a higher level of integration between space and ground data and services is desired. If the countries affected by earthquakes want to benefit from space-based data, rescue personnel must be trained in the data use.

The main purpose of managing ice hazard is to allow ships transiting icy waters to navigate safely and to support maritime rescue operations. High-resolution products derived from several satellites are being used to manage ice hazard. SAR systems are the best alternative to high-resolution optical data. The currently functional agreement between the Canadian and the U.S. ice services provides a good model of cooperation among nations, but the funding arrangements are not easy to make. Research and development will be crucial as the next generation of satellites becomes operational.

Oil spill is a common technological disaster. The concerned working group emphasized the importance of fusion of airborne, sea state, and satellite data to manage oil spills. The requirement is for daily coverage and revisit at short intervals to monitor oil spill. Data reception, processing, and delivery should be as automated as possible. It was suggested that a fund for supporting the use of space data in oil spill disasters could be established by oil companies, oil transporters, and the governments of major oil importing and exporting countries.

The action team then went on to establish implications and characteristics of an integrated system and ended up with a set of recommendations in a formal document submitted to the UN General Assembly, "Implementation of the Recommendations of the Third United Nations Conference on the Exploration and Peaceful Uses of Outer Space (UNISPACE III): Final Report of the Action Team on Disaster Management. United Nations General Assembly, A/AC.105/C1/L-273, 22 December 2003." The most prominent of these recommendations was the creation of an international space coordination body for disaster management, nominally identified as the Disaster Management International Space Coordination Organization (DMISCO) that would provide the necessary means to optimize the efficiency of services for disaster management. The concept of DMISCO would be based on a disaster management space support system for all stakeholders: users of various backgrounds (civil protection agencies, lending institutions, emergency response units, and national authorities; value-added centers and companies; and space data providers in the public and private sectors). The implementation of the concept would rely on existing space and ground resources and would cover the entire disaster management cycle. The DMISCO concept received approval in principle of the UN General Assembly, and on the advice of the CEOS Scientific and Technical Subcommittee, an ad hoc expert group was formed in early 2005 to work toward implementing the DMISCO concept. The group has already submitted its final report for CEOS consideration of the implementation options.

20.5 About the charter

The International Charter "Space and Major Disasters" was established to promote cooperation among space agencies and space system operators with regard to the use of space technologies for managing crises arising from natural or technological disasters. Essentially, the charter is about the provision of space-based data at no cost to the user, where these data and the derived information or services would help in crisis management. The charter membership is on a voluntary basis and does not involve exchange of funds. The overall charter administration is the responsibility of a board on which all the member agencies are represented. The board is assisted by an executive secretariat with routine charter administration and operations.

The most important aspect of the International Charter is the entities that are entitled to request the charter data. Unlike the historical use of satellite remote sensing data, either by commercial or R&D groups, the charter data are available to those who are actually engaged in crisis management. They are called associated bodies and, for the purpose of the charter, are the institutions or services responsible for rescue and civil protection, defense, and security under the authority of a state whose jurisdiction covers an agency or an operator that is a member of the charter. The member states of the ESA are included, as are those of any other international organization that is party to the charter, such as the Disaster Monitoring Constellation (DMC). Similar bodies of nonmember countries have the choice of requesting the charter activation in the event of crisis in their countries, either by applying directly to the charter board, or through the associated bodies. Last, the charter partners can act on their own if they judge that there is a need for space-based response to a disaster occurrence.

The charter member agencies use their best efforts to fulfill their membership obligations in mobilizing their space and ground facilities to contribute disaster-related data or information, which the associated bodies can use or deliver to end users dealing with the crisis. The contribution may be in the form of satellite image data, data products, or refined information. In other cases, these may be services related to data acquisition, positioning, telecommunications, and TV broadcasting under the control of the member agencies.

The crisis situation is identified by the authorities of the country affected by the disaster and by at least one associated body, which alerts the charter executive secretariat as soon as possible following the advent of the disaster having the potential of creating a crisis situation. In response, the charter authorities pool their resources together in terms of coordinated spacecraft tasking, data archive searches, data mergers, value adding, and data deliveries.

In implementing the charter, the member agencies and operators are expected to maintain information for their mutual use on assets that are under their direct management and under other public or private sector control but to which they have access. The information may pertain to space mission and orbital characteristics, operational conditions, programming procedures, data availabilities in the archives, and products and services provided by ground systems, including the data processing facilities. The experience gained with the charter functioning is extensively documented and constantly analyzed to develop crisis scenarios and improve operational procedures. The charter functionaries keep themselves abreast of new technological developments.

In addition to the associated bodies, there are other organizations called co-operating bodies, such as the European Union and the UN agencies, with which the charter maintains a cooperative relationship. The co-operating bodies can work with the associated bodies to obtain the charter data and facilitate its access to the end users.

The charter membership is open to all the space agencies and space system operators, as long as they can commit to contributing their space and ground resources for data and information helpful in managing disaster crisis. Requests for membership are received by the charter board that formulates its response within 180 days of the reception of the request and studies it in light of the extent of the requester's contribution and commitment to the charter functions and objectives. In accepting a new membership, the board makes sure that there will be no disruption or structural modifications in the deployment of the systems already in place.

20.5.1 History and operations

The charter idea came into being at the third UN space conference UNISPACE III held in Vienna in July 1999. In the face of increasing destruction and damage to life and property caused by natural disasters and conscious of the benefits that space technologies can bring to rescue and relief efforts, the ESA and the CNES set out to establish the text of the charter, which they themselves signed on 20 June 2000, while inviting other space agencies to do the same (Bessis et al., 2003). The CSA was the first to come onboard and sign the charter on 20 October 2000. These three founding space agencies then went on to establish the architecture essential for implementing the charter.

The charter implementation required first and foremost the identification of its functions and the corresponding procedures for their operation. It also required defining the charter rules and policies. In the face of these tasks, the founding agencies established a high-level Charter Implementation Plan and a Charter Policies and Procedures document. The latter document describes the process and the conditions that apply to become a charter member, an associated body, or a co-operating body. The document also contains definitions of what constitutes an emergency that will qualify for a charter activation and acquisition of data and information. According to these definitions, an "emergency" is an urgent need for space services associated with a unique and important event, when something unanticipated suddenly occurs and requires prompt action, beyond normal procedures to prevent or limit injury to person or damage to property. As a consequence of their unpredictability, the satellite coverage for such events cannot take place in the usual manner and as part of the normal data acquisition planning. Two kinds of emergencies were recognized, an environmental emergency, which is a sudden uncontrolled event resulting from human activities or natural causes, with considerable potential for harm to the environment, public health, or general welfare, and which demands timely and appropriate action to restore control and counter the (potential) effects. A nonenvironmental emergency is a sudden state of affairs that will have harmful consequences unless acted on immediately and for which governments may take on special powers or role. Military actions are excluded from the charter scope and intervention. The charter implementation is for emergency response and crisis management. It does not cover requests related to the use of the charter for research and development, rehabilitation, and sustained monitoring.

The various functions needed to implement the charter are a board, an executive secretariat, and four key players, namely, a group of authorized users (AUs), an on-duty operator (ODO), a pool of emergency on-call officers (ECOs), and a selection of project managers (PMs).

The board is the overall charter authority that negotiates new membership and interacts with other international organizations, such as the co-operating bodies of the charter. It is the responsibility of the board to provide safeguards against unauthorized distribution of the charter data. The technical and administrative functions are performed under the supervision of the executive secretariat, including the exchange of similar functions with the member agencies, the associated and co-operating bodies, and the end users. In the course of exercising its responsibilities, the executive secretariat establishes and approves operational procedures, internal and external reports, interface documentation, and the activities related to information and promotion.

The first of the four functional units is a group of AUs, who alone are entitled to activate the charter. In the context of the charter, the AUs are from the same civil protection, defense, and security institutions or services that were earlier defined as "associated bodies." A list of these AUs is kept up to date with their full contact information for identity purposes.

As a space agency becomes the charter member, it automatically acquires the privilege to nominate one or more AUs. On acceptance, the AU receives the phone/fax number of the 24-hour charter "call center," as well as a form, the user request form (URF), for completing it with information on the disaster for which the charter assistance is required, and localization of the affected area. The URF is faxed to the next functional unit called the ODO, which is a centralized 24 hours a day call-receiving unit staffed by the ESA's Earth Observation Centre ESRIN in Frascati, Italy. The ODO's role is to check the identity of the caller as being a bona fide AU. The ODO interfaces with the next functional unit, the ECO. The information exchange that takes place at the ODO-ECO interface relates to contact details about the AU and the type and location of the disaster, as described in further detail later.

The ECO is the most critical function in the charter operations. It is at this level that the charter unveils its attribute of a unified system of space data acquisition and delivery to users. The ECO's primary function is to carry out on an emergency basis multisatellite data acquisition and delivery planning by liaising with the mission planning staff of the member agencies. Like the ODO, this is also a 24 hour a day, 7 day a week function, and is performed by the member agencies on a weekly rotation. A monthly roster of ECO duty assignment is issued by the executive secretariat in advance. In addition to its planning function, the ECO is responsible for updating the status of the space resources belonging to the ECO's member agency. This status update includes periods of satellite outages, ground station unavailability, and other factors that should be considered in developing a draft acquisition/archive plan (AAP). When a charter activation call is received and accepted, the ECO on duty is required to compile a "dossier" with all relevant information on the disaster, communications with the AU, the URF, and the AAP. The latter is in reality a collection of data acquisition forms of the individual agencies. The forms are filled, among others, with geographic coordinates of the coverage area, date and time of acquisitions, orbits, sensor/imaging modes, processing levels, and data destinations. The contextual and job content knowledge requirements for the ECO functions relate to sensor types, orbit characteristics, outages, data downlink options, coverage, resolution, data ordering time lines, product standardization, and agency allocation/contribution. As such, the ECO is able to use the tools provided by each member agency to plan new requests and undertake archive searches. The ECO also possesses a general knowledge of the processing levels and product types associated with the various sensors, and is familiar with the charter provisions and the level of commitment of each member agency. Generally speaking, the ECO has sufficient understanding of imaging, telecommunications, meteorological, data collection, and positioning satellite systems. The ECO's work is facilitated by means of pre-existing guidelines and checklists for processing a charter activation call.

Once the "dossier" is complete, the ECO passes it on to the next functional unit, a PM. Unlike the ODO and the ECO, the PM is available during the normal working hours only and is selected by the executive secretariat based on the following criteria: geographic location of the disaster occurrence, disaster type, sensor(s) used to cover the disaster, availability of potential PMs in a member agency, uniform distribution of PM appointments among the member agencies, and value-added processing proposal received from a member agency. In fact, each agency provides a list of PMs in advance to the executive secretariat, along with the information on the PM's field of expertise. The PM is the overall project authority and is the one that coordinates the various activities and interactions occurring in the course of a charter activation. The PM has expert knowledge of remote sensing satellites and their supporting ground segments; data delivery networks; remote sensing data applications, particularly disaster management; civil protection organizations; remote sensing data value adding; and project management. The PM also has a good understanding of the predefined scenarios. To complement the information on the disaster and to know more specifically about the needs, the PM may directly contact the AU and the end users.

As an agency's charter membership is accepted, it is required to undergo cross-training and rehearsals with the existing member agencies in order to qualify for its integration with the charter operations. Likewise, apart from its contribution in the form of space data or services, the new member is expected to be in a position to provide the personnel and other operating resources that are needed as a result of an emergency notification. The purpose of the cross-training and rehearsals is precisely to ensure that these personnel and resources can act properly in the mobilization of space and ground equipment, such as the data acquisition planning tools and application software. The successful completion of the training and the rehearsals is also a manifestation on the part of the new member that it is able to furnish the ECO and the PM functions. During the period the new member's operational integration is taking place, it is entitled to nominate a technical point of contact, which is later replaced by the member's representation in the charter executive secretariat. At the beginning of the training and rehearsals, a detailed test plan is established under the supervision of a technical team, and the outcome of its implementation is recorded in a test report that becomes the basis of new member's operational certification.

After their mutual training, the three charter founding agencies, namely, the ESA, the CNES, and the CSA declared the charter operational and open for business as of 1 November 2000. The charter has been activated to date more than 100 times to cover disasters of natural causes or human actions worldwide.

NOAA and the Indian Space Research Organization (ISRO) were the first to show interest in joining the charter after the start of its operations and became members in September 2001, followed by the Argentinean Comisión Nacional de Actividades Espaciales (CONAE) in July 2003. In February 2005, the JAXA formally signed the charter. In fact, the U.S. NOAA membership was a representation of the concerned U.S. government agencies; however, in April 2005, the U.S. Geological Survey (USGS) was brought directly into the charter's fold, and has since been active in the charter implementation. The latest member to join the charter was the DMC, which was built by the UK company Surrey Satellite Technology Ltd. and comprises five satellites owned by the UK, Nigeria, Algeria, Turkey, and China.

Figure 20.16 depicts the charter operational cycle commencing with the reception of a disaster coverage request from an AU and ending with the data and data product delivery to the end users (Mahmood et al., 2004). The operational cycle can be explained as follows. Following a disaster of significant potential of damage to life and property, the AU collects the relevant information that it would have received from the end users to complete the URF and faxes it to the ODO. A backup electronic copy of the URF is also submitted. The AU then calls the ODO to identify the fax. The ODO acknowledges receipt of the faxed form, and confirms its legibility and completeness with the AU.

The ODO checks the identity of the caller as being an AU by consulting a list already provided to the ODO by the executive secretariat. The ODO makes sure that the caller's phone, fax, and e-mail addresses are properly recorded. On receiving a call, the ODO checks the information required in the URF that has been faxed or additionally e-mailed to the ODO by the AU. In the case of problems in reading the faxed document, the ODO refills the form over the phone and reconfirms its contents with the AU. Next, the ODO calls the ECO on duty for the week in the concerned space agency and leaves the ODO's name and contact numbers. The ECO returns the ODO's call within 20 minutes of its arrival. If the ODO does not receive a response from the ECO within 1 hour, the ECO is paged again. As part of the information exchange, the ODO provides the ECO with the URF. The ODO logs the date and time of the successful contact with the ECO. At the same time, the ODO copies the URF to the executive secretariat and to the charter webmaster, who creates a

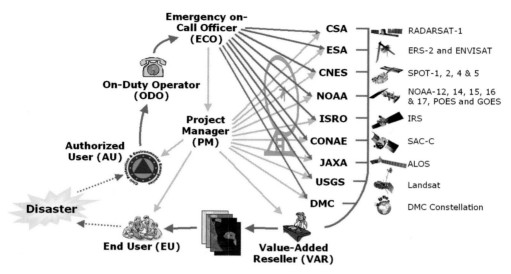

Figure 20.16 The operational cycle of the International Charter from the declaration of a disaster to the delivery of data and information products to the end users.

temporary folder on the charter ftp site for use by those involved in the charter activation, including the ECO, the PM, the member agency mission planners, and often the processing facility personnel.

On receiving a call from the ODO, the ECO ascertains the validity of the disaster relief request by checking again that the requesting organization is on the list of the AUs. If the request is from an organization that is not on the list, the ECO may contact the organization for further inquiry. The ECO may proceed with developing a draft acquisition plan if the disaster is deemed to be of serious proportions. The ECO has the standing guidelines from the executive secretariat as to whether to entertain a request from a non-AU. At this point, the ECO is required to contact its executive secretariat representative to proceed with the activation or stop the process.

The ECO consults the guidelines provided by the executive secretariat for accepting a charter activation request and keeps the executive secretariat informed of the requests that are not accepted. The ECO develops the AAP based on the information obtained from the user, the guidelines for covering the disaster, the data commitment of each member agency, and the ECO's own analysis of the disaster event. After consulting the disaster scenarios available in the charter procedures and confirming details about the event through media reports, the ECO establishes the AAP. In creating the AAP, the ECO takes into account the following:

- Suitability of the space sensor for covering the disaster; this decision should be based on the sensor type, its availability, coverage, resolution, and product type, using operational information provided by the space system operators
- Need for new or archive data
- Member agency's data contribution limit
- AUs needs in terms of the nature of data and information, such as telemedicine services, communication and broadcasting, and others; some of these services, however, are not yet available to the charter operations

The AAP specifies the name of the satellite(s), type of sensor(s), imaging parameters (if applicable), coverage frequency, ground resolution, acquisition period, type of processing, delivery time, and delivery medium and address.

The ECO communicates with the member agency mission planning staff with regard to the draft AAP and coordinates its finalization in light of the availability of satellite and archive resources, and data acquisition planning constraints and conflicts. The ECO remains in the consultation mode with the AU, and where practical, presents to the AU a summary of the requested acquisitions and the actions taken on the AU's request. After duly informing the AU, the ECO places the data orders using the agency-prescribed emergency request forms (ERFs). The draft AAP is thus finalized and executed, and is incorporated in the dossier, along with other details about the disaster. The ECO then passes the dossier to the next functional unit, which is the PM, with copies going to the member agency staff that performs ECO function and the executive secretariat.

The ECO contacts the PM during the PM's regular business hours, who returns ECO's call using the ECO's pager or emergency number, and leaves a message with the PM's name and contacts. The ECO then responds to the PM within 20 minutes of receiving the message. If the PM does not get a response from the ECO within 1 hour, the ECO is called again. As part of the information exchange between these two entities, the ECO provides the PM with the dossier, which the ECO also places in the temporary call folder opened on the charter ftp site.

On its designation, the PM studies the dossier in detail to acquire in-depth knowledge of the disaster and the data requirements. The designated PM may require help from other PMs about the information on local data delivery channels, value-added resellers (VARS), and data application potential. The PM may eventually complement the dossier with further actions for the maximum utilization of space facilities, such as the use of meteorological, telecommunication, data collection and positioning satellites, and enhanced data processing, and for that purpose follows the relevant PM scenarios that are developed and recorded on a regular basis. The scenarios are incorporated in the Charter Implementation Plan, and these are meant to guide the PM about the data acquisition possibilities and characteristics of the participating satellites/sensors. These scenarios provide general information on the satellite tasking conflicts and planning constraints and specific choices with regard to each disaster type.

In the case of earthquakes, for example, the purpose of use is identified to be the assessment of damage to buildings, bridges, roads, dams, and other critical infrastructure. The satellite data products are base and change maps. High-resolution optical sensors (2–5 m) are the most appropriate, but radar sensors of highest resolution may also be tasked for complementary data sets.

The required information for flood, as for all the other disaster types, is the time and location of the occurrence, the extent of damage, and the expected evolution, but the knowledge of estimated flood peak time is most important. The standard deliverables are flood extent and change detection maps that can be generated in the case of small-scale flooding with multispectral optical data; however, radar sensors are very effective for general viewing of the inundated areas, particularly when cloud cover hampers the use of optical sensors. For the coverage of very large areas ($>1,000$ km^2) and synoptic meteorological conditions, the scenarios recommend sensors such as NOAA AVHRR.

For landslides, damage assessment and recovery operations remain the main purpose of use, and the damage extent and change detection maps the desired deliverables. High-resolution optical sensors are best for this disaster, although comparably high-resolution radar imagers with optimum angles of incidence can reveal further details on the landslide scars.

Hotspot detection and damage assessment in support of recovery and safety operations is the purpose of using space technologies for lava flows and ash location. The choice of sensor varies with the type of volcanic hazard. Deformation and lava flows can be imaged best with multispectral and panchromatic and radar high-resolution sensors, hotspots, and ash plumes with infrared, midinfrared, and visible channels on the charter's environmental satellite payloads.

Fire detection, monitoring, and behavior are what need to be carried out with the satellite surveillance system. Support to burnt area emergency rehabilitation is a specific purpose of use related to wildfire disaster. A special information requirement is the anticipated development of the fires based on factors such as weather data and forecasts, terrain, and fuel conditions. In addition, knowledge of neighboring forested areas, grasslands, and population centers is useful. The deliverables are base maps, smoke plume reports, maps of hotspots, and burnt areas. The base maps can be prepared with fine to medium resolution radar and optical satellite data. For smoke plumes and hotspot detection, payload channels for night and daytime imaging on the environmental operational NOAA satellites are selected. For fires, the charter functionaries have in the past sought data from specialized sensors outside the charter group of satellites, such as BIRD, to complement MODIS Land Rapid Response System data (from Terra and Aqua satellites) and ATSR.

The purpose of using the charter data for sea ice hazard is specifically for ice extent, classification, and concentration in order to support rescue and recovery operations. The common deliverables are ice type and extent maps. For ice detection, visual, infrared, and radar data in wide swath are planned under the scenario guidelines. For ice classification, in contrast, radar sensors are primarily chosen.

Data acquisition planning over multiple days is required to support the oil slick containment and monitoring process. Space sensor data are interpreted for slick location and tracking. Charter radar sensors, preferably with VV polarization and wider swaths, are by far the most commonly used. In the case of the maritime disasters of oil slicks and sea ice, archive data are not a requirement, although they are critical for the land disaster applications for change detection and for which reference imagery is essential.

The PM interacts closely with the agency mission planners, VARs, data processing facilities, and the executive secretariat to ensure fast data, information, and service delivery directly to the requestor or through the concerned civil protection authorities, or any other channels of quick turnaround. A best product catalogue is maintained to aid the PM in the selection of value-added products. With regard to the remote sensing satellites, the commitments of the charter member agencies are limited to providing satellite imagery of a predetermined processing level only. In actual practice, however, the member agencies have generously supported the development of value-added and information products that are needed by the end user. One of the most common products is a "map," which for present purposes can be defined as a set of layers of geographic information, vector or raster in nature, and which may ultimately (although not necessarily) be printed on paper. Some of these layers are derived from the space EO data products; others are obtained from external sources. The following components of the catalogue are specific to the disaster type that is being confronted and will be reflected in the image products that are shown later in the Case Histories section.

- **Damage themes:** These maps are obtained by adding a set of extra layers on top of the selected reference map, and these layers correspond to the type of damage described in the disaster scenarios referred to previously. The following are a few examples recorded in the best product catalogue.
- **Flooded areas:** In flood extent maps, the inundated areas are represented as colored polygons. The underlying flooded surface is sometimes represented in lighter colors.

Burnt surfaces: In the case of charter activation in the Var region of southern France, burnt areas were expressed as polygons in dark blue coloration.

Earthquake-damaged urban zones: Earthquake-affected area maps derived from the space EO data are the most difficult to draw because of limitations of ground resolution of the imagers currently included in the charter satellite constellation. Better results are achieved by means of photointerpretation of very high-resolution imagery. A good example is the charter activation for the 2003 earthquake in Algiers, for which the damage zones were spotted on the imagery.

Oil spill extent and location: The SAR sensors have been extensively used so far to recognize oil slicks, which can be easily confused with wind shadows, both having a dampening effect on radar brightness.

Lava flows: Change detection has been possible by means of radar imagery of sufficiently high resolution pre- and postdating the disaster. A good example is the Nyiragongo eruption in the Democratic Republic of Congo in 2002. In this case, a classification scheme was applied and the results were superimposed on 10-m resolution optical data. Further details are provided in the Case Histories section.

Volcanic debris: Radar images are also used for delineating the distribution of pyroclastics and volcanic ash deposits. A good example from the past is Montserrat volcanic eruption, which was covered by the charter.

Additional themes: Apart from the damage themes, maps are also needed for the emplacement of facilities for the population displaced by a disaster, such as refugee camps. In the case of Nyiragongo volcanic eruption, space imagery was used to locate areas not under the threat of lavas flowing from the mountain slopes to move people to safer grounds. In the aftermath of the Algerian earthquake, the locations of the tent sites for the homeless were recognized on satellite images. The area selection for camps and other such sites can be done by means of colored polygons as for the damage themes.

Reference data sets: Topographic maps are the basic products of traditional land surveys, which are used as the underlays for the previous theme maps. In the case of oil spill detection, these are marine maps. As a minimum, the reference maps should show map scale, geographic grid, and the characteristics of the projection applied, depending on the cartographic projection system used in the affected country, political and administrative boundaries, annotations, rail/road and hydrographic network and its classification, location of utilities (hospitals, fire brigade posts, service centers, airports, helipads, food markets, etc.)

In some cases, such maps are not readily available during emergencies, or they are too old to be reliable. Instead, geospatial maps may be used; these are obtained with a raster background derived from optical or radar data and by using pseudo- or true colors. For some charter activations, vector layers from the digital chart of the world were added to provide additional information. Where available, isoclines from the digital terrain model may be superimposed on the map. In some other cases of the charter activation, a hybrid background was used, where the reference topographic map was updated with information derived from a recent satellite image of the area, in order to update a newly built roadway. Land use classification is another option for selecting the background.

Geographic extent: A typical map scale is 1:50,000, but the scale may vary from 1:10,000 to 1:250,000, depending on the available space imagery and the extent of the area of interest. The overall situation of a large area (scale: LARGE) is over 100×100 km^2, synoptic viewing maps (scale: MEDIUM) are typically 50×50 km^2, and detailed maps (scale: SMALL) are 5×5 km^2.

Presentation: The information derived with the help of the satellite imagery is presented in several different ways, as two-dimensional traditional maps, as three-dimensional (3D) models, and as fly-by views, in which case the information is draped over a 3D model of the area, simulating an aerial tour of the disaster affected area.

Format: Several formats have been used for recording the products. Among the digital formats, the commonly used JPEG format with built-in compression scheme is preferred. In addition to maintaining the best product catalogue, the charter service delivery standards are also pursued.

Delivery time: For damage assessment maps, the target is to provide maps within 12 hours following acquisition of the image. In the case of the charter activation over floods in the Gard region of France, the total time elapsed from the activation request reception to data product delivery was 38 hours, including 18 hours of value adding.

Delivery mode: Data delivery through e-mailing has been the most frequently used means of communication. In the case of the Algerian earthquake, electronic e-mailing was used to communicate with the local authorities. However, limitation on file size often impedes this communication mode. A dedicated or user-provided FTP site has been successfully used in a number of cases, but this requires the end user to access to Internet resources, which can be difficult during remote field operations.

Disaster size and scope: The charter is tailored to relatively large events, where major losses of life and property are feared. Generally speaking, the size of the affected area should not be less than about 1,000 km^2; however, in reality, the charter activation requests have been on much more localized sites.

Data sources: Although the charter assets are confined to those of its member agencies, specialized or high-resolution data from outside resources have been procured on a member's own initiative with the PM's and value-added service provider's assistance. The national institutes (IGN, France; IGM, Italy; and Water Institute, Argentina), some of which have regularly provided PM services, have used their own resources to generate ancillary data for the charter activations.

Cost: There is no interagency transfer of funds, but the member space agencies do cover the costs of data acquisition and processing, as well as funding for value-added work. As a matter of practice, the member agency that has provided the PM services also takes on the responsibility for developing value-added products and information.

Most of the charter activations described in the Case Histories section are included in the best product catalogue and are made available to the PMs for consultation.

20.5.2 A constellation of sensors and satellites

The sensors aboard the satellites of the member space agencies are available to the charter operations. The main characteristics of these satellite sensors are described in this section for each agency.

European remote sensing satellite (ERS)

The two European Remote Sensing Satellites, ERS-1 and ERS-2, were primarily active microwave imagers, but also carried aboard their platform a radar altimeter and other instruments for SST and wind measurements. ERS-1 was launched in 1991, and ERS-2 in 1995; the two were built as identical twins, except with an extra instrument on ERS-2 designed to monitor ozone levels in the atmosphere. The active microwave instrument combined the functions of a SAR and a wind scatterometer. The SAR operates in imaging mode for acquisition of wide-swath, all-weather images over land and ocean. In wave mode, the SAR produces image frames, roughly 5 × 5 km, for the derivation of the length and direction of ocean waves. The wind scatterometer uses three antennas for the generation of sea surface wind velocity. The radar altimeter provides accurate measurements of sea surface elevation, significant wave heights, various ice parameters, and an estimate of sea surface wind speed. The ATSR combines an infrared radiometer and a microwave sounder for the measurement of SST. Other instruments are related to the accurate determination of spacecraft position and orbit. The additional GOME (global ozone monitoring experiment) aboard ERS-2 is an absorption spectrometer, which measures the presence of ozone, trace gases, and aerosols in the stratosphere and troposphere.

ERS-1 was launched in a near-polar elliptical sun synchronous orbit with a mean altitude of 785 km and a local nodal crossing in descending path at 10:30 am. The ERS-1 mission also included a number of orbit maneuvers to achieve a 3-day, 35-day, and 168-day repeat cycle to meet various ground coverage requirements. Shortly after the launch of ERS-2, the two spacecrafts were linked in a tandem mission, which lasted for 9 months. During this time, the increased frequency of coverage and the exact repeat of the orbital track of one satellite 24 hours apart from the other satellite allowed the collection of a wealth of data (Figure 20.17) for global interferometric and other applications. In March 2000, a computer and gyro control failure led to the ERS-1 satellite finally ending its operations. The ERS-1 archives remain a major data source for the charter, along with the continuation of new acquisitions that can be planned with ERS-2.

Environment Satellite (ENVISAT)

In March 2002, the ESA launched ENVISAT, an advanced polar-orbiting EO satellite for atmospheric, ocean, land, and ice measurements. It flies in a sun-synchronous polar orbit at an altitude of 800 km. The repeat cycle is the same as the nominal cycle for ERS-1 and ERS-2. Its wide-swath sensors enable complete coverage of the globe within 1 to 3 days.

The ENVISAT payload consists of a variety of instruments, operating in a broad electromagnetic spectral range from microwave to ultraviolet, in order to make measurements of the atmosphere and of the surface of Earth. There are two radar instruments, three spectrometers of different types, two different radiometers (broad and narrow band), high-resolution interferometer for long-term observation, and two instruments for range measurements. The instrument most commonly used for the charter operations is the advanced synthetic aperture radar (ASAR). The medium resolution imaging spectrometer (MRIS) is occasionally used.

Satellite Pour l'Observation de la Terre (SPOT)

SPOT of the French space agency, CNES, has been in operation for 20 years as a major high-resolution earth imaging system. Figure 20.18 is a graph showing the major milestones in the life of the SPOT series of remote sensing satellites. Five satellites in the SPOT family have been launched thus far. The high-resolution visible and infrared instrument (HRVIR) of SPOT-1, launched on 22 February 1986, acquired its first image of the surface of the Earth 2 days later with a spatial resolution of 10 to 20 m. SPOT 2 was launched in 1990. Identical to its predecessor, SPOT-2 was the first in the series to carry the DORIS precision positioning instrument. SPOT-3 followed the others in 1993, and in addition to DORIS, it also carried the American passenger payload POAM II used for measuring atmospheric ozone at the poles. Currently, SPOT-2 is still in operation, SPOT-1 entered on-orbit storage in 2002, and SPOT-3 failed in 1996, but after having completed its nominal mission life.

SPOT-4 was launched on 24 March 1998, and features significant advancements over the earlier SPOT satellites. These advancements are in the form of longer 5-year design life, enhanced imaging telescope features and onboard recording capability. A spectral band in the short-wave infrared (SWIR, 1.55–1.75 μm) was added, and the panchromatic band (0.49–0.69 μm) was replaced by the B2 band (0.61–0.68 μm), which acquired images on request, with 10 to 20 m ground resolution. SPOT-4 carries a special optical payload called VEGETATION, which has a spatial resolution of 1 km and a wide imaging swath. VEGETATION operates in the spectral bands of the HRVIR instrument mentioned previously, the SWIR band, and a new blue band B0 (0.43–0.47 μm). The HRVIR and the VEGETATION instruments thus use the same spectral bands and the geometric

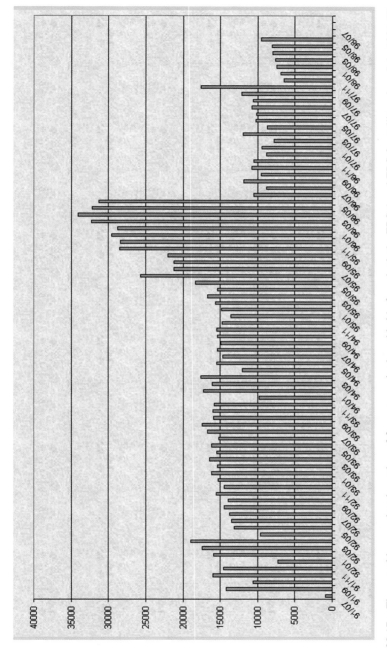

Figure 20.17 The monthly cumulated number of frames acquired worldwide from both the ERS-1 and ERS-2 satellites since July 1991. The frames acquired by more than one station are counted only once.

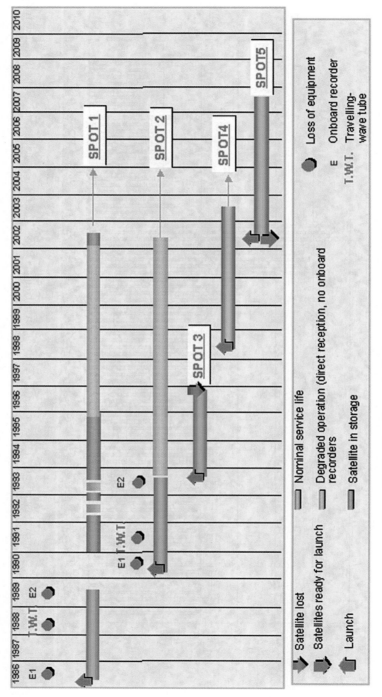

Figure 20.18 Time lines and milestones of the SPOT satellite programs.

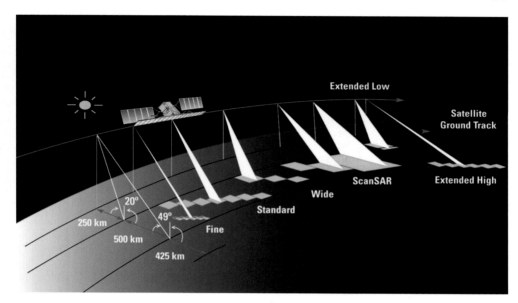

Figure 20.19   RADARSAT-1 synthetic aperture radar (SAR) imaging modes.

reference system, which means that the image data from the two are highly complementary and can easily be coregistered. These instruments are used for tracking wildfires.

The development of the SPOT-5 system was based on the considerations related to data continuity and increased image quality. While retaining the main imaging characteristics of the earlier missions, SPOT-5 made significant headway by offering a ground resolution of 10 m in its multispectral and 2.5 to 5 m in the panchromatic and infrared modes. SPOT-5 has a fore- and aft-pointing, high-resolution camera for stereo imaging of the same area and for generating digital elevation models required for image rectifications. SPOT-5 also carries a VEGETATION-2 instrument identical to that of SPOT-4.

RADARSAT-1

When the satellite was launched in November 1995 by the CSA, with a design life of 5 years, RADARSAT-1 became the first remote sensing program of its kind in several respects. This was Canada's first EO satellite and an orbital radar system with multimode imaging capability in terms of viewing angles and image resolutions. Currently, it is in the twelveth year of operations. A summary of RADARSAT-1 mission performance has been provided in Mamen and Mahmood (2003). The imaging device of the satellite is a SAR, which operates at C-band frequency of 5.3 GHz, with horizontal polarization for both transmission of the radiofrequency pulses and reception of the signals returned from the ground. The main feature of the advanced RADARSAT-1 technology is its antenna, with its electronic beam steering, beam shaping, and rapid beam switching capabilities (Luscombe et al., 1993). The various beam positions and their naming order are depicted in Figure 20.19.

There are seven Standard (S1 to S7) modes, three Wide (W1 to W3) modes, five basic Fine (F1 to F5) modes, one Extended Low (EL) and six Extended High (EH1 to EH6) modes, two ScanSAR Wide (SCWA and SCWB) modes, and two ScanSAR Narrow (SCNA and SCNB) modes, each related to an angle of the incident beam with the local perpendicular, the width of the illuminated swath, and the ground resolution. The ScanSAR is, in fact, a

Table 20.1. *Parameters for RADARSAT-1 SAR beam modes*

Beam Mode	Beam Position	Incidence Angle Range	Resolution Range (m)	Resolution Azimuth (m)	Nominal Coverage (km)	Effective Number of Looks
Fine	F1n	36–40	8.3	8.4	50 × 50	1
	F1	37–40	8.3	8.4	50 × 50	1
	F1f	37–40	8.3	8.4	50 × 50	1
	F2n	39–42	7.9	8.4	50 × 50	1
	F2	39–42	7.9	8.4	50 × 50	1
	F2f	40–43	7.9	8.4	50 × 50	1
	F3n	41–44	7.6	8.4	50 × 50	1
	F3	41–44	7.6	8.4	50 × 50	1
	F3f	42–44	7.6	8.4	50 × 50	1
	F4n	43–46	7.3	8.4	50 × 50	1
	F4	43–46	7.3	8.4	50 × 50	1
	F4f	44–46	7.3	8.4	50 × 50	1
	F5n	45–47	7.1	8.4	50 × 50	1
	F5	45–47	7.1	8.4	50 × 50	1
	F5f	46–48	7.1	8.4	50 × 50	1
Standard	S1	20–27	24.2	27.0	100 × 100	3.1
	S2	24–31	20.4	27.0	100 × 100	3.1
	S3	30–37	25.3	27.0	100 × 100	3.1
	S4	34–40	23.4	27.0	100 × 100	3.1
	S5	36–42	22.1	27.0	100 × 100	3.1
	S6	41–46	20.3	27.0	100 × 100	3.1
	S7	45–49	19.1	27.0	100 × 100	3.1
Wide	W1	20–31	33.8	27.0	150 × 150	3.1
	W2	31–39	24.6	27.0	150 × 150	3.1
	W3	39–45	20.8	27.0	150 × 150	3.1
ScanSAR Narrow	A: Composite[1]	20–39	81.5–43.8	47.8–53.8	300 × 300	3.5
	B: Composite[2]	31–46	54.4–38.4	71.1–78.8	300 × 300	3.5
ScanSAR Wide	A: Composite[3]	20–49	162.7–73.3	162.8–76.6	500 × 500	7
	B: Composite[4]	20–46	93.1–117.5	93.1–117.6	500 × 500	7
Extended High	EH1	49–52	18.0	27.0	75 × 75	3.1
	EH2	50–53	17.7	27.0	75 × 75	3.1
	EH3	52–55	17.3	27.0	75 × 75	3.1
	EH4	54–57	16.8	27.0	75 × 75	3.1
	EH5	56–58	16.6	27.0	75 × 75	3.1
	EH6	57–59	16.4	27.0	75 × 75	3.1
Extended Low	EL1	10–23	39.1	27.0	170 × 170	3.1

[1] ScanSAR narrow A: W1 + W2.
[2] ScanSAR narrow B: W2 + S5 + S6.
[3] ScanSAR wide A: W1 + W2 + W3 + S7.
[4] ScanSAR wide B: W1 + W2 + S5 + S6.
Adapted from Mahmood et al. (1996).

composite beam combining Standard and Wide beams. The spatial resolution varies from about 10 m for the Fine beam in an image swath of 50 km across to 100 m for the ScanSAR Wide beam in a 500-km-wide swath. Table 20.1 gives these details, as well as information on the number of looks, which refers to the number of range lines averaged along track for image processing for each beam mode. A network of data receiving stations and onboard tape recorders ensures a truly global coverage of RADARSAT-1 day and night, regardless of the weather conditions. The selection of imaging mode is made according to the disaster situation, whether the data requirement is for synoptic viewing of a large-scale disaster, such as floods, or the need is for a close-up of urban damage caused by an earthquake.

Indian remote-sensing satellites (IRS)

Indian satellites have been providing space-based data and services for well more than two decades. Each successive mission has been a continuous improvement in data quality and delivery. The first generation IRS-1A and IRS-1B, launched on 17 March 1988 and 29 August 1991, respectively, had onboard an LISS and a WiFS camera of moderate resolution. The second-generation satellite IRS-1C launched on 28 December 1995 carried a solid-state pushbroom panchromatic camera (PAN, operating in 0.5–0.75 μm spectral range with a 5.8-m pixel and a swath width of 70 km). The improved LISS instrument on this satellite provided higher resolution images than its predecessors. The WiFS of IRS-1C had a spatial resolution of 188 m and a swath width of 895 km. IRS-P3 was a versatile satellite mission that adopted time sharing between Earth and stellar orientation. Launched in 1996, it carried a multispectral opto-electric scanner operating in visible near-infrared short-wave region, with a resolution of 500 to 1,500 m and a swath 200 km wide. This satellite covered a wide range of applications, primarily related to oceans, but to some degree also to land and atmosphere.

IRS-1D was the next one in the second generation of IRS series with the same payloads as that of IRS-1C. The first Indian satellite mission dedicated to ocean application was IRS-P4. It carried an ocean color monitor operating in eight spectral bands with 360-m resolution, and a multispectral scanning microwave radiometer (MSMR) with very large area viewing (1,360-km-wide swath and 40–120-km resolution). IRS-P6 (RESOURCESAT-1) has a pushbroom camera and a high-resolution (5.8-m) LISS-4 instrument, as well as a medium resolution (23.5-m) LISS-2 instrument. There is also an advanced AWiFS of 70-m resolution. IRS-P5 (CARTOSAT-1) is India's first cartographic mission; it provides stereoscopic imagery of Earth's landmass. DEMs can be generated from the stereoscopic data, which are required for making topographic maps. The stereo imagery is created with two PAN cameras mounted on the platform with a nadir tilt of +26 degrees and −5 degrees. The platform is steerable across a track by applying a roll bias. Any place in the world can be covered in a period of 5 to 6 days. The upcoming CARTOSAT-2 satellite is India's highest-resolution satellite with better than 1-m pixel in a 10-km-wide swath. The operational design life is 5 years, and the CARTOSAT-2 imagery is slated for use, particularly for flood hazards.

NOAA satellites and information service

The operational satellite system made available to the charter by NOAA is composed of GEOS and POES environmental satellites.

The GOES satellites are meant for continuous monitoring of the same position on the Earth's surface. The geosynchronous plane is about 35,800 km above the surface of the Earth, high enough to allow the satellites a full-disc view of the Earth. Because of their "fixed" position with regard to the rotation of the Earth, they provide a constant surveillance opportunity to take stock of severe weather conditions such as ocean storms and the resulting floods.

The POES satellite system offers, like the other satellites of the charter constellation, the advantage of daily global coverage because of the polar orbits roughly fourteen times a day. Currently in orbit are both a morning satellite and an afternoon satellite, which provide global coverage four times a day. The POES system includes the AVHRR and the TIROS

Table 20.2. *Landsat mission dates and sensors*

Satellite	Launched	Decommissioned	Sensors
Landsat 1	23 July 1972	6 January 1978	MSS/RBV
Landsat 2	22 January 1975	25 February 1982	MSS/RBV
Landsat 3	5 March 1978	31 March 1983	MSS/TM
Landsat 4	16 July 1982	15 June 2001	MSS/TM
Landsat 5	1 March 1984	Operational	MSS/TM
Landsat 6	5 October 1993	Did not achieve orbit	ETM
Landsat 7	15 April 1999	Operational	ETM+

operational vertical sounder. Because of the polar orbiting nature of the POES series of satellites, these are able to provide data on a broad range of natural disasters, such as volcanic eruptions, forest fires, and coastal floods.

Landsat

Landsat represents the world's longest continuously acquired collection of space-based land remote sensing data. In the mid-1980s, stimulated by the success in planetary exploration using unmanned remote sensing satellites, the U.S. Department of the Interior, NASA, the U.S. Department of Agriculture, and others embarked on an ambitious initiative to develop and launch the first civilian EO satellite program. On 23 July 1972, NASA launched the first in a series of Landsat satellites designed to provide repetitive global coverage of Earth's landmass. Originally named "ERTS" for earth resources technology satellite, it has continued to provide high-quality, moderate-resolution data depicting land and coastal signatures. As a result of subsequent satellites, launched throughout the 1980s and 1990s, there is a continuous set of Landsat data from mid-1972 until the present time (Table 20.2).

There are three main sensor types on the Landsat satellites, as shown in Table 20.2. The return beam video was essentially a television camera, but it did not achieve the same level of popularity as of the multispectral scanner (MSS) sensor, which was accompanied by a thematic mapper (TM) on Landsat 3 to 5. The sensor onboard Landsat 6 was the enhanced thematic mapper (ETM), and Landsat 7 carried an ETM+ version. The resolution of the MSS sensor was approximately 80 m, with four bands of spectral coverage ranging from the visible green to the NIR wavelengths. The MSS sensor on Landsat 3 included a fifth band in the thermal IR. Landsat 4 and 5 carried both MSS and the newly developed TM sensors. The MSS onboard Landsat 4 and 5 were identical to the one on Landsat 3. The TM sensor of Landsat 4 and 5 included several additional bands in the SWIR, as well as improved spatial resolution of 120 m for the thermal IR band and 30-m resolution for the other six bands. The currently operating Landsat 5 and 7 have a 16-day orbit cycle, but the two orbits are offset by 8 days, thus allowing 8-day repeat coverage. Landsat 5 and 7 are also designed to collect data over a 185-km-wide swath. The ETM+ on board Landsat 7 gives 30-m resolution in the visible and IR bands, 60 m in the thermal band, and 15 m in the panchromatic band.

SAC-C

Argentina's SAC-C satellite is part of an international "morning constellation" of EO satellites that also includes the U.S. satellites Landsat 7, EO 1, and Terra. Operated by the Argentinean space agency CONAE (Comisión Nacional de Actividades Espaciales), the satellite has several instruments onboard, which were built in partnership with the space

20.5 About the charter

Table 20.3. *DMC satellites specifications*

Satellite	Launch	Expected Life (Years)	Sensors Sensors	Operational Condition
ALSAT-1	2002	5	32-m SLIM6 MS	Operational
NigeriaSat-1	2003	5	32-m SLIM6 MS	Operational
BILSAT	2003	5	26-m MS/12-m pan	Operational
UK-DMC	2003	5	32-m SLIM6 MS	Operational
Beijing-1	2005	5	32-m MS/ 4-m pan	Commissioning being completed

organizations of the United States, Argentina, Denmark, Italy, France, and Japan. The primary instrument of interest to the charter is the multispectral medium resolution scanner that operates in visible IR bands. Its spatial resolution is 175 m, and the width of its imaging swath is 360 km. An associated high-resolution technological camera is used to selectively obtain 35-m resolution within an imaging swath of 90 km.

Disaster monitoring constellation (DMC)

DMC is a consortium of small satellites built and operated on behalf of government and private sector organizations. The constellation operations are aimed at broad area coverage with rapid revisit and are particularly suited to space data provision for disaster response. The Disaster Monitoring Constellation International Imaging Ltd. (DMCII) of the United Kingdom is a partner in the consortium and provides support and coordination for joint operations of the constellation for the Charter. Other partners of the consortium are the Centre National des Techniques Spatiales of Algeria, the owners and operators of the ALSAT-1 satellite of the constellation; the National Space Research and Development Agency of Nigeria, which owns and operates the NigeriaSat-1; the Information Technologies and Electronics Research Institute of Turkey, which owns and operates BILSAT; the Surrey Satellite Technology Ltd., the owners and operators of the UK DMC satellite; and, finally, the Beijing Landview Mapping Information established with the support of the Chinese government and the Ministry of Science and Technologies, which owns and operates the DMC+4 or the BEIJING-1 satellite. Further details on the individual satellites are given in Table 20.3.

The Standard Disaster Monitoring Constellation imager is a six-channel, Surrey Linear Imager (SLIM6) and is built by the SSTL, UK. The sensor is an evolution of the previous multispectral camera flown on various SSTL missions. The SLIM6 design provides for a nadir-viewing, three-band multispectral scanning camera capable of providing moderate-resolution images of the Earth's surface.

Advanced Land Observing Satellite (ALOS)

JAXA's ALOS satellite has disaster monitoring as one of its main applications. It dwells on the earlier EO technologies of the Japanese Earth resource satellite-1 (JERS-1) and the advanced Earth observing satellite. ALOS is carrying three sensors: a panchromatic remote sensing instrument for stereo mapping called PRISM, the advanced visible and near-infrared radiometer type 2 (AVNIR-2), and the phased array type L-band synthetic aperture radar (PALSAR). The PRISM instrument is a 2.5-m resolution panchromatic sensor with a 70-km swath width. PRISM has three optical systems for forward, nadir, and backward viewing. PRISM images of disaster areas can be acquired every 46 days. AVNIR-2 has a 10-m resolution four-band (visible through near infrared) sensor with a 70-km imaging swath. PALSAR enables day and night, all-weather imaging, as do other radar imagers included in the charter constellation. It provides 10-m ground resolution, again in a 70-km imaging swath in its fine mode. The resolution degrades to 100 m in the 350-km-wide ScanSAR mode swath. Generally, the images of a disaster-affected area can be acquired every 2 to 3 days by either AVNIR or PALSAR. In addition to ALOS data, JAXA has also committed to providing to the charter archived JERS-1 data to be used for the sake of predisaster reference imagery.

20.5.3 Mission summaries

The nature of satellite mission management varies from agency to agency. In the case of some missions, satellite tasking is part of the routine access to data products for operational purposes, such as for the NOAA satellites; in other cases, elaborate mission planning and conflict management in response to specific data requests are involved, a case in point being the Canadian RADARSAT-1 satellite. The following is a high-level account of mission operations organization and constraints related to spacecraft and ground segment tasking for data acquisition at the various member agencies.

European Space Agency (ESA)

The ESA has operated the three charter constellation satellites, namely, ERS-1, alone from July 1991 to 1995, and in tandem with ERS-2 for a period of time until ERS-1 end of mission in March 2000, and ENVISAT since March 2002. The ERS mission operations are conducted from ESRIN (European Space Research Institute) in Frascati, Italy, and the European Space Operations Centre in Darmstadt, Germany. In the absence of onboard storage facility, ERS-1 and -2 tandem mission data were downlinked to twenty-seven national satellite reception stations covering most of the world's landmass. ERS raw and fast delivery products are sent to distributed data processing and archiving facilities from the main Kiruna ground station. The archiving facilities now contain 14 years of ERS data. The ENVISAT flight operations segment comprises a number of command and control centers responsible for telemetry; command and control, including payload planning; and for providing interfaces with the other ground segment entities. For routine operations, the Kiruna ground station is used for satellite tracking and command and control functions. Outside the Kiruna mask, the satellite is monitored using four or five passes over the Svalbard station. The payload data control segment is used to acquire, process, store, and distribute the ENVISAT data to its end users. Nominally, the Kiruna ground station and a communication satellite Artemis are used for data reception and relay. The Svalbard station is used as a backup. In addition to telemetry data, 250 gigabytes of ENVISAT data products are generated daily at ESA.

French Centre National d'Etudes Spatiales (CNES)

The operation of the SPOT family of satellites is shared between the CNES and Spot Image, which is the licensee of SPOT data distribution internationally. The satellite control and command, including the execution of the data acquisition plan, is under the direct responsibility of the CNES, whereas Spot Image, working from its main facility in Toulouse, establishes the daily acquisition plan, receives image data, processes the telemetric data to update the image catalogue, and generates products for delivery. The main data reception station is located on the Spot Image premises in Toulouse. The telemetry, tracking and command (TT&C) functions are performed by the CNES from the Issus-Aussagel station near Toulouse and the Kiruna station in northern Sweden. The Pretoria station in South Africa is used as a backup. CNES also runs an image quality system by developing data processing parameters used by the Pre-Processing and Archiving Centre and by ground receiving stations. The payload tasking center is operated by Spot Image to process daily customer orders received from data distributors or direct data downlink stations after the necessary conflict management. In view of the optical nature of the SPOT payload, the daily data acquisition plan prepared by Spot Image, in addition to resolving conflicts and technical constraints between the various customer and data reception station requests, takes into account the forecasts furnished by the French weather service to estimate the amount of cloud cover affecting the imagery to be acquired. In this regard, Spot Image offers several levels of service, depending on the amount of cloud cover, the number and priority of acquisition attempts, and the image quality in terms of the cloud cover. The data price is set according to the level of service chosen by the customer.

Canadian Space Agency (CSA)

RADARSAT-1 satellite is operated by the CSA from its headquarters in Saint-Hubert, Quebec, which houses mission planning and mission management, the mission control center (MCC), and two of the five customer order desks. A backup TT&C facility is located in Saskatoon, Saskatchewan, in central Canada. RADARSAT-1 data are captured by direct downlink to three receiving stations in North America, called partners' stations, in Gatineau (Quebec), Prince Albert (Saskatchewan), and Fairbanks (Alaska), called the Alaska SAR Facility. In addition, there are thirty-two other data receiving facilities (DRFs) spread across the world that are certified for RADARSAT-1 data reception and processing. Twenty-two of these also act as data archiving facilities. Any region of the world left out of a DRF reception mask can be covered by means of the onboard tape recorder. The tape playback only takes place over the Canadian data receiving stations. RADARSAT-1 data can be requested through an order desk network. There are five of these order desks, one each for the Canadian government, save the Canadian Ice Services, the U.S. government, which is a program partner, the CSA for its baseline data acquisitions and image calibration activities, and MacDonald Dettwiler and Associates Geospacial Services, previously known as Radarsat International, for commercial data distribution. The fifth order desk is dedicated to the Canadian Ice Services, who are the principal operational users of RADARSAT-1 data. The data requests from these order desks are received by the CSA for planning, which is done after a rigorous conflict management process to assign planning priorities. After spacecraft health and safety, these priorities are ordered in decreasing importance as follows: data to assist in emergencies, data for calibrations, data for time-critical use, and data for non time-critical use. Once the payload command data are ready, these are passed on to the MCC for uplink. The RADARSAT mission management office is open year-round, and on-call support is available 24 hours a day to accommodate emergency data acquisition requests. In fact, these requests can be accepted for planning as late as 31 hours before their execution by the satellite.

Indian Space Research Organization (ISRO)

With the experience gained through pilot studies, ISRO has developed for the Indian space resources, the IRS and INSAT series of satellites, the concept of space-based observation, and the communication system for disaster management (DMS). India provides the charter data from IRS 1C, IRS 1D, IRS P4, IRS P5, and IRS P6 satellites. These are operated from the ISRO Telemetry, Tracking and Command network (ISTRAC), with ground stations in Bangalore, Lucknow, Port Blair, Sriharikota, and Thiruvananthapuram. IRS P5 and IRS P6 satellites are also equipped with onboard recorders to image any part of the world, and therefore are not limited to ground station visibility for data acquisition. In addition, ISTRAC has foreign TT&C stations in Mauritius, Bearslake (Russia), Biak (Indonesia), and Brunei. A multimission Spacecraft Control Centre is located in Bangalore. ISRO also operates the Local User Terminal/Mission Control Centre under the international Satellite-Aided Search and Rescue Program. The Earth Station in Shadnagar near Hyderabad acquires data within its visibility mask. This covers data over India and parts of China, Bhutan, Tadzikstan, Afghanistan, Iran, Oman, Sumatra, Thailand, Mayanmar, and Bangladesh. The IRS data coverage from the International Ground Stations in Fairbanks, Norman, Tehran, Beijing, Moscow, Thailand, and Korea provide additional coverage. The National Remote Sensing Agency Data Centre in Hyderabad has been providing users with the necessary tools for data selection. The web-based software packages guide users in their submission of data requests according to the formats appropriate for data acquisition planning.

U.S. Geological Survey (USGS)

The USGS and the NASA support the Landsat Project. NASA developed and launched the spacecraft, whereas the USGS handles the flight operations, maintenance and management of all ground data reception, processing, archiving, product generation, and distribution. The Landsat ground stations located in Sioux Falls, South Dakota, and Alice Springs, Australia, receive both the X-band payload (science) data and the spacecraft housekeeping S-band data. The S-band facility at these stations also provides for TT&C services. The housekeeping data are sent real time to mission operations centers (MOCs) for immediate spacecraft health and safety monitoring. The science data are sent to Landsat 7 Processing System or the Landsat Archive Conversion System to process Landsat 7 or 5 data, respectively. Both the two systems are located at the USGS Earth Resources Observation and Science centre in Sioux Falls. The NASA Tracking and Data Relay Satellite System (TDRSS) ground segment, the White Sands Ground Terminal, located near Las Cruces, New Mexico, controls the TDRSS and receives/transmits data from/to customers' low earth orbit satellites through TDRSS. The ground sites in Poker Flat, Alaska, and Svalbard, Norway, are used as backup sites in times when extra ground resources are required to fulfill mission objectives. Landsat 7 and Landsat 5 follow the worldwide reference system of paths and rows, with a temporal offset of orbit cycles, as mentioned previously. Landsat is typically scheduled to collect approximately 250 scenes per day in accordance with the long-term acquisition plan (LTAP). All scheduled LTAP scenes populate the USGS archives. The scenes will generally become available for search and order within 3 to 24 hours after downlink for the data received at the Sioux Falls station, and between 3 to 5 days for the data captured by the other stations listed previously. In addition to LTAP (USGS-archived) scenes, members of the international ground station network also collect Landsat 7 data falling within their reception masks. Landsat 5 does not have an onboard recorder, which means that all sensed data must be downlinked directly to a ground station. All scenes received at the Sioux Falls station are archived there and are available within 24 hours.

Comisión Nacional de Actividades Espaciales (CONAE)

The MCC, located at Teofilo Tabanera Space Centre near Cordoba, is responsible for planning, commanding, and monitoring the Argentinean satellites. MCC is part of the national space center that also houses the TT&C facility, the archiving facility, the satellite testing and integration facility, and the Institute for Advanced Space Studies. SAC-C, which is part of the morning constellation of satellites along with EO-1, Landsat 7 and Terra, is supported by the NASA Integrated Services Network (NISN) for real-time telemetry and commands from the constellation's ground tracking stations. The X-band image data are

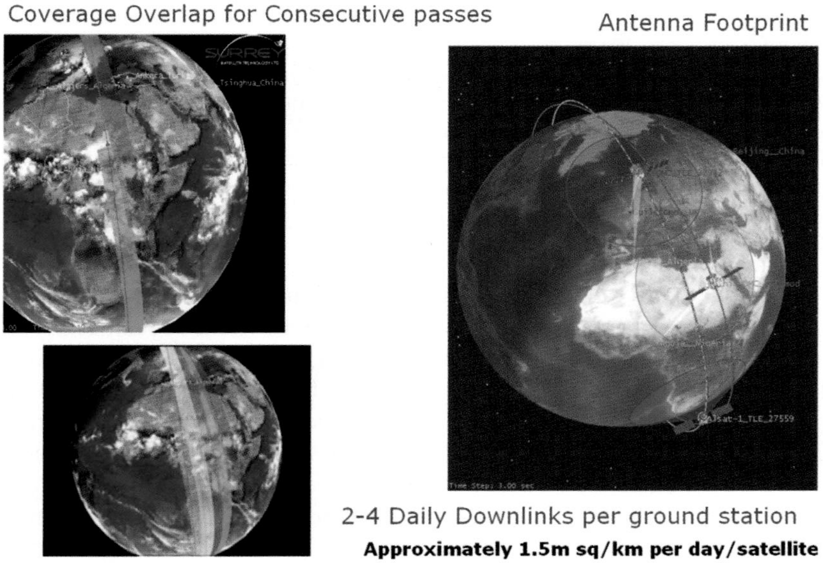

20.5 *About the charter*

Coverage Overlap for Consecutive passes

Antenna Footprint

2-4 Daily Downlinks per ground station
Approximately 1.5m sq/km per day/satellite

Figure 20.20 Satellite coverage and downlink of the disaster monitoring constellation (DMC).

received at the Cordoba Ground Station Estacion Terrena Cordoba. The MCC at Cordoba performs the overall functions linked to mission planning and scheduling, spacecraft and instrument command and control management and health monitoring, trending analysis, and orbit/attitude control. The image data distribution is also carried out from the station to science and other users. Post-pass tracking, data from the NISN is transmitted to Goddard Space Flight Centre for processing, tracking, and distributing data from the ground stations. The spacecraft ephemeris data are then fed back to the CONAE mission control.

Disaster Monitoring Constellation (DMC)

The high revisit frequency of the DMC is made possible by the fact that the orbits of the individual spacecraft are not frozen but phased to within $+/-3.5$ degrees of nominal 90-degree spacing around the polar, sun-synchronous, circular orbit plane of the four spacecraft currently in operation, namely, ALSAT-1, BILSAT, NigeriaSat-1, and UK-DMC. There is a coverage overlap between successive passes, as shown in Figure 20.20 with the reception masks of the four tracking antennas in the UK, Turkey, Algeria, and Nigeria. Furthermore, these four satellites have solid-state recorders for onboard data storage and later playback to a ground station. The orbits of these four satellites constitute orbit plane 1. The Chinese Beijing-1 satellite would constitute plane 2, with phased equidistant around the orbit plane at 90 degrees.

Japan Aerospace Exploration Agency (JAXA)

The three sensors onboard ALOS are capable of generating a high data volume, and this required the development of appropriate data handling and transmitting technology. The ALOS data handling system is therefore designed to perform three important functions:

- **Data compression:** Takes place onboard the spacecraft for the two optical payloads, PRISM and AVNIR. Image data on land surface are easy to compress with almost no degradation for later transmission.

- **Data accumulation:** The onboard storage of the compressed data is needed until a downlink opportunity is available. The recorder onboard has a capacity of 96 Gb.
- **Data transmission:** There are two methods of data transmission, either through a relay satellite (DRTS) in geostationary orbit or by direct downlink to a ground station. In the first case, ALOS sends the data to DRTS, and the transponder onboard amplifies the data before they are transmitted to the ground stations. The ALOS visibility of the geostationary satellite is longer than the visibility mask of a ground station; therefore, this data relay method has obvious advantages. However, the direct transmission replaces the DRTS transmission when ALOS is not within sight of the geostationary satellite.

The worldwide ALOS data processing and archiving follows a node concept. There are four data nodes: at ESA covering the European and African region, at JAXA for the Asian region, at NOAA/ASF for the North and South American region, and at Geoscience Australia for the Oceania region. The node concept offers advantages in terms of increased capacity for ALOS data processing and archiving, accelerated scientific and practical use of ALOS data, increased international co-operation on data validation and science studies, and enhanced service delivery for the ALOS users. The order of priority in data acquisition planning is as follows: satellite emergency operations, housekeeping operations, disaster area monitoring, calibration/validation, basic observation, sole Japanese governmental use, sole data node use, sole research purpose use, and all other data requests.

20.5.4 Applicable policies

The charter membership is on a voluntary, best-effort basis, and the parties or partner agencies are not under any legal obligations with regard to membership. Although the AUs with the right to activate the charter are required to abide by the provisions of the charter by accepting the conditions of the charter data use laid down in a nondisclosure agreement, the satellite data they receive are subject to individual data policies of the data providers. The policies for the various charter member agencies are summarized in the next sections.

European Space Agency (ESA)

The objectives of the ESA ERS data policy are based on the use of ERS data to achieve a balanced development of science, public utilization, and commercial applications. The creation of a revenue stream is of less importance to the agency.

The ESA data policy provisions were originally contained in the document ESA/PB-EO (90)57 Revision 6 that was approved by the ESA Program Board for Earth Observation (PB-EO) in May 1994. The currently observed data policies take into account the ENVISAT data policy (ESA/PB-EO [97] 57 rev.3), which was approved by the board in February 1998.

The agency on behalf of the participating states retains title to and ownership of all primary data originating from ERS payloads, together with any derived products generated under the ESA contract, as well as other products, to the extent that the contribution of ERS is substantial and recognizable. This entitlement is protected with the applicable copyright and intellectual property laws.

> **Data acquisition:** The ERS data are received from a suite of instruments, which include AMI (active microwave instrumentation) operating in SAR image, wind, wave, or wind/wave mode; ASTR; GOME (on ERS-2 only); MWR (microwave radiometer); RA (radar altimeter); and PRARE (precise range and range-rate equipment). AMI-SAR operates in high-bit rate mode. All other instruments operate in low rate mode. The ESA, through its receiving stations and agreements with national and foreign stations, undertakes acquisition, processing, and near real-time distribution of ERS products as part of the ESA services.

The ERS successor ENVISAT mission carries three instruments for atmospheric measurements: global ozone monitoring by occultation of stars (GOMOS), Michelson interferometer for passive atmospheric sounding, and scanning imaging absorption spectrometer for atmospheric chartography. These instruments measure aerosols and the greenhouse and trace gases causing ozone depletion. GOMOS data are processed in Sodankylä, Finland, at the Processing and Archiving Centre of the Arctic Research Centre, which is part of the Finish Meteorological Institute, an original coproposer of the GOMOS instrument. The other instruments on board, such as the radar altimeter and the advanced along-track spectroradiometer measure ocean currents with high accuracy. The MERIS measures ocean color and is well suited to give information on bioproduction in the sea. In addition to these instruments, the ENVISAT Advanced SAR is bringing valuable improvements in monitoring sea ice after the ERS radar imagers. The ASAR, along with the MWRs and the Doppler orbitography and radiopositioning integrated by satellite, also collects useful land data.

The national stations are those set up by nationally registered private or public entities in participating states to acquire ERS SAR data and generate products for distribution to users. These include fixed stations and mobile receivers located within or outside the national territories of the participating states. The national stations of the participating states have the right to receive low-rate ERS SAR data.

The foreign stations are owned and operated by organizations legally established in nonparticipating states. The data acquisition by these stations is through agreements signed with the ESA Council.

Data use: The conditions attached to the distribution of ERS data depends on the use of the data. The following two categories of use are defined.

- **Category 1 use:** Research and applications development in support of the mission objectives, including research on long-term issues of Earth system science, research and development in preparation for future operational use, certification of receiving stations as part of the ESA functions and ESA internal use. The identification of category 1 use generally requires approval by PB-EO on the basis of an appropriate peer review process.
- **Category 2 use:** All other uses that do not fall into category 1 use, including operational and commercial use.

Data distribution: The ERS data are made available on an open and nondiscriminatory basis, in accordance with the United Nations Principles on Remote Sensing of the Earth from Space (United Nations Resolution 41/65, 3 December 1986). The ESA retains the overall policy and programmatic responsibility for distribution of ERS data for category 1 use and is the direct provider of the service from its own facilities, whenever feasible. The responsibility of category 2 use data distribution is with the distribution entities, which are either selected through a submission process or are appointed directly, primarily from among satellite operating and data processing facilities in the participating states. ESA grants nonexclusive licenses to the distribution entities for the distribution of category 2 use data over well-defined areas and for a period of 3 years.

The distribution entities establish distribution schemes and negotiate third-party agreements, contracts, and sublicenses for further distribution of ERS data and data products. The data distribution by third parties complies with the copyrights and intellectual property rights vested in the ESA. The third parties are prohibited from further distribution of the data without the consent of the distribution entities. When the ERS data have been substantially modified to create products beyond the scope of the product levels (raw, level 0, level 1b, and level 2) certified by the ESA, the property and the distribution rights may be claimed by the concerned entity.

Data pricing: The ESA makes the data available at marginal cost to users, both for category 1 use and category 2 use.

For category 1 use, the ESA fixes the price for ERS products. The price is set at or near the cost of reproduction of the data. The PB-EO can authorize the ESA on possible price waivers for category 1 use for specific projects and for specified quantities of data. The ERS announcement of opportunity (AO) projects continue to benefit from these price waivers.

For category 2 use, the ESA fixes the price of ERS data products and services, which it provides to the distribution entities. The price is set at a level comparable to the price for category 1 use. In turn, the distribution entities are allowed to set prices for ERS products and services, which are at or above the price that the ESA charges to the distribution entities.

Data archiving: The ERS data archives are stored by the ESA for 10 years following the end of the satellite mission. The ESA also has the right to obtain a copy of all ERS data acquired at national and foreign stations. If a station is unable to keep the long-term archives, the station is required to at least offer to the ESA to take over the archives. The ESA provides fair and nondiscriminatory access to its ERS archives to all users following the distribution modalities described previously.

Centre National d'Etudes Spatiales (CNES)

The French space agency, CNES, has assigned the responsibility for commercial operation of SPOT to Spot Image, which is the distributor of SPOT imagery worldwide. The SPOT program was developed by the French space agency CNES in partnership with Sweden and Belgium. The European Union and Italy are also partners on the VEGETATION program. Through its worldwide network of subsidiaries, ground receiving stations, and distributors, Spot Image offers users a wealth of experience and know-how acquired since the first SPOT data appeared on the market in 1986.

Partners of the SPOT VEGETATION program have decided to take significant measures toward a new no-cost distribution policy. As the first concrete action of this renewed data policy, the VEGETATION TEN-DAY synthesis archive is freely accessible through the website of the program. This free access to the high-quality VEGETATION archive is an open invitation to scientists and to the industry to develop new algorithms, models, and applications; to better cope with the user community requirements; and to provide decision makers with high-level information on the state of global natural resources. Indeed, since its successful launch onboard the SPOT 4 satellite in March 1998, the VEGETATION instrument is dedicated to the daily observation of terrestrial ecosystems and the biosphere, particularly for addressing global change and environment issues. The decision of the VEGETATION program to launch VEGETATION 2 onboard SPOT 5 further guarantees the continuation of the VEGETATION mission until 2008.

Canadian Space Agency (CSA)

Canada's RADARSAT-1 is the first radar (SAR) imaging satellite intended to meet the operational needs of remote sensing data users. The RADARSAT-1 program also has specified commercial interests. To satisfy these interests, the CSA, as owner of the intellectual property for the RADARSAT-1 data, has negotiated a master license agreement with MDA GSI, previously Radarsat International, a private company fully owned by MacDonald Dettwiler and Associates (MDA) of Richmond, British Columbia, for the commercial distribution of the data. RADARSAT-1 is at the same time a program of international partnership between the Canadian and the U.S. governments, and has benefited from the Canadian provincial governments' participation and contribution. These program partners, contributors, and participants are entitled to data allocations pursuant to the various agreements among the parties. The RADARSAT-1 program partners, contributors, and participants receive the data at the cost of processing and distribution under their respective data allocations, and they use the data in fulfillment of their departmental mandates.

The successor RADARSAT-2 mission, slated for launch in 2007, will mark a complete shift from public sector ownership of the satellite to a private sector enterprise. The intellectual property rights for all RADARSAT-2 SAR data are reserved solely for MDA, the builder of the spacecraft, regardless of the form or location of the data. MDA has the sole right to identify, qualify, and appoint commercial distributors. The Canadian government, having contributed to the development of the RADARSAT-2 mission, is entitled to preferential terms and conditions for receiving RADARSAT-2 data and data products.

Data acquisition: RADARSAT-1 data are acquired for users on an open nondiscriminatory basis in compliance with the UN resolution referred to previously. There are five levels of data acquisition planning priority:

- Data for spacecraft health and safety
- Data for emergency acquisition
- Data for calibration purposes
- Data for time-critical uses
- Data for non–time-critical uses

RADARSAT-1 data may be obtained either through the existing data archives or as new data using the satellite resource (i.e., the imaging time of the SAR payload). The imaging time is shared between the Canadian and the U.S. governments and MDA. The same sharing principles apply to real-time data downlink and onboard data tape recording. Given the multimode imaging capabilities of the spacecraft and a variety of data requesters, governmental as well as commercial, the data acquisition planning process requires a high level of conflict resolution at the RADARSAT-1 mission management office.

Data use: The use of RADARSAT-1 data for internal governmental use by the program partners is the choice and the privilege of the partners, provided the data are not sold, given, or otherwise made available to third parties, except for research studies and application development approved individually or jointly by the partners and supported through AOs. The main AO implemented so far under RADARSAT-1 program is called applications development and research opportunity (ADRO) and its follow-on phases.

Data distribution: A widespread RADARSAT-1 data distribution is made possible through a network of order desks, each dedicated to entertaining data requests of a particular user group. The MDA order desk looks after the RADARSAT-1 data needs of commercial users. For these customers, the primary mode of operation for the provision of data worldwide is through local distributors licensed by MDA for various regions of the world. The distributors place all data and data product orders directly and exclusively with MDA and satisfy the data demands within their respective territories. Value-added product developers and vendors are free to develop products from RADARSAT-1 data for sale and distribution if these do not contain recognizable RADARSAT-1 imagery; otherwise, MDA's consent is required under the royalty agreements that it has with the CSA. In the case of the data accessed through government allocation, value-added products containing retrievable RADARSAT-1 imagery may be developed and distributed, but only on behalf of the government or agency concerned and in a manner that does not compete with MDA's rights as the sole commercial distributor of RADARSAT-1 data. The data made available to government departments are for internal use only in support of departmental mandates. Government contractors and subcontractors having access to the data from government allocation can develop products only on behalf of the client department; they cannot distribute or use the data for their own purposes. All RADARSAT-1 data and data products should clearly display the CSA copyright inscription: © Canadian Space Agency/Agence Spatiale Canadienne, (year of reception).

Data pricing: MDA periodically revises the commercial price list and terms of sale. The prices are determined by the level of processing and imaging beam mode selected, and some incremental charges are made, depending on the nature of data request and the means of delivery. The use of local distributor does not affect the price, nor does the means (direct downlink or onboard tape recorder) and source (current or archive) of data acquisition. The Canadian federal and provincial government departments can make use of a national master standing offer to order data and data products from MDA at government prices. The departments should request RADARSAT-1 data and data products through the government order desk. The departments receiving data and data products under the standing offer and benefiting from the government price thereof are restricted to noncommercial use of these data and data products pursuant to the user license agreement.

Data archiving: All RADARSAT-1 data downlinked to a ground facility are archived. These data can be accessed through the order desks. Data less than 6 months old are considered to be "current," and special conditions to their access apply, for accounting purposes, if accessed through the U.S. order desk. The Canadian Centre for Remote Sensing operates the Canadian Data Archiving Facility for RADARSAT-1. In addition, every foreign licensed facility is required to either archive data or render a copy to the Canadian archive. The data received at the Canadian data receiving facilities are kept as raw signal data in the archive and can be ordered in any of the applicable standard data products. The CSA, NASA, and NOAA as program partners and MDA have unlimited access to all archive data. Other users can access the archives through MDA.

Indian Space Research Organization (ISRO)

Remote sensing data is an important source of information for managing the nation's natural resources, supporting and monitoring developmental activities at the local level, and disaster response, as well as for tracking weather patterns, the state of the ocean, and the environment in general. The IRS data are acquired and distributed to government, private sector, and educational institutions for a variety of applications and to the global market. The government-approved comprehensive remote sensing data policy covers the acquisition and distribution of satellite remote sensing data from Indian and foreign satellites for civilian use in India. The Department of Space (DoS) is the nodal agency for implementing the policy. The following two aspects of the policy implementation need to be emphasized:

- Pursuant to the policy, government permission is required for operating the remote sensing satellite from India and also for the distribution of satellite images in India. The National Remote Sensing Agency has been identified as the national acquisition and distribution agency for any foreign satellite data within India and has been authorized to enter into agreements for any foreign satellite data distribution in India. Antrix Corporation of DoS is identified to license capacities outside India.
- The policy safeguards are meant to ensure that images of sensitive areas are screened so national security interests are duly protected. In particular, the policy streamlines the distribution of

high-resolution data to government users; private users involved in developmental activities with government; and other private, public, academic, or foreign users.

NRSA is also responsible for archiving Indian and foreign satellite data (Landsat, SPOT, ERS, Metsat). The important elements of the data archival policy are that all good data from these satellites should be preserved for 5 years on high-density tapes and digital linear tapes (DLTs). Older data are transcribed onto new media, such as DLT/CD-R, for four cycles per annum in nonmonsoonal period. The four cycles will be February–March, April–May, October–November, and December–January. If the data type for a given period is available from more than one sensor, the data type of better quality and continuity will be archived. Therefore, the users need to be aware that data older than 5 years and of periods other than those listed here are purged.

NOAA, NASA, and USGS

NOAA, NASA, and USGS have responsibility over the U.S. government space EO mission or data holdings. In general, the unclassified government programs operate according to the principles outlined in the U.S. Office of Management and Budget Circular A-130 (OMB), which specifies open and nondiscriminatory access to EO data. Following this government directive, the data prices should be set at a level no higher than the price that would allow the recovery of the cost of data dissemination. The original collection or processing costs are to be included. There are exceptions to the data pricing guidelines, depending on the statutes or an agency's mission. When information is collected for the benefit of a specific group, beyond what is useful for the general public, the government can charge more than the cost of dissemination. The U.S. government may also impose restricted access to data under certain circumstances, where there are conflicting interests.

NOAA: Data policy is consistent with the OMB Circular A-130 specifications for EO data. There are no copyright or other restrictions on distribution. NOAA data are accessible through the NOAA National data centers for the cost of reproduction. The primary data access is achieved through the NOAA Satellite Active Archives (SAA) of the Office of Satellite Data Processing and Distribution and the NOAA data centers (National Climatic Data Centre, National Geophysical Data Centre, and National Oceanographic Data Centre).

Generally speaking, users may obtain, at no charge, small, near real-time satellite data acquisitions through the SAA. The SAA delivers data in none other than electronic form. Currently, there are no fees associated with the SAA's online services. NOAA's only condition for its free data is the recognition of the data source by including, in all publications, articles, and papers resulting from the use of these data, the following citation: "The NOAA Satellite Active Archive."

For large retrospective acquisitions, users may obtain data from the data centers by paying a fee calculated to cover the costs of reproduction and transmission. Real-time access to all NOAA geostationary satellite data is also provided to U.S. and international users through direct broadcast from the satellites. In this way, any user with a receiving antenna may obtain regional imagery and sounding data. Direct broadcast is NOAA's preferred means of regional transmission because it provides the most efficient utilization of data from the satellites for the largest number of users. It is NOAA policy, based on the U.S. Department of Commerce guidelines, to acquire advance payment on the sale of mission information to non–U.S. government organizations or individuals.

NASA: Like NOAA, NASA-funded missions and instruments have an open data policy, offering data to all users at no more than the marginal cost of filling a user's request. The direct broadcast architecture described previously is also used by NASA as part of its EO system. Commercial entities are welcome to acquire data from NASA and provide value-added services to the community.

There is no single approach to the release of data from NASA missions. For each new mission, a data release plan is determined early in the mission. The plan is consistent with the general NASA policy of quick release, as well as with the nature of the mission and its data stream. The mission specific plan is then included in the AO.

The primary source of EO data from NASA is from the distributed active archive centers (DAACs). These centers process, archive, and/or distribute EO system and other NASA Earth science data, and provide full support to users of these data.

USGS—Landsat: To maximize the value of the Landsat program to the American public, nonenhanced Landsat 4 through Landsat 5 data are made available at the cost of fulfilling user request to global environment change researchers and other researchers who are funded by the U.S. government. Nonenhanced Landsat 7 data are made available to all users at the cost of fulfilling user requests.

The EROS Data Centre (EDC) was originally assigned the responsibility of receiving, processing, and distributing data collected and transmitted to the center by Landsat satellites. Since its opening, the EDC has stored, processed, and distributed a variety of cartographic, satellite, and aircraft data. The cartographic data collection is referred to U.S. GeoData. As part of its data archiving role, the EDC operates the National Satellite Land Remote Sensing Data Archive, a legislatively mandated program designed to maintain a high-quality database of space-acquired images of Earth for use in studying global change and other related issues. In addition to Landsat and NOAA's AVHRR data, the planned archive holdings of the EDC include data from NASA's MODIS instrument (part of the Mission to Planet Earth), ASTER (a cooperative effort between NASA and the Japan Ministry of International Trade and Industry), the SRTM, and DeClass II.

The data access is through a variety of information management systems (MIS) and interfaces, among which are EOS Data Gateway and Earth Explorer. The gateway is a NASA worldwide web-based system, and the Earth Explorer is a data entry and access tool.

Comisión Nacional de Actividades Espaciales (CONAE)

Currently, CONAE supplies, free of charge, SAC-C data to users for covering emergencies caused by natural and manmade disasters. This includes data acquisition; processing; value adding, if requested; and delivery. CONAE channels the data and information deliveries through local authorities dealing with the disaster. The policy has been extended to all the countries included in the footprint of ETC. The same data policy provisions apply to the data delivered under the charter, for both the real-time downlinked and the onboard recorded data.

Disaster Monitoring Constellation (DMC)

The DMC is founded on principles that align closely with that of the charter and considers disaster response as a key activity from the conception of the constellation; in fact, 5% of the spacecraft capacity of the constellation is free for daily imaging of disaster areas.

Japan Aerospace Exploration Agency (JAXA)

JAXA's data distribution, use, and pricing policy applies to all data obtained from the EO satellites and sensors owned by it. The data distribution is divided into two categories from the viewpoint of the said data policy, one for research and development use contributing to JAXA's R&D activities, and one for all other purposes of use.

The first category includes the following type of activities:

- Technology development embraces proof of concept and applications. The data use in this case is for technology evaluation of the space segment or technology demonstration of remote sensing data analysis in support of disaster monitoring or Earth's environment monitoring by space means. The results of this data use are exploited for the development of the space and ground segments of the future systems.
- Application activities related to operational support to disaster management, including the charter, and to natural resource management and environmental monitoring.
- Earth science studies to understand and resolve global water circulation and climate change issues.
- Other activities with regard to JAXA's public relationing and education.

JAXA will distribute data, by means of individual agreements, directly to R&D users, such as administrative agencies, educational institutions, and research establishments. For the other category of data use, JAXA will distribute data commercially to general users through a private enterprise. JAXA selects the private commercial distributor and retains all intellectual property rights to the data.

JAXA's pricing policy likewise applies differentially to the previously defined categories of use. For the R&D use, JAXA will set the price at the marginal cost of processing and reproduction. It may also distribute limited amounts of data free of charge if the users' contribution toward JAXA's activities is considered significant, for public relations and educational purposes. For all other uses, data prices are set by the commercial distributor, but JAXA will see to it that the prices are affordable and the data quality is maintained. If private companies or other agencies as subdistributors develop new products based on JAXA's original data, these entities will set their own prices for the products developed. In these cases, JAXA will charge a royalty for the original data used for the product development.

In addition to the previous general data policy, there will be data policy specific to satellite data or mission, including a detailed distribution scheme and pricing and conditions of data use.

20.5.5 Performance update

In its first 6 years, from 1 November 2000, when it was declared operational, to 3 November 2006, the charter was activated 112 times all across the world. Figure 20.21 shows the worldwide distribution of the charter activations according to the disaster type. The disasters have been classified into three broad categories (Table 20.4). Those related to the processes of solid Earth include earthquakes, landslides, and volcanic eruptions. Disasters caused by weather-related phenomena are ocean storms/hurricanes, ice and snow hazard, floods and ocean waves, forest fires, drought, fog/haze, and epidemic occurrence. Drought is considered a weather-related disaster that does not have a realistically defined response phase, and is therefore not admissible to the charter support. Likewise, an epidemic occurrence is also excluded for the lack of a direct connection between the disaster and the application of satellite remote sensing. The third category relates to technological or nonenvironmental disasters, such as oil spills and hazardous material spread.

Each disaster type is assigned a different color symbol on the world map (Figure 20.21). By far, the most common charter intervention has been for floods, primarily fluvial, but in some cases coastal floods also. In the period mentioned previously, 54 of the 112 activations were for floods, followed by earthquakes, 17 in all. There has been a steady increase from year to year in the number of activations, understandably as a result of increasing charter

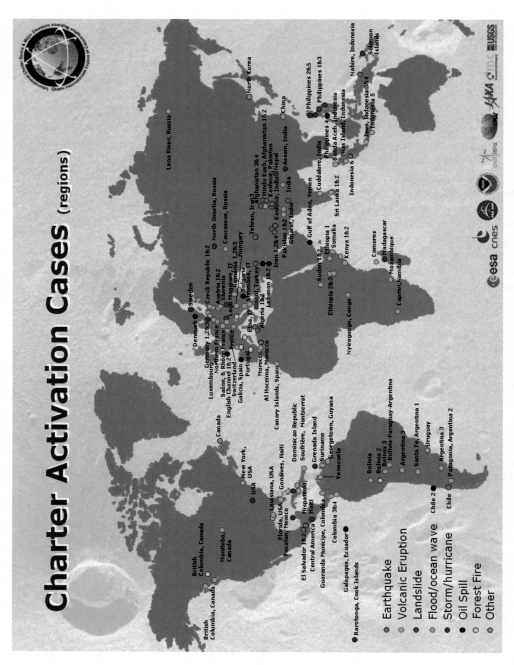

Figure 20.21 World map of the charter activation locations and disaster types.

20.5 About the charter

Table 20.4. *Charter activation cases by year and disaster type from 1 November 2000 to 3 November 2006*

		2000	2001	2002	2003	2004	2005	2006	Subtotal	Total
Solid Earth	Earthquake		3	1	3	5	3	2	17	
	Landslide	1		2	2			1	6	31
	Volcanic eruption		1	1	2	2	1	1	8	
Weather/ atmospheric	Storm/ hurricane			1	2	3	3		9	
	Ice/ snow hazard								0	
	Flood/ ocean wave		4	8	4	9	16	13	54	71 112
	Forest fire				5	1	2		8	
	Drought								0	
	Fog/haze								0	
	Epidemic								0	
Technological	Oil spill		3	2				4	9	
	Hazardous material epidemic					1			1	10
	Total/year	1	11	15	18	21	25	21		

membership and the accompanying AUs. The charter co-operating body, namely, UN OOSA, has been one of the most frequent requesters of the charter data on behalf of the various regional and humanitarian assistance agencies of the UN. A higher number of the charter activations in Europe can be explained by the number of AUs located in that part of the world. In fact, all countries of the European Union have been allowed to nominate an AU.

The histograms included in Figure 20.22 are meant to show the distribution of the data units (images) for the various disaster types. For earthquakes, the high-resolution SPOT sensor was the sensor of choice to carry out change detection analysis for assessing the urban damage. The use of archive data for this disaster was almost as important as the acquisition of new images for the change detection analysis. In the case of volcanic eruptions, on the contrary, radar imagers were used as often, if not more, as the optical sensors. As explained elsewhere, the radar brightness is particularly sensitive to surface roughness and degree of lava solidification, making radar sensors useful tools for mapping lava flows. The availability of reference archive images would depend on the length of the archive duration for a given satellite mission. Consequently, older satellite missions such as ERS and RADARSAT-1 would offer a higher likelihood of finding an archive match of a new radar acquisition than a newer mission such as ENIVSAT. The data types for landslide are similar to earthquake, although in this case the high spatial resolution requirement is more pressing than the spectral resolution. In the hilly terrains where landslides occur, the interpretation of radar images is rendered difficult because of the shadowing, foreshortening, and, in extreme cases, layover effects; however, radar images were used just as commonly for landslide mapping as the optical images. Judging from the pattern obtained in Figure 20.22, the optical and radar imagers are equally effective for flood monitoring, and in this case, the main driver

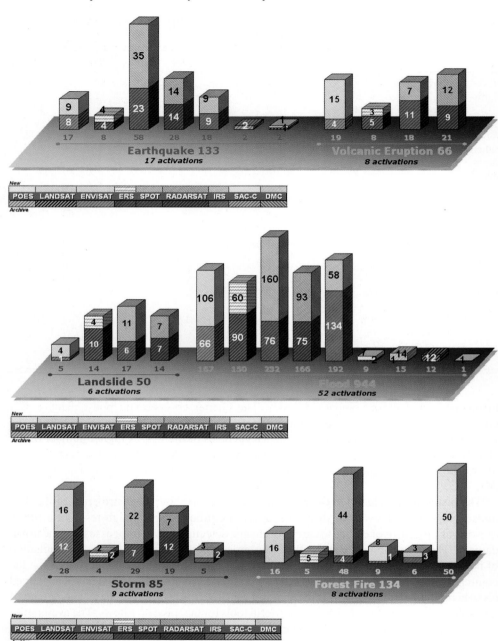

Figure 20.22 Distribution of satellite data units from the archives and new acquisitions used for different disaster types covered by the charter.

for disaster coverage is the rapidity and revisit. The main response phase requirement in the case of forest fire disaster is the burning fire location; therefore, only optical sensors equipped with their thermal and infrared bands appear on the chart. As for the fires, only newly acquired, postdisaster images from optical and radar sensors have been used for oil spill disaster, with the exception of a few cases where the damage to the coastline needed to be monitored with the reference predisaster data. SAR data are always useful in view of the dampening effect of oil on the sea surface roughness causing darker image signature. The dark signature may also be caused by wind shadows indicative of calmer waters. Therefore, caution needs to be exercised in interpreting SAR images for locating oil slicks.

20.6 Disaster coverage

The ECO-prepared Archive Acquisition Plan (AAP) is executed by the mission planning staff of the individual agencies whose satellites would have been identified for the disaster coverage. This section describes the handling of a charter request by the agency mission planners, along with the activation and the end user satisfaction criteria.

20.6.1 Activation criteria

The 112 activations reported here (Table 20.4) were a result of 134 calls received by the ODO. This shows that not every call materialized into charter activation. In some cases, the calls were not from a bona fide AU, but in other cases the request did not qualify in terms of the appropriateness of space technologies to disaster response. The decision to accept a call in the first place lies with the ECO. In cases where the call is judged to be unacceptable by the ECO, the latter contacts the executive secretariat for a decision. To help an ECO with its evaluation of a request for the charter data, pre-established activation criteria are applied by the ECO.

The determination of the scale of the disaster as being "major" is basically an AU prerogative. No casualty figure, either in terms of loss of life or damage to property, has been set to qualify a disaster as "major" and thus requiring the charter intervention. Therefore, when a call for charter activation is received from an AU, it is assumed that the disaster event is admissible. Nevertheless, in the following cases, an AU call for charter activation may be questioned by the ECO, examined by the executive secretariat, and arbitrated by the board, as the case may be.

- Non emergency situations
  (a) Oil spill monitoring and cleanup operations
  (b) Ice monitoring operations, except for specific events
- Emergencies falling out of the charter scope
  (a) Wars or armed conflicts
  (b) Humanitarian actions not linked to a particular disaster event
  (c) Search and rescue support not linked to a specific disaster
- Emergencies with doubtful or no benefit from space assets
  (a) Drought
  (b) Routine epidemiological outbreaks
- Calls exceeding the emergency period
  (a) Commonly, a charter activation requested more than 10 days after the first occurrence of the crisis is not accepted

The duration of an activation is generally limited to a maximum of 15 days from the date of activation. To add, the requests may not be accepted if the size and scope of the disaster is such that the available sensor resolutions are not appropriate for disaster coverage.

20.6.2 Data acquisition planning

As a space agency's charter membership application is accepted, it is required to submit a plan to describe the manner in which it foresees the implementation of the charter requirements within its organization. An agency charter implementation plan is therefore a prerequisite to its integration with the charter operations. The document also contains information on a member agency's contribution to the charter, in terms of satellite data and services, as well as any other operational resources and their deployment.

The ESA, being one of the charter's founding members and initially the lead agency having developed the charter concept with CNES, designed its implementation document for the purpose of describing the primary roles of the various functionaries and the ones that would be under the ESA's responsibility. In particular, the ESA implementation plan takes into account the resolution of issues related to the operation of ERS for emergency data acquisitions.

CNES, apart from committing resources to the charter implementation functions, ensured the availability of its own space facilities and those under control of private or public operators. The necessary contractual arrangements were completed with Spot Image in order to have access to its products and services, which included priority data acquisition planning and rush delivery. CNES put special emphasis on cooperation with the industry for generating timely satellite data products and delivering these to the end users by means of all the technical resources available: telecommunications, data collection, and navigation.

At the CSA, a request for RADARSAT-1 data by the ECO triggers a chain of events that can have an impact on several data users. The CSA on-call staff receives the charter request, analyzes the accompanying documentation for completeness and consistency with the charter procedures, and gathers the information needed to evaluate the impact of the request on the data acquisition planning process, including any potential conflicts with other user requests. After the necessary approvals from the operation planning manager, the charter request is assigned priority 2, as for all emergency data requests. Data requests can be submitted at any point up to a limit of 29 hours prior to execution of the plan for regular requests and 31 hours for emergency requests; in the latter case, a significant amount of additional paperwork needs to be completed. The plan uplinked to the spacecraft covers a 24-hour period and spans from 19:00 UTC to 19:00 UTC the following day. The charter acquisitions are planned with near real-time processing and delivery. All stages of the process are properly documented, and all interested parties are informed of the status of their requests for RADARSAT-1 data.

The charter data acquisition planning process is relatively long and complex when compared to the nominal operations. Furthermore, if the charter request is conflicting with already planned requests, the lower priority requests are abandoned, some of which may be for time-critical data. Therefore, every effort is made by the charter authorities and the ECOs to see that proper activation criteria have been applied to incoming charter requests if these are of questionable nature.

In addition to providing a charter board member, who is the DMS director, and other functionaries (executive secretariat member, ECOs, and PMs), ISRO has also constituted an International Charter Review Committee with experts from various centers dealing with data acquisition and analysis to constantly review the mechanism for providing charter support. ISRO has built up expertise in data handling and value adding within its various centers, as well as with the industry to offer it to the charter operations.

The National Environmental Satellite, Data, and Information Service (NESDIS) Senior Advisor for Systems and Services leads NOAA's participation in the charter. The Office of Satellite Data Processing and Distribution provides the bulk of the facilities, data, products, and services. The Satellite Analysis Branch (SAB) shift supervisor is NOAA's primary point of contact for an ECO or a PM from a non-NOAA charter participating agency. The shift supervisor can provide operational products that are produced by the SAB and experimental products produced in the NESDIS. Products are created routinely and are available on the Internet. Working with the Spacecraft Operations Control Center (SOCC) scheduler, the SAB shift supervisor initiates tasking of the NOAA POES satellite to record and deliver 1-km AVHRR local area coverage, over an area requested by a charter AU. Conflict resolution is performed by the SOCC scheduler and the ECO or PM. The SAB shift supervisor can facilitate the process, for instance, by providing viable, scientifically sound alternatives. NOAA level IB (charter level IA) data are available through the NOAA Comprehensive Large Array data Stewardship System (CLASS). Archived AVHRR data are also available through CLASS. The ECO or PM establishes an account on CLASS to search for and download data via FTP. Other products are available from CLASS that may support the emergency response. The products are listed on the NOAA Data Request Submission Procedures Form. The NESDIS Interagency and International Affairs Office plays a role in coordinating the U.S. government activities for the charter.

The USGS EROS disaster response coordinators have, 24 hours a day, 365 days a year, the responsibility for coordinating USGS EROS operational support for emergency remote sensing requirements. They coordinate data acquisitions from the long-term archive, new satellite and airborne acquisitions, generation of value-added products, data distribution, and archiving. They act as the on-call persons for USGS and their designated staff is the first point of contact for the ODO, if the weekly ECO assignment is USGS, and for the ECO or the PM as the USGS mission planning contact. Working with the Landsat data acquisition manager and the Landsat 7 flight operations team, the designated staff initiates tasking of the Landsat 7 satellite to record and deliver 28.5 m Enhanced Thematic Mapper+ (ETM+) data over an area requested by an AU. Conflict resolution is carried out by the Landsat data acquisition manager, the Landsat 7 flight operations team, and the designated staff. The latter facilitates the process in the same way as does the NOAA SAB shift supervisor. The designated staff also provides access to the Landsat 5 and 7 archives located at USGS EROS.

The emergency responsible (EMR) is the CONAE's point of contact that has been assigned the responsibility for the development, implementation, and follow-up of the Argentinean space agency's participation in the charter. The application group of CONAE coordinates activities under this program to support the creation of value-added products. Considering the urgency for data availability to meet a charter request, CONAE has integrated the charter procedure with its existing procedures for planning, acquisition, product generation, and delivery for national emergencies. An emergency request by the ECO is addressed to the CONAE EMR by phone. A fax should follow with the complete relevant information. All incoming SAC-C data requests at the time of the charter activation are processed following emergency request acceptance criteria. CONAE works closely with other national institutions (SIFEM, INTA) to fulfill its roles and responsibilities under the charter, particularly in terms of the PM services and value adding.

As mentioned previously, the DMC consortium considers disaster response as a key activity from the conception of the constellation; therefore, every effort is made to ensure that requests from the charter are routed to the closest point of satellite tasking without delay. The consortium has provided an initial point of contact for all charter activations so as to coordinate the deployment of DMC assets for data supply. The programming requests for satellite data acquisition are sent to DMCII by the ECO. The requests are checked for completeness, feasibility, and availability of data providing coverage of the area of interest in the DMC archives. Data from the archives are supplied where they can justify the programming request or can furnish supplementary value to the request, such as in the form of predisaster reference imagery. The feasibility of programming requests depends on technical constraints such as potential conflicts with other programming requests and climatic conditions in the area of interest. For the sake of standardization and minimizing the effort, DMC accepts requests using the ERF that can be mailed electronically by the ECO. DMCII will then forward each programming request to all DMC satellite operations teams and make direct requests into DMC satellite mission planning systems. Individual DMC operations teams are responsible for generating commands within various satellite and ground station schedules to ensure data acquisition and downlink. Command schedules are generated automatically by the satellite mission planning system or manually using standalone software tools. After the acquisition of data and processing, the PM is informed of the successful data delivery by DMCII.

The programming conflicts of ALOS are managed by JAXA/Disaster Management Support Systems Office in liaison with the ECO and the PM. JAXA has prepared guidelines that can be used by the ECOs to select the most appropriate of the three ALOS instruments and the imaging mode for a given disaster type. The highest planning priority, barring spacecraft health and safety, is assigned to emergency requests received under the charter.

20.6.3 Reporting and user feedback

From the moment of its nomination, the PM is in charge of a charter activation, which is treated as a project. With the exception of the charter annual report prepared by the executive secretariat and posted on the charter website (www.disasterscharter.org) for public viewing, all reporting requirements lie with the PM. There are two fundamental reports that the PM is required to submit to the executive secretariat, namely, the preliminary and the final project manager report of a given charter activation.

The PM builds the preliminary report of the event using a standard template and based on the "dossier" forwarded to the PM by the ECO, as well as the PM's later interaction with the different users. This report is submitted to the executive secretariat within 1 month from the date of the charter activation, and it contains the following information:

- Details of the initial request, organizations, and contact names
- Geographic location of the affected area
- Data and documents provided to the AUs with indication of the sources of data
- Proposed additional actions (e.g., fast delivery, special data or information products)
- General information gathered from news sources
- Estimation of the cost of the emergency situation using information given by the providers of data and information
- Suggested improvements in the conducted disaster scenarios and/or implementation of the scenarios, based in part on user feedback

With input from all participating bodies concerned with the disaster, the PM writes up a final operation report following a recommended table of contents for submitting the report to the executive secretariat within 3 months from the date of the charter activation.

The most important aspect of the PM reports is the user feedback. Two of the preliminary report items relate exclusively to the feedback in order to record user satisfaction with the data delivery process and to measure the effectiveness of space technology applications through the charter for disaster management. The Case Histories section, therefore, takes into account user feedback received for the selected charter activations.

20.7 Case histories

Since its inception, the charter has covered the majority of the major disasters that struck various parts of the world. A few of these have been selected to describe how the charter operations were conducted and what type of satellite data products were delivered to provide space-based information in emergency situations. More important, the end user feedback has been included in the descriptions to gauge customer satisfaction of the space-based disaster management.

20.7.1 Nyiragongo volcanic eruption

The Nyiragongo volcanic eruption took place on 17 January 2002 and was reported by the Belgian Civil Protection Agency for activating the charter 4 days later. The ECO function for the week was being performed by CNES, which also provided the PM services.

According to the initial reports on the disaster, the lava flowed into the city of Goma, killing more than 100 people and destroying more than 12,000 homes, forcing hundreds of thousands of people to flee. After preliminary consultation with the end user of the Musée Royal d'Afrique Centrale in Tervuren, Belgium, production of maps of the affected region was started. Four satellites were tasked for data acquisition. These were SPOT-2, SPOT-4, RADARSAT-1, and ERS-2. The commercial data distributors, specifically Spot Image and RSI, had already programmed acquisitions over the region by the time the charter was triggered, thereby quickly providing posteruption images for 21 and 22 January. An ERS-tandem mission derived DEM was provided by the Institute of Geophysics and Planetology, University of Hawaii. When the eruption began on Thursday, 17 January, people had already fled their homes. By the weekend, however, most of the residents of Goma had returned and needed to be sheltered. Map products prepared with the charter data were to be used in setting up camps for the refugees in zones not at risk of lava encroachment; therefore, a final deadline of Sunday, 27 January, was set for the delivery of image products.

Due to the very complex topography of the area, the PM decided to contract the French Institut Géographique National to orthoregister the images with sufficient accuracy. The data (images, maps, and DEM) were perfectly registered in the cartographic projection chosen by the end user. The French value-added reseller SERTIT was contracted to do photointerpretation and final map production, and the Institut de Physique du Globe de Paris (IPGP) from Paris-VI University was asked to provide its geological expertise and help with terrain measurements. As new data became available, they were deposited on a CNES FTP server, where they could be immediately downloaded and analyzed by the end user, SERTIT, and IPGP. This allowed the end user to get an early start on land use analysis using the original imagery, while the other organizations worked on the image products. The first value-added maps were delivered on 24 January 2002, 3 days before the deadline set by the end user.

Figure 20.23 Ground truthing (*left*) and validation of lava flows delineated on RADARSAT-1 SAR image (*right*), Nyiragongo volcanic eruption, January 2002.

Three types of product were provided to the end user:

- Land use map with identification of the lava flow
- Lava flow location (updated with automatic change detection), road network extraction, and location of refugees' camps
- Damage caused by the lava flow to urban areas

The land use maps together with the detection of the lava flow were used in order to find possible locations for the refugee camps. Most of the lava flow was well detected, but there were areas of false alarm, and the western lobe of the lava flow escaped detection on the satellite data-derived map (Figure 20.23). The discrepancy may have been caused by the rate of solidification of the lava surfaces, which affected the radar backscatter. It was determined that the surface of the lava flows had, in some cases, a very rough texture, while in other cases the surfaces were smooth. Furthermore, the lava flowed along the streets without affecting the buildings, making it difficult to detect it with the 10-m resolution radar images that were used for the purpose (Figure 20.24). Clearly then, for this kind of event, even 10-m resolution imagery may not be good enough. More precise lava detection could have been obtained had another type of sensor been available together with SAR. The high-resolution SPOT data could not be used due to cloud cover.

The end user found that the map production delay was acceptable for this kind of emergency, but it would be too long in the case of an earthquake where lives must be saved and the information on field conditions is required much faster than 3 days, which was the time taken in the case of Nyiragongo volcanic eruption to deliver the first charter products. The end user considered the information contained in the maps provided by the charter to be useful for the relief efforts; however, a user guide or image viewer software would have been helpful in reading the information quickly.

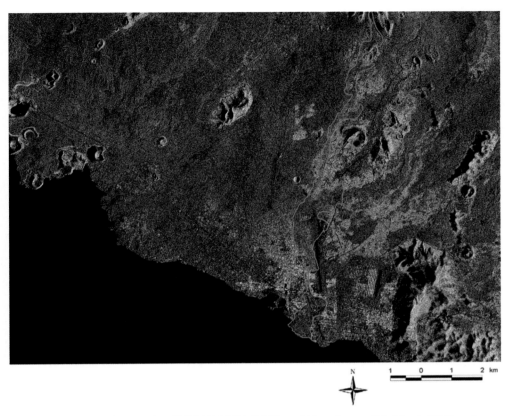

Figure 20.24 Raw radar (RADARSAT-1 SAR) multidate image (22 March 2001 and 21 January 2002) composite to map lava flow surfaces (red outline) by means of change detection, Nyiragongo volcanic eruption, January 2002.

20.7.2 Southern Manitoba flood

Heavy rainfall began on 9 June 2002, in southern Manitoba, Canada, and northern Minnesota, United States, causing the highest rain-generated flood event on record, with river levels cresting at 5.2 m and some areas recording up to 30 cm of total rainfall. Large areas of farmland were flooded, causing concerns over the contamination of river and groundwater due to hog farm runoff. The Canadian AU, namely the Office of Critical Infrastructure Protection and Emergency Preparedness, triggered the charter on 13 June 2002, to help the Water Branch of Manitoba Conservation and Manitoba Emergency Management. The ESA was the ECO for the week, and the PM services were delivered by staff of the CSA.

All five RADARSAT-1 scenes that were requested were provided to the value-added reseller hired by the CSA for product generation and information extraction. These were two archive and three newly acquired scenes postdating the event. A total of nine SPOT images of the flood-affected area were also obtained; however, five of these image frames could not be used due to excessive cloud cover. The remaining four frames were provided directly to the end user. The value-added reseller, Vantage Point International, a company based in Ottawa, delivered the first products only 3 days after the charter activation, along with a document for interpreting the satellite data products. The products were delivered in digital format to facilitate their distribution and integration with data from other sources.

The main feature of these products was the visualization of the flooded surfaces by using pre- and postflood orthorectified RADARSAT-1 imagery (Figure 20.25). The information derived from these products was handed over to the end user on the morning of 18 June 2002. Although the entire process went rapidly and smoothly and the information was found useful, it was not fully used in the field because the water had started to recede too fast and the need for locating flooded areas was removed. Contrary to the 1997 Red River flooding, which unfolded over an extended period of time, this 2002 event was more of a flash flood. The use of the charter-provided information was held more appropriate for flood forecasting and public awareness.

According to the end user, the data for the June 2002 event was considered useful for determining flooded areas for the purpose of hydrologic modeling. This included water ponding and areas of overland flow, both of which are important for modeling runoff volumes and stream flow. The water is also an ultimate test of the topographical detail of watersheds, which is not commonly available from land surveys. The flooded areas and areas of very moist soil are used in conjunction with rainfall data, runoff data, soil type, vegetation, and topography for hydrologic modeling.

Another application of the data was to determine the extent of over-bank flows on rivers for flood routing purposes. This feeds into the design studies and future operational use. A further use of the satellite data is to highlight inadequacies of the drainage network, which is a very topical issue. Moreover, there is a water quality component because flooding of major hog and livestock operations poses a potentially serious problem to the environment and to the safety of drinking water. It is interesting to note that there had been a strong proliferation of livestock operations in southeastern Manitoba in the 6 years preceding the flood event being reported. Manitoba Emergency Management used the data mainly to determine the extent and location of flooding. This is required in making important decisions with respect to the eligibility for flood compensation. It is also highly important to define the magnitude of the flood for the purpose of applying for the federal financial assistance for flood damages and recovery costs. In this regard, the remotely sensed data are effective in recognizing flooding on quarter sections of land and even smaller parcels, and in discerning standing water and moist soils. The land use information also applies to the types of lands and crops that are affected by the flood.

In this case, although the data were not used as expected, it was demonstrated that the information extraction could be done rapidly and the products could be delivered to the end user in a relatively short time. The value-added reseller used a generic algorithm to digitally map the extent of the flooded area, as shown in Figure 20.25. The digital maps were produced in the well-known ArcView format so these could be transferred and displayed rapidly by the end user. Both the end user and the AU were very much impressed by the efficiency and quality of the products delivered and decided to include space-based information in their future organizational practices.

20.7.3 Galicia oil spill

This is a case of a technological disaster covered by the charter. The *Prestige* tanker started leaking fuel off the coast of Galicia, Spain, on Thursday, 14 November 2002, when it encountered a violent storm about 150 miles out in the Atlantic Ocean. For several days, it was pulled away from the shore, but the crippled tanker carrying more than 67,000 tons of oil split in half off the northwest coast of Spain on Tuesday, 19 November, threatening it with one of the worst environmental disasters in history. The rear section of the *Prestige* sank early in the day, taking many of the oil tanks with it. By this time, it had already leaked at least 7,000 tons of oil.

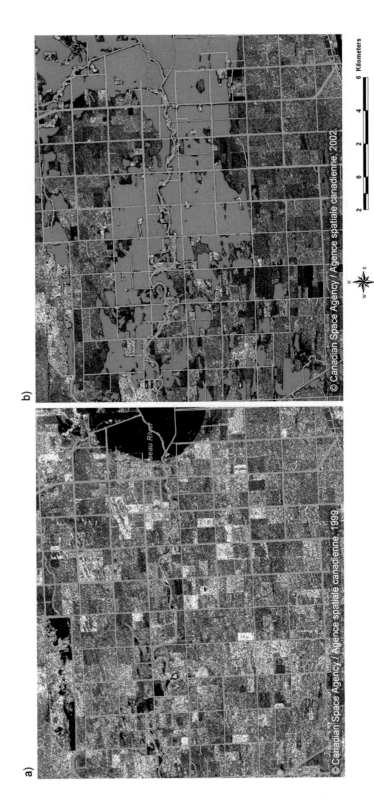

Figure 20.25 Derivation of flooded areas with (a) pre- and (b) postevent RADARSAT-1 imagery, Southern Manitoba flood, June 2002. Flooded surface is shown in red.

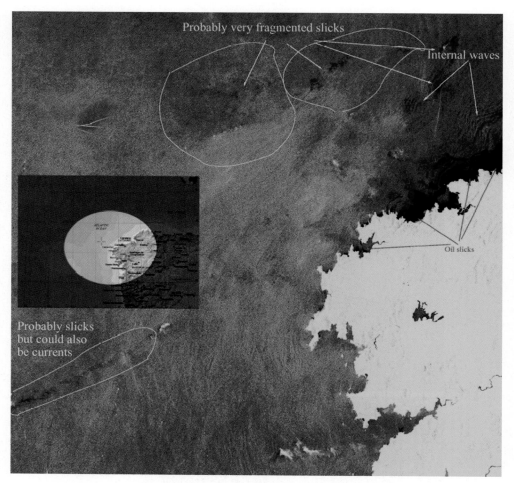

Figure 20.26 Galicia oil spill detection on RADARSAT-1 image (18 November 2002).

The charter activation request came from the General Directorate for Environment of the European Commission, and the end users were the Spanish authorities of the Department of Public Administration. The ESA was the ECO for the week, and CNES furnished the PM services. Initially, SPOT satellites were identified by the ECO for data acquisition, but these requests were canceled by the PM due to weather conditions, and because optical data was not seen as the most appropriate for oil spill detection. Therefore, all available radar satellites were tasked, including ENVISAT, which was still in its commissioning phase. ESA provided a series of ENVISAT ASAR images for this event. Two RADARSAT-1 images were also provided, one acquired on 18 November and the other on 22 November. Additional RADARSAT-1 images were obtained through NOAA's National Oceanic Service (NOS) on the direct request of the Spanish authorities. These images were received from the Scotland ground receiving station on 24 and 26 November, and were processed by NOS. CNES/QTIS (Qualité et Traitement de l'Imagerie Spatiale) provided the radar intensity images, enhanced quicklooks, and value-added products to the users. They provided the RADARSAT-1 images on a reference map giving the position and shape of the detected slicks (Figure 20.26). These products were distributed with a brief annotation giving the level of confidence to the detection.

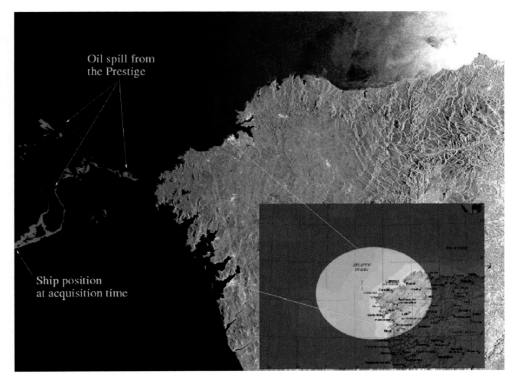

Figure 20.27   Galicia oil spill detection on ENVISAT ASAR image (17 November 2002).

The AU itself interpreted raw ENVISAT ASAR images (Figure 20.27) and noticed that VV polarization was more effective than HH polarization of RADARSAT-1 in oil spill detection due to higher backscatter of the sea surface, thus accentuating the dampening effects of oil on the surface. Furthermore, wide swath modes were chosen because of the size of the area covered and spill displacement, depending on the wind conditions.

The end users highly appreciated the conduct of the charter operations and the co-operation among the various parties involved in the activation. The end user feedback also helped the charter authorities to improve the data communication channels for faster delivery. In proof of the quality of the charter product and services, the Joint Research Centre of the European Commission and the end users proposed a more formal framework and expertise for the image analysis and interpretation of the charter data at the European level.

20.7.4 South Asian Tsunami

On 26 December 2004, an earthquake of magnitude 9.0 on the Richter scale struck off the west coast of Sumatra in the Indonesian Archipelago at 06:29 IST (00:59 GMT) at 3.4°N, 95.7°E. The focal depth of the earthquake was 10 km, and the epicenter was located 250 km south-southeast of Banda Aceh and 320 km west of Medan, the capital of North Sumatra Province. This earthquake triggered a devastating tsunami that killed hundreds of thousands of people in different countries. Indonesia was the worst affected, followed by Sri Lanka, India, Thailand, Somalia, Maldives, Malaysia, and Myanmar.

Three separate charter activations covered this huge disaster, indeed one of humanity's worst. ISRO and NOAA carried the ECO duties for these activations. The ISRO Disaster

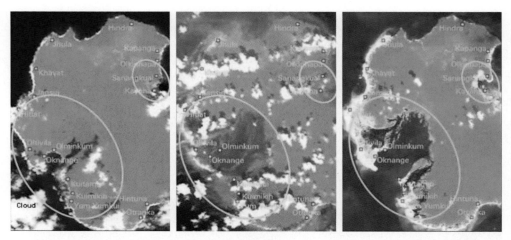

Figure 20.28 Three IRS color-coded images of Katchall Island acquired before (21 December 2004), during (26 December 2004), and after (4 January 2005) the tsunami show the changes induced by the disaster.

Management System was the first AU to call the charter at approximately 12:30 UTC to cover two locations of the country, on the Nicobar and Anadman islands and along the coast of the Tamil Nadu State. The second charter activation call originated from the French AU, namely, the civil defense and security directorate, for Sri Lanka. The Canadian executive secretariat lead had already issued, by 16:00 UTC on 26 December, the PM nominations for the two activations from the Indian National Remote Sensing Agency (NRSA) and CNES for the Indian and the Sri Lankan activations, respectively. By 17:00 UTC, the first satellite data acquisition requests, consisting of two new and two archive RADARSAT-1 frames, had been submitted for planning by the CSA.

At 22:00 UTC, the charter was activated a third time by OOSA on behalf of UN OCHA to monitor the disaster effects on Indonesia and the neighboring countries. The UN staff performed the PM services for the third charter activation. A French value-added reseller, the Indian NRSA and the UN OSAT, along with the USGS EROS center, generated a variety of satellite data products, which are displayed and described in Figures 20.28 to 20.33. More than 200 ENVISAT, IRS, RADARSAT-1, and SPOT scenes were planned, in addition to Landsat and high-resolution commercial data contributions (IKONOS, QuickBird).

Cloud cover in the days following the tsunami prevented the acquisition of optical imagery over Sri Lanka until 12 January 2005, so the first products were all produced with radar imagery. RADARSAT-1 was used to image the southeastern coastline of Sri Lanka, and ENVISAT acquired imagery over its southern portion. IRS images of the northeastern and SPOT images of the southern and southeastern parts were later provided.

Image products provided included extracts of satellite imagery in real and in natural colors, space maps on various scales, before and after images, damage maps, and 3D overview videos. A population distribution map was also provided. These products were transmitted to the end users via FTP. The NRSA data center, ISRO, provided IRS data on CDs.

The procured satellite data comprised of different sensors with different spatial resolutions ranging from 360 to 2.5 m. The coarse resolution (56-m) satellite data, such as of the AWiFS sensor onboard IRS P6 with a temporal coverage frequency of 5 days, provided an opportunity to cover a large extent of the affected areas in a very short span of time. Although the sensor resolution is too coarse for identifying the damage in detail, it

20.7 *Case histories* 527

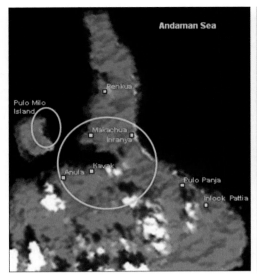
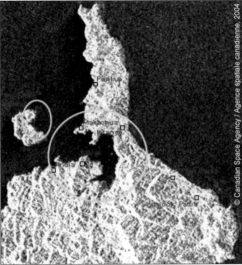

Figure 20.29   A closeup view of the coast of Nicobar Island with pretsunami, 21 December 2004, optical IRS (*left*) and posttsunami, 31 December 2004, cloudfree, RADARSAT-1 image (*right*) to show the damage to the coastline.

provided an overview of the areas affected by the tsunami. Wherever possible, medium- and high-resolution data were used for analysis.

Furthermore, for some of the worst affected areas, high-resolution sensors such as PAN and LISS IV MX and SPOT-MLA were programmed. Wherever optical data were cloudy, microwave data from ENVISAT and RADARSAT-1 allowed the analysts to obtain clear images of the affected areas. Efforts were made to analyze in almost real time the satellite data acquired. Pre-event satellite data sets, mostly from the month of December, were procured and analyzed alongside the postevent satellite data sets for damage assessment. The drastic changes to the coastlines and disappearance of land are depicted in Figure 20.30.

20.7.5 *French forest fires*

Southern France experienced some major forest fires in 2003 and 2005. The charter was activated three times in 2003 and once in 2005 for these forest fires. The affected areas included the Var region, the island of Corsica, and the Maritime Alps. More than 40,000 hectares of land were devastated by the fires. The 2005 activation was for fires affecting the French Riviera. Here, only the 2003 event is described as a case history of this disaster type.

The fires started in mid-July 2003, and three separate charter activation calls, on July 19, July 25, and September 2, were placed by the French AU (COGIC, Ministry of Interior) to obtain help for the French Civil Protection and local authorities in the Var region. The calls arrived during the ECO duty periods of NOAA, ESA, and CONAE. The PM services for all three activations were delivered by the CNES staff.

The end user requested maps showing the perimeter of the burnt area, as well as before and after space maps, in Lambert 2 extended map projection. CNES/QTIS provided the various types of image products to the end users so they could identify which products were most useful for this disaster type. A typical color-coded burnt area map prepared with the

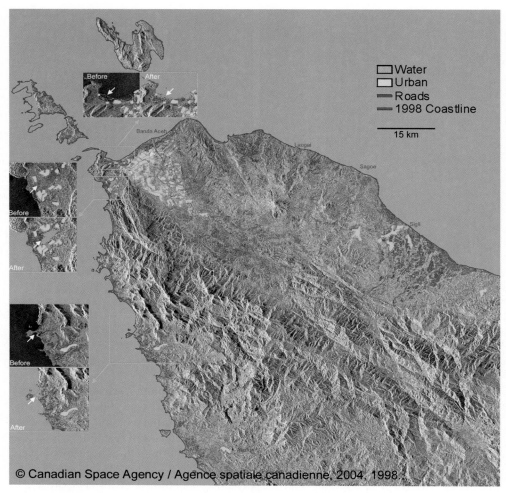

Figure 20.30 RADARSAT-1 before (9 April 1998) and after (31 December 2004) image interpretation to show the damage caused by the tsunami to coastline and urban areas of northern Sumatra.

orthorectified SPOT imagery of the land cover is presented in Figure 20.34. These products were transmitted to the end users via FTP, while quicklook images were sent via e-mail. Delays in delivering final products varied between 6 hours for simple satellite images to 24 to 36 hours for more complex image products. The charter intervention for this disaster was a sure case of the benefits of space-based information to the local authorities, first responders, and frontline forces in emergency situations.

20.7.6 Hurricane Katrina

Hurricane Katrina caused considerable damage, as well as extensive flooding, in the Mississippi Delta and in the states of Louisiana, Mississippi, and Alabama, leaving thousands of people without shelter, electricity, and drinking water. Thousands of people took refuge in the New Orleans Super Dome stadium. The charter coverage of the disaster was by means of two separate activations, one requested by the U.S. AU on 31 August 2005, and the other

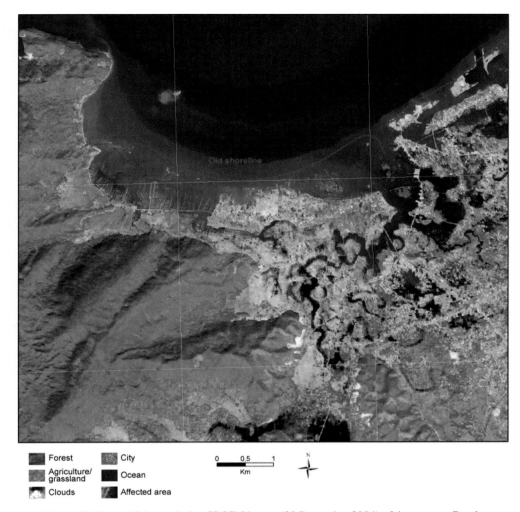

Figure 20.31 A high-resolution SPOT-5 image (30 December 2004) of the western Banda Aceh region taken after the tsunami, with changes in the land cover and the coastline. The pretsunami coastline is shown in red.

by the French AU on 2 September 2005. The latter request was specifically for the cities of LaFayette, Mobile, Jackson, and Baton Rouge. The ISRO ECO handled the two activation calls, and the PM tasks were carried out, respectively, by the USGS and CNES staff. The main end users in the case of this disaster were the U.S. Federal Emergency Management Agency (FEMA) and the French Civil Protection, who were planning to send relief teams to the affected region.

Twenty-five space maps were produced covering the New Orleans flooded areas, Biloxi and the surrounding areas, and the coastal islands south of Mobile. These maps were provided in addition to raw radar imagery (Figure 20.35). Some good quality, value-added products were produced by SERTIT (Figure 20.36), which were all made available to the AUs of both activations in time, as well as to the German Civil Security (ZKI). The maps were used by the German teams deployed in the affected areas, as well as by the central German authorities, who used the maps to localize the teams at work. The space maps were highly valued by the AUs.

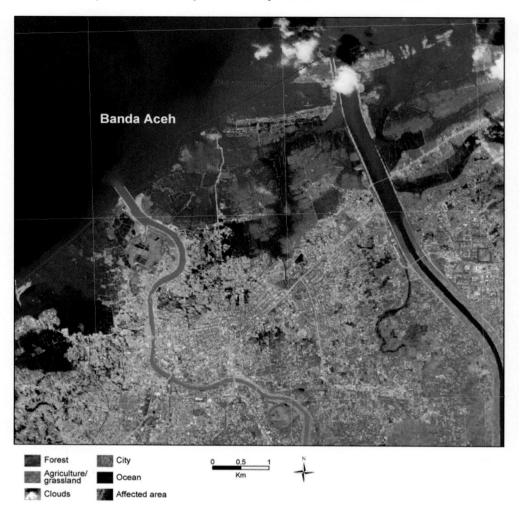

Figure 20.32 A high-resolution SPOT-5 image (30 December 2004) of the Banda Aceh city area to show the devastation caused by the tsunami. The pretsunami coastline is shown in red.

Feedback from some of the responding agencies indicated that the data were used in many ways to assist in the emergency response efforts. Some of these included identifying damaged buildings, tracking water levels in the national parks, tracking water levels in New Orleans, performing Red Cross damage assessment planning, assessing port damage, assessing coastal highways, determining areas of washout, looking for stranded and wrecked vessels, locating helicopter landing sites, assessing petroleum/hazardous chemical areas, and creating search and rescue maps. The end user exploited the data to monitor the draining of water from New Orleans, in conjunction with other data layers to assist in search and rescue efforts.

From the charter implementation and logistical viewpoint, because of the large number of responding agencies involved, it was difficult for the PM to gather the required feedback about the degree of usefulness of the generated data products. The PM simply did not have access to the agencies' internal systems for copies of their products. It was also unclear

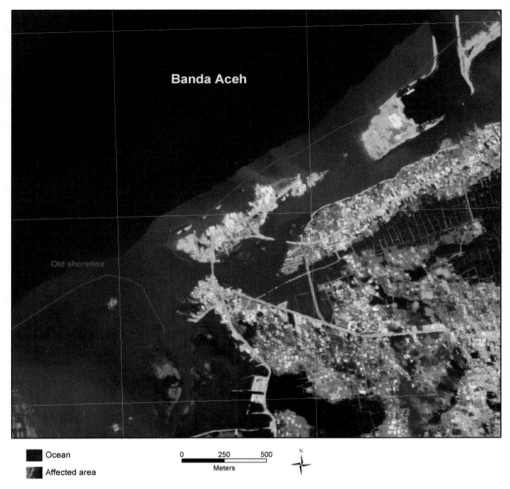

Figure 20.33 A SPOT-5 (30 December 2004) closeup of western Banda Aceh to show the tsunami damage. The pretsunami coastline is shown in red.

how many products were actually used in the decision support systems for providing input to the relief planning process. Some of the other problems recorded as lessons learnt were related to the sheer size and number of multisource data sets. The format problems were to be addressed by designing and implementing a preprocessing system that would convert images to standard processing levels and formats. The image quality problems were to be addressed by making better web browsing tools available for use in the field.

20.7.7 Kashmir Earthquake

An earthquake of magnitude 7.6 on the Richter scale shook northern India and Pakistan at 9.30 am local Indian time on 8 October 2005. The preliminary location of the earthquake epicenter was close to Muzuffarabad in Pakistani-administered Kashmir and west of the Uri sector of the Baramulla district in India. The severe shaking was felt in most of northern India and Pakistan. Massive destruction was reported in both Indian and Pakistani Kashmir, particularly around the epicenter zone near Muzaffarabad. The disaster was handled under

Figure 20.34　Burnt area (in blue) map of the Var region of France prepared with orthorectified SPOT image (3 September 2003).

two separate charter activations by the CSA ECO. The activation call from the Indian AU covered the regions on the Indian side of the line of control (LoC). Both the French AU and the UN OOSA requested the charter data for the Pakistani side of the LoC. The Indian NRSA staff performed PM functions for their part, and an ESA staff took over as the PM for the Pakistani side. The main data requirement was high-resolution imagery to detect damage to towns and villages, roads, communication lines, and other infrastructures. Given the hilly nature of the terrain, there was particular concern about landslide damage. In fact, some new landslides were detected on the satellite imagery; others were old landslides reactivated by the earthquake (Figures 20.37 and 20.38). Fine mode RADARSAT-1 SAR and very high-resolution optical data from commercial satellite systems, such as IKONOS and Quickbird, were sought by the users. Topographic map updates using SRTM, Landsat, and IKONOS data were prepared by the French value-added reseller SERTIT for logistical support and damage assessment purposes.

20.7.8 Philippines landslide

A huge landslide on the island of Leyte on the west point of the islands of the Philippines caused at least 200 deaths and 1,500 people were reported missing, according to the Red Cross, at the time the charter PM report was written. The local authorities had only officially announced a total of 23 dead. Some 500 houses and a primary school were buried under the mud, which, after 2 weeks of continuous rain, came from a mountainous escarpment. The governor of Leyte Province said that it was feared that 500 houses and a primary school full of pupils were buried in Ginsahugan, a village that had approximately 2,500

Figure 20.35 A raw RADARSAT-1 image (5 September 2005) of downtown New Orleans showing radar corner reflection effects caused by buildings and surrounding water surface. RADARSAT-1 data © Canadian Space Agency, 2005.

inhabitants. Two other villages were also affected, and about 3,000 people were evacuated to a state building. According to the ambassador of the United States, the American Army was sending a ship to bring aid to the affected people.

UNOSAT-UNITAR, on behalf of the UN OOSA, triggered the charter on 17 February 2006, with the main objectives of providing reference, crisis, and change maps of the affected area to the rescue teams working in the field. Both the ECO and the PM duties were performed by the ESA staff.

Although multisatellite data acquisitions were planned, either these could not be materialized because of technical problems or planning conflicts or excessive cloud cover precluded information extraction in the case of most of the optical (SPOT, IRS, and DMC) data. The area affected by the landslide is located in a tropical zone where dense cloud cover prevents the use of optical sensors. The first and the only exploitable optical crisis image came from ALOS AVNIR-2 on 20 February. This was combined with SPOT-5 archive data to generate the first crisis product about 3 days after the charter activation. Fourteen attempts by different optical sensors were made before these image data were obtained.

■ Flooded areas ▨ Nonflooded areas 0 1 2 N
▨ Flooded streets ■ Rivers, lake   km

Figure 20.36   Classification of flooded surfaces in urban Louisiana area following the passage of Hurricane Katrina (SPOT-5, 2 September 2005).

This activation was, therefore, the first JAXA ALOS (DAICHI) contribution to a charter emergency. JAXA constructively contributed to this activation by performing a change detection study based on DAICHI PALSAR and JERS-1. The resulting product was the first SAR derived for this activation and gave an excellent overview of the landslide-affected area (Figure 20.39). RADARSAT-1 Fine beam mode scenes, together with ENVISAT ASAR IMP, allowed the European value-added service provider RESPOND to produce the final product of this activation with appropriate vector overlays by means of the archive SPOT image maps.

The products provided were useful for the coordination of the operation, particularly the early products, due to their rapid availability. The later products were appreciated for their quality, but they arrived a little too late to be of practical use for the search and rescue operation. These products were still considered useful, however, for risk assessment and reduction planning. The products were even used by the government agency responsible for coordination in its reporting to the president of the Philippines. Several hard copies of the satellite data products were printed and distributed to field teams early on in the disaster occurrence.

20.7.9 Central Europe floods

The flooding of large swaths of the Czech Republic, Slovakia, Romania, Hungary, and Germany was caused by swollen rivers due to the combined effect of heavy snow melt and rainfall in early spring 2006. In Germany, the water level along the Elbe River, in some regions, topped the level of the century flood in 2002. In Hitzacker (Germany), the water level reached a height of 7.63 m, which is 10 cm above the century flood level. The historic part of the city was flooded, and all inhabitants had to be evacuated. The charter was activated by the German AU. The activation was handled by the ESA ECO and by a PM from the DLR.

The data provided included ERS-2, RADARSAT-1, DMC, and IRS scenes. The first space map was produced about 48 hours after the charter was activated (Figure 20.40).

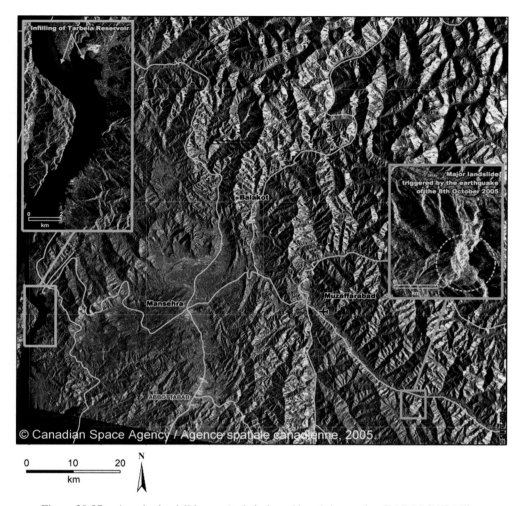

Figure 20.37 A major landslide scar (red circle and inset) detected on RADARSAT-1 Fine beam mode resolution imagery (predisaster, 20 June 2005; postdisaster, 18 October 2005) of the earthquake-affected region of Pakistani Kashmir. RADARSAT-1 data © Canadian Space Agency, 2005.

There were some data delivery problems, most notably a 6-day delay between acquisition and delivery of ERS-2 data. The processing and transfer was sped up after the PM contacted the ESA help desk.

DLR produced eighteen maps of the Elbe River from the Czech border to Hamburg. Several areas of special interest to end users were mapped with radar data to overcome cloudy weather conditions. DLR analyzed the satellite data and extracted a flood mask of the inundated areas. The flood mask was put on a topographic map, and in the case of one of the scenes, on a Landsat image background, to provide information to the local users. The entire Elbe River was mapped on 1:50,000 scale, and individual maps were updated after several days in order to adjust the relief activities. Near real-time products were delivered to the end user; these included satellite-derived maps and shape files of the extracted flood masks.

In the Czech Republic, at least four people, including two children, perished in the floods. In the city of Usti Nad Labem, 120 people were evacuated, in addition to about 300 people

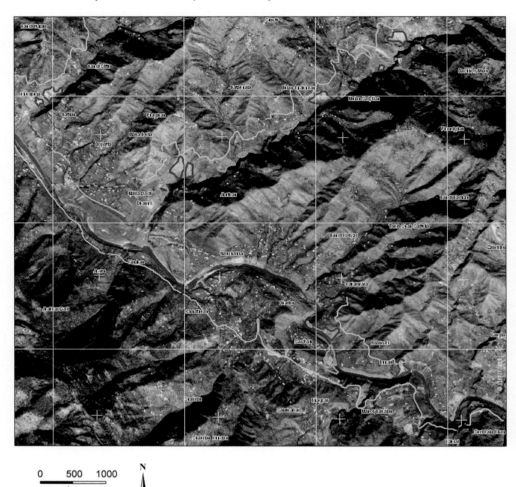

Figure 20.38 Mapping of reactivated and newly formed landslide surfaces in the earthquake-affected region of Pakistani Kashmir. Newly detected landslides outlined in orange; reactivated landslides outlined in red. (Predisaster—SPOT-4, 21 September 2005; postdisaster—SPOT-5, 11 October 2005).

who were cleared from their homes near the river embankment, and 150 who had to leave their homes around 20 km upstream near Litomerice. The situation was stable in the two worst-hit Czech regions, south Moravia and south Bohemia, where around 2,000 people had already been evacuated from the south Moravian town of Znojmo and its surroundings.

On 4 April 2006, the European Commission, DG ENV, Civil Protection unit A.5, triggered the charter with the main objectives of providing reference, crisis, and flood assessment maps of the affected areas. The call for triggering the charter was received by the CNES ECO. Another DLR staff also provided the PM services for this activation. The data provided were from ENVISAT, RADARSAT-1, ERS-2, SPOT, IRS, DMC, SAC-C, and Landsat.

Despite communication problems, the PM was in regular contact with the AU and the end user to get more information on the local situation. All products created by the European

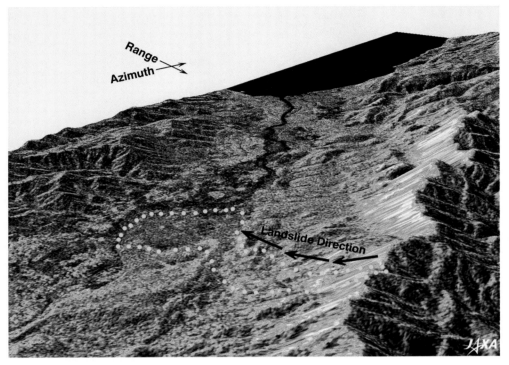

Figure 20.39 Bird's eye view of the northwest side of the landslide area in Leyte Island, the Philippines (predisaster—JERS-1 SAR, 2 February 1996; postdisaster—ALOS PALSAR, 24 February 2006).

GMES RISK-EOS Consortium (SERTIT) were orthorectified, geocoded, and projected to UTM zone 33 N. The first crisis product was made available via FTP on 5 April 2006. Original raster data were provided again via FTP to the Czech end user. The data quality and the foolproof data transfer was very much appreciated by the Czech authorities.

Due to violent rainfall and melting snow, the water levels of the Morava River (March River) in Austria quadrupled in comparison to its normal discharge. On 2 April 2006, a dam burst on the March River, separating Austria from Slovakia and flooding the Austrian village of Duernkrut. Roughly 200 people were evacuated from the village, parts of which were submerged in 1 m of water.

The Regional Alarm Centre, Lower Austria, triggered the charter on 7 April 2006, with the main objectives of providing reference, crisis, and change maps of the affected area. The same CNES ECO, as in the case of the previous activation, managed the charter call and processed the request for help with the collaboration of an ESA PM.

ERS-2, ENVISAT, RADARSAT-1, SAC-C, Landsat, and SPOT-4 data were provided. The value-added reseller SERTIT carried out timely and quality data analysis, producing no less than eleven image products, the first of which was available by 11 April 2006. The AU contributed with a large amount of support material, including field photos, aerial photos, and locality maps. The end user was completely satisfied with the charter data and service delivery.

In Hungary, the most critical situation shifted from the Danube River, where the water level had reached its highest mark in 100 years, to the Tisza River (18 April 2006). In particular, the cities of Szolnok and Szeged were affected, where the Hungarian civil

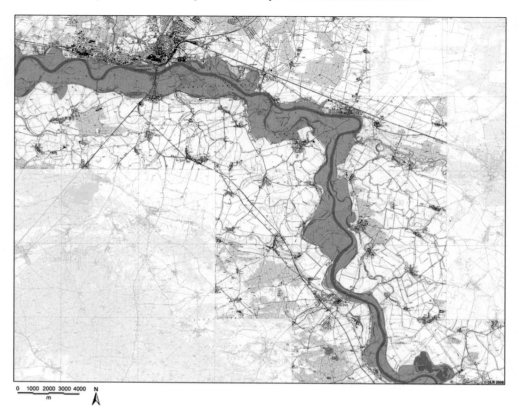

Figure 20.40 Topographic map with satellite data-derived flood mask (ERS, 4 April 2006), Elbe River, Germany.

protection tried to reinforce the dams with sand bags. In total, about fourty settlements were endangered by the flood.

The call for the charter help was placed by the European AU, Monitoring and Information Center, EC, DG Environment. The satellite data provided for the activation included RADARSAT-1, IRS, and Terra/Aqua. The USGS was performing the ECO duties for the week, and the PM nomination came from DLR. The AU and the PM coordinated their efforts extensively in compiling information on the disaster event and the satellite data acquisition planning. The overall scope of the call (i.e., the region of interest and the time period) was relatively small, and no operational problems were encountered in the course of the activation.

DLR produced six maps of the Tisza River and the Koeroes River in Hungary. Several areas of special interest to the end user (large cities along the rivers) were mapped with radar data due to the cloudy weather conditions. DLR analyzed the satellite data and extracted a flood mask of the inundated areas. The flood mask was integrated with the archived Landsat images with additional GIS and land use information.

The whole Tisza River/Koeroes River area was covered with maps on 1:75,000 scale, and the maps were updated after several days in order to help adjust the relief activities. In addition, shape files of the flood masks were sent to the users as e-mail attachments.

The MIC AU commented that the products were of good quality but needed to be delivered more rapidly, so an extension for events of longer duration was suggested for consideration in the charter operational procedures.

In Romania, large areas along the Danube River were flooded for a period of 4 to 5 weeks. In this time interval, several villages were inundated due to dam breaks along the Danube River, at Bistret, for example. In addition, large parts of the Danube Delta were flooded. The situation in Romania remained critical over several weeks due to the fact that peaking of the water level was long and rain set in periodically. The charter intervention was requested by the same AU as for the Hungarian region, and the same DLR staff was appointed as PM. The ECO weekly functions had, however, been taken over by CONAE in the meantime. SPOT, RADARSAT-1, ERS-2, ENVISAT, Landsat, IRS, DMC, and MODIS satellites were tasked.

The first map was produced by DLR on 19 April 2006, over the Bistret region. ROSA (Romanian space agency) was assigned access to the charter data and performed the value-added work. ROSA subsequently created the rest of the products for this activation. The Romanian governmental end user had prior experience with the charter data and working with the European RISK-EOS (DLR) service. ROSA has an extensive database for Romania and has learnt to produce RISK-EOS–adapted maps (in content and layout). There were frequent communications between the PM and ROSA to discuss data and product delivery, and ROSA maintained a close relationship with the end user, including the rescue teams in the field.

The data and products provided for this activation helped relief organizations and crisis teams in evacuating people, locating water pumps, reacting to dam breaks, building camps for evacuated people, and making forecasts for regions downstream. ROSA was pleased with the "good and fruitful cooperation" and the resulting products.

20.8 Concluding remarks

Management of major disasters using satellite EO data is being increasingly recognized as a priority application in both space and disaster relief communities. In view of their usefulness, space technologies have been popularly factored in the overall disaster planning by several governments.

The International Charter "Space and Major Disasters" is the world's primary source of space-based data and information in situations of crisis caused by natural and technological disasters. The data and information products delivered under the charter have helped crisis management authorities and relief organizations in the provision of base and damage assessment maps for organizing logistical support, in identifying safe areas for relocating the affected populations, in forecasting and monitoring disaster impact, and in promoting cooperation among the various stakeholders of disaster management teams.

The quality of the charter data and information is improving as new generations of satellites of higher spatial and spectral resolution and enhanced imaging parameters are added to the charter constellation, as the charter operations are improved and upgraded with user feedback, and as new members with their resources join the charter, also resulting in higher coverage frequency of disasters and faster data deliveries to the first responders and other end users. This chapter is meant to describe the baseline operations and provide a summary of the first 6 years of the implementation of the charter. The status of the charter performance and operations will continue to be reported in future activities of the international space community.

Acknowledgments

Dr. Ahmed Mahmood is the Canadian member of the charter executive secretariat. He wants to thank all of his colleagues who represent the other charter member agencies, namely, the European Space Agency (ESA), the French Centre National d'Etudes Spatiales (CNES), the Indian Space Research Organization (ISRO), the U.S. National Oceanic and Atmospheric Administration (NOAA), the U.S. Geological Survey (USGS), Comisión Nacional de Actividades Espaciales (CONAE) of Argentina, the U.K. Disaster Monitoring Constellation Imaging Ltd (DMCII), and the Japan Aerospace Exploration Agency (JAXA), for their generous support in providing information, directly or through the many charter documents, without which this contribution would not have been possible. Thanks are also due to Ms. Amy McGuire, Consultant SED Systems at the Satellite Operations Directorate, Canadian Space Agency, for her help with information compilation, illustrations, and database maintenance.

The intellectual property rights for the image data and products are reserved for the concerned satellite data providers and their suppliers, and in this regard copyright restrictions may apply. For full details regarding copyright restrictions and credits, the charter web management (www.disasterscharter.org) may be consulted.

Published with the permission of HER MAJESTY THE QUEEN IN RIGHT OF CANADA, © Government of Canada 2008.

References

Becker, F., and Li, Z. (1990) "Towards a Local Split Window Method Over Land Surface," *Int. J. Remote Sensing* **11**, pp. 369–393.

Bessis, J.-L., Bequignon, J., and Mahmood, A. (2003) "The International Charter 'Space and Major Disasters' Initiative," *Acta Astronautica* **54**, pp. 183–190.

Dech, S. (2005) "The Earth Surface," in *Utilization of Space Today and Tomorrow*, eds. B. Feuerbacher and H. Stoewer, pp. 53–90, Springer-Verlag, New York, New York.

Deutsch, M., Strong, A. E., and Estes, J. E. (1977) "Use of Landsat Data for the Detection of Marine Oil Slicks," in *Ninth Annual Offshore Technology Conference*, pp. 311–318, Houston, Texas, 2–5 May.

Elachi, C. (1987) *Introduction to the Physics and Techniques of Remote Sensing*, pp. 273–300, John Wiley & Sons, New York, New York.

Elachi, C. (1988) *Spaceborne Radar Remote Sensing: Applications and Techniques*, pp. 199–208, IEEE Press, New York, New York.

Fitch, J. P. (1988) *Synthetic Aperture Radar*, pp. 33–83, Springer-Verlag, New York, New York.

Franceschetti, G. F., and Lanari, R. (1999) it Synthetic Aperture Radar Processing, pp. 24–54, CRC Press, Boca Raton, Florida.

Hein, A. (2004) *Processing of SAR Data: Fundamentals, Signal Processing and Interferometry*, pp. 197–251, Springer-Verlag, New York, New York.

Jain, A. K., Duin, R. P. W., and Mao, J. (2000) "Statistical Pattern Recognition: a Review," *IEEE Transactions on Pattern Analysis and Machine Intelligence* **22**, no. 1, pp. 4–37.

Luscombe, A. P., Furguson, I., Shepherd, N., Zimick, D. G., and Naraine, P. (1993) "The RADARSAT Synthetic Aperture Radar Development," *Canadian J. Remote Sensing* **19**, no. 4, pp. 298–310.

Mahmood, A., Carboni, S., and Parashar, S. (1996) "Potential Use of RADARSAT in Geological Remote Sensing," in *11th Thematic Conference and Workshop on Applied Geologic Remote Sensing*, pp. I-475–I-484, Las Vegas, Nevada, 27–29 February.

Mahmood, A., Crawford, J. P., Michaud, R., and Jezek, K. C. (1998) "Mapping the World with Remote Sensing," *EOS, Transactions, American Geophysical Union* **79**, no. 2, pp. 17–23.

Mahmood, A., Cubero-Castan, E., Bequignon, J., Lauritson, L., Soma, P., and Platzeck, G. (2004) "Disaster Response in Africa by the International Charter," *Proc. 5th African Association of Remote Sensing of the Environment Conference*, 5 pages, Nairobi, Kenya, 17–22 October.

Mamen, R., and Mahmood, A. (2003) "RADARSAT-1 Accomplishments Beyond its Nominal Mission," *54th International Astronautical Congress* (IAC), 9 pages, Bremen, Germany, 29 September–3 October.

Massom, R. A., and Lubin, D. (2006) *Polar Remote Sensing: Ice Sheets*, pp. 39–136, Springer-Verlag, New York, New York.

Mather, P. M. (2004) *Computer Processing of Remotely Sensed Data*, pp. 107–137, John Wiley and Sons, New York, New York.

Mesev, V., Gorte, B., and Longley, P. A. (2001) "Modified Maximum-Likelihood Classification Algorithms and their Applications to Urban remote Sensing, ", in *Remote Sensing and Urban Analysis*, eds. J. Donnay, M. J. Barnsley and P. A. Longley, pp. 69–94, Taylor & Francis, New York, New York.

Mogil, H. M. (2001) "The Skilled Interpretation of Weather Satellite Images: Learning to See Patterns and Not Just Cues," in *Interpreting Remote Sensing Imagery*, eds. R. Hoffman and A. B. Markman, pp. 235–255, Lewis Publisher, Chantilly, Virginia.

Price, J. C. (1984) "Land Surface Temperature Measurements from Split Window Channels of the NOAA 7 Advanced Very High Resolution Radiometer," *J. Geophys. Res.* **89**, pp. 7231–7237.

Pierce, L., Kellndorfer, F., and Ulaby, F. (1996) "Practical SAR Orthorectification," *Int. Geosci. & Remote Sensing Symp.* **4**, pp. 2329–2331.

Rajesh, K., Jawahar, C. V., Sengupta, S., and Sinha, S. (2001) "Performance Analysis of Textural Features for Characterization and Classification of SAR Images," *Int. J. Rem. Sensing* **22**, no. 8, pp. 1555–1569.

Richards, M. (2005) *Fundamentals of Radar Signal Processing*, pp. 462–482, McGrow-Hill, New York, New York.

Richards, P. B. (1982) "Space Technology Contributions to Emergency and Disaster Management," *Adv. Earth Oriented Appl. Space Techn.* **1**, no. 4, pp. 215–221.

Richards, J. A., and Jia, X. (1999) *Remote Sensing Digital Image Analysis: An Introduction*, pp. 109–134, Springer, New York, New York.

Robinov, C. J. (1975) "Worldwide Disaster Warning and Assessment with Earth Resources Technology Satellites," USGS Project Report (IR), NC-47, Washington, D.C.

Rose, L. A., and Krumpe, P. F. (1984) "Use of Satellite Data in International Disaster Management: The View from the U.S. Department of State," *Eighteenth International Symposium on Remote Sensing of Environment*, pp. 301–306, Paris, France.

Toutin, T. (2003) "Review article: Geometric Processing of Remote Sensing Images: Models, Algorithms and Methods," *International Journal of Remote Sensing* **25**, pp. 1893–1924.

Tuceryan, M., and Jain, A. K. (1998) "Texture Analysis," *Handbook of Pattern Recognition and Computer Vision*, eds. C. H. Chen, L. F. Pau and P. S. P. Wang, pp. 207–248, World Scientific, London, United Kingdom.

Ulaby, F. T., Jakob, J., and Zyl, V. (1990) "Wave Properties and Polarization," in *Radar Polarimetry for Geoscience Applications*, ed. F. T. Ulaby and C. Elachi, pp. 4–40, Artech House, Norwood, Massachusetts.

Ulaby, F. T., Moore, R. K., and Fung, A. K. (1981) *Microwave Remote Sensing: Active and Passive. vol. I, Fundamentals and Radiometry*, pp. 186–251, Artech House, Norwood, Massachusetts.

Ulaby, F. T., Moore, R. K., and Fung, A. K. (1986) "Microwave Remote Sensing: Active and Passive. vol. II, Radar Remote Sensing and Surface Scattering and Emission Theory," pp. 630–652, Artech House, Norwood, Massachusetts.

Venkatachary, K. V., Rangrajan, S., Nageswara, P. P., and Srivastava, S. K. (1999) "Remote Sensing Applications for Natural Disaster Management: Retrospective and Perspective," *Proceedings ISRS National Symposium of Remote Sensing Applications for Natural Resources, Retrospective and Perspective*, pp. 23–34, Bangalore, India.

Verblya, D. L. (1995) *Satellite Remote Sensing of Natural Resources*, pp. 107–156, CRC Lewis Publisher, Boca Raton, Florida.

Walter, L. S. (1999) "Satellite Usage in Natural Disaster Management," *Earth Observation Magazine*, pp. 42–44, April.

Yuan, D, Elvidge, D. C., and Lunetta, R. S. (1998) "Survey of Multispectral Methods for Land Cover Change Analysis," *Remote Sensing Change Detection: Environmental Monitoring Methods and Applications*, eds. R. S. Lunetta and C. D. Elvidge, pp. 21–39, Ann Arbor Press, Ann Arbor, Michigan.

Zhang, G. P. (2000) "Neural Networks for Classification: a Survey," *IEEE Transactions on Systems, Man and Cybernetics. Part C: Applications and Reviews* **30**, no. 4, pp. 451–462.

Zhou, G., and Jezek, K. (2004) "Satellite Navigation Parameter-Assisted Orthorectification for over 60N Latitude Satellite Imagery," *Photogrammetric Engineering and Remote Sensing* **70**, no. 9, pp. 1201–1209.

21

Weather satellite measurements: their use for prediction

William L. Smith

Severe weather can cause significant loss of life and property. Floods, droughts, tropical storms and hurricanes, severe thunderstorms and tornados, extreme hot or cold temperatures, extreme snowfall, and air pollution are major weather-related disasters facing humankind. In this chapter, we review how weather satellite measurements can be used to observe and predict some of these weather-related disasters. Included is a brief discussion of the importance of satellite data for the Global Earth Observation System of Systems (GEOSS), which is a major international initiative to improve environmental predictions. Finally, a new geostationary satellite instrument concept is presented that, when implemented on the international system of geostationary satellites, would provide a major source of the atmospheric state data for the GEOSS.

21.1 Introduction

Weather-related disasters are a primary natural cause of life and property loss. As pointed out in a study by Lott and Ross (2006), the United States alone sustained more than $500 billion in overall inflation-adjusted damages and costs due to these events during the 1980 to 2005 period. Hurricane Katrina alone caused more than $100 billion in damage and more than 1,300 deaths. Within the United States, there are usually several hundred deaths each year due to severe thunderstorms and tornados, tropical storms and hurricanes, and heavy flooding. When a severe drought/heat wave occurs, there can be tens of billions of dollars in damage and several thousand deaths. On a worldwide basis, these statistics are even more startling, particularly for nations in north Africa that suffer from severe drought and nations in south Asia that often suffer from severe flooding.

The statistics and sociological impacts of such natural disasters are covered in detail elsewhere in this book. This chapter highlights a few of the ways that weather satellites can help observe and predict some of these weather-related disasters. It is not an exhaustive summary, but it does provide references to more detailed papers on the subject. A few examples of the use of current weather satellites are presented, and the promise of future environmental satellite systems for improving the observing and prediction capability is discussed.

21.2 Weather satellite measurements

Environmental satellites generally possess two types of instruments for making weather observations: (1) visible, infrared, and microwave "imagers"; and (2) infrared and microwave "sounders." Imagers generally differ from sounders in terms of their spatial and spectral resolution. Imagers traditionally possess high horizontal resolution but relatively poor spectral resolution, whereas the sounders traditionally possess relatively high spectral resolution at the expense of relatively high spatial resolution. Visible and infrared imagers are used to observe clouds and associated storm systems; land surface properties such as temperature and vegetation cover and health, sea surface temperature, and aerosol concentration; and upper-level water vapor patterns related to the large-scale circulation of the atmosphere. Microwave imagers are used to observe precipitation, sea state and ocean surface winds, and the liquid and vapor water content of the atmosphere.

Sounders observe the radiation from the atmosphere in narrow spectral channels of an atmospheric absorption band. The radiation observed as a function of the wavelength of measurement can be interpreted in terms of the temperature or the absorbing gas mixing ratio as a function of altitude. The radiance to space from a cloudfree atmosphere observed in spectral regions of strong molecular absorption will originate from the upper levels of the atmosphere because the molecules within the upper atmosphere will absorb the radiation from lower levels. In contrast, radiation to space observed in spectral regions of weak molecular absorption will originate from the lower levels of the atmosphere because the molecules in the upper atmosphere will be transparent to the radiation emitted from below. If the absorbing gas molecular concentration is known (e.g., oxygen and carbon dioxide), then the spectral distribution of radiance can be related to the vertical temperature profile. If the temperature profile is known, then the spectral distribution of the radiance across a water vapor absorption band can be related to the vertical profile of water vapor mixing ratio. Generally, both sets of spectral radiance measurements are interpreted in a manner that allows one to infer both the temperature and the water vapor profile simultaneously. Infrared radiances generally provide higher vertical resolution than microwave measurements, but microwave measurements have the ability to sense through nonprecipitating cloud. The vertical resolution achievable with sounding instruments depends on the spectral resolution and the number of spectral channels observed (Smith, 1991).

There are typically two types of satellites used for global and regional weather observations. Polar orbiting satellites, generally orbiting at an altitude of about 1,000 km, with a period of about 90 minutes, are used to obtain global coverage but with a frequency of only twice per day for a single satellite. Polar orbiting weather satellites are usually inserted into a sun-synchronous orbit so that the satellite passes over every point on the Earth at roughly the same local time. The relatively low orbit of a polar orbiting satellite allows meteorological data to be collected by instruments at a relatively high spatial resolution. The high spatial resolution combined with global coverage allows polar orbiting systems to provide real-time environmental data for initializing global weather forecast models.

Geostationary satellites, orbiting around the equator at an altitude of 36,000 km to match the rotation velocity of the Earth, remain parked over fixed location of the equator. From this geographically "fixed" position, imagers and sounders can achieve very high temporal resolution as needed to watch storm systems evolve and move. However, the geographic coverage from a single geostationary satellite is limited to about 60 degrees of great circle arc (i.e., 60 degrees of latitude along the satellites subpoint longitude). It is the imagery from the geostationary satellites, along with ground-based radar data, that are generally used to provide the public with warnings of severe thunderstorms and tornados, as well as the landfall position and time of tropical storms and hurricanes. Also, by observing cloud and water vapor movement across the Earth from a geostationary satellite, wind velocities can be estimated. Generally, the temperature and water vapor profile information from the polar satellites and the "wind" determinations from the geostationary satellites provide the basic atmospheric state observations used with other surface-based and balloon-borne atmospheric profile observations to initialize global numerical weather prediction models.

Figure 21.1 Visible cloud (*left panel*), infrared cloud (*upper right panel*), and water vapor (*lower right panel*) observed with the METEOSAT second-generation satellite.

An example of visible and infrared cloud and water vapor imagery obtained from the European geostationary satellite, METEOSAT, is provided in Figure 21.1.

In addition to the operational geostationary and polar orbiting satellites, there are a large number of research satellites contributing unique information about the global environment. For example, the innovative U.S.–Taiwan COSMIC constellation of six low-orbiting satellites now provides high vertical resolution atmospheric data daily in real time at thousands of points on Earth for both research and operational weather forecasting. The temperature profile is derived from a vertical array of bending angles, dependent on the density profile of the atmosphere, as observed by satellite receivers of radio signals sent by the U.S. global positioning system (Anthes et al., 2000), as the receiver satellites occult the Earth's atmosphere. The National Aeronautics and Space Administration's (NASA's) A-train consists of six satellites flying in formation that provide unique active (i.e., Radar and Lidar) and passive infrared and microwave spectrometer and solar reflectance and polarization measurements of clouds and aerosols (Anderson et al., 2005), as well as atmospheric chemistry (Santee et al., 2005) and improved quality atmospheric temperature and moisture profile data (Chahine et al., 2006) that are improving weather forecasts (Le Marshall et al., 2006). These are but two examples of numerous research satellites, of many nations, that are obtaining measurements of the Earth's environment.

21.3 Global Earth Observation System of Systems

The GEOSS was initiated on 16 February 2005, when sixty-one countries agreed to cooperate and build, over the next 10 years, a system that will revolutionize our understanding of the Earth's environment and enable us to predict its future state. Member countries of the Group on Earth Observations at the Third Observation Summit held in Brussels agreed to a 10-year implementation plan for the GEOSS. Nearly forty international organizations also support the emerging global network. The GEOSS project will help all nations involved produce and manage their information in a way that benefits the environment and humanity (see www.epa.gov/geoss/).

The GEOSS focuses on nine societal benefit areas:

- Improve weather forecasting.
- Reduce loss of life and property from disasters.
- Protect and monitor our ocean resource.
- Understand, assess, predict, mitigate, and adapt to climate variability and change.
- Support sustainable agriculture and forestry and combat land degradation.
- Understand the effect of environmental factors on human health and well-being.
- Develop the capacity to make ecological forecasts.
- Protect and monitor water resources.
- Monitor and manage energy resources.

The challenge to achieving the benefits of the GEOSS is the ability to ingest data from all environmental monitoring sensors around the globe to create an accurate global model of the Earth system, where the observations and the model can communicate with each other. Observations from moored and free-floating buoys, land-based environmental stations, and research and operational environmental satellites orbiting the globe will need to be continuously assimilated into a sophisticated model of the global environment. The model must contain the physics of hydrological processes, on all scales, the physics of surface/atmosphere energy exchange processes, a coupling of chemistry and dynamics, as well as accurate numerical procedures for performing both long-term climate and shorter-term numerical weather forecasts. The time and space gaps in the observations will be filled in by short-term predictions provided by the numerical model, so that, at any instant in time, the environmental conditions will be known at every location on Earth. Both global and regional scale models will be able to be initialized with the globally available environmental conditions in order to provide long- and short-term predictions of weather and other environmental phenomena (e.g., air pollution) for any point on the globe. The global model fields will also be able to be used to enable monthly, seasonal, and annual climate predictions on a worldwide basis. Figure 21.2 shows a schematic of the conceptual data system to be used for the GEOSS.

21.4 The current and planned space component

The current space component of the global observing system consists of polar, geostationary, and low inclination satellites from many different nations, as shown in Figure 21.3. Throughout the next decade, China, Europe, Japan, Korea, Russia, and the United States will maintain geostationary and polar orbiting operational environmental satellites. In addition to these operational satellites, there will be many research and development satellites that will contribute unique information about the environment to the GEOSS. These satellites will carry advanced sensors for monitoring the state of the atmosphere, as well as for observing the state of the sea and the land surface. Atmospheric temperature, moisture, clouds, aerosols, ozone, carbon monoxide, and other trace gases will be observed with high spatial resolution from low Earth orbiting satellites and with high temporal resolution from geostationary satellites. Measurements of sea surface wind, temperature,

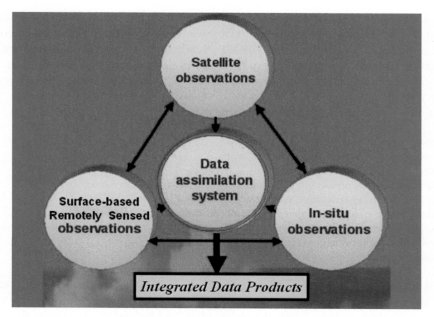

Figure 21.2 An example schematic of a data system to be used for Global Earth Observation System of Systems (GEOSS).

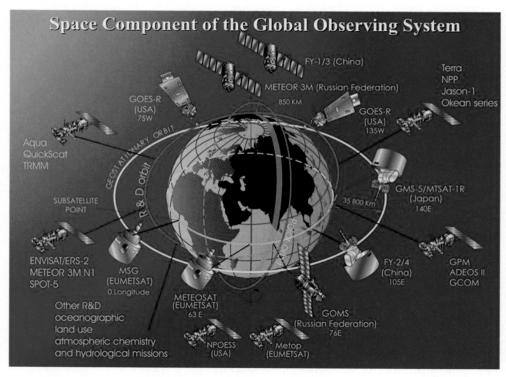

Figure 21.3 Space component of the global observing system (www.wmo.ch/web/www/images/GOS/Figure%20II-10%20Satellites.jpg).

and topography, as well as land surface temperature, soil moisture, and vegetation state will also be observed. The environmental applications include meteorology, climate, ocean, land, and ecology. The data will be communicated in a common format in order to easily blend the observations from the satellites at different altitudes and from different nations.

The National Polar-orbiting Operational Satellite System (NPOESS) provides an example of the satellite measurement capability expected during the next decade. The environmental data record to sensor mapping, shown in Figure 21.4, provides a summary of the measurement capability planned for the NPOESS system. The sensor acronyms are as follows:

- VIIRS (visible/infrared imager/radiometer suite) obtains visible and infrared radiometric data of the Earth's atmosphere, ocean, and land surfaces.
- CMIS (conical microwave imager/sounder) provides global microwave radiometry and sounding data to produce microwave imagery and other meteorological and oceanographic data.
- CrIS (cross-track infrared sounder) observes the spectrum of Earth's radiation with high spectral resolution to determine the vertical distribution of temperature, and moisture in the atmosphere in-between and above clouds.
- OMPS (ozone mapping and profiler suite) observes the vertical and horizontal distribution of ozone in the Earth's atmosphere.
- SESS (space environment sensor suite) observes the neutral and charged particles, electron and magnetic fields, and optical signatures of aurora.
- APS (aerosol polarimeter sensor) provides aerosol and cloud parameters using multispectral photo polarimetry.
- ATMS (advanced technology microwave sounder), in conjunction with CrIS, provides global observations of temperature and moisture profiles for the clouded, as well as clear, atmosphere.
- ERBS (Earth radiation budget sensor) measures Earth radiation budget parameters using instruments similar to the ERBE and CERES heritage instruments.
- ALT (radar alttimeter) is used to measure sea surface topography ocean surface topography to an accuracy of 4.2 cm.
- TSIS (total solar irradiance sensor) is a total solar irradiance monitor plus a 0.2 to 2 microns solar spectral irradiance monitor.

The NPOESS Preparatory Program (NPP) satellite and two NPOESS satellites will compliment the operational European polar satellite METOP to provide a total of three satellites in sun-synchronous orbit with four different equator-crossing times (i.e., NPP at 10:30, C1 at 13:30, C2 at 17:30, and METOP at 21:30 ascending nodes). The first of the METOP series, METOP-A, was launched on 7 October 2006, carrying the highly advanced infrared atmospheric sounding interferometer with more than 8,000 spectral channels for meteorological and chemistry sounding of the Earth's atmosphere. The NPP and the NPOESS C1 and C2 satellites will be launched in 2009, 2013, and 2016, respectively. Replacement of the METOP and NPOESS satellites is planned to take place in 4- to 5-year increments, depending on the lifetime of the METOP and NPOESS spacecraft. Consequently, beginning in 2016, the polar orbiting environmental satellite system will provide high vertical resolution sounding observations over the entire Earth every 4 hours for their assimilation into global numerical weather prediction (NWP) models. Thus, the NPOESS and METOP systems will play their most important roles in the GEOSS during its initial years of operation.

21.5 Vegetation index

Polar orbiting satellites collect imagery worldwide with daytime and nighttime coverage of most areas of the globe. The normalized difference vegetation index (*NDVI*) is widely

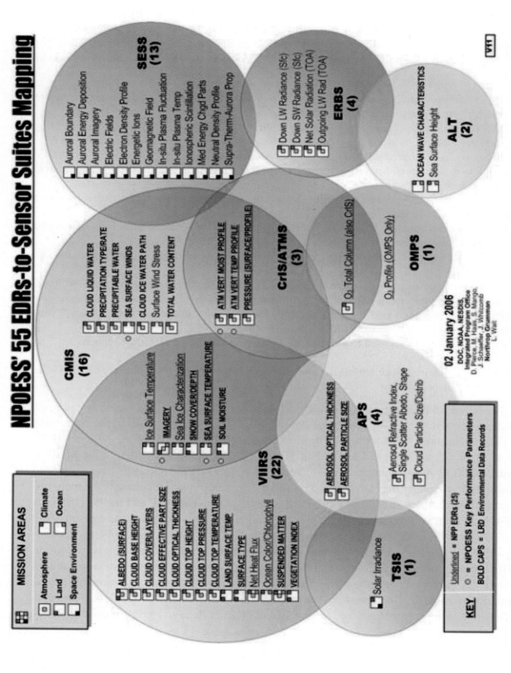

Figure 21.4 The mapping of NPOESS sensor to environmental data record (EDR) (www.ipo.noaa.gov/Science/sensorToERD.html).

used for vegetation monitoring (Kogan, 1998, 2000). The *NDVI* is calculated as

$$NDVI = \frac{(NIR - RED)}{(NIR + RED)} \tag{21.1}$$

where *NIR* is the reflectance radiated in the near-infrared waveband (e.g., 0.725–1.0 microns), and *RED* is the reflectance radiated in the visible red waveband (e.g., 0.58–0.68 microns) of the satellite radiometer. This ratio provides an indicator such that the higher the *NDVI*, the greater the level of photosynthetic activity in the vegetation. It has been extensively used for vegetation monitoring, crop yield assessment, and drought detection. It has been demonstrated that a time series of *NDVI* derived from satellite data is a useful tool for monitoring vegetation condition on a regional and global scale.

Figure 21.5 shows a product of *NDVI*, the stress index, provided by the NESDIS STAR website (www.orbit.nesdis.noaa.gov/star/droughtdetection.php). These products came from the operational National Oceanic and Atmospheric Administration (NOAA) satellites based on channels 1 and 2 of the advanced very high-resolution radiometer (AVHRR). For 2005, severe drought conditions are observed in parts of Kenya, Ethiopia, and Somalia for the sixth year in a row. These conditions cause starvation, water shortages, widespread crop losses, and disease outbreaks. As shown, the 2005 drought gripped the region known as the Horn of Africa in January and continued to impact areas of eastern Kenya, southeastern Ethiopia, and northern and central Somalia. At stake was the harvest during the agricultural season, March through May, which normally provides enough food to sustain the population through the fall when the next harvest begins.

Future imaging instruments (e.g., Visible/Infrared Imager/Radiometer Suite (VIIRS)) to be flown on future NPOESS satellites will have spatial resolutions better than 500 m, as compared to the 1-km resolution of the AVHRR. Thus, the NDVI will become more useful for crop health monitoring as a result of the higher spatial, as well as temporal, density of the measurements to be obtained during the next decade.

21.6 Flash floods

The use of satellite-based precipitation estimation techniques is important for providing spatially complete fields of rainfall, particularly needed for forecasting flash flood disasters. There are three types of satellite instruments that are used to estimate precipitation. They are the geostationary infrared sensor, the polar orbiting Microwave sensor, and the tropical rainfall measurement mission (TRMM) precipitation radar. The Infrared sensors aboard the geostationary satellites are the ones most useful for predicting flash flood disasters because they provide rainfall estimates with high temporal resolution. The geostationary infrared radiometer detects radiation within the infrared (IR) window wavelengths that provide a measure of the temperature of opaque cloud tops, which is related to the cloud height. The rainfall estimate is based on the assumption that the colder the convective cloud temperatures, the higher the vertical extent of the cloud, and therefore, the cloud must be producing more rainfall. The microwave sensors aboard NOAA, Defense Meteorological Satellite Program (DMSP), and TRMM satellites provide rainfall estimates based on a more physically based relationship than that used for the IR sensor estimation. The radiation received by the microwave sensor is actually emitted and attenuated by cloud liquid water droplets and scattered by suspended ice particles. Rainfall estimation is based on relationships between the actual rainfall rate and the brightness temperature measured at certain spectral frequencies of the microwave radiometer. The TRMM precipitation radar is an active sensor that measures the rainfall intensity from the measured strength of the backscattered radar

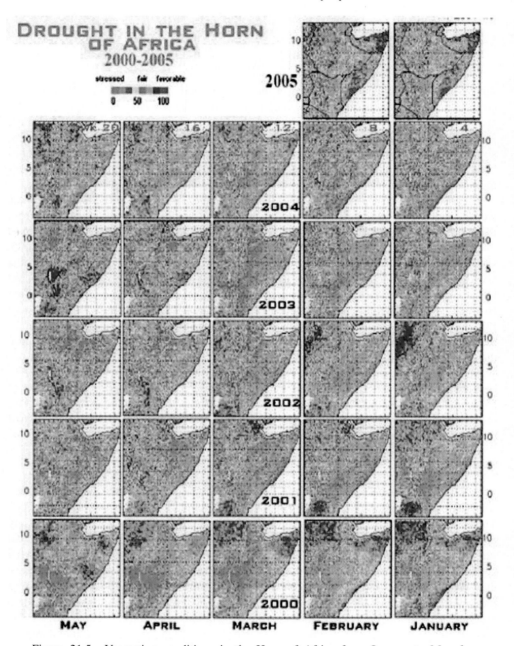

Figure 21.5 Vegetation conditions in the Horn of Africa from January to May for the years 2000 to 2005. The normalized difference vegetation index (*NDVI*) indications of drought conditions are indicated by red shading. (www.orbit.nesdis.noaa.gov/star/droughtdetection.php).

pulse, similar to ground-based radar systems. The best method for obtaining a rainfall estimate at any instant of time is to base quantitative rainfall estimates on all three types of satellite instrument measurements. The timeliest geostationary infrared measurement is used to provide relative updates to the more physically based polar satellite microwave estimates and the low inclination TRMM satellite radar measurements, when available for the region of interest.

Figure 21.6 shows an example of a combined IR and microwave (MW) rainfall technique called SCaMPR (self-calibrating multivariate precipitation retrieval) developed by Kuligowsky (2002). The SCaMPR uses predictor data from three geostationary satellite IR channels (i.e., 6.7 μm water vapor, 11 μm long-wave window, and 12 μm split window channel brightness temperatures), together with precipitation estimates from polar orbiting satellite advanced microwave sounding unit (AMSU) water vapor profile sounders (i.e., AMSU-B) and liquid water and water vapor imagers (i.e., the special sensor microwave imager [SSM/I]). The rain rate calibration is performed by fitting a nonlinear function of geostationary satellite brightness temperatures to the rainfall rate provided by nearly coincident microwave imager and/or sounder observations. These relationships are applied to the high temporal resolution geostationary satellite data until a new simultaneous set of polar satellite MW and geostationary satellite IR data become available.

As shown in Figure 21.6, the SCaMPR product compares well with the radar/rain gauge measurements over the region where these two different types of measurement are available. The advantage of the geostationary satellite product is the superior geographic and temporal coverage achieved in regions void of rain gauge and radar rain rate data. The technique can be applied globally using the international system of geostationary satellites to provide flash flood warnings anywhere in the world.

21.7 Severe thunderstorms and hurricanes

Sounding and imaging instruments on geostationary satellites are used to forecast where thunderstorms will develop and where hurricanes will landfall (Menzel and Purdom, 1994). The current sounder on the U.S. geostationary operational environmental satellite (GOES) is an eighteen-channel IR filter wheel device whose spectral channels are shown in Figure 21.7. The blue bands denote spectral channels used to observe the temperature profile of the atmosphere, the red bands observe atmospheric water vapor, and the green bands are channels used to observe clouds and the Earth's surface. The yellow band is used to provide a total ozone concentration measurement. Although the vertical resolution of these sounding measurements is relatively low (3–5 km, depending on altitude), the hourly resolution data are useful for monitoring the thermodynamic stability of the atmosphere, related to the temperature and moisture concentration of the lower and middle troposphere. By monitoring atmospheric stability, it is possible to predict where thunderstorms and tornados might occur.

Figure 21.8 shows an example of how these data are used to broadly isolate where severe thunderstorms and tornados will develop. The colors denote different values of the atmospheric stability expressed in terms of the lifted index (LI), which is computed from the temperature and moisture profile of the lower half of the troposphere. The red region is where the LI is less than -10, which indicates very unstable air. As can be seen, the LI values observed at 1800 UTC portray where the thunderstorms will develop almost 5 hours later at 2300 UTC. Unfortunately, the vertical resolution of the current instrument is not good enough to pinpoint geographically where the severest weather will develop. However, future high vertical resolution hyperspectral sounders on geostationary satellites,

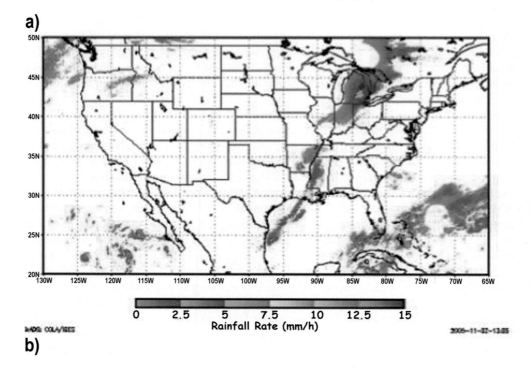
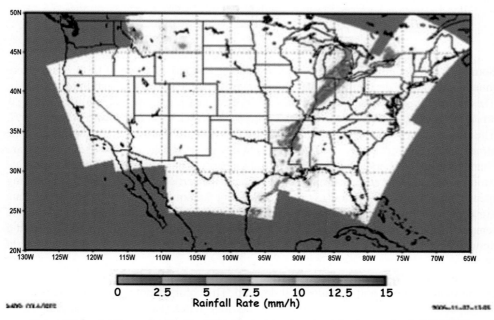

Figure 21.6 (a) Example of self-calibrating multivariate precipitation retrieval (SCaMPR) total rainfall field. (b) Corresponding stage IV radar/rain gauge field. For 0800–0900, 1 November 2005. Adapted from Kuligowsky et al. (2006).

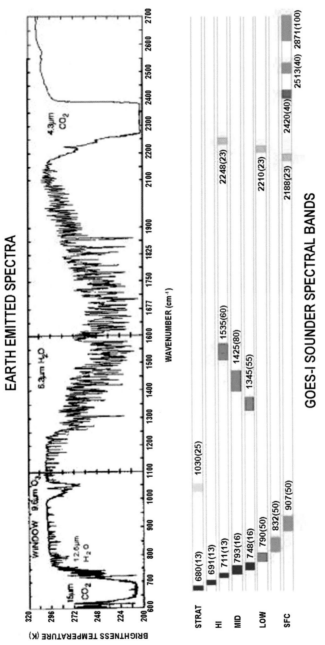

Figure 21.7 Spectral bands of the current U.S. geostationary satellite sounding radiometer.

Figure 21.8 Atmospheric stability (colored regions) in cloudfree atmospheric regions expressed in terms of the lifted index (LI) as derived from geostationary operational environmental satellite (GOES) sounder measurements on 3 May 1999. The noncolored regions are regions of cloudiness where the LI cannot be derived. Courtesy of the Cooperative Institute for Meteorological Satellite Studies, CIMSS, at the University of Wisconsin–Madison.

with a 1–2 km vertical resolution, similar to that now being provided by the atmospheric infrared radiation sounder (AIRS) on the Aqua satellite (Chahine et al., 2006), are expected to provide much higher resolution and more reliable atmospheric stability information for pinpointing the location and timing of severe thunderstorm development. This improvement is discussed in more detail at the end of this chapter.

The geostationary satellite is also used to observe cloud and water vapor winds (Le Marshall, 1998; Stewart et al., 1985; Velden, 1996). These winds are used to initialize large-scale NWP models and to provide steering currents for hurricane track forecasting.

As an example, Figure 21.9 shows the cloud and water vapor winds calculated, at CIMSS, 36 hours before Hurricane Katrina made landfall near New Orleans, Louisiana. The wind barbs overlay an IR image of the upper tropospheric water vapor radiance observed with the GOES imager. Figure 21.10 shows the hurricane steering currents produced from these cloud and water vapor wind vectors. The best estimate of the direction of hurricane movement is given by the axis formed between the southwesterly and the northeasterly wind flow. As can be seen, this axis intersects New Orleans at the point of landfall, which occurred about 24 hours after these steering currents were produced. The use of the geostationary satellite to observe the position and movement of tropical storms and hurricanes enable early warnings to be provided for securing property and evacuating people who are endangered by such storms. Although it is hard to estimate how many lives are saved as a result of geostationary satellite imagery, one can estimate the cost savings from the fact that it costs the American

Figure 21.9   Water vapor winds computed for 00 GMT, 28 August 2005. Courtesy of the CIMSS.

Figure 21.10   Steering currents for Hurricane Katrina computed from the water vapor winds about 26 hours before landfall. Courtesy of the CIMSS.

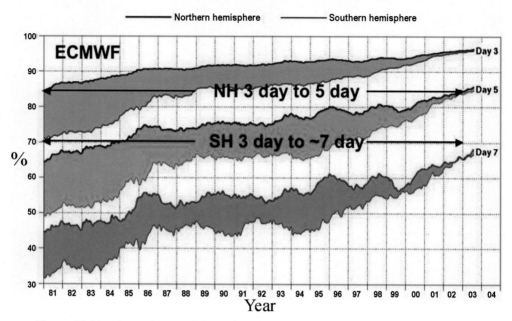

Figure 21.11 Anomaly correlation of the 500 hPa geopotential height for the 2004 European Centre for Medium range Weather Forecasts (ECMWF) model run from a re-analysis of data between 1980 and 2004. Courtesy of ECMWF.

taxpayer about $1 million per mile of coastline warned. Because the average 36-hour landfall prediction error is about 100 miles, each hurricane costs the American taxpayer about $200 million in preparedness. Improved forecasts due to improved model physics or better satellite wind data (e.g., from the advanced geostationary satellite instrument to be discussed next) are expected to cut these errors in half, saving about $100 million per storm, as well as reduce the loss of life as a result of the more dependable landfall forecasts.

21.8 Improvements in the satellite observing system

Improvements since the Global Weather Experiment: Since the global weather experiment observational phase (1 December 1978–30 November 1979), there have been enormous improvements in NWP as a result of improvements in the resolution and physics of the computer models and the manner in which satellite data are assimilated into the forecast system (Bengtsson et al., 2004). Figure 21.11 shows that the NWP has greatly improved as a result of assimilating satellite data into the NWP operation (Uppala et al., 2005). Shown is the anomaly correlation of patterns of 500 hPa geopotential height, which provides a good measure of the large-scale weather systems. The anomaly is the difference between the forecast geopotential height field and an analysis of the actual geopotential field for a given forecast range (3-, 5-, and 7-day forecasts are shown here). Thus, an anomaly correlation of 100 would indicate that the forecast weather pattern is in perfect agreement with the observed weather pattern. An anomaly correlation below 60 is considered to be a useless forecast. In any event, the improvements in the forecast shown here are based on the latest European Centre for Medium range Weather Forecasting (ECMWF) model applied to the data available for the year shown. Thus, the improvements shown here are primarily due to the increase in the density and accuracy of the satellite data used to generate the initial analysis used for the forecast. As can be seen, the forecasts for the Southern Hemisphere, for which satellite data dominates the initial analysis, have become as good as Northern Hemisphere forecasts. The improvements in satellite data between 1980 and 2004 have improved forecasting range by about 2 days in the Northern Hemisphere and by about 4 days in the Southern Hemisphere. As is shown later, another big improvement in the forecasts has resulted since 2004 due to the use of hyperspectral AIRS sounding data.

Hyperspectral resolution sounders: Data from the first of the new series of hyperspectral sounding instruments was orbited on the Aqua satellite on 2 May 2002 (Aumann et. al., 2003). Data from this instrument (Figure 21.12), called AIRS, are being used to observe greenhouse and pollutant gases, as well as to provide high vertical resolution thermodynamic profile data leading to significant improvements in weather prediction (Chahine et al., 2006; Le Marshall et al., 2006). The improvement in the forecast (Figure 21.13) results from the increased vertical resolution provided by the hyperspectral resolution sounder (Smith, 1991). It is noteworthy that a strong positive impact of AIRS data on extended-range forecasts results even though only a relatively small volume of the available clear sky radiance measurements are used. The impact of hyperspectral sounding measurements is expected to be much larger, particularly for short-range forecasts, once cloudy sky radiance measurements are assimilated into the weather analysis/forecast operation. Interferometer spectrometers launched on operational satellites (i.e., METOP IASI, NPP/NPOESS CrIS) will enable the assimilation of cloudy, as well as clear sky, sounding information, which are expected to lead to greater improvements in the accuracy of extended-range numerical forecasts than those already gained from the assimilation of the experimental AIRS clear sky radiance data.

The use of hyperspectral sounding capability to detect and track pollutant gases impacting air quality has also been demonstrated with AIRS data (McMillan et al., 2005, 2006). As shown in Figure 21.12, the AIRS measures the absorption of Earth-emitted IR radiance within the 2180 to 2220 cm^{-1} region of the spectrum with very low brightness temperature noise (i.e., NEdT <0.1 K). These carbon monoxide (CO) radiance measurements, along with the temperature and water vapor profiles derived from the AIRS measurements, are used to estimate the tropospheric profile of CO. It should be noted that the vertical resolution of this measurement is on the order of 5 km, so the profile produced is a vertically smoothed representation of the true CO profile of the atmosphere.

Figure 21.14 shows the results in terms of a CO mixing ratio for the midtroposphere (i.e., 500 mb) for two different days during which there was a strong transport of CO, produced by Alaskan wildfires, eastward across North America to the northeast coast of the United States. As can be seen, the CO cloud disperses as it crosses Canada, with one branch of the cloud propagating southward across Wisconsin, while the main cloud propagates toward the heavily populated regions of southeast Canada and the New England states of the United States. With a single polar orbiting satellite, such maps of toxic atmospheric pollutants can be obtained every 12 hours, and these will prove useful for large-scale, extended-range air quality forecasts. Improved measurements of CO and the ability to monitor other greenhouse gases (e.g., methane) are now being obtained with the high spectral resolution and continuous spectral coverage (3.5–15 μm) infrared atmospheric sounding interferometer (IASI) (Blumstein et al., 2004; Chalon et al., 2001), which was orbited 7 October 2006 aboard the operational METOP satellite.

To obtain higher spatial resolution shorter time scale thermodynamic profile measurements for the forecasting and monitoring of severe weather, as well as for observing

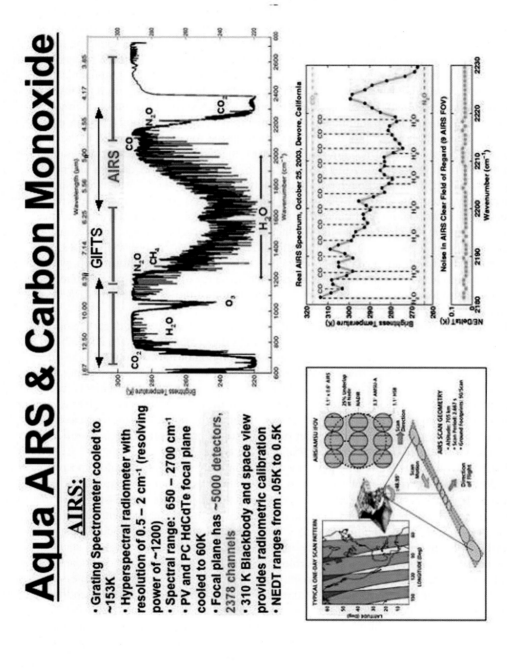

Figure 21.12 The atmospheric infrared radiation sounder (AIRS) instrument radiometric, spectral, and spatial sampling characteristics. A measured spectrum of carbon monoxide (CO) emission is shown in the lower right-hand portion of this figure.

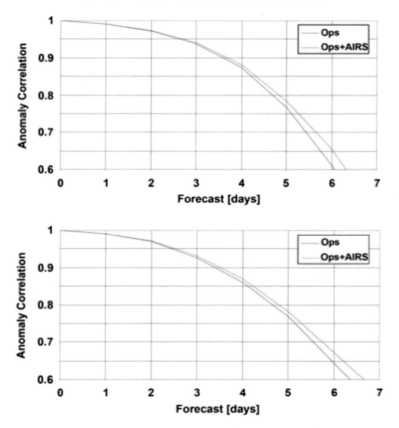

Figure 21.13 (a) Forecast anomaly correlation for the global forecast system with (ops.+atmospheric infrared radiation sounder [AIRS]) and without (Ops) AIRS data in the Southern Hemisphere. (b) Forecast anomaly correlation for the global forecast system with (ops.+AIRS) and without (Ops) AIRS data in the Northern Hemisphere.

the transport of gaseous pollutants, the hyperspectral sounding capability will be implemented on future geostationary satellites. The measurements desired are to be made using a device called the Geostationary Imaging Fourier Transform Spectrometer (GIFTS), shown as an artist conception in Figure 21.15. The GIFTS (Smith et al., 2001a) uses three large area format focal plane detector arrays to observe the Earth and atmosphere in the visible region of the spectrum, with 1-km spatial resolution, and the spectrum of IR radiation, with 0.6 cm^{-1} resolution, in two spectral bands, 680 to 1,150 cm^{-1} (8.7–14.7 μm) and 1,650 to 2,250 cm^{-1} (4.4–6.1 μm), with 4-km footprint size. The field of regard for the GIFTS 512 × 512 visible detector element array and the two 128 × 128 infrared detector element arrays is 524 × 524 km at nadir. A sampling time of 11 seconds is required to achieve the highest spectral resolution (0.6 cm^{-1}) IR radiance measurements. As can be seen from the spectrum shown in Figure 21.12, the GIFTS is able to sample, with high spectral resolution, (1) the long- and short-wave IR radiance emitted by CO_2 and N_2O, used for deriving the temperature profile; (2) the important long-wave IR window spectrum (8.7–12.7 μm) used to observe clouds, the surface characteristics, dust, volcanic aerosol, etc.; (3) the 9.6-μm O_3 emission used to measure both tropospheric and stratospheric ozone concentration; (4) water vapor radiance emission from several hundred water vapor lines of varying strength,

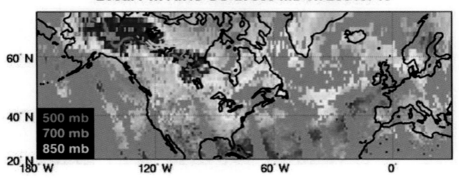

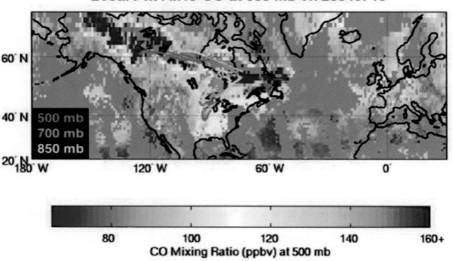

Figure 21.14 Aqua satellite atmospheric infrared radiation sounder (AIRS) instrument measurements of a carbon monoxide (CO) cloud created by Alaskan forest fires. The two images show how the cloud has moved from Alaska to the southeast coast of Canada. The trajectories observed at 500, 700, and 850 mb are also shown. Courtesy of W. McMillan, University of Maryland Baltimore Campus (UMBC).

used to measure the water vapor profile of the atmosphere; and (5) the emission at dozens of lines in the 4.7-μm CO band at much higher spectral resolution than the AIRS (0.6 vs. 2.5 cm^{-1}), enabling more higher vertical resolution retrievals of CO profiles. It is remarkable that the new and revolutionary focal plane detector array and associated detector readout technology used in the GIFTS instrument, enable more than 80,000 high vertical resolution soundings of temperature and water vapor to be observed every minute. Because of the GIFTS capability to image the time variation of the three-dimensional distribution of water vapor with high vertical resolution, wind profiles can be measured by tracking the motion of water vapor features at six to eight distinct altitudes of the troposphere.

Figure 21.16 shows the spectral variation of radiance produced by the absorption and scattering of radiation by dust aerosol. The strong spectral variation between of

21.8 *Improvements in the satellite observing system* 561

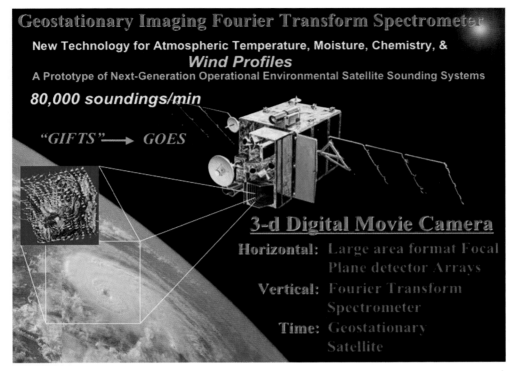

Figure 21.15 An artist conception of a geostationary imaging Fourier transform spectrometer (GIFTS) instrument viewing a hurricane from a geostationary satellite.

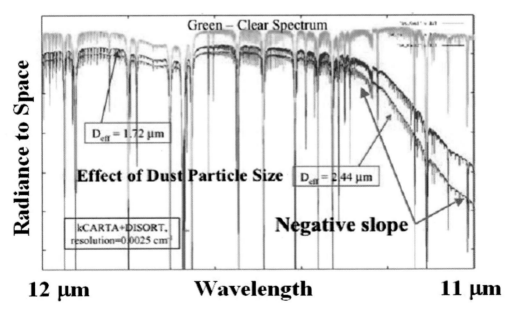

Figure 21.16 The spectral variation of radiance for a clear atmosphere and for two different dust particle sizes.

Figure 21.17 A dust front is shown by the boundary between light purple shading (the dust cloud) and white to blue background (the African desert surface and the Atlantic Ocean) as delineated by the negative difference between the 11-μm and 12-μm channel brightness temperatures measured with the METEOSAT instrument on the European METEOSAT second-generation satellite. Courtesy of EUMETSAT.

radiance between 11 and 11.5 μm is characteristic of dust attenuation, and this variation is unique in that variations of opposite slope are produced by water vapor and cloud attenuation.

The ability to observe and track a dust front from a geostationary satellite with an instrument capable of sensing the unique inverse slope of the spectral radiance is demonstrated in Figure 21.17. A time sequence of these images shows the propagation of this dust cloud from the northeast toward the southwest. The dust cloud is delineated from the African surface and the Atlantic Ocean by the negative differences between the 11-μm and 12-μm channel brightness temperatures measured with the multichannel radiometer on the European geostationary satellite. In the case of the GIFTS, the height of the dust cloud will also be measurable from the depth of the absorption lines, which will be observed within the 10- to 11.5-μm region. The height of the dust layer is important for both air pollution forecasting and for diverting aircraft over, or around, this hazard to safe flight.

The unique GIFTS measurement concept to observe water vapor with high vertical resolution and to provide wind profiles for improved weather prediction has been demonstrated using similar radiance spectral measurements obtained with the NPOESS Airborne Sounding Testbed Interferometer (NAST-I) flying aboard NASA's high-altitude ER-2 aircraft (Fig. 21.18). The spatial detail of the retrieved humidity distribution, shown in Fig. 21.18, is particularly noteworthy. The fine scale vertical details shown in the radiosonde validation data are displayed by the NAST-I cross-section of soundings.

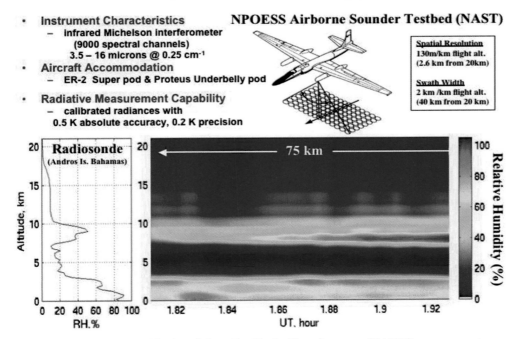

Figure 21.18 NPOESS Aircraft Sounding Testbed Interferometer (NAST-I) measurement characteristics and a 150-km vertical cross-section of water vapor measured near Andros Island, Bahamas. A mean of four different radiosondes launched from Andros Island within 2 hours of the overpass time is shown for comparison.

The primary measurement objective of GIFTS is to obtain high spatial and temporal resolution temperature, moisture, and wind profiles as needed for the prediction of localized severe weather. The wind profiles to be specified from GIFTS data will also be very useful for the initialization of global NWP models, particularly over oceanic regions where there is a void of wind profile data. The polar orbiting satellites currently provide temperature and moisture profile data over the oceans of the world. The wind profiling technique involves retrieving the temperature and water vapor profile from the radiance spectra observed for each field of view of the instrument and at multiple time steps of the measurement for the same area of the Earth. The result is high vertical resolution (1–2 km) temperature and water vapor mixing ratio profiles obtained using rapid profile retrieval algorithms (Smith et al., 2005; Zhou et al., 2002). For GIFTS, the profiles are obtained on a 4-km grid and then converted to relative humidity profiles. Images of the horizontal distribution of relative humidity for atmospheric levels, vertically separated by approximately 2 km, are constructed for each spatial scan (see Figure 21.19 for an example of the three-dimensional relative humidity to be retrieved from GIFTS data). The sampling period will range from minutes to an hour, depending on the spectral resolution and the area coverage selected for the measurement. Successive images of clouds and the relative humidity for each atmospheric level are then animated to reveal the motion of small-scale thermodynamic features of the atmosphere (Smith et al., 2001b; Velden et al., 2004). Automated correlation feature tracking programs (Nieman et al., 1997; Velden et al., 1997) are then used to compute the speed and direction of movement of these small-scale features, providing a measure of the wind velocity distribution at each atmospheric level. The net result is a dense grid of temperature, moisture, and wind profiles that can also be used for atmospheric analyses and operational weather prediction.

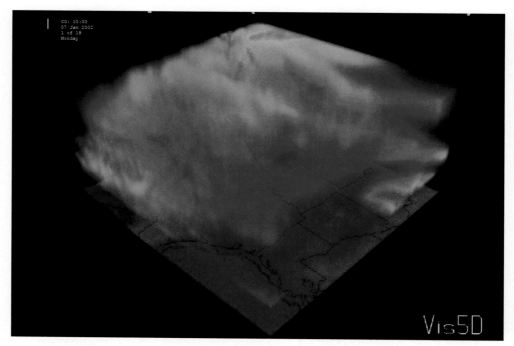

Figure 21.19 An example of the three-dimensional water vapor distribution to be observed with the Geostationary Imaging Fourier Transform Spectrometer (GIFTS) using a regional sample of 3 × 3 focal plane array frames, each covering an area of about 512 km on a side. Courtesy of CIMSS.

This wind measurements concept has been tested using simulated and aircraft data (Velden et al., 2004). GIFTS data were simulated using the GIFTS forward radiative transfer model applied to the temperature and water vapor mixing ratio fields produced by the MM5 (Mesoscale Meteorological Model, version 5) forecast model for 12 June 2002 over the southern Great Plains of the United States. Although generally clear sky conditions were prevalent prior to convective initiation, a cloud mask was added to eliminate data below the level of any detectable cloud tops in all of the moisture fields (therefore, no vectors are attempted beneath clouds). Three time steps at 30-minute intervals were used to track moisture gradients on pressure-resolved horizontal planes at 50-mb increments from 1,000 to 350 mb. A visualization of the resultant wind fields is given in Figure 21.20. The resultant wind field coverage is a significant improvement over what is currently achievable from geosynchronous satellites, especially given the profile nature of the GIFTS wind information.

The ability to observe wind profiles by tracking the motion of water vapor at different altitudes has also been demonstrated using the airborne NAST-I data obtained during a NASA ER-2 aircraft mission off the coast of California on 11 February 2003. About eight race tracks were flown, each taking about 30 minutes, which allowed water vapor imagery to be produced from the relative humidity profiles. A Doppler Wind Lidar (DWL) was also flown on a Twin Otter aircraft (G.D. Emmitt, personal communication) 2 November 2003 beneath the NAST-I, to validate the accuracy of the NAST-I produced winds.

Figure 21.21 shows an example comparison of the wind profile derived from three consecutive frames of NAST-I water vapor imagery and the mean Doppler LIDAR wind observed during the 1.5-hour measurement period. As can be seen, there is excellent agreement in the wind direction and reasonable agreement in the wind speed for these two totally independent measures of the velocity profile. The comparison is limited to the lower

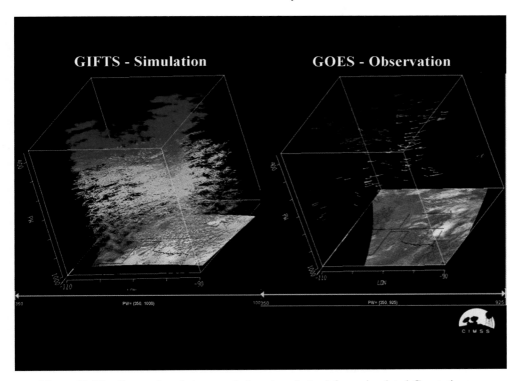

Figure 21.20 Comparison between wind vectors derived from simulated Geostationary Imaging Fourier Transform Spectrometer (GIFTS) hyperspectral sounding data and those available from the current geostationary operational environmental satellite (GOES) satellite over the Great Plains states on 12 June 2002. Courtesy of CIMSS.

troposphere due to limitations in the signal to noise of the Doppler LIDAR above 3 km altitude. In the case of the radiometrically derived wind profiles, the accuracy is expected to increase with increasing altitude throughout the troposphere.

Aside from the ability to produce wind profiles for NWP and hurricane track and landfall location and time predictions, the GIFTS high vertical and temporal resolution soundings over land areas will enable revolutionary improvements in severe thunderstorm and tornado forecasts. This is because GIFTS will enable the detection of areas of decreasing atmospheric stability where there is considerable moisture convergence in the lowest levels of the atmosphere. As a consequence, the location where intense convective storms will develop will be observable from the otherwise invisible moisture fields long before clouds and precipitation occur when the storms can be seen in visible cloud imagery and on radar. It is estimated that the GIFTS will be able to identify the location where severe convective storms will develop as much as an hour before they occur, thereby providing significant warning time to decrease death, injury, and property loss due to these storms.

21.9 Summary

Satellite data play a key role in the understanding, modeling, and prediction of weather phenomena. The international GEOSS will provide Earth data for predicting environmental phenomena, including natural hazards, which cause loss of life and property. Hyperspectral remote sensing spectrometers being implemented on polar orbiting satellites and

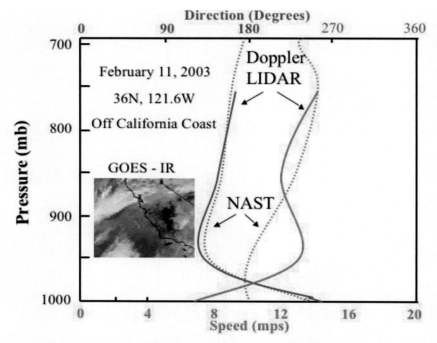

Figure 21.21 An example comparison of the wind profile derived from three consecutive frames of NPOESS aircraft sounding testbed-interferometer water vapor imagery and the mean Doppler LIDAR wind observed during the 1.5-hour NAST-I measurement period. The red X on the insert shows the geographical location of the wind profile.

high-resolution imaging spectrometers being implemented on geostationary satellites will provide four-dimensional thermodynamic and wind sounding data that will revolutionize the observation, modeling, and prediction of hazardous weather systems on a global scale. Thus, the world population will benefit greatly through an improved quality of life as a result of the technology advances being made in the global environmental satellite observing system during the next decade, and beyond.

Acknowledgments

I want to acknowledge the support of the Integration Program Office for the National Polar-orbiting Operational Satellite System (NPOESS) and the National Aeronautics and Space Administration (NASA) Langley Research Center for their support of the preparation of this chapter.

References

Anderson, T. L., Charlson, R. J., Bellouin, N., Boucher, O., Chin, M., Christopher, S. A., Haywood, J., Kaufman, Y. J., Kinne, S., Ogren, J. A., Remer, L. A., Takemura, T., Tanré, D., Torres, O., Trepte, C. R., Wielicki, B. A., Winker, D. M., and Yu, H. (2005) "An 'A-Train' Strategy for Quantifying Direct Climate Forcing by Anthropogenic Aerosols," *Bull. Amer. Meteorol. Soc.* **86**, pp. 1795–1809.

Anthes, R. A., Rocken, C., and Kuo, Y. H. (2000) "Application of COSMIC to Meteorology and Climate," *Terrestrial, Atmospheric, and Oceanic Science* **11**, pp. 115–156.

Aumann, H. H., Chahine, M. T., Gautier, C., Goldberg, M. D., Kalnay, E., McMillin, L. M., Revercomb, H., Rosenkranz, P. W., Smith, W. L., Staelin, D. H., Strow, L. L., and Susskind, J.

(2003) "AIRS/AMSU/HSB on the Aqua Mission: Design, Science Objectives, Data Products, and Processing Systems," *IEEE Transactions on Geoscience and Remote Sensing* **41**, pp. 253–264.

Bengtsson, L., Doos, B. R., Uppala, S., Tennebaum, J., Coughlan, M. J., Newson, R. L., Smith, N. R., Zillman, J. W., Shapiro, M. A., Thorpe, A. J., Anthes, R. A., and Cornord, S. G. (2004) "The Global Weather Experiment: The 25th Anniversary," *World Meteorological Organization Bulletin* **53**, pp. 191–230.

Blumstein, D., Chalon, G., Carlier, T., Buil, C., Hébert, P., Maciaszek, T., Ponce, G., Phulpin, T., Tournier, B., Siméoni, D., and Tournier, B. (2004) "IASI Instrument: Technical Overview and Measured Performances," SPIE Conference, Denver, Colorado, SPIE 2004-5543-22, August.

Chahine, M. T., Pagano, T. S., Aumann, H. H., Atlas, R., Barnet, C., Blaisdell, J., Chen, L., Divakarla, M., Fetzer, E. J., Goldberg, M., Gautier, C., Granger, S., Hannon, S., Irion, F. W., Kakar, R., Kalnay, E., Lambrigtsen, B .H., Lee, S. -Y, Le Marshall, J., McMillan, W. W., McMillin, L., Olsen, E. T., Revercomb, H., Rosenkranz, P., Smith, W. L., Staelin, D., Strow, L. L., Susskind, J., Tobin, D., Wolf, W., and Zhou, L. (2006) "AIRS: Improving Weather Forecasting and Providing New Data on Greenhouse Gases," *Bulletin of the American Meteorological Society* **87**, pp. 911–926.

Chalon, G., Cayla, F., and Diebel, D. (2001) "IASI: An Advanced Sounder for Operational Meteorology," *Proceedings of the 52nd Congress of IAF*, Toulouse, France, 1–5 October.

Kogan, F. N. (1998) "A Typical Pattern of Vegetation Conditions in Southern Africa During El Niño Years Detected from AVHRR Data Using Three-Channel Numerical Index," *Int. J. Rem. Sens.* **19**, pp. 3688–3694.

Kogan, F. N. (2000) "Satellite-Based Sensitivity of World Ecosystems to El Niño/La Niña," *Rem. Sens. Environ.* **74**, pp. 445–462.

Kuligosky, R. J. (2002) "A Self-Calibrating GOES Rainfall Algorithm for Short-Term Rainfall Estimates," *J. Hydrometeorology* **3**, pp. 112–113.

Kuligowsky, R. J., Qiu, S., and Im, J.-S. (2006) "Global Application of the Self-Calibrating Multivariate Precipitation Retrieval (SCaMPR)," *20th Conference on Hypdrology*, Atlanta, Georgia, January.

Le Marshall, J. (1998) "Cloud and Water Vapour Motion Vectors in Tropical Cyclone Track Forecasting—A Review," *Meteorology and Atmospheric Physics* **65**, pp. 141–151.

Le Marshall, J., Jung, J., Derber, J., Chahine, M., Treadon, R., Lord, S. J., Goldberg, M., Wolf, W., Liu, H. C., Joiner, J., Woollen, J., Todling, R., van Delst, P., and Tahara, Y. (2006) "Improving Global Analysis and Forecasting with AIRS," *Bull. Amer. Meteorol. Soc.* **87**, pp. 891–894.

Lott, N., and Ross, T. (2006) "Tracking and Evaluating U.S. Billion Dollar Weather Disasters, 1980–2005," *The 86th AMS Annual Meeting, Atlanta Georgia*, 28 January–3 February, www.ncdc.noaa.gov/oa/reports/billionz.html.

McMillan, W. W., Warner, J. X., Comer, M. M., Maddy, E., Chu, A., Sparling, L., Eloranta, E., Hoff, R., Sachse, G., Barnet, C., Razenkov, I., and Wolf, W. (2006) "AIRS Views of Transport from 12–22 July 2004 Alaskan/Canadian Fires: Correlation of AIRS CO and MODIS AOD and Comparison of AIRS CO Retrievals with DC-8 In Situ Measurements During INTEX-NA/ICARTT," *J. Geophys. Res.*

McMillan, W. W., Barnet, C., Strow, L., Chahine, M. T., McCourt, M. L., Novelli, P. C., Korontzi, S., Maddy, E. S., and Datta, S. (2005) "Daily Global Maps of Carbon Monoxide: First Views from NASA's Atmospheric Infrared Sounder," *Geophys. Res. Lett.*, doi:10.1029/2004GL012821.

Menzel, W. P., and Purdom, J. F. W. (1994) "Introducing GOES-I: The First of a New Generation of Geostationary Operational Environmental Satellites," *Bull. Amer. Meteorol. Soc.* **75**, pp. 757–781.

Nieman, S., Menzel, W. P., Hayden, C. M., Gray, D., Wanzong, S., Velden, C. S., and Daniels, J. (1997) "Fully Automated Cloud-Drift Winds in NESDIS Operations," *Bull. Amer. Meteorol. Soc.* **78**, pp. 1121–1133.

Santee, M. L., Manney, G. L., Livesey, N. J., Froidevaux, L., MacKenzie, I. A., Pumphrey, H. C., Read, W. G., Schwartz, M. J., Waters, J. W., and Harwood, R. S. (2005) "Polar Processing and Development of the 2004 Antarctic Ozone Hole: First Results from MLS on Aura," *Geophys. Res. Lett.* **32**, L12817, 10.1029/2005GL022582.

Smith, W. L. (1991) "Atmospheric Soundings from Satellites—False Expectation or the Key to Improved Weather Prediction?," *Quart. J. Roy. Meteorol. Soc.* **117**, pp. 267–297.

Smith, W. L., Harrison, F. W., Revercomb, H. E., and Bingham, G. E. (2001a) "Geostationary Imaging Fourier Transform Spectrometer (GIFTS)—The New Millenium Earth Observing-3 Mission," *Proceedings of the International Radiation Symposium: Current Problems in Atmospheric Radiation*, St. Petersburg, Russia, 24–29 July 2000, pp. 81–84.

Smith, W. L., Harrison, W., Hinton, D., Parsons, V., Larar, A., Revercomb, H., Huang, A., Velden, C., Menzel, P., Peterson, R., Bingham, G., and Huppi, R. (2001b) "GIFTS—A System for Wind Profiling from Geostationary Satellites," *Proceedings of the 5th International Winds Workshop*, Lorne, Australia, 28 February–3 March, European Organization for the Exploitation of Meteorological Satellites, Darmstadt, Germany, pp. 253–258.

Smith, W. L., Zhou, D. K., Larar, A. M., Mango, S. A., Knuteson, R. B., Revercomb, H. E., and Smith, W. L., Jr. (2005) "The NPOESS Airborne Testbed Interferometer—Remotely Sensed Surface and Atmospheric Conditions During CLAMS," *J. Atmos. Sci.* **62**, pp. 1118–1134.

Stewart, T. R., Smith, W. L., and Hayden, C. M. (1985) "A Note on Water-Vapor Wind Tracking Using VAS Data on McIDAS," *Bull. Amer. Meteorol. Soc.* **66**, pp. 1111–1115.

Uppala, S. M., Kållberg, P. W., Simmons, A. J., Andrae, U., da Costa Bechtold, V., Fiorino, M., Gibson, J. K., Haseler, J., Hernandez, A., Kelly, G. A., Li, X., Onogi, K., Saarinen, S., Sokka, N., Allan, R. P., Andersson, E., Arpe, K., Balmaseda, M. A., Beljaars, A. C. M., van de Berg, L., Bidlot, J., Bormann, N., Caires, S., Chevallier, F., Dethof, A., Dragosavac, M., Fisher, M., Fuentes, M., Hagemann, S., Hòlm, E., Hoskins, B. J., Isaksen, L., Janssen, P. A. E. M., Jenne, R., McNally, A. P., Mahfouf, J.-F., Morcrette, J.-J., Rayner, N. A., Saunders, R. W., Simon, P., Sterl, A., Trenberth, K. E., Untch, A., Vasiljevic, D., Viterbo, P., and Woollen, J. (2005) "The ERA-40 Re-Analysis," *Quart. J. Roy. Meteorol. Soc.* **131**, pp. 2961–3012.

Velden, C. S. (1996) "Winds Derived from Geostationary Satellite Moisture Channel Observations: Applications and Impact on Numerical Weather Prediction," *Meteorology and Atmospheric Physics* **60**, pp. 37–46.

Velden, C. S., Dengel, G., Dengel, R., Huang, A. H.-L., Stettner, D., Revercomb, H., Knuteson, R., and Smith, W. L. (2004) "Determination of Wind Vectors by Tracking Features on Sequential Moisture Analyses Derived from Hyperspectral IR Satellite Soundings," *Proceedings of the 13th Conference on Satellite Meteorology and Oceanography*, Norfolk, VA, 20–23 September, Paper No. P1.10.

Velden, C. S., Hayden, C. M., Nieman, S., Menzel, W. P., Wanzong, S., and Goerss, J. (1997) "Upper Tropospheric Winds Derived from Geostationary Satellite Water Vapor Observations," *Bull. Amer. Meteorol. Soc.* **78**, pp. 173–195.

Zhou, D. K., Smith, W. L., Li, J., Howell, H. B., Cantwell, G. W., Larar, A. M., Knuteson, R. O., Tobin, D. C., Revercomb, H. E., and Mango, S. A. (2002) "Thermodynamic Product Retrieval Methodology for NAST I and Validation," *Applied Optics* **41**, pp. 957–967.

Epilogue
Mohamed Gad-el-Hak

ALONG THE PARIS STREETS, the death-carts rumble, hollow and harsh. Six tumbrils carry the day's wine to La Guillotine. All the devouring and insatiate Monsters imagined since imagination could record itself, are fused in the one realisation, Guillotine. And yet there is not in France, with its rich variety of soil and climate, a blade, a leaf, a root, a sprig, a peppercorn, which will grow to maturity under conditions more certain than those that have produced this horror. Crush humanity out of shape once more, under similar hammers, and it will twist itself into the same tortured forms. Sow the same seed of rapacious license and oppression over again, and it will surely yield the same fruit according to its kind.
(From A Tale of Two Cities *by Charles Dickens)*

> The body's delicate: the tempest in my mind
> Doth from my senses take all feeling else
> Save what beats there. Filial ingratitude!
> Is it not as this mouth should tear this hand
> For lifting food to't? But I will punish home:
> No, I will weep no more. In such a night
> To shut me out! Pour on; I will endure.
> In such a night as this! O Regan, Goneril!
> Your old kind father, whose frank heart gave all,—
> O, that way madness lies; let me shun that;
> No more of that.
> *(From William Shakespeare's* King Lear*)*

What have we accomplished in this book? In a sentence, we provided a broad coverage of the subject of prediction, prevention, control, and mitigation of large-scale disasters. We began by defining a large-scale disaster as an event that adversely affects a large community and/or ecosystem. Such disaster taxes the resources of local and central governments, and typically forces a community to diverge substantially from its normal social structure. That broad definition allowed us to include all types of natural and manmade calamities, including some unconventional—although unfortunately not infrequent—ones such as pollution, energy crisis, global warming, dictatorship, genocide, and ethnic cleansing. We then went on to introduce a universal metric by which all types of disasters can be measured. The proposed scale is logarithmic (i.e., the severity of the calamity increases by an order of magnitude as we move up the scale), and considers the number of people and the extent of the geographic area adversely affected by the catastrophe. The defined scope gives officials a quantitative, dynamic measure of the magnitude of the disaster as it evolves so proper response can be mobilized and adjusted as warranted.

The book was written mostly from a scientific viewpoint and includes chapters on scientific principles and modeling fundamentals for predicting disaster's evolution, issues in disaster relief logistics, perspectives on medical response, root causes of disasters, energy crisis, seawater irrigation, anthropogenic hazards, tsunamis, climate change, water resources, arid land meteorology, numerical weather predictions, climate models, the International Charter, and weather satellite measurements.

Disasters can be modeled as nonlinear dynamic systems, which has implication on the ability to predict and control the calamity under consideration. In contrast to natural disasters, manmade ones are, in general, somewhat easier to control but more difficult to predict. The war on terrorism is a case in point. Who could predict the behavior of a crazed suicide bomber? However, a civilized society spends its valuable resources in the form of intelligence gathering, internal security, border control, and selective/mandatory screening to prevent (control) such devious behavior, whose dynamics obviously cannot be distilled into a differential equation to be solved. Yet, even in certain disastrous situations that depend on human's behavior, predictions can sometimes be made; crowd dynamics being a prime example where the behavior of a crowd in an emergency can to some degree be modeled and anticipated so proper escape or evacuation routes can be designed.

The tragedy of the numerous manmade disasters all around us is that they are all preventable, at least in principle. We cannot prevent a hurricane, at least not yet, but global warming trends could be slowed down by using less fossil fuel and seeking alternative energy sources. Conflict resolution strategies can be employed between nations to avert wars, and so on. Speaking of wars, the Iraqi–American poet Dunya Mikhail, lamenting on those too many manmade disasters, calls the present period "the tsunamical age." A bit more humanity, common sense, selflessness, and moderation and a bit less greed, meanness, selfishness, and zealotry, and the world will be a better place for having fewer manmade disasters.

Modeling disasters as nonlinear dynamic systems carries with it some good and some bad aspects. On the one hand—the bad one—such systems are capable of chaotic behavior and are therefore characterized by extreme sensitivity to initial conditions. Under those circumstances, reliable long-term predictions are not possible. The weather is an example of such initial value problem, while, paradoxically, very long-term predictions of boundary value problems such as the climate become feasible because the system's dependence on initial conditions diminishes asymptotically. Of course, climate predictions are mere trends and cannot by any stretch of the imagination foretell the temperature or chance of rain, say, on a particular date years from now. Such *weather* predictions are confined to a theoretical maximum of 20 days, while climate trends are given in years, decades, and centuries.

On the other hand—the potentially good one—it is conceivable, at least in principle, to exploit the extreme sensitivity to initial conditions in such a way as to achieve significant desired change in a future state by applying small perturbations to the present state. Such chaos control strategies have been demonstrated in simple systems but never yet in a complex, infinite-degrees-of-freedom one such as, for example, the weather. The payoff is tremendous, however. Just think, for instance, of the ability to weaken or even prevent the formation of a future hurricane, or to reverse the devastating effects of global warming for future generations to enjoy an even better environment than the one we enjoy today.

For now, humans can do little to stop most natural large-scale disasters, but timely, efficient response to the calamities can reduce the damage and mitigate at least some of the suffering and despair. The last *annus horribilis*, in particular, has shown the importance of being prepared for large-scale disasters, and how the world can get together to help alleviate the resulting pain and suffering. In its own small way, this book will hopefully better prepare scientists, engineers, physicians, first responders, and, above all, politicians to deal with manmade and natural disasters.

Once disaster strikes, mitigating its adverse effects becomes the primary concern. Issues such as how to save lives, take care of the survivors' needs, and protect properties from any further damage are at the forefront. Dislocated people need shelter, water, food, and medicine. Both the physical and mental health of the survivors, as well as relatives of the deceased, can be severely jeopardized. Looting, price gouging, and other law-breaking activities need to be contained, minimized, or eliminated. Hospitals need to prioritize and even ration treatments, especially in the face of the practical fact that the less seriously injured tend to arrive at emergency rooms first, perhaps because they transported themselves there. Roads need to be operable and free of landslides, debris, and traffic jams for the unhindered flow of first responders and supplies to the stricken area, and evacuees and ambulances from the same. Buildings, bridges, and roads need to be rebuilt or repaired, and power, potable water, and sewage need to be restored.

From an evolutionary points of view, disasters bring out the best in us, except when there is a profound sense of injustice. It almost has to be that way. Humans survived ice ages, famines, infections, and so on, not because we were strong or fast, but because in a state of extreme calamity, we tend to be resourceful and cooperative. Commenting on a unique museum's exhibition, Caroline Ash writes in *Science* (vol. 317, p. 1869, 28 September 2007), "Although the perpetual threat of disaster makes us fear the unexpected, our imaginations also prepare us to manage disorder and suffering. Consequently, given some period of peace after a catastrophe, societies rapidly regroup and act not only to assist their own members but also to help others, often far distant and quite anonymous, to rebuild their infrastructure and their faiths. The objects that help humans to be resilient are the subject of the exhibition *Scénario Catastrophe*, . . . , currently at the Musée d'ethnographic de Genève."

There is much in common when it comes to preparing for the occasional calamity of whatever type and managing the resulting mess. Open lines of communication among all concerned, efficient command structure, supplies, material and financial resources, and fairness of the system are needed to overcome the adversity and return the community to normal. However, we must be prepared for the rare but inevitable disaster.

What better way to end this book than with the folksy words of wisdom from the respected broadcast journalist Tedd Koppel. In two National Public Radio commentaries delivered on 3 and 18 August 2006, he affirmed that Americans (and by extension everyone else, I presume) are not ready for major disasters. Koppel lamented, "Preparing for a disaster is neither rocket science nor brain surgery. It is making sure that people at the grass-roots level know what to do. There are some very simple things we could be doing that would cover a variety of catastrophes." The British-born Koppel ended with this enlightening story, "I was a little boy in London during the Second World War and one of my earliest memories is of my middle-aged father and a neighbor going off every night to patrol our street, carrying a garbage can lid and a broom. The Germans were dropping incendiary bombs back then. These would land on a rooftop, burn through and then set the house on fire. But one man could sweep them off the roof without much risk, and a second man with a garbage can lid could smother the bomb and render it harmless. Simple. It could be handled at the neighborhood level by a couple of middle-aged civilians. But they had to know what to do. And someone had to tell them." One hopes that certain officials who listened to Mr. Koppel and perhaps even read this book would tell us what to do. And that includes rocket scientists, brain surgeons, and even respected journalists.

Index

Abbe, Cleveland, 430
Advanced Research WRF (ARW), 303
aerosol, 221
    parameterization of, 305
Al-Salam Boccaccio ferry, sinking of, 56
anthropocene, geological period, 9
atmospheric mesoscale motions, 295
atmospheric predictability, 449

biomass, 213
bird flu, 59
boundary layer theory, 20
butterfly effect, 27, 319

carbon dioxide, 183
    emission mitigation, 194
casualty triage, 153
cellular automata, 85
chaos control, 28
    OGY strategy, 28
chemical accident, Bhopal, 77
climate, 318, 329
    change, 331, 363
    components of, 71
    definition of, 72, 330
    extreme events, 329, 368
    feedback, 331
    flooding, 346
    heat waves, 347
    length scales, 71
    long-term simulation, 72
    model, 330
    modeling of, 72, 186
    outlook, 331
    prediction, 320, 331
    projection, 331
    scenario, 331
    system, 330
    time scales, 71
    variability, 331
    warming, 347
    wind storm, *see* hurricanes
climate change
    precipitation, 373
climate change, impact on water cycle, 363
cloud physics, 450
coarse-grained methods, 84
Community Climate Model, 72
Community Climate System Model (CCSM), 367

community involvement, 174
complexities, flow, 23
compressibility, 17
constitutive relations, 15
contingency estimating, 165
contingency planning, 165
continuum methods, 88
cooperative regional capacity augmentation, 171
coordination issues, 130
Critical Path Method (CPM), 283
Crutzen, Paul, 9
cumulus parameterization, 450

data collection systems (DCS), 455
day-to-day prediction, 449
desert climate, 377
desert weather, 377
desertification, 396
    anthropogenic contributions, 401
    definition of, 396
    degrees of, 398
    detection, 417
    extent of, 399
    mapping, 417
    natural contributions, 409
    physical-process feedbacks, 414
disaster
    conceptual scale, 8
    coordination issues, 130
    corruption, 139
    discrimination, 139
    ethical issues, 139
    evaluation, 158
    funding issues, 126
    information technology, 136
    local resources, 137
    logistical challenges, 123
    management, 31, 149, 454
    management of information, 129
    metric, 2, 7, 8
    military use, 140
    mitigation, 151
    needs assessment, 127
    network design, 132
    operational issues, 132
    personnel issues, 134
    phases of, 150
    political issues, 140
    procurement, 127

disaster (cont.)
  psychological definition of, 156
  relief issues, 122
  relief standardization, 132
  response failure, 159
  scope, 2, 8, 64
  space-based management, 479
  stakeholders, 124
  stressors, 135
  supply chain issues, 123
  transportation infrastructure, 132
Disaster Management Support Group (DMSC), 479
disaster management, remote sensing, 454
disaster prediction strategies, 448
Disaster Severity Scale (DSS), 162
disaster, definition of, 147
disaster, health impact, 148
disaster-related weather, 447
disasters, characteristics of, 148
disasters, classification of, 148
disasters, in literature, 33
disasters, inevitability, 147
disasters, modeling of, 12
disasters, sociology of, 33
diurnal cycle, 366
Doppler Bandwidth (DB), 468
dry deposition, 304
dust storm, *see* sand and dust storms
  hazards, 385
dust storms, 379
  haboob, 388
  mitigation, 390
dynamical climate prediction, 322
dynamical systems, 26
dynamical systems theory, 26

earthquake, 25
  detection, 285
  Sumatra, *see* Sumatra earthquake
earthquake, 1999 Turkey, 136
earthquake, Izmit, 41
earthquake, Kashmir, 49
earthquake, San Francisco, 34
Eckert number, 17
El Niñ, 322
Emergency Management System (EMS), 149
energy
  balance, 184
  consumption, 179
  fossil fuel, 179
energy crisis, 61
equations of motion
  for a fluid, 14
  turbulent flows, 22
ethical issues, 139
Euler equations, 17
explicitly resolved convective systems, 311
extreme weather event, 330

Fisher, Henry, 33
FitzRoy, Robert, 428
flooding, 346
flooding, Central Europe, 534

forecast probability distribution function, 324
fossil fuel, 179
fully-compressible coupled models, 312
fully-coupled online modeling, 303
funding issues, 126

Geostationary Operational Environmental Satellite (GOES), 455
global climate modeling, 72
global climate models, 74
Global Disaster Information Network (GDIN), 480
Global Earth Observation System of Systems (GEOSS), 30, 453
global warming, 184, 347
government role, 173
Grashof number, 17
Group on Earth Observations (GEO), 454

Hajj stampede, 54
health care capacity, 163
  Medical Severity Factor, 164
  Medical Severity Index, 164
heat waves, 347
heterogeneous multiscale method, 93
hospital
  command and control, 172
  management, 172
Hospital Emergency Incident Command System (HEICS), 155
Hospital Treatment Capacity (HTC), 166
humanitarian relief logistics, 121
Hurricane Katrina, 45, 352
Hurricane Rita, 352
Hurricane Wilma, 49
hurricanes, 351
Hyatt Regency walkway collapse, 38

Incident Command System (ICS), 149
incompressible flow equations, 16
influenza pandemic, 59
information technology, 136
inhalation anesthesia, 157
Integrated Global Observing Strategy (IGOS), 454
International Charter, 454, 479
International Charter, Space and Major Disasters, 482
Izmit earthquake, 41

Kashmir earthquake, 49
Katrina, Hurricane, 45, 352
Kolmogorov, 294

large-scale disaster
  definition of, 1, 6, 94
  time scale of, 2, 7
large-scale disasters
  books on, 2
  control of, 10
  coordinating and collaborating, 121
  examples of, 34
  facets of, 10
  journals on, 2
  logistics, 121
  multiscale modeling, 81
  origin of, 5

prediction of, 10
web sites for, 3
lattice Boltzmann method, 90
Lawson's Report, 26
Lawson, Andrew, 26
limits to predictability, 74
local resources, 137
logistical challenges, 123
logistics, humanitarian relief, 121
Lorenz' attractor, 319
Lorenz, Edward, 319

Mach number, 19
management of information, 129
manmade disasters
    causes of, 2, 7
    examples of, 2, 7
Margules, Max, 432
Mass Casualty Incident (MCI), 152
mass trauma casualty predictor, 166
mathematical homogenization, 86
Medical Assistance Chain (MAC), 163, 166
medical disaster, 149
Medical Rescue Capacity (MRC), 165
medical response to disaster, 152
    roles of specialists, 156
Medical Severity Factor, 164
Medical Severity Index, 164
Medical Transport Capacity (MTC), 165
medium-range prediction, 448
mesoscale, 293
meteorology
    Bergen school, 433
    history of, 428
    non divergent barotropic vorticity equation, 436
    prediction, 439
    theoretical, 429
Mikhail, Dunya, 12
military use in disaster relief, 140
model physics, 450
molecular dynamics, 83
Monte Carlo methods, 84
multi-casualty incident (MCI), 149
multiscale modeling, 81

National Disaster Medical System (NDMS), 161
natural disasters
    examples of, 2, 6
    human impact, 97
Navier–Stokes equations, 15
needs assessment, 127
network design, 132
neural networks, 86
nonlinear dynamical systems, 26
nonlinear dynamical systems theory, 26
Normalized Difference Vegetation Index (NDV), 456
Numerical Weather Prediction (NWP), 440
numerical weather prediction model, 447

off-site patient care, 173
offline modeling, 306
offline simulations, 308
online modeling systems, 302
operational issues, 132

Pacific tsunami, 43
personnel issues, 134
photolysis frequencies, 306
political issues, 140
Prandtl number, 18
Prandtl's boundary layer theory, 20
precipitation, 363
preparedness for disaster, 150, 173
procurement, 127
Péclet number, 17

quasi-equilibrium assumption, 14

radiation, 450
regional anesthesia, 157
regional climate models, 74, 329
relief issues, 122
remote sensing
    change detection, 475
remote sensing, disaster management, 454
Reynolds number, 17, 296
Reynolds-averaged equations, 22
Richardson, Lewis Fry, 13, 434
root causes of disasters
    case studies, 98
Rossby, Carl-Gustaf, 437

Sahara, 213
San Francisco earthquake, 34
sand and dust storms, 221
    characteristics of, 381
    China case study, 249
    Egypt case study, 223
    electrical effects, 384
    geographic favorability, 386
    physics of, 381
sand storm, *see* sand and dust storms, *see* sand and dust storms, *see* sand and dust storms
sand storms, 379
satellite data, 542
satellite imaging
    geophysical parameter retrieval, 473
    image interpretation, 473
    image processing, 470
    information contents, 473
Scénario Catastrophe, 571
seawater irrigation, 216
seawater/saline agriculture, 215
September 11, 42
severe weather in arid lands, 378
short-range prediction, 448
space-based management, of disasters, 479
stampede, Hajj, 54
Standardized Emergency Management System (SEMS), 149
standards of care, modification of, 170
stratosphere, 293
Sumatra earthquake, 267
Sumatra tsunami, 269
    effects on ecosystem, 278
    impact on Sri Lanka, 271
    physical impacts, 272
    wave observations, 270

supply chain issues, 123
surge capability, 162
surge capacity, 162
   community-based patient care, 162
   health care facility-based, 162
   public health, 162
synthetic aperture, 467
Synthetic Aperture Radar (SAR), 455

thermodynamic equilibrium, 14
total intravenous anaesthesia (TIVA), 157
toxic agents
   modeling of their release, 77
   release of, 77
transport equations, 14
transportation infrastructure, 132
triage
   definition of, 153
   mass critical care, 171
   officers, 154
   priorities, 164
   Simple Triage and Rapid Tagging (START), 154
tsunami, 258
   causes of, 262
   Critical Path Method (CPM), 283
   definition of, 258
   detection, 285
   ecological impact, 265
   hydrodynamics, 263
   planning, 283
   Sumatra, *see* Sumatra tsunami
   warning center, 286
   warning system, 282
tsunami, Pacific, 43

turbulence, 21
   computer requirement, 22
   equations of motion, 22
turbulent flows, 21

UN Humanitarian Information Center, 129
UN Joint Logistics Center, 131
United Nations Environment Programme (UNEP), 480
United Nations Geographic Information Working Group (UNGIWG), 480
United Nations Institute for Training and Research (UNITAR) (UNITAR), 480
United Nations Office for the Coordination of Humanitarian Affairs (UNOCHA), 480
United Nations Operational Satellite Applications Programme (UNOSAT), 480

very-short-range prediction, 449
von Neumann, John, 437

walkway collapse, Hyatt Regency, 38
water cycle, 363
weapons of mass destruction (WMD), 171
weather
   atmospheric prediction, 440
   changes, 192
   debris flows, 394
   definition of, 72
   flood, 392
   forecasting, history of, 427
   modeling of, 303
   prediction, 427
   rainstorms, 391
Weather Research and Forecast (WRF), 303
weather-chemistry modeling, 312
Wilma, Hurricane, 49